Biology 11

Biology *11*

Authors

Dr. Bob Ritter
Edmonton Catholic School Board

Christine Adam-Carr
Ottawa-Carleton Catholic District School Board

Douglas Fraser
District School Board Ontario North East

Program Consultant

Maurice Di Giuseppe
Toronto Catholic District School Board

NELSON
THOMSON LEARNING

Australia • Canada • Mexico • Singapore • Spain • United Kingdom • United States

NELSON

THOMSON LEARNING

Nelson Biology 11

Dr. Bob Ritter
Christine Adam-Carr
Douglas Fraser

Director of Publishing
David Steele

Publisher
Kevin Martindale

Program Manager
Eileen Jung
First Folio Resource Group, Inc.

Developmental Editors
Catherine Oh
Michael Pidgeon
Jim Dawson
Janice Palmer
Shirley Barrett

Editorial Assistant
Matthew Roberts

Senior Managing Editor
Nicola Balfour

Senior Production Editor
Rosalyn Steiner

Copy Editor
Susan Till

Proofreader
Bonnie Di Malta

Production Coordinator
Sharon Latta Paterson

Creative Director
Angela Cluer

Art Management
Allan Moon
Kyle Gell

Interior Design
Peter Papayanakis
Ken Phipps
Peggy Rhodes

Cover Design
Katherine Strain

Cover Image
Stephen Dalton/Photo Researchers

Composition
AMID Studios
Kyle Gell Design

Photo Research and Permissions
Mary Rose MacLachlan

Printer
Transcontinental Printing Inc.

**National Library of Canada
Cataloguing in Publication Data**

Ritter, Bob, 1950–
Nelson biology 11

Includes index.
ISBN 0-17-612100-5

1. Biology.
I. Adam-Carr, Christine, 1963– .
II. Fraser, Douglas, 1957– . III. Title.

QH308.7.R57 2001   570
C2001-930491-9

Acknowledgments
The authors wish to thank Doug Jones, Matthew Calder, and Kristy
McGowan for their assistance in the development of this book.

Reviewers

Contents

Unit 1

Cellular Functions

Unit 1 Are You Ready?   4

Chapter 1   Cell Biology   6

1.1   The Cell Theory   8
1.2   Overview of Cell Structure   10
1.3   Looking at Cells and the Cytoplasmic Organelles   12
1.4   Special Structures of Plants   19
      ACTIVITY 1.4.1: Plant and Animal Cells   21
1.5   Cell Research in Medicine   23
      Chapter 1 Summary   27
      Chapter 1 Review   28

Chapter 2   Chemistry of Life   30

2.1   Nutrients   32
      ACTIVITY 2.1.1: Investigating 3-D Molecules   33
2.2   Carbohydrates   34
2.3   Lipids   37
2.4   Proteins   41
      INVESTIGATION 2.4.1: Identifying Nutrients   42
2.5   Nucleic Acids   45
2.6   The Living Cell Membrane   47
2.7   Passive Transport   50
      ACTIVITY 2.7.1: Observing Diffusion and Osmosis   53
      INVESTIGATION 2.7.1: Factors Affecting the Rate
      of Osmosis   55
2.8   Active Transport   56
2.9   Energy Flow in Photosynthesis and Cellular
      Respiration   58
      Career   61
2.10 Cellular Respiration in Plants and Animals   62
2.11 Types of Respiration   64
      INVESTIGATION 2.11.1: Yeast Fermentation   66
      Chapter 2 Summary   69
      Chapter 2 Review   70

Unit 1 Performance Task: Factors Affecting
      Cell Membrane Permeability   72
Unit 1 Review   74

Unit 2

Genetic Continuity

Unit 2 Are You Ready?   80

Chapter 3   Cell Division   82

3.1   Principles of Cell Division   84
3.2   The Cell Cycle   86
3.3   A Cell Clock   90
      ACTIVITY 3.3.1: Frequency of Cell Division   91
3.4   Cloning   94
3.5   Cancer   100
3.6   Meiosis   103
3.7   Comparing Mitosis and Meiosis   108
      ACTIVITY 3.7.1: Comparing Mitosis and Meiosis   110
3.8   Reproduction and Cell Division   112
3.9   Abnormal Meiosis: Nondisjunction   116
      Chapter 3 Summary   123
      Chapter 3 Review   124

Chapter 4   Genes and Heredity   128

4.1   Early Beliefs and Mendel   130
4.2   Single-Trait Inheritance   134
4.3   Selective Breeding   140
4.4   Multiple Alleles   143
4.5   Incomplete Dominance and Codominance   144
4.6   Dihybrid Crosses   150
      ACTIVITY 4.6.1: Genetics of Corn   155
      Chapter 4 Summary   157
      Chapter 4 Review   158

Chapter 5 The Source of Heredity   160

5.1	Early Developments in Genetics	162
5.2	Development of the Chromosomal Theory	163
5.3	Morgan's Experiments and Sex Linkage	163
	ACTIVITY 5.3.1: Sex-Linkage in Fruit Flies	168
5.4	Looking Inside the Chromosome	172
5.5	Discovering the Structure of DNA	174
5.6	The Structure of DNA	175
5.7	Genes That Change Position	177
5.8	Gene Research and Technologies	180
5.9	DNA Fingerprinting	182
	Career	184
5.10	Gene Therapy	185
	Chapter 5 Summary	187
	Chapter 5 Review	188

Unit 2 Performance Task: Investigating Human Traits	190
Unit 2 Review	192

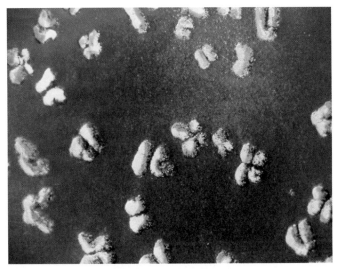

Unit 3

Internal Systems and Regulation

Unit 3 Are You Ready?	198

Chapter 6 Digestion and Nutrition   202

6.1	Organs and Organ Systems	204
6.2	Importance of Digestion	208
6.3	Ingestion	210
	ACTIVITY 6.3.1: The Effect of the Enzyme Amylase on Starch Digestion	213
6.4	The Stomach and Digestion	214

6.5	The Small Intestine and Pancreas	216
	INVESTIGATION 6.5.1: The Effect of pH and Temperature on Starch Digestion	219
6.6	The Liver and Gall Bladder	222
6.7	Absorption of Materials	224
6.8	Homeostasis and Control Systems	226
6.9	Defining Food Energy	230
6.10	Essential Nutrients	234
	Chapter 6 Summary	237
	Chapter 6 Review	238

Chapter 7 Circulation and Blood   240

7.1	The Importance of a Circulatory System	242
7.2	Components of Blood	244
7.3	Blood Groups	246
7.4	Blood Vessels	250
7.5	The Mammalian Heart	256
7.6	Setting the Heart's Tempo	259
7.7	Heart Sounds	261
7.8	Blood Pressure	264
7.9	Variation in Cardiovascular Output	266
	INVESTIGATION 7.9.1: Effects of Posture on Blood Pressure and Pulse Rate	269
	INVESTIGATION 7.9.2: Effects of Exercise on Blood Pressure and Pulse Rate	270
7.10	Capillary Fluid Exchange	272
7.11	The Lymphatic System	274
	Chapter 7 Summary	277
	Chapter 7 Review	278

Chapter 8 Respiratory System   280

8.1	The Importance of the Respiratory System	282
8.2	The Mammalian Respiratory System	285
8.3	Gas Exchange and Transport	290
8.4	Disorders of the Respiratory System	294
	ACTIVITY 8.4.1: Determining Lung Volume	296
	Career	298
	INVESTIGATION 8.4.1: The Effects of Exercise on Lung Volumes	299
8.5	Organ Transplants in Canada	303
8.6	The Effect of Psychoactive Drugs on Homeostatic Adjustment	304
8.7	Digestive, Circulatory, and Respiratory Systems of the Fetal Pig	307
	ACTIVITY 8.7.1: Examining the Systems of a Fetal Pig	308
	Chapter 8 Summary	313
	Chapter 8 Review	314

Unit 3 Performance Task: Determining Fitness Level	316
Unit 3 Review	318

Chapter 11 Invertebrates 406

11.1	The Animal Kingdom	408
11.2	The Simplest Animals	412
11.3	Worms	417
	ACTIVITY 11.3.1: Earthworm Dissection	424
11.4	Mollusks and Echinoderms	428
11.5	Arthropods	435
	Chapter 11 Summary	443
	Chapter 11 Review	444

Chapter 12 Chordates 446

12.1	Phylum Chordata	448
12.2	Fish	452
12.3	Amphibians	455
12.4	Reptiles	457
12.5	Birds	461
12.6	Mammals	465
	ACTIVITY 12.6.1: Using Models for Limb Movement	472
	Career	479
	Chapter 12 Summary	480
	Chapter 12 Review	481

Unit 4 Performance Task: Cladistics 482

Unit 4 Review 484

Unit 4

Diversity of Living Things

√

Unit 4 Are You Ready?	324

Chapter 9 Taxonomy and the World of Microorganisms and Viruses 326

9.1	Taxonomic Systems	328
	ACTIVITY 9.1.1: Using a Classification Key	332
9.2	Viruses	334
9.3	Kingdoms Archaebacteria and Eubacteria	340
	INVESTIGATION 9.3.1: Effects of Antiseptics	346
9.4	Kingdom Protista	349
	ACTIVITY 9.4.1: Examining Protists	357
	Chapter 9 Summary	360
	Chapter 9 Review	361

Chapter 10 Fungi and Plants 362

10.1	Kingdom Fungi	364
10.2	Life Cycle of Fungi	366
	INVESTIGATION 10.2.1: Monitoring Bread Mould Growth	367
10.3	Importance of Fungi	369
10.4	Kingdom Plantae	374
10.5	The Evolution of Terrestrial Plants	375
10.6	Alternation of Generations	378
10.7	Mosses	381
10.8	Ferns	386
10.9	Seed Plants	390
	Chapter 10 Summary	403
	Chapter 10 Review	404

Unit 5

Plants: Anatomy, Growth, and Functions

Unit 5 Are You Ready?	490

Chapter 13 Plants: Form and Function — 492

13.1	Vascular Plant Structure and Function	494
13.2	Plant Tissues	497
13.3	Leaves	501
	ACTIVITY 13.3.1: Leaf Adaptations	507
13.4	Roots	512
	ACTIVITY 13.4.1: Root Anatomy	516
13.5	Stems	518
	ACTIVITY 13.5.1: Stem Anatomy	523
13.6	Transport in Plants	525
	INVESTIGATION 13.6.1: Water Movement in Stems and Leaves	527
13.7	Reproduction	528
	ACTIVITY 13.7.1: Monocot and Dicot Seeds	531
13.8	External Factors Affecting Plant Growth	534
	INVESTIGATION 13.8.1: The Effect of Light on Plant Growth	535
13.9	Internal Factors Affecting Plant Growth	540
	Chapter 13 Summary	543
	Chapter 13 Review	544

Chapter 14 The Importance of Plants — 546

14.1	Succession	548
14.2	Plants and Biodiversity	554
14.3	Fertilizers	557
	INVESTIGATION 14.3.1: Assessing Synthetic Fertilizers	560
14.4	Food Plants	562
14.5	Fuels, Fibres, and Wood Products	569
14.6	Medicinal and Nonmedicinal Chemical Plant Products	574
	Career	577
14.7	Pest Management	579
14.8	A Sustainable Future	584
	Chapter 14 Summary	587
	Chapter 14 Review	588

Unit 5 Performance Task: Plants That Changed the World	590
Unit 5 Review	592

Appendixes	596
Appendix A: Skills Handbook	598
Appendix B: Safety Skills	620
Appendix C: Reference	630
Appendix D: Answers	636
Glossary	640
Index	651
Credits	658

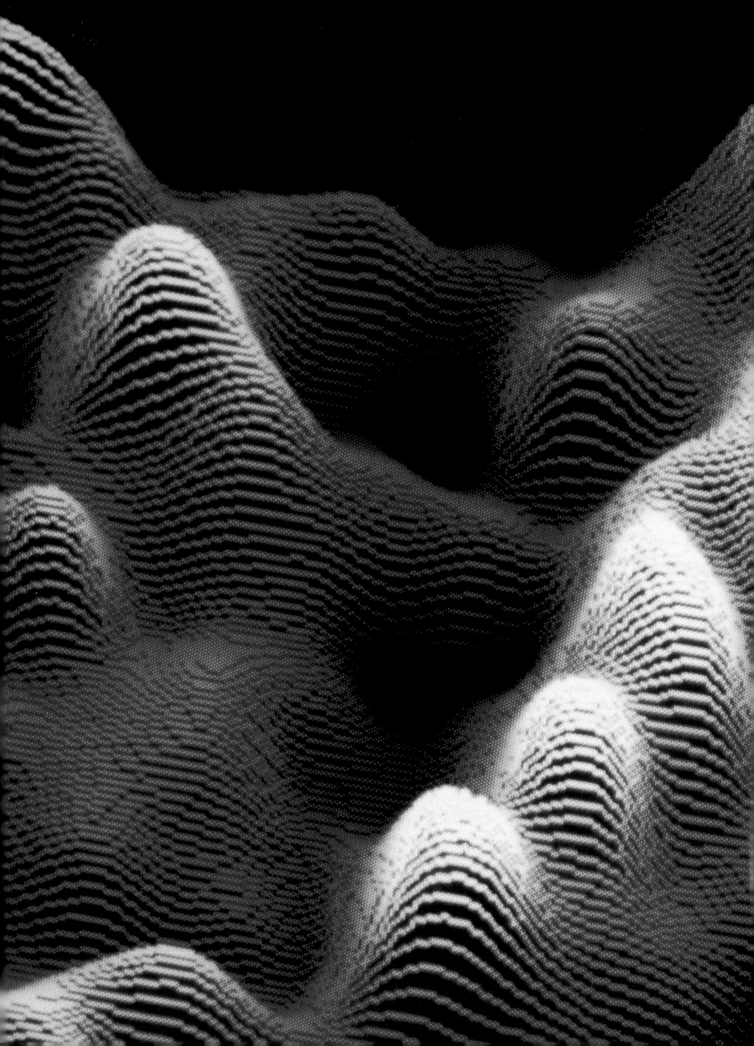

There are many variations in reproduction among the plantlike protists belonging to the three phyla Chlorophyta, Phaeophyta, and Rhodophyta. Variations range from complex life cycles that include both sexual and asexual reproduction, to simple asexual reproduction by fragmentation—the organism's body simply breaks apart. Most commonly, reproduction is asexual by binary fission. *Spirogyra* also reproduce by conjugation (**Figure 7**).

Animal-like Protists (Protozoa)

In contrast to plantlike protists, all protozoa are heterotrophs. Generally, they must move about to obtain food. Most protozoa are said to be holozoic because they engulf their food (bacteria and other microorganisms). Others, called saprozoic organisms, absorb predigested or soluble nutrients directly through the cell membranes.

The major criterion used in classifying the four phyla of protozoa is their type of locomotion (**Table 3**). Other criteria, such as the types of organelles, life cycle, mode of reproduction, nutrition, and whether the organism is free-living or parasitic, are also important in the modern identification and classification at the lower taxonomic levels.

Protozoans rival bacteria in both population numbers and number of species. They occupy a diverse range of moist habitats. Their range in size is considerable, with some small protozoan blood parasites measuring only 2 μm (micrometres) long. A small mammalian red blood cell may contain as many as a dozen of these protozoans. On the other hand, some foraminiferans can grow to as large as 5 cm in diameter.

A tube develops, connecting the two cells together.

The contents of the left-hand cell move into the right-hand cell.

The nucleus of the left-hand cell fuses with the nucleus of the right-hand cell.

The cytoplasm rounds off and develops a thick wall, becoming a zygospore.

Figure 7
Conjugation of *Spirogyra*, a filamentous green algae that also reproduces asexually by binary fission and by fragmentation

Table 3: Animal-like Protists

	Phylum	Descriptions
	1. Sarcodina (e.g., *Amoeba*, foraminiferans, radiolarians)	Most sarcodines are free-living forms that thrive in fresh water, salt water, and soil. A few parasitic species are found in the intestines of animals. Many are motile, with adult forms possessing pseudopods ("false feet") for locomotion. Movement results from repeated cytoplasmic extension and retraction of the pseudopods. The amoeba feeds by having its pseudopods flow around and engulf food particles.
	2. Mastigophora (e.g., flagellated protozoans)	Zooflagellates move by means of flagella and are found in both fresh and salt water. Most are parasitic and cause disease in animals. Reproduction is asexual, by longitudinal fission. There are no reported cases of sexual reproduction. Flagellates also form cysts, which is the way many parasitic forms are spread from host to host.
	3. Ciliophora (ciliated protozoans, e.g., *Paramecium*)	Ciliates, in contrast to flagellates, are considered to be the most complex and advanced of the protozoans. They are characterized by the presence of hairlike structures called cilia. Cilia are synchronized for swimming in the free-moving organisms like paramecia. Typically, reproduction is asexual by binary fission; however, the organisms may also reproduce sexually by conjugation.
	4. Sporozoa (e.g., *Plasmodium*)	Sporozoans lack means of independent locomotion. Sporozoans, and their relatives, are exclusively parasitic, and depend entirely upon the body fluids of their hosts for movement. They are characterized by a sporelike stage. Sporozoa display complex life cycles.

cysts: cells that have a hardened protective covering on top of the cell membrane

ectoplasm: the thin, semirigid (gelled) layer of the cytoplasm under the plasma membrane

endoplasm: the fluid part of the cytoplasm that fills the inside of the cell. The endoplasm is responsible for an amoeba's shape as it moves.

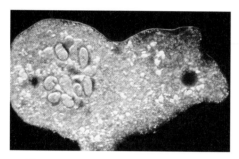

Figure 8
Amoeba with ingested paramecia

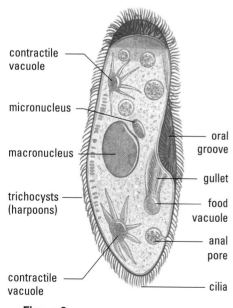

Figure 9
The paramecium is sometimes called the "slipper animal" because of its shape.

Reproduction in protozoans is usually asexual, by binary fission. Under adverse conditions, protozoans may form resting cells called **cysts**. If favourable conditions return, the organisms may emerge from their cysts.

Phylum Sarcodina

Except for a few parasitic species found primarily in animal intestines, most sarcodines are free-living forms. On the basis of pseudopods—fingerlike projections used for locomotion and for food capture—the sarcodines are divided into four groups: amoebas, foraminiferans, radiolarians, and heliozoans.

One of the largest yet least complex protozoans is the amoeba (**Figure 8**). The amoebas, which do not have shells, are exceptional organisms to view under the microscope and clearly demonstrate the movement of the sarcodines.

An amoeba moves by repeatedly extending and retracting its pseudopods. Its cytoplasm has two layers: **ectoplasm**, a thin, semirigid layer under the plasma membrane, and **endoplasm**, a more fluid part that fills the inside of the cell. The continuous movement of the endoplasm causes the amoeba to change shape constantly as it moves. Can you understand why Linnaeus named the amoeba *Chaos chaos* when he saw the organism for the first time?

An amoeba feeds by phagocytosis; its pseudopods simply flow around and engulf food particles. The food eventually becomes enclosed in a food vacuole, where the food is digested. Water taken in with the food, along with extra water that enters the organism through diffusion, collects in a contractile vacuole. As this vacuole fills, it contracts and the water is discharged through a pore in the plasma membrane.

Reproduction in the amoeba is by binary fission. Once the amoeba splits, the two organisms then grow to their full size and may split again. Under proper conditions, this could occur once a day. How many amoebas would be produced from an original cell in a month of 30 days, assuming all survived and reproduced?

Phylum Mastigophora

Mastigophora, animal-like flagellates, are protozoa that have one or more flagella. Free-living types live mainly in freshwater or marine habitats, while the parasitic types live inside other organisms, including humans.

One fairly common internal parasite is *Giardia lamblia*. The parasite usually causes stomach upsets and diarrhea, but can have severe effects on some people. Cattle and wild animals often have the parasite. The parasite passes to humans when they drink water or eat food that has been tainted by the feces of infected animals. Water withdrawn straight from streams should be boiled to kill the parasite.

Another common parasite, *Trichomonas vaginalis*, is passed through sexual intercourse and can result in infection in the urinary and reproductive tracts. African sleeping sickness and Chagas disease are other disorders caused by parasites, both of which are transmitted by insects.

Phylum Ciliophora

Ciliates possess hairlike structures called cilia, which in some cases cover the entire surface of the organism. While similar in function to flagella, cilia are usually shorter and more abundant. In free-moving organisms, such as *Paramecium*, cilia are synchronized for swimming, like oars or the arms of a swimmer. In ciliates that are attached to some surface, such as *Vorticella* and *Stentor*, the cilia attach the microorganism to the surface.

Ciliates are the most advanced of the protozoans. They live in both freshwater and marine habitats. The most commonly studied ciliate, the paramecium, illustrates many of the group's complex features (**Figure 9**). The oral groove contains a mouth that leads into a cavity called the gullet. Specialized cilia in the gullet sweep

The following labels appear on Figure 9: contractile vacuole, micronucleus, macronucleus, trichocysts (harpoons), contractile vacuole, oral groove, gullet, food vacuole, anal pore, cilia.

bacteria and other food particles into the cavity. From the gullet, food enters a food vacuole where digestion takes place. Wastes are expelled from the food vacuole through an anal pore. Similar to other protists, paramecia collect excess water in a contractile vacuole, from which it is expelled into the environment.

A paramecium has two nuclei: a large one called the macronucleus, which controls most of the cell's activities, and a smaller one called the micronucleus, which is involved in reproduction. Reproduction is usually asexual by binary fission. Periodically, the organism may undergo a form of sexual reproduction called conjugation. During this process, a pair of paramecia join to combine their genetic material. Following this process, the daughter cells undergo fission.

The paramecium also has an interesting defence mechanism. Hundreds of poison-laden barbs can be discharged either to drive away predators or to capture prey. These structures are called trichocysts.

Phylum Sporozoa

Sporozoans are unique among the protozoans because they lack any means of independent locomotion; they have no pseudopods, cilia, or flagella (except for a few types that have flagella while maturing). They are exclusively parasitic and depend entirely on the body fluids of their hosts for movement. Also, they have fewer organelles and specialized structures than other protozoans.

The complex life cycles of sporozoans contrast with their relatively simple structure. Sporozoans are characterized by having reproductive cells that can produce a new organism without fertilization (**spores**) during their life cycle. At this early stage of development, they are called sporozoites. Sometimes two or more hosts are involved, in which case the parasite reproduces sexually in one host and asexually in the other. Insects are frequently responsible for transmitting sporozoans from one host to the next and are referred to as insect vectors. One sporozoan, *Plasmodium*, is particularly notorious as it causes malaria in humans and some other mammals and birds (**Figure 10**).

spores: reproductive cells that can produce a new organism without fertilization. Spores have a haploid number of chromosomes. Fungal spores have thick, resistant outer coverings to protect them.

Within the mosquito, the gametocytes mature into gametes and fuse to form zygotes. The zygotes develop into sporozoites.

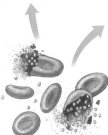

Mosquito bites an infected human and sucks up blood containing gametocytes of *Plasmodium vivax*, a sporozoan parasite.

When the mosquito bites another human, the sporozoites enter the blood stream and move to the liver.

Some merozoites return to the liver; others reinfect red blood cells. Both cause repeated malarial episodes.

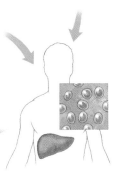

Merozoites enter the bloodstream and infect red blood cells. There they reproduce asexually. The red blood cells eventually burst, sending the merozoites and male and female gametocytes into the bloodstream.

Merozoites are produced in the liver by asexual reproduction of the sporozoites.

Figure 10
The life cycle of the malaria-causing sporozoan, *Plasmodium vivax*

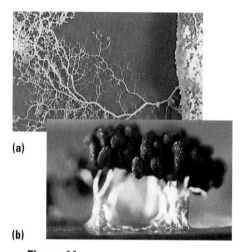

(a)

(b)

Figure 11
(a) A plasmodial slime mould
(b) Fruiting body of the slime mould, where
spores are produced

Figure 12
Dr. G.F. Bennett

Fungilike Protists

Fungilike protists, also referred to as slime moulds (**Figure 11**), are placed in the phylum Gymnomycota. Slime moulds prefer cool, shady, moist places and are usually found under fallen leaves or on rotting logs. The name is derived from the slimy trail left behind as the mould moves over the ground.

During some stage of their life cycle, slime moulds resemble protozoans and become amoebalike or have flagella. At other times, they produce spores much like fungi do. Unlike other primitive organisms, the slime moulds do not always remain as single-celled organisms. *Dictyostelium* spends some of its life cycle as a single-celled amoebalike organism; however, these single cells can converge into a large slimy mass plasmodium. The plasmodial mass itself begins to act like a single organism. It extends into a sluglike form and begins to creep, feeding on organic matter as it goes. When it runs into an object, it retracts and slithers around it. The coordinated movements are slow—sometimes only a few millimetres a day.

Biologists recognize that becoming multicellular is a tremendous advancement. As individual cells begin to work together, the groundwork for cell specialization is laid. In more advanced organisms, found in the higher kingdoms, the specialized cells develop structures suited for specific tasks.

Research in Canada: Dr. G.F. Bennett

Much of the research into sporozoan-caused diseases like malaria involves studying nonhuman species such as birds and small mammals. This work is often coordinated in special laboratories or centres that have access to a wide range of expertise, equipment, and information. One such centre is the International Reference Centre for Avian Haematozoa (IRCAH), located at Memorial University in Newfoundland. As the words "avian haematozoa" imply, the centre specializes in the study of blood parasites of birds.

From 1969 until the 1990s, Dr. G.F. Bennett (**Figure 12**), a Canadian entomologist born and raised in India, directed the work at the facility, originally established under the sponsorship of the World Health Organization. The centre has accumulated over 120 000 indexed reference collections of bird blood parasites and contains articles representing 99% of the total material on the subject. Bennett and IRCAH attracted some of the world's leading malariologists to conduct research at IRCAH. Bennett researched bird-biting insects in Ontario, studied simian (monkey) malaria in Malaysia, and also investigated blood-sucking cattle ticks in Australia. He has personally acted as an experimental host for two species of simian malaria.

Bennett was dedicated to clarifying the connection between bird blood parasites, their vectors, and hosts. This required a greater understanding of the biology and life cycles of a number of sporozoan diseases.

Other projects at IRCAH include studies on the environmental impact of diseases on birds and the development of biological control agents for insect pests.

Practice

Understanding Concepts

1. Name the three distinct groups that make up Protista.
2. What is phytoplankton?
3. Which group of protists is called protozoans?
4. What are some plantlike and animal-like characteristics of *Euglena*?

5. What does an amoeba have in common with a human white blood cell?

6. How do sporozoans differ from all other protozoans?

7. Describe movement in an amoeba, a paramecium, and a euglenid.

8. Why is conjugation considered a form of sexual reproduction?

9. What ecological role do algae play?

10. Why are fungilike protists also called slime moulds?

Making Connections

11. How do you think birds in Canada would become infected with malaria?

Exploring

12. Eutrophication is a process where a body of water becomes enriched in dissolved nutrients such as phosphates.
    (a) Runoff from agricultural land is often rich in nutrients. Do you think an increase in the natural level of nutrients would be beneficial for a pond or lake?
    (b) Research to find out how algae are involved in eutrophication. What are algal blooms?
    Follow the links for Nelson Biology 11, 9.4.

    GO TO www.science.nelson.com

    (c) How does eutrophication ultimately result in low oxygen levels in the body of water?

Activity 9.4.1

Examining Protists

In this activity, you will examine, under a light microscope, the cell structure of some common protists using prepared slides of *Paramecium*, *Spirogyra*, as well as some other examples. You will also observe live paramecia move, feed, and respond to their environment.

Paramecium, a ciliated protozoan, is one of the most complex of the single-celled protists. It reproduces asexually by binary fission, and periodically reproduces sexually by conjugation. *Spirogyra*, a filamentous green alga, reproduces asexually by fragmentation from spring until autumn, and sexually by conjugation during the colder months. Zygospores formed during the cold season remain dormant until the following spring.

Materials

lab apron
light microscope
paramecia culture
10% methyl cellulose
    solution or glycerine
prepared slides of
    Paramecium, *Spirogyra*, and
    any of the following protists:
    Euglena, Amoeba, Volvox

medicine dropper
microscope slides and cover slip
stained yeast culture (congo red)

Table 4: Observations of the Paramecium

Feature	Description and/or sketch
anterior/posterior end	
general movement pattern	
reaction to obstacles	
leading end during locomotion	
cilia length and movement	
yeast cells in gullet	
food vacuoles, number and location	
trichocysts	
contractile vacuoles	

Procedure

Part 1: Examining Protists

1. Put on your apron.
2. Place each prepared slide on the stage of the microscope. Using medium-power or high-power magnification, draw one example of each type of organism you observe. For each drawing, label all the visible structures (e.g., macronucleus, micronucleus, contractile vacuole, food vacuole, cilia, oral groove, anal pore).
3. Calculate and record the length and width of each protist.
4. Describe reproduction in the paramecium if it is observed.

Part 2: Observing Paramecia

5. Using the medicine dropper and microscope slide, prepare a wet mount of the paramecia culture. Add a cover slip.
6. In your notebook, prepare a table similar to **Table 4**. Record your observations in your table as you observe the movement of paramecia under low- and medium-power magnification.
7. Prepare another wet mount by mixing one drop of methyl cellulose solution (or smearing a small amount of glycerine) with an equal-sized drop of culture fluid on a slide. Before applying a cover slip, add a drop of the stained yeast culture to the mixture. Under low-power magnification, examine the slide for cilia movement, yeast cells, and food vacuoles. Add your observations to the table.

Analysis

Part 1: Examining Protists

(a) Compare the protists that you examined with another group. Were the same protists approximately the same shape and size? If not, what might account for the differences?
(b) What clues to functions are provided in the names of the structures you observed (e.g., oral groove, contractile vacuole, trichocysts, chloroplasts)?

Part 2: Observing Paramecia

(c) From your examination, summarize how a paramecium ingests and digests food, and eliminates wastes.
(d) From your examination, summarize how a paramecium moves.
(e) Why was methyl cellulose solution (or glycerine) added to the slide?
(f) From your examination, what might be the function of trichocysts?
(g) Was reproduction evident? If so, describe the stage(s) that you viewed.

Evaluation and Synthesis

(h) *Spirogyra* is similar to plants because the cells contain chloroplasts and have cell walls. Why isn't it classified as part of the plant kingdom?
(i) *Euglena* does not have cell walls. Why is it considered to be plantlike?
(j) Describe the difference between conjugating cells and nonconjugating cells in *Spirogyra* and in *Paramecium*.
(k) What might happen to the contractile vacuoles in a paramecium if the organism was transferred from the culture medium into seawater? Explain.

SUMMARY Kingdom Protista

1. Kingdom Protista contains plantlike, animal-like, and fungilike protists. The group contains organisms that do not fit into the other five kingdoms.

2. Protists are eukaryotes and obtain energy in a variety of ways.

3. Animal-like protists are distinguished by their modes of locomotion.

4. Plantlike protists vary from simple one-celled forms to large multicellular organisms.

5. Fungilike protists exhibit complex life cycles and exist in various cellular forms.

6. Reproduction in protists may be sexual or asexual, with many having complex reproductive cycles.

7. Algae are primary food producers and the source of biological energy for most aquatic food webs; they supply about two thirds of the world's oxygen.

Section 9.4 Questions

Understanding Concepts

1. How can bacteria and protists be distinguished from one another?

2. How can protists be distinguished from the higher kingdoms of plants, animals, and fungi?

3. What are the basic similarities and differences between cilia and flagella?

4. Describe asexual and sexual reproduction in paramecium. How do the end products differ genetically from one another?

5. When do some amoeba form cysts, and what is the function of a cyst?

6. What does the term plasmodium refer to when it is used to describe
   (a) a slime mould?   (b) an animal-like protist?

7. What is the difference between an insect vector and a viral vector?

Making Connections

8. What would happen in a pond if water pollution killed the algae?

Exploring

9. In 1999, researchers fertilized a small part of the ocean near Antarctica with iron. They found that the iron stimulated the growth of algae that consume carbon dioxide, a major greenhouse gas.
   (a) Find out more about the research and describe the procedures and findings.
   Follow the links for Nelson Biology 11, 9.4.

   GO TO www.science.nelson.com

   (b) Is this a viable solution to global warming?

Key Expectations

Throughout this chapter, you have had opportunities to do the following:

- Use characteristics of organisms and the principles and nomenclature of taxonomy to classify organisms (9.1).
- Define the fundamental principles of taxonomy and phylogeny (9.1).
- Demonstrate, through applying classification techniques and terminology, the usefulness of the system of scientific nomenclature in the field of taxonomy (9.1).
- Classify various organisms in a pond, following the principles of taxonomy (9.1).
- Explain how species are categorized and named according to structure and/or evolutionary history (9.1, 9.3, 9.4).
- Locate, select, analyze, and integrate information on topics under study, using appropriate library and electronic research tools, including Internet sites (9.1, 9.3, 9.4).
- Classify representative organisms from each of the kingdoms according to their nutritional pattern, type of reproduction, habitat, and general structures (9.1, 9.3, 9.4).
- Use a microscope effectively and accurately for making observations (9.1, 9.3, 9.4).
- Compare and contrast the structure and function of different types of prokaryotic and eukaryotic cells (9.1, 9.4).
- Give examples of how viruses and bacteria are used in biotechnology (9.2, 9.3).
- Describe selected anatomical and physiological characteristics of representative organisms from each life kingdom and a representative virus (9.2, 9.3, 9.4).
- Compare and contrast the life cycles of representative organisms from each life kingdom and a representative virus (9.2, 9.3, 9.4).
- Describe the importance of the resistance to infection by "new" microorganisms, and the resistance of bacteria to antibiotics (9.3).
- Explain the importance of sexual reproduction (including the process of meiosis) to variability within a population (9.4).
- Identify and describe Canadian contributions to research and technological developments in biotechnology (9.4).

Key Terms

antibiotics	lysis
Archaebacteria	lysogeny
bacteriophages	Monera
binomial nomenclature	obligate aerobes
capsid	obligate anaerobes
cysts	phylogeny
dichotomous key	Protista
ectoplasm	species
endoplasm	spores
endospores	taxa
Eubacteria	taxonomy
facultative anaerobes	vaccines
host range	viruses

Make a Summary

In this chapter, you studied the principles of taxonomy and classified microorganisms. To summarize your learning, create a spider key that shows the differences between viruses and living organisms, and also shows the differences between prokaryotic and eukaryotic cells. Label the spider key with as many of the key terms as possible. Check other spider keys and use appropriate designs to make your sketch clear.

Reflect on your Learning

Revisit your answers to the Reflect on Your Learning questions at the beginning of this chapter.

- How has your thinking changed?
- What new questions do you have?

Understanding Concepts

1. How was the introduction of classification keys a major contribution to taxonomy?

2. Develop an argument to support or refute the following statement: Viruses are nonliving.

3. How are viruses like cellular organisms? How are they different?

4. In many viral diseases (e.g., smallpox, mumps, influenza), illness occurs shortly after exposure to the virus. In others (e.g., AIDS), the victim may not show symptoms for many years following the initial infection. How would you explain the difference in these situations?

5. Why do lysogenic viruses, such as retroviruses, hold greater hope for curing some human genetic disorders than lytic viruses do?

6. Why are antibiotics not prescribed for viral diseases such as the common cold?

7. Certain bacteria are credited with establishing the oxygen component of Earth's atmosphere. Explain how the evolution of other life forms might have been different without these bacteria.

8. Identify situations in which a bacteria-free environment is desirable and those situations in which it would not be desirable.

9. One of the most common diseases of the human respiratory system is pneumonia, an inflammation of the lungs usually caused by bacteria. Pneumonia can also be caused by viruses. Viral pneumonia is more difficult to treat. What physical characteristic of viruses accounts for this?

10. Why are organisms belonging to Eubacteria considered to have evolved earlier than those belonging to Protista?

11. Explain why some unicellular protists are considered to be complex organisms. Give examples.

12. Identify the disadvantage(s) of a sporozoan such as *Plasmodium* being totally parasitic.

13. What might an antibiotic do to the beneficial bacteria in your intestine? How can this be prevented?

14. Why are pathogenic bacteria that form tough, resistant resting cells feared more than bacteria that form none?

15. What reasons might account for the relatively late discovery of fossil bacteria?

16. What adaptations in paramecia might suggest that they are more complex than amoebas?

17. How might plantlike protists be used to benefit humans? List some possibilities that you have learned about.

Applying Inquiry Skills

18. Botulism is a deadly form of food poisoning caused by bacteria (*Clostridium botulinum*) that grow in anaerobic conditions. Toxins produced from these deadly bacteria can paralyze muscles and lead to death. A student notices that a can has become misshapen. When opened, a foul-smelling gas is released from the can. The student concludes that the botulism-producing bacteria may be present, but reasons that just exposing it to the air will cause it to die. After heating the food thoroughly, the student begins eating. Do you agree with the reasoning? Explain why or why not.

19. Could you accept the hypothesis that viruses were the precursors of life on this planet? Provide reasons.

20. Flagellated protozoans in the termite intestine enable the insect to digest the cellulose in wood. Investigate the symbiotic relationships that aid plant-eating animals, such as termites and cattle, in digesting cellulose.

21. Conduct a detailed analysis of a typical bacterium and a protistan cell. Consider their similarities and differences. Include cell size, cell type, organelles, metabolism, and reproduction.

Making Connections

22. Preventing and controlling bacterial growth is a major concern within the food industry. Suggest some advantages and disadvantages of the following food preservation methods: freezing, canning, dehydration, pickling, and salting. Can you name some other methods?

23. Explain the difficulty in proving that viruses are associated with cancer in humans.

24. If a virus was found to reproduce outside a cell, how would this affect current scientific thinking about viruses?

25. Tuberculosis, a disease that was once believed to be almost eliminated, has seen a dramatic increase in recent years. To make matters worse, many of the new strains of this bacteria are resistant to most conventional drugs.
    (a) Speculate about why new strains are showing increased resistance to drugs.
    (b) Suggest a plan that would help bring the numbers of these bacteria into check.

Exploring

26. Conduct an Internet or library search to find out what causes red tides. How do red tides affect organisms in the ocean and the organisms that eat them?
Follow the links for Nelson Biology 11, Chapter 9 Review.

 GO TO   www.science.nelson.com

Fungi and Plants

In this chapter, you will be able to

- describe selected anatomical and physiological characteristics of plants;
- compare and contrast the life cycles of fungi and various plants;
- describe plant adaptations to different environmental conditions;
- explain the importance of sexual reproduction to variability within a population;
- demonstrate an understanding of the connection between biodiversity and species survival;
- demonstrate, by applying classification techniques and terminology, the usefulness of the system of scientific nomenclature in the field of taxonomy.

It is a problem to provide food for an ever-increasing human population, so scientists are trying to make sure that there are enough plants to feed us all. The world's need for food will nearly double by the year 2030, but cultivated land usage is near its limit already. The world's farmers rely on technological and biological advancements in agriculture to keep pace. Grain yields have nearly tripled since the 1950s, with much of the expansion coming from already existing farmland. Fertilizers, irrigation, pesticides, and many other agricultural advancements have so far enabled us to keep pace with population growth. However, many experts question whether this can continue much longer.

Scientists, more than ever, are selecting plants that produce maximum growth and nutrient content to maintain increased food production. Consumer preferences also require that the crops be appealing. Tomatoes are selected for a thin dark-red outer skin and a symmetrical shape. Oranges must be large, brilliant orange in colour, provide a heavy aroma, and not show blemishes. By selecting plants with desired characteristics, scientists have been able to produce both more food and food with greater appeal. However, in doing so, they have also reduced the genetic variety of crops. All tomatoes begin to look the same and fewer types of oranges are produced. It is no accident that all McIntosh apples in grocery stores have a striking resemblance.

Historically, humans relied on about 200 different species of plants. Today we rely heavily on about 20 different species—wheat, rice, corn, and potatoes being the most common. By raising crops that are genetically uniform, farmers could be risking our food supply. Plants selected for rapid growth and desired aesthetic qualities may not be ideally suited to withstand disease and other negative environmental conditions. For example, the corn blight fungus devastated crops in the southwestern United States in 1970. If there had been greater genetic variability in the corn crop, the effects might not have been as widespread. The problem was solved by switching to many varieties of corn with different genetic makeups. Often crops that are not resistant to pests or diseases are sprayed with chemicals to protect them (**Figure 1**). However, many people fear eating food treated with chemicals.

Reflect on your Learning

1. Make a list of ways in which people rely on plants.
2. How has selecting plants for food affected biological diversity?
3. Describe how environmental conditions affect plants that grow in your area.
4. What structural adaptations must aquatic plants have in order to survive on land?

Figure 1
Improved agricultural technologies have greatly increased the food crop yield, but at what cost?

Try This
Activity

Studying Plant Distribution

- On your way to or from school, take a close look at the plants in flowerbeds along the north side and along the south side of any buildings.
- Identify the plants found. Consult a book on plants or use other resources, if necessary. Write a brief description of the plants; include height, colour, and shapes of the blooms and leaves. Take a photograph or make a sketch for reference.
- Compare the plants in the north- and south-facing flowerbeds:
  (a) What is the total number of shrubs? flowering plants?
  (b) How many types of shrubs were there? flowering plants?
  (c) What are the differences between the different types of shrubs? types of flowering plants?
  (d) If a particular flowering plant is found in both flowerbeds, compare their general state of health, the number of flowers per plant, the size of the flowers, and the stage of flowering.
  (e) What is the soil colour and texture of the flowerbeds?
  (f) Which flowerbed had the most grass or other weeds?

DID YOU KNOW ?

In plant biology, the category phylum is called a division.

10.1 Kingdom Fungi

Fungi were once classified as members of the plant kingdom. However, enough differences exist between plants and fungi to place them in separate kingdoms. **Table 1** summarizes some of their similarities and differences.

Fossils resembling fungi date back about 900 million years. By about 300 million years ago, representatives of all modern fungi had evolved. Well before the Cambrian period (570 million years ago), fungi took a different evolutionary pathway than plants. They adapted to a heterotrophic way of life. The fact that fungi depend on other organisms for nutrients is a major difference between plants and fungi.

Fungi are adapted for two main functions: absorption of nutrients, and reproduction. Digestion is extracellular—nutrients are digested externally before being absorbed. In multicellular forms, nutrient absorption takes place in the **mycelium** (plural: mycelia), a mesh of microscopic branching filaments that are usually on or below the surface of the substrate. Each of these filaments is called

mycelium: a collective term for the branching filaments that make up the part of a fungus not involved in sexual reproduction

Table 1: Similarities and Differences Between Plants and Fungi

Plants	Fungi
Similarities	
cells are eukaryotic	
numerous organelles	
have cell walls	
most anchored in soil or other **substrate**	
reproduction can be asexual, sexual, or both	
stationary	
Differences	
have one nucleus per cell	often have many nuclei per cell
most are autotrophs	are heterotrophs
starch is the main storage molecule	have few or no storage molecules
most have roots	have no roots
have cellulose in cell walls	often have **chitin** in cell walls
some reproduce by seed	none reproduce by seed

substrate: a surface in or on which an organism grows or is attached

chitin: a nitrogenous polysaccharide of long fibrous molecules

Figure 1

Some fungi have hyphae with cross-walls. Most hyphae have cell walls that are reinforced with chitin. It forms structures of considerable strength.
(a) Hyphae with cross-walls
(b) Hyphae without cross-walls
(c) Mycelium showing many interlocking hyphae

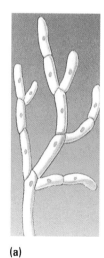

(a)   **(b)**   **(c)**

a **hypha** (plural: hyphae) (**Figure 1**). Often the only visible parts of a fungus are its reproductive structures, which display a wide variety of sizes, shapes, and colours (see **Table 1**).

hypha: one of the filaments of the mycelium

saprophytes: organisms that obtain nutrients from dead or nonliving organic matter

Classification of Fungi

Classification schemes for fungi may differ slightly depending on the source of information. **Table 2** lists some of the divisions of the major classes, their common names, habitats, descriptions, and methods of reproduction.

Table 2: Classification of Fungi

Division: Mastigomycota
Main characteristics: produce flagellated (motile) spores; cellulose cell walls

Class	Examples	Habitat	Description	Reproduction
	chytrids, water moulds	mainly aquatic, some terrestrial	• hyphae with cross-walls • some **saprophytes**, some parasites	• asexual and sexual

Division: Amastigomycota
Main Characteristics: produce nonmotile spores; chitin in cell walls

Class	Examples	Habitat	Description	Reproduction
Zygomycetes	common moulds (bread moulds, dung moulds)	mainly terrestrial (soil, decaying plant matter)	• few single-cell forms • mainly multicellular • cross-walls lacking in hyphae • some saprophytes	• asexual and sexual
Ascomycetes (sac fungi)	yeast, morels, truffles	terrestrial and aquatic	• few single-cell forms • mainly multicellular • cross-walls in hyphae • many are pathogens	• asexual and sexual
Basidiomycetes (club fungi)	mushrooms, shelf fungi	mainly terrestrial	• mainly multicellular • cross-walls in hyphae • many are pathogens	• sexual; asexual spores absent

Division: Deuteromycota
Main characteristics: known as imperfect fungi

Class	Examples	Habitat	Description	Reproduction
	parasitic fungi, *Penicillum,* and fungi that cause diseases (e.g., athlete's foot and yeast infections)	mainly terrestrial	• some resemble sac fungi • others resemble club fungi	• asexual by spore formation • sexual reproduction unknown

DID YOU KNOW?

The fungus *Armillaria ostoyae* is believed to be the largest organism in the world. The largest one found so far is in eastern Oregon and covers about 9 km². Because this species kills trees, it often can be found by searching for areas of dead trees.

sporangia: the reproductive structures in which spores are produced

germinate: to grow or sprout; refers specifically to the embryo inside a plant seed

vegetative: describes any part of a fungus or plant that is not involved with sexual reproduction

dikaryotic: describes cells that contain two haploid nuclei each of which came from a separate parent

10.2 Life Cycle of Fungi

A life cycle describes the development of an organism from a single cell to its reproductive stage. The cycle is complete when the organism produces the next generation. There is a wide variety of fungal life cycles. Fungi reproduce both asexually and sexually but always produce spores as reproductive cells. Spores have a haploid chromosome number and are produced in specialized structures called **sporangia** (singular: sporangium). Spores are usually dispersed by air currents. If spores are in a suitable environment, they **germinate**. Some fungi may also reproduce asexually by fragmentation, which is the breaking apart of portions of the mycelium.

Field Mushroom

Field mushrooms grow wild in fields, lawns, and gardens (**Figure 1**). They are the type of mushrooms most commonly cultivated for human consumption. The **vegetative** part of the mushroom, the mycelium, consisting of many hyphae, is usually just beneath the surface of the soil.

The reproductive phase begins when compatible haploid parent hyphae fuse and become **dikaryotic**. These new hyphae develop enlargements that increase in size until they break through the soil's surface and can be seen as

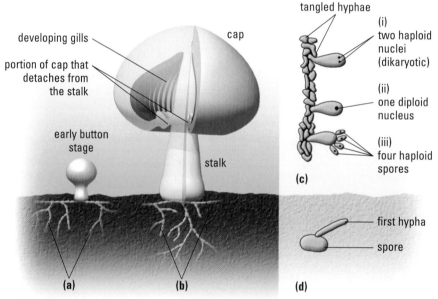

Figure 2
General life cycle of a field mushroom.
(a) An early button consisting of dikaryotic tissue created by the fusion of the haploid hyphae, not the nuclei, of two separate parents. **(b)** A longitudinal section through an immature field mushroom. It consists entirely of dikaryotic tissue because of the fusion of the haploid hyphae, not the nuclei, of two separate parents. When the lower rim of the cap detaches from the stalk, the cap opens like an umbrella to allow the gills to be seen. **(c)** A highly magnified view of a small section of one side of one gill consisting of tangled hyphae. Three extensions are shown at sequential stages of development: (i) an extension with two haploid nuclei, (ii) an extension with one diploid nucleus resulting from the fusion of the two haploid nuclei, (iii) an extension with four haploid spores resulting from meiosis of the diploid nucleus. **(d)** An enlarged view of a germinating haploid spore showing the first hypha protruding from the ruptured protective spore coat

Figure 1
Field mushrooms showing caps and stalks. Spore-producing cells line the dark mature gills on the undersurface of the mushroom cap.

small white spheres called buttons (**Figure 2**). As the buttons grow and mature, they form a stalk and a cap that remains spherical. Within this cap, many changes occur. Thin membranous gills are formed and radiate out from the stalk. The gills consist of many tangled modified hyphae. At first the gills are pink but darken as they grow and mature. As stated, all this tissue is dikaryotic. Extending from the gills are specialized extensions. In a process considered to be a sexual union, the two nuclei fuse in some of these extensions. The new diploid nucleus then undergoes meiosis to produce four haploid spores.

The underside of the cap detaches from the stalk and eventually looks like an opening umbrella. The maturing gills are now visible. As the spores mature, they are ejected into the spaces between the gills and downward toward the ground. However, these spores are small enough to be carried away by air currents even before they land on the ground. Spores are discharged from any one cap over a period of several days. It is estimated that over two billion spores may be produced from one average field mushroom. If a spore lands in a favourable environment, its protective coat will eventually split and the haploid cell will divide by mitosis to produce a new hypha. It will continue to grow and branch into a complex mycelium (**Figure 3**).

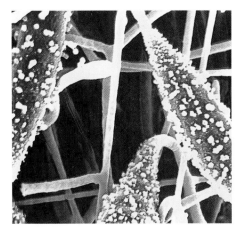

Figure 3
A scanning electron micrograph of a culture of the fungus Microsporum canis. The filamentous structures are hyphae.

Investigation 10.2.1

Monitoring Bread Mould Growth

Moulds usually reproduce by means of spores. Under certain conditions, mould will grow on bread and can be observed after several days. For this investigation, you and your group will prepare mould cultures to study their various structures.

Question

Which environmental conditions promote or inhibit the growth of mould on bread?

Design

Eight sealed containers, each containing a cube of bread, are subjected to various conditions. Observations of mould growth are recorded in a table and the possible types of mould grown are identified.

Materials

lab apron
eye dropper
paper towel
microscope
water
small metric ruler
cardboard box with lid
marker or ink pen
cubes of various types of bread
  (labelled by type)

petri dishes, or baby food jars with lids
masking tape
hand lens (or magnifying glass)
prepared microscope slides
  of bread mould
incubator or warm radiator
refrigerator or other cooling
  appliance or cool location
lamp or battery-operated lamp

INQUIRY SKILLS

○ Questioning   ◉ Recording
○ Hypothesizing   ◉ Analyzing
○ Predicting   ◉ Evaluating
○ Planning   ◉ Communicating
◉ Conducting

Any student allergic to airborne particles should inform the teacher immediately.

Procedure

1. Put on your lab apron.

2. Line the bottom of eight petri dishes or jars with one layer of barely moist paper towel.

3. Place one cube of bread on the paper towel. Use the same type of bread for all containers and record that type.

4. Label each container with your group number and the type of bread used. Also number your containers from 1 to 8.

5. Use an eyedropper to moisten (but not soak) the bread cube in containers 1, 2, 3, and 4 with water. Pour off any water that may have pooled on the bottom of these containers.

6. Allow all the bread cubes to stand for 20 to 30 min in the open containers.

7. After 20 to 30 min, cover the petri dish and tape the top and bottom of the petri dishes together or put the lids on the jars. DO NOT OPEN these containers again. Try to avoid bumping, tipping, or upsetting them as you place them according to the instructions in step 8.

8. Store all containers for 4 to 7 days as follows:
   (a) Store containers 1 and 5 in a warm incubator with a light source or near a warm radiator with a lamp turned on for at least 10 h per day.
   (b) Store containers 2 and 6 in a warm incubator with no light source or near a warm radiator in a closed cardboard box so they receive no light at all.
   (c) Store containers 3 and 7 in a refrigerator with a battery-operated lamp or a cool place with exposure to light for at least 10 h per day.
   (d) Store containers 4 and 8 in a refrigerator or other cool place in a closed cardboard box so that they receive no light at all.

9. Make daily observations without opening the containers. Try to avoid bumping, tipping, or upsetting the containers as you take them to and from their storage site. Be sure to return the containers to the same storage site. Record your observations in a table in your notebook. For each container, include the approximate number of "spots" or colonies and also the approximate diameters of one or two of the larger colonies on each bread cube.

10. When the mould becomes quite dark or different colours are noticed, examine the bread with a hand lens WITHOUT OPENING the containers. Identify the following types of hyphae: sporangiophores, which develop sporangia at their tips; stolons, which run horizontal to the bread surface; and rhizoids, which develop at the swellings along the stolons and anchor the fungus to the bread (**Figure 4**).

11. WITHOUT OPENING the containers, make drawings of what you observe. Draw each bread cube, without its container, and indicate the location of any of the types of mould listed below.

    Note: Mould colours may sometimes indicate different mould types. For example, common black bread mould is *Rhizopus nigricans,* blue-green mould is *Penicillium,* brown mould is *Aspergillus,* and grey mould is *Mucor.* Also draw enlarged views of any of the parts you identified in step 10.

12. Obtain a prepared microscope slide of bread mould. Examine the slide under the microscope. Make drawings of parts listed in step 10, plus the spores.

13. When the experiment is finished, give the closed containers to the teacher for proper disposal.

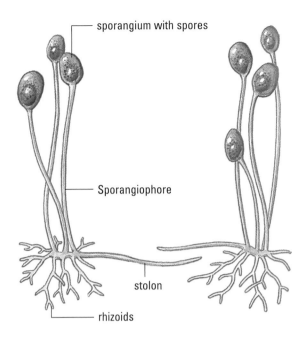

sporangium with spores

Sporangiophore

stolon

rhizoids

Figure 4
Main structures of bread mould

Analysis

(a) Using the colony count of the colony diameter measurement, determine the growth rate of each mould over several days.
(b) Using your observations about the mould colours, identify the types of mould that grew on your bread.
(c) List your numbered containers in order from the one with the most mould to the one with the least at the end of the observation period.
(d) Which conditions favour mould growth?
(e) Which conditions seem to inhibit mould growth?
(f) What is the control in this investigation?

Evaluation

(g) Why was it necessary to keep the containers sealed shut? Explain what results you might have observed or what results might have been different if the containers had not been kept shut.
(h) Compare your results with a group who used a different type of bread. Suggest reasons for any differences.
(i) What conclusion(s) could you make if no mould formed in any the containers?
(j) List possible sources of error. Suggest improvements to the design.

10.3 Importance of Fungi

Many of the roles of fungi appear to be destructive, at least from a human viewpoint. However, fungi are vital to other organisms and to the proper functioning of ecosystems through their role as decomposers. Fungi are part of nature's recycling system. Fungi, along with Eubacteria, transform complex organic substances into raw materials that other fungi and plants use for growth and development. Plants also use these recycled materials to produce new supplies of energy-rich organic compounds (**Figure 1**).

Figure 1
Fungi are important decomposers because their mycelia, often invisible, break down and absorb nutrients from their organic substrate.

Some fungi benefit humans. Yeast is used to make bread, wine, and beer; *Penicillium* produces an antibiotic; *Aspergillus* is used to flavour soft drinks. Other fungi, such as mushrooms, morels, and truffles, are sought-after food items. Certain fungal species are potentially useful in decomposing harmful pollutants. Research is continuing into ways of using fungi to break down some complex hazardous chemicals in toxic dumpsites and waste-water treatment plants.

Fungi and plants have helped to shape the biosphere as it exists today. Together they have paved the way for other terrestrial organisms.

Fungus Symbionts

Many fungi are involved in a **symbiotic relationship** with another organism. For example, **mycorrhizae** (singular: mycorrhiza) are symbiotic relationships between the hyphae of certain fungi and the roots of many specific plants. In this symbiotic relationship, the extensive fungal hyphae help the plant absorb nutrients, such as phosphorus. In return, the fungi obtain nutrients such as sugars that the plant has made during photosynthesis. Some plant seeds will not germinate in the absence of the mycorrhizal fungi. For that reason, environmental biologists often monitor mycorrhizal fungal growth in contaminated soils to see whether plant growth can occur in these soils. It has even been suggested that the evolution of land plants has been dependent on the presence of mycorrhizae.

Lichen is the name given to a combination of green algae or cyanobacteria and a fungus growing together in a symbiotic relationship. Lichens familiar to Canadians are reindeer moss and British soldiers (**Figure 2**). In the arctic tundra and boreal forests of northern Canada, lichens are an important source of food for caribou and other animals. They can be found on soil, rocks, and trees. Lichens are important in plant succession. By being able to establish themselves on rocks and in barren areas, lichens help form basic soil material.

Lichens are really two organisms in one. The photosynthetic partner is generally a cyanobacterium or a green alga, and the fungus is usually a sac fungus. In a lichen relationship, the fungal mycelium surrounds the photosynthetic cells and provides them with carbon dioxide and water for photosynthesis. The mycelium also lends structural support to the entire organism. The autotrophic organisms share the carbohydrates they manufacture with the fungi. Lichens reproduce asexually by fragmentation.

Lichens: Air-Quality Monitors

Since the Industrial Revolution, researchers have recognized the value of lichens in detecting air pollutants. Unlike plants which absorb water that has been filtered through the ground, lichens absorb water directly from the air. Thus, they absorb more dissolved toxic substances than do plants, which allows

symbiotic relationship: a relationship between two organisms in which both partners benefit from the interaction. Some of these relationships may be necessary for the survival of the partners while others benefit both partners but are not necessary.

mycorrhizae: symbiotic relationships between the hyphae of certain fungi and the roots of many specific plants

lichen: a combination of green algae or cyanobacteria and a fungus growing together in a symbiotic relationship

(a) (b)

Figure 2
Lichens familiar to Canadians are **(a)** reindeer moss and **(b)** British soldiers.

them to be sensitive indicators of pollutants. However, as was the case with the radioactive fallout from the Chernobyl nuclear power-plant disaster, identifying contaminated areas usually occurs after the environmental devastation has taken place. Often the detection method is based on which lichens survive and which ones die.

More recently, lichenologists (scientists who study lichens) have been attempting to determine the potential of using lichens for monitoring air quality to indicate problems in their early stages. Scientists are also hopeful that lichens may serve as indicators for a wide range of airborne contaminants.

Try This Activity

Collecting and Examining Lichens

Materials: ruler, small plastic bags, hand lens, prepared slide of lichen, microscope

- Use the edge of a ruler to carefully scrape lichens from various trees and rocks. Try to collect the scrapings from separate lichens in separate small plastic bags.
- Examine the lichen pieces under a hand lens and classify them according to **Table 1**.
- Examine a prepared slide of a lichen using a light microscope.
  (a) How are you able to distinguish the photosynthetic partners from the fungi?

Table 1: Appearance of Lichens

Lichen type	Description	
crustose	flat or crusty, forms a matlike structure on rocks or tree bark	
foliose	leaflike lobes, spreads outward, has a paperlike appearance	
fruticose	raised structure with stalks, multiple branching threads, and may hang from trees	

Frontiers of Technology: Fungal Mimicry

Modern technology and new scientific knowledge are being applied to the study of fungal species that display mimicry. The results of the studies might offer information that could benefit humans.

The fungus *Monilinia* has consistently devastated the valuable blueberry fruit crop. The organism—which damages leaves and shoots and mummifies blueberries into hard, white, inedible kernels—causes more than $100 million in annual losses worldwide. Traditionally, it was thought that the fungal disease spread from one bush to another by airborne spores. However, current research, including a closer and more careful examination of the diseased bushes, has uncovered some surprises as to how the spores are dispersed.

Visiting flies and bees have been observed to lick the fungal spores off the blueberry leaves as if they were feeding on flowers. Recently, scientists using ultraviolet light, which is visible to insects, discovered that both lesions caused by the fungus and blueberry flowers appear as bright points against a dark background. Additional research showed that the sugars produced by the lesions are the same as those found in the nectar of the blueberry flowers. It also showed that the fungus exudes the same tealike odour as the flowers. As the insects feed on what they sense as being nectar, they accidentally collect fungal spores, which they then carry to healthy plants that they visit next. With this new knowledge about the relationship between the insects, the fungal lesions, and the blueberry flowers, scientists are developing techniques to prevent the spread of the fungus without altering the role of the insects as pollinators.

Practice

Understanding Concepts

1. What are some similarities and differences between fungi and plants?
2. Describe the roles of fungi as decomposers and as parasites. Give examples.
3. For each of the following, provide a definition and describe the function:
   (a) mycelium   (b) hyphae
   (c) spores   (d) sporangia
4. What is the dikaryotic stage of a fungus life cycle?
5. Provide the common names used to describe three classes of fungi and give an example of each.
6. Describe the role of fungi in each of the following:
   (a) mycorrhizae   (b) lichens
7. Refer to **Figure 2** in section 10.2 to answer the following questions about the life cycle of field mushrooms:
   (a) Exactly where do the haploid spores form and by what process?
   (b) Describe when the dikaryotic stage begins.

Research in Canada: Dr. Faye Murrin

Dr. Faye Murrin of Memorial University in Newfoundland is a cell biologist and mycologist (fungus specialist). Using such techniques as fluorescence microscopy, in which fluorescent dyes are used to tag specific molecules, and electron microscopy, which can magnify images up to two hundred thousand times, Dr. Murrin investigates the functions and mechanisms of the cytoskeleton in cells. The cytoskeleton, a hallmark of eukaryotic cell evolution, is a versatile network of fine protein filaments and microtubules. The cytoskeleton controls cell shape, maintains intracellular organization, and is involved in cell movement.

Dr. Murrin uses the fungus *Entomophaga*, which produces an insect pathogen, as a model system in which to study the cytoskeleton. An understanding of *Entomophaga* can help explain cytoskeleton function during infection. *Entomophaga* is also a potential biological control agent and an alternative to chemical sprays for combating forest pests such as the spruce budworm and hemlock looper.

Dr. Murrin has also set her sights on large fungi and is collaborating on a study of the potential for wild mushroom harvest as a small-scale industry in Newfoundland, a venture that has proved successful on Canada's west coast.

SUMMARY Fungi

1. Fungi do not photosynthesize.
2. Fungi can reproduce sexually and asexually but always produce spores.
   - Spores are haploid and are produced in sporangia.
   - If a spore germinates, it produces hyphae.
3. Fungi are very important to ecosystems because of their role as decomposers.
4. Fungi may have symbiotic relationships with plant roots to form myccorhizae.
5. In mycorrhizae, fungal hyphae help the plant absorb nutrients; the plant shares its carbohydrates with the fungus.
6. Fungi may also have symbiotic relationships with cyanobacteria or green algae to form lichens.
7. In lichens, fungal mycelia provide structural support and carbon dioxide and water to the autotrophic partner; the autotrophic partner shares carbohydrates it has manufactured with fungi.
8. Lichens are important in natural succession.

Sections 10.1–10.3 Questions

Understanding Concepts

1. What defining features are missing that prevent fungi from being included in the plant kingdom?
2. Predict what would happen to an ecosystem if fungi began to disappear.
3. In what ways may fungi be useful to other living things?
4. In what ways may fungi be harmful to other living things?
5. What role do lichens play in the establishment of plants on barren lands?
6. Explain why lichens are often able to survive in environments considered to be too severe for plants.

Making Connections

7. Why do bakers add mould inhibitors to baked goods?
8. What environmental conditions would you recommend to prevent mould growth on foods?
9. (a) What features are most desirable in fresh food items such as lettuce, carrots, apples, and oranges?
   (b) What connection exists between the size, shape, and colour and the taste or nutritive value of fresh foods?
   (c) What dangers might exist if we select fresh food items based on a very narrow set of characteristics such as size, shape, and colour?

10.4 Kingdom Plantae

Plants are organisms that lack mobility. Their cells are eukaryotic, have numerous organelles, and have cell walls that contain cellulose. Most plants have the ability to synthesize carbohydrates through the process of photosynthesis. To review the similarities and differences between plants and fungi, reread **Table 1** in section 10.1.

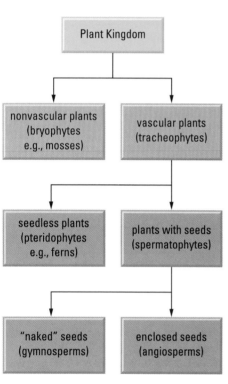

Figure 2
The major divisions of Kingdom Plantae

nonvascular: to be without the conductive tissues found in vascular plants. Nonvascular plants are also referred to as bryophytes.

vascular: describes the system of conductive tissue (xylem and phloem) found in plants to transport water and dissolved materials throughout a plant. Vascular plants are referred to as tracheophytes.

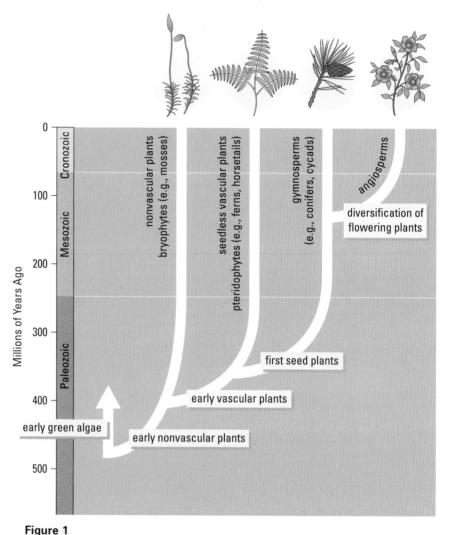

Figure 1
Phylogeny of plants. Of the three algal groups, early green algae gave rise to early **nonvascular** plants which, in turn, gave rise to present nonvascular plants, such as mosses and liverworts, and early **vascular** plants. Early vascular plants gave rise to current vascular plants and the first seed plants. The first seed plants gave rise to the gymnosperms. Flowering plants, the angiosperms, have arisen most recently of all the plants.

The history of life on Earth has been dramatically influenced by the success of plants as they have diversified and established themselves in the terrestrial environment. **Figure 1** illustrates the relationship among the major divisions of plants and the increasing success of Earth's major plant taxa that has taken place over geological time. The three hundred thousand to five hundred thousand species of identified plants display such diversity that no single species can be cited as a typical example of the kingdom. Consequently, scientists have devised a number of classification schemes for plants. Refer to **Figure 2** to review the major divisions considered in this text.

Practice

Understanding Concepts

1. List five characteristics shared by plants.
2. What does it mean if a plant is nonvascular?
3. What is a simpler name for tracheophytes?
4. What are two related functions of vascular tissue?
5. Name the two types of vascular tissue.
6. What is the difference between gymnosperms and angiosperms? Give one specific example of each.

10.5 The Evolution of Terrestrial Plants

As the environment changes, populations that are suited to their new surroundings survive. Those that are not suited become extinct. This simple fact demonstrates how important it is to have a variety of different organisms in any natural community. It is reasonable to assume that organisms existing in large numbers at any given time are well suited to their environment (if they were not, there would not be so many of them in the population). We can also assume that environmental conditions change with time. On a very large time scale, ice ages alternate with temperate periods, while on a smaller scale, a particular species may have to contend with an infestation of disease-causing organisms. A change in the environment will require a species to change its characteristics and pass newly acquired traits on to its offspring. However, we know that individuals cannot change their inheritable characteristics at will (curling straight hair or tanning pale skin will not produce children with curly hair or tanned skin). Long-term survival is only possible if, by chance, there exists a group of individuals in the population with the characteristics needed to survive in the new environment. Thus, having individuals with a variety of inheritable traits in a population increases the likelihood that the species will survive changes in environmental conditions. A population of identical individuals (i.e., clones) stands little chance of survival in the face of environmental change. If a thick coat of hair is needed to survive the next ice age, then a population comprised entirely of hairless individuals would certainly become extinct. Extinction would not happen immediately, but over a number of generations as the poorly adapted individuals' numbers dwindle and the adapted individuals' numbers increase. Evolution is a very slow process, taking hundreds, even thousands of generations to become noticeable.

The colonization of the world by terrestrial fungi and plants more than 400 million years ago represents one of the major events in the evolutionary history of life (**Figure 1**). Before this time, our world looked quite different than it does today. Most of our planet was covered by water, and any dry land was barren and inhospitable. Organisms existed only in seas and freshwater ponds, where they were protected from the harsh conditions of the earth's atmosphere. These conditions prevailed for almost 90% of the time that life has existed on our planet.

The long, slow evolutionary transition from aquatic life to life on land took millions of years of environmental changes, accompanied by the adaptation of species to the changing terrestrial conditions. Again, remember that there was no intentional movement to drier conditions. Members of many species simply got washed up during storms or regular tidal movement. Those species that had the

Figure 1
A model and diagram of one of the earliest known terrestrial plants called *Cooksonia*. It grew along lakeshores more than 415 million years ago. Its branched, upright stem, less than 7 cm tall, contained vascular tissue but lacked leaves. The tips of some of its branches looked similar to fungi because they bore sporangia.

innate ability to survive, even with great difficulty, began to colonize the land. Those species that could not survive did not.

The earliest species that washed ashore may also have included animal species, but since animals are heterotrophs, they did not survive because they had no source of food. Plants led the way to all terrestrial life because they are autotrophs and are not dependent on other life forms for nutrients. Plants transformed the landscape. As some plants died, they contributed to the formation of richer soil. As other plants lived and grew tall, they shaded the soil, thus changing the microclimate by lowering the temperature and raising the humidity under them. The roots of these plants helped to stabilize the soil by reducing erosion by wind and water. The plants also created habitat and nutritional opportunities for herbivorous animals and their predators. Early plant life, while enjoying many advantages offered by a land existence, also encountered a series of challenges.

Many plants that are successful in aquatic environments simply cannot survive on land. What makes success on land so difficult? To be successful on land, plants must have structures to help keep them upright in air because air does not provide the buoyant support of water. Since they are not constantly bathed by water bringing nutrients and removing wastes, they must have conducting tissue to transport various materials through their bodies that are often quite large (**Figure 2**). Finally, successful land plants must have a method of reproduction that does not depend on water.

As previously discussed, in an aquatic environment, water and nutrients bathe the plant. To be successful on land, plants must have ways to actively obtain, transport, and distribute water, dissolved minerals, and products of photosynthesis to the entire plant body—something that was accomplished by the earliest forms of vascular tissue. The earliest successful terrestrial plants did not have well-developed conducting vessels, but they did have enough special tissue that materials could be transported throughout their bodies, even if somewhat inefficiently. The earliest terrestrial plants did not have complex structures such as roots, stems, and leaves, but they did manage to survive. Gradually, those species and members within a species that had the best conducting systems became the most successful terrestrial plants on land. The gradual evolution of roots, stems, and leaves helped plants overcome some of the problems of life on land. One early development was the evolution of tiny rootlike structures called **rhizoids**. Roots and stems are more recent developments than rhizoids and the holdfast and other analogous parts of seaweeds (**Figure 3**). Leaves gradually evolved as the primary organs for photosynthesis, allowing plants to make carbohydrates more efficiently. A vascular system consists mainly of **xylem** tissue (mostly dead cells that form microscopic

rhizoids: hairlike structures that function like tiny roots and are found on the lower surfaces of certain parts of mosses, ferns, and other small organisms. They probably evolved before true roots with vascular tissue.

xylem: a vascular tissue in plants that carries water and dissolved materials up from the roots to the other plant parts

Figure 2
Ferns were among the first terrestrial plants.

(a) seaweed holdfast   **(b)** terrestial plant root

Figure 3
Structures for anchoring plants

tubes to convey water and dissolved materials up from the roots) and **phloem** tissue (mainly living cells that transport sugars produced by photosynthesis throughout the plant, including down to the root for storage). Bryophytes do not have specialized xylem and phloem tissue and are considered nonvascular plants. Bryophytes tend to be small organisms and many can only survive in very moist environments. They do have some tissue capable of transporting dissolved materials for very short distances, but it is not nearly organized enough to be considered true vascular tissue.

Being exposed to air also means that terrestrial plants must prevent excessive water loss by evaporation and must maintain proper exchange of gases between the plant's tissues and the air. The evolution of a waxy covering, called a **cuticle**, over the stem and leaves assists many plants in retaining water. The gradual development of microscopic pores called **stomata** (singular: stoma), particularly in leaves, permits a controlled exchange of gases between the plant and the atmosphere.

Water helps transport the gametes and disperse the "young" of the next generation of aquatic plants. Terrestrial plants can only be successful if they have evolved other methods to transport their gametes and disperse their "young." Terrestrial plants have evolved a wide range of reproductive strategies or features including nectar, colourful flowers, pollen, fruit, spores, and seeds with protective coats.

In aquatic environments, environmental conditions do change but their range of change is relatively small compared to the huge fluctuations in environmental conditions that occur on land. To be successful on land, plants must have evolved structures or processes that allow them to live and reproduce through a vast range of extreme conditions.

The evolution of various terrestrial ecosystems has depended largely on the evolution of a great diversity of plant groups that managed not only to survive but also to thrive.

Figure 2 in section 10.4 indicates how biologists currently divide the plant kingdom. The remaining part of this chapter will help you understand how all these groups have evolved ways to solve the two biggest problems encountered by plants on land: transportation of materials throughout their bodies and reproductive strategies that are independent of an aquatic environment. It should be noted that there are still representative species of bryophytes, pteridophytes, and spermatophytes that live in aquatic environments. However, the focus of our studies will be those that are successful on land.

phloem: a vascular tissue that transports sugars, which were synthesized in the leaves, throughout the plant and down to the root for storage

cuticle: in plants, a layer of noncellular material secreted by epidermal cells designed to protect cells from drying out

stomata: pores in the epidermis of plants, particularly in leaves. They permit the exchange of gases between the plant and atmosphere while at the same time helping to prevent excessive water loss.

The Evolution of Terrestrial Plants

1. Different species have different abilities to cope with different environmental conditions.

2. Different members of the same species have different abilities to cope with different environmental conditions.

3. As plants began to survive on land, they altered the conditions by slowly creating soil and microclimates. Plants provided food for animals that were also evolving systems to cope with terrestrial life.

4. Some plant modifications:
   • supportive structures—stems and roots/rhizoids
   • anchoring structures—roots/rhizoids
   • transport systems to carry nutrients and water—vascular system
   • means to reduce water loss—cuticle
   • means to regulate gas exchange and reduce water loss—stomata
   • reproductive strategies for dry conditions—spores, pollen, nectar, colourful flowers, fruit, and seeds

Practice

Understanding Concepts

1. Describe environmental conditions on Earth before plants colonized land.

2. What were the two biggest problems encountered by the first plants that washed up on land?

3. List the three possible outcomes for any plant that washed ashore.

4. Explain why the cuticle and stomata are important for plant survival on land.

5. List six reproductive strategies or features that helped plants survive terrestrial life.

6. What is the difference between rhizoids and roots?

7. Why did plant life have to evolve on land before animal life?

8. Explain why diversity of species is critical for the success of any natural community, not just on land.

Making Connections

9. Suggest the major reason why potted plants often have difficulty surviving modern houses and apartments, even if they get plenty of light.

10. Suggest two types of geographical locations on Earth where many plants have difficulty surviving and explain why.

10.6 Alternation of Generations

Recall that growth and reproduction in living organisms involves mitosis and meiosis. In animals, diploid cells undergo mitosis to produce diploid cells. Meiosis, on the other hand, reduces the chromosome number by half, producing four haploid cells from one diploid parent cell. The haploid cells do not normally undergo further cell division. In plants as in animals, diploid cells undergo mitosis to produce additional diploid cells, and as in animals, certain diploid cells

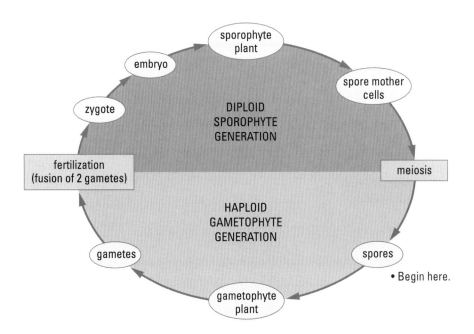

• Begin here.

Figure 1
Generalized life cycle with alternation of generations. In reality, the two generations are not usually the same size nor do they last the same length of time. Since it is a cycle, there is no specific beginning or end. The labels here will apply to all other plant life cycles in this chapter.

divide by meiosis to produce haploid cells. However, unlike animal cells, certain haploid plant cells divide by mitosis to produce more haploid cells.

The life cycle for all plants includes both haploid and diploid stages or generations. One of these generations produces spores and the other produces gametes. However, this is not an "either/or" situation. Both generations are necessary. Each of these stages or generations leads to the other. One complete life cycle is referred to as an **alternation of generations** and does not have a true beginning or end.

For this generalized introduction we will start with the spores. As you read about the life cycle, follow the names and the direction of the arrows in **Figure 1**. Pay particular attention to whether the part is haploid or diploid. Plant life cycles may seem complex but basically they are all the same.

Spores are haploid and are the first cells of the **gametophyte** generation. They grow by mitosis to produce all the haploid cells of various structures of the gametophyte plant. Eventually, specialized parts of these haploid plants produce haploid gametes. Two of these gametes (also known as sex cells), an egg and a sperm, fuse during the process of fertilization to form a diploid zygote. This zygote is the first cell of the **sporophyte** generation. The zygote grows by mitosis to form the embryo. The embryo matures into the mature diploid sporophyte plant. Eventually, **spore mother cells** are produced. These cells are also diploid because they were produced by mitosis. The spore mother cells are the last cells of the sporophyte generation because they undergo meiosis to produce haploid spores, which is where the story started. It is the two processes of fertilization and meiosis which change the chromosome number and cause the switch from one generation to the next.

What these cells and plant parts look like varies from one plant type to another, but the pattern of life is the same. Reread this description as you look carefully at **Figure 1** again.

Remember that the alternation of generations is sexual reproduction. There are male and female gametes. When fertilizaton takes place, more is happening than simply restoring the diploid number. Genetic material from two separate parents is combined to provide for limitless variety. It is this diversity within a species that allows it to survive a variety of environmental situations.

alternation of generations: refers to the complete life cycle of a plant, where the haploid stage produces gametes, and the diploid stage produces spores

gametophyte: refers to a stage in a plant's life cycle in which cells have haploid nuclei. This stage begins with the haploid spores. During this stage, the sex cells (gametes) are produced by mitosis.

sporophyte: refers to a stage in a plant's life cycle in which cells have diploid nuclei. This stage arises from the union of two haploid gametes. During this stage, spores are produced by meiosis.

spore mother cells: the last sporophyte cells. They undergo meiosis to produce the haploid spores.

Understanding Concepts

1. To which generation do spore mother cells belong?
2. To which generation do spores belong?
3. What process brings about the change from the gametophyte to the sporophyte generation?

SUMMARY **Alternation of Generations**

1. Plants have a life cycle in which a haploid form alternates with a diploid form.
2. The haploid form is called a gametophyte because it produces gametes.
3. The diploid form is called a sporophyte, which is a result of the union of two gametes.
4. Some cells of the sporophyte undergo meiosis to produce haploid spores.
5. Each spore develops into a haploid gametophyte.
6. The life cycle of a haploid stage followed by a diploid stage is referred to as an alternation of generations.

Sections 10.4–10.6 Questions

Understanding Concepts

1. What gave rise to the early vascular plants? Approximately when did this part of evolution occur?
2. What was the last major group of plants to evolve?
3. Explain why diversity within any one species is critical for the success of that species, on land and in water.
4. What are the two major divisions of the plant kingdom and what is the major difference between them?
5. Some people argue that the sporophyte generation is dependent on the gametophyte generation. Others argue that the reverse is true. Explain why neither idea is correct.

Making Connections

6. Suggest reasons why humans know more about animals than about plants.
7. Suggest reasons why humans know more about angiosperms than other types of plants.
8. Why might humans be surprised to discover that oak trees are angiosperms?
9. Why might fossil evidence for early plants be harder to find than for early animals?
10. Find at least two examples of plants that are not autotrophic. Also find out where and how they live.
11. Jewellery made with pieces of amber is very popular. What is amber? How is it formed? Where is it found? List some of the colours of amber that have been found.
12. What is petrified wood? How is it formed?

10.7 Mosses

Mosses form the largest and most familiar group of bryophytes. Mosses, like other bryophytes, are mostly restricted to swampy regions and other relatively moist environments. During a walk through the woods, you have likely encountered mosses in the form of low-growing, tightly knit clumps of green, velvety carpets. They also form a thin mat on rocks and tree trunks in very moist areas (**Figure 1**).

Mosses have evolved many of the same features that are found in other successful terrestrial plants, but they lack one important evolutionary development: a vascular system. Like multicellular algae and fungi, mosses depend on simple diffusion, osmosis, and active transport to move nutrients, wastes, and water to and from their cells. As a result, there are strict limits on their overall size. Unlike multicellular algae and fungi, the moss plant body of the gametophyte generation possesses structures that superficially resemble leaves, stems, and roots but are more simple structures that lack true xylem and phloem, the components of a real vascular system. What they do have are elongated cells in the middle of their stalks and thin leafy appendages that carry out photosynthesis. In place of the complex roots found in more recently evolved plants, mosses have simple rootlike filaments called rhizoids, which have only a limited capacity for absorption and anchorage.

Mosses and other bryophytes are pioneer plants. They are often the first to establish themselves on exposed rock or soil surfaces where there are no other plants. As the moss plants die and decompose, they create organic matter to act as soil and nutrients for other types of plants. Indeed, mosses pave the way for other terrestrial plants.

There are at least 1170 species of mosses in North America alone. It is necessary to use a hand lens and a microscope at different times during their life cycles to identify many of the species. Some species grow in many different geographical locations; others are found only in a limited area. To add to the confusion, many plants that are called mosses are not really mosses at all. For example, Spanish moss, often found dangling from trees in the southern states, is not a moss. In fact, it is a flowering plant that produces seeds. No moss ever has flowers or seeds. Some lichens are also mistakenly called moss. For example, reindeer moss is really a lichen. There is even a group of flowering plants called the moss pinks, which are referred to in this way partly because they grow in very compact low clumps.

(a)   (b)   (c)   (d)   (e)

Figure 1
Mosses are usually small plants growing in clumps or mats, often resembling thick green velvet **(a)** and **(b)**. Moss often grows with liverworts, another bryophyte **(c)**. Sometimes moss grows with fungi **(d)**. When moss grows on decaying tree stumps, it is often with a variety of lichens **(e)**.

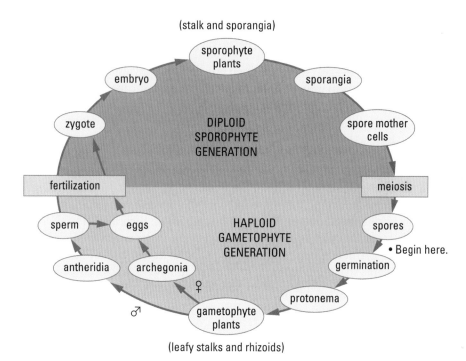

(stalk and sporangia)

Figure 2

Alternation of generations of a moss. Starting with the spores at the right side of the diagram, follow the cycle clockwise and note the labels carefully as you read the detailed description in the main text.

gemmae: the small clumps of haploid photosynthetic cells produced in little cup-shaped structures on the gametophyte plant. They are dispersed by splashes of rain to grow into other gametophyte plants.

protonema: the young gametophyte of a moss in the early stages after the germination of the spore

antheridium: the sex organ that produces male gametes in mosses and ferns

archegonium: the sex organ that produces female gametes in mosses and ferns

Life Cycle of Mosses

Some mosses and other bryophytes occasionally reproduce asexually by fragmentation or by the release of **gemmae** (singular: gemma) from a small structure which may appear on a gametophyte plant. The full life cycles of mosses and other bryophytes involve an alternation of generations as discussed in section 10.6. As mentioned there, the basic cycle is always the same. However, what the structures look like varies greatly from one species of moss to another. Follow **Figures 2** and **3** carefully as you read about the life cycle.

The tiny haploid spores that have been produced by meiosis are the first cells of the gametophyte generation. As they are released, they are usually carried by the wind or sometimes by water. If a spore lands in an environment suitable for its growth, the protective covering splits and the spore germinates. The single cell divides rapidly by mitosis. The resulting haploid plant is called a **protonema** and resembles a filamentous green alga. This tiny gametophyte plant continues to grow. Rhizoids are produced from its lower surface to help anchor it. In time, the protonema produces little buds that grow into larger gametophyte plants, which may grow upright or along the surface of the soil or rock. As these plants mature, sex organs are formed. In the case of upright plants, these structures are surrounded by the leaflets at the top of the stalks (**Figure 4**).

The male sex organ, called an **antheridium** (plural: antheridia) is tiny and shaped somewhat like an elongated balloon. Several of these organs are clustered at the top of one plant. Inside these organs, male gametes (sperm cells) are produced by mitosis. The female sex organ, called an **archegonium** (plural: archegonia), is also tiny and shaped like a bowling pin. Inside each of these organs, a female gamete (egg) is produced by mitosis. These sex cells are haploid, as are all the cells of this gametophyte generation.

When the sperm are mature, they are released from the antheridia. The mature eggs remain in the archegonia, which now produce a very sticky material. Transfer of the sperm from the male plants to the female plants can only occur if there is water. However, the plants are usually very close together and the water required can be as little as dew drops. Once in the vicinity of the archegonia, the

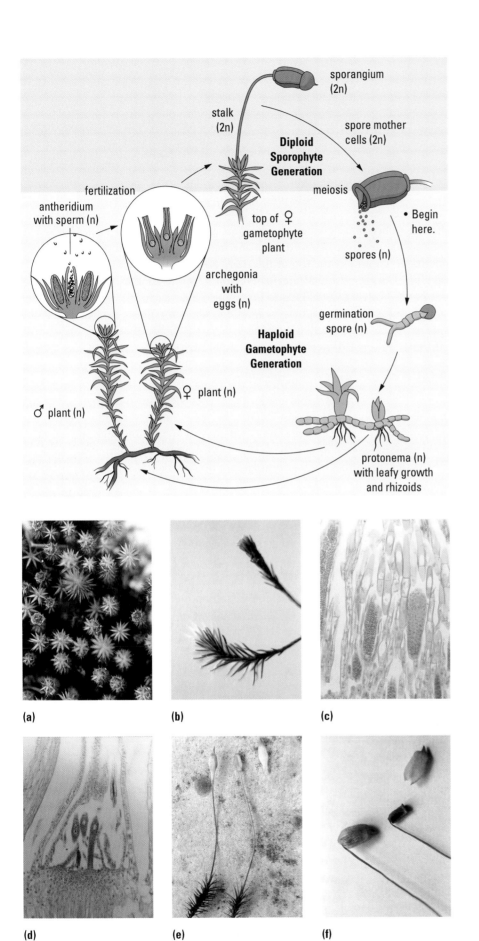

Figure 3

Alternation of generations of a moss. Starting with the spores at the right side of the diagram, follow the cycle clockwise and note the labels carefully. Compare these details with the text and with the simpler diagram, **Figure 2**.

(a)   **(b)**   **(c)**

(d)   **(e)**   **(f)**

Figure 4

Note: Not all the photographs are of the same moss species. The top of the male (cup-shaped) and female (star-shaped) gametophytes **(a)**; male and female gametophytes **(b)**; microscopic view through antheridia **(c)** and archegonia **(d)**; sporophyte growing out of the female gametophyte with "hat" still on the sporangium, and then removed **(e)**; closer view of sporangium, with and without its "hat" **(f)**

Figure 5
Because dead mosses are able to hold large amounts of water, they are used extensively by gardeners and in plant nurseries to condition the soil or keep roots moist during shipment. Moss is frequently used in recreational fishing to keep earthworms and other bait alive.

sperm cells are attracted to the sticky material and swim down the neck of the archegonia. Only one sperm fuses with the waiting egg. This fertilization marks the beginning of the new sporophyte generation.

The first cell of this new sporophyte generation is the diploid zygote that grows rapidly by mitosis to form the diploid embryo—still in the archegonium. The embryo continues to grow into the new sporophyte plant, remaining embedded in the archegonium, which supplies nearly all the nutrients for the growing sporophyte plant. As the embryo grows, it becomes visible as a thin, brown stalk rising out of the top of the female gametophyte plant. Sometimes the brown stalk seems to be wearing a "hat." This "hat" is the top of the old archegonium, which was torn off and rides up as the stalk grows. The stalk's total height is often equal to the height of the gametophyte plant supporting it. Gradually the top of the stalk, under the "hat," enlarges into a sporangium, inside of which are many diploid spore mother cells. Each spore mother cell undergoes meiosis to form four haploid spores, which are the first cells of the next gametophyte generation. The sporangia open and eject the spores, which are carried away by air currents. The stalk height increases the efficiency of spore dispersal.

Peat

Mosses play an important role in a variety of ecosystems but are also of significant commercial value to humans (**Figure 5**). One moss, sphagnum, once was extremely abundant. Dead portions of these vast sphagnum bogs are collected, dried, and sold as peat moss to be used in gardening and agricultural applications.

As the dead sphagnum accumulates, it becomes quite dense and can be cut and dug up in brick-shaped pieces. Dried peat has been used as a fuel source in

Examining Peat Moss

Peat moss is used in gardens and in potting mixes. Why?

Materials: 50 mL of peat moss, small beaker or glass of water, pH indicator paper, balance

- Determine the dry mass of the peat moss.
- Allow the peat moss to absorb water by soaking it overnight in the water.
- Pour off and collect the excess water.
- Measure the mass of the saturated moss.
- Calculate the ratio of final mass:dry mass.
- Using a piece of pH indicator paper, measure the pH of the decanted water.

(a) By what process do mosses appear able to obtain water from the environment?

(b) What beneficial effects would the addition of peat moss have on garden soil?

(c) What are some potential problems associated with using too much peat moss in garden soil?

(d) What are some ecological and economic benefits of peat moss? Follow the links for Nelson Biology 11, 10.7.

GO TO www.science.nelson.com

homes in Northern Europe and Asia for years. Even in Newfoundland, peat has been used as fuel for centuries. Newfoundland's peat resource is ranked as eighth in the world. Over millions of years, large peat deposits became compressed and chemically altered to form coal. Some people are encouraging an increased use of peat for home heating as well as for fuel for large industries such as pulp and paper mills. These proposed new uses for peat, along with an increase in serious gardeners, are creating large-scale demands on Earth's supply of peat.

Peat's tremendous absorption capability and antiseptic properties, because of its low pH, have also facilitated its use in "mopping up" oil spills, as surgical dressings for wounds during World War I, as natural diapers for babies in past rural and Native settlements, and in current feminine hygiene products.

SUMMARY Mosses

1. Mosses, the most common bryophytes, are nonvascular plants.
2. Transportation of water and nutrients occurs through elongated cells.
3. Both sporophyte and gametophyte generations occur on the same plant.
4. Only the gametophyte plants are photosynthetic.
5. Haploid spores are produced by meiosis in the sporangium.
6. A spore germinates and grows by mitosis into a protonema.
7. A protonema eventually grows into a mature gametophyte plant with sex organs (antheridia or archegonia).
8. Sperm transfer can only occur in moist conditions.
9. A zygote develops into a mature sporophyte.

Practice

Understanding Concepts

1. (a) Describe the general appearance, structures, habitat, and growth habits of bryophytes.
   (b) In what ways are bryophytes similar to vascular plants?
   (c) In what ways are they less evolved?
2. How are mosses able to transport materials throughout their body?
3. Draw a diagram to describe the stages of the moss life cycle. Use as many of the terms you learned as possible.
4. (a) In which generation of mosses are the cells haploid?
   (b) In which are the cells diploid?
   (c) Where in the life cycle does the reduction in chromosome number occur and by what process?
   (d) Where is the diploid condition restored and by what process?
5. Describe a moss protonema.
6. Explain at which point in the moss life cycle it is particularly important that the plants form mats of thousands of plants.

Applying Inquiry Skills

7. After performing some soil tests, a gardener determines that her soil is very sandy and dry and is slightly alkaline with a pH of 7.4. What action would you recommend to rectify this problem? Explain.

8. Peat is continually forming from dead plants of sphagnum moss. Suggest why promoting its use as an alternative to petrochemical fuels would be advantageous. Suggest why some environmentally concerned people are boycotting the purchase of peat moss.

Reflecting

9. Why is it important for humans to gain deeper understanding of the great diversity of plants and their wide range of characteristics?

10.8 Ferns

Ferns are tracheophytes, which are plants with a vascular system. The earliest vascular plants to evolve were the pteridophytes, of which ferns are the most well-known. The feature that enables most people to readily recognize a fern is the leafy portion of the plant called the **frond**. The fronds of ferns are a common sight in many forested areas. In fact, nearly everything we see of ferns is the fronds. A distinctive feature of many ferns is that their fronds mature from the base to the tip by gradually unfurling (**Figure 1**). The stems are called **rhizomes** and are usually horizontal at the surface of the soil. Ferns do not have flowers or seeds.

Ferns are extremely diverse in habitat, form, and reproductive method (**Figure 2**). In size alone, they range from 2 mm tall to huge tree ferns that grow to over 15 m in height. Some ferns are twining vines while others grow on or just under the surface of ponds. However, most ferns are plants living in shaded, moist areas in temperate or tropical regions. Their numbers diminish with increasing latitude and decreasing moisture. Dry, cold areas have few, if any, ferns. The greatest diversity of ferns is displayed in the tropical rain forests, where one hundred or more species may be identified in only a few hectares. Also, many

frond: the leaf portion of a fern

rhizomes: the barely visible stems of ferns

Figure 1
A few types of ferns are consumed by people in many parts of the world, including eastern Canada. Fiddlehead greens, the tightly furled fern fronds, are considered a delicacy and are collected in the wild or actually cultivated for food. You can find them in specialty shops in late spring and may even be able to find them in the frozen food section of your supermarket. Ferns are also used extensively as decorative indoor and outdoor plants and as part of many cut floral arrangements.

(a)   (b)   (c)

(d)   (e)   (f)

Figure 2
Various ferns: **(a)** holly fern; **(b)** maidenhair spleenwort; **(c)** hart's tongue fern; **(d)** walking fern; **(e)** royal fern; **(f)** common bracken

ferns grow on the trunks and branches of trees and are known as **epiphytes**. As with mosses, common names are not always meaningful. The asparagus fern and artillery fern are not ferns at all.

Ferns play an important role in ecological succession. Like mosses, ferns are pioneer plants. They grow on relatively bare, exposed rock surfaces and in open bogs and marshes. They are even able to grow in areas where there has been a forest fire or where there is a base of volcanic lava or ash. Although most ferns thrive in shaded, moist areas, the bracken fern, the best-known genus worldwide, is frequently found growing in old fields where it is warm and sunny. As the ferns grow year after year, they contribute to the organic material. Thus, they create environmental conditions that other plants require. These new conditions often have a negative effect on their own ability to survive. In addition, they often cannot compete with the plants that are now successful in the area. This sequence is the basis for all ecological succession.

There is a group of plants called fern allies—the club mosses, horsetails, and whisk ferns. They are also vascular, seedless plants and it is thought that they appeared on Earth at about the same time as ferns. However, they are not closely related.

Life Cycle of Ferns

A few ferns occasionally reproduce asexually by fragmentation of the rhizome. Some fern gametophytes can reproduce asexually by producing gemmae. The walking fern, found along the Niagara Escarpment, got its name because it "walks." It produces new plants asexually when the tip of a frond stays in contact with the soil and produces its own rhizome and fronds. The full life cycle of ferns involves an alternation of generations. As mentioned in section 10.6, the basic cycle is always the same. However, what the structures look like varies greatly among fern species. Follow **Figures 3** and **4** (page 388) carefully as you read the details that follow.

The tiny haploid spores that have been produced by meiosis are the first cells of the gametophyte generation. As they are released, they are usually carried by the wind. If a spore lands in an environment suitable for its growth, the protective covering splits and the spore germinates. The single cell divides rapidly by mitosis. The resulting haploid plant is called a **prothallus** (or prothallium). This thin, green, heart-shaped gametophyte is about the size of the fingernail on your baby finger. A cluster of rhizoids grow from the underside of the prothallus. Also on the underside of the prothallus are found spherical antheridia, in which sperm are produced by mitosis, or flask-shaped archegonia, in which eggs are produced by mitosis. Whether the male and female sex organs are on the same or separate gametophyte plants depends on the fern species.

Like mosses, mature fern sperm are released from the antheridia but require moisture to help transfer them to the female sex organs, the archegonia. Following fertilization, the diploid zygote, the first cell of the sporophyte generation, grows by mitosis into an embryo that continues to grow. The tiny immature sporophyte plant produces small roots to absorb water and minerals and a tiny frond that can photosynthesize even before it has fully unfurled and reached its full size. A rhizome is also produced, which grows laterally and produces more fronds and roots. The prothallus withers and dies.

The mature sporophyte frond bears clusters of sporangia on its lower surface. One of these clusters is called a **sorus** (plural: sori)(**Figure 5**, page 389). In some fern species, the sporangia develop on special separate fronds with a distinctive form and colour. Many species of fern can be identified by the distinctive patterns formed by the sori. Diploid spore mother cells are produced by mitosis inside each

epiphytes: plants that grow on the stems and branches of other plants but continue to photosynthesize

prothallus: the gametophyte plant of ferns

sorus: a cluster of sporangia on a fern frond

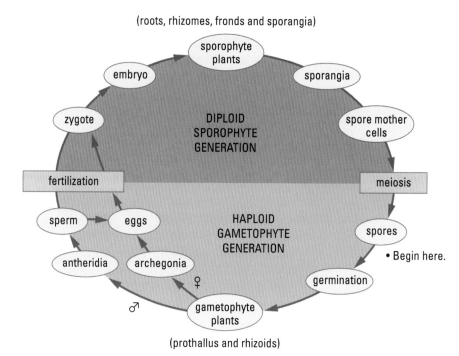

(roots, rhizomes, fronds and sporangia)

(prothallus and rhizoids)

Figure 3
Alternation of generations of a fern. Starting with the spores at the right side of the diagram, follow the cycle clockwise and note the labels carefully as you read the detailed description in the main text.

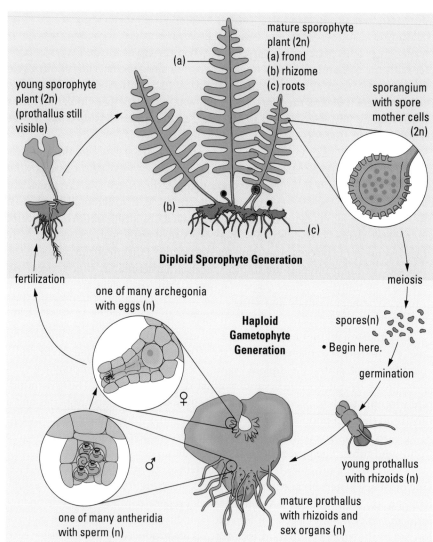

Figure 4
Alternation of generations of a fern. Starting with the spores at the right side of the diagram, follow the cycle clockwise and note the labels carefully as you read the detailed description in the main text and review the simpler diagram in **Figure 3**.

Figure 5
The sori found on the undersurface of fern fronds produce thousands of dustlike spores.

sporangium. Each of these undergoes meiosis to form four haploid spores, the first cells of the next gametophyte generation. When the spores are mature, the sporangia use a variety of mechanisms to eject the spores. One frond alone can release many thousands of spores, which will be carried away in the wind.

SUMMARY **Ferns**

1. As tracheophytes, ferns have a vascular system.
2. As pteridophytes, ferns do not have flowers or seeds.
3. The gametophyte plant, the prothallus, has rhizoids. It also produces eggs and sperm by mitosis in the archegonia and antheridia, respectively.
4. Sperm transfer can only occur in moist conditions.
5. Fertilization produces a zygote, which is the first cell of the diploid sporophyte generation.
6. The prothallus withers once the sporophyte plant matures.
7. Plants of both generations can photosynthesize.
8. The familiar part of the fern, by which we identify it, is the frond (leaf) of the sporophyte generation.
9. Sporangia, found in clusters called sori on the undersides of fronds, contain spore mother cells, which undergo meiosis to produce haploid spores.
10. Spores are the first cells of the gametophyte generation.
11. Spores germinate to produce a haploid prothallus.

Practice

Understanding Concepts

1. (a) What is the major difference between bryophytes and tracheophytes?
   (b) Discuss the advantages provided to ferns by this difference.
2. What is a pioneer plant and why are bryophytes and ferns considered pioneer plants?
3. Although most ferns are terrestrial, during what part of their life cycle do they depend on water?
4. Draw a diagram to describe the stages of a fern life cycle. Use as many of the terms you learned as possible.
5. (a) What generation of ferns do we normally see growing in the woods?
   (b) What specific part of that generation is the most noticeable?

6. Where exactly in the fern life cycle does the diploid number get restored?

7. Describe a fern prothallus.

8. Why is it significant that both generations of a fern are photosynthetic?

Making Connections

9. The vascular tissue of ferns is strengthened by a special chemical called lignin. What extra advantage do ferns have because of this chemical?

10. In the 1990s, Ontario experienced a number of record-breaking hot, dry summers.

   (a) What impact might this unusual weather have on moss and fern populations?

   (b) What broader long-term effects might these moss and fern changes have on the natural communities?

11. A group of people were volunteering for a planting event to help regenerate a wooded area. During the process of sorting the plants, one careless person broke a number of the fern rhizomes. To hide his carelessness, he put all the broken pieces into a garbage bag. A friend saw what he had done, pulled out all the bits, and planted them. Explain why you think this second person was, or was not, wasting her time and energy.

10.9 Seed Plants

Seed plants are called spermatophytes, a subdivision of the tracheophytes or vascular plants. They have roots, stems, and leaves exhibiting a huge variety of textures, sizes, shapes, odours, and colours. Spermatophytes are the most recently evolved plant group. They have reproductive structures that are not dependent on water. The major reproductive strategy that sets spermatophytes apart from all other plants is the seed, which contains a plant embryo, a partially developed plant. A seed provides protection for the embryo. Spermatophytes are the most widely distributed and complex group of plants on Earth today. The ancestors of the more than 270 000 known seed plants first appeared about 370 million years ago during the Devonian period.

(a) (b) (c) (d)

Figure 1
Examples of gymnosperms **(a)** and **(b)**, and angiosperms **(c)** and **(d)**. White cedar **(a)** and white pine and eastern hemlock **(b)** are all native to Ontario. White cedar can grow in a variety of locations. White pine and eastern hemlock often grow together and require more specific conditions than white cedar. Elm trees **(c)** and poison ivy **(d)** are both angiosperms, even if their flowers are not obvious.

Some spermatophytes can reproduce asexually by a variety of mechanisms, but their main method of reproduction is alternation of generations. Seed plants have separate male and female gametophyte tissue. Most seed plants are autotrophic but some are saprophytic or parasitic. The majority of seed plants live on dry land, but others, such as pitcher plants, orchids, and a few trees, thrive in wetlands. Other seed plants, such as duckweed, water lilies, and cattails, only do well when they are submerged or partially submerged in water. Seed plants range in size from the very small duckweeds to the giant redwoods and eucalyptus trees, which can grow more than 100 m tall. Vascular tissue, strengthened by special chemicals, allows spermatophytes to reach heights that are unattainable by nonvascular plants.

Whether or not the seeds are enclosed is the main criterion for distinguishing the two major groups of spermatophytes, **gymnosperms** and **angiosperms** (**Figure 1**). Gymnosperms produce unprotected, or "naked," seeds in conelike structures and are often referred to as conifers. Angiosperms produce seeds that are enclosed and protected inside a fruit, which is formed by various flower parts. All seeds ensure the survival of embryos by reducing excessive water loss.

gymnosperms: spermatophytes that produce "naked" seeds, usually inside cones

angiosperms: spermatophytes that produce seeds enclosed in fruit formed by certain flower parts

Gymnosperms

Gymnosperms include the pines, spruces, junipers, cedars, and other cone-bearing plants (**Figure 2**). Most gymnosperms have evolved thin, needlelike leaves, which help the plants resist the harshness of hot, dry summers, cold winters, and moderate rainfall. The needles are covered by a hard, waxy cuticle, which helps the plant retain moisture. Gymnosperms also have evolved roots that extend over a wide surface area, rather than penetrating deep into the soil. These roots anchor the tree even in locations where soil is scarce.

Gymnosperms form the basis of many large ecosystems by anchoring the soil and providing habitat and food for many animal species. Coniferous forests are also an important source of raw materials for a variety of commercial and industrial products. Economically, conifers are extremely important. They provide about 85% of all the wood used in building (spruce and cedar) and furniture (pine) construction. The pulp and paper industry uses millions of tonnes of

DID YOU KNOW ?

Graceful conifers were once worshipped by ancient Germanic tribes. These trees became the inspiration for the Yuletide song "O Tannenbaum" ("O Christmas Tree").

(a)

(b)

Figure 2
(a) Conifers are the best known gymnosperms, as seen here in northern Ontario.
(b) Many conifers are tall trees capable of withstanding weather extremes. Conifers are adapted to grow in areas with very little soil.

Figure 3
The bark of the Pacific yew was once the prime source of taxol, a cancer drug.

conifers annually, mostly for the production of newsprint and other paper products. Conifers provide people with other useful products—varnishes, turpentine, disinfectants, fuels, and medicines such as taxol. Conifers—and forests, in general—also help control flooding by absorbing rainwater through their roots and by anchoring topsoil, preventing erosion by water or wind.

Although many conifers grow back relatively quickly following a forest fire or timber harvest, some, such as the California redwood, Ontario white pine, and Newfoundland red pine, have been decimated by over-harvesting.

Frontiers of Technology: Taxol Research

Clinical tests in the 1980s of a new cancer drug called taxol sparked excitement among cancer researchers. Since that time, this "natural drug" has shown considerable promise in the treatment of a variety of cancers, particularly ovarian and breast cancer. In about 30% of women whose advanced ovarian cancers do not respond to other therapies, taxol has been shown to shrink the tumours significantly.

Initially, the prime source for taxol was the bark of the yew tree, *Taxus brevifolia*, a small understorey tree found in the old growth forests of the Pacific Northwest (**Figure 3**). Unfortunately, both the tree and the drug were in limited supply. Estimates suggested that it would take up to three of the 150- to 200-year-old trees to treat a single patient.

Driven by the high cost of this drug therapy (approximately $1000 per treatment cycle) and environmental concerns, laboratories around the world raced to develop synthetic and natural alternatives. Bristol-Myers Squibb quickly found a synthetic production method that started with the needles of European and Asian yews rather than the bark of the scarce Pacific yew. Needles are plentiful and harvesting them isn't fatal to trees. Furthermore, the taxol yield is high. This process was so successful that the drug quickly went from being very scarce to being the world's best-selling cancer drug!

Scientists at the University of Portland have recently extracted the same active ingredient found in taxol, paclitexal, from hazelnuts while researchers at Montana State University have discovered that taxol is also produced by a fungus, *Taxomyces andreanae*. Interestingly, the fungus was discovered in the phloem of the Pacific yew tree. Although the amount of taxol from hazelnuts (1/10 that of yew needles) and from fungus is low, improved culturing techniques and genetic engineering are expected to improve the yields and lower production costs.

Life Cycle of Gymnosperms

The pine tree will provide a general example of a gymnosperm's life cycle (**Figure 4**). A pine tree is the diploid sporophyte plant. In the spring, each tree produces two types of cones, neither of which looks like the woody, brown cones you have seen. The male cones, sometimes called pollen cones, are quite small and delicate and are found in clusters. Each male cone consists of many scales, each one with two sacs. In each sac, diploid microspore mother cells undergo meiosis to form four haploid microspores. Each of these develops into a haploid **pollen** grain, which is the male gametophyte. The female cones, sometimes called seed cones, are also quite small and somewhat sticky. They are often a pinkish-purple colour and are found singly or in groups of two or three. Each cone consists of many scales. On the upper side of each scale are two **ovules**. In each ovule, the megaspore mother cell undergoes meiosis but only one survives as a haploid megaspore, the female gametophyte.

When the pollen grains are mature, the tiny sacs of the male cones disintegrate and millions of dry pollen grains are released. The pollen grains have little flaps or wings that allow them to be carried easily by the wind. A parked car

pollen: the grains that contain the haploid male gametophyte in seed plants

ovules: the plant structures that contain the megaspore mother cell and, later, the single haploid megaspore, which is the female gametophyte

under a pine tree during this pollen release will accumulate a layer of yellow dust. The remnants of the male cones gradually dry up and fall off the tree.

The female cones are held by the tree such that the tip is pointing upward and the scales angle downward (see **Figure 5(d)**, page 394). When ripe, airborne pollen lands on the female cones of the same or a different pine tree, the sticky sap and angles of the scales ensure that the pollen moves toward the ovules. After **pollination**, the female cones become greenish, grow very quickly in size, and reorient themselves so that the tips are pointing downward (see **Figure 5(f)**, page 394). In some gymnosperms, fertilization may occur right away, but in pines it usually takes a year before fertilization occurs.

Fertilization, the union of the microspore and megaspore nuclei, produces the diploid zygote, which is the first cell of the next sporophyte generation. The zygote grows by mitosis to produce the diploid embryo, which remains inside the ovule. Now that there is an embryo, the ovule becomes a **seed**. It develops a seed coat which protects the embryo until there are suitable conditions for germination. Some gymnosperm seeds may develop within a few months, but most species take two to four years to mature. During this seed development, the female cones become brown and take on a woody texture. As they dry, the scales separate and the seeds fall out. Eventually the empty female cones fall to the ground. If conditions are suitable, the seeds germinate. After germination the tiny plant is called a seedling. Gymnosperms usually have to grow for many years before they produce male and female cones.

Note that the gymnosperm gametophyte generation is extremely tiny, both in size and duration, but it still creates variety in the next generation of plants to ensure survival under many different circumstances.

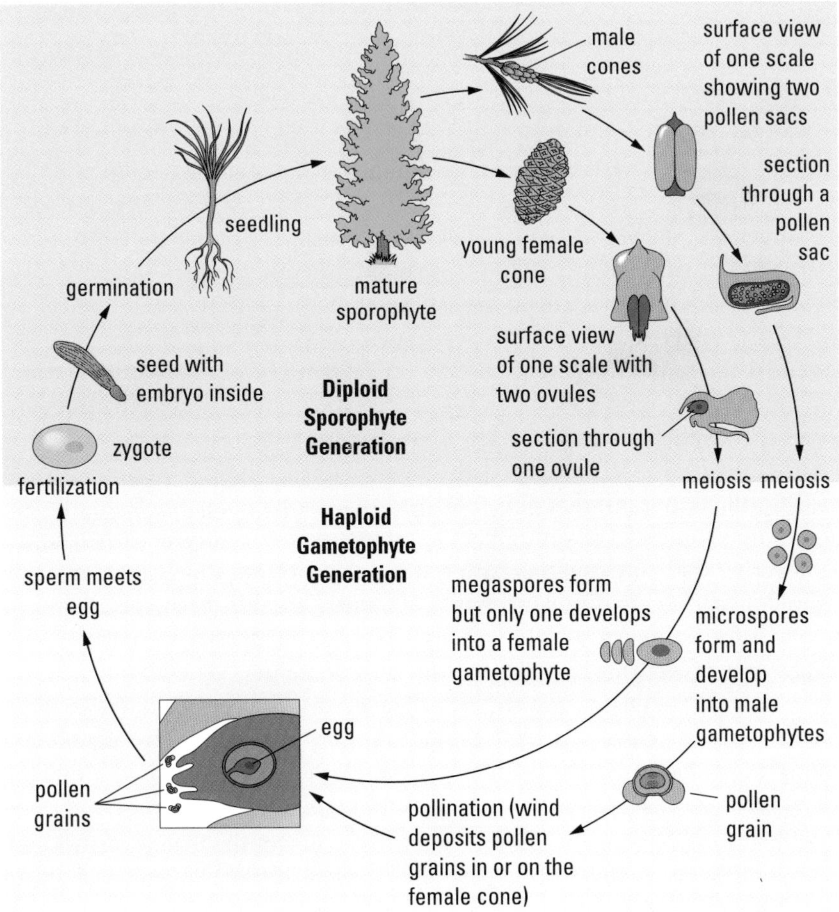

pollination: the transfer of pollen from the pollen-producing organs to the organs containing the female gametophyte

seed: an ovule after fertilization, containing an embryo which developed from the zygote

Figure 4
Alternation of generations of a pine tree. Starting with the mature diploid sporophyte tree at the top of the diagram, follow the cycle clockwise and note the labels carefully as you read the detailed description in the main text. Remember that even though the gametophyte generation is small, it is very important for maintaining diversity within the species. See **Figure 1** in section 10.6 if you need to review a general alternation of generations cycle.

(a)   (b)   (c)   (d)

(e)   (f)   (g)   (h)

Figure 5

Note: Not all the photos are of the same pine species.

(a) Mature male cones below new needle growth
(b) Microscopic section through two pollen sacs
(c) Pollen grains
(d) New female cones above new needle growth
(e) Microscopic section through one scale and ovule
(f) Female cones after pollination
(g) Year-old female cones below new needle growth and above last year's needle growth
(h) Two-year-old female cones open to release seeds.

Angiosperms

There are over two hundred and fifty thousand known species of angiosperms—more species of plants than in all the other plant divisions combined. New species are discovered almost daily. Angiosperms are all tracheophytes. Many can reproduce asexually by a broad variety of mechanisms, but they all reproduce by alternation of generations. The seeds they produce are enclosed in fruit formed by various flower parts. The sexual phase of alternation of generations allows genetic material to be recombined as a result of fertilization. Even the young produced by the same two parents can be very different from each other. Genetic variability within a species allows the species to survive various environmental forces it encounters. Young that are all identical may be resistant to attack from a particular disease. However, it is also possible that if they are not resistant, the species will disappear if that disease strikes. Genetic variability increases the possibility that some of the young have resistance and will survive. It is true that plants other than angiosperms have genetic diversity, but flowering plants have evolved a reproductive strategy that is extremely reliable. Each species has the potential for limitless diversity and also produces seeds that are nature's most reliable way of securing dispersal and survival of the next generation.

Many of the plants you see are angiosperms. Trilliums, trout lilies, raspberries, thistles, clovers, asters, goldenrod, and apple trees are all obvious examples of angiosperms. What many people do not realize is that grasses, wheat, corn, beans, plus trees such as elms, maples, and birches, are all flowering plants. In these last examples, the flowers are easily missed because they are not large and colourful, and the fruit is easily missed because it is brown and thin instead of fleshy.

Plant Diversity

Find out which plant groups are in your vicinity by sampling the plants in a specific area and then identifying and classifying them. Note that sampling does not necessarily mean collecting. Sampling also includes recording observations, photographing, and making sketches.

- Go to a natural area such as a pond, marsh, forest, or natural section of a local park.
- Use wooden stakes and string or yarn to outline several plots of equal size. Make one plot for each group in the class.
- In groups of two or more, survey the plants and fungi in your sampling plot. Photograph or sketch representative specimens and take careful notes on their appearance and any other information that will help you to identify them later. Also note how many individuals of each species are found in your plot.
- When you return to the classroom, refer to field guides, electronic media, or call your local naturalists' club to help identify each plant found.

  (a) What kinds of areas do you think would show the most diversity among the various plots and within each plot?
  (b) Why would you make the plots larger if there appears to be low diversity?
  (c) What species were in your plot? Compare the species found in your plot with those found in the plots of other groups. If all the plots contained generally the same species, suggest why. If the plots showed quite a lot of differences, suggest why. If some plots showed diversity and some did not, suggest why.
  (d) What taxonomic groups are represented in the area?

(a) roadside flowers

(b) male and female flowers on separate willow trees

(c) fragrant water lily

(d) common cattail flowers, male above female, on the same stalk

Figure 6
Various angiosperms

(a) black-eyed Susan

(b) common milkweed

(c) sand cherry

(d) wood lily

(e) yellow lady's slipper

(f) Indian paintbrush

Figure 7
Various flowers

Flowers

Angiosperms are flowering plants. No other division of plants has flowers. Flowers come in all shapes, sizes, designs, colours, and scents (**Figure 7**). These characteristics do not define a flower. The flower is the reproductive centre of the plant. Flowers, like the cones of gymnosperms, belong to the sporophyte generation, but within them meiosis occurs to produce the haploid tissue, which will take part in fertilization.

The flower of a typical angiosperm, the wood lily, contains both male and female reproductive organs and is called a perfect flower. Although the common parts of a perfect flower are shown in **Figure 8**, there is incredible variation in the

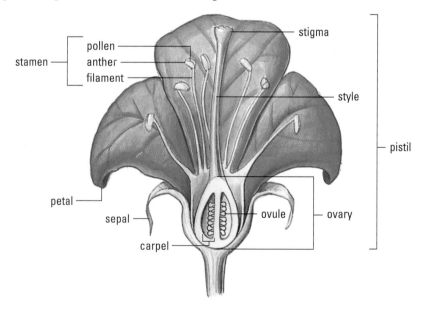

Figure 8
Features common to a typical perfect flower

shape, number, and arrangement of these parts (**Figure 11**, page 399). Some species of angiosperms have separate male and female flowers, and these are called imperfect flowers. Some species, such as pumpkins, have these imperfect flowers on the same plant. Other species, such as willows, have their imperfect flowers on separate plants.

The stamen is the male part of the flower. It consists of a thin stalk, called the filament, and an anther, in which pollen grains are formed. The pistil is the female part of the flower. It consists of three main parts: stigma, style, and ovary. The term carpel refers to a simple ovary or one subunit of a more complex ovary.

Sepals are often small, green, leaflike structures that enclose the developing flower. Petals are often large and colourful, but sometimes it is the sepals that are large and coloured. In many flowers, there is a tiny swelling deep inside the flower that contains a sugary liquid called nectar. The bright colours and sweet liquid often attract insects and some small birds, such as hummingbirds.

Monocots and Dicots

Angiosperms are grouped into two classes: the monocotyledons (monocots), which have one seed leaf, and dicotyledons (dicots), which have two seed leaves (**Figure 9**). A **cotyledon** is a seed leaf that stores carbohydrates for the young sporophyte and often becomes the first leaf to appear as the seed germinates. Common monocots are water lilies, onions, orchids, and grasses, including crop plants such as wheat, corn, barley, and rye. Dicots, the larger of the two groups, include maples, oaks, cacti, and crop plants such as peas, beans, potatoes, and cabbage.

cotyledon: a seed leaf that stores carbohydrates for the seedling and often is the first photosynthetic organ of a young seedling

Life Cycle of Angiosperms

The entire angiosperm plant, including the roots, stem, leaves, and flowers, belongs to the diploid sporophyte generation. In the appropriate season, the flower bud opens and the petals unfurl. The reproductive parts are revealed and,

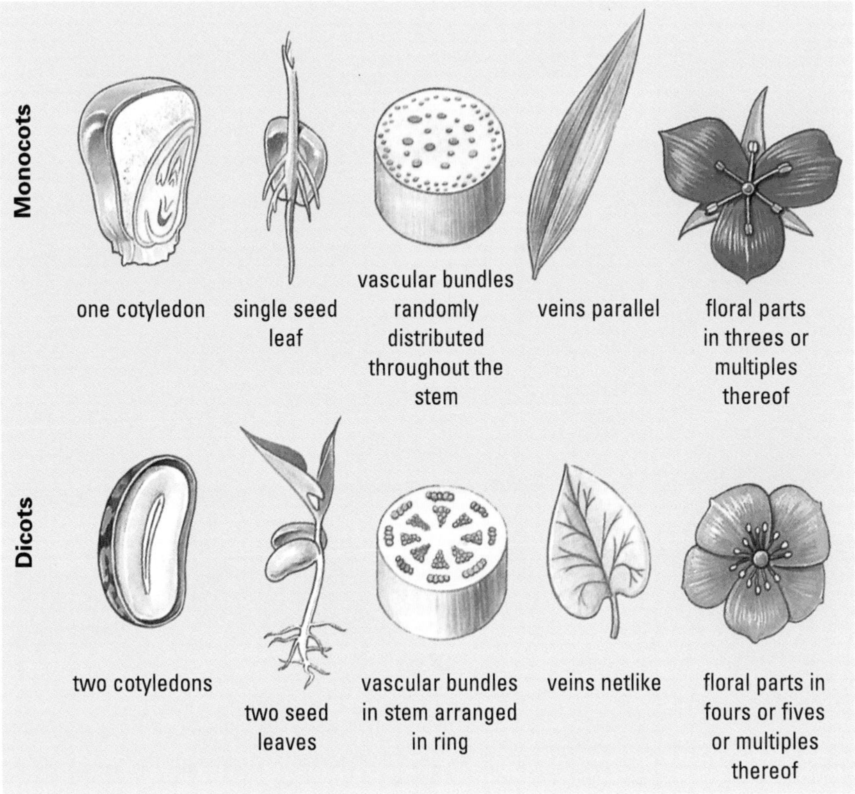

Monocots

| one cotyledon | single seed leaf | vascular bundles randomly distributed throughout the stem | veins parallel | floral parts in threes or multiples thereof |

Dicots

| two cotyledons | two seed leaves | vascular bundles in stem arranged in ring | veins netlike | floral parts in fours or fives or multiples thereof |

Figure 9
The differences between a monocotyledon and a dicotyledon

in a few days, they mature (**Figure 10**). The filament of the stamen elongates and the anther enlarges. Each anther consists of several chambers in which diploid microspore mother cells are located. Each of these undergoes meiosis to form four haploid microspores, or male gametophytes. Each will develop into a mature pollen grain. When the pollen grains reach maturity, the anther chambers split, and as they curl inside out, the pollen grains appear to be coating the outside of the anthers. The pollen of some species is quite sticky, while in others it is like dry powder. The pollen of some plants has tiny wings, while in others the surface has distinctive ridges and grooves. During this stamen development, the style of the pistil also elongates and the stigma enlarges slightly and secretes a sticky, sometimes scented, substance that covers its surface. At the bottom of the pistil, the ovary also enlarges. Inside are one or more ovules. Within each ovule, the diploid megaspore mother cell undergoes meiosis and forms four haploid megaspores, but only one survives as the female gametophyte.

Pollination is usually carried out by wind or insects, but for some angiosperms, pollination is aided by birds or bats. The transfer of pollen from the anther to the stigma on the same flower or another flower on the same plant is called self-pollination. When pollen is transferred to a flower on a different plant, it is called cross-pollination. The pollen grains tend to adhere to the sticky stigmas. Part of the pollen makes its way down through the style tissue and the sperm eventually reaches the egg in the ovule.

Fertilization, the fusion of microspore and megaspore nuclei, produces the diploid zygote, which is the first cell of the next sporophyte generation. The

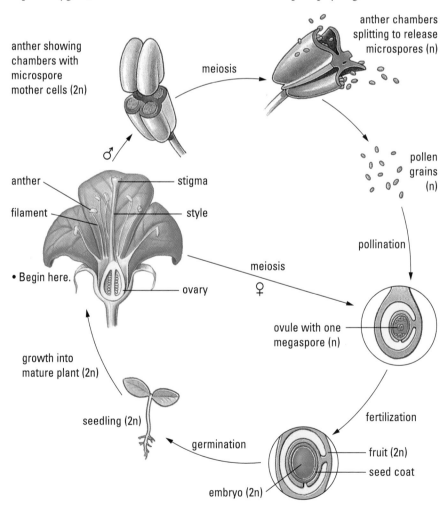

Figure 10

Alternation of generations of a typical flowering plant. Starting with the mature diploid sporophyte flower at the left of the diagram, follow the cycle clockwise and note the labels carefully. Remember that even though the gametophyte generation is small, it is very important for maintaining diversity within the species.

(a) anthers of a wood lily

(b) pale green ovary is almost hidden by stamen filaments (wood lily)

(c) pitcher plant with many short stamens under an enlarged stigma

(d) stamens deep within a Jack-in-the-pulpit flower

(e) pistils deep within a Jack-in-the-pulpit flower

Figure 11
Variations of flower parts

zygote grows by mitosis to form an embryo, which remains inside the ovule. Now that there is an embryo, the ovule is called a seed and has its own protective seed coat. Besides the embryo, the seed also contains some special tissue that will provide nourishment to the developing embryo during seed germination and to the seedling until photosynthetic leaves become functional. While these changes are taking place inside the ovule, the ovary and perhaps some surrounding tissue are developing into a fruit. The fruit may be fleshy or quite hard and dry. The fruit provides protection for the seeds and often helps secure dispersal of the seeds. During seed and fruit development, the other flower parts often become dry and blow away or may stay attached to the fruit as withered bits of tissue.

After some time elapses, the mature fruit falls or is carried away by animals. Under suitable conditions, the fruit decomposes, the seed coat splits, and germination occurs. The embryo grows very quickly and is now called a seedling. In some angiosperms, the seedling will grow big enough to produce its own flowers and seeds within a few months. In other angiosperms, the plant has to grow many years before it produces its own flowers.

There are many angiosperms that also reproduce asexually. The methods vary widely. Humans have also intervened in plant reproduction by devising methods of reproducing angiosperms by asexual or vegetative means that the plants could not actually do themselves. Humans also intervene in the pollination process, either by manually transferring the pollen from one flower to another or by preventing pollen from reaching the stigmas of the flowers.

Practice

Understanding Concepts

1. To what major division of the plant kingdom do the biggest terrestrial plants belong? What are the two subdivisions of this large group?

2. To what generation do most of the spermatophyte plant parts belong?

3. Describe the male and female sporophyte parts of gymnosperms and angiosperms.

4. Compare the seeds of gymnosperms and angiosperms.

5. Compare pollination in gymnosperms and angiosperms.

6. Compare fertilization in gymnosperms and angiosperms.

7. What are the two classes of angiosperms? How do they differ?

Alternative Farming Methods

Conventional farms tend to share some characteristics such as high reliance on technology, including irrigation and the use of pesticides; operating on a large scale; growing a single crop continuously on the same plot of land; and using genetically uniform high-yielding crops. These conventional methods have greatly increased the food yield worldwide. The World Bank estimates that between 70% and 90% of the recent increases in food production are the result of conventional agriculture techniques rather than an increase in the amount of land that is farmed. When discussing food production in a country or the whole world, we must not only look at the amount produced but also examine the amount produced per capita and the amount produced per unit input of energy.

Alternative farming, a growing movement in agriculture, has an approach completely different from the conventional farming that is based on social and economic goals. An important component of alternative farming is the use of organic farming methods. Some of the primary concerns of organic farming are

(a)   (b)

Figure 12
Conventional farms **(a)** have much lower species diversity than organic farms **(b)**, but seem to have higher yields when based only on the amounts produced.

DECISION-MAKING SKILLS

- Define the Issue
- ○ Identify Alternatives
- Research
- Analyze the Issue
- Defend a Decision
- Evaluate

Explore an **Issue**

Debate: Alternative Farming

You have learned that conventional farming methods may or may not be the best method of farming. Alternative methods are being advocated. Different farming methods involve different philosophies, social and economic considerations, and have different ecological impacts.

Statement

Conventional farmers should be persuaded to use alternative farming methods.

- In your group, research the issue. Learn more about both conventional and alternative farming methods. Search for information in newspapers, periodicals, CD-ROMs, and on the Internet. Follow the links for Nelson Biology 11, 10.9.

GO TO www.science.nelson.com

- Write a list of points and counterpoints that your group considered. You might consider these factors:
  - the effects on species and genetic diversity
  - environmental problems
  - yields and how they are determined
  - social, economic, or environmental circumstances that make one farming method viable relative to another
- Decide whether your group agrees or disagrees with the statement.
- Prepare to defend your group's position in a class discussion or debate.

retaining as much genetic and species diversity as possible in the farmed area, and eliminating the use of chemical products. To meet these goals, organic farmers grow several species with wide genetic variability on smaller plots instead of growing a single genetic strain of a single species on a large plot of land. Plants are chosen to complement one another. For example, some types of plants, such as legumes, can enrich the land and improve the growth of other plants. Some plants naturally exude chemicals that deter pests and weeds. The organic farmer chooses a diverse combination of plants that help one another grow and that eliminate the need for chemical additives.

SUMMARY Seed Plants

Gymnosperms

1. Gymnosperms are cone-bearing plants.
   - Most have needles covered by a waxy protective cuticle.
   - Seeds are "naked," or not enclosed.
   - They are important to construction, furniture, and pulp and paper industries.

2. There are two types of cones: male and female.

3. In male cones, microspore mother cells undergo meiosis to produce haploid pollen grains, the male gametophytes.

4. Female cones have ovules in which a megaspore mother cell undergoes meiosis to produces only one megaspore, or female gametophyte.

5. Pollen gets trapped on sticky sap secreted by the female cone.

6. After fertilization, the diploid zygote develops into an embryo, which remains in the ripened ovule, now called a seed.

7. Mature "naked" seeds fall out of the female cones.

8. It may be several years before the seedling grows into a plant mature enough to produce its own cones.

Angiosperms

1. Angiosperms are flowering plants.
   - Flowers have male parts, female parts, or both.
   - Monocots have one cotyledon and dicots have two cotyledons.

2. Within the anther chambers, microspore mother cells undergo meiosis to produce haploid pollen grains, or male gametophytes.

3. Within the ovule, a megaspore mother cell undergoes meiosis to produce only one megaspore, or female gametophyte.

4. Pollination is aided by wind, insects, birds, and bats.

5. Pollen gets trapped by the sticky substance on the stigma.

6. Self-pollination involves one plant only; cross-pollination involves two separate plants.

7. After fertilization, the diploid zygote grows into an embryo, which remains in the ripened ovule, now called a seed.

8. As the seeds develop, the ovary and other parts develop into the fruit enclosing the seeds.

9. Fruit offers protection for the seeds and may aid in the dispersal of the seeds.

10. Seedlings may produce mature plants within months or perhaps not for several years.

Understanding Concepts

1. Discuss the ways in which spermatophytes are suited for terrestrial life.

2. Why are mosses said to have achieved only a partial adaptation to terrestrial conditions?

3. How are spermatophytes better suited to dry habitats than mosses and ferns?

4. Describe some characteristics of gymnosperms that allow them to live in extreme conditions.

5. Why is it significant that the gametophyte plant is green and the sporophyte is often brown in many mosses?

6. How is a fern gametophyte different from a moss gametophyte?

7. How is a fern sporophyte different from a moss sporophyte?

8. **Figure 13** shows the reproductive structures of a species of pine tree.

   (a) Are the cones shown male or female? Explain your answer.

   (b) How can you verify that your answer to part (a) is correct? What evidence would you be seeking?

Figure 13
Young cones on an Austrian pine

9. Describe some methods of pollination and how they would differ for plants in different situations.

10. (a) How does sexual reproduction contribute to genetic diversity?

    (b) How does this diversity contribute to species survival?

11. What problems might occur if seeds were all dropped directly beneath the parent plant or were dispersed too far from each other?

Making Connections

12. Explain how the cutting of a diverse forest ecosystem to create fields for planting only one crop could create environmental problems.

13. Why are gymnosperms considered economically and ecologically important?

Reflecting

14. What methods do you find most effective at helping you to remember how organisms are classified?

15. As you learn more about each group of plants or fungi

    (a) how does the information about previous groups add to your understanding of the new group?

    (b) how does the information about a new group help with the understanding of previous groups?

    (c) which patterns of characteristics or development among the groups can you see?

    (d) how can you use these patterns to help you remember information?

Key Expectations

Throughout this chapter, you have had opportunities to do the following:

- Use appropriate sampling procedures to collect various organisms in an ecosystem (10.9).
- Select appropriate instruments and use them effectively and accurately in collecting observations and data (10.0, 10.2, 10.9).
- Communicate the procedures and results of investigations and research (10.0, 10.2, 10.7, 10.9).
- Describe anatomical and physiological characteristics of fungi and plants (10.1, 10.4, 10.6, 10.7, 10.8, 10.9).
- Describe plant adaptations to different environmental conditions (10.1, 10.4, 10.5, 10.7, 10.8, 10.9).
- Compare and contrast the life cycles of fungi and various plants (10.2, 10.5, 10.7, 10.8, 10.9).
- Describe the importance of fungi and plant groups to ecosystems and to humans (10.3, 10.4, 10.5, 10.7, 10.8, 10.9).
- Describe Canadian contributions to research about fungi or plants (10.3).
- Describe technologies employing fungi or plants that benefit agricultural or medicinal research (10.3, 10.9).
- Explain how nonvascular plants function without a specialized vascular system (10.4, 10.5, 10.6).
- Explain the importance of sexual reproduction to variability within a population (10.5, 10.6, 10.7, 10.8, 10.9).
- Demonstrate an understanding of the connection between biodiversity and species survival (10.5, 10.7, 10.8, 10.9).

Key Terms

alternation of generations
angiosperms
antheridium
archegonium
chitin
cotyledon
cuticle
dikaryotic
epiphytes
frond
gametophyte
gemmae
germinate
gymnosperms
hypha
lichen
mycelium
mycorrhizae
nonvascular

ovules
phloem
pollen
pollination
prothallus
protonema
rhizoids
rhizomes
saprophytes
seed
sorus
sporangia
spore mother cells
sporophyte
stomata
substrate
symbiotic relationship
vascular
vegetative
xylem

Make a Summary

In this chapter, you studied the principles of taxonomy and you classified plants. To summarize your learning, create a spider key, chart, table, or set of Venn diagrams that shows the differences among mosses, ferns, gymnosperms, and angiosperms. Use as many of the key terms as possible.

Reflect on your Learning

Revisit your answers to the Reflect on Your Learning questions at the beginning of this chapter.

- How has your thinking changed?
- What new questions do you have?

Understanding Concepts

1. Given your knowledge of fungi and plants, develop a rationale for
   (a) placing fungi and plants in two separate kingdoms;
   (b) keeping both in the same kingdom.

2. The sentences below contain errors or are incomplete. Write a complete and correct version of each statement.
   (a) Mosses live in dry coastal habitats.
   (b) Bryophytes rely on diffusion and vascular bundles to obtain and transfer water.
   (c) The major advance of the bryophytes was the development of roots.
   (d) Mosses are uncommon plants of little use to humans.
   (e) Moss pollen is carried to the egg cell through rainwater or dew.

3. The vegetative body or mycelium of a fungus may be present but unnoticed. Why?

4. How does a vascular system relate to plant height?

5. In this chapter, you have studied representatives from the major divisions of terrestrial plants. Which would you consider to be the most complex and the least complex organisms? Explain your answers.

6. One definition of cuticle is that it is a special layer on the surface of leaves and green stems. Explain why you would or would not expect to find a cuticle on non-green stems and around root tissues?

7. What is the basis for classifying both angiosperms and gymnosperms as spermatophytes?

8. What characteristics of a flower can help you determine its method of pollination?

9. What advantages do seed-producing plants have over plants that produce only spores?

10. Explain why lichens are sometimes called "two for the price of one" organisms.

11. What is the difference between rhizoids and rhizomes?

12. What are mycorrhizae? Explain their importance to the health of our planet.

13. Describe one use that humans have for each of the following: fungi, mosses, and ferns.

14. List the plant features that allow them to be successful on land.

15. Not all flowering plants are terrestrial. Provide three specific examples of flowers that are not terrestrial.

16. Not all flowers are colourful. Provide three specific examples of flowers that lack colour.

17. What is the difference between a perfect flower and an imperfect flower?

18. The moss sporophyte generation is often dependent on the gametophyte generation. Explain why.

19. Explain why plant diversity in a flower or vegetable garden is very important.

20. Describe how any group of plants can alter the soil and microclimate where they are living.

21. Discuss the advantages that wind-pollinated plants gain by having flowers with very small sepals and petals.

Applying Inquiry Skills

22. A student gathers two samples of "mosses." Sample A is composed of very long, silvery-grey threadlike masses and is found clinging to trees and on the surface of other plants. Sample B is composed of small green plants less than 2 cm in height and is found close to the ground. Both samples absorb water when placed in a beaker of water. One sample is a true moss and the other is not.
    (a) Which sample is the true moss? Why?
    (b) What might the other sample be? Why?

23. The leader of a nature hike points to a leafy green organism about 1 m tall and with brown spots under its leaves. He identifies it as a true moss.
    (a) Do you agree or disagree with the leader? Why?
    (b) Suggest what this organism might be and explain why.

Making Connections

24. What would happen to the other organisms in a pond if the aquatic plants suddenly disappeared?

25. The vascular system of tracheophytes is sometimes compared to the circulatory system of humans. Explain why you would agree or disagree with this comparison.

26. The ancient fern forests have all disappeared. What are some possible reasons for their disappearance?

27. Most alien (non-native) plants were introduced into Canada for food, fibre, herbal, or ornamental purposes. In agriculture, approximately five hundred introduced plants have become pests, such as yellow rocket (*Barbarea vulgaris*) and the Canada thistle (*Cirsium arvense*).
    (a) Make a list of problems that can be caused by these plants.
    (b) Why might they be so difficult to control?
    (c) Many retired Canadians who move into other parts of the world take plants from Canada with them to re-create the feeling of being at home. Describe one problem with this practice.

28. (a) In your notebook, or using a spreadsheet or database, create a table with 23 rows and 6 columns.

Label columns 1 to 6 with these headings: Feature/Process, Fungi, Mosses, Ferns, Gymnosperms, and Angiosperms, respectively. Starting at the second row, place these terms in the first column: phloem, frond, tracheophyte, xylem, meiosis, flowers, bryophyte, archegonium, gills, seeds, rhizoids, pollen, prothallus, antheridium, hyphae, ovule, gemmae, protonema, spermatophyte, spores, rhizomes, fruit.

(b) Place a check mark in the cells to indicate the features that are found or the process that is carried out for each type of plant.

29. **Figure 1** shows part of the underside of a fern frond. In this case, some of the "dots" appear white because they have a little white membrane covering them. Other "dots" appear brown because their membrane has dried and dropped away.
(a) What are the "dots" called?
(b) What are the components of each "dot" called?

Figure 2 is a photomicrograph of two of these components.
(c) What are the first cells within this structure called?
(d) What process takes place in each of these cells?
(e) What are the products of that process called?
(f) What will happen when these products are released and land in a suitable location?

Figure 1
Close-up photo of a partial fern frond

Figure 2
Photomicrograph of a fern structure

30. **Figure 3** shows two kinds of plants on this rock.
(a) What are the two plants shown? Describe briefly where they are in the picture and how you have identified them.
(b) What other type of organism is present?
(c) Name and describe each of the different kinds of the organism identified in part (b).
(d) Which of the organisms identified in parts (a) and (b) probably existed on this rock longer than the others?
(e) How did this early organism pave the way for the others?

Figure 3
A rock covered by living organisms

Exploring

31. Many plants called mosses, such as Irish moss and Spanish moss, are not mosses at all. Conduct research to uncover other so-called mosses that actually belong to other plant groups. Why are these plants not true mosses? How and why were these plants named mosses? Follow the links for Nelson Biology 11, Chapter 10 Review.

GO TO www.science.nelson.com

32. Find out more about the rosy periwinkle, a plant which produced chemicals that have medical benefits for people with Hodgkin's disease, a form of leukemia or white blood cell cancer. Explain how it was discovered, how it works in the body, how rare the plant is and how difficult the chemicals are to extract or synthesize. Follow the links for Nelson Biology 11, Chapter 10 Review.

GO TO www.science.nelson.com

33. Earth's atmosphere once contained far less oxygen than it does now. Provide a possible explanation of why oxygen levels have increased from the early days of the planet.

Invertebrates

Symmetry is often found in nature—in the shapes of leaves, the petal arrangements of flowers, and even the hexagonal facets that make up the compound eye of insects. The concept of symmetry is the same in biology as it is in mathematics:

- A shape has bilateral symmetry (bi meaning two) if one half of a shape is the mirror image of the other half. For example, a bee is bilaterally symmetric.
- A shape has radial symmetry if the shape of one part is repeated a number of times about a central axis (**Figure 1**). For example, a bee's honeycomb cell is radially symmetric (**Figure 2**).

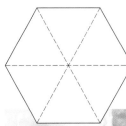

Figure 1
In mathematics, the regular hexagon is said to have radial symmetry of order 6.

Figure 2
The bee and honeycomb are examples of symmetry in nature.

In this chapter, you will be able to

- explain how invertebrates are classified;
- describe the anatomy and physiology of different invertebrates;
- compare the life cycles of different invertebrates;
- explain the success of arthropods in a wide variety of environments;
- examine the internal features of representatives of the invertebrates (worms and insects);
- appreciate the ecological role played by various invertebrates.

Reflect on your Learning

1. All of the animals in **Figure 3** are invertebrates. What do they have in common that classifies them as invertebrates?

2. List some other animals that you think are invertebrates, and some that you think are vertebrates.

3. (a) Do you think it is reasonable to say that vertebrates (fish, amphibians, reptiles, birds, and mammals) have bilateral symmetry? Why?
   (b) Why do you think that animals with bilateral symmetry tend to be more complex than those with radial symmetry?

sand dollar

flatworm

millipede

spider

octopus

sea anemone

Figure 3
A variety of organisms

Try This
Activity

Symmetry in Nature

Examine the organisms in **Figure 3**. Which of the organisms display bilateral symmetry and which display radial symmetry?

- Sketch the body shape of each organism and draw the line or central point of symmetry.
- Draw a diagram of a butterfly and a daisy and describe the symmetry of each.

11.1 The Animal Kingdom

What characteristics make an organism a part of the animal kingdom? Like plants, animals are multicellular and eukaryotic. However, animals differ significantly from plants because their cells have no cell walls, only a cell membrane, and they are heterotrophic—they cannot make their own food. Animals obtain their food from plants or other animals, then digest that food, circulate its nutrients throughout the body for growth and energy supply, and dispose of it as metabolic waste. They must coordinate their activities, avoid predators and other hazards, grow, and reproduce.

Scientists who study **zoology** divide this complex kingdom into two broad groups: **invertebrates** and **vertebrates**. The invertebrates do not have a backbone. They are the only large taxonomic group defined by the lack of a characteristic rather than by the presence of a common feature. The vertebrates, to which humans belong, have a **notochord** for at least part of their life cycle.

The following are some of the major characteristics used to classify animals:

- body organization—Does the animal have tissues, or tissues organized into organs, or organ systems?
- number of body layers—Does the animal have two or three layers?
- body symmetry—Does the animal have radial or bilateral symmetry?
- digestive tract or gut—Does the animal's gut have only one opening or does it have two openings: a mouth for food intake and an anus for expulsion of body waste (**Figure 1**)?
- **coelom** or body cavity—Does the animal have a true body cavity, is it partially formed, or is it absent?

zoology: the study of animal life

invertebrates: multicellular, eukaryotic heterotrophs that do not have a notochord

vertebrates: multicellular, eukaryotic heterotrophs that have a notochord at some stage in their life

notochord: a skeletal rod of connective tissue that runs lengthwise along the dorsal surface and beneath the nerve cord. Notochords are present at some time during vertebrate development.

coelom: the fluid-filled space inside the body, lined with a layer of cells called the peritoneum

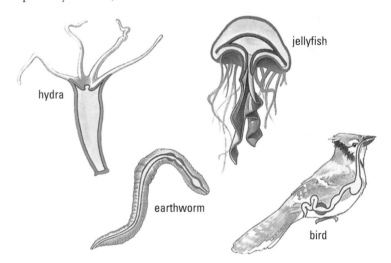

Figure 1
Hydra and jellyfish have a saclike gut with a single opening. More complex animals, such as earthworms and birds, have a tubelike gut with a separate mouth and anus. This arrangement enables the one-way movement of food through the gut, and permits greater specialization in a region, such as for grinding or chewing, chemical digestion, and absorption of nutrients.

Number of Body Layers

One important characteristic that separates certain animal phyla from others is the number of **germ layers** their members possess. (In this case germ refers to the beginning of something, as in germination.) Germ layers appear very early in the development of most young organisms and give rise to specific tissues in the adult. There are three different types of germ layers: ectoderm (*ecto* meaning outer and *derm* meaning skin or layer), endoderm (*endo* meaning inner), and mesoderm (*meso* meaning middle).

Cells from the ectoderm will form skin and the nervous system. In more complex animals, feathers, scales, hair, and nails also come from the ectoderm.

germ layers: layers of cells in the embryo that give rise to specific tissues in the adult

Cells from the endoderm will form the lining of the gut. In most animals, there is a germ layer between the ectoderm and endoderm: the mesoderm, which gives rise to organs of the circulatory, reproductive, excretory, and muscular systems.

The simplest animals, such as the sponges and jellyfish, have no mesoderm; therefore, they lack structures such as a circulatory system. With only an ectoderm and endoderm, each cell of the animal is exposed to seawater. A circulatory system is unnecessary because nutrients are received directly from the water; and an excretory system is unnecessary because wastes are discharged into the water. By comparing many different animals, zoologists have found that as their complexity increases, a larger mesoderm layer develops and structures become more elaborate (**Figure 2(c)**). Because the middle layer of cells is no longer directly connected to the external environment, an internal transport system becomes important for carrying nutrients throughout the body and for removing wastes.

Body Cavities

Another characteristic useful for classifying animals is the presence or absence of a body cavity called a coelom. The coelom is located between the body wall and the gut, and contains and protects the internal organs. The coelom develops from the embryo's mesoderm in all vertebrates and higher invertebrates, and has a definite lining of cells called the **peritoneum**. The peritoneum not only lines the inner surface of the body wall but also surrounds the internal organs and holds them in place (**Figure 2(c)**). Less complex invertebrates may lack a coelom (**Figure 2(a)**), or may have an intermediate structure called a **pseudocoelom**, which is a fluid-filled space of variable shape and has no peritoneum (**Figure 2(b)**).

peritoneum: a covering membrane that lines the body cavity and covers the internal organs

pseudocoelom: a fluid-filled cavity that lacks the mesodermal lining of a true coelom

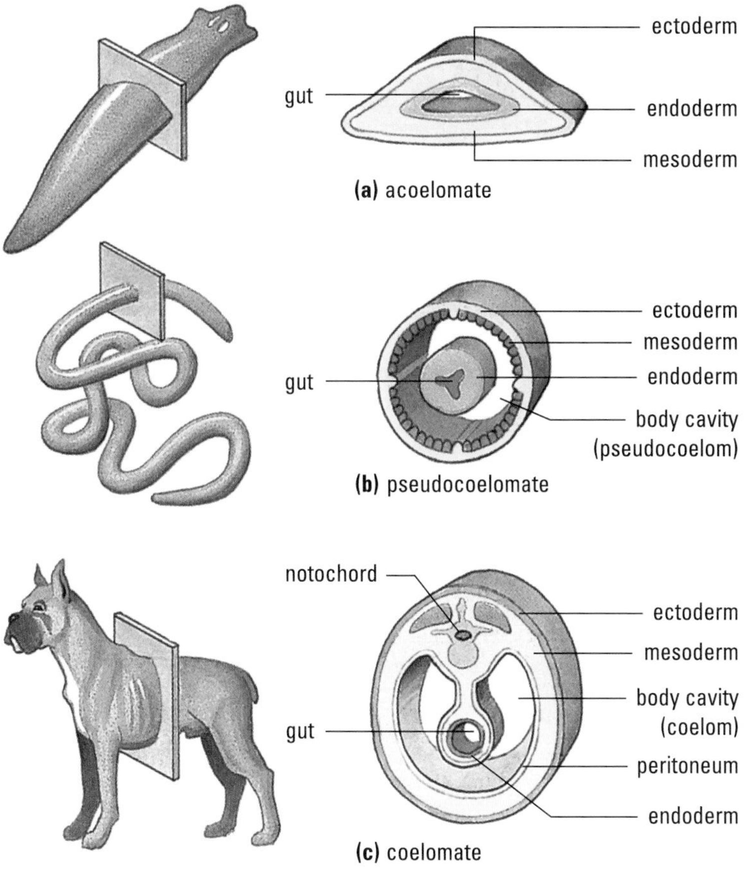

(a) acoelomate

(b) pseudocoelomate

(c) coelomate

Figure 2
(a) Acoelomates, such as flatworms, lack a body cavity.
(b) Pseudocoelomates, such as roundworms, do not have a continuous peritoneal lining.
(c) Coelomates, such as the higher invertebrates and vertebrates, have a true coelom lined with a continuous peritoneum.

The development of a coelom was probably associated, at least in part, with an increase in animal size. As a solid structure increases in volume, the exchange of gases, food materials, and waste is less easily accomplished, and the folding of different organs to increase the surface area is somewhat restricted. Having a body cavity allows space for internal organs such as lungs, heart, stomach, and intestines to expand and contract and to slide by each other as the animal moves. The fluid of the cavity may further aid in waste removal and in the circulation of food materials and oxygen.

Symmetry

How is body symmetry related to an animal's lifestyle and brain development? Animals that display radial symmetry, such as hydra, jellyfish, and starfish, are not well suited to rapid locomotion (**Figure 3**). One explanation for the slower movement can be traced to the fact that no one region always leads. Only animals that display bilateral symmetry have a true head region. Because the head, or anterior region, tends to enter a new environment first, nerve cells are usually concentrated in this area. **Cephalization** (from the Latin *cephalicus*, meaning head) enables the rapid processing of stimuli such as food or danger. Not surprisingly, bilateral symmetry is advantageous for animals that actively move forward (**Figure 4**).

cephalization: the concentration of nerve tissue and receptors at the anterior end of an animal's body

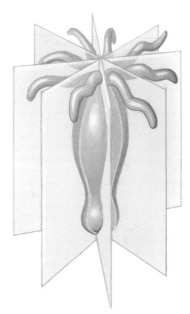

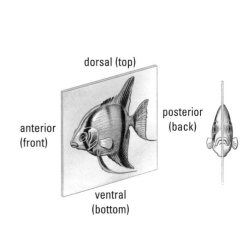

Figure 3
Planes of radial symmetry in a hydra

Figure 4
Plane of bilateral symmetry in a fish

Classifying Invertebrates

Numbering over one million species, the invertebrates comprise over 95% of all described animal species and affect life on Earth in countless ways. Their diversity ranges from a microscopic mite in house dust to the large graceful Portuguese man-of-war drifting in the ocean, from a termite digesting wood to a leech ingesting blood, from a **sessile** sponge found on the bottom of the ocean to a **motile** butterfly that migrates thousands of kilometres.

Some of the 30 or so phyla in the animal kingdom are described in **Table 1**, arranged from least to most complex. All but one of these phyla include invertebrates. Scientists group all vertebrates in only one phylum.

sessile: not capable of independent movement. Sessile animals remain fixed in one place throughout their adult lives.

motile: capable of movement. Motile animals are able to move from place to place by expending cellular energy.

Table 1: Summary of Major Phyla of the Animal Kingdom

Phylum (common name)	Representative members	Description	Approx. no. of species
1. Porifera (sponges)	giant sponge vase, redbeard sponge	• sessile • irregular shape • no mouth or digestive cavity • marine and freshwater	5 000
2. Cnidaria	jellyfish, hydra	• sessile or motile • medusoid form and polyp form in life cycle of some organisms • stinging nematocysts • radial symmetry • marine, with a few freshwater	10 000
3. Platyhelminthes (flatworms)	turbellarians, flukes, tapeworms	• free-living in marine or fresh water, or parasitic • body flattened dorsoventrally • mouth but no anus	19 000
4. Nematoda (roundworms)	hookworm, pinworm, vinegar eel	• cylindrical, slender, tapered at either end • free-living or parasitic • all habitats	20 000+
5. Rotifera (wheel animals)	rotifers	• anterior end ringed with cilia, posterior end tapering to a "foot" • mostly freshwater, with some marine • microscopic size	1 500
6. Annelida (segmented worms)	earthworms, leeches, polychaetes	• segmented body • terrestrial and aquatic • mouth and anus	12 000+
7. Mollusca (mollusks)	snails, clams, squids	• muscular foot • shell present in many forms • all habitats	100 000+
8. Arthropoda (arthropods)	insects, crab, mites, ticks, spiders, centipedes	• segmented body, some segments may be fused; jointed appendages; external skeleton • all habitats and modes of life, including parasitism	1 000 000+
9. Echinodermata	starfish, sea cucumbers, sea urchins	• adults have pentamerous (five-sided) radial symmetry • marine	7 000
10. Chordata (chordates)	fish, amphibians, reptiles, birds, mammals	• notochord at some time in life history • all habitats	7 000

Practice

Understanding Concepts

1. What are three features shared by all organisms in the animal kingdom?
2. How many germ layers are present in the embryos of complex animals? How are the germ layers different from one another?
3. Define coelom and describe one advantage provided by having a coelom.
4. What type of body symmetry do humans possess? Explain.

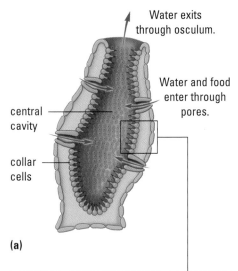

Water exits through osculum.

Water and food enter through pores.

central cavity

collar cells

(a)

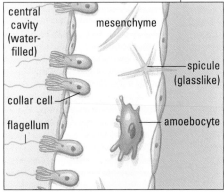

central cavity (water-filled)

mesenchyme

spicule (glasslike)

collar cell

flagellum

amoebocyte

(b)

Figure 1
(a) The body plan of a sponge
(b) Section through a sponge body wall

hermaphroditic: sharing both male and female sex cells or organs

larva: an intermediate form that an organism goes through to achieve its adult form. Tadpoles are larval frogs; caterpillars are larval butterflies or moths.

metamorphosis: a series of stages that an organism goes through, from egg to adult. The intermediate forms are quite often different from the final form.

5. How is body symmetry related to an animal's lifestyle and to brain development?

6. Name two animals that are sessile, and two that are motile.

Making Connections

7. If sensory receptors were concentrated near an animal's posterior instead of its head, the animal would sense environmental information after its entire body was exposed. Explain why this is unfavourable.

11.2 The Simplest Animals

Phylum Porifera: Animals Without Tissues

Sponges were once thought to be plants because they did not have any of the characteristics of animals. By the early 1800s this view changed. Sponges are considered to be the simplest and most primitive of animals. While they have some specialized cells, these cells are not arranged into tissues or organs. Without tissue development, the sponge has many limitations. It has no nervous system or brain to coordinate activities, and no muscle tissue, so it is incapable of independent movement.

Despite their limitations, sponges have existed for over 500 million years. In part, their success is due to a system of feeding that brings food to them. Collar cells line the inside of the body cavity (**Figure 1(a)**). Their flagella help to create currents that move water and small particles of food into the sponge through many small pores, and move wastes out via the osculum, or open end of the sponge. Digestion occurs in the collar cells or in special cells called amoebocytes that wander around in the gelatinous mixture called mesenchyme (**Figure 1(b)**). The mesenchyme also contains spicules and/or spongin fibres—nonliving material that provides the frame for the sponge body.

Reproduction

Most sponges are **hermaphroditic**, producing both male and female gametes; however, only one type of gamete (egg or sperm) is produced at a time, so that sperm will not fertilize eggs from the same sponge. The fertilized egg develops into a free-swimming **larva**, which soon attaches to the bottom and undergoes **metamorphosis**. A motile stage is common in sessile aquatic invertebrates, allowing them to distribute to new areas. Sponges can also reproduce asexually by budding or branching, and have considerable regenerative capabilities. Occasionally, part of the sponge may be broken off, and some pieces will form new individuals.

Ecological Role of Sponges

Sponges are an important food source for some species of snails, sea stars, and fish. Also, sponges serve as shelter and homes for many smaller invertebrates that live in or on the hollow spaces inside their bodies. In one study, a zoologist counted 16 000 small shrimp living inside the body of a loggerhead sponge. Other sponges have a symbiotic relationship with photosynthetic organisms, such as bacteria and plantlike protists. In this relationship, the microorganisms provide food and oxygen for the sponge and remove wastes.

Special amoebocytes found in a group called the boring sponges play an important ecological role in marine habitats by helping to clean up the ocean floor. They release chemicals into the water that break up old shells and pieces of coral, thereby recycling calcium in the ocean.

Phylum Cnidaria: Tissue Development

The word Cnidaria (the C is silent) comes from the Greek word *knidae*, meaning nettle or stinging hairs. Cnidarians include hydra, jellyfish, sea anemones, and coral (**Figure 2**), and are found exclusively in aquatic ecosystems. Most live in marine environments but some species are found in freshwater ecosystems, hydra being the most common freshwater cnidarian.

Figure 2
Coral are small organisms that secrete a wall of calcium carbonate around themselves. These skeletons grow together, forming groups or colonies, with new generations building on top of older generations. Often, these coral homes grow into a variety of shapes and colours, forming coral reefs.

All cnidarians are radially or biradially symmetric, have tentacles, and possess true tissues. However, like sponges, they have only two germ layers: ectoderm and endoderm. Between the two layers is a jellylike material called mesoglea. Cnidarians have specialized nerve, muscle, and digestive tissues. A nerve net encircles their body. If hydra or sea anemones are touched, they respond by flattening out and pulling their tentacles inward. This type of movement helps obtain food and indicates coordination of nerves and muscles—something not seen in the primitive sponge. (Further comparison of the two phyla is given in **Table 1**, page 416.)

The body structure can be in one of two forms: polyp or medusa (plural: medusae), which are illustrated in **Figure 3**. The polyp form has a sessile tubelike body in which the oral end is directed upward and the opposite end is attached

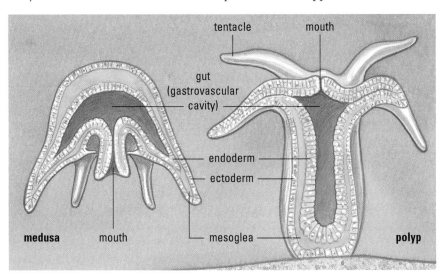

Figure 3
The body plan of a medusa and a polyp

(a)   (b)   (c)

Figure 4
(a) Hydra are polyps that often reproduce by budding.
(b) The medusa of a jellyfish has an extremely thick mesoglea.
(c) The Portuguese man-of-war is a complex colony that has many polyps and medusae joined together.
The different forms are specialized for different functions (e.g., feeding, defence, and reproduction).

to the ocean floor, as shown in **Figure 4(a)**. The motile medusa resembles a bell or umbrella, with the convex side upward and the mouth located in the centre of the concave undersurface, as shown in **Figure 4(b)**. Some cnidarians, such as hydra, sea anemones, and coral, exhibit only the polyp form. Some floating bell-shaped colonies have both medusoid and polyp form, such as *Physalia*, the Portuguese man-of-war, as pictured in **Figure 4(c)**.

Feeding

Cnidarians capture food such as small fish by using specialized stinging cells that contain **nematocysts** (**Figure 5**). The nematocysts have coiled, hollow, threadlike tubes that can shoot out rapidly and penetrate the skin of a prey or predator. Some inject toxic material, and if sufficient nematocysts are triggered, the target can be paralyzed. Thousands of the stinging cells can be found along the tentacles of a jellyfish.

Captured food is digested in the gastrovascular cavity or gut, which has a single opening. Food enters the same opening from which wastes are expelled. Because the entry to the gut can be opened and closed by muscles, cnidarians can ingest larger animals than sponges can.

Reproduction

Some cnidarians such as *Obelia* display both medusa and polyp forms in their life (**Figure 6**). In *Obelia*, a single polyp reproduces asexually by budding to form

nematocysts: stinging capsules that aid in the capture of prey

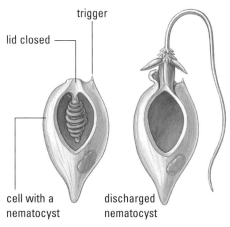

Figure 5
This nematocyst has a bristlelike trigger.

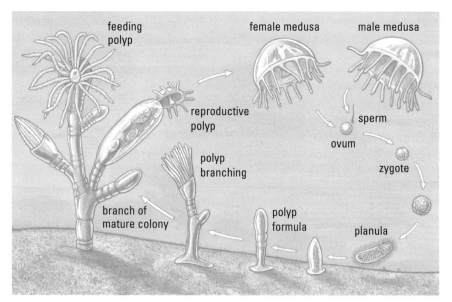

Figure 6
Life cycle of *Obelia*, which has a free-swimming medusa as well as a polyp form. There is regular alternation of sexual and asexual generations.

a colony. In a mature colony, there are two types of polyps: feeding polyps with tentacles, and reproductive polyps that are specialized to produce medusae. All medusae reproduce sexually by forming and releasing eggs or sperm into the surrounding water. Following fertilization, the resulting zygote rapidly develops into a motile ciliated larva called a planula. The planula is elongated and biradially symmetrical, but without a gut or mouth. Following several hours to several days, the planula attaches to an object and becomes a polyp, which then develops into a colony, and the cycle begins again.

In hydra, there is no motile medusoid form. Hydras usually reproduce asexually by budding, where the buds begin as outgrowths of the body wall, or by regenerating whole bodies from pieces of the organism that have broken off. (It is this feature that gives hydra its name, which means "many headed." In Greek mythology, Hydra is the many-headed monster that grows two heads to replace each one that is cut off.)

During unfavourable conditions, hydras reproduce sexually by forming eggs and/or sperm. Following fertilization, the zygote develops and then becomes enclosed by a protective casing, called a cyst. The cyst drops off the parent and lies dormant, generally until the next spring, when the young hydras hatch and develop into adults.

Ecological Role of Cnidaria

Cnidarians, like sponges, play an important role in ocean ecology and add to the great diversity of marine life. Many fish, turtles, and other small animals feed or live in the protection of sea anemones and coral reefs. In the case of sea anemones, fish and other organisms feed on the scraps of food trapped in their tentacles; in turn, these organisms are believed to help clean the sea anemone and provide protection against certain predators.

The fish and shellfish protected by coral reefs serve as important sources of protein for people who live in these areas. Coral reefs also protect land along coastal areas from damage caused by rough seas and large storm waves. However, water pollution, caused by the dumping of sewage and by oil spills, is becoming a serious threat to the survival of many coral reef environments. Once the living coral is killed, waves and currents break apart the coral skeleton, not only destroying important habitats but also exposing much of the land surface to destructive wave action.

DID YOU KNOW ?

Cnidarians are common along Canada's three shorelines. The purple-coloured Vellela occasionally floats into popular beaches on maritime shores. When a swimmer brushes against its tentacles, the nematocysts are triggered and release their toxin, which causes swelling in humans. The jellyfish Chironex, found off the coast of Australia, produces toxins so powerful, they can be life threatening for humans. In Australia, its sting has claimed twice as many victims as shark attacks have.

Practice

Understanding Concepts

1. Why did scientists, up to the early nineteenth century, consider sponges to be plants?
2. Name the two body forms found in cnidaria.
3. How do tentacles and nematocysts help sea anemones obtain food?
4. What are two methods of asexual reproduction found in hydras?
5. Why are sponges considered to be more primitive than cnidarians?
6. Name one important ecological role played by sponges and one by cnidaria.

Exploring

7. Research to find out more about the new coral reef reserve established in Hawaii.
   Follow the links for Nelson Biology 11, 11.2.

GO TO www.science.nelson.com

8. Research to find out how natural sponges are harvested and processed into soft pads useful for washing. Are most store-bought sponge pads made of natural sponge? Explain your answer.

On the Internet, follow the links for Nelson Biology 11, 11.2.

GO TO www.science.nelson.com

SUMMARY **The Animal Kingdom and the Simplest Animals**

Table 1 summarizes the main similarities and differences between the two phyla of invertebrates discussed so far.

Table 1: Sponges and Cnidarians

	Porifera	Cnidaria
Representative	sponge	hydra, jellyfish, coral colonies
Habitat	predominantly marine, but also freshwater	predominantly marine, but also freshwater
Body plan	asymmetrical, two germ layers	radial symmetry, bell-shaped medusae, and tube-shaped polyp; two germ layers
Reproduction	asexual—budding; sexual—egg and sperm	asexual—budding; sexual—egg and sperm
Transport and digestion	none—incoming current supplies food, and excurrent, or outgoing current, carries away wastes	body cavity functions in transport and digestion
Nervous system	no nerves	nerve tissues, but no central collection of nerves that would resemble a brain

1. The majority of animals are invertebrates, grouped together because they lack a backbone.

2. Most of the more complex animals are bilaterally symmetrical and form a type of tissue called mesoderm during the embryo stage. The mesoderm gives rise to the coelom (body cavity).

3. The presence and form of a coelom, the type of symmetry, and the complexity of body systems help separate the animals into groups.

4. The Porifera, or sponges, are the simplest animals because they lack true tissues or organs.

5. Cnidaria possess two different body plans: polyp and medusa.

6. Some cnidarians show an alternation of sexual and asexual generations in their life cycle.

Sections 11.1–11.2 Questions

Understanding Concepts

1. What main characteristics are used to classify animals into different phyla?

2. Why is a gut with openings at both ends an improvement over one with only a single opening?

3. Explain how a concentration of nerve tissue in the anterior portion of an animal's body can be related to bilateral symmetry. Why would the nerve tissue not be concentrated near the posterior end?

4. Why is it necessary for larger animals to have a digestive system?

5. Describe the similarities and differences between the two basic cnidarian body forms. Give examples of animals that possess each form.

6. What is the major difference between the way poriferans and cnidarians digest food?

7. In certain protista groups, colonies are simple aggregations of unicellular organisms that have become stuck together. If an individual cell is separated from the others, it has the ability to reproduce and build a new colony. Why are cnidarian floating colonies such as *Physalia* considered to be more complex than protista colonies like *Volvox*?

8. Why do you think nematocysts are vital structures for hydras?

Making Connections

9. Why do you think so many species live in aquatic habitats?

11.3 Worms

Worm is a general term that includes invertebrates from the three main groups of animals classified by a true body cavity (**Figure 2(b)** in section 11.1). Flatworms are acoelomates because they lack a coelom. Roundworms are pseudocoelomates because they do not have a peritoneum covering the irregular body cavity. Segmented worms, such as earthworms, are coelomates, with a true coelom.

Phylum Platyhelminthes: Bilateral Symmetry and the Primitive Brain

The Platyhelminthes, or flatworms, are the most primitive animals to show bilateral symmetry and cephalization. Flatworms also have mesoderm—a tissue not found in simpler organisms—as well as true organs and rudimentary organ systems for digestion and excretion. The phylum includes both free-living and parasitic organisms.

Free-Living Flatworms

The best-known free-living flatworms are the Planaria, which occur in moist or submerged habitats in both marine and fresh water. Most are small, less than 1 cm in length, but some terrestrial forms may reach 60 cm in length. Like cnidarians, they have a "blind" digestive system, with the mouth as the only entrance or exit. Unlike cnidarians, they have a rudimentary excretory system

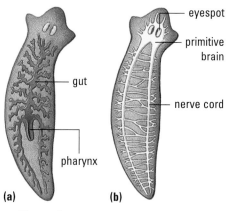

Figure 1
Planaria, a typical flatworm
(a) The digestive system includes a pharynx, a muscular organ that elongates out of the mouth and sucks food in.
(b) The nervous system includes a simple brain, a frontal sense organ, and two eye spots in the head region.

DID YOU KNOW ?

The planarian's powers of regeneration are so great that if a cut is made at the tip of its head, and the cut is kept open, it will develop two heads. Repeated cutting can produce a worm with many heads, each with two eye spots!

scolex: the knoblike head of a tapeworm

proglottids: the segmentlike divisions of a tapeworm's body

consisting of a network of fine tubules running throughout the body and opening to the outside via tiny pores (**Figure 1(a)**).

Nerve tissue is concentrated in the head area, resembling a primitive brain. Two nerve cords, each made up of many nerve cells, arise from the brain and run along the ventral side of the body toward the tail (**Figure 1(b)**). This system coordinates body movements and receives input from simple sensory cells. Flatworms can avoid light and respond to chemical substances, which enable them to sense and move toward potential food. They can even learn their way out of a simple maze!

Free-living flatworms are hermaphroditic. A reproductive system is present only during the breeding season and degenerates at other times of the year. After two individuals mate, fertilization occurs and then capsules containing a small number of eggs and thousands of nutritive yolk cells are discharged. The capsules fasten to objects in the water and, in two to three weeks, they hatch into juveniles, which resemble the adults. In addition to sexual reproduction, some planarians reproduce asexually by regenerating complete worms from small fragments.

Parasitic Flatworms

Flukes and tapeworms look quite different from their free-living relatives, and can be found both inside and outside a host animal's body. While the largest tapeworm, that of a sperm whale, may reach 30 m, the smallest are only a few millimetres long. Human tapeworms occasionally reach lengths of 7 m.

A parasitic flatworm that lives within its host requires little sensory information, and sensory receptors are reduced or absent. Some species do not have their own digestive system because they feed off the food digested by their host or the nutrients carried in the host's blood. Other species ingest tissues of their host and have a somewhat reduced digestive system. The reduced or absent digestive system allows more room for a highly developed reproductive system. Parasitic flatworms are capable of producing hundreds of thousands of eggs.

Life Cycle of a Tapeworm

Tapeworms are intestinal parasites of vertebrates, including dogs and humans. The parasite is protected from the digestive enzymes and immune response of the host by a nonliving cuticle which is secreted by ectodermal cells. Also, a tapeworm has a structure with suckers and/or hooks called the **scolex**, which enables it to attach itself to the intestinal wall of a host. Behind the scolex are mature segments called **proglottids**, which contain a full set of sex organs. These individual hermaphroditic units can mate with each other by the worm doubling back on itself. By the time the segments reach the posterior of the worm, they are full of eggs. Proglottids at the extreme end drop off and pass out with the host's feces, taking the eggs with them. There they become available to infect an intermediate host, such as a fish, snail, or rabbit.

Figure 2 shows cattle as intermediate hosts and humans as final hosts. In the muscle tissues of the intermediate host, the tapeworm embryo forms a cyst about the size of a pea. No further development occurs, so the tapeworm does not harm the intermediate host. However, when the cyst is consumed by humans, usually in improperly cooked meat, a young tapeworm emerges, attaches itself to the wall of the human intestine and, over time, grows to full size. Mature adult tapeworms, which can live for several years, use up a substantial portion of food in a human's digestive system, thereby weakening the final host.

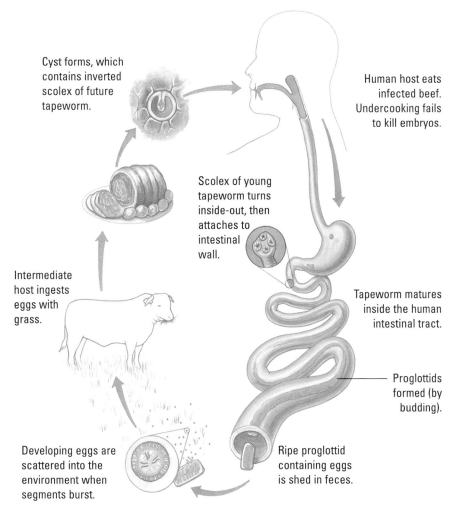

Cyst forms, which contains inverted scolex of future tapeworm.

Human host eats infected beef. Undercooking fails to kill embryos.

Scolex of young tapeworm turns inside-out, then attaches to intestinal wall.

Intermediate host ingests eggs with grass.

Tapeworm matures inside the human intestinal tract.

Proglottids formed (by budding).

Developing eggs are scattered into the environment when segments burst.

Ripe proglottid containing eggs is shed in feces.

Figure 2
The life cycle of the beef tapeworm *Taenia saginata*

Phylum Nematoda: Complete Digestive Tracts

Most nematodes, or roundworms, are free-living, small, and harmless, but parasitic nematodes can cause major health problems in both plants and animals. The total number of described species in the phylum varies from 20 000 to 80 000, but this likely represents only about 10% of the total number. Free-living species are often present in enormous numbers; a decomposing apple on the ground in an orchard has yielded 90 000 roundworms!

Nematodes are the simplest organisms to have a complete digestive tract, with a mouth at one end and an anus at the other. The development of an anus means that food can be digested step by step as it moves through the system, and the removal of waste does not interfere with the intake of food. This is an important advancement toward increased complexity.

A roundworm's nervous system has an anterior ganglion, or aggregation of nerve cells, and lateral or ventral nerve cords extending along the body. There is no respiratory or circulatory system, due in part to the organism's small size. Oxygen and carbon dioxide are exchanged by simple diffusion.

Parasitic roundworms attack virtually all groups of plants and animals, making them one of the most dreaded of the parasitic animal groups. These roundworms display all degrees of parasitism. Potato cyst nematodes, for example, can have debilitating effects on potato crops. Hookworms, heartworms,

DID YOU KNOW ?

The golden nematode (*Globodera rostochiensis*) is the most common North American potato cyst nematode, and has been subjected to stringent control measures. In Newfoundland, in order to keep the organism from spreading to the mainland, you are not allowed to take any soil-containing materials from the island (including your house plants, if you are moving). Cars leaving by ferry are thoroughly washed at the ferry terminal to remove any soil, much to the amusement of tourists not familiar with the reason for their free car wash!

and pinworms commonly infect domesticated animals and humans. However, some parasitic nematodes can be helpful to humans; for example, when microscopic nematodes are sprayed on a lawn, they enter host pests, such as white grubs, earwigs, or slugs, and can kill the pests within 72 h.

Life Cycle of Ascaris

Ascaris is a parasitic roundworm that can cause blockage in the intestines (**Figure 3**). Unlike some of its relatives, which may have two or three hosts in their life cycle, the ascaris inhabits various organs of a single host, such as a human, cow, dog, cat, or other mammal. In the host's intestine, adult worms mate, fertilization occurs, and then the female lays its eggs, releasing up to about 200 000 eggs per day. The eggs then pass out with the feces. A new infection occurs when a human or other animal consumes food or water contaminated with the eggs. The eggs hatch in the small intestine of the new host. The newly hatched larvae burrow through the intestinal wall, enter the blood stream, and are carried to the liver, heart, and then the lungs. They continue to grow in the lungs and after about 10 days, the young worms migrate up the windpipe and are swallowed and transported back to the intestine.

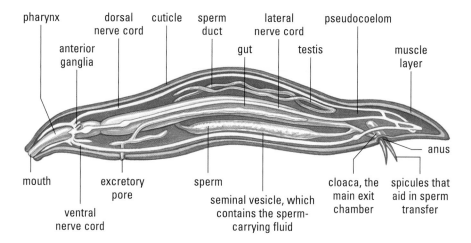

Figure 3
The anatomy of Ascaris, showing the male roundworm

Practice

Understanding Concepts

1. What is a proglottid?
2. What are the three body layers found in flatworms and other complex animals?
3. What is the function of the cuticle in tapeworms?
4. Why do intestinal parasites, such as tapeworms, have poorly developed digestive systems? Does this provide an evolutionary advantage? Explain.
5. Describe the life cycle of a tapeworm and the life cycle of Ascaris.
6. How does the presence of an intermediate host ensure that a tapeworm is carried from one host to the next?

Making Connections

7. A new method of controlling white lawn grubs involves the use of nematodes. The microscopic nematodes are sprayed on a lawn, where they enter the host insect (white grubs, earwigs, slugs, etc.)

and release a bacteria which destroys the grub larvae within 72 h. Discuss the advantages and disadvantages of using nematodes as a form of insecticide.

Exploring

8. Find out about biological forms of pest control other than the one mentioned in question 7. Select one organism and describe how it can be used as a pesticide. On the Internet, follow the links for Nelson Biology 11, 11.3.

 GO TO www.science.nelson.com

Phylum Annelida: Segmentation and a Coelom

The Annelids, or segmented worms, include well-known earthworms and red worms, used in backyard composters, as well as polychaetes and leeches. Annelids are the simplest animals that possess a coelom; consequently, they can be classified with the higher invertebrates. They also have organs and systems for circulation, digestion, reproduction, excretion, and coordination.

Annelids exhibit another characteristic found only in higher invertebrates and vertebrates—they have segmented bodies, which means that the body is composed of many structurally similar units. **Segmentation** is particularly advantageous for locomotion, because it allows flexibility of movement. A segmented organism can move individual parts at different times, whereas an unsegmented organism must usually move its entire body all at once.

segmentation: the repetition of body units that contain some similar structures

The Earthworm: A Representative Annelid

Ecologically, earthworms (**Figure 4**) are of tremendous importance to agriculture. They burrow in the soil, mixing and churning it, and thereby increasing aeration and drainage. They also help break down organic matter, making it available as nutrients for crop plants.

The earthworm's body is well suited to burrowing; it has approximately one hundred segments that form a bilaterally symmetrical, elongated tube tapered at both ends. Each segment is separated from the next by a septum, or wall, formed by a double layer of peritoneum. A combination of muscles around the body and along the length of the body allows individual segments to contract. The contractions move along the length of the worm in a wave—a process called peristalsis. Externally, bristles, known as setae, project from the lower surface of each segment. These bristles can be extended to anchor the worm in its burrow, and to assist in locomotion.

Earthworms eat soil. Soil, containing dead and decaying plant and animal matter, enters the mouth and is sent backward to the crop by contractions of the muscular pharynx. Here it is stored until it moves into a grinding organ called the gizzard. In the gizzard, organic matter is ground into small pieces, with the help of sand contained in the soil. Soil particles and undigested matter are then passed along the intestine, which extends the remaining length of the body, and are expelled through the anus. This material, called castings, helps fertilize the surrounding soil.

In the intestine, gases and molecules of digested food are absorbed into the watery fluid called blood and then carried to all the body cells via an extensive network of blood vessels. The branches of the vessels become progressively smaller, until they develop into capillary beds. There is no central heart, but thickened, muscular blood vessels in the anterior region act as pumps to help move the

DID YOU KNOW ?

While most North Americans are familiar with earthworms from their gardens or as bait used by anglers, they would be startled to encounter the giant earthworm of Australia. Specimens of these have measured over 3 m long.

clitellum

Figure 4
An easily recognized annelid, the earthworm. Partway along the body is a thickened band covering some of the segments, called the clitellum. This structure is permanently present in earthworms, but in other segmented worms it may form only during the breeding season.

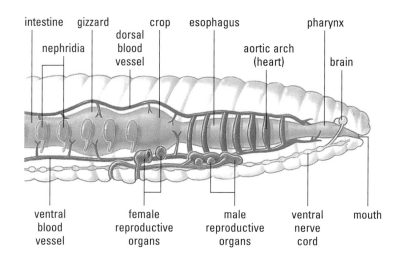

intestine   gizzard   crop   esophagus   pharynx

nephridia   dorsal   aortic arch   brain
             blood   (heart)
             vessel

ventral   female   male   ventral   mouth
blood   reproductive   reproductive   nerve
vessel   organs   organs   cord

Figure 5
The circulatory, excretory, nervous, and reproductive systems of an annelid

nephridia: open-ended tubules that function in excretion

clitellum: a smooth swollen band found about one third of the way along the body of some annelids. It secretes a protective covering for the eggs.

Figure 6
A polychaete, showing the fleshy projections on each side of each segment. These parapodia ("side feet") function in locomotion and gas exchange, and are armed with bristles for protection and traction.

blood. This entire system, known as a closed circulatory system, is not seen in the lower invertebrates. All the systems of an annelid are illustrated in **Figure 5**.

The annelid's excretory system is very much linked to its segmentation. Two open-ended tubules called **nephridia** (singular: nephridium) are found in virtually every segment. Each nephridium opens into the coelom in one segment, then passes through the septum into the next posterior segment. In that segment, the nephridium ends in an exit pore (nephridiopore) in the body wall. Waste materials are excreted through this pore to the outside.

A large dorsal ganglion (brain) is found in the annelid's head region. A ventral nerve cord runs the length of the worm, with small ganglia giving rise to nerves in every segment. There is a concentration of sensory cells at the anterior end, and along the body are scattered cells that appear receptive to chemical, mechanical, and light stimulation.

Earthworms are hermaphroditic, but fertilization occurs between two individuals. When the eggs are laid, the **clitellum** secretes a cocoon, which envelops and protects them. The young develop in the cocoon and leave it as fully formed, but smaller, worms.

Polychaetes and Leeches

Other types of annelids include polychaetes and leeches. Polychaetes are a diverse group of marine worms (**Figure 6**). Some species burrow into the ocean bottom; others live beneath stones and shells or build tubes to live in. Some are free-living, while others are parasitic.

Leeches are found in lakes, slow-moving streams, ponds and marshes, and on moist vegetation in humid environments such as jungles. About 75% of the known species of leeches are blood-sucking, external parasites (**Figure 7**). The leech attaches to its host and cuts through the skin to create a flow of blood. The leech's salivary glands produce a chemical, hirudin, which prevents the host's blood from clotting. Digestion is very slow, and leeches can tolerate long periods (over one year) of fasting between meals.

Figure 7
Feeding leeches. The best-known leeches are those used in medicine. The use of leeches for medical purposes goes back to 100 A.D., and was prevalent in the 16th to 19th centuries, when "blood letting" was a common treatment.

Frontiers of Technology: Biosurgery—Nature's Scalpels

Treating infected wounds is becoming more and more difficult. Many of the bacteria are becoming antibiotic resistant, and we are running out of treatment options. In New York State and in Great Britain, doctors are rediscovering new techniques in biosurgery by applying new scientific understanding to an old practice. What is biosurgery? It involves the use of living organisms in surgical treatments. This can include using maggots (the larval stage of flies) to cleanse wounds and leeches to reduce pooled blood pockets. Both of these techniques serve to speed the healing process.

In the case of wound cleansing, doctors working with African patients noticed that patients arriving with maggots in their wounds were generally healthier than patients with similar but nonmaggot-infested wounds. The maggot-riddled wounds were found to be significantly cleaner than expected. The maggots eat dead flesh and bacteria, leaving clean healthy tissue that heals more quickly. Such maggots are now being bred for use in hospitals and home-care, but only after more traditional methods of wound treatment fail. It is promising better relief for diabetic patients who often have circulation problems that can lead to gangrene in their feet. In the past, amputation was often the best way to stop the spread of the gangrene.

In the case of leeches, they play a critical role in helping to restore circulation in reattached fingers, toes, and limbs. Arteries and veins need to be reattached to restore blood circulation to the area as quickly as possible to prevent tissue damage. Occasionally, the veins are damaged beyond repair, and blood begins to pool in the newly reattached body part, often resulting in the loss of the limb. However, carefully applied leeches can painlessly suck up this stagnant blood. Leeches are well suited to their new task of draining blood from reattached body parts. Leech saliva contains a variety of chemicals, including a natural anesthetic, an anticlotting agent (hirudin), and a vasodilator (which causes blood vessels to open up). This combination allows the patient to feel no pain and promotes the circulation of blood critical to the healing of the reattached part. Several leeches later, the reattached body part has healed sufficiently to function and circulate blood correctly on its own. In many cases, the finger, toe, or limb would have been lost without the help of the leeches.

Practice

Understanding Concepts

1. What are setae?
2. What parts of an earthworm's anatomy are involved in its locomotion?
3. How does the earthworm carry out respiration?
4. What is the function of the clitellum?
5. Why is the earthworm considered an important member of the terrestrial ecosystem?
6. In which phyla do we first see three distinct body layers, ganglia, and a coelom?

Making Connections

7. (a) What special properties do the chemicals in a leech's saliva have?
   (b) Why are these medically useful?
8. Describe some possible consequences if earthworms did not exist.

Earthworm Dissection

In this activity, you will examine the external and internal anatomy of an earthworm. You will number each earthworm segment in sequence from anterior to posterior. Identifying particular segments will help you locate various organs. Use the worm map in **Figure 8** and draw diagrams to keep track of what you observe as you work through the dissection.

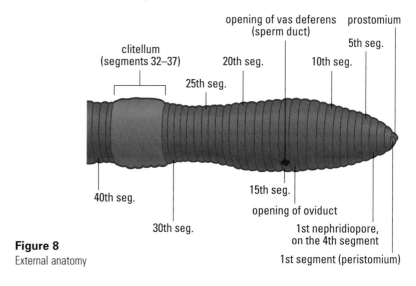

opening of vas deferens (sperm duct)
prostomium
5th seg.
clitellum (segments 32–37)
20th seg.
10th seg.
25th seg.
40th seg.
15th seg.
30th seg.
opening of oviduct
1st nephridiopore, on the 4th segment
1st segment (peristomium)

Figure 8
External anatomy

Materials

lab apron
latex or vinyl gloves
hand lens (dissecting microscope)
dissecting pins and forceps
probes or hard round toothpicks

eye glasses or goggles
earthworm (frozen or preserved)
dissecting pan
dissecting scissors or scalpel
ruler

Procedure

External Anatomy

1. Put on your lab apron, gloves, and eye glasses or goggles.

2. Place a moistened paper towel in the bottom of the dissecting pan to prevent the earthworm from drying out. Add water to the pan periodically to keep the animal moist.

3. Place an earthworm in the dissecting pan. Locate the anterior and posterior ends and the dorsal and ventral surfaces. Identify the following structures: mouth (surrounded by a hoodlike structure called the prostomium); anus (small hole in last segment); clitellum; and setae (on the ventral surface).

4. Record the number of body segments and examine the ventral surface of segment number 14, which contains the oviduct openings, and segment number 15, which contains the sperm duct openings. You may also be able to observe the nephridial pores, which are found in almost every segment.

 CAUTION: If preserved earthworms are used, they should be kept under running water for several hours before use to remove any preservative and eliminate any fumes.

Internal Anatomy

5. Stretch your earthworm out with the dorsal side up and the anterior end facing away from you. Place pins in the second and second-last segments to hold the worm in place.

6. Make a *very shallow* cut into the surface of the worm because the outer covering is extremely thin. Starting at the posterior end, carefully cut through the skin and muscle of your worm just to the midline along the dorsal surface.

7. As you cut, spread the edges apart carefully by cutting through the thin membranes (septa) that separate each segment. Hold the body wall back with forceps. After every 15 segments, place a pin (at a 45° angle) through the body wall to hold it in place. Repeat the cutting and pinning until you reach the anterior end. Describe the body cavity (coelom) of the earthworm.

 CAUTION: Use scalpels with care. Do not use rusty or dull ones. Do not keep a scalpel in your hand if you are not using it. If you have to do anything other than cut in your specimen, put the scalpel aside, away from the edge of your bench. Do the same with the probes.

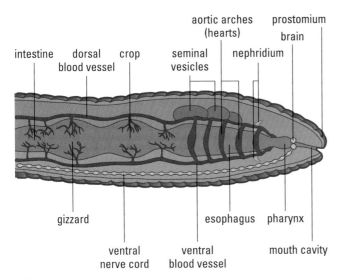

Figure 9
Digestive system (bilateral section)

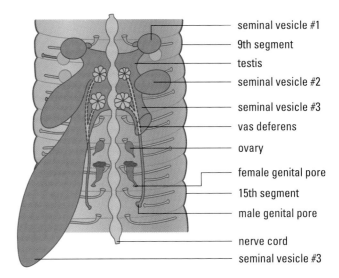

Figure 10
Reproductive system

8. Use **Figures 9** and **10** to help you locate and identify these features of the following body systems:

   Digestive System: pharynx, esophagus, crop, gizzard, intestine
   • Measure and record the length of the intestine and the digestive tract in millimetres.

   Circulatory System: dorsal and ventral blood vessels, five "hearts"
   • Examine and describe the circulatory system.

   Reproductive System: seminal vesicles (segments 9–12), sperm ducts (segment 15), oviducts (segment 14), and ovaries and testes
   • Some of these structures are small and hard to identify. Refer to the worm map. Examine and describe the reproductive system.

   Nervous System: "brain" (ganglion mass), ventral nerve cord
   • Because of its ventral location, you may need to remove part of the intestine to see the nerve cord beneath. Examine and describe the worm's nerve cord.

   Excretory System: paired organs called nephridia
   • These organs are small and located against the lateral walls in each segment except the first three and the last segment. Examine and describe the nephridia.

9. When you are finished, wrap your worm in paper towel and dispose of the tissue according to the teacher's instructions.

10. Wash the work area and your hands thoroughly.

Analysis

(a) Compare the number of segments of your earthworm with those of a classmate's. Explain any differences in the number of body segments.

(b) Why is the earthworm's body plan often called a "tube within a tube"?

(c) Approximately what percentage of the digestive tract is intestine?

(d) What are the functions of the crop and gizzard?

(e) Why are the "hearts" not considered true hearts?

(f) Is the circulatory system open or closed? Explain.

(g) What is the function of the seminal vesicles?

(h) Explain why an earthworm is not self-fertilizing.

(i) How does the position of the worm's nerve cord compare with your nerve cord?

(j) What is the function of the nephridium?

(k) How does the earthworm obtain oxygen if no respiratory system exists?

Synthesis

(l) How does the nervous system of the earthworm demonstrate greater specialization than that of the more primitive invertebrates shown in **Figure 11**?

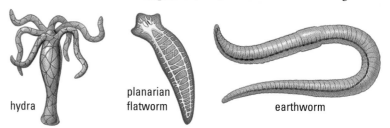

hydra

planarian
flatworm

earthworm

Figure 11

(m) How is the increased complexity of the earthworm's digestive system related to segmentation?

(n) Using **Table 1** as a guide, compare the features and traits of a representative invertebrate and vertebrate. (Refer to Chapter 8, Activity 8.7.1: Examining the Systems of a Fetal Pig to help you.)

SUMMARY **Worms**

Table 1 summarizes the main similarities and differences among the three phyla of worms.

1. Platyhelminthes, or flatworms, have no coelom but they show bilateral symmetry, have mesoderm, true organs, and a primitive brain.

2. Nematodes, or roundworms, are more advanced because they have a pseudocoelom, a nervous system, and a complete digestive tract with a mouth and an anus.

3. Parasitic worms show a variety of adaptations to a parasitic lifestyle.

4. Annelids are segmented worms and possess a true coelom.

5. Segmentation is an important evolutionary advantage because it allows for greater specialization.

Table 1: The Worms

	Platyhelminthes (flatworms)	Nematoda (roundworms)	Annelida (segmented worms)
Representative	tapeworm, planarian, fluke	ascaris, hookworm	earthworm, leech
habitat	• aquatic and terrestrial • many are parasites	• aquatic and terrestrial • some are parasites	• aquatic and terrestrial • few are parasites
body plan	• bilateral symmetry • three germ layers • acoelomate	• bilateral symmetry • three germ layers • pseudocoelomate	• bilateral symmetry • three germ layers • true coelom
reproduction	• asexual reproduction by fission • sexual reproduction—hermaphroditic cross-fertilization	• sexual reproduction—separate sexes	• sexual reproduction—hermaphroditic with cross-fertilization
circulation	• none except diffusion	• none except diffusion	• five pairs of aortic arches and two large blood vessels run along the dorsal and ventral surfaces
digestion	• single opening functions as a mouth and anus • digestive organs present but reduced in parasite species	• separate mouth and anus • digestive organs present	• separate mouth and anus • digestive organs present
nervous system	• primitive brain • two longitudinal nerve cords	• primitive brain • dorsal and ventral nerve cords	• more advanced brain • large ventral nerve cord with many peripheral ganglia

Section 11.3 Questions

Understanding Concepts

1. What type of body symmetry is characterized by the possession of a head? Explain.

2. Describe the benefits of segmentation.

3. Worms encompass three different phyla. Why are they not grouped in the same phylum?

4. What two major advances do flatworms have over cnidarians?

5. (a) What special advantage does a circulatory system provide the earthworm?
   (b) Explain how sponges and cnidaria have survived for approximately 500 million years without benefit of a circulatory system.

6. **Figure 12** represents a phylogenetic tree showing relationships that have evolved over 570 million years. Copy the diagram into your notebook and attach the following labels: radial symmetry, bilateral symmetry, acoelomates, pseudocoelomates, coelomates, animals without tissues, animals with tissues, animals with organs.

7. Why do you think annelids are referred to as "true" worms? Provide an example of a "nontrue" worm to help you explain.

(continued)

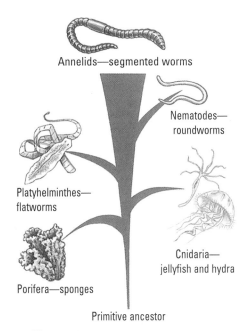

Annelids—segmented worms

Nematodes—roundworms

Platyhelminthes—flatworms

Cnidaria—jellyfish and hydra

Porifera—sponges

Primitive ancestor

Figure 12
Phylogenetic tree of invertebrates

Exploring

8. Heartworm is a roundworm disease in dogs caused by the bite of the female mosquito. Research this disease.
   (a) Describe the generalized life cycle of this annelid.
   (b) Describe the effects that this parasite has on the host.
   On the Internet, follow the links for Nelson Biology 11, 11.3.

GO TO www.science.nelson.com

11.4 Mollusks and Echinoderms

Phylum Mollusca

The phylum Mollusca contains roughly one hundred thousand species and is the second largest in the animal kingdom, next to the arthropods. Mollusks show extreme diversification in appearance, and are grouped into seven classes. The three largest classes are the gastropods (e.g., snails, slugs), bivalves (e.g, clams, oysters), and cephalopods (e.g., squids, octopuses).

Despite their diversity, all mollusks have a coelom, a complete digestive tract, and are bilaterally symmetrical at least during some part of their lives. For example, a snail shows bilateral symmetry as a larva (**Figure 1(a)**). When it becomes an adult, its body becomes coiled like its shell so its shape becomes asymmetrical (**Figure 1(c)**). Another commonality is that all mollusk eggs follow a similar pattern of development: the mouth develops from the first opening formed, and the anus develops from a second opening formed later in the embryo. Consequently, mollusks are called protostomes (meaning "first mouth").

Most mollusks go through a stage of development when they are free-swimming larvae, called trochophore larvae. All mollusks, except bivalves, have a distinct head on which various sensory organs are located, and have a radula, which is a rasping, tonguelike organ with hard teeth that is used to scrape or cut up food. Scientists speculate that bivalves lost these basic traits as they became specialized at filtering food from the water.

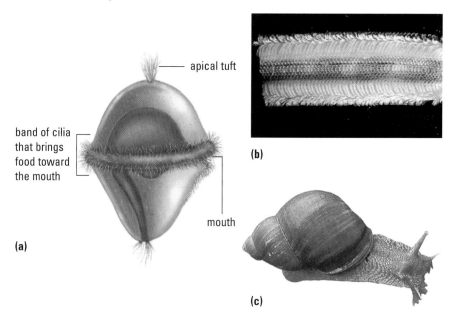

apical tuft

band of cilia that brings food toward the mouth

mouth

(a)

(b)

(c)

Figure 1

(a) Tiny trochophore larva of a snail

(b) Close-up of a radula

(c) The land snail is typical of terrestrial gastropods. Its shell is sealed except for a small hole through which air passes directly into the many blood vessels found at the tissue surface.

Figure 2 illustrates variations of the same basic body plan shared by the three major groups of mollusks: gastropods, bivalves, and cephalopods. The following are the most conspicuous common features of this body plan:

- foot—This muscular foot functions in locomotion. In cephalopods, the foot has become modified into a head and appendages.
- mantle—This tissue hangs down like a cloak around some or all of the body.
- shell—It can be an internal or external structure, consisting largely of calcium carbonate.
- gills—Gills are specialized respiratory structures that exist only in aquatic species and that arise as outgrowths of the mantle wall.
- visceral mass—It comprises the internal organs consisting of the gut, kidneys, heart, and reproductive organs.

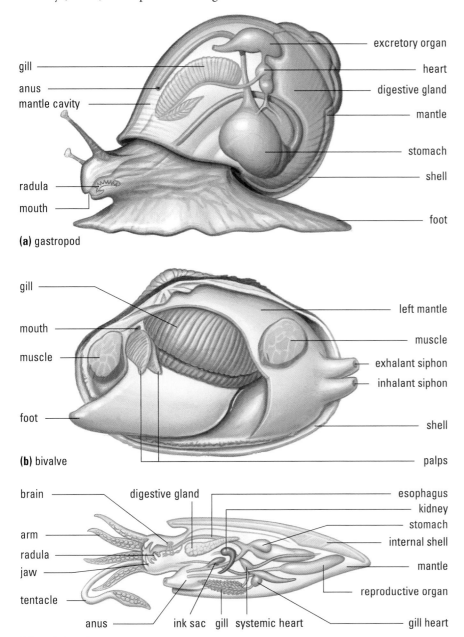

(a) gastropod

(b) bivalve

(c) cephalopod

Figure 2
(a) Body plan of a gastropod: an aquatic snail, which has gills
(b) Body plan of a bivalve: a clam, showing its shell lifted off
(c) Body plan of a cephalopod: a squid, which has arms and tentacles

Figure 3
Some aquatic gastropods have lost their shell and are soft bodied. The common name for these gastropods is nudibranchs.

sedentary: used to sitting still much of the time; moving little and rarely

Ecological Role of Mollusks

Mollusks inhabit virtually all known aquatic and terrestrial ecosystems and are critical links in many food chains. They are sources of food for fish, starfish, birds, humans, and a variety of other mammals. Their empty shells also provide shelter and homes for other invertebrates. Herbivorous mollusks feed on plant matter that is scraped from rocks and shells; terrestrial gastropods, such as snails and slugs, are garden pests and can cause damage to agricultural crops. Mollusks are often intermediate hosts for various parasitic flukes. Predatory mollusks, such as squids and octopuses, capture fish and invertebrates with powerful tentacles and suckers. Bivalves such as clams, oysters, and mussels, are filter feeders, filtering food such as phytoplankton from the water using feathery gills. In the process, dangerous and toxic pollutants become concentrated in the tissues of bivalves. While this feature is useful to scientists for the purpose of environmental monitoring, organisms that consume these mollusks may become sick and die.

Class Gastropoda

Gastropods (meaning "stomach footed") make up the largest group of mollusks and include snails, slugs, whelks, limpets, periwinkles, conches, and sea hares (**Figure 3**). The most familiar forms have a heavy, coiled shell. The foot has a flat, hard plate, which fills the opening to the shell and protects the soft body when the foot is withdrawn. A carnivorous gastropod uses its radula to drill a hole in the shell of a prey, and then extends its tubelike mouth into the hole to tear and ingest pieces of food.

Class Bivalvia

Bivalves include clams, mussels, oysters, and scallops (**Figure 4**). They have a two-part shell which is pulled together by two strong muscles. Most bivalves are **sedentary** because of their heavy shells. They have a complex filter-feeding mechanism that involves the use of large gills covered with cilia. Water circulating through the body carries many small particles of food, including single-celled algae, small animals, and dead organic material. As the water passes across the gills, the particles are filtered out and embedded in mucus. Strings of mucus and food are then moved toward the mouth by the cilia. Digestion occurs in the stomach, and undigested material moves along the intestine to exit via the anus, which is situated near the exhalant siphon.

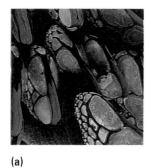 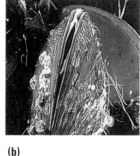 

(a)   **(b)**   **(c)**

Figure 4
The bivalves are familiar to humans as an important source of food.
(a) Mussels secrete mucous threads that harden on contact with water and that attach these animals to rocks and to one another.
(b) Oysters are often grown on farms, not only for food but also to produce pearls that are used in jewellery.
(c) Scallops have numerous stalked blue eyes, each with a lens, on the fringed mantle along the shell edge.

Class Cephalopoda

Cephalopods (meaning "head footed") are the most complex mollusks, and include squids, nautiluses (**Figure 5**), and octopods. Octopods (*octo* meaning eight) have eight equal arms, all equipped with suckers. Squids have ten appendages, including two tentacles that are longer and have suckers on expanded tips. The tentacles trap and grasp prey organisms, which are then drawn toward the mouth. Nautiloids have a cluster of many slender, suckerless arms.

In squids, the body is streamlined and holds its shape with an internal remnant of a shell. In octopods, the shell is absent and the body is very flexible. Only the nautilus has a complete external shell. In both octopods and squids, digestion occurs in the muscular stomach, aided by enzymes from the digestive gland. Nutrients are absorbed in the stomach, with some additional absorption in a straight or coiled intestine. Undigested material is excreted into the mantle cavity.

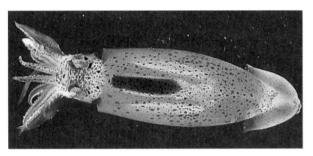
(a)

(b)

Figure 5
(a) A squid. Both squids and octopods use water jets, powered by contraction of the mantle, for fast swimming.
(b) Nautiloids swim with the aid of their numerous arms.

A cephalopod's predatory nature, which involves the capture of motile prey, requires efficient respiratory exchange and circulation. The circulatory system of cephalopods is closed and capable of maintaining a high flow rate and blood pressure. The large, extensively folded gills have a separate "gill heart" to pump blood back to a "systemic heart," which pumps it out to the body.

The cephalopod's nervous system is also highly developed, with a large brain that contains several ganglia. In the octopus, the brain has ten lobes, and the animal is a capable learner. Octopods have been trained to solve simple problems and have the ability to remember. Squids and octopods have highly developed eyes that function in much the same way as human eyes do. Complex eyes are crucial to the survival of these predatory animals, which depend on vision to find their prey.

Cephalopods are **dioecious**, and fertilization is internal. In squids, the fertilized eggs are deposited in gelatinous capsules, and both the male and female die after mating. In octopods, the fertilized eggs are attached to the roof of the cave or crevice in which the female lives, and she guards and cleans them for several months until they hatch. During this time, the female eats little or nothing.

dioecious: describes organisms in which the male and female reproductive organs or gonads are carried by separate individuals

Frontiers of Technology: Mussels and Superglue

Have you ever wondered how certain shellfish such as mussels cling so tenaciously to underwater objects? Mussels stick to substrates by means of a byssus (a mass of filamentous threads) tipped with a strong, waterproof adhesive protein made in the foot of the animal. The "glue" secreted by these marine mussels is strong, hardens quickly, and sticks underwater.

A major goal of biotechnology is to duplicate naturally occurring substances for use by humans. No commercial glue works well in wet, saline environments. Until recently, mussel glue extractions required a laborious process that yielded just 1 g of glue from every 3000 mussels. In an attempt to decrease the harvesting

Figure 6
The sea cucumber is a sand burrower and plankton feeder.

endoskeleton: an internal skeleton

Figure 8
The sea urchin exhibits a spiny skin typical of echinoderms. The harvesting of sea urchins is a fairly significant industry in many coastal areas. In 1998, approximately 1540 tonnes (t) of sea urchins were landed in California, with a wholesale value of $8.2 million (US).

of large numbers of marine mussels, scientists turned to tools and techniques of biotechnology and genetic engineering.

In initial research, the protein that makes up the sticky substance was identified. Researchers have since isolated and created a gene that carries the code for the protein. When this modified gene is inserted into yeast or bacteria, it directs the production of large quantities of the glue, which sets faster, bonds more strongly, and can be broken down easily, if required.

This superglue adheres to almost any kind of surface and, as a result, synthetic versions of the protein are being developed for medical, dental, and industrial uses. In the area of dentistry, the adhesive is ideal as a sealant for filling cavities, bonding teeth, and forming a strong bond between gums and teeth. Surgical applications include coating sutured tissues to prevent infections, joining small broken bones or tendons, and acting as an ophthalmic glue for cornea and retina repairs. Industrial uses include making underwater repairs to hulls of ships to prevent corrosion and leaking.

Phylum Echinodermata

Echinoderms (*echinos* meaning spiny or bristly) are relatively large aquatic animals that are sessile (e.g., sea lilies), sand burrowing (e.g., sand dollars), or slow crawling (e.g., starfish). The body structure shows pentamerous radial symmetry, *penta* referring to the five similar body segments arranged around a central core. The larval stages, however, are bilaterally symmetric; it appears that echinoderms have a bilateral ancestor and developed radial symmetry secondarily. The development of echinoderm eggs follows a unique pattern found in only one other major phylum, Chordata, suggesting that echinoderms and chordates share a common ancestor. The anus forms from the primary opening and the mouth forms from a second opening later in development; consequently, echinoderms and chordates are deuterostomes (meaning "second mouth").

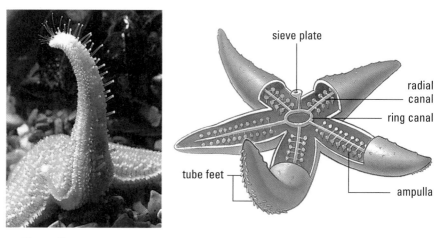

Figure 7
Many echinoderms can regenerate missing parts of their bodies. A starfish can regenerate an entire arm. Every arm contains identical structures, including the male or female sex organ.

The major unifying characteristic of echinoderms is their water vascular system, which is a unique system of interconnecting, fluid-filled tubes (**Figure 7**). Locomotion and feeding in echinoderms are accomplished by the water vascular system, which operates much like a hydraulic pump. The skeleton may be loosely connected to allow flexible movement, as in starfish, or fused to form a rigid skeleton, as in sea urchins (**Figure 8**). Since the skeleton is covered with a thin layer of epidermis, it is a true **endoskeleton**.

Ecological Role of Echinoderms

Echinoderms exist as bottom dwellers in all oceans, from shoreline waters to great depths. Many tend to live in aggregations of enormous numbers and play vital roles within the environments they inhabit. For example, some echinoderms recycle nutrients in the ecosystem by feeding on dead and decaying materials. Starfish, as important marine carnivores, control populations of other animals. In certain coastal zones, the presence of sea urchins is essential to the control and distribution of algae. But due to their voracious appetites, large starfish populations can cause disasters, such as wiping out entire oyster beds in a short time, and extensively damaging coral reefs, as seen recently in the Pacific Ocean. Sea urchins compete with gastropods for kelp.

While echinoderms are virtually the only major phylum in the animal kingdom without parasitic species, they are the favourite hosts for an enormous number of parasites from other animal groups. Among the worst enemies of echinoderms are certain parasitic snails that either penetrate the body wall, leaving a hole from which they feed, or take up residence inside the body of the host.

Birds also take a heavy toll on certain echinoderms, such as shoreline sea urchins, by breaking the hard skeleton and feasting on the soft internal organs.

Explore an Issue

Role Play: Zebra Mussel Conference

DECISION-MAKING SKILLS

- Define the Issue
- Identify Alternatives
- Research
- Analyze the Issue
- Defend a Decision
- Evaluate

When species are introduced, either deliberately or accidentally, into biological communities where they have never been before, they can create enormous ecological and economic damage. This is the case with zebra mussels, the small oblong mussels with distinctive black and white striped markings. Sometime in 1986, zebra mussels entered the Great Lakes and, since then, they have spread dramatically throughout the entire Great Lakes system and connecting waterways.

Question

What can be done to stop the zebra mussel invasion in eastern Canada?

- In your group, research one of the following aspects of the situation:
  (a) the life cycle and feeding habitats of zebra mussels (e.g., Why are zebra mussels so successful?);
  (b) the source of the zebra mussels and the history or time line of their spread in eastern North America;
  (c) the ecological impacts (Has their presence created additional endangered species in Canada?);
  (d) the economic impact of the zebra mussel invasion on various industries;
  (e) control strategies and their success.
- Search for information in newspapers, periodicals, CD-ROMs, and on the Internet.
  Follow the links for Nelson Biology 11, 11.4.

GO TO www.science.nelson.com

- Write a report, outlining the information your group considered.
- Prepare to share your group's report in a class discussion, and suggest a solution, or solutions, to the question.
- Apply the class information to suggest ways of avoiding, or at least reducing, accidental introduction of new species into established biological communities.

Understanding Concepts

1. List five members of the phylum Mollusca, and two members of the phylum Echinodermata.

2. What are the main parts of the mollusk body plan?

3. What is the function of the mantle?

4. How does the mollusk's gas exchange system compare with that of the segmented worm?

5. Which phylum has
   (a) a trochophore larva?
   (b) a planula larva?

6. List three factors that suggest that cephalopods are the most advanced mollusks.

7. What type of symmetry do adult echinoderms possess?

8. What is the function of the water vascular system in echinoderms?

SUMMARY Mollusks and Echinoderms

1. All mollusks possess the same early development of the embryo, and have a larval stage before they mature into adults.

2. Adult mollusks vary greatly in their appearance but they share a common body plan.

3. All echinoderms are spiny-skinned animals that live in marine habitats.

4. Echinoderms possess a different type of early embryonic development from mollusks. The adult forms possess radial symmetry with five similar body segments.

Section 11.4 Questions

Understanding Concepts

1. Describe the common body plan of mollusks.

2. Why are cephalopods included in the phylum Mollusca even though they look so different from snails, clams, and most of the other mollusks?

3. Echinoderms and mollusks share similar environments.
   (a) Why are clams, snails, squids, and octopuses placed in the same phylum?
   (b) Why are the sea urchins and starfish not classified with the mollusks?

4. Most of the shelled mollusks are relatively small, whereas cephalopods can become very large. Suggest a reason for this.

5. What makes mollusks similar to annelids but different from the early invertebrates?

6. How might an organism's size be related to the complexity of its digestive tract?

7. Echinoderms and chordates are both deuterostomes, yet they have different kinds of body symmetry. Provide a possible explanation.

11.5 Arthropods

Phylum Arthropoda is by far the largest phylum in the animal kingdom, containing roughly one million species or 90% of all known animal species. Arthropods include insects, as well as centipedes, scorpions, spiders, mites, ticks, crabs, lobsters, and barnacles. They can be found in all habitats, including the Arctic ice, deserts, the deep ocean, the mountains, prairie ponds, and the skies overhead.

Arthropods evolved from a segmented ancestor; however, in most cases, segments have been lost or fused together. Generally, the fused segments form a head, thorax, and abdomen, each specialized to perform certain functions for the whole animal. In many groups, the head and thorax are fused into a cephalothorax. Located on the fused segments are highly specialized jointed appendages that may be sensory or adapted for food manipulation. Features of various arthropods are shown in **Figure 1**.

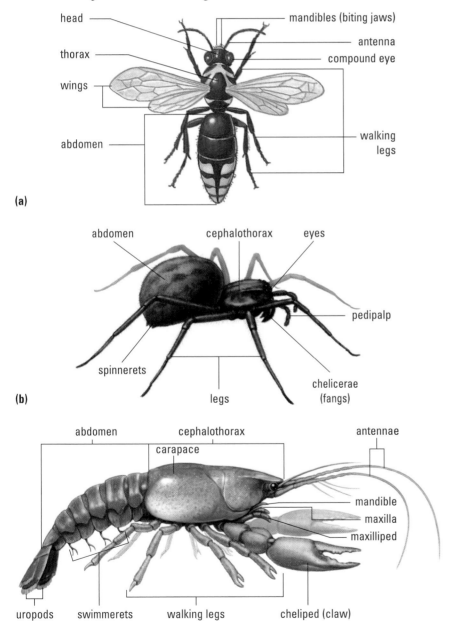

(a)

(b)

(c)

Figure 1
Features of an arthropod
(a) A wasp. The frontmost appendages on insects and crustaceans are the antennae (singular: antenna), used for sensory function.
(b) A spider. The frontmost appendages are used for feeding.
(c) A crayfish. Crustaceans have many specialized appendages—some for food handling, some for defence, some for locomotion.

One of the key features separating arthropods from all other animals is the strong external skeleton, called the **exoskeleton** or cuticle. It is composed of chitin, a nonliving material, that may be thick and heavy, as on a lobster's claws, or thin and light, as on a grasshopper's wings. The innermost layer is thin, relatively flexible, and is continuous across the joints, protecting the arthropods while allowing them to move. Without jointed appendages, the exoskeleton would be too cumbersome for locomotion.

The exoskeleton protects internal muscles and blood vessels. In humans, by contrast, the muscles cover the internal skeleton, making muscles and blood vessels more susceptible to injury; however, the muscles also cushion the skeleton, making it less vulnerable to impact. Also, your skeleton contains living bone and so it grows with you, unlike the arthropod's skeleton.

The exoskeleton cannot grow along with the organism, so an arthropod grows by shedding its cuticle, or moulting. In moulting, the inner layers of the cuticle dissolve and are recycled to make a new and larger cuticle beneath the old one. This new cuticle is initially flexible and has many folds. The animal swells as it takes in water (or air, in some cases). The pressure created by the swelling splits the remaining old cuticle, which is shed, and expands the new cuticle. Once it has stretched, the new cuticle begins to harden. An arthropod is very vulnerable during moulting, as it lacks the protection of its cuticle, and may not even have enough rigidity in its body to escape a predator. Once the cuticle hardens, the excess water or air that swelled the body is expelled to provide growth space. The crayfish moults seasonally and is also capable of regenerating lost appendages (**Figure 2**).

Figure 2
Crayfish are arthropods that live in freshwater habitats and moult seasonally. They can regenerate lost appendages.

Other Arthropod Characteristics

The nervous system of primitive arthropods resembles that of annelids. More advanced arthropods, especially those with complex social behaviours such as crabs, show an increase in brain size and complexity, allowing for an increased ability to sense and respond to changes in their environment. This allows for greater control over the entire body.

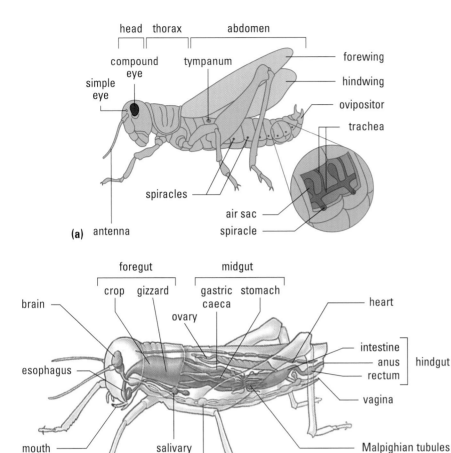

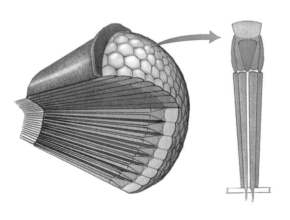

Figure 3
A representative arthropod—the grasshopper
(a) External anatomy
(b) Internal anatomy

The external and internal anatomy of a representative arthropod are shown in **Figure 3**.

Various types of sensory organs exist. The eyes may be simple and consist of only a few **photoreceptors**, or may have thousands of receptors capable of forming a crude image (**Figure 4**). Antennae are touch-sensitive and may be important at light levels that are too low for eyes to function well. Primitive "ears," composed of a flexible membrane stretched across an opening in the cuticle, may also be present. Projections or pits scattered around the exoskeleton are sensitive to mechanical and chemical stimuli and may be important in prey detection.

photoreceptors: sensory receptors that detect light energy

Figure 4
Insects and many crustaceans have a compound eye, which is composed of many light-sensing units. Each unit provides a dot of light, and the image is made up of an array of combined dots. Compound eyes have no mechanism to change focus, so the image they perceive may be very coarse. However, they are efficient at detecting movement.

Arthropods have an open circulatory system, with blood bathing the tissues in spaces or sinuses. The system of open spaces, called a **hemocoel**, replaces the coelom as the major body cavity. A coelom exists but is much reduced. A heart and arteries deliver blood to the body and sinuses. The blood may be oxygenated by gills in aquatic forms, or by organs called book lungs in some terrestrial forms. In many groups—the insects, in particular—the blood serves only a secondary role in delivering oxygen to the cells and removing carbon dioxide. These groups have a tracheal system, consisting of open tubes connecting to the outside by spiracles, which are openings in the cuticle. The extensively branched and cuticle-lined tracheal system reaches throughout the body and connects cells directly to the atmosphere.

All arthropods have a digestive system consisting of three basic parts:

1. foregut—from the mouth to the esophagus and then through any preliminary digestive organs (e.g., crop, gizzard in insects)

2. midgut—the stomach, where digestion of food is completed

3. hindgut—from the intestine through to the anus

Many terrestrial arthropods have specialized excretory organs called Malpighian tubules. These elongated organs join to the beginning of the intestine and secrete fluid into the hindgut, where reabsorption of fluids and dissolved materials occurs. This system enables the organism to retain much of the water contained in its food.

Most arthropod species are dioecious and have internal fertilization. The eggs of some species develop directly into a form that resembles the adult, with no intervening larval stage (e.g., millipedes, spiders, scorpions, some beetles). In others, during the nymph stage (**Figure 5**), the free-living stage after the egg, it is similar to the adult but differs in some ways—perhaps by the number of appendages or the absence of sex organs (e.g., grasshoppers, dragonflies, cicadas, cockroaches). Many groups have a larval stage, when they resemble the adult only superficially or not at all; such forms undergo metamorphosis (e.g., moths, flies, ants, lobsters).

Figure 5
Some cicadas live for 17 years, spending 16 years as nymphs, living in the ground, feeding on roots. In the last year of its life, the nymph usually crawls up on a tree for its moult to the adult. The newly emerged adult, which lives for only 4 to 6 weeks, does not eat.

DID YOU KNOW ?

For its size, spider silk is five times as strong as steel. A spider gene that codes for silk production has been transferred to bacteria to allow for industrial production of this strong fibre.

Arthropod Diversity

Arthropods are divided into four subphyla: Trilobitomorpha, Chelicerata, Crustacea, and Uniramia. Trilobites, abundant 65 million years ago, are now extinct. About 4000 species have been identified from fossil records (**Figure 6**). The other three groups are diverse, with many living members.

Chelicerates have no antennae, and the first pair of head appendages are feeding appendages. Arachnida, the best-known class in this subphylum, includes spiders, scorpions, ticks, mites, and several smaller groups. There are over 70 000 species, about half of which are spiders (**Figure 7**).

Figure 6
Trilobites were segmented, with jointed appendages, and their body was marked by two longitudinal furrows into three lobes—hence, their name.

Figure 7
Spiders capture and paralyze prey by injecting poison and then digestive fluid, which turns the tissue of the prey into a liquid to be sucked up by the spider.

(a)

(b)

Figure 8
(a) A rock crab is a typical crustacean.
(b) Barnacles are the only crustaceans that are sessile, but their motile larvae are similar to those of other crustaceans. Many species attach themselves to ocean docks as well as the bottom of oceangoing ships and submarines. Considerable effort and money is spent cleaning these animals off.

Crustaceans (**Figure 8**) include well-known, economically important taxa such as lobster, crab, and shrimp, as well as tiny crustaceans such as krill, copepods, and water fleas (*Daphnia*). Typical crustaceans have two pairs of antennae, three pairs of feeding appendages grouped around the mouth, and a variety of modified trunk appendages for swimming or walking. Most species have a free-swimming larval stage.

The remainder of the arthropods are in the subphylum Uniramia. Arthropods in this group are characterized by uniramous or unbranched appendages, and have one pair of antennae and a hardened head capsule. There are two main divisions: myriapods and insects. Myriapods include centipedes and millipedes, neither of which has the number of legs their name implies. Their bodies are strongly segmented. Centipedes (**Figure 9**) have one pair of walking legs on almost every body segment, while millipedes have two pairs of walking legs on almost every segment.

Class Insecta is the most diverse group of organisms on Earth and accounts for over 750 000 species, which is about 85% of all species of living things (**Figure 10**). Insects can be distinguished from other arthropods by three pairs of legs. The legs may be used for walking, but some may be modified for grasping prey, as in the praying mantis. Also, all insects have a single pair of antennae and one pair of compound eyes. There may be one or two pairs of wings, or no wings. The last segment of the abdomen consists of only reproductive structures for mating or egg laying (**Figure 11**).

Figure 9
This centipede's body is quite flat and is well suited for slipping into small crevices. Millipedes have rounded bodies.

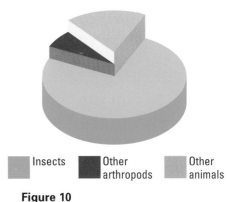

■ Insects ■ Other arthropods ■ Other animals

Figure 10
Approximate proportions of animal species. Because many insects have not been classified, the proportions are only estimates.

(a)

(b)

(c)

(d)

Figure 11
There are 30 orders of insects, but most belong to (a) Coleoptera: beetles; (b) Diptera: flies; (c) Hymenoptera: ants, bees, wasps, and sawflies; or (d) Lepidoptera: moths and butterflies.

Feeding habits may be predicted by looking at an insect's mouth. Some insects have mouthparts designed for biting (e.g., scorpionflies, horseflies, blackflies), or for chewing (e.g., beetles, earwigs, grasshoppers, crickets). Others have structures specially adapted for piercing and then sucking animal or plant juices (e.g., mosquitoes, fleas, aphids, chinch bugs), or for chewing and sucking (e.g., bees and wasps), or for sponging up liquified food (e.g., houseflies). Butterflies and moths suck nectar from flowers with their long tubular mouthparts.

Ecological Role of Arthropods

Arthropods are found nearly everywhere and are the direct or indirect source of food for countless organisms. Insects such as bees and butterflies pollinate many of the world's plants, including vital fruit and vegetable crops, and they play an integral role in the food webs that include humans. Crustaceans such as lobster, crab, and shrimp are an important food source for humans, as well as other animals. Aquatic food webs are based on tiny crustaceans such as copepods and larvae and shrimplike krill (**Figure 12**), which are a direct food source for both large and small animals. Antarctic krill are extremely important, being the main food supply for baleen whales, penguins, and seals.

On the negative side, insects compete with humans for food, claiming over half of the harvest in some areas of the world. Billions of dollars are spent annually on insecticides, which act with varying degrees of success to reduce the harmful effects of insect pests. Also, costly damage often results from the work of arthropods, such as barnacles, termites, and carpenter ants. Insect vectors, such as ticks, fleas, and mosquitoes, transmit serious diseases such as malaria and sleeping sickness as they feed on the blood of humans, agricultural livestock, or other organisms. Natural control of arthropods takes several forms: parasitic arthropods help control the numbers of other arthropods by laying their eggs on the host's body, and many arthropod predators kill pest organisms by the millions.

Figure 12
In polar seas, the springtime turnover of water stimulates a growth explosion of phytoplankton, which in turn stimulates a growth explosion of krill. Krill live in great swarms that are often several hundred metres in diameter and 40 m thick. An average-sized blue whale may consume 3 t at one feeding.

SUMMARY **Arthropods**

1. The most diverse phylum of animals is Arthropoda, characterized by an exoskeleton and jointed appendages.

2. Arthropods include spiders, crustaceans, and insects, which play a vital role in the survival of all life on Earth.

3. In arthropods, some segments are fused and are specialized to perform certain functions for the whole animal.

Practice

Understanding Concepts

1. Describe at least two features that all arthropods have in common.
2. Why is Arthropoda often called the most successful phylum?
3. Why is moulting so important to arthropods?
4. What is a tracheal respiration system? What are its advantages?
5. What is the advantage of metamorphosis in insects?
6. What body features help the crayfish adapt to its environment?
7. List some of the harmful effects arthropods can have.
8. List some useful products which are the result of arthropod activity.

Reflecting

9. Describe what it would be like to have an exoskeleton. For example, imagine the difficulties you would experience if you were fitted with a suit of armour at birth.

10. How do you think diversity and abundance contribute to the success of a phylum such as Arthropoda?

Classifying Invertebrates

Use the information in **Table 1** to help you classify each of the following invertebrates according to phylum and class, if possible.

(a) has six legs that are found on the middle body segment, the thorax

(b) has many tube feet on each of five arms arranged around a central core

(c) has eight appendages and a flexible body with no skeleton

(d) has a closed circulatory system and an asymmetrical external shell

(e) has eight legs and an exoskeleton

(f) has an external shell, a closed circulatory system, and many tentacles

(g) has a heavy two-part shell and gills for gas exchange

(h) has many jointed appendages and gills for gas exchange

Table 1: Mollusks, Echinoderms, and Arthropods

	Mollusks	**Echinoderms**	**Arthropods**
Representative	• clam, oyster, octopus	• starfish, sea urchin, sand dollar	• insect, crayfish, spider
habitat	• aquatic and terrestrial; a few are parasites	• aquatic	• ubiquitous (almost everywhere)
body plan	• bilateral symmetry • three germ layers • true coelom	• adults—pentamerous radial symmetry • larvae—bilateral symmetry • three germ layers • true coelom	• bilateral symmetry • three germ layers • true coelom • various stages of metamorphosis
reproduction	• sexual reproduction—hermaphroditic with cross-fertilization, or separate sexes	• sexual reproduction—separate sexes, regeneration	• sexual reproduction—separate sexes, at some time during life cycle
circulation	• closed circulatory system	• open circulatory system (a series of blood channels without walls)	• open circulatory system • heart located in hemocoel • blood pigments do not carry oxygen
digestion	• radula • filter feeding	• one-way digestive tube	• one-way digestive tube
nervous system	• primitive brain • light receptors	• oral and radial nerve cords	• definite brain with two ventral nerve cords
respiration	• gills • mantle	• diffusion	• tracheal system • gills

Figure 13
Dr. Gail Anderson

Research in Canada: Dr. Gail Anderson

Forensic entomology is the study of insects associated with a corpse, usually a murder victim, to determine when the victim died and if the body was moved after death. One of the few forensic entomologists in Canada is Dr. Gail Anderson (**Figure 13**) of Simon Fraser University.

There are two main ways of using insects to determine time of death. One is to study the successional waves of insects on the body. As decomposition takes place, many different species of insects are attracted to the remains during the various stages of decomposition. Knowing the insect fauna common to a particular region and the times at which they will colonize carrion (dead and decaying flesh), Dr. Anderson can tell from the remains roughly when death took place.

The second method is to study maggot age and development. The insects analyzed in this method are blowflies (Calliphoridae), which are the first to arrive on a corpse and lay their eggs. Blowfly development follows a predictable cycle; therefore, Dr. Anderson can work out time of death simply by identifying the age of the insects present. For example, if the new generation of insects on the body are 10 days old, it means the victim has been dead for 10 days.

Dr. Anderson and her students are now creating a provincial database of carrion insects by geographic region, season, and body location (in the sun or shade; above ground, below ground, or in water). They are also starting to look at using insects to determine the presence of drugs such as cocaine and valium in the victim's system at the time of death. With all of this information, the RCMP in British Columbia—and eventually across Canada—will be able to use forensic entomology in any homicide investigation.

Section 11.5 Questions

Understanding Concepts

1. What are some features that have made arthropods so successful?

2. What is an exoskeleton? Name three functions of the exoskeleton.

3. What were some features that enabled arthropods to adapt to terrestrial life?

4. What are some of the positive effects and negative effects which arthropods have on humans?

5. Compare the circulatory system of an earthworm with that of a grasshopper.

Making Connections

6. Why are arthropods vital to the survival of life on Earth?

7. In the tropics, most types of arthropods exist in greater numbers and have a wider variety of species. What are possible reasons for this?

Exploring

8. Research the income generated from lobster, oyster, mussel, and clam fisheries in Canada over several years. Has the income increased, decreased, or remained steady? Why?
Follow the links for Nelson Biology 11, 11.5.

GO TO www.science.nelson.com

Key Expectations

Throughout this chapter, you have had opportunities to do the following:

- Demonstrate an understanding of the connection between biodiversity and species survival (11.1, 11.2, 11.3, 11.4, 11.5).
- Describe selected anatomical and physiological characteristics of representative organisms from each life kingdom (11.2, 11.3, 11.4, 11.5).
- Compare and contrast the life cycles of representative organisms from each life kingdom (11.2, 11.3, 11.4, 11.5).
- Demonstrate, through applying classification techniques and terminology, the usefulness of the system of scientific nomenclature in the field of taxonomy (11.1, 11.5).
- Locate, select, analyze, and integrate information on topics under study, working independently and as part of a team, and using appropriate library and electronic research tools, including Internet sites (11.2, 11.3, 11.4).
- Select appropriate instruments and use them effectively and accurately in collecting observations and data (11.3).
- Compare the anatomy of different organisms—vertebrate and invertebrate—by carrying out a dissection or using a computer-generated dissection (11.3).
- Explain the importance of sexual reproduction to variability within a population (11.5).
- Classify representative organisms from a kingdom (11.5).

Key Terms

cephalization	nematocysts
clitellum	nephridia
coelom	notochord
dioecious	peritoneum
endoskeleton	photoreceptors
exoskeleton	proglottids
germ layers	pseudocoelom
hemocoel	scolex
hermaphroditic	sedentary
invertebrates	segmentation
larva	sessile
metamorphosis	vertebrates
motile	zoology

Make a Summary

Make a chart comparing the phyla you studied in the animal kingdom. Include headings such as the following in your chart: Phylum name, Meaning of name, Habitat—terrestrial or aquatic (marine or freshwater), Physical similarities of representative organisms, Style of feeding, Tissue organization, Nervous system, Digestive system, Circulatory system, Gas exchange system, Reproduction, Life cycle characteristics, Benefits to other animals (including humans), Ecological role.

Reflect on your Learning

Revisit your answers to the Reflect on Your Learning questions at the beginning of this chapter.

- How has your thinking changed?
- What new questions do you have?

Understanding Concepts

1. In which phylum or phyla of the less complex invertebrates do we first see three distinct body layers, ganglia, and a coelom?

2. How has the feeding process in sponges contributed to their success? Compare this process to the one in cnidarians.

3. What evolutionary adaptations are exhibited by flatworms and roundworms?

4. What traits distinguish earthworms from other animals?

5. The tube-within-a-tube body plan is found in nematodes. Explain why this arrangement, found in all higher animals, is a more advanced body plan.

6. What are some advantages and disadvantages of having an exoskeleton? Describe how one disadvantage is overcome.

7. The term coelom is of the same origin as coelenterate and pseudocoel. What is the meaning of coelom and how does it relate to the other two terms?

8. What are the advantages to organisms that possess bilateral symmetry? segmentation? a gut with two openings?

9. At what time in the life history of an arthropod would you expect to find it hiding in a secluded place? Explain.

10. Many people think of spiders as insects. Explain why spiders are classified separately from insects.

11. Describe four areas of specialization that are unique to arthropods.

12. Give evidence that supports the theory that arthropods evolved from annelids.

13. How does diversity and abundance contribute to the success of a phylum such as Arthropoda?

14. Hermaphroditic organisms such as sponges produce either only eggs or only sperm at a time, ensuring that sperm will not fertilize eggs from the same organism. Why is this important?

15. Animals that display radial symmetry are found only in water. Why is this type of symmetry not well suited to life on land?

16. Why is the digestive system of roundworms considered to be more advanced than that of flatworms?

17. Why would cephalization be of greater advantage to annelids than to echinoderms?

18. Why would having a highly motile larval stage be an advantage to the slow-moving sea urchin?

19. Several organisms examined in this chapter can regenerate whole organisms or appendages from one piece. List those which can regenerate in this way. What is the advantage of being able to do this?

20. For certain clams, such as the giant clams of the South Pacific, at least part of their food is furnished by algae that live in amoeboid cells in the blood. How do the algae obtain light for photosynthesis?

21. A guiding biological principle is that an organism's structure is related to its function. Using one of the phyla studied in this chapter, explain how this principle is illustrated.

22. The body structures of multicelled animals, such as those studied in this chapter, are said to be specialized. What does this mean (in a scientific sense)?

23. A fossilized coral reef has been found between Ontario and Ohio. It is believed to be over 400 million years old. What can be inferred about the history of this region?

Applying Inquiry Skills

24. Examine the images of different insects and their mouths in **Figure 1**. Indicate which of the following functions each insect's mouth is adapted for: siphoning, puncturing, or chewing.

grasshopper

butterfly

mosquito

Figure 1
Mouths of several insects

25. While walking along a seashore, you come across four organisms. Each of the organisms matches one of the following descriptions:
   - hard outer shell in two major sections, long tail, underside also hard, with a soft inner body structure; bilaterally symmetrical; 20 cm in diameter
   - long, narrow, smooth, soft tube; openings visible at both ends; bilaterally symmetrical; 2 mm in length
   - a fossilized structure with an exoskeleton, segments containing four legs each; bilaterally symmetrical; 1 cm in length
   - a soft, round, clear structure; radially symmetrical; 15 cm in diameter.

   How would you classify these organisms? What other investigations could you complete to help you with your classification?

26. A planarian is able to respond to simple environmental stimuli. Name the organ systems involved in responding to stimuli. Design a lab investigation to examine how a planaria would react to light.

27. Compare hydra, grasshoppers, and earthworms according to their appendages and locomotion, digestion and digestive organs, gas exchange, and their ability to respond to their environment.

Making Connections

28. List the similarities and differences between animals and plants. How do these differences affect the lifestyles of organisms in each kingdom?

29. What abilities are lost as animal cells become increasingly specialized?

30. Why might scientists be interested in the way sponges (**Figure 2**) grow entire bodies from fragments? Why do sponge farmers throw pieces of sponge back into the water?

Figure 2

31. Green hydra contains algae that give the hydra its colour. The algae found in the green hydra are indistinguishable from the common unicellular algae, called chlorella. What do you think the evolutionary significance of this is?

32. Green hydra (**Figure 3**) grows more quickly in the light than in the dark. What is the benefit of algae and green hydra existing together?

Figure 3

33. As animals made the transition to life on land, they faced two challenges—the first was to avoid drying out and the second was to develop a means to obtain oxygen. Which invertebrates made the transition successfully and what adaptations have they evolved to cope with life on land?

34. Summarize the relationship between humans and invertebrates. Make a list of invertebrates that are helpful and a second list of those that are harmful. In what ways are we harmful to invertebrates?

   Follow the links for Nelson Biology 11, Chapter 11 Review.

   GO TO www.science.nelson.com

Exploring

35. Parallels are often drawn between insect societies and human societies. Research this topic and describe the similarities and differences between the two. Report your findings to the class.

   Follow the links for Nelson Biology 11, Chapter 11 Review.

   GO TO www.science.nelson.com

36. Spiders show tremendous variety in their behaviour. Research one of the following areas of spider behaviour:
   - spiders that do not spin webs
   - hunting spiders
   - courtship displays
   - varying size in male and female spiders within a species

   Report on your findings, identifying examples of the spiders in your report.

   Follow the links for Nelson Biology 11, Chapter 11 Review.

   GO TO www.science.nelson.com

In this chapter, you will be able to

- describe, compare, and classify the basic body structures of animals with backbones, such as fish, amphibians, reptiles, birds, and mammals;
- explain the success of vertebrates as a result of their adaptability;
- examine the adaptive features of representatives of the vertebrates;
- appreciate the ecological role played by various vertebrates.

Chordates

In 1798, scientists from the London Museum of Natural History huddled around a strange-looking beast. As they poked at its flat, ducklike bill and examined its webbed feet, they first concluded that the duck-billed platypus (**Figure 1**) was not a real animal, but rather a cruel joke fashioned to embarrass the prestigious group of scientists.

The platypus looks like no other animal, although it has individual features similar to those of other animals. Like other mammals, it has hair and produces milk to feed its young. The flat, furry tail is reminiscent of a beaver's and the webbed feet are similar to an otter's. However, like reptiles and birds, the platypus has a single opening, called a cloaca, which functions for both reproduction and excretion. Thin-shelled eggs, laid by the platypus, provide another link to reptiles and birds. The tiny young that hatch from the eggs look much more like tiny, pink embryos than the young of reptiles, birds, or most mammals (**Figure 2**). The only other animals that produce immature young that resemble those of the platypus are pouched marsupials, such as kangaroos and wallabies.

The platypus is one of nature's success stories because its structures are well adapted for its lifestyle and habitat. At night, the platypus swims in rivers and lagoons in search of mollusks. The wide, highly sensitive bill is ideally suited for locating and scooping the mollusks from the mud. The dense, oily fur keeps cold water from stealing heat from the relatively small body. A thick layer of fat immediately below the skin provides a second layer of insulation and is an important store of energy. Conserving body heat is essential because the platypus remains submerged in cool, watery habitats all night. An animal with poor insulation would soon become chilled and sluggish, eventually succumbing to hypothermia. The wide, long tail works like a rudder, steering the body through the currents, while the webbed feet act as paddles.

During the day, the small creatures hide from larger predators in underground burrows. The clawed hind feet are excellent for digging. The tiny, poorly developed eyes are no mistake; a creature that spends most of the daytime underground would have no use for large eyes, which could become irritated by soil. Instead, more than 80 000 sense organs are located in the bill. Some of these sensory nerves detect wave motion created by prey that attempt to escape, or weak electrical impulses created by the flicking of a shrimp's tail or legs.

Reflect on your Learning

1. In what ways do mammals, birds, reptiles, amphibians, and fish differ from insects and other invertebrates?

2. Make a list of similarities and differences among mammals, birds, reptiles, amphibians, and fish.

3. Use the example of the platypus to explain how classification systems are artificial creations of human thought rather than something that really exists in nature.

4. Although streams and lagoons exist in many areas of the world, the platypus is found only in remote areas of Australia. Propose an explanation that accounts for the fact that the platypus is not found in other parts of the world with similar habitats.

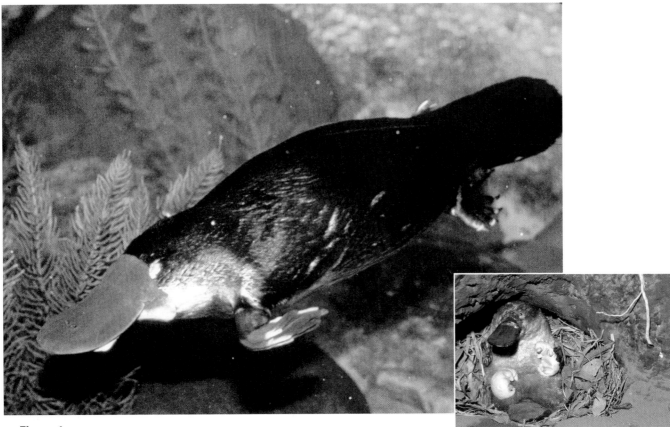

Figure 1
The platypus has a collection of parts normally found on different kinds of animals. This odd mix of features makes the platypus successful in a particular environment.

Figure 2
Platypus young

 Try This Activity

Designing an Animal

Invent an animal adapted for living in a habitat that has long, cold winters (**Figure 3**). Assume that 50 cm of snow will fall over the winter months and that the average winter temperature is –20°C. Summers are short and often have temperatures below freezing during the night. Give reasons for your design.

The following checklist might be useful:
- Is your animal a predator? (Must it sneak up on its prey?)
- Do larger animals prey on the animal that you designed? (Must it hide from other animals?)
- Does it have feathers, scales, or fur for insulation?
- Does it live on top of the snow or under the snow cover?
- Does it ever enter the water? If it has fur, will the fur get wet?
- Does it need large eyes? (The reflected light off the snow is very bright.)

aid in listening for predators

result in body heat loss

provides camouflage during the winter

aid walking on the snow

Figure 3
Some adaptations of the snow hare

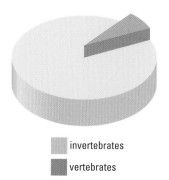

Figure 1
While invertebrates account for more than 1 million different species, chordates consist of about 47 000 species.

invertebrates
vertebrates

12.1 Phylum Chordata

When asked to picture an animal in their mind, most people would imagine a chordate since this phylum contains, besides humans, such familiar animals as cats, dogs, and horses. Although chordates are often considered to be the most familiar and interesting of all the animals, they make up only about 5% of all known animal species and are one of the most recent groups of animals (**Figure 1**).

Classification of Chordates

Chordates form a distinct phylum, containing all those organisms that at some time in their development have a notochord, a hollow dorsal nerve tube, a tail that extends beyond the anus, and a muscular tube called a pharynx. The pharynx functions in feeding and/or respiration. At some time in their embryonic development, chordates have slits in the wall of the pharynx.

Phylum Chordata can be divided into three subphyla: Cephalochordata (such as lancelets), Urochordata (such as tunicates), and Vertebrata (such as fish, amphibians, reptiles, birds, and mammals), as shown in **Table 1**. Members of Cephalochordata (**Figure 2**) and Urochordata (**Figure 3**) are filter-feeding marine organisms that have notochords but do not have backbones or vertebral columns. These organisms may represent a transition between the invertebrates and the chordates; therefore, studying them provides clues about the origins of the chordates.

Table 1: Classification in the Phylum Chordata

Classification	Example
subphylum Cephalochordata	lancelets
subphylum Urochordata	tunicates
subphylum Vertebrata	
class Agnatha	jawless fish (such as lamprey)
class Placodermi	jawed armoured fish (extinct)
class Chondrichthyes	cartilaginous fish (such as sharks)
class Osteichthyes	bony fish (such as perch)
class Amphibia	amphibians (such as frogs)
class Reptilia	reptiles (such as alligators)
class Aves	birds (such as blue jays)
class Mammalia	mammals (such as dogs)

Figure 2
Lancelets belong to the subphylum Cephalochordata. These filter feeders spend much of their time in the sand on the ocean's floor.

Figure 3
The Urochordata are diverse and abundant. About 90% of urochordates are tunicates.

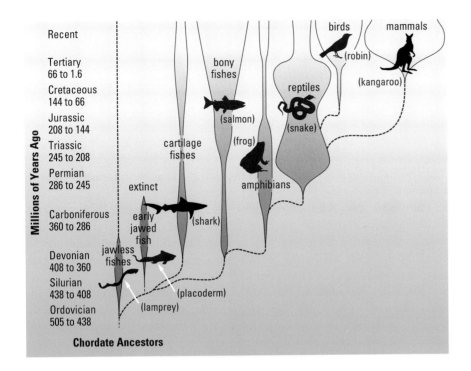

Figure 4
An outline of the evolutionary record of the vertebrates. The geological periods and their relative times are indicated. The width of each vertebrate class indicates the number of species in any geological period.

Approximately 95% of chordates belong to the subphylum Vertebrata (**Figure 4**). All vertebrates have vertebrae (singular: vertebra), which are hollow, cartilaginous, or bony structures that surround and protect the dorsal nerve cord. Other features shared by vertebrates include an endoskeleton, a large brain protected by a skull, an advanced nervous system, a complex heart and circulatory system, a special outer covering, one or two pairs of appendages, and a large coelom, which contains the vital organs. Birds and mammals are warm-blooded organisms (**endotherms**), whereas all other vertebrates are cold-blooded organisms (**ectotherms**). In cold-blooded forms, the body temperature and, hence, metabolic rate vary with changes in the surroundings. Conversely, the body temperature of warm-blooded animals remains constant regardless of external changes in temperature. They are able to vary their metabolic rate. The ability to regulate their temperature makes warm-blooded animals much more independent of their environment.

endotherms: organisms that are able to maintain a constant body temperature. They are also described as homeothermic.

ectotherms: organisms that are not able to maintain a constant body temperature. They are also described as poikilothermic.

Practice

Understanding Concepts

1. Name the three subphyla of the phylum Chordata, and give an example of each.
2. In what ways are tunicates and lancelets different from other chordates?
3. List eight general features of the vertebrates.
4. What is the difference between an endothermic animal and an ectothermic animal?

Ecological Role of Vertebrates

Vertebrates, the most important group of chordates, are found almost everywhere—in the ocean, freshwater lakes, and streams; on the land; and in the air. They include fish, amphibians, reptiles, birds, and mammals.

Figure 5
Vertebrates can be found almost everywhere and they have a major impact on our environment.

Fish are a vital part of many living systems and an important food source for large numbers of aquatic animals, birds, and mammals, including humans. They are one of the most important sources of protein for people. As consumers, fish help to control the populations of the organisms they eat.

Amphibians play a pivotal role in wetlands ecology by preying on insects and devouring algae and dead plants that can fill up ponds and streams. They are also a source of food for reptiles (snakes and turtles), fish, birds, and mammals.

Reptiles play an important role in the environment and agriculture. As predators, they keep a variety of potential pests in check. For example, snakes that feed on small mammals, insects, and other invertebrates are extremely beneficial to farmers. Reptiles also provide food for other animals.

Birds also play an important role in maintaining a balance in the environment. By eating enormous quantities of insects, they help keep insect populations under control. Predatory birds feed on rats and mice and keep them in check. Many birds have roles as pollinators and as agents of seed dispersal, thereby serving important functions in fields and forests.

Mammals, including humans, are also influenced by and alter the environment in many ways. Small burrowing mammals, such as moles and ground squirrels, aerate soils. Beavers create dramatic environmental changes when they build dams and create ponds (**Figure 5**). Larger herbivorous mammals consume enormous quantities of the earth's plant material before becoming food for carnivorous mammals. Some mammals such as rats, squirrels, and dogs transmit disease. Humans play a very significant role in living systems and have a major impact on the environment. They influence the environment in many positive ways, but they also have a negative impact by creating pollution and causing untold damage to the biosphere.

SUMMARY Phylum Chordata

1. The phylum Chordata contains the most complex and advanced living organisms.

2. All chordates, at some stage in their life cycle, have a notochord, a hollow dorsal nerve tube, a tail, and a pharynx.

3. Phylum Chordata can be divided into three subphyla: Cephalochordata, Urochordata, and Vertebrata. Members of the first two subphyla are filter-feeding marine organisms without a vertebral column.

4. The majority of chordates are vertebrates. The vertebrates include fish, amphibians, reptiles, birds, and mammals. Vertebrates have a cartilaginous or bony vertebral column, an endoskeleton, a large brain protected by a skull, an advanced nervous system, a complex heart and circulatory system, a special outer covering, one or two pairs of appendages, and a large coelom.

5. Birds and mammals are endotherms; all other chordates are ectotherms.

Cladistics

Cladistics is an alternative method available for analyzing and predicting the phylogeny (evolutionary relationships) of organisms. Cladistics is based on an analysis of shared characteristics. Since these characteristics are changing over time, cladists (people who study cladistics), distinguish between recent, derived (**apomorphic**) characteristics and older, primitive (**plesiomorphic**) characteristics. They then arrange organisms based on their hypothesized degree of sharing of apomorphic characteristics rather than of plesiomorphic characteristics. For

cladistics: a system which classifies organisms based on the presence or absence of shared derived characteristics

apomorphic: in cladistics, this term describes a recent characteristic that is derived from, but no longer the same as, an ancestral characteristic. Derived characteristics arose at some time after the first "splitting" of a member from the group, and therefore derived characteristics differ among the members of the group.

plesiomorphic: in cladistics, this term describes a primitive characteristic that is thought to be ancestral to all members of the group under consideration. Primitive characteristics are widespread and cannot be used to distinguish between members of such a group.

example, a recently derived feature is the single toe of horses. The original primitive state of five-digit appendages, exhibited by monkeys, turtles, and crocodiles, indicates an ancient but common ancestor. The fact that monkeys and turtles both have five digits does not demonstrate close kinship. Cladists propose that more closely related organisms will share a greater number of unique derived features which were not present in their distant ancestors.

Cladograms are branching diagrams used to display these hypothesized relationships, usually in the form of a "tree." Examine the simple cladogram in **Figure 6** for six organisms. Each organism is placed at the end of a line. In this case, the organisms are all modern species. The lines represent the lineages leading to each end group. Following these lines down the cladogram represents going back in time through a continuum of ancestors. These lines join to other lines at nodes, such as A, B, C, and D. A node represents the point in time at which a common ancestral species "splits" to give rise to two different lineages. For example, node A represents an organism that is a common ancestor to all organisms on this cladogram. This ancestral species split, giving rise to the lineage leading to the perch, as well as to the lineage leading to all other organisms in the cladogram. The ancestor at node B, however, is not an ancestor to the perch but is a more recent organism that is a common ancestor to both the birds and mammals. Note that the perch, chimp, and human species did not exist at time A. Node A simply represents the point in time at which we shared a common ancestor with the modern perch.

The D node is further back in time than the C node. Analysis of this cladogram therefore suggests that humans and chimps share the most recent common ancestor, followed by robins and crows. Also, note that birds are not more closely related to perch than humans, chimps, or horses. Rather, these organisms must be equally closely related to perch as they all share the same common ancestor at node A.

Like many of the biological sciences, the explosion in genetic information has revolutionized cladistics. Cladists are now able to compare shared derived genetic codes rather than just outward traits. This has allowed them to find new compelling and convincing evidence of many phylogenetic relationships.

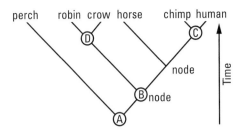

Figure 6
Cladograms depict proposed evolutionary relationships.

SUMMARY Cladistics

1. Cladistics is a system that classifies organisms based on the presence or absence of shared derived characteristics.

2. Apomorphic or derived features are characteristics that originated within members of a group and therefore differ among them. Derived features are useful in cladistic classification.

3. Plesiomorphic or primitive features are those that are ancestral to all members of a group. Primitive features are not useful in cladistic classification.

Section 12.1 Questions

Understanding Concepts

1. What are five ways in which vertebrates play important ecological roles in nature?

2. How are the lancelets and tunicates a transition between the invertebrates and the chordates?

(continued)

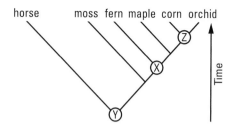

Figure 7
Sample cladogram

Figure 1
The lamprey has caused great damage, especially to the populations of whitefish and lake trout in the Great Lakes. Normally found in marine water, the lamprey has invaded freshwater lakes by swimming up the St. Lawrence Seaway.

DID YOU KNOW ?

Niagara Falls was a natural barrier for the sea lamprey. The completion of the Welland Canal in 1932 provided the lamprey direct access to the Great Lakes beyond the Falls.

Figure 2
Sharks are one of the planet's most successful creatures. They have changed little in the past 400 million years.

3. Explain the following statement: All vertebrates are chordates, but not all chordates are vertebrates.

4. Name the major groups of vertebrates and describe their most significant distinguishing characteristics.

5. Give the possible evolutionary sequence in which the classes of vertebrates first appeared. Explain your logic.

6. What advantages do endotherms have over ectotherms?

7. Examine the sample cladogram in **Figure 7** and answer the following questions:
   (a) Which two species share the most recent ancestor? Explain.
   (b) Species X is an ancestor to which organisms on the cladogram?
   (c) Is the maple or corn more closely related to the fern? Explain.
   (d) Would you expect the species at position Y to resemble a horse? Explain.
   (e) What does this cladogram suggest to you about the phylogenetic relationship between modern animals and plants?

12.2 Fish

Fish, the most numerous and widespread of the vertebrates, are divided into three groups: jawless fish, cartilaginous fish, and bony fish. All fish possess gills for exchange of oxygen and carbon dioxide with the water. Of the nearly 22 000 species of fish, 59% live in salt water, 40% in fresh water, and the remaining 1% move regularly between the two habitats.

Jawless Fish: Class Agnatha

The jawless fish are covered with a slimy skin and lack paired fins. They have soft, eel-like bodies with a notochord and a skeleton made up of cartilage, not bone (i.e., cartilaginous). Agnatha is made up of about 70 species of lampreys and hagfish.

Lampreys are widely known for the damage they have caused to commercial fisheries, particularly in the Great Lakes (**Figure 1**). They spawn in streams where the larvae live for some years before metamorphosing to adults and migrating to a lake or the sea. The mouth of a lamprey is suckerlike and lined with hard teeth. The adult attaches itself to the body of its prey, tears a hole in the flesh, and sucks out blood and body fluids.

Hagfish, which are bottom dwellers, differ from lampreys in a number of ways. They are scavengers, feeding on dead or dying animals. The mouth of a hagfish is rather inconspicuous, being surrounded by fleshy tentacles or "feelers," but is equipped with a row of toothlike plates that can tear the flesh from its prey.

Cartilaginous Fish: Class Chondrichthyes

This class includes the sharks, skates, and rays—fish with a cartilaginous skeleton. The cartilaginous fish have two important advantages over the jawless fish: biting jaws and paired fins. The best-known cartilaginous fish are the sharks. The largest shark, the whale shark, is a filter feeder that feeds on plankton. Sharks most often prey on other fish or bottom-dwelling animals such as mollusks or lobsters.

The shark's streamlined body and paired fins enhance its stability and manoeuvrability (**Figure 2**). The paired gills (usually five to seven pairs per side) are efficient structures for the exchange of oxygen and carbon dioxide. Sharp, triangular teeth, used to capture and tear off chunks of prey, ring the mouth. These teeth are repeatedly shed and replaced. The shark's paired nostrils connect to sensitive organs of smell that are used to locate prey. Another sensory structure, the **lateral line** (a series of pits along the side of the body), allows the shark to sense movement in the water. A shark's body is covered with **placoid scales**, giving the skin a texture like sandpaper. Male and female sharks remain separate, except when mating. Fertilization is internal, and embryo development takes place inside the body of the female for nine months to two years. Some shark species lay egg cases containing well-developed embryos, while others give birth to litters of 2 to 60 live young. Young sharks are born fully formed and are able to exist independently.

Skates and rays, although similar to sharks in many ways, have flattened bodies adapted to living on the bottom of the ocean (**Figure 3**). This distinctive feature is a result of enlarged pectoral fins, which extend onto the side of the head. Because skates and rays often bury their mouths in the ocean bottom, they take in water through modified gill slits called spiracles.

Bony Fish: Class Osteichthyes

Bony fish are the most numerous vertebrates. Representatives include trout, salmon, cod, perch, sole, eel, halibut, and tuna (**Figure 4**). These fish, and nearly all living species of bony fish, belong to the largest of the three subclasses, the ray-finned fish. Ray-finned fish exhibit a wide range of diversity and live in almost all aquatic habitats. All have gills in a common chamber protected by a plate called an **operculum**. The surface of the body is covered with flat, flexible scales. Blood is pumped through a two-chambered heart. Fertilization is usually external, with the female shedding numerous eggs into the water. The male deposits a sperm-laden fluid called milt over the eggs.

lateral line: a line of sensory cells along each side of a fish's body

placoid scales: toothlike scales composed of dentine, within a pulp cavity

Figure 3
The skate is a close relative of the shark.

operculum: also called a gill cover. It is a bony plate covering the gill chamber.

(a)

(b)

(c)

(d)

Figure 4
Representatives of the bony fish include **(a)** Atlantic cod; **(b)** brown trout; **(c)** seahorse; and **(d)** halibut.

Primitive fish, such as sharks, will sink if they stop moving because they have no **swim bladder** (**Figure 5**, page 454). Bony fish do not sink as they are equipped with a small, balloonlike structure, which allows them to float at different levels. The swim bladder, which is filled with oxygen, nitrogen, and carbon dioxide, helps the fish maintain buoyancy and permits it to change depth with ease. By adding or removing gas, the fish can adjust its buoyancy. **Table 1**, page 454, shows the function of the swim bladder and other organs and structures of fish.

swim bladder: a small, balloonlike structure filled with gases which helps fish maintain buoyancy and allows them to float at different depths. It is also called an air bladder.

Figure 5

The development of a swim bladder is an important evolutionary step.

Labels: caudal fin, muscle, dorsal fin, kidney, spinal cord, brain, olfactory bulb, heart, liver, gall bladder, pectoral fin, stomach, intestine, gonad, swim bladder, urinary bladder

Table 1: Some Fish Organs and Structures

Organ/structure	Function
heart	pumps blood containing dissolved gases to the tissues of the body and to the swim bladder
blood vessels	carry gases to and from the swim bladder
swim bladder	fills with gases from the blood, making the fish more buoyant
caudal (tail) fin	propels the fish
dorsal fin	stabilizes the fish, keeping it upright
pectoral fins	used for movement and provide stability, preventing the fish from rolling over
olfactory bulb	organ involved in sense of smell

When air from the blood enters the swim bladder, the fish rises in the water. As air is pushed from the swim bladder back into the blood, the fish sinks. When a fish is sick, it can lose muscular control over the swim bladder. Often, this causes additional air to move into the swim bladder and the fish floats sideways on the surface of the water.

The other two subclasses of bony fish, lungfish and lobe-finned fish (**Figure 6**), have both gills and lunglike air sacs that enable them to remove oxygen from both air and water.

Figure 6

The lobe-finned fish were thought to be extinct until a specimen of the coelacanth, *Latimeria,* was captured off the coast of South Africa in 1938. Many scientists believe the coelacanth resembles the early land animals.

Practice

Understanding Concepts

1. Provide one example of an animal that belongs to each of the following groups: jawless fish, cartilaginous fish, and bony fish.
2. Describe the environmental impact caused by the lamprey invading the Great Lakes.
3. Indicate two advances of sharks and rays over the jawless fish.
4. What are placoid scales?
5. Why must sharks keep moving?
6. Why is the lateral line of sharks important to their survival?
7. What is the function of a swim bladder?

SUMMARY Fish

1. Fish are the most numerous and widespread of the vertebrates.
   - They are divided into jawless fish, cartilaginous fish, and bony fish.
   - All fish have gills.
2. The earliest vertebrates were aquatic creatures that lacked jaws.
   - They have a notochord and a cartilaginous skeleton.
   - The two groups of living jawless fish are lampreys and hagfish.
3. Cartilaginous fish are jawed.
   - They have a skeleton made of cartilage.
   - This group includes sharks, skates, and rays.
4. All other jawed fish are bony fish.
   - Bony fish are the most numerous.
   - Bony fish have swim bladders, which help maintain buoyancy and allow the fish to float at different levels.

Section 12.2 Questions

Understanding Concepts

1. How do bony fish differ from cartilaginous fish?
2. Describe the characteristics of the three major groups of living fish that would help you classify a fish you've found.
3. Why might sharks be considered one of the most successful animals on Earth?
4. Why is the swim bladder considered to be an important evolutionary advancement?

Making Connections

5. The completion of the Welland Canal allowed ships to bypass Niagara Falls, but it also enabled the lamprey to invade the western Great Lakes. What might have been done differently?
6. Why do you think there are so few species of jawless fish living today?

12.3 Amphibians

About 350 million years ago, a lineage of lobe-finned fish were the first vertebrates to make the transition from water to land, about 100 million years after the first plants made the same transition. These fish had specialized characteristics that allowed for terrestrial living, including limblike fins for crawling and primitive lungs that enabled them to breathe air for at least short periods. Their descendants, which could stay out of water for longer periods, were the first amphibians.

Class Amphibia: Two Lives

There are about twenty-five hundred species in three orders:

1. legless amphibians—include the wormlike caecilians
2. tailless amphibians—include frogs and toads
3. tailed amphibians—include salamanders and newts

Frogs and toads are the most familiar amphibians (**Figure 1**).

(a)

(b)

(c)

(d)

Figure 1
Amphibians include **(a)** caecilians; **(b)** salamanders; **(c)** toads; and **(d)** frogs.

Amphibians can be considered to have two lives. Amphibians have external fertilization and their eggs are laid in fresh water. The eggs hatch as tadpoles. At this tadpole stage, most amphibians are herbivores. They exchange gases through gills, like fish. Tadpoles then change into adult forms that reside on land. Most adult amphibians are carnivores. Adult amphibians have a three-chambered heart, unlike fish (**Figure 2**). Blood from the body enters the right atrium and then flows into the single ventricle. Blood from the lungs enters the left atrium and then flows into the ventricle. This causes the ventricle to have a mixture of oxygenated blood from the lungs and deoxygenated blood from the body. The ventricle contracts and pumps blood to both the body and the lungs. This is not an efficient method of circulation. However, amphibians can exchange gases directly through their moist skin as well as through their lungs. The lungs have a simple structure, with little internal folding to increase the surface area for the exchange of gases.

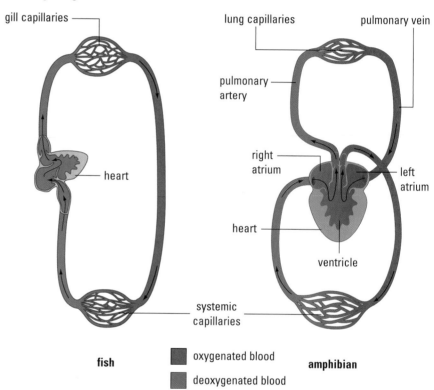

Figure 2
Comparison of fish and amphibian circulatory pathways

Ecological Role of Amphibians

Because amphibians must lay eggs in water, their range is restricted to wetter areas. It has also been suggested that amphibians and, particularly, frogs might be uniquely sensitive to environmental contaminants. Frogs are found in both water and on dry land, and can absorb gases through their skin. They also eat a wide variety of foods, including both plant and animal material. Over the past decade, scientists in many areas of the world have noticed a decline in the population levels of amphibians, and some have speculated that the decline is in response to a general degradation of the environment. Others, however, do not feel that there is any one cause for the decline, but that there have been declines in a number of populations in response to a range of factors and that it is only coincidental that several populations have been affected at the same time.

Amphibians

1. The ancestral land vertebrate was likely a lobe-finned fish with lungs and limblike fins that enabled it to live on land for short periods. Its descendants were the first amphibians.

2. The three orders are the legless amphibians, the tailless amphibians, and the tailed amphibians.

3. Amphibians have two lives.
   - Reproduction occurs in water. They hatch from eggs and, as tadpoles, are herbivores and exchange gases through gills.
   - When they become adults, they live on land, are carnivores, and exchange gases through their skin and lungs.

12.4 Reptiles

Reptiles appeared about 310 million years ago. They are the first fully terrestrial vertebrates. Even reptiles that spend a great deal of time in water do not have to return to water to lay eggs.

Class Reptilia

There were few competitors when reptiles "invaded" the land and, as a result, they had little trouble increasing in numbers and variety and are now found in almost all habitats in the warmer parts of the world. At one time, they were even found in the air, but large flying reptiles have been extinct for millions of years.

The approximately seven thousand species of living reptiles belong to three subclasses (**Figure 1**):

1. Anapsida (turtles and tortoises)
2. Lepidosaura (lizards, snakes, and the tuatara)
3. Archosaura (crocodiles, alligators, and dinosaurs)

Life in a Terrestrial Environment

To live a fully terrestrial existence, terrestrial vertebrates have needed many special characteristics to enable them to efficiently handle the varied geographical, atmospheric, and other environmental conditions found in a land environment. This is in contrast to the relatively uniform conditions encountered by fish and some species of aquatic amphibians.

To live on land, organisms need a skin that reduces water loss. A dry, waterproof skin accomplishes this, but it is unable to exchange gases. To exchange gases while living on land, organisms need efficient lungs and circulation. Reptiles have lungs that have internal folding, which increases the surface area to exchange gases (**Figure 2**). Most reptilian hearts have three chambers; however, the septum

(a)

(b)

(c)

Figure 1
Representative reptiles: (**a**) tortoise; (**b**) snake; and (**c**) crocodile. All reptiles have scales and a bony skeleton.

Figure 2
The development of lungs in vertebrates
(**a**) Many amphibians rely on their moist skin for respiration. Their lungs are simple sacs with only a few ridges to increase the surface area.
(**b**) The lungs of reptiles are more developed. They are divided into many partitions which are further subdivided into smaller chambers, thereby greatly increasing the surface area.
(**c**) In mammals, the entire lung is filled with thousands of saclike structures that provide a tremendously large area for gas exchange.

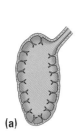
(a)

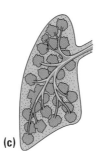
(b)

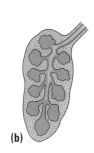
(c)

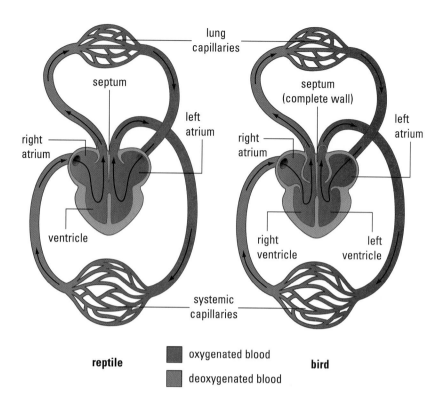

lung capillaries

septum

right atrium

left atrium

ventricle

systemic capillaries

reptile

septum (complete wall)

right atrium

left atrium

right ventricle

left ventricle

oxygenated blood

deoxygenated blood

bird

Figure 3
The reptilian heart is more efficient than the amphibian heart. The partial division of the ventricle by a septum results in deoxygenated blood mixing with the oxygenated blood. In birds and mammals, the heart is fully partitioned into two halves.

amniotic eggs: eggs that have leathery or calcified shells that surround internal membranes, including the amnion

amnion: a membrane which surrounds the embryo and holds the amniotic fluid

allantois: the embryonic membrane that functions in respiration and in the storage of metabolic wastes in reptiles, birds, and some mammals

chorion: the outer membrane around a developing embryo

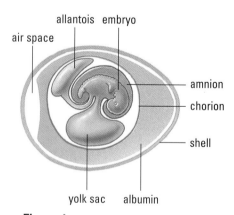

air space

allantois embryo

amnion

chorion

shell

yolk sac albumin

Figure 4
The amniotic egg allowed reptiles to lay eggs on land and colonize dry habitats.

continues quite far into the single ventricle. This functions to separate oxygenated and deoxygenated blood so that the ventricle pumps blood with a higher concentration of oxygen to the body (**Figure 3**).

The most significant advance of reptiles over amphibians was the development of the **amniotic egg**, a key factor that allowed vertebrates to inhabit the land (**Figure 4**). The amniotic egg provides a self-contained environment for the embryo. It has allowed reptiles to spread into some of the driest areas on Earth. It is also a characteristic of birds and some mammals. The egg generally has a leathery or calcified shell. Inside the shell are four membranes that nourish and protect the developing embryo. The **amnion** is a fluid-filled sac that holds and cushions the embryo, allowing the embryo to move freely and protecting it from changes in temperature. The yolk sac contains the yolk, or stored food, upon which the embryo feeds. The **allantois** holds the wastes produced by the embryo and, along with the amnion, provides for the exchange of oxygen and carbon dioxide through the shell. The **chorion** lines the inside of the shell and encloses the other three membranes.

Limbs and a lightweight skeleton are also needed to live on land. Limbs probably evolved from the lobe-finned fish. The overall structure of limbs, and the number, relative sizes, and positioning of the bones developed in complexity and efficiency from the amphibians and reptiles through to the mammals and birds. Two fundamental changes occurred simultaneously in the limbs. First, the strengthening of the bones and muscles of the limbs allowed an animal to drag its body from one place to another. Over time, repositioning of the limbs to support the body above the ground produced highly efficient movement with minimum damage to the body proper. Second, the knee joints changed from a side to a forward position, greatly increasing forward push and speed. In mammals, the characteristic four-footed gait resulted in a longer stride and increased speed.

In any species of vertebrate, the precise limb structure is a reflection of the ecological niche that it occupies (**Figure 5**). For example, snakes have lost functional

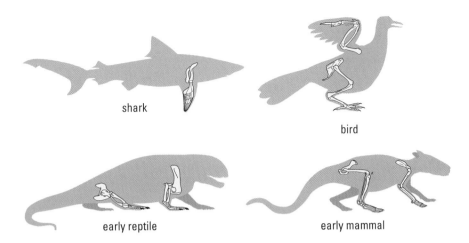

shark

bird

early reptile

early mammal

Figure 5
The change in positions of the limbs has resulted in animals that can support their bodies above the ground.

limbs and move by pushing themselves with body undulations. Vertebrates such as salamanders, which live a semiaquatic existence, retain the simpler sideways limb structures.

Early Reptiles

Dinosaurs dominated life on Earth for approximately 160 million years and then, over a period of time, became extinct. One dinosaur, *Apatosaurus,* was probably most like today's reptiles. It was a herbivore and among the largest of the dinosaurs. A familiar group is *Stegosaurus,* characterized by bony plates on its back and a relatively small head. In addition, there were the giant forms such as the carnivorous *Tyrannosaurus rex,* which stood erect and walked on its hind legs, and many other forms that are known from fossil records. Much has still to be learned about these ancient reptiles (**Figure 6**). The reason for their disappearance is still debated.

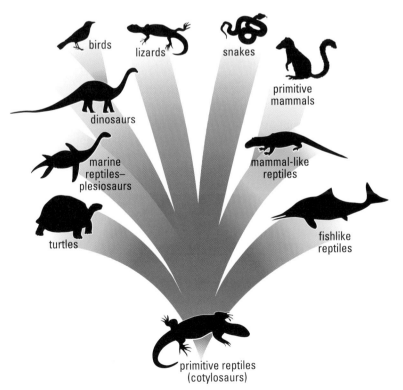

birds

lizards

snakes

primitive mammals

dinosaurs

marine reptiles— plesiosaurs

mammal-like reptiles

turtles

fishlike reptiles

primitive reptiles (cotylosaurs)

Figure 6
The reptiles probably developed from cotylosaurs.

Understanding Concepts

1. List the three major groups of living amphibians.
2. What are the major characteristics of amphibia? How do they represent the transition in habitat from water to land?
3. Reptiles were the first vertebrates able to reproduce on land. What specialization allowed them to do this?
4. Many reptiles have returned to living most of their lives in the water. Name some reptiles that are primarily aquatic.

Making Connections

5. The caecilians are wormlike amphibians. What characteristics make them different from earthworms?
6. All terrestrial vertebrates obtain oxygen directly from the air. However, many biologists believe that this ability came before the first vertebrates existed on land. Describe a possible sequence of events in the development of air breathing.

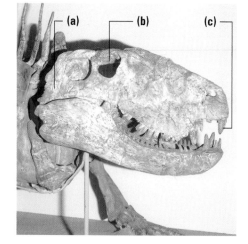

Figure 7
Skull of dimetrodon
(a) The large area for muscle attachment indicates a strong jaw.
(b) The forward-set eyes provide depth perception allowing the animal to estimate distances.
(c) The sharp pointed teeth are adapted for puncturing and holding food.

Frontiers of Technology: Using Computer Simulations to Reconstruct the Past

Have you ever wondered how scientists know so much about dinosaurs? No human ever saw a dinosaur eat or run. These huge lizards disappeared from our planet about 65 million years ago and the human species *Homo sapiens* did not exist before 500 000 years ago, according to the fossil record. If humans and dinosaurs never coexisted, where does all our information come from?

Skeletons of dinosaurs have been reconstructed. The skeleton provides indirect evidence of how a dinosaur might have lived. Evidence from the skull in **Figure 7** indicates that this dinosaur might have been a meat eater. No one ever saw this dinosaur eating meat; the evidence to support this conclusion came from examining the skull shape and the structure of the teeth.

Scientists often make models to help them understand how living things work. Models can be small-scale structures that simulate what is found in nature. For example, a scientist might reconstruct the climatic conditions of 65 million years ago to uncover what might have happened to the dinosaurs. Another type of model could be nonliving structures that work in a similar fashion. For example, the human heart is often understood from the model of a pump.

Recently, scientists have begun using computers to make mathematical models. Unlike the structural models, these models only exist as numbers.

SUMMARY Reptiles

1. Reptiles were the first group to be able to complete their life cycle out of water. When reptiles invaded land, they increased greatly in number and variety since there were few competitors.
2. Subclasses are Anapsida (turtles and tortoises), Lepidosaura (lizards, snakes, and the tuatara), and Archosaura (crocodiles, alligators, and dinosaurs).
3. Reptiles needed to adapt to land.
   - They developed a dry, waterproof skin, and more efficient lungs and circulation.

- The amniotic egg was the most significant development. The amniotic egg provides a self-contained environment for the embryo, allowing reptiles to live in very dry areas.
- Limbs also became increasingly efficient.

4. Dinosaurs ruled Earth for millions of years. The reason for their disappearance is still debated.

Sections 12.3–12.4 Questions

Understanding Concepts

1. Describe two adaptations of fish that allowed for movement onto the land.
2. Compare amphibians and reptiles to explain adaptations for living on land.
3. The word amphibia is derived from two words: "amphi" meaning two and "bios" meaning life. Explain why the name "two lives" is particularly appropriate for frogs and salamanders.
4. What significant advancement in reptiles allowed them greater access to the land than amphibians?
5. Describe the components of the amniotic egg.
6. Indicate two significant changes that must have occurred in the development of limbs for fish and early amphibians for movement on land to be possible.
7. What challenge is faced by terrestrial vertebrates in obtaining oxygen from the atmosphere rather than from the water?
8. How does the development of the amphibian respiratory system over its life cycle parallel the changes that occur as animals move from the water to the land?

Making Connections

9. If it is true that amphibians are especially sensitive to environmental contaminants, how might biological diversity in a given population of amphibians help them survive?

Reflecting

10. Reptiles have been hunted as pests or for their soft, leathery skin. Some naturalists have argued that because reptiles are deemed less attractive than seals or birds, they generate less sympathy.
    (a) Should reptiles be hunted? Justify your answer.
    (b) Is the statement about beauty and sympathy true? Give your reasons.
    (c) Do crocodiles and snakes serve important roles in ecosystems? Explain.

12.5 Birds

Birds are believed to have descended from bipedal, crocodilelike reptiles that existed about 160 million years ago (**Figure 1**, page 462). *Archaeopteryx*, a fossil with traits similar to both reptiles and modern birds (**Figure 2**, page 462), may represent a transition between the two groups. Its reptilian characteristics included thick bones, teeth, and a long bony tail; the most obvious avian characteristics were the elongated, winglike hand and the presence of feathers. Feathers, which evolved from the scales of the bird's reptilian ancestor, are the one trait that is unique to birds. Feathers provide insulation and protect the underlying skin.

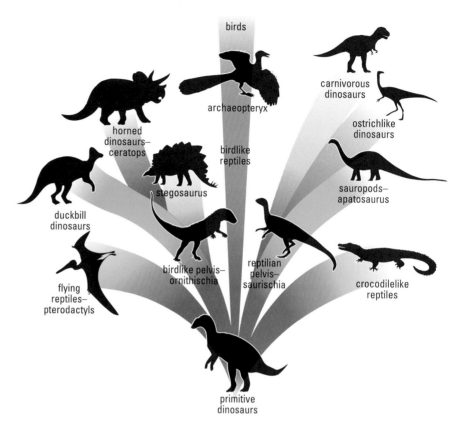

Figure 1

Reptiles were the dominant animals during the Jurassic and Triassic eras, about 60 million years ago. Birds and mammals are believed to have evolved from primitive reptile stock, each following a distinct pathway. Reptiles and birds are very similar.

birds

archaeopteryx

carnivorous dinosaurs

ostrichlike dinosaurs

horned dinosaurs— ceratops

birdlike reptiles

stegosaurus

sauropods— apatosaurus

duckbill dinosaurs

birdlike pelvis— ornithischia

reptilian pelvis— saurischia

crocodilelike reptiles

flying reptiles— pterodactyls

primitive dinosaurs

Figure 2

A 140-million-year-old fossil of *Archaeopteryx*

Class Aves

Birds are one of the best-studied animal groups, and most of the approximately nine thousand living species and subspecies have been described by scientists. Our fascination with birds is undoubtedly linked to their often musical songs and calls, and to their ability to fly, which continues to stimulate our imagination.

type	examples	type of feet	type of beak
birds of prey	hawks owls eagles	grasping	tearing
wading birds	herons sandpipers cranes	wading	spearing
swimming birds	ducks swans gulls	webbed	filtering or straining
perching birds	cardinals chickadees robins	perching	cracking seeds
nonperching birds	woodpeckers nuthatches turkeys	clinging	chiseling

Figure 3

Five common groups of birds

Birds are found in all habitats, and they show a wide range in plumage and in behaviour patterns, many of which are related to reproduction and to care of the young.

Phylogenetic relationships among birds are poorly understood due to gaps in the fossil record and due to the many cases in which unrelated species have adapted to similar environments in similar ways. However, the end result is clear: Aves is a spectacular and diverse class of animals. Structural characteristics are the prime features used to identify and classify birds. Two of the most familiar structures, beaks and feet, delineate five common groups of birds (**Figure 3**).

Characteristics for Flight

In class Aves, specialization has occurred in response to the pressures of flight. The most obvious characteristic for flight is feathers. The flight (contour) feathers of birds are lightweight with hollow main shafts (quills). The barbs that branch out from the shaft are held together by hooks, keeping the feathers flat and aerodynamic. Feather attachment in the wings provides the curved surface that is required for lift and flight (**Figure 4**).

Birds have lungs, and many, such as the pigeon, also have air sacs among the internal organs and inside the bones. Air is inhaled through the mouth and nose, down the trachea, and into the lungs and connecting air sacs. During exhalation, carbon dioxide is expelled, and some of the air in the air sacs moves into the lungs. This arrangement permits an almost continuous supply of fresh air and allows the bird to breathe in the thin air found at higher altitudes. The series of air sacs provides a particularly efficient gas exchange system, meeting the high energy requirements of flight, and also increasing the buoyancy of the bird in the air—a distinct flight advantage (**Figure 5**).

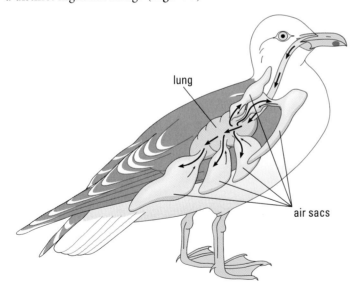

Any reduction in body mass aids in flight. A system of hollow but strong bones in a rigid framework provides strength without weight; only the neck and tail retain flexibility. Some bones are actually joined to the lungs. The air that enters the bones has been warmed up as it passes through the body, increasing buoyancy of the bird. In flying birds, the **sternum** is quite large to allow attachment of the large powerful flight muscles (**Figure 6**, page 464).

Flight also requires efficient circulation to ensure that the muscles are well supplied with oxygen. A four-chambered heart is at the centre of the bird's circulatory system. The complete separation of oxygen-carrying arterial blood

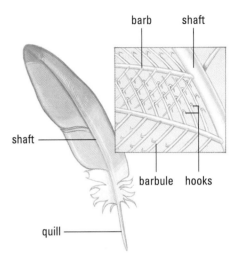

Figure 4
The flight feathers consist of a hollow shaft to which are attached small branches called barbs and barbules. Down feathers are hair-like and are designed for insulation.

Figure 5
The respiratory system of a bird is the most efficient of all chordates.

sternum: the breast bone found at the front of the chest in birds and mammals

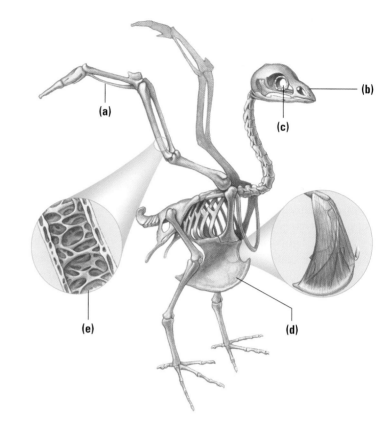

Figure 6

Many of the secrets of bird flight can be found within the skeleton. Birds carry little excess weight.

(a) Long limbs provide a large gliding surface.

(b) Birds use light beaks instead of heavy teeth for biting and tearing their food.

(c) A large socket holds the well-developed eye. Flight requires excellent vision.

(d) A large breastbone provides an attachment place for very big, strong flight muscles.

(e) Lightweight bones have many round air spaces.

from carbon-dioxide-carrying venous blood is a marked improvement over the reptilian three-chambered heart. It eliminates the mixing of oxygenated and deoxygenated blood, allowing higher levels of oxygen in the blood and a very efficient metabolism, which helps the bird maintain a constant temperature.

Birds show a number of other characteristics that allow for their aerial mode of life (**Figure 7**). They lack urinary bladders, which is another weight-reducing

Figure 7

The bird's body is adapted for flight in many ways. Small feathers make the bird's body smooth and aerodynamic; large feathers provide lift and control. Special organs help improve circulation of oxygen and compensate for the lack of teeth.

(a) Small contour feathers provide a stream-lined body shape.

(b) Long, broad primary feathers provide a large surface area to catch air and provide maximum lift.

(c) The hollow quill provides maximum strength with very little weight.

(d) The interlocking projections provide a large, flat airfoil that provides lift.

(e) The long tail feathers act as a rudder, controlling the direction of flight.

(f) Small pebbles held in the gizzard grind the food.

(g) Air sacs attached to the lung help the bird take in a large volume of air.

(h) The crop stores food.

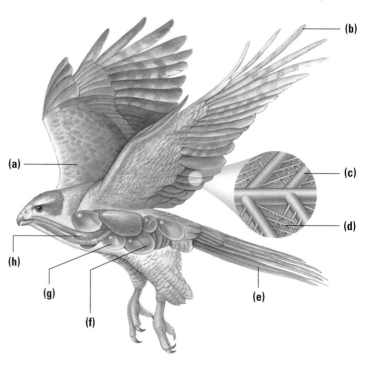

adaptation. Urine is replaced by uric acid, a semisolid, which is excreted with the feces. Digestion is rapid and efficient. Birds must eat large amounts to provide the energy for flight and maintain a constant body temperature. Birds also possess good hearing and acute colour vision. Many birds of prey can see great distances and often see better at night than most people can see during daylight.

SUMMARY **Birds**

1. Early in the history of reptiles there was great diversification, during which time the ancestors of all the living reptile groups, as well as birds and mammals, evolved.

2. Birds, like mammals, are warm-blooded; however, unlike mammals, all birds have feathers and lay eggs.

3. Birds display many characteristics for flight, such as feathers, hollow quills and bones, very efficient lungs with air sacs, an efficient heart, good vision, and rapid digestion.

Practice

Understanding Concepts

1. List three characteristics that allow birds to fly.
2. Why have birds evolved feathers?
3. Why have some biologists argued that birds are really reptiles?
4. Refer to **Figure 3**, page 462, and design a bird according to each of the following descriptions:
   (a) eats fish and hunts in deep, muddy swamps
   (b) is a predator that feeds on smaller birds
   (c) is a small bird that feeds on nuts and seeds and lives in conifers

12.6 Mammals

The forerunners of modern mammals were probably reptiles called therapsids, which lived about 240 million years ago and had teeth, a skull, and limbs similar to those of present-day mammals. Mammals remained small and uncommon until the dinosaurs died out some 65 million years ago. At that time, with little competition for food and living space, mammals did not have much difficulty surviving and reproducing. Consequently, conditions were right for the development of a wide range of new species.

Class Mammalia

From the tiny hognose bat that weighs less than 1.5 g, to the massive blue whale that can exceed 100 tonnes (t), the 4500 species of mammals show great diversity. Like other vertebrates, mammals have a bony skeleton that protects delicate organs of the body. The main axis runs along the backbone, separating the right and left sides, demonstrating bilateral symmetry.

Monotremes and Marsupials

Only the early mammals are egg layers, referred to as **monotremes**. The reptilian ancestors of monotremes were different from the reptilian ancestors of all other mammals, including the marsupials (**Figure 1**). The eggs of monotremes are

(a)

(b)

Figure 1
(a) The odd-looking duck-billed platypus is an egg-laying mammal that lives in Australia.
(b) Spiny anteater: Only two species of this unique monotreme exist in the world today.

monotremes: mammals that reproduce by laying eggs

marsupials: mammals that give birth to partially developed embryos that continue further development in the mother's pouch

placental: a type of mammal that has all of the embryo development within the uterus of the female

(a)

(b)

Figure 2

(a) Native to British Columbia, the opposum, or possum, is the only Canadian marsupial. Over the past few years, it has moved into the southern parts of Ontario and Quebec, which now have resident populations.

(b) The koala bear is a threatened species in Australia. It is slow moving and spends most of its time in eucalyptus trees, feeding on the leaves.

prehensile: capable of grasping

incubated outside the body, and the young are nourished from mammary glands that secrete milk directly into the fur. The three living species are the duck-billed platypus of Australia and two species of spiny anteaters of Australia and New Guinea.

Marsupials are usually born in an extremely immature condition and undergo further development in a pouch on the ventral side of the mother. The hairless, partially formed young crawl into the mother's pouch and attach to her nippled mammary glands, where they remain for the duration of the feeding period.

With the exception of the opossums of North, Central, and South America, modern marsupials are confined almost entirely to Australia and the neighbouring islands. The best-known Australian marsupials are kangaroos, wallabies, and koala bears. The North American opossum is an omnivore (**Figure 2**). Many Australian marsupials are herbivores. The Tasmanian devil and the extinct Tasmanian wolf represent marsupial carnivores.

Placental Mammals

Approximately 95% of mammals, including humans, are **placental** mammals. The placenta develops in the uterus to facilitate the exchange of materials between mother and young. As the embryo develops, an umbilical cord forms that connects it to the placenta. The umbilical cord contains the veins and arteries that act as the lifeline between the mother and the developing embryo. The placenta contains capillaries inside strands of tissue called villi. The capillaries come close to, but never make direct contact with, the mother's blood. Oxygen and nutrients diffuse from the mother's blood into the villi and capillaries of the embryo's blood system. Carbon dioxide and wastes return from the embryo to the mother in much the same way.

Placental mammals provide varying degrees of maternal care. For cats, rabbits, and humans, which are born naked and helpless, the young are dependent on the mother for long periods. In the case of horses (which can stand shortly after birth) and whales (which swim beside their mother immediately after being born), the period of dependence is shorter. Besides nursing, placental mammals provide other types of care for their young. Adults protect the young from predators and teach them survival skills such as searching for food and seeking protection from the elements and from other animals. The gestation period—the span of time during which an embryo is carried in the uterus—also varies among species of placental mammals.

There may be as many as 24 separate orders of placental mammals, including fossil forms. The major orders are summarized in **Table 1**, pages 468–469, and one particularly important group, the primates, will be discussed in detail.

Order Primates

The classification of the order Primates is complex and being reviewed. **Figure 3** shows one of the present classification systems for some examples of primates.

Like their ancestors, most primates are tree dwellers. Lemurs, lorises, and related forms are classified as a suborder, Prosimians. The Prosimians are largely arboreal (tree dwelling) and are most active at night. Lemurs have a foxlike head with a pointed muzzle; their tails are long but not **prehensile**. Tarsiers are small, lemurlike creatures with long ears, large and protruding eyes, and elongated heels.

Monkeys, apes, and humans are referred to as the suborder Anthropoids and are further classified into major groups called the New World monkeys, Old World monkeys, and hominoids (apes and humans).

New World monkeys, found in North and South America, are tree-dwelling. In many species, the long tail is prehensile and can be used to grasp objects. The

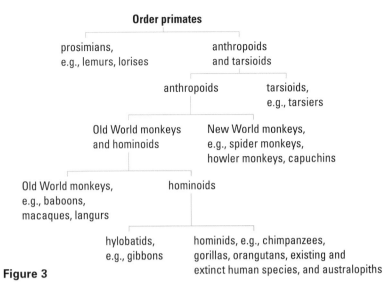

Order primates

prosimians, e.g., lemurs, lorises

anthropoids and tarsioids

anthropoids

tarsioids, e.g., tarsiers

Old World monkeys and hominoids

New World monkeys, e.g., spider monkeys, howler monkeys, capuchins

Old World monkeys, e.g., baboons, macaques, langurs

hominoids

hylobatids, e.g., gibbons

hominids, e.g., chimpanzees, gorillas, orangutans, existing and extinct human species, and australopiths

Figure 3

nostrils are widely separated and open to the sides. The nostrils of Old World monkeys, found in Asia and Africa, are close together and face forward. Their tails are not prehensile and they possess sitting pads, often brightly coloured, that act as a cushion when the monkey is in the sitting position.

While the name "ape" is sometimes used to designate any monkey, it is usually applied to the hylobatids and all nonhuman hominids. The great apes are those that closely resemble humans. Apes lack tails, live in forests, and their natural walking position is on all four limbs. While most are vegetarians, the chimpanzees do eat some meat and are capable of using simple tools.

The human, *Homo sapiens*, is a member of the **hominid** family. A highly complex brain, the specialization of parts of the skeleton and other body structures, the ability to make and use tools, and the fact that they are endothermic are major reasons why humans have adapted to almost every climatic condition. The development of an opposable or gripping thumb was one of the most significant adaptions.

hominid: a family consisting of existing and extinct human species and close relatives

Practice

Understanding Concepts

1. What are the identifying characteristics of a mammal?
2. Why are therapsids considered to be the link between reptiles and mammals?
3. List the characteristics that would help you identify a monotreme and a marsupial.
4. What is a prehensile tail? In what group is it found?
5. Why are rodents often called gnawing animals?
6. What are some of the distinguishing features of New and Old World monkeys?

General Characteristics

A typical mammal is a warm-blooded, air-breathing, four-legged vertebrate. Its skin is covered with hair (in the form of bristles, wool, scales, or fur) and is equipped with sweat glands. Hair and sweat glands help mammals to control their body temperature. Teeth may be of four different types and are fixed into

Table 1: The Major Orders of Placental Mammals

Order	Representative members	Characteristics	
1. Insectivora	insect eaters: moles, shrews, European hedgehog	• small, with elongated snouts for burrowing and snuffling out food • mostly nocturnal • moles have poor vision but keen smell and hearing • some shrews may eat twice their body weight each day	1.
2. Chiroptera	bats	• forelimbs modified for flight • approximately one thousand species • many are nocturnal and use echolocation to navigate and find food • can hibernate through unfavourable periods	2.
3. Edentata (or Xenarthra)	toothless mammals: armadillos, anteaters, sloths	• lack teeth or have only poorly developed molar teeth • anteaters have sharp claws and a long, sticky tongue for collecting ants	3.
4. Rodentia	gnawing animals: rats, mice, beavers, chipmunks, voles, porcupines, guinea pigs	• chisel-shaped incisor teeth, two in front of both upper and lower jaw, for chewing wood and other hard material • incisors grow continuously • no canine teeth	4.
5. Lagomorpha	rodentlike mammals: rabbits, pikas, hares	• teeth similar to rodents, but with four incisors in upper jaw (second pair behind the first) • short, stubby tails	5.

Order	Representative members	Characteristics	
6. Cetacea	fully aquatic mammals: whales, dolphins, porpoises	• thick layer of fat or blubber beneath skin • pectoral appendages modified as flippers • tail flattened horizontally • breathe at surface by means of modified nostrils on top of head	6.
7. Carnivora	flesh eaters: cats, dogs, bears, seals, sea lions, mink	• three pairs of small incisors in both upper and lower jaws • large canines for killing and cutting prey • digitigrade foot type (walking on toes) except for aquatic forms, which have webbed feet	7.
8. Perissodactyla	odd-toed hoofed animals: horses, donkeys, zebras, tapirs, rhinoceri	• herbivores • third toe is largest • walk on nails or hooves • usually large • some have nose and upper lip drawn out into a short snout	8.
9. Artiodactyla	even-toed hoofed animals: cattle, sheep, goats, camels, llamas, deer	• herbivores • walk on the nails of their third and fourth toes	9.
10. Proboscidea	trunked mammals: elephants, and related extinct mammoths and mastodons	• nose and upper lip extended to form long, prehensile snout or trunk, with nostrils at its free end	10.
11. Primates	monkeys, lemurs, apes, tarsiers, humans	• highly specialized nervous system, with complex brain	11.

sockets in the jawbone. The pattern and arrangement of teeth reveal whether an animal is an omnivore, carnivore, insectivore, or herbivore.

Mammals have fleshy lips, a diaphragm separating the lungs from the general body cavity, a four-chambered heart, a middle ear, and a well-developed brain that enables them to learn a variety of behaviour patterns. Except for monotremes such as the duck-billed platypus, all are **viviparous** (bear live young). The embryo is surrounded by two membranes and is connected to a placenta by an umbilical cord. The most distinctive mammalian characteristic is the presence of mammary glands on female adults, which secrete milk to nourish the developing young.

Mammalian limbs are usually pentadactyl (with five toes); variations are a result of adaptations to running. For some mammals, the central toes have become more important and the total number is reduced (**Figure 4**). Other limb adaptations include those used for climbing (ape, monkey), leaping (rabbit, kangaroo), burrowing (mole), flying (bat), and swimming (dolphin). In upright mammals, limiting the hind limbs for movement freed up the forelimbs for other specialized functions such as grasping and manipulating objects. Regardless of specific adaptations, the number and organization of the limb bones have remained relatively constant.

viviparous: describes animals that give birth to live young

Figure 4
The different ways in which the foot is placed on the ground can be used to identify the group of mammals.
(a) Plantigrade: The whole foot is placed on the ground.
(b) Digitigrade: The heels are raised off the ground, and walking occurs using the digits to create the gait.
(c) Unguligrade: The running is efficiently accomplished on the tips of the toes, which are covered by hooves.

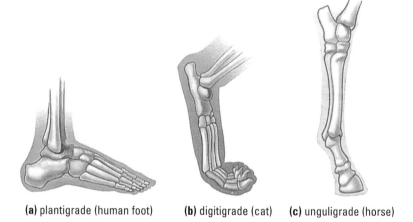

(a) plantigrade (human foot)   **(b)** digitigrade (cat)   **(c)** unguligrade (horse)

Characteristics for Keeping Warm

A dense coat of hair is found on most terrestrial mammals and is involved in the regulation of the animals' body heat. Both the longer and outer guard hairs and the short underhairs provide excellent insulation, helping the animal prevent heat loss. In areas of extreme cold, the outer guard hairs are often hollow, giving them a superior insulating value. In hot climates, hair can protect animals from absorbing too much heat. Coat coloration can also allow an animal to blend into its environmental background. This is advantageous for both predators and prey (**Figure 5**).

In some mammals, the hair has become specialized. The guard hairs on the porcupine are represented by quills, a well-known defence mechanism. In pronghorns, the horns are covered with a skin containing specialized hairs. The whiskers found on many animals are hairs that provide them with an increased sensitivity to touch. The whiskers on a deer mouse inform the animal of the size of an opening it may wish to pass through.

Hair contains the protein keratin which is produced in the individual hair follicles and fills the dead cell of the growing hair. Keratin also makes up parts of other animal structures which are, therefore, hair related. These include antlers, horns, toes, hooves, fingernails, and claws.

Figure 5
The arctic fox in winter illustrates the importance of hair, which insulates as well as provides the animal with camouflage.

Characteristics for Feeding

The most visible characteristics for feeding in mammals are the types and arrangement of teeth in their mouths. Teeth are of four basic categories: incisors, canines, premolars (bicuspids), and molars (tricuspids). They appear in the above order from the front to the back of the mouth on both sides and in the upper and lower jaw. The teeth in both jaws match up, unlike those of the reptile. This helps in chewing. **Figure 6** compares the teeth of a wolf and a cow.

(a) **(b)**

Figure 6
(a) The incisors and canine teeth of the wolf are specialized for holding and tearing meat.
(b) The cow uses its incisors to tear up grass and its molars to grind the grass.

Each type of tooth is specialized. The incisors are used for biting and tearing. In elephants, the upper incisors have developed into tusks. The canines of mammals, which can be quite large, assist a predator in holding and piercing its prey. In the walrus, they grow up to 90 cm long and are also called tusks (**Figure 7**). The premolars and molars of mammals are generally used for chopping and grinding food.

Many animals can be identified by a dental formula. In this type of formula, one half of the jaw (from the front to the back on one side) is represented by the specific number of each tooth type. The top part of the fraction represents half of the upper jaw and the bottom represents half of the lower jaw. In humans, the formula is $\frac{2123}{2123}$, indicating the presence, in the half jaw, of 2 incisors, 1 canine, 2 premolars, and 3 molars. In gnawing rodents like the beaver, the second number in the formula has a zero value, as the canine is not present in these mammals.

In animals that rip up grasses (cattle) or browse on buds and leaves (elk), the dominant teeth are the incisors and molars. Their digestive systems are specialized to break down foods with high cellulose content. These ruminants have a multiple stomach system consisting of four chambers. The food enters the first two chambers, the rumens, which are filled with bacteria that produce special cellulose-splitting enzymes. After a period of time, the animal ruminates or chews its cud by regurgitating some of the partially digested food. This results in more mechanical breakdown of the food as well as enzymes being mixed with it. Then the food is swallowed a second time and enters the last two chambers for further chemical digestion before passing into the intestine.

Figure 7
The long canine teeth of the walrus are referred to as tusks.

Characteristics for Running

The endoskeletons of mammals confer three advantages over organisms with exoskeletons: mammals can grow continuously without a moulting phase; they do not have a vulnerable period while the new exoskeleton forms; and they do not have to divert their metabolic energy to produce the new skeleton. In combination with well-developed muscles and bones, particularly where they are jointed, the vertebrate skeleton is highly flexible. The structure of the limbs permits both ease of motion and speed. The pectoral and pelvic girdles have been modified for support and to accommodate the location of the limbs directly under the body.

The limbs of the various orders of mammals have adapted to their specific environment. For example, members of the cat family have developed grasping claws on their feet, used for climbing and holding down prey. The snowshoe hare has elongated hind legs specialized for travelling on snow and for hopping. Grazing animals, often found in prairielike conditions, have modified toes designed for speed—their best means of avoiding a predator. The human foot is designed to support bipedal motion.

Activity 12.6.1

Creating Models for Limb Movement

Your limbs work as levers, which come in three different classes (**Figures 8**, **9**, and **10**). Muscles provide the force to move the levers. The fixed point around which a lever turns is the fulcrum. In most situations, the joints act as the fulcrum. In this activity, you will study movement by building levers to represent limbs.

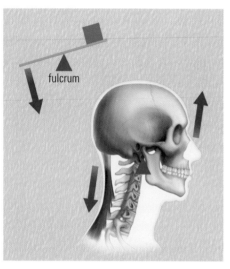

Figure 8
First-class levers apply a downward force to move a mass upward. For example, muscles at the back of the neck pull downward to move the head upward. The atlas bone, which attaches the vertebral column (backbone) to the skull, acts as the fulcrum.

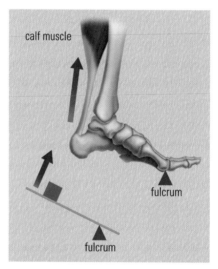

Figure 9
In second-class levers, the mass being moved is positioned between the fulcrum and the force applied. For example, the calf muscle pulls the leg upward.

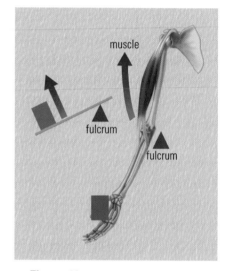

Figure 10
Third-class levers apply force between the mass being moved and the fulcrum. For example, the elbow acts as a fulcrum. The load rests on the hand, and the bicep muscle, shown in the diagram, pulls the arm upward.

Materials

Geostrips; a flat, plastic Mechano set; very stiff cardboard; or foam board or wide popsicle sticks with holes cut out, as shown in **Figure 13**
bristol board or another type of thick but pliable cardboard
brads or butterfly tacks
elastic bands
adhesive tape
books
scissor
ruler

years. They often use their flipperlike front limbs for walking on the ocean floor while searching for food. They do not fear humans and have a pleasant disposition. Regrettably, this has led to a serious reduction in their numbers.

The beaver, muskrat, and otter are other mammals whose modifications fit their natural aquatic environment.

Research in Canada: Kristine Webber

Kristine Webber studies coyotes—not rural coyotes, but urban coyotes. As a graduate student in the Department of Animal Sciences at the University of British Columbia, Webber became interested in this novel aspect of "problem wildlife" when concerned pet owners started calling the veterinary clinic where she worked to report coyote sightings. As she began to learn the extent of their presence in Vancouver (and in many North American cities, including Toronto, New York, and Los Angeles), she decided to make the terrier-sized *Canis latrans* the subject of her master's degree research.

Now head of Vancouver's "Urban Coyote Project," Webber is involved in measuring the size of the urban coyote population, learning more about its basic ecology, and educating the public on how to coexist with the animal. Standard ways of dealing with the perceived invasion by coyotes into urban areas have been aimed at eradicating or relocating the animals. Webber describes both these solutions as expensive and ultimately ineffective. Working with wildlife control agencies in the Vancouver area, she is hoping instead to develop an urban coyote management plan that focuses on controlling the population and allaying public concerns.

Already Webber's study is turning up some interesting results about what she calls "a fascinating creature that has adapted extremely well to an urban environment." She has found that coyotes travel throughout the city, even venturing away from parks and golf courses right into the downtown. They appear to be active during the day as well as at night. Seventy percent of their diet is small mammals, such as rats, voles, field mice, and rabbits. The rest is made up of fruits, vegetables, garbage, and—only occasionally—domestic pets.

Animals and Humans

Our history of living with wildlife has not been a good one. Wolves once roamed across North America but, today, they seek refuge in the remotest areas. Many ranchers curse the protected wolves that leave the national parks to hunt their cattle. Before the wolves were protected, the answer to such confrontations was to eliminate the menace. As a result, the wolf has been completely eliminated from some areas without a recorded case of an attack on humans.

Removing large predators is not without its costs. Small rodent and hare populations often increase dramatically once a predator is removed. Exploding populations of these smaller animals often result in crop damage, damage to housing, or the outbreak of contagious diseases.

As urban centres continue to grow, wildlife areas are being destroyed or changed. Garbage dumps have attracted bears. The promise of free food not only draws the bears into areas where humans congregate, but reduces the bears' fear of people. A startling increase in the number of bear problems over the past few years can be traced to the fact that bear and human populations are overlapping in many more locations. The disruption of wildlife areas has also meant a decline in the natural food source for most of the larger predators. To survive, they must seek a new food source.

Case Study: Tracing Evolutionary Trends in the Vertebrate System

In this case study, you will compare the structure of vertebrate organs of the circulatory, respiratory, and excretory systems, and identify trends in evolutionary development.

Procedure/Analysis

1. **Figure 17** shows the heart in fish, amphibians, and mammals.
   (a) What is the major evolutionary trend in the vertebrate heart?

→ oxygenated blood

→ deoxygenated blood

→ mixed oxygenated/deoxygenated blood

▬ atrium

▬ ventricle

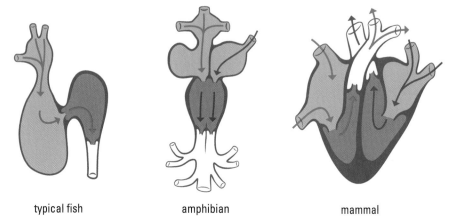

typical fish   amphibian   mammal

Figure 17

2. Use **Figure 18** to explain this statement:
   Vertebrates that use gills for respiration are said to have a single-loop circulatory system, whereas those with lungs have a double-loop system.
   (b) The fluid pressure of the blood passing through the systemic capillaries in a single-loop system is greatly reduced. Why do you think this is so? What disadvantage would this have for the animal?

Single-loop circulatory system   **Double-loop circulatory system**

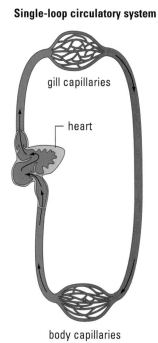
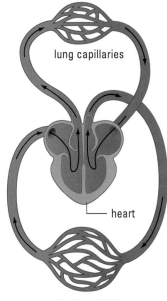

gill capillaries   lung capillaries

heart   heart

body capillaries   body capillaries

Figure 18

(c) Compare the number and arrangement of chambers in the hearts of fish, amphibians, reptiles, birds and mammals.

(d) Look at the type of blood that flows through the systemic capillaries in the three-chambered and four-chambered heart. What advantages do you think the four-chambered heart has over the three-chambered heart?

3. **Figure 19** shows the brain in fish, amphibians, birds, and mammals.

(e) What is the major evolutionary trend in the vertebrate brain?

(f) Compare the relative sizes of the various parts of the brain in the four vertebrate representatives. Are there any differences?

(g) The cerebellum (or midbrain) coordinates movement and controls balance. Account for the differences in the sizes of the cerebellum in the four brains.

(h) The cerebrum is the "thinking" part of the brain. Account for the varying sizes of the cerebrum in the four brains.

(i) Compared to other animals, humans have poorly developed senses. Use the comparative sizes of the brains to explain this.

Synthesis

(j) Explain why endothermic vertebrates have a greater requirement for a four-chambered heart than fish or frogs.

(k) In less intricate vertebrates, the brain stem is the main centre of the brain. In more complex forms, specific areas of the brain have become more intricate and have increased in size.

  (i) Which area of the complex vertebrate brain is involved in coordinating motor activities? Explain.

  (ii) Which part of the mammalian brain might be associated with the ability to learn from experience? Explain.

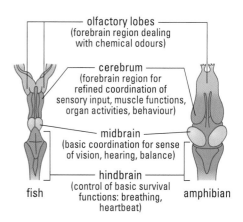

olfactory lobes
(forebrain region dealing with chemical odours)

cerebrum
(forebrain region for refined coordination of sensory input, muscle functions, organ activities, behaviour)

midbrain
(basic coordination for sense of vision, hearing, balance)

hindbrain
(control of basic survival functions: breathing, heartbeat)

fish   amphibian

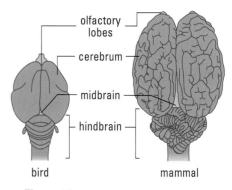

olfactory lobes

cerebrum

midbrain

hindbrain

bird   mammal

Figure 19

Explore an **Issue**

Take a Stand: Controlling Wildlife

The sharing of land by humans and animals is not a simple relationship. Animals may benefit humans by controlling other populations within the region. The extent of animal control and protection is an ongoing issue associated with urban development.

Statement

Large predatory animals must be protected in more areas. Human populations cannot expand into areas presently occupied by other animals.

- In your group, research the issue. Learn more about confrontations between humans and wildlife, diseases that wildlife carries which affect humans, and the environmental impact of controlling wildlife.
- Search for information in newspapers, periodicals, CD-ROMs, and on the Internet.
  Follow the links for Nelson Biology 11, 12.6.

GO TO www.science.nelson.com

- Write a list of points and counterpoints which your group considered.
- Decide whether your group agrees or disagrees with the statement.
- Prepare to defend your group's position in a class discussion.

DECISION-MAKING SKILLS

- Define the Issue
- Identify Alternatives
- Research
- Analyze the Issue
- Defend a Decision
- Evaluate

SUMMARY Mammals

1. Mammals diversified and became very numerous only after the death of the dinosaurs.

2. Only the early mammals, the monotremes, lay eggs. Marsupials bear live, immature young that complete their development in a pouch attached to the mother. Most mammals are placentals; in placentals, all embryo development occurs within the uterus.

3. Typically, mammals are warm-blooded, air breathing, and four legged.
   - They have hair, sweat glands, teeth, fleshy lips, a diaphragm, a four-chambered heart, and a middle ear.
   - Females have mammary glands, which secrete milk to nourish the young.
   - An endoskeleton means that mammals can grow continuously without moulting.
   - A complex brain and nervous system enables mammals to learn from experience and adapt readily to environmental changes.

Sections 12.5–12.6 Questions

Understanding Concepts

1. Based on fossil records, which developed earliest, birds or mammals?

2. What advantages does a four-chambered heart have over a three-chambered heart?

3. Describe some of the specializations for flight in birds.

4. How do monotremes and marsupials differ from placental mammals?

5. What is the function of the placenta? How is it adapted for that function?

6. Mammals reproduce sexually. How would this characteristic have helped mammals to diversify and expand all over the world?

Pest Control Officer

Pest control officers evaluate and correct problem situations involving animal pests in a variety of environments, ranging from industrial complexes and housing, to tree management in cities and forests. The pests may be rodents, but the majority of work involves insects, so qualifications for an entomologist are often required. It is essential for any pest control officer to understand the life cycle and physiology of the pests, as well as the environmental effects of chemicals and other treatments for eliminating or discouraging pests.

Veterinary Technician

Veterinary technicians perform many of the same tasks for a veterinarian that nurses and other professionals perform for physicians. The majority are employed in companion animal practice, but opportunities exist in related areas such as biomedical facilities, veterinary supply sales, zoos, and animal control centres.

Practice

Making Connections

7. Identify several careers that require knowledge about microorganisms, plants, or animals.

8. Survey newspapers or conduct a web search to identify career opportunities in these areas.

9. Select one of the careers that most interests you. Which high school subjects will you need for this career?

10. Investigate and list features that appeal to you about this career. Also make a list of features that you find less attractive about this career.

11. What are some technology applications that aid biological research? Follow the links for Nelson Biology 11, 12.6.

GO TO www.science.nelson.com

Research Scientist

Research scientists investigate a wide range of subject areas related to microorganisms, plants, or animals. For example, mycologists study fungi, ornithologists study birds, icthyologists study fish, and parasitologists study parasites. Research may be conducted at colleges, universities, or diagnostic laboratories, and involves extensive use of computers—whether for using the Internet to keep up with new developments in the field and communicating results, or to collect, measure, and analyze information in digital libraries.

Key Expectations

Throughout this chapter, you have had opportunities to do the following:

- Describe selected anatomical and physiological characteristics of representative organisms from the animal kingdom (12.1–12.6).
- Demonstrate, through classification techniques and terminology, the usefulness of the system of scientific nomenclature in the field of taxonomy (12.1, 12.2, 12.4).
- Classify representative organisms from the animal kingdom (12.1, 12.2, 12.4, 12.6).
- Compare and contrast the life cycles of representative organisms from the animal kingdom (12.2, 12.3, 12.4, 12.5, 12.6).
- Demonstrate an understanding of the connection between biodiversity and species survival (12.3, 12.4, 12.6).
- Explain the importance of sexual reproduction to variability in a population (12.4).
- Select appropriate instruments and use them effectively and accurately in collecting observations and data (12.6).
- Locate, select, analyze, and integrate information on topics under study, working independently and as part of a team, and using appropriate library and electronic research tools, including Internet sites (12.6).
- Select and use appropriate modes of representation to communicate scientific ideas, plans, and experimental results (12.6).
- Identify and describe science- and technology-based careers related to the subject area under study (12.6).
- Use techniques of sampling and classification to illustrate the fundamental principles of taxonomy (Unit 4 Performance Task).

Key Terms

allantois	marsupials
amnion	monotremes
amniotic eggs	operculum
apomorphic	placental
chorion	placoid scales
cladistics	plesiomorphic
ectotherms	prehensile
endotherms	sternum
hominid	swim bladder
lateral line	viviparous

Make a
Summary

In this chapter, you studied the diversity and characteristics of chordates. To summarize your learning, create a phylogenetic tree of chordates with representatives from each class. Label the tree with as many of the key terms as possible. Check other phylogenetic trees and use appropriate designs to make your sketch clear.

Reflect on your Learning

Revisit your answers to the Reflect on Your Learning questions at the beginning of this chapter.

- How has your thinking changed?
- What new questions do you have?

Understanding Concepts

1. What characteristics have allowed the bony fish to be successful in so many different environments?

2. How did the amniotic egg allow reptiles to dominate the terrestrial environment?

3. Describe two adaptations that have occurred in vertebrates to permit movement on land.

4. How do many amphibians compensate for a less-developed lung?

5. How is the surface area for gas exchange increased in the mammalian lung?

6. What three adaptations, which are lacking in reptiles, enable birds to fly?

7. How does the circulatory system of a reptile differ from that of a bird or mammal?

8. Provide examples that explain how the development of the limb structure of vertebrates is related to their ecological niche.

9. Compare the lungs of amphibians to those of reptiles and mammals.

10. What adaptations in lung development provide more active animals with greater amounts of oxygen?

11. Indicate three advantages of the endoskeleton of chordates over the exoskeletons of insects and crustaceans.

12. Indicate three advantages of the exoskeletons of insects and crustaceans over the endoskeleton of chordates.

Applying Inquiry Skills

13. Each of the birds in **Figure 1** has special characteristics.
    (a) Which bird or birds are specialized for eating meat? List special characteristics for birds that eat meat.
    (b) The hummingbird uses a long tongue to get nectar from flowers. What other characteristics does the hummingbird have that make it successful at feeding?
    (c) Identify what might be the special feeding characteristics of the pelican.
    (d) Swallows are insect eaters. Would you expect that swallows catch insects in the air or scoop them from lakeshores? Give your reasons.

(a) owl   **(b)** hummingbird   **(c)** pelican   **(d)** swallow   **(e)** hawk

Figure 1

14. Bearing live young occurs in several vertebrate classes. Biologists believe that this type of reproduction has probably occurred independently in each group. Name some live-bearing vertebrates. What do you think the advantage is of carrying the young inside the body instead of laying an egg?

Making Connections

15. Drawing upon your knowledge of bird flight, indicate what special characteristics humans would need to fly (unassisted by technological devices such as airplanes or hot-air balloons). Draw a picture which illustrates some of the characteristics described.

16. Recently ecologists have been surprised to discover rapid declines in the population of some amphibians. Although many hypotheses have been suggested, no comprehensive explanation has been widely accepted. One hypothesis proposes that amphibians are ecological indicators and that their decline may signal deepening environmental problems. Why might scientists believe that amphibians are ecological indicators? How would you go about testing such a hypothesis?

17. Many animals are considered to be in danger of becoming extinct. Strict laws have been established in many countries to protect endangered species, and many people are involved in programs to preserve important habitats such as the tropical rain forests. Comment on this statement: Humans have a responsibility to preserve the diversity of species found on Earth.

18. Although the transition from water to land is a major step in animal evolution, some terrestrial animals (e.g., whales) later returned to the sea. Discuss the adaptations required to return to an aquatic existence.

19. How might the upright stance and opposable thumb in humans have influenced technology and cultural development?

Exploring

20. There has been considerable controversy over the development of game farms and fish farms in some parts of Canada. Research this issue and discuss both the benefits and potential dangers of such farms.

21. The negative effect of the lamprey on other species in the Great Lakes is well documented. Conduct an Internet or library search to find out about other invasions, such as the introduction of the horse to Sable Island, Nova Scotia.
    Follow the links for Nelson Biology 11, Chapter 12 Review.

 www.science.nelson.com

Cladistics

In this activity, you will apply your classification skills to build a cladogram.

How to Make a Cladogram

1. Study **Table 1**. The plus sign (+) indicates that the animal has the trait, while the minus sign (−) indicates that the trait is missing.

Table 1

Animal	Bony skeleton	Paired appendages	Enlarged cranium	Mammary glands	Small canines	Total # of derived traits
lamprey	−	−	−	−	−	0
dog	+	+	−	+	−	3
frog	+	+	−	−	−	2
shark	−	+	−	−	−	1
chimp	+	+	+	+	−	4
human	+	+	+	+	+	5

Note: When constructing a cladogram, the characteristics chosen should be derived, meaning that they have evolved more recently than the common ancestor to all organisms. For example, the "presence of a tail" would be a poor choice since the ancestor to all the organisms listed is thought to have possessed a tail. Therefore, the "absence" or "reduction of the tail" would be a derived trait that could have been considered for the cladogram.

2. Using graph paper, begin to construct your cladogram. The organism with the fewest of the derived traits examined (in this case, the lamprey) is the first to split from the group. This first organism is often referred to as the "out group." Draw a large *V*. The base of the *V* represents your first node and is the common ancestor that split into the two lineages. "Lamprey" is placed at the top end of one line, as shown in **Figure 1(a)**.

3. Next, determine which organism or organisms have the next fewest derived characteristics that are shared with all remaining group members. In this example, the shark only shares a single trait with all remaining members. Place this organism at the end of a new line that splits from the group just above the base of the *V*. **Figure 1(b)** illustrates this step.

4. Continue repeating step 3, adding new nodes and lines as each organism that possesses fewer shared derived characteristics splits from the group. Refer to **Figures 1(c)** and **(d)**. Note that it is useful and informative to include the derived traits in their relative positions on the cladogram.

5. A somewhat more complex example is provided in **Table 2**. Note that organisms may split off in groups rather than individually if they share additional derived traits with each other rather than with the entire group.

Figure 2 is the cladogram corresponding to the information in **Table 2**. Circled numbers correspond to specific traits.

(a)

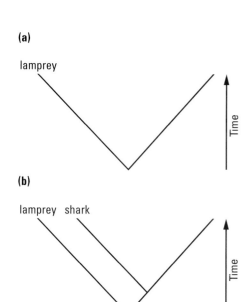

(b)

(c)

(d)

Figure 1
Steps in building a cladogram

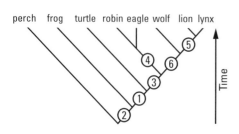

Figure 2

Table 2

Animal	Trait 1	Trait 2	Trait 3	Trait 4	Trait 5	Trait 6	Total # of derived traits
perch	–	–	–	–	–	–	0
lion	+	+	+	–	+	+	5
lynx	+	+	+	–	+	+	5
turtle	+	+	–	–	–	–	2
wolf	+	+	+	–	–	+	4
robin	+	+	+	+	–	–	4
eagle	+	+	+	+	–	–	4
frog	–	+	–	–	–	–	1

Questions

How similar are the animals to each other, based on number of shared derived traits? What is their phylogenetic history?

Materials

graph paper   colouring pencils

Procedure

1. Complete the materials list, if necessary.

2. Choose 12 different animals from the lists provided by your teacher. Select a variety of traits and use them to compare your animals. Your traits should all be significant characteristics that are derived and clearly distinctive. For example, for separating hummingbirds from bumblebees, the feature "has feathers" is a more fundamental trait than "can fly."

3. Display your data in a table and use it to create a cladogram.

Analysis

(a) Which animals are most similar to each other in the number of shared adaptive traits? Which animals share the most recent common ancestor? Confirm your conclusions using text or Internet resources.

(b) Which animals are least similar (most distantly related) to each other?

(c) Compare your results with a classmate's for the same pair of animals. Are they different? If so, why?

Evaluation

(d) Suggest some ways to improve the results of your study.

Synthesis

(e) A more recent method for classifying organisms is by examining the genetic code of different species. A recent study examined DNA sequences from a wide variety of mammals. The researchers found that hippopotamuses and whales were the only living mammals to share a common derived DNA sequence. What does this suggest about the relationship between hippopotamuses and whales? Does this information allow you to determine whether or not hippos or whales are more closely related to elephants? Explain.

Unit 4   Review

Understanding Concepts

1. For Linnaeus's system of classification, the most specific group is given which of the following names?
   (a) genus   (b) family   (c) class   (d) species

2. The smaller groups that a kingdom is divided into are called which of the following?
   (a) phyla   (b) families   (c) species   (d) colonies

3. Some kinds of differences among individuals of a species, such as differences in shape, colour, and size, are known as which of the following?
   (a) structural adaptations   (b) variations
   (c) natural selection   (d) diversity

4. How many genera are included in **Table 1**?
   (a) 2   (b) 3   (c) 4   (d) 5

Table 1: Cat Names

Common name	Genus/species
bobcat	*Lynx rufus*
lynx	*Lynx canadensis*
black-footed wildcat	*Felis nigripes*
mountain lion	*Puma concolor*
African lion	*Panthera leo*
European wildcat	*Felis sylvestris*
tiger	*Panthera tigris*
jaguar	*Panthera onca*

5. How many species are included in **Table 1**?
   (a) 2   (b) 4   (c) 6   (d) 8

6. The scientific name for the common house cat is *Felis domesticus*. Which cats in **Table 1** are most closely related to the common house cat?

7. Using the example provided by the cats in **Table 1**, explain one advantage of scientific names over common names.

8. The classification system used in this textbook identifies six kingdoms: Archaebacteria, Eubacteria, Protista, Fungi, Plantae, and Animalia. Identify the correct kingdom for each of the following organisms. Give reasons to support your answer.
   (a) pine tree   (b) moss   (c) fungi
   (d) mushroom   (e) ant   (f) *Euglena*
   (g) cyanobacteria (h) yeast   (i) *E. coli*
   (j) sponge   (k) worm   (l) fern

9. Approximately two thousand years ago, Aristotle classified organisms into two groups: plants and animals.
   (a) Why is this classification scheme limited?

(b) Aristotle lived long before the first microscopes. If he had been able to view organisms under a microscope, how might his system of classification have changed? What are some examples of organisms that cannot be classified according to his system?

10. Indicate whether each of the following statements about viruses is true or false. For each false statement, rewrite it to make it true.
    (a) Viruses are not living organisms.
    (b) Viruses are tiny cells.
    (c) Viruses do not contain genetic information.
    (d) Viruses do not contain a cell membrane, nucleus, or cytoplasm.
    (e) Viruses contain mitochondria and ribosomes.
    (f) Viruses do not require food or produce wastes if they are found outside of a host cell.
    (g) The genetic material from viruses can direct the host cell to produce virus proteins.

11. Indicate whether each of the following statements about bacteria is true or false. For each false statement, rewrite it to make it true.
    (a) Bacteria are not living organisms.
    (b) Bacteria do not contain a nuclear membrane.
    (c) Bacteria do not contain genetic information.
    (d) Bacteria do not contain a cell membrane, or cytoplasm.
    (e) Bacteria only grow in an environment rich in oxygen.
    (f) Bacteria may be parasitic or free-living organisms.
    (g) Some bacteria contain chlorophyll.

12. In what way are amoeba different from bacteria or cyanobacteria?

13. A biologist finds an unclassified organism that has the following characteristics:
    - long, filamentous hyphae
    - cell wall surrounds the cell membrane
    - multicellular—many nuclei appear to be in a single cell
    - eukaryotic cells with mitochondria and ribosomes
    - reproduction may be asexual or sexual
    - no chloroplasts are found
    - grows well in dark, moist places

    To which kingdom does this organism likely belong?
    (a) Archaebacteria (b) Protista   (c) Fungi
    (d) Plantae   (e) Animalia   (f) Eubacteria

14. Lichens are often the first organisms that appear in an area following a flood or fire. Explain why lichens are well-suited to growing on barren land.

15. The process shown in **Figure 1** is
    (a) conjugation, a form of sexual reproduction;

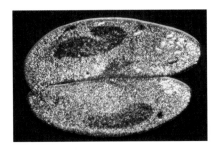

Figure 1
Paramecium

sponge
(phylum Porifera)   jellyfish
(phylum Cnidaria)   sea anemone
(phylum Cnidaria)

tapeworm
(phylum
Platyhelminthes)   pinworm
(phylum
Aschelminthes)   earthworm
(phylum Annelida)

starfish
(phylum Echinodermata)   squid
(phylum Mollusca)

Figure 2
Invertebrates

(b) conjugation, a form of asexual reproduction;
(c) spore formation, becoming multicellular;
(d) spore formation, becoming unicellular.

16. The reason why the process shown in **Figure 1** allows diversity is because
(a) sexual reproduction allows the combination of new genes;
(b) asexual reproduction allows the combination of new genes;
(c) spore formation allows the most successful genes to survive;
(d) spore formation allows the least successful genes to survive.

17. Use **Table 2** to compare mosses, ferns, and angiosperms. Copy the table into your notebook and place a check mark in the appropriate cells.

Table 2: Characteristics of Three Plant Groups

Characteristic	Mosses	Ferns	Angiosperms
vascular system			
spores			
seeds			
true roots, stems, and leaves			
water required for reproduction			
gametophyte dominant			
sporophyte dominant			

18. List two similarities and two differences among angiosperms and gymnosperms.

19. Which of the animals in **Figure 2** demonstrate radial symmetry?
(a) sponge and pinworm
(b) squid and earthworm
(c) jellyfish and starfish
(d) tapeworm and earthworm

20. Which of the animals in **Figure 2** has no true tissues?
(a) sponge   (b) tapeworm
(c) earthworm   (d) starfish

21. Which of the animals in **Figure 2** demonstrate the development of a true brain?
(a) sponge and tapeworm
(b) sea anemone and starfish
(c) tapeworm and earthworm
(d) none of the animals listed

22. Which of the animals in **Figure 2** has a segmented body plan?
(a) sponge   (b) jellyfish
(c) sea anemone   (d) earthworm

23. Which phylum has no mouth or digestive cavity?
(a) Porifera   (b) Cnidaria
(c) Platyhelminthes   (d) Nematoda

24. Which phylum alternates between polyp and medusoid body forms?
(a) Porifera   (b) Cnidaria
(c) Platyhelminthes   (d) Nematoda

25. Flatworms are
(a) not able to respond to light;
(b) radially symmetrical;
(c) strictly marine;
(d) hermaphroditic.

26. Nematodes have
    (a) a circulatory system;
    (b) no parasitic species;
    (c) a digestive system with one opening;
    (d) a simple brain.

27. The body cavity in earthworms is known as a
    (a) gastrovascular cavity; (b) coelom;
    (c) gut;   (d) peritoneum.

28. Which of the following has a complete digestive system?
    (a) leech   (b) planarium
    (c) fluke   (d) tapeworm

29. Echinoderms are
    (a) marine organisms;
    (b) terrestrial organisms;
    (c) freshwater organisms;
    (d) both marine and freshwater organisms.

30. Digestion and absorption of food in mollusks takes place in the
    (a) crop;
    (b) radula;
    (c) stomach and digestive glands;
    (d) intestine.

31. Which group of mollusks has the most well-developed nervous system?
    (a) chitons
    (b) squids, octopuses, and nautiluses
    (c) clams, mussels, oysters, and scallops
    (d) snails and whelks

32. Which of the following is part of the insect order?
    (a) crayfish   (b) trilobites
    (c) centipedes   (d) butterfly

33. Indicate the kingdom to which each of the following organisms belongs:
    (a) bread mould
    (b) fir tree
    (c) nematode

34. Summarize the ecological role played by various invertebrate phyla.

Applying Inquiry Skills

35. It is not always easy to identify whether or not something is living. Many times incorrect conclusions are drawn because they were based on limited information or evidence. A number of different conclusions were drawn by students in a grade 11 classroom.
    • Examine each of the following statements and state whether or not you would agree with it.
    • Indicate how you would find further information to check whether or not the statements are correct.

(a) Statement: Hair is alive.
    Evidence: On a shampoo commercial, I heard that hair needs vitamins. I think hair must be living for a short time even after you cut it.

(b) Statement: Yogurt must contain living things.
    Evidence: If you leave yogurt for a long time a smelly gas collects at the top. The yogurt also changes form if it is left too long. I examined yogurt under the microscope and found little rod-shaped structures inside of it. When the yogurt got smelly, I found more rod-shaped things.

(c) Statement: Fat droplets must be living.
    Evidence: I watched two fat droplets dance across the water after I did the dishes. When two fat droplets touch they form one larger droplet. It's like they're growing.

(d) Statement: There are livings things in sauerkraut.
    Evidence: You need living cells to make the sauerkraut sour. If you boil the cabbage until all the living things are killed and then seal the sauerkraut container, it will never get sour.

36. Construct a dichotomous key to classify the leaves in **Figure 3**.

basswood   striped maple   sugar maple   sassafras

elm   hickory   beech

birch   white pine   persimmon

white oak   white spruce   hemlock

Figure 3
Leaves

37. Use **Figure 4** to identify the bird that has special characteristics for each of the following:
    (a) wading
    (b) swimming
    (c) walking on snow
    (d) grasping prey
    (e) perching on a branch

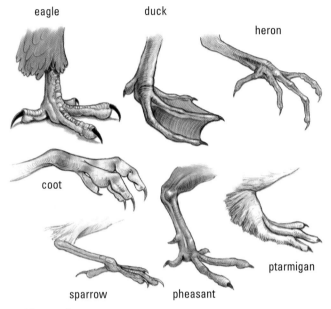

eagle   duck   heron   coot   sparrow   pheasant   ptarmigan

Figure 4
Birds' feet

38. Build a bird suited for a long migratory flight. What features will you want to include? Draw a picture of the bird and list the specialized characteristics that you have included.

39. Place the following words in the correct place on the concept map (**Figure 5**): cell wall, single cell, cyanobacteria, fern, autotrophic, no chlorophyll, yeast, movement, methanogen. (*Note:* There is more than one correct word for some boxes and some words may be used more than once.)

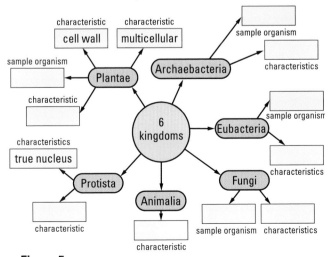

characteristic — cell wall
characteristic — multicellular
sample organism
sample organism — Plantae
characteristic
Archaebacteria
sample organism
characteristics
true nucleus
characteristics
6 kingdoms
Eubacteria
sample organism
characteristics
Protista
Animalia
Fungi
characteristic
characteristic
sample organism   characteristics

Figure 5

40. (a) Compare the circulatory systems of representative organisms from phylum Annelida and phylum Arthropoda. Which organism is more complex? Explain your reasoning.
    (b) Would your answer change if you compared the respiratory systems of the representative organisms? Explain.

Making Connections

41. Humans and insects are often said to be the dominant animals on Earth, and yet there is only one species of humans but over seven hundred thousand species of insects. How would you explain this phenomenon?

42. What is the link between widespread geographical distribution and biological success?

43. Besides parasitism, invertebrates exhibit various other symbiotic relationships. Use library resources to find information on these additional relationships.

44. In many areas of northeastern North America, Lyme disease is becoming a serious problem. Using science journals and magazines, trace the history of the disease and what scientists are doing to fight it.

Exploring

45. Indoor composting is an alternative for people who cannot compost outdoors. Red wigglers (a relative of the earthworm) are commonly used in the composting process. Research and report on how to set up an indoor composter using these red worms.
    Follow the links for Nelson Biology 11, Unit 4 Review.
    GO TO www.science.nelson.com

46. Around the world, many countries are fighting against diseases which infect their livestock. Research and report on the infectious agents responsible for mad cow disease and foot-and-mouth disease.
    Follow the links for Nelson Biology 11, Unit 4 Review.
    GO TO www.science.nelson.com

3

Internal Systems and Regulation

Dr. C.K. Govind is a professor of animal physiology at the University of Toronto.

"In first-year courses, students learn about the internal systems that keep an organism alive. In advanced courses, current research is examined. In my research, we are trying to answer the question, 'What determines where the major, or crusher, claw develops—on the right or left side of the body?' We found that greater use of one of the paired claws during a critical period in early development induces that claw to become a crusher, while the other becomes a minor, or cutter, claw. Thus the interaction of nature and nurture determines claw type in lobsters. Such research is also useful in helping us understand critical periods in human brain development for language, speech, and other abilities."

**Dr. C.K. Govind,
University of Toronto**

Overall Expectations

In this unit, you will be able to

- describe and explain the major processes, mechanisms, and systems by which plants and animals maintain their internal environment;
- describe and explain the roles of the respiratory, circulatory, and digestive systems in internal regulation;
- investigate internal regulation in plants and animals;
- evaluate the impact of personal lifestyle decisions on human health;
- explain how concern for human health has advanced the development of technologies related to the regulation of internal systems.

Are You Ready?

Knowledge and Understanding

1. Add the missing labels to the diagram of the heart (**Figure 1**). The arrows show the direction of blood flow. Use the following labels: aorta, arteries to body, veins from lung, right atrium, left ventricle.

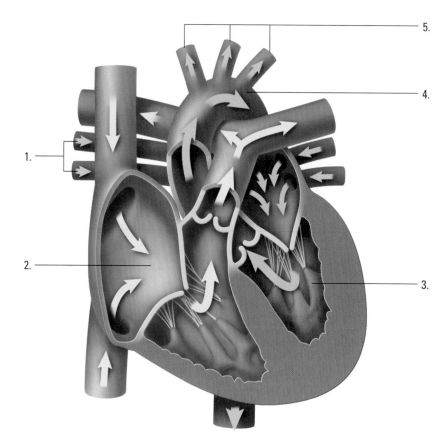

Figure 1
Structure of the human heart

2. In **Figure 2**, which gases do x and y represent? Explain the importance of each of these gases.

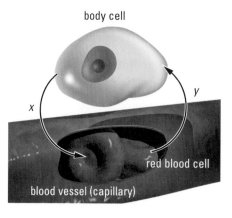

Figure 2
Movement of gases into and out of cells

3. In **Table 1**, match each structure in column A with its function in column B.

Table 1

Column A	Column B
1. gills	(a) carries blood away from the heart
2. lung	(b) stores fibres and other wastes
3. artery	(c) respiratory organ of crayfish and tadpoles
4. vein	(d) produces digestive enzymes for carbohydrates, fats, and proteins
5. heart	(e) a small muscular tube that connects the mouth with the stomach
6. stomach	(f) the major area of food digestion and absorption in the gut
7. esophagus	(g) the respiratory organ of mammals and birds
8. small intestine	(h) a blood vessel that carries blood to the heart
9. pancreas	(i) the initial site of protein digestion
10. large intestine	(j) the organ that pumps blood

4. Predict the effect of exercise on each of the following and give reasons for each of your predictions.
   - (a) heart rate
   - (b) blood pressure
   - (c) respiratory rate
   - (d) ability to digest foods

5. Blood samples were taken from four different people. The samples were placed in a centrifuge and spun to separate the different components of the blood. In **Figure 3,** subject 1 is considered to be in excellent health. Note the following:

- Red blood cells (RBC) carry oxygen.
- White blood cells (WBC) are part of the immune system and defend against foreign invaders.
- Plasma contains water, proteins, and sugar.

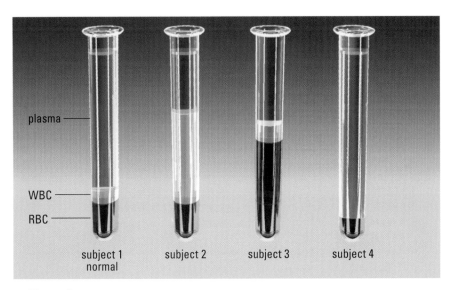

Figure 3

(a) Which subject might have suffered a blood loss? Explain your answer.
(b) Which subject might live at a high altitude? Explain your answer.
(c) Which subject might have a severe infection? Explain your answer.

Inquiry and Communication

The experiment shown in **Figure 4** is conducted in a high school laboratory. A synthetic form of saliva with amylase enzyme is used.

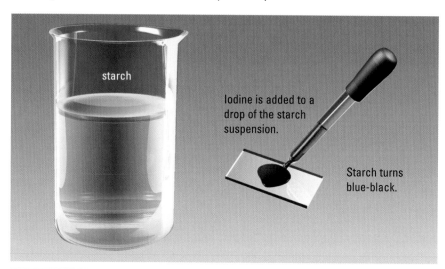

starch

Iodine is added to a drop of the starch suspension.

Starch turns blue-black.

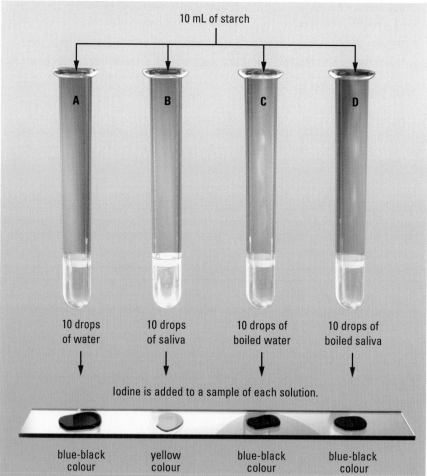

10 mL of starch

| A | B | C | D |

| 10 drops of water | 10 drops of saliva | 10 drops of boiled water | 10 drops of boiled saliva |

Iodine is added to a sample of each solution.

| blue-black colour | yellow colour | blue-black colour | blue-black colour |

Figure 4

6. Identify the question the experiment is designed to investigate.
   (a) What is the effect of iodine on starch?
   (b) What is the effect of amylase on starch digestion?
   (c) What is the effect of temperature on starch digestion?
   (d) What is the effect of water on starch digestion?

7. Which test tubes serve as controls for the experiment?
   (a) test tube A
   (b) test tube B
   (c) test tube C
   (d) test tube D

8. What is the purpose of the iodine in the experiment?
   (a) starch indicator
   (b) stain
   (c) starch enzyme
   (d) control

9. Identify the dependent variable in the experiment.
   (a) time
   (b) starch digestion
   (c) temperature
   (d) colour of iodine

10. Identify the independent variable(s) in the experiment.
    (a) time
    (b) temperature
    (c) solution type
    (d) colour of iodine

11. A living animal cell is placed in a 2% solution of glucose. Each of the dots shown in **Figure 5** represents a glucose molecule. Identify the processes that occur from 0 h to 1 h.
    (a) cellular respiration
    (b) diffusion
    (c) photosynthesis
    (d) endocytosis

time = 0 h   time = 0.5 h   time = 1 h

Figure 5

Technical Skills and Safety

12. The following are materials needed for a worm dissection. Where appropriate, indicate the safety precautions that you should follow when handling each item.

preserved earthworm	dissecting pan
hand lens (magnifying glass)	dissecting pins
dissecting scissors, scalpel, or single-edged razor blade	probes

6

Digestion and Nutrition

In this chapter, you will be able to

- describe the process of digestion in humans;
- compare how different organisms obtain nutrients from their environment, and identify their anatomical structures;
- identify nutrients and explain how they maintain health of the human body;
- identify lifestyle activities that could improve or impair health.

You have probably noticed that the foods of various cultures differ not only in flavour, but also in the types of ingredients used. For example, in South India, most people are vegetarian; in many parts of East Asia, little or no dairy products are used; beans are an important part of South American cuisine; while in Tibet, a large part of the diet consists of meat and dairy products.

A variety of diets result in different nutrition-related problems in each culture. The high rate of heart disease in North America is due in part to the large amounts of fat consumed. People in East Asian countries, in contrast, have very low rates of heart disease; stomach cancer is the bigger problem, possibly because of the preservatives used in the foods. Recently, more North Americans have decided that a vegetarian diet is healthier. Some worry that such a diet doesn't provide enough of certain types of nutrients, such as protein or iron. But supporters believe that it isn't necessary to consume as much of these nutrients as has traditionally been the practice in North America, and that vegetarianism benefits health.

How much of each type of nutrient do you really need? What are the nutritional properties of different diets that lead to the different diseases? What can you do to improve the nutritional value of your diet?

Try This Activity

Canada's Food Guide to Healthy Eating

- On the Internet, find the Web site address for Health Canada's Food Guide to Healthy Eating.
  Follow the links for Nelson Biology 11, 6.0.

  GO TO www.science.nelson.com

  (a) What recommendations does the food guide give? Why do you think it is recommended you eat large amounts of some foods and smaller amounts of others?
  (b) According to the guide, which foods should be eaten in larger quantities? Which foods should be eaten in smaller quantities?
  (c) Write down everything you might eat on a typical day. Score yourself using the Healthy Eating Scorecard.
  (d) Research the typical daily diet of a person from a country outside of North America. Compare your diet to it.

Reflect on your Learning

1. Make a list of essential nutrients that must be included in every diet.

2. Identify a diet-related disorder and the cause of the disorder.

Figure 1
Typical meals from **(a)** North America, **(b)** Japan, **(c)** China, **(d)** South America

6.1 Organs and Organ Systems

organs: structures composed of different tissues specialized to carry out specific functions

Organs are groups of different tissues specialized to carry out particular functions. The stomach is an excellent example of a complex organ (**Figure 1**). The stomach's outer structure is covered by epithelial tissue. Next, there are three layers of smooth muscle that run in different directions so that muscle contractions and relaxations can churn the food within. Inside the muscle layers is the submucosa, made up of loose connective tissue with many blood and lymph vessels. The innermost mucosa contains cells that secrete gastric juices and mucus. Nerve cells regulate and synchronize the contractions that release partially digested material into the small intestine.

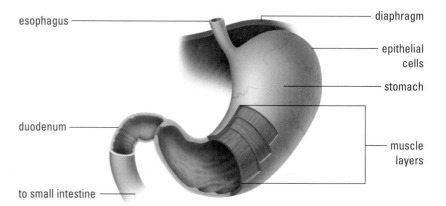

Figure 1
Stomach structure

Your hands and kidneys are further examples of complex organs. Although each organ is composed of a variety of different tissues, the tissues act together to accomplish a common goal.

organ system: a group of organs that have related functions. Organ systems often interact.

An **organ system** is a group of organs that have related functions. The digestive system is made up of the esophagus, stomach, small intestine, large intestine, and other associated organs, such as the liver and pancreas. **Table 1** outlines the levels of cell organization for the body's important organ systems.

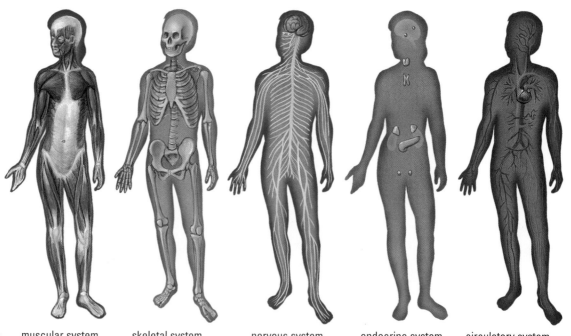

Figure 2
Human organ systems

muscular system   skeletal system   nervous system   endocrine system   circulatory system

Table 1: Levels of Cell Organization

Organ system	Organs	Tissues
nervous	brain, spinal cord, eye, ear, peripheral nerves	epithelial, nerve, connective
excretory	kidney, bladder, ureter, urethra	epithelial, nerve, connective, muscle
circulatory	heart, blood vessels	epithelial, nerve, connective, muscle
digestive	esophagus, stomach, intestines	epithelial, nerve, connective, muscle
reproductive	both sexes: glands male: testes, vas deferens female: ovary, uterus, fallopian tubes	epithelial, nerve, connective, muscle
respiratory	lungs, windpipe	epithelial, nerve, connective, muscle
endocrine	pancreas, adrenal glands, pituitary	epithelial, nerve, connective

Organ systems interact closely with each other. For example, the body's digestive system would not be able to function properly if the circulatory system did not allow for the transport of materials and the respiratory system did not provide adequate gas exchange.

Some organs are classified according to anatomy rather than function. Consider the kidneys and the large intestine, which are both involved in waste removal from the body. The kidneys, considered part of the excretory system, remove wastes from the blood, while the large intestine, part of the digestive system, concentrates and stores undigested matter while also absorbing water and vitamins. **Figure 2** shows an overview of human organ systems.

Monitoring Organs

The first X-ray photograph was taken in 1895 when the German physicist Wilhelm Roentgen used X rays to take a picture of his wife's hand. These high-energy electromagnetic waves pass through soft tissue, like muscle, but are absorbed by denser bone.

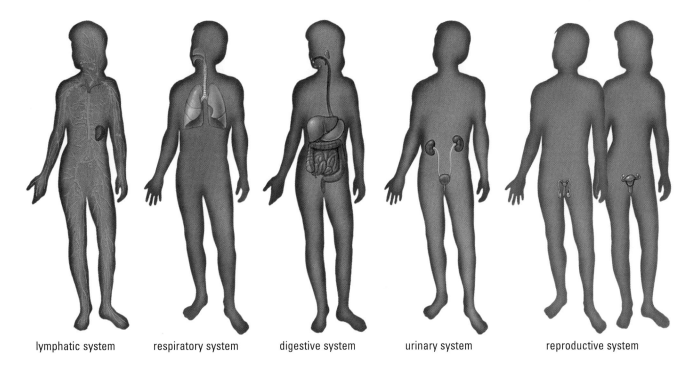

| lymphatic system | respiratory system | digestive system | urinary system | reproductive system |

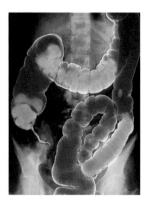

Figure 3
The penetrating properties of X rays, combined with stains, are used to monitor the state of internal organs. This X ray shows the large intestine.

computerized axial tomography (CAT) scan: a procedure in which an X-ray machine takes many pictures of an object from different angles; a computer then reassembles the images to allow viewing of the object in cross section and in 3-D

nuclear imaging: a medical imaging technique that uses radionuclides to view organs and tissues of the body

radionuclides: the nucleii of unstable atoms that emit rays of energy. In nuclear imaging techniques, the energy emitted by radionuclides injected into the body is scanned to produce a picture.

nuclear magnetic resonance (NMR) technology: a technique to determine the behaviour of the nucleus of an atom. In magnetic resonance imaging, NMR technology is used to produce a picture of the internal structures of the human body.

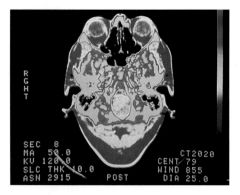

Figure 4
A CAT scan of a cross section through the top of the skull shows the eyeballs at the top.

In 1898, Walter Cannon was the first person to use X rays to view soft tissue. Cannon developed a stain that contained bismuth, a nontoxic mineral opaque to X rays. When a patient drank the bismuth mixture, the X rays were unable to penetrate the stained gastrointestinal organs, resulting in a white image of the organs on a black background (**Figure 3**). For the first time, organ structures could be observed without surgery. Today, barium is used instead of bismuth.

Today, scientists have combined X-ray technology with computer technology to view body organs in even more complex ways. In the **computerized axial tomography (CAT) scan** procedure, an X-ray machine rotates around the patient, taking hundreds of individual pictures from numerous angles. The images are stored in a computer along with their location and angle. The computer can reassemble the pictures to provide thin cross-sectional views, and it can organize the pictures to permit three-dimensional imaging. The organ can then be viewed section by section (**Figure 4**). The CAT scan is so accurate that it can detect abrasions as small as one millimetre. The scanner can also distinguish between gases, liquids, and solid tissues, and is able to identify tumours embedded in the brain or liver. CAT scans are particularly useful as a diagnostic tool for assessing head injuries involving blood clots, which can be life threatening.

Frontiers of Technology: Nuclear Medicine

Nuclear imaging is a valuable diagnostic tool that allows doctors to view a beating heart or detect bone cancer without resorting to surgery. Unlike the CAT scan, which uses external radiation to produce an image, nuclear imaging measures the radiation emitted from within the body and provides information about the function of the organ as opposed to its structure.

Nuclear imaging employs **radionuclides** (sometimes referred to as radioisotopes) to identify organs, much like X rays use opaque dyes. Radionuclides are the nucleii of unstable atoms that emit rays of energy. Radionuclides injected into the body collect in the target organs. A scanner, called a gamma camera, records the energy released from the radionuclides and produces a picture (**Figure 5**).

Different radionuclides are used to identify different organs. A thallium isotope is very valuable for heart imaging. Large amounts of the isotope collect in areas of damage, called infarctions, and produce a "hot spot." If the camera detects the hot spot, the physician knows that the damage has occurred within the past five days, the maximum amount of time in which the infarction will collect the radionuclides.

Another technique, **nuclear magnetic resonance (NMR) technology** complements the CAT scan. NMR works by subjecting the nucleus of a specific atom to a combination of magnetic forces and radio waves to determine whether or not the nucleus behaves normally. Because NMR does not use any external source of radiation, it is, theoretically, safer than the CAT scan. The use of NMR techniques for medical purposes is called magnetic resonance imaging (MRI). MRI uses the spinning motions of the atomic nuclei to produce a map of the internal structure of human body tissue and to reveal how the organs are functioning (**Figure 6**). MRI technology is an excellent tool for determining if tumours are cancerous. Despite equipment costs that can run up to $1.5 million, MRI technology is being introduced throughout the country. The University of Western Ontario's Robarts Research Institute is in the forefront of medical imaging and the development of new techniques in 3-D ultrasound and functional MRI.

| SUMMARY | Organ Systems and Imaging Technologies |

1. Organ systems often interact. They can be classified by anatomy or by function.

2. X rays are high-energy electromagnetic waves that pass through soft tissue like muscle, but are absorbed by denser bone.
   • Wilhelm Roentgen was the first to use X rays, in 1895.
   • In 1898, Walter Cannon used bismuth with X rays to make the gastrointestinal tract visible.

3. CAT scans provide information about structure, including soft tissue.
   • Many X rays are taken from many angles and reassembled to make three-dimensional images.
   • CAT scans can distinguish between gases, liquids, and solid tissues and are useful for detecting blood clots.

4. Nuclear imaging techniques provide information about function.
   • Radionuclides injected into the body collect in target organs. The radiation emitted is scanned to produce a picture.

5. Nuclear magnetic resonance (NMR) technology provides information about structure and function.
   • A specific atom is subjected to a combination of magnetic forces and radio waves to determine whether or not the nucleus behaves normally.
   • Magnetic Resonance Imaging (MRI) is the medical application of NMR techniques.

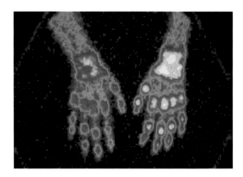

Figure 5
Hands of a person with extensive rheumatoid arthritis. The arthritic joints appear as brighter areas. The image records the distribution and intensity of gamma rays emitted by a tiny amount of radionuclide injected into the patient.

Practice

Understanding Concepts

1. Differentiate between a tissue, an organ, and an organ system. Provide examples of each in your explanation.

2. Why do different classification schemes for organ systems exist?

3. Explain how your hand is a complex organ. Hint: What types of tissues does it contain?

Making Connections

4. Discuss the advantages and disadvantages associated with X rays, CAT scans, nuclear imaging, and MRI techniques.

5. The cost of nuclear medicine, CAT scans, and artificial body parts is extremely high and places a heavy financial burden on the health-care system. In your opinion, can we continue to support such expensive research projects? State your reasons.

6. Research career opportunities in the field of medical imaging at your local hospital and report on an aspect that you find interesting.

7. Look in books or follow the links for Nelson Biology 11, 6.1, to locate photographs of X rays, nuclear imaging, and MRIs. Which structures can you identify? Which imaging techniques are most appropriate for different types of structures (e.g., soft tissue, bone)?

GO TO www.science.nelson.com

Reflecting

8. You have learned about a number of technologies used in medicine, each of which has advantages and disadvantages. At what point do you think that the advantages overcome the disadvantages? What factors do you think influence how acceptable these technologies are to you?

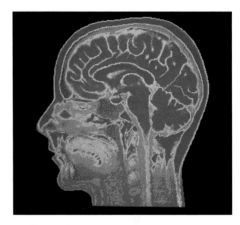

Figure 6
Magnetic resonance image of a section of a normal 42-year-old female's head, showing structures of the brain, spine, and facial tissues

6.2 Importance of Digestion

Unlike plants, which make their own food, heterotrophs must consume organic compounds to survive. These organic compounds, called nutrients, are digested in the gastrointestinal tract, absorbed, and transported by the circulatory system to the cells of the body. Once inside the cells, the nutrients supply the body with energy, or the raw materials for the synthesis of essential chemical compounds used for growth, maintenance, and tissue repair.

The digestive system is responsible for the breakdown of large, complex organic materials into smaller components that are utilized by the tissues of the body. Every organ system depends on the digestive system for nutrients, but the digestive system also depends on other organ systems. Muscles and bones permit the ingestion of foods. The circulatory system transports oxygen and other needed materials to the digestive organs and carries the absorbed foods to the tissues of the body. The nervous and endocrine systems coordinate and regulate the actions of the digestive organs. In many respects, the study of the digestive organs is a study of the interacting body systems.

In this chapter, you will study four components of digestion:

1. ingestion—the taking in of nutrients
2. digestion—the breakdown of complex organic molecules into smaller components by enzymes
3. absorption—the transport of digested nutrients to the tissues of the body
4. egestion—the removal of waste food materials from the body

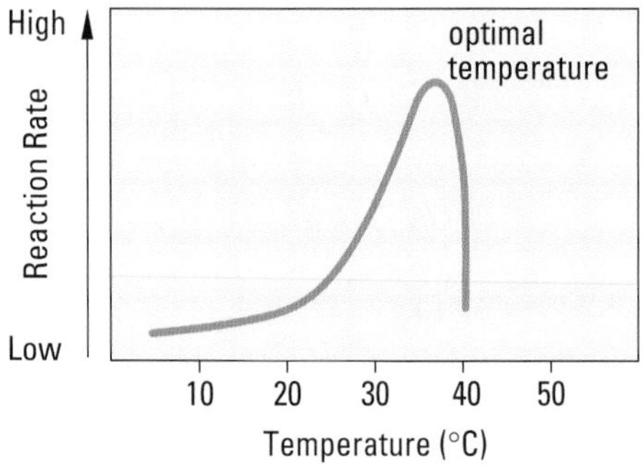

Digestive Enzymes

The rate at which an enzyme functions best to break down complex molecules is affected by two factors, temperature and pH.

In general, as the temperature increases, more energy is added, and the greater the enzyme activity. However, for most human enzymes, their efficiency peaks at about 37°C (**Figure 1**) and then drops as more energy is added. Recall, that enzymes are proteins that when subjected to high temperatures change shape. Once the protein is denatured, it is no longer active.

Enzymes function best within certain pH ranges. The enzyme pepsin, shown in blue in **Figure 2**, works best at a low pH. It is found in the acidic environment of the stomach. The enzyme amylase, shown in grey, works best in neutral pH environments, such as the mouth. The enzyme trypsin, shown in green, works best in basic environments, such as the small intestine.

Digestion in Simpler Organisms

The simplest form of digestion occurs in organisms such as protozoa, which digest their meals inside food vacuoles within cells. An amoeba, a single-celled organism, engulfs its food by phagocytosis. It extends pseudopods to engulf the food, and a vacuole is formed inside the cell (**Figure 3**). These vacuoles fuse with lysosomes in the cell. (Recall from Chapter 1 that lysosomes are vesicles formed by the Golgi apparatus.) Lysosomes contain **hydrolytic enzymes**, which use molecules of water to break down food.

Many simple organisms, such as hydra, have a digestive sac with a single opening. This pouch, or **gastrovascular cavity**, encloses part of the external environment and allows food storage and digestion to take place. Digestive

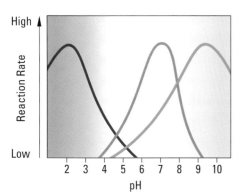

Figure 1
Effect of temperature on enzyme reaction

Figure 2
The enzymes indicated by the blue, grey, and green lines are most effective at a pH of 2.5, 7.0, and 9.0, respectively.

hydrolytic enzymes: enzymes that use water to break down molecules

gastrovascular cavity: a digestive compartment, usually with a single opening that functions as both mouth and anus

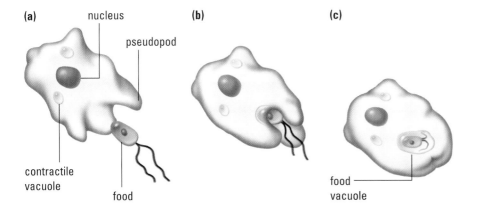

Figure 3
Intracellular digestion by amoeba. The amoeba moves and feeds at the same time.
(a) The amoeba wraps itself around food.
(b) It slowly engulfs food particles by extending its pseudopods.
(c) The membrane around the food forms a food vacuole. Digestion takes place inside the vacuole.

enzymes are released into the cavity, breaking down the larger food molecules. Then the smaller food molecules are engulfed by cells that line the digestive cavity, and digestion continues within the cytoplasm of the cells. Wastes that remain in the digestive cavity are then expelled from the same opening which ingested the food (**Figure 4**).

Digestion in More Complex Organisms

More complex animals digest food along digestive tracts, called alimentary canals. These long tubes have a separate opening for a mouth and an anus. Because food moves along the canal in one direction, the canal can be organized into specialized regions that enable the breakdown and absorption of food in a stepwise process. As illustrated by the earthworm and bird, food ingested through the mouth travels through the muscular **pharynx** and into the **esophagus.** Then, depending upon the species, it is held in the stomach or **crop**. Animals that don't grind food with teeth have a muscular **gizzard** that physically breaks down food particles. Pulverized food from the gizzard moves into the intestine, where hydrolytic enzymes complete the stage of chemical digestion. Then nutrients are absorbed across the lining of the intestine into the blood. The circulatory system carries the digested nutrients to the cells of the body. Undigested wastes are removed through the anus (**Figure 5** and **Figure 6**, page 210).

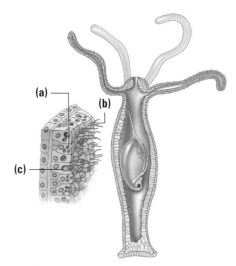

Figure 4
The hydra, a small pond organism, has two cell layers and little cell specialization. Prey is stuffed into the gastrovascular cavity.
(a) Cells release digestive enzymes that begin to break down food.
(b) Other cells with flagella keep the contents of the cavity moving.
(c) Still other cells engulf the food particles, much as an amoeba would.

pharynx: a muscular section of the digestive tract. Air and/or food passes through this muscular tube.

esophagus: a tube that carries food from the mouth to the stomach

crop: a receptacle for storing undigested food

gizzard: a muscular chamber designed to physically break down food

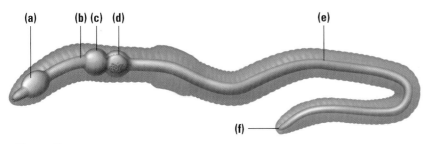

Figure 5
The earthworm has specialized cells in specialized areas of its digestive system.
(a) A muscular pharynx pulls food and soil into the mouth.
(b) Muscles in the wall of the esophagus push food to the crop.
(c) The crop stores and moistens food.
(d) The gizzard is a muscular chamber. Small particles of sand and gravel in the gizzard aid in the breakdown of food.
(e) In the intestine, food is broken down chemically and absorbed.
(f) Waste is eliminated through the anus.

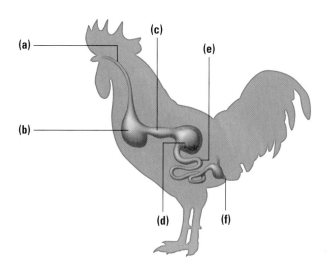

Figure 6

The digestive system of a bird

(a) The esophagus carries food to the crop.

(b) The crop stores food.

(c) The gizzard has thick muscular walls. Birds swallow small pebbles to help grind the food.

(d) The stomach begins chemical digestion.

(e) The intestine absorbs digested food.

(f) Waste is removed through the anus.

SUMMARY **Importance of Digestion**

1. Digestive systems enable the absorption of nutrients needed for growth, maintenance, and repair. The digestive system and other organ systems are interdependent. The four components of digestion are ingestion, digestion, absorption, and egestion.

2. In protozoa, digestion occurs in intracellular vacuoles.

3. Some simple organisms have a single gastrovascular cavity with one opening, which both stores food and digests it.

4. Digestive tracts allow the separation of processes into different regions.

Practice

Understanding Concepts

1. Define ingestion, digestion, absorption, and egestion.

2. Compare the digestive systems of the amoeba, bird, and earthworm.

3. How does the presence of an alimentary canal alter the organization of the digestive organs within the body?

6.3 Ingestion

The digestive tract of adult humans, normally 6.5 m to 9 m long, stores and breaks down organic molecules into simpler components. Physical digestion begins in the mouth, where food is chewed and formed into a *bolus* (the Greek word for ball) by the tongue. **Figure 1** shows the entire digestive system.

Saliva

amylase: an enzyme that breaks down complex carbohydrates

Saliva, the watery fluids produced by the salivary glands, contains **amylase** enzymes, which break down starches (complex carbohydrates) to simpler carbohydrates. Saliva lubricates the food so it can be swallowed, dissolves food particles, and makes it possible to taste what is being eaten.

The way that we discern flavour is that food particles dissolved in solution penetrate the cells of the taste buds located on the tongue and cheeks. Different types of receptors respond to specific flavours. For example, the taste buds are

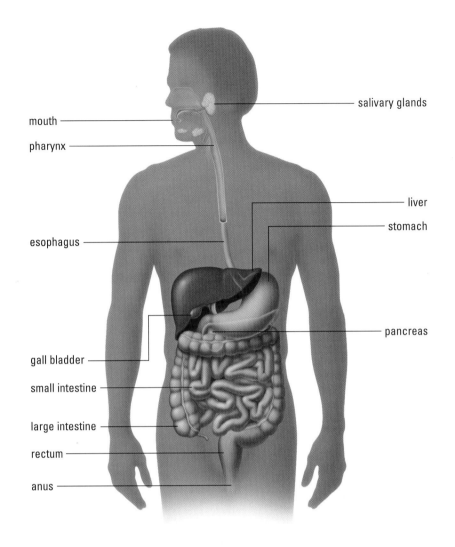

mouth

pharynx

esophagus

gall bladder

small intestine

large intestine

rectum

anus

salivary glands

liver

stomach

pancreas

Figure 1
The human digestive system and accessory organs

equipped with receptors (**Figure 2**) that have a specific geometry that permit the identification of sweet tastes from carbohydrates. Nerve cells for taste are stimulated when receptor sites are filled by chemical compounds with a complementary shape. You can find out the significance of dissolving foods by drying your tongue and then placing a few grains of sugar or salt on it. You will not detect any flavour until the crystals dissolve.

Teeth

The teeth are important structures for physical digestion. Eight chisel-shaped teeth at the front of your mouth, called incisors, are specialized for cutting. The incisors are bordered by sharp, dagger-shaped canine teeth that are specialized for tearing. Next to the canine teeth are the premolars. These broad, flattened teeth are specialized for grinding. The molars are found next to the premolars. These teeth tend to be even broader and have cusps that are even more flattened. They are designed for crushing food. The last set of molars are the wisdom teeth, so called because they usually do not emerge until about 16 to 20 years of age. Often these molars are troublesome and must be removed because there is not enough space for them to grow in.

Each tooth has two divisions: the root and the crown. An enamel crown covers the tooth with calcium compounds and forms the hardest substance in the body. Immediately inside the enamel is dentin, a bonelike substance, which is

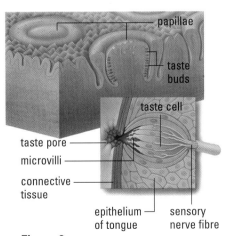

papillae

taste buds

taste cell

taste pore

microvilli

connective tissue

epithelium of tongue

sensory nerve fibre

Figure 2
Taste buds are located along the tongue.

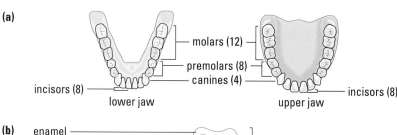

(a)

molars (12)
premolars (8)
canines (4)
incisors (8)
lower jaw
incisors (8)
upper jaw

(b)

enamel
dentin
pulp cavity (contains nerves and blood vessels)
root canal
periodontal membrane
bone
crown
gingiva (gum)
root

Figure 3
(a) Human teeth and
(b) a human molar. Each tooth has a root and an enamel crown. Enamel, which is formed of calcium compounds, is the hardest substance in the body.

part of the root structure. The dentin encases the pulp cavity, which contains nerves and blood vessels (**Figure 3**).

Tooth decay is caused by bacteria living off nutrients that cling to the teeth. These harmful microbes produce corrosive acids that erode a tooth's structure. Infections can also spread to the periodontal membrane, which anchors the teeth to the jawbone. As an infection progresses, the periodontal tissue is slowly destroyed and the teeth loosen.

Esophagus

peristalsis: rhythmic, wavelike contraction of smooth muscle that moves food along the gastrointestinal tract

Once swallowed, food travels from the mouth to the stomach by way of the esophagus. The bolus of food stretches the walls of the esophagus, activating smooth muscles that set up waves of rhythmic contractions called **peristalsis**. Peristaltic contractions, which are involuntary, move food along the gastrointestinal tract (**Figure 4**). The only points at which food is moved voluntarily along the tract is during swallowing and during the last phase, egestion. Peristaltic action will move food or fluids from the esophagus down to the stomach even if you stand on your head.

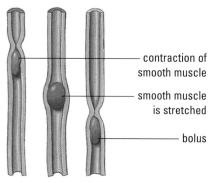

contraction of smooth muscle
smooth muscle is stretched
bolus

phase 1   phase 2   phase 3

Figure 4
Rhythmic contractions of the smooth muscle move food along the digestive tract.

| SUMMARY | **Ingestion** |

1. Saliva is important in digestion because it
   - contains amylase enzymes that initiate carbohydrate breakdown;
   - lubricates the food passage;
   - dissolves food particles;
   - activates the taste buds.
2. Teeth are necessary for biting, tearing, grinding, and crushing food into smaller particles.
3. After food is swallowed, movement through the esophagus is regulated by peristalsis, contractions of smooth muscle.

Practice

Understanding Concepts
1. What are the functions of saliva?
2. How does chewing assist in the digestion of food?
3. What are amylase enzymes and why are they necessary?

Activity 6.3.1

The Effect of the Enzyme Amylase on Starch Digestion

Starch is a polysaccharide made of a large number of glucose molecules that are bonded together. Recall from Chapter 2 that glucose molecules are bonded together by dehydration synthesis to form starch. In this activity, you will use the enzyme amylase to break the bonds between the glucose molecules in starch and release simpler sugars.

Materials

apron	goggles
4 test tubes	4 rubber stoppers for test tubes
1% cornstarch suspension	5% amylase solution
Lugol's solution	Benedict's reagent
hot plate	test-tube rack
250-mL beaker	10-mL graduated cylinder or pipette
tap water	labelling materials
test-tube holder	timer or watch
utility stand	ring pipette with suction device clamp

 Benedict's reagent and Lugol's solution are toxic and can cause an itchy rash. Avoid skin and eye contact. Wash all splashes off your skin and clothing thoroughly. If you get any chemical in your eyes, rinse for at least 15 min and inform your teacher.

Procedure

1. Put on your apron and goggles.
2. Label the test tubes from 1 to 4. Set up a water bath (**Figure 5**).
3. Pour 3 mL of the cornstarch suspension into each of test tubes 1 and 2.
4. Add 2 drops of Lugol's solution to test tube 1. Record your observations.
5. Add 5 mL of Benedict's reagent to test tube 2 and place it in a hot water bath at 100°C. Record your observations after 5 min. Do not let the test tube sit in the hot water bath for more than 5 min.
6. Pour 10 mL of cornstarch suspension into test tube 3. Add 5 drops of the amylase solution to the test tube. Place a stopper in the test tube and shake while pushing on the stopper with your index finger.
7. Let test tube 3 sit in the rack for 20 min. Record any observations during the 20 min.
8. After the 20-minute wait, pour half of the contents of test tube 3 into test tube 4.
9. Add 2 drops of Lugol's solution to test tube 3. Record your observations.
10. Add 5 mL of Benedict's reagent to test tube 4 and place it in a hot water bath at 100°C. Repeat the rest of step 5.

Figure 5

Analysis

(a) What was indicated by the result of the Lugol's solution test in step 4?
(b) What was indicated by the result of the Benedict's reagent test in step 5?
(c) What was indicated by the result of the Lugol's solution test in step 9?
(d) What was indicated by the result of the Benedict's reagent test in step 10?
(e) Explain any differences that are found in the results of steps 9 and 10.

sphincters: constrictor muscles that surround a tubelike structure

mucus: a protein produced by a layer of epithelial cells known as a mucous membrane

pepsin: a protein-digesting enzyme produced by the stomach

ulcer: a lesion along the surface of an organ

capillary: a blood vessel that connects arteries and veins. Capillaries are the sites of fluid and gas exchange.

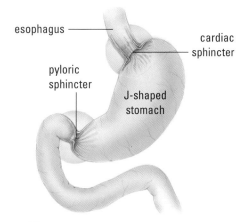

Figure 1
The stomach is the initial area of protein digestion. The cardiac sphincter controls the movement of food from the esophagus to the stomach. The pyloric sphincter controls the movement of food from the stomach to the small intestine.

Sections 6.1–6.3 Questions

Understanding Concepts

1. How are the digestive system and other organ systems interdependent?

2. What advantage is gained from digesting food along a digestive tract rather than in a single gastrovascular cavity?

3. How do toothless animals such as birds break down food particles? Suggest advantages of being able to swallow food without needing to chew.

4. Differentiate between physical and chemical digestion. Provide examples of each.

5. Is the movement of food through your digestive system voluntary or involuntary? What mechanisms are responsible for moving food along the alimentary tract?

6. The type of teeth that a mammal has is matched to diet. Keeping in mind the function of different types of teeth, name an animal that would have well-developed (a) canines, and (b) molars and premolars.

6.4 The Stomach and Digestion

The stomach is the site of food storage and initial protein digestion. The movement of food to and from the stomach is regulated by circular muscles called **sphincters**. Sphincters act like the drawstrings on a bag. Contraction of the cardiac sphincter closes the opening to the stomach located nearer the heart, while its relaxation allows food to enter. A second sphincter, the pyloric sphincter, regulates the movement of foods and stomach acids to the small intestine (**Figure 1**).

The J-shaped stomach has numerous ridges that allow it to expand so that it can store about 1.5 L of food. Millions of cells line the inner wall of the stomach. These cells secrete the various stomach fluids, called gastric fluids or gastric juice, that aid digestion. Approximately 500 mL of these fluids are produced following a large meal. Gastric fluid includes mucus, hydrochloric acid (HCl), pepsinogens, and other substances. **Mucus** provides a protective coating. Hydrochloric acid kills many harmful substances that are ingested with food. It also converts pepsinogen into its active form, **pepsin**, which is a protein-digesting enzyme. Pepsin breaks the long amino acid chains in proteins into shorter chains, called polypeptides.

The pH inside the stomach normally ranges between 2.0 and 3.0, but may approach pH 1.0. Acids with a pH of 2.0 can dissolve fibres in a rug! It is the high acidity (low pH) of hydrochloric acid that makes it effective at killing pathogens and allows pepsin to do its work. How does the stomach safely store these strong chemicals, both of which dissolve the proteins that make up cells? A layer of alkaline mucus protects the stomach lining from being digested. Pepsinogen moves through the cell membrane and mucous lining, is activated by HCl, and becomes pepsin. The pepsin breaks down the proteins in the food, but not the proteins of the stomach's cells because these proteins are protected by the mucous layer.

Ulcers

When the protective mucous lining of the stomach breaks down, the cell membrane is exposed to the HCl and pepsin. The destruction of the cell membrane leads to a peptic **ulcer**. Beneath the thin layer of cells is a rich **capillary** network.

As the acids irritate the cells of the stomach lining, there is an increase in blood flow and acid secretions. With this increased blood flow and acid secretion, more tissue is burned, the allergic reaction becomes even stronger, and the cycle continues. Eventually the blood vessels begin to break down.

Most ulcers have been linked to a bacterium called *Heliobacter pylori (H. pylori)* shown in **Figure 2**. Although diet, stress, and other factors may still play a part in the development of ulcers, this harmful microbe has changed how stomach ulcers are commonly treated. Dr. Barry Marshall, an Australian physician, is credited with changing the minds of many in the medical community. Prior to Dr. Marshall's research, scientists believed that microbes were unable to withstand the highly acidic conditions of the stomach. However, *H. pylori* can survive in this harsh environment and often only powerful antibiotics can kill it.

Dr. Marshall's research, now conducted in the United States, is attempting to prove a possible link between the microbe and some forms of stomach cancer. A simple breath test for the presence of *H. pylori* is now widely available.

Frontiers of Technology: Ulcers and Lasers

By 1960, American physicist Theodore Maiman had built the first laser. Laser beams have many medical applications. They can be used to remove damaged tissues such as those created by stomach ulcers. The laser beam is thinner than most scalpels and provides the added advantage of sealing small blood vessels. In addition, the laser may reduce the need for surgery.

A device called an **endoscope** can be fitted with a light-emitting glass fibre and then positioned inside a patient's body (**Figure 3**). The endoscope can be used to view such things as stomach ulcers. Tiny forceps, fitted in the endoscope, can even extract small pieces of tissue for a biopsy.

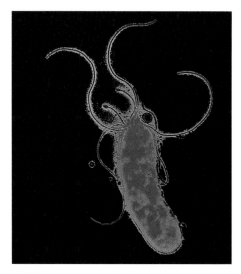

Figure 2
Heliobacter pylori

endoscope: an instrument to view the interior of the body

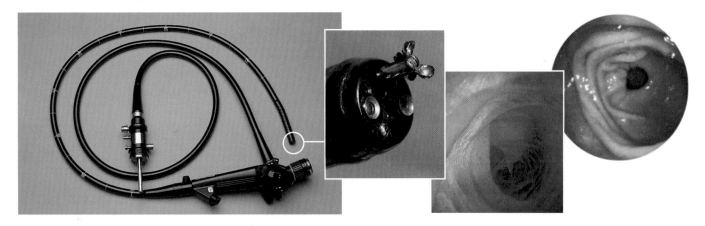

Figure 3
The endoscope can be used to provide a view of the stomach.

SUMMARY **Digestion in the Stomach**

1. Sphincter muscles regulate the movement of food into and out of the stomach.
2. Digestive fluids in the stomach include hydrochloric acid (HCl), pepsinogens, and mucus.
   - HCl kills pathogens and helps convert pepsinogen into pepsin.
   - Pepsin digests proteins.
   - Mucus protects the stomach from the above two fluids.
3. Ulcers are caused by the breakdown of the mucous lining in the stomach, exposing the stomach to stomach acids. Ulcers are linked to *Heliobacter pylori*.

DID YOU KNOW ?

The stomach capacity of a newborn human baby can be as little as 60 mL. An adult stomach has a maximum capacity of about 1.5 L, while the stomach of a cow is divided into 4 compartments and may hold up to 300 L.

Understanding Concepts

1. How is movement of food into and out of the stomach regulated?

2. What substances make up gastric fluid?

3. What is the function of the mucous layer that lines the stomach?

4. What is an endoscope and why is it useful?

5. List and discuss two factors that affect enzyme activity. Provide two examples.

Making Connections

6. Find out about the different kinds of ulcers. Learn about the risk factors, symptoms, and treatments.
Follow the links for Nelson Biology 11, 6.4.

GO TO www.science.nelson.com

6.5 The Small Intestine and Pancreas

Most digestion takes place in the small intestine, so named because of its narrow diameter. In humans, the small intestine is up to 7 m in length, but only 2.5 cm in diameter (**Figure 1**). The large intestine, by comparison, is only 1.5 m in length, but 7.6 cm in diameter. In mammals, the length of the small intestine is related to diet. Meats are relatively easy to digest, while plant materials are more difficult to digest. Accordingly, carnivores such as wolves and lions have short small intestines while herbivores, such as rabbits, have long small intestines. Omnivores, such as raccoons, pigs, bears, and humans have small intestines that are of intermediate length, allowing them to digest both types of food.

The majority of digestion occurs in the first 25 to 30 cm of the small intestine, an area known as the **duodenum**. The second and third components of the small intestine are called the jejunum and ileum. The three segments are differentiated by cell shape.

As you already know, food moves from the stomach to the small intestine. Partially digested foods reach the small intestine already soaked in HCl and pepsin. How are the cells of the small intestine protected? To answer this question, you must look beyond the small intestine to the pancreas.

duodenum: the first segment of the small intestine

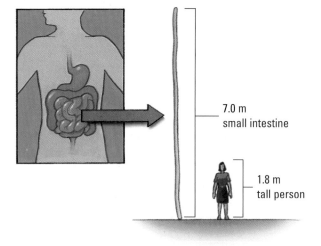

Figure 1
A comparison of the length of the small intestine with the height of a tall person

7.0 m
small intestine

1.8 m
tall person

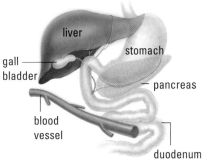
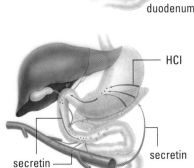

liver
gall bladder
stomach
pancreas
blood vessel
duodenum

HCl
secretin
secretin

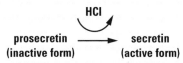

- HCl enters the duodenum from the stomach.
- HCl stimulates the conversion of pro- secretin into secretin

HCl

prosecretin $\longrightarrow$ **secretin**
(inactive form) **(active form)**

- Secretin is aborbed into the blood vessels.

- Secretin is carried by the circulatory system to the pancreas.
- Here secretin acts as a chemical messenger, stimulating the release of pancreatic fluids.
- The HCO_3^- ions, released by the pan- creas, neutralize the HCl from the stomach. The neutralization of acids protects the lining of the duodenum.

Figure 2
The function of secretin

secretin: a hormone that stimulates pancreatic and bile secretions

When acids enter the small intestine, a chemical called prosecretin is converted into **secretin** (**Figure 2**). Secretin is absorbed into the bloodstream and carried to the pancreas, where it signals the release of a solution containing bicarbonate ions. Bicarbonate ions (HCO_3^-) are carried to the small intestine, where they neutralize the HCl in gastric fluid and raise the pH from about 2.5 to 9.0. The now basic pH inactivates pepsin. Thus, the small intestine is protected from stomach acids by the release of secretin.

The pancreatic secretions also contain enzymes that promote the breakdown of the three major components of foods: proteins, carbohydrates, and lipids. A protein-digesting enzyme, called trypsinogen, is released from the pancreas. Once trypsinogen reaches the small intestine, an enzyme called **enterokinase** converts the inactive trypsinogen into **trypsin**, which acts on the partially digested proteins. Trypsin breaks down long-chain polypeptides into shorter-chain peptides. A second group of enzymes, the **erepsins**, are released from the pancreas and small intestine. They complete protein digestion by breaking the bonds between short-chain peptides, releasing individual amino acids (**Figure 3**).

enterokinase: an enzyme of the small intestine that converts trypsinogen to trypsin

trypsin: a protein-digesting enzyme

erepsins: enzymes that complete protein digestion by converting small-chain peptides to amino acids

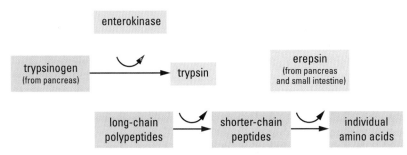

enterokinase

trypsinogen (from pancreas) $\longrightarrow$ trypsin

erepsin (from pancreas and small intestine)

long-chain polypeptides $\longrightarrow$ shorter-chain peptides $\longrightarrow$ individual amino acids

Figure 3
Breakdown of proteins in the small intestine

The pancreas also releases amylase enzymes, which continue the digestion of carbohydrates begun in the mouth by salivary amylase. The intermediate-size chains are broken down into disaccharides. The small intestine releases disaccharide enzymes, called disaccharidases, which complete the digestion of carbohydrates. (Note that the suffix "ase" is used to identify enzymes. For example, amylase is the enzyme that breaks down amylose, and disaccharidases break down disaccharides into monosaccharides.)

Lipases are enzymes released from the pancreas that break down lipids (fats). There are two different types of lipid-digesting enzymes. Pancreatic lipase, the most common, breaks down fats into fatty acids and glycerol. Phospholipase acts on phospholipids.

For a summary of the enzymes in the small intestine, where they are produced, and the reactions that take place, see **Table 1**.

Table 1: Digestion in the Small Intestine

Enzyme name	Produced by	Reaction
lipase	pancreas	fat droplets + H_2O —> glycerol + fatty acids
trypsin	pancreas	protein + H_2O —> peptides
erepsin	pancreas, small intestine	peptides + H_2O —> amino acids
pancreatic amylases	pancreas	starch + H_2O —> maltose
maltase	small intestine	maltose + H_2O —> glucose

SUMMARY **The Small Intestine and Pancreas**

1. Most digestion occurs in the duodenum.

2. When acids enter the small intestine, prosecretin is converted to secretin. This causes the pancreas to signal the release of bicarbonate ions, which help neutralize HCl and inactivate pepsin.

3. Pancreas secretions (such as trypsinogen and erepsins) play a large role in protein digestion.
   - The pancreas also secretes amylase enzymes, which continue the digestion of carbohydrates that was started in the mouth.
   - The pancreas also releases lipases, which are lipid-digesting enzymes.

4. The small intestine secretes disaccharidases, which complete the digestion of carbohydrates.

Practice

Understanding Concepts

1. How are the cells of the small intestine protected from stomach acids? Explain the mechanism and the chemicals involved.

2. What enzymes secreted by the pancreas promote digestion?

3. Explain the chemicals and processes involved in protein digestion and carbohydrate digestion. Why are carbohydrates not digested in the stomach?

4. List the lipid-digesting enzymes secreted from the pancreas. Do these enzymes allow for complete breakdown of lipids?

5. How is the duodenum protected against stomach acids? Why does pepsin not remain active in the duodenum?

6. In cases of extreme obesity, a section of the small intestine may be removed. What effect do you think this procedure has on the patient?

○ Questioning   ● Recording
● Hypothesizing   ● Analyzing
● Predicting   ○ Evaluating
○ Planning   ○ Communicating
● Conducting

Investigation 6.5.1

The Effect of pH and Temperature on Starch Digestion

Very little starch is broken down in the mouth. The low pH of the digestive fluids in the stomach halts digestion of carbohydrates such as starch until the carbohydrates leave the stomach and enter the small intestine.

Questions

What is the best pH for starch digestion? What is the best temperature for starch digestion?

Hypothesis/Prediction

Based on what you learned previously, form a hypothesis about the ideal pH level for starch digestion. Also predict whether a very cold temperature or very warm conditions would promote the most complete breakdown of starch.

Design

A cornstarch suspension will be mixed with an enzyme solution at different pH levels and at different temperatures to see which acidity level and which temperature result in the most complete breakdown of starch. The efficiency can be measured by how much sugar is produced. Benedict's reagent is used to indicate the presence of maltose, a disaccharide. Refer to Activity 2.4.1 for a guide on the nutrients tests.

Materials

apron	goggles
10 test tubes	test-tube rack
1% cornstarch suspension	5% amylase solutions at pH 2.0, 7.0, and 12.0
Benedict's reagent	hot plate
ice cubes	thermometers
two 250-mL beakers	utility stand
ring clamp	tap water
25-mL graduated cylinder	labelling materials
eyedropper	timer or watch
rubber stoppers for test tubes	glass stirring rod

Procedure

Part 1: The Effect of pH on Starch Digestion

1. Copy **Table 2** in your notebook and complete it as you perform each step in the activity.

Table 2

Test tube	Appearance after 20 min without heating	Appearance after 5 min in hot water bath at 100°C
1. cornstarch suspension only	?	?
2. cornstarch suspension and amylase at pH 2.0	?	?
3. cornstarch suspension and amylase at pH 7.0	?	?
4. cornstarch suspension and amylase at pH 12.0	?	?

2. Put on your apron and goggles.

3. Label 4 test tubes from 1 to 4. Set up a water bath as shown in **Figure 4**.

4. Place 15 mL of the 1% cornstarch suspension into each of the 4 test tubes.

5. Add 5 drops of the pH 2.0 amylase solution to test tube 2. Add 5 drops of the pH 7.0 amylase solution to test tube 3. Add 5 drops of the pH 12.0 amylase solution to test tube 4. Put a rubber stopper in each test tube and shake (**Figure 5**).

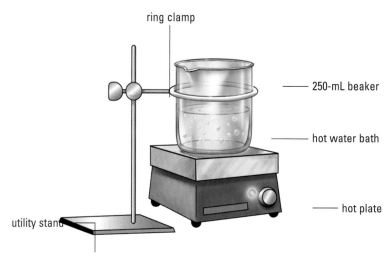

Figure 4

6. Let the test tubes sit for 20 min. Record your observations about each test tube. A colour change from blue to yellow to orange indicates maltose.

 Benedict's reagent is toxic and can cause an itchy rash. Avoid skin and eye contact. Wash all splashes off your skin and clothing thoroughly. If you get any chemical in your eyes, rinse for at least 15 min and inform your teacher.

7. Add 5 mL of Benedict's reagent to each of the 4 test tubes and place them in the hot water bath at 100°C. If you use the same cylinder as in Step 4, make sure to rinse and dry it first. Record your observations after 5 min. Do not let the test tubes sit in the hot water bath for more than 5 min (**Figure 4**).

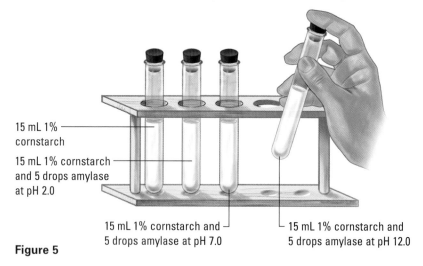

15 mL 1% cornstarch

15 mL 1% cornstarch and 5 drops amylase at pH 2.0

15 mL 1% cornstarch and 5 drops amylase at pH 7.0

15 mL 1% cornstarch and 5 drops amylase at pH 12.0

Figure 5

Analysis

(a) In which test tube did starch digestion occur? How could you tell?

(b) What is the function of test tube 1?

(c) At what pH does amylase work best to control starch digestion?

Part 2: The Effect of Temperature on Starch Digestion

8. Copy **Table 3** in your notebook and add 5 more rows, numbered up to 6. Complete the table as you perform each step in the activity.

Table 3

Test tube	Temperature (°C)	Appearance after 20 min without heating	Appearance after 5 min at 100°C
1			

9. Label 6 test tubes from 1 to 6.

10. Place 15 mL of cornstarch suspension in each test tube.

11. Add 5 drops of amylase solution at pH 7.0 to test tubes 1, 3, and 5.

12. Place test tubes 1 and 2 in the hot water bath and heat until the cornstarch suspension reaches 50°C. Do not heat the contents of the test tubes above 50°C (**Figure 6**).

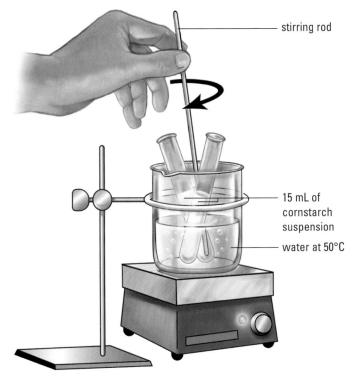

— stirring rod

— 15 mL of cornstarch suspension

— water at 50°C

Figure 6

13. Place test tubes 3 and 4 in a beaker of ice water. Let the cornstarch suspensions chill to a temperature between 0°C and 5°C. Stirring the water with a stirring rod may speed the cooling process.

14. Keep test tubes 5 and 6 at room temperature. Record the temperature of the cornstarch suspension. Record your observations about each test tube.

15. Let all test tubes stand for 20 min. Maintain temperature conditions for the test tubes.

16. Add 5 mL of Benedict's reagent to each test tube and place them in a hot water bath at 100°C for 5 min. Record your observations in the table.

Analysis

(a) What would overheating have done to the contents of test tubes 1 and 2? What happens to the ability of the enzyme to convert starch to sugar at the tested temperatures?

(b) What was the function of test tubes 2, 4, and 6?

(c) At what temperature did amylase work best to convert starch to sugar?

Evaluation

(d) Identify possible sources of error, and indicate how you could improve the procedure.

Synthesis

(e) How are the conditions in the experiment similar to the conditions in the digestive system? How are they different?

Sections 6.4–6.5 Questions

Understanding Concepts

1. State the functions of the enzymes amylase and pepsin.

2. What causes stomach ulcers?

3. In stomach cells, protein-digesting enzymes are stored in the inactive form. Once the enzymes leave the stomach, an acid in the stomach changes the shape of the inactive enzyme, making it active. The active enzyme begins to digest proteins. Why must protein-digesting enzymes be stored in the inactive form?

4. Where would you expect to find digestive enzymes that function best at a pH of 2.0? at a pH of about 7.0?

5. Why does the low pH of the stomach stop the starch digestion that begins in the mouth? What is the advantage to the body of this delay?

6. Would a mouth with a pH of 5.0 have more or less tooth decay than a mouth with a pH of 7.0? Why?

7. Why is the digestion of proteins more complex than the digestion of fats or carbohydrates?

Making Connections

8. Heartburn, or acid indigestion, occurs when stomach acids back up into the esophagus, burning its lining. Antacids can be taken to reduce the burning sensation. How might using antacids to mask the pain of heartburn inadvertently lead to more serious problems?

6.6 The Liver and Gall Bladder

bile salts: the components of bile that break down large fat globules

The liver continually produces a fluid called bile. Bile contains **bile salts**, which speed up fat digestion. When the stomach is empty, bile is stored and concentrated in the gall bladder.

When there are fats in the small intestine, the hormone cholecystokinin (CCK) is released. CCK is carried in the blood to the gall bladder (**Figure 1**) and triggers the gall bladder to release bile salts. Once inside the small intestine, the bile salts emulsify, or break down, large fat globules. The breakdown of fat globules into

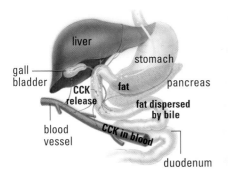

- Fats enter the duodenum and stimulate the release of the hormone CCK.
- CCK is carried by the bloodstream to the gall bladder.
- CCK stimulates the release of bile salts from the gall bladder.
- Bile emulsifies fats.

Figure 1
The function of cholecystokinin

smaller droplets is physical digestion, not chemical digestion, since chemical bonds are not broken. The physical digestion prepares the fat for chemical digestion by increasing the exposed surface area on which fat-digesting enzymes, such as pancreatic lipase, can work.

Liver and Gall Bladder Problems

The production and concentration of bile can result in certain problems. Cholesterol, an insoluble component of bile, acts as a binding agent for the salt crystals found in bile. The crystals precipitate and form larger crystals called **gallstones**. Gallstones can block the bile duct, impairing fat digestion and causing considerable pain.

Bile also contains pigments. The liver breaks down hemoglobin from red blood cells and stores the products in the gall bladder for removal. The characteristic brown colour of feces results from hemoglobin breakdown. Any obstruction of the bile duct or accelerated destruction of red blood cells can cause **jaundice**, turning skin and other tissues yellow.

The liver also stores glycogen and vitamins A, B$_{12}$, and D. In addition, the liver is able to **detoxify** many substances in the body. Harmful chemicals are made soluble and can be dissolved in the blood and eliminated in the urine. One of the more common poisons is alcohol, which the liver breaks down to usable materials and wastes. Alcohol, like many other harmful agents, can destroy liver tissue if consumed in large quantities. Damaged liver cells are replaced by connective tissue and fat, which are not able to carry out normal liver duties. This condition, which can also result from nutritional deprivation or infection, is referred to as **cirrhosis** of the liver.

Table 1 outlines the various functions of the liver.

gallstones: crystals of bile salts that form in the gall bladder

jaundice: the yellowish discoloration of the skin and other tissues brought about by the collection of bile pigments in the blood

detoxify: to remove the effects of a poison

cirrhosis: a chronic inflammation of liver tissue characterized by an increase of non-functioning fibrous tissue and fat

Table 1: Liver Functions

synthesis	• produces bile salts, which are stored in the gall bladder and which emulsify fats, breaking them into smaller droplets • manufactures blood protein from amino acids, which are found in the blood
breakdown/conversion	• removes the highly toxic nitrogen group from amino acids, forming urea (the main solid component of urine) • converts the toxic component of hemoglobin, allowing it to be excreted with bile salts
storage	• converts glucose into glycogen and glycogen to glucose to maintain a constant blood sugar level
detoxification	• converts harmful compounds, such as alcohol, to less harmful products

DID YOU KNOW ?

The liver was once considered to be the centre of emotions. The term lily-livered, meaning cowardly, implies inadequate blood flow to the liver.

SUMMARY **The Liver and Gall Bladder**

1. The liver produces bile, which contains bile salts. Bile salts emulsify fats, forming small droplets and providing greater surface area upon which fat-digesting enzymes work.

2. Gallstones are formed from salt crystals and cholesterol in the gall bladder.

3. Jaundice is a yellowing of the skin caused by the accumulation of bile salts in the blood.

4. The liver stores glycogen and vitamins and detoxifies many harmful substances.

5. If liver tissue is destroyed, the liver cells are replaced by connective tissue and fat; this is called cirrhosis of the liver.

Try This
Activity

Bile Salts and Dissolving Fats

Materials: eyedropper, test tube, vegetable oil, bile salts or liquid soap, hand lens (magnifying glass), test tube stopper

- Fill a test tube one-quarter full of water.
- Add 10 drops of vegetable oil. Record the location and appearance of the oil in the test tube.
- Shake the test tube (with stopper) and immediately examine its contents with the hand lens. Record your observations.
- Let the test tube stand for 2 to 3 min. Observe any changes.
- Add about 5 drops of liquid soap or a pinch of bile salts to the test tube.
- Shake the test tube (with stopper). Immediately examine with the hand lens and record your observations.

(a) What effect did the liquid soap or bile salts have on the oil?

Practice

Understanding Concepts

1. What are the components of bile? Where is bile created and where is it stored?

2. Explain the importance of bile salts in digestion.

3. Why doesn't fat dissolve in water?

4. Why is the liver important in processing toxins in the body? What happens if the level of toxins is very high?

6.7 Absorption of Materials

Chemical digestion is complete by the time food reaches the large intestine. The **colon**, the largest part of the large intestine, must store wastes long enough to reabsorb water out of the wastes. Some inorganic salts, minerals, and vitamins are also absorbed with the water. The stomach absorbs some water, specific vitamins, some medicines, and alcohol.

colon: the largest segment of the large intestine, where water reabsorption occurs

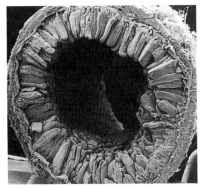

Villi in the mammalian intestine

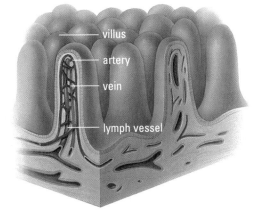

villus
artery
vein
lymph vessel

Figure 1
Villi greatly increase the surface area for absorption of nutrients. Each villus has blood capillaries and lymph vessels.

The large intestine houses bacteria, such as *E. coli*, which are essential to life, and use waste materials to synthesize vitamins B and K. Cellulose, the long-chain carbohydrate characteristic of plant cell walls, reaches the large intestine undigested. Although cellulose cannot be broken down by humans, it serves an important function: Cellulose provides bulk. As wastes build up in the large intestine, receptors in the wall of the intestine provide information to the central nervous system which, in turn, prompts a bowel movement. The bowel movement ensures the removal of potentially toxic wastes from the body. Individuals who do not eat sufficient amounts of cellulose (roughage or fibre), have fewer bowel movements. This means that wastes and toxins remain in their bodies for longer periods of time. Scientists have determined that cancer of the colon can be related to diet. Individuals who eat mostly processed, highly refined foods are more likely to develop cancer of the colon.

Medical reports that state fibre may also reduce cholesterol levels have been embraced by many food manufacturers. Although the actual benefits of bran, a form of roughage, in reducing cholesterol can be debated, its value in providing a balanced diet is undeniable.

Although some water and vitamins are absorbed in the large intestine and the stomach, most absorption takes place within the small intestine. Long fingerlike tubes called **villi** (singular: villus) greatly increase the surface area of the small intestine (**Figure 1**). One estimate suggests that villi account for a tenfold increase in surface area for absorption. The cells that make up the lining of each villus have **microvilli**, which are fine, threadlike extensions of the membrane that further increase the surface for absorption (**Figure 2**).

villi: small fingerlike projections that extend into the small intestine which increase surface area for absorption

microvilli: microscopic fingerlike outward projections of the cell membrane

Microvilli on the surface of an epithelial cell

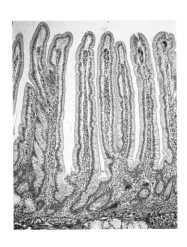

Figure 2

lacteals: small vessels that provide the products of fat digestion access to your circulatory system

Each villus is supplied with a capillary network which intertwines with lymph vessels called **lacteals** that transport materials. Some nutrients are absorbed by diffusion, but some nutrients are actively transported from the digestive tract. Carbohydrates and amino acids are absorbed into the capillary networks; fats are absorbed into the lacteals.

SUMMARY ## The Large Intestine and Absorption

1. Chemical digestion is complete by the time food reaches the large intestine.
2. The colon stores wastes long enough to reabsorb water.
3. The large intestine houses bacteria that use waste materials to synthesize vitamins.
4. Most absorption takes place in the small intestine.
5. Surfaces of the small intestine have villi; cells that line each villus have microvilli. Together, villi and microvilli increase the surface area available for absorption.

Practice

Understanding Concepts

1. What is the function of the colon in the digestive system?
2. Why is cellulose considered to be an important part of your diet?
3. Describe what the inside of the small intestine looks like and how this structure increases the efficiency of its operation.

6.8 Homeostasis and Control Systems

Your body works best at an internal temperature of 37°C, with a 0.1% blood sugar level, and at a blood pH level of 7.35. However, the external environment does not always provide the ideal conditions for life. Air temperatures in Canada can fluctuate between –40°C and 40°C. Foods rarely contain 0.1% glucose and rarely have a pH of 7.35, and you place different demands on your body when you take part in various activities, such as playing sports or digesting a large meal. Your body systems must adjust to these variations to maintain a reasonably constant internal environment.

The term homeostasis is often used to describe the body's ability to adjust to a fluctuating internal and external environment. The word is derived from the Greek words *homoios,* which means similar or like, and *stasis,* which means standing still. The term is appropriate because in homeostasis the body maintains a constant balance, or steady state, despite environmental fluctuations.

The system of active balance requires constant monitoring or feedback about body conditions (**Figure 1**). Special receptors located in the body sense information about blood sugar, body temperature, oxygen levels, and many other body conditions. When homeostasis is disrupted, a monitor sends a signal to the coordinating centre (the brain) where the normal limits are set. The brain relays the information to the appropriate regulator, which helps restore balance. For example, an increase in heart rate during exercise or the release of glucose from the liver to restore blood sugar levels are adjustments made by regulators.

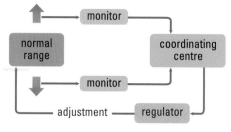

Figure 1
A control system

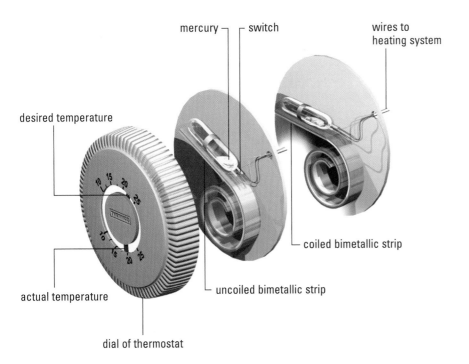

mercury — switch
wires to heating system

desired temperature

coiled bimetallic strip

actual temperature

uncoiled bimetallic strip

dial of thermostat

Figure 2
When the temperature in a room falls below a desired value, the bimetallic strip of the thermostat uncoils. This action causes a drop in mercury to close a switch and start the heating system. After the temperature has risen to the desired level, the strip coils up and the mercury opens the switch, shutting off the heat.

Homeostatic systems are feedback systems that respond to changes in the internal and external environments. As shown in **Figure 2,** a thermostat is an example of a regulator that ensures a stable temperature.

The regulation of blood sugar level provides an example of how a homeostatic system works. Chemical reactions help ensure that the concentration of sugar in the blood remains at an appropriate level. Eating a meal causes the blood sugar level to rise. This stimulates chemical receptors that cause the pancreas to release the hormone insulin. Insulin acts on the cells of the liver and muscles, making them more permeable to glucose. In the liver, excess glucose, which cannot be stored by the body, is converted to glycogen, which can be stored.

A decrease in blood sugar level activates receptors that stimulate the pancreas to release a second hormone called glucagon. It promotes the conversion of glycogen into glucose, which the body can use for its immediate needs. The relationship between insulin and glucagon allows the body to maintain the appropriate amount of sugar in the blood.

Homeostatic Control of Digestion

Homeostatic controls over digestion act before food is absorbed. The control is exerted by the nervous and hormonal systems and the nerves in the stomach wall. Seeing, smelling, or tasting food will produce gastric secretions even before there is any food in the stomach. Swallowing motions also stimulate production of gastric juices, regardless of whether food is actually swallowed.

Hormones play a large role in the control mechanism. For example, secretin, which you learned about in section 6.5, is released when acids from the stomach move into the small intestine along with foods. Secretin is absorbed into the blood and travels to the pancreas where it initiates the release of substances that raise the pH of the small intestine. Another hormone, called **gastrin**, is produced when partially digested proteins are present in the stomach (**Figure 3**, page 228). Gastrin stimulates the release of gastric juices to digest proteins.

The speed at which the digestive system processes food is also under homeostatic control. When food enters the stomach, nerves in the stomach wall cause the muscles to contract and gastric fluids to be secreted. A large meal will activate

DID YOU KNOW ?

Sweating is an important homeostatic mechanism. Evaporation of sweat requires heat. When sweat evaporates from the skin, it takes heat with it, cooling the body. Following exercise, sweat is produced all over the surface of your body. Fear or nervousness produces sweat mainly on the palms and soles of your feet.

gastrin: a digestive hormone secreted by the stomach that stimulates the release of gastric juices to digest proteins

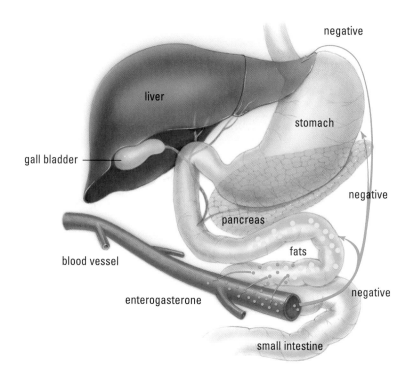

Figure 3
Undigested protein in the stomach stimulates the release of a stomach hormone called gastrin. Gastrin is carried in the blood to the parietal cells of the stomach, where it signals for increased HCl production.

receptors, causing more forceful stomach contractions and faster emptying. If the meal is fatty, the small intestine secretes a digestive hormone (enterogasterone) that slows peristaltic movements, allowing time for fat digestion and absorption.

Practice

Understanding Concepts

1. Define homeostasis.
2. Use the example of a thermostat to explain homeostasis.
3. Explain the function of gastrin.
4. Copy and complete **Figure 4** to explain homeostatic feedback mechanisms for temperature control. The following are the missing words: muscles, sweating, shivering, adjustment, adjustment, sweat glands, and evaporation.

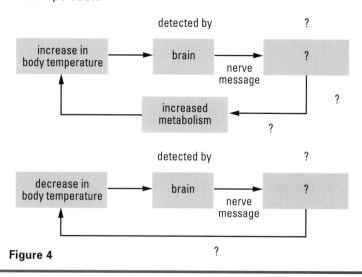

Figure 4

Digestion and Substances Involved in Digestion

Table 1: Digestion

Organ	Function
mouth	chewing of food and digestion of starch
stomach	storage of food and the initial digestion of proteins
small intestine	digestion of carbohydrates, proteins, lipids; the most absorption of nutrients
pancreas	production of digestive enzymes that act on foods in the small intestine; production of the hormone insulin which regulates blood sugar; storage of HCO_3^- ions that neutralize stomach acids in the small intestine
large intestine	absorption of water and storage of undigested food

Table 2: Substances Involved in Digestion

Organ	Secretion	Function
mouth, salivary glands	salivary amylase	initiates the breakdown of polysaccharides to simple carbohydrates and disaccharides
stomach	hydrochloric acid	converts pepsinogen to pepsin; kills microbes
	pepsinogen	when converted to pepsin, initiates the digestion of proteins
	mucus	protects the stomach from pepsin and HCl
	rennin	coagulates proteins in milk
pancreas, small intestine	bicarbonate	neutralize HCl from the stomach
	trypsinogen	when activated to trypsin, converts long-chain peptides (many amino acids) into short-chain peptides
	lipase	breaks down fats to glycerol and fatty acids
	erepsin	completes the breakdown of proteins
	disaccharidases (e.g., maltase)	break down disaccharides (e.g., maltose) into monosaccharides
liver	bile	emulsifies fat
gall bladder		stores concentrated bile from the liver
large intestine	mucus	helps movement of food

Sections 6.7–6.8 Questions

Understanding Concepts

1. What important physical change must fats undergo before chemical change can take place? Where and how does this physical change occur?

2. Explain the mechanism that triggers the release of bile salts.

(continued)

3. Are nutrients absorbed passively or actively in the digestive tract? Into what are carbohydrates, amino acids, and fats absorbed?

4. Sketch the two ways in which absorbed nutrients leave the intestine and get to body cells.

5. What are some signals that trigger the secretion of digestive fluids even in the absence of food in the stomach?

6. How do hormones help regulate digestion?

7. What are gallstones and what causes them?

8. What is jaundice? Why does this condition produce a yellowing of the skin?

9. Why might a diet that includes a large proportion of fried foods cause an increase in body mass?

10. What kind of dietary changes would a person without a gall bladder need to make? Why?

Exploring

11. Research the latest techniques used in the removal of gallstones. Follow the links for Nelson Biology 11, 6.8.

GO TO   www.science.nelson.com

6.9 Defining Food Energy

Energy is a term that is often used but is not easy to describe. You can't see energy, smell it, or find its mass. You only know it is present when it changes forms. In your body, chemical energy can be converted by muscle cells into kinetic energy. Every time you lift a book you use energy. However, you use energy even when you don't move. Breathing, thinking, digestion, and waste removal also require energy. About half of the food energy that you take in is used to help your body stay at a constant temperature. Any drastic drop in food energy would affect not only your activity level but also your body temperature.

A Unit for Measuring Energy

In Canada, energy is measured in joules (J) or kilojoules (kJ), a unit developed by the British scientist James Joule. In some countries, such as the United States, energy in food is measured in calories. One calorie is equal to 4.18 J. The word calorie is still frequently used to refer to food energy. For example, you will find some foods labelled as calorie reduced.

Your lowest energy need is called your **basal metabolic rate** (BMR). This describes the amount of energy needed to keep you alive, even in the most relaxed state. See **Table 1** for the amount of energy required daily by humans of varying ages and lifestyles.

Balancing Energy

The balance between energy input and energy output is delicate. To maintain a balance, the energy in the food taken in should equal the energy output (**Figure 1**). The energy output is used for growth, exercise, and normal metabolic reactions.

If you increase your food intake or decrease your energy output, an imbalance will be created. For example, as growth begins to slow, the body does not need to output as much energy.

What happens to an individual if the energy input continually exceeds the energy output? The person will gain mass since the excess energy will be converted

basal metabolic rate: the minimum amount of energy that a resting animal requires to maintain life processes

Table 1: Daily Energy Requirements

Description of person	Energy requirement (kJ per day)
newborn	2 000
child (age 2–3)	6 000
teenage girl	9 500
teenage boy	12 000
office worker	11 000
heavy manual worker	15 000

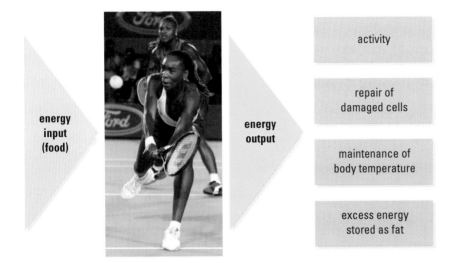

Figure 1

into fat or other energy-storage compounds. The energy that goes into the body is never lost; it only changes forms. This helps explain why many people tend to gain mass as they age.

Factors That Affect Metabolism

Individuals who receive fewer than 9200 kJ of energy per day are said to be starving. However, different people have different energy requirements. Larger people, with a greater number of cells, have greater energy needs. The more active people are, the greater are their energy requirements. Also, different individuals, even if they have the same body mass, can have different rates at which metabolism occurs in their bodies. Metabolism is the sum of all the chemical reactions that occur within the body cells (**Figure 2**). This includes those reactions that convert food energy into other forms of energy. Have you ever noticed how one person can eat huge lunches and never have a change in body mass and another person, the same size and with the same activity level, must eat small lunches to maintain the same body mass?

A chemical messenger produced in the thyroid gland regulates the rate at which food energy is converted into other forms of energy by the cells of the body. If the release of the chemical messenger from the thyroid gland increases, metabolism increases. A portion of the energy from metabolism is converted into heat energy to warm the body, and the rest escapes. A person with a low metabolic rate uses food energy very efficiently. Little food energy is converted into heat and therefore less energy escapes from the body. Instead, the food energy is

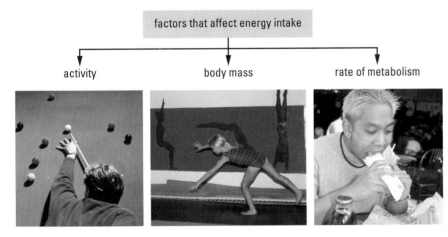

Figure 2

Table 2: Energy Factors for Various Activities

Type of activity	Energy factor (kJ/kg/h)
sleeping	4.1
sitting	5.2
writing	6.0
standing	6.3
singing	7.1
using a computer keyboard, playing cards	9.0
washing car, cooking	10.5
playing piano	11.2
walking (3.2 km/h)	11.6
bowling	13.6
bicycling (13 km/h)	15.8
walking (4.8 km/h)	16.2
walking (6.4 km/h)	20.6
badminton	21.5
cycling (15.3 km/h)	25.8
hiking, fast dancing	27.0
tennis, downhill skiing	36.2
climbing stairs, running (8.8 km/h)	37.5
cycling (20.9 km/h)	40.5
cross-country skiing	42.0
swimming crawl (45.7 m/s)	49.1
handball	49.5
running (12.9 km/h)	62.0
competitive cross-country skiing	73.6

converted into the energy-storage compound—fat. A person with a high metabolic rate converts a large portion of the food energy into heat, less food energy is converted into fat, and the person remains relatively thinner.

Calculating Energy Requirements

It takes more energy to sit than it does to sleep and more energy to stand than to sit. The muscles in your back must contract to keep you erect. Muscle activity requires energy. Any increase in the activity level of your body's cells requires more energy. **Table 2** provides estimates of the amount of energy consumed for different activities. The estimates are taken for moderate activity.

Sample Problem

How much energy would a 75-kg person use if he or she slept all day? The energy factor for sleeping is 4.1 kJ/kg/h. The total energy required can be calculated by multiplying the energy factor by the body mass in kilograms by the amount of time spent on that activity.

Solution

$$\text{Energy required for 1 day} = \text{energy factor} \times \text{body mass} \times \text{time}$$
$$= 4.1 \text{ kJ/kg/h} \times 75 \text{ kg} \times 24 \text{ h}$$
$$= 7.4 \times 10^3 \text{ kJ}$$

The total energy required for a 75-kg person to sleep for one day is 7.4×10^3 kJ.

Dieting

It is now well documented that people who are overweight are more prone to certain diseases such as arteriosclerosis and diabetes. Less well known is the fact that being underweight can also cause problems, such as fatigue and increased susceptibility to illness and injury.

In recent decades, society has promoted thinner ideal body types for women. Thus, many women diet as a way of life, even if their weight is in the proper range. Constant dieting can be dangerous for health. A mid 1990s report from the U.S. National Institute of Health indicated that 95% of people who diet will regain the lost weight. Many follow an endless cycle of dieting and gaining, dieting and gaining. This constant weight fluctuation may be more dangerous to health than keeping a greater, but stable, body mass. The study also revealed that weight is regained more quickly after each dieting phase. This dangerous practice subjects the body to repeated stress. Some weight-loss plans include appetite suppressants such as amphetamines or laxatives, which are both very dangerous if abused.

A slightly different problem exists for many young males. They may wish to emulate the muscular heroes they see in movies, sports, and advertising. However, most adolescent boys do not have muscular bodies. At this age many are just starting to grow in height. Muscle development becomes most accelerated only once height growth has slowed. In an effort to look larger, many men use protein diets, which are aimed at increasing muscle development by increasing food intake beyond the body's needs. For some people, anabolic steroids have become a quick but dangerous way to gain muscle mass. Negative side effects include mood swings and prematurely halted growth.

Practice

Understanding Concepts

1. Define basal metabolic rate.

2. Why is it necessary to maintain a balance between food intake and energy needs?

3. List factors that affect energy output.

4. Explain the role of the thyroid gland in metabolic processes.

5. What is the relationship between metabolic rate and heat energy?

6. Most people cannot work intensely for a long time. However, well-trained cross-country skiers can operate at close to peak performance for 3 h or more. Calculate the total amount of energy consumed in a 3-hour period of competitive cross-country skiing. (Assume the cross-country skier has a mass of 50 kg.)

7. How long would you have to bowl to use the same amount of energy as the competitive cross-country skier referred to in question 6?

Explore an Issue

Take a Stand: Dangerous Diets

Figure 3 shows the stereotypes of "ideal" body shapes for particular periods.

Statement: The advertisements of the diet food industry must be closely regulated. They should show bodies of different shapes, not just the stereotypical "ideal" figures.

- In your group, define the issue—is the problem the diet food industry itself, or is it social standards of ideal body types?
- Research the issue. Search for information in newspapers, periodicals, CD-ROMs, and on the Internet.
  Follow the links for Nelson Biology 11, 6.9.

  GO TO www.science.nelson.com

  You might consider these aspects:
  - the role of genetics in influencing weight
  - the prevalence of eating disorders such as anorexia nervosa and bulimia
  - factors that you think contribute to society's image of the ideal body
  - whether or not all diets are dangerous
  - whether, in the absence of a diet-food industry, there would be other weight-reduction options available for people whose obesity is dangerous to their health

- Identify the perspectives of each of the opposing positions.
- Develop and reflect on your opinion.
- Brainstorm suggestions for a solution.
- Write a position paper summarizing your stand.

DECISION-MAKING SKILLS

- ● Define the Issue
- ○ Identify Alternatives
- ● Research
- ● Analyze the Issue
- ● Defend a Decision
- ○ Evaluate

Figure 3

SUMMARY   Food Energy

1. The basal metabolic rate (BMR) is the minimum amount of energy that a resting animal requires to maintain life processes.

2. In Canada, energy is measured in joules.

3. Larger people, active people, and people with high metabolic rates need more energy to sustain activity.

4. Food intake must provide enough energy for growth, exercise, and normal metabolic reactions.

5. A low metabolism means high efficiency. Little food energy escapes as heat, and more food energy is stored as fat. A high metabolism means low efficiency. More food energy escapes as heat and less is stored as fat.

6.10 Essential Nutrients

In addition to providing energy, food must also contain certain essential nutrients that cannot be made in the body. Of the 20 amino acids needed to produce proteins that make up cell structures, 8 cannot be synthesized. These are the essential amino acids that must be obtained from food. Nutrients identified as essential for one species may not be essential for another.

One great danger of dieting or changing diets to remove a food group is that one or more of the essential amino acids may not be ingested. Common sources of essential amino acids are meats and animal products such as cheeses and eggs. A vegetarian must obtain the essential amino acids by eating specific combinations of plant foods that complement each other. For example, corn provides the essential amino acid methionine, which is missing in beans, while beans contain the essential amino acid lysine, which is missing in corn.

Fats, often given a bad name because of their link with weight gain, are essential in the diet of humans. Fats help the body absorb many important **vitamins**, serve as body insulation under the skin, protect some of the delicate organs of the body, provide a protective coating around nerves, and are an important component of the cell membrane. Your diet must include appropriate amounts of the essential fatty acids needed for making and repairing cell membranes.

vitamins: organic molecules needed in trace amounts for normal growth and metabolic processes

Vitamins help your body change food into energy. **Table 1** shows some of these essential vitamins. Vitamins can be classified as water soluble, such as vitamin C and the B-group vitamins, or fat soluble, such as vitamins A, D, E, and K. Water-soluble vitamins cannot be stored in your body and must be consumed daily because when not absorbed immediately, they are excreted in the urine.

Table 1: Some Vitamins Important for the Body

Vitamin	Sources	Needed for	Deficiency symptoms	Fat or water soluble
A	green vegetables, yellow vegetables, carrots, tomatoes	good vision, normal growth of bones and teeth, healthy skin	poor vision, night blindness, kidney problems	fat
B_1	pork, liver, peas, soybeans, grains, vegetables	proper functioning of heart, nerves, muscles	poor appetite, nerve problems, beriberi	water
B_2	lean meat, eggs, milk, liver, fish, poultry, leafy vegetables	healthy skin and hair, good vision, growth, reproduction	poor growth, hair problems, poor vision	water
C	citrus fruits, potatoes	maintaining cells and tissues	low resistance to infections	water
D	fish oils, eggs, milk	strong teeth and bones, growth	weak teeth and bones	fat
E	leafy vegetables, grains, vegetable oils, liver	forming red blood cells	no symptoms	fat
K	leafy vegetables, liver, potatoes	assisting blood clotting, healthy bones	hemorrhaging	fat

Fat-soluble vitamins can be stored in the fatty tissues of your body. If you take too many of these vitamins, they can build up in your body and become harmful. For example, accumulating vitamin D, which helps regulate calcium and phosphate levels for your teeth and bones, can cause serious problems. Excessive levels of vitamin D have been linked to calcium deposits in the kidney and blood vessels and slower growth in some children.

Some vitamins can be created in the body. For example, vitamin K can be found in food but is also produced by bacteria in our intestines. Vitamin D is found only in fish liver oils, butter, and milk; however, ultraviolet radiation changes a chemical in your skin into vitamin D.

Several **minerals** are also needed in small amounts. Some important minerals and their functions are listed in **Table 2**. Like vitamins, minerals help you obtain the necessary energy from the foods you eat and help maintain normal body functioning.

DID YOU KNOW ?

In countries that have limited sunlight during most or part of the season, vitamin D is often added to commercially produced milk. In sunny countries such as Mexico, this vitamin is not added.

minerals: elements (such as copper, iron, calcium, potassium, etc.) required by the body, often in trace amounts. Minerals are inorganic.

Table 2: Some Minerals Important for the Body

Mineral	Sources	Needed for	Deficiency symptoms
calcium	milk, cheese, grains, beans, hard water	growth and maintenance of bones and teeth, aids blood clotting	soft bones and teeth, osteoporosis
iodine	seafood, table salt	proper working of thyroid gland	swollen thyroid gland, goitre
iron	green vegetables, liver, whole-wheat bread, grains, nuts	transport of oxygen through the body	lack of energy, anemia
phosphorus	meats, fish, dairy products, grains	growth and maintenance of bones and teeth, some cell reactions	poor development of bones and teeth
potassium	meats, grains, milk, fruits, green vegetables	needed to make proteins	weak muscles
sodium	table salt, vegetables, canned meat	regulates movement of water between cells and blood	dehydration

Antioxidants

A great deal of excitement is being generated by a group of vitamins: C, E, and beta carotene (the chemical parent of vitamin A). These chemicals are known as **antioxidants**. Early research suggests that these chemicals are able to lessen the danger of a group of harmful molecules known as oxygen-free radicals.

Free radicals are created in your body by exposure to sunlight, X rays, ozone, tobacco smoke, car exhaust, and other environmental pollutants. They damage the genetic information in your body cells, causing mutations. The altered instructions found in the genetic information can cause the cell to divide at uncontrolled rates or even die. Scientists believe that these free radicals play a major role in the development of cancer, heart disease, lung disease, and even cataracts (a condition that makes the lens of the eye opaque). By taking these chemicals "out of commission," you would live longer and experience better health. By eating certain vegetables, you can provide your body with the antioxidants vitamins A and C (**Figure 1**).

antioxidants: chemicals that reduce the danger of oxygen-free radicals. Vitamin C is a common antioxidant.

Figure 1
Vegetables such as cabbage, Brussels sprouts, and broccoli are rich in vitamins A and C.

SUMMARY   Essential Nutrients

1. The essential amino acids are the eight amino acids that cannot be synthesized by your body and must be ingested.
   - A common source of essential amino acids is animal products.
   - The essential amino acids can be obtained in vegetarian diets by eating complementary foods.

2. Vitamins are either water soluble or fat soluble.
   - Water-soluble vitamins dissolve in the blood and are excreted in urine. Thus they cannot be stored and must be consumed daily.
   - Fat-soluble vitamins can be stored in fatty tissue. Excessive amounts of fat-soluble vitamins can build up in the body and cause problems.
3. Antioxidants can make free radicals less toxic.
   - Free radicals are created in the body in response to things like sunlight, X rays, pollution, etc.
   - Free radicals may contribute to cancer and other disorders.

Sections 6.9–6.10 Questions

Understanding Concepts

1. Young children tend to eat less than adults despite being more active. Why do young children require less food energy?

2. Is it possible for two people who are equally active to maintain the same body mass yet eat different amounts of food? Explain your answer.

3. Refer to **Table 2** in section 6.9.
   (a) Calculate the amount of energy consumed by a 75-kg person sitting for 15 h. How much energy is consumed by the same person sitting and using a computer keyboard for 15 h?
   (b) Calculate the amount of energy required by a 55-kg person standing for 4.5 h.
   (c) How much energy would you use for 25 min of continuous swimming? Show your work.
   (d) List the activities you participate in on a typical day. Estimate how much time you spend at each activity, and then calculate your energy demands in one day.

4. What are amino acids? What are essential amino acids?

5. Explain how both vegetarians and nonvegetarians obtain all necessary nutrients.

6. Even though our bodies can convert carbohydrates into fat for storage, some fat in the diet is necessary. Explain why.

7. Explain why vitamins should be included in your diet.

8. Name one important vitamin and indicate a source of that vitamin.

9. Name one important mineral and explain why it must be included in your diet.

10. Why is it more dangerous to take in large quantities of fat-soluble vitamins than water-soluble vitamins?

11. (a) Identify two lifestyle choices that could lead to a vitamin deficiency.
    (b) Describe the deficiency symptons that could result.

Making Connections

12. Cod liver oil is high in vitamin A. Research why some people think children need cod liver oil supplements, especially during the winter months.
    Follow the links for Nelson Biology 11, 6.10.
    GO TO www.science.nelson.com

13. Why is vitamin A added to 2% and skim milk but not to whole milk?

14. What are free radicals and antioxidants? Explain their possible link to cancer and other disorders.

Key Expectations

Throughout this chapter, you have had opportunities to do the following:

- Identify examples of technologies that have enhanced scientific understanding of internal systems (6.1, 6.4).
- Provide examples of Canadian contributions to the development of technology for examining internal systems (6.1, 6.4).
- Analyze and explain how societal needs have led to scientific and technological developments related to internal systems (6.1, 6.4).
- Compare the anatomy of different organisms: vertebrate and/or invertebrate (6.2).
- Explain the role of transport systems in the transport of substances in an organism (6.2, 6.3, 6.7).
- Describe the importance of nutrients and digestion in providing substances needed for energy and growth (6.4, 6.7, 6.9, 6.10).
- Select appropriate instruments and use them effectively and accurately in collecting observations and data (6.3, 6.5, 6.6).
- Select and use appropriate modes of representation to communicate scientific ideas, plans, and experimental results (6.3, 6.5).
- Present informed opinions about how scientific knowledge of internal systems influences personal choices concerning nutrition and lifestyle (6.5, 6.6, 6.8, 6.9, 6.10).
- Select and integrate information about internal systems from various print and electronic sources, or from several parts of the same source (6.4, 6.6, 6.9).
- Locate, select, analyze, and integrate information on topics under study, working independently and as part of a team, and using appropriate library and electronic research tools, including Internet sites (6.4, 6.6, 6.9).
- Express the result of any calculation involving experimental data to the appropriate number of decimal places or significant figures (6.9, 6.10).

Key Terms

amylase	lacteals
antioxidants	lipases
basal metabolic rate	microvilli
bile salts	minerals
capillary	mucus
cirrhosis	nuclear imaging
colon	nuclear magnetic
computerized axial	resonance (NMR)
tomography (CAT) scan	technology
crop	organs
detoxify	organ system
duodenum	pepsin
endoscope	peristalsis
enterokinase	pharynx
erepsins	radionuclides
esophagus	secretin
gallstones	sphincters
gastrin	trypsin
gastrovascular cavity	ulcer
gizzard	villi
hydrolytic enzymes	vitamins
jaundice	

Make a
Summary

In this chapter, you studied the importance of digestion in providing substances needed for energy and growth. To summarize your learning, create a concept map that shows how the digestive system breaks down foods and absorbs nutrients. Use as many of the key terms as possible. Check other concept maps to help you make your sketch clear.

Reflect on your Learning

Revisit your answers to the Reflect on Your Learning questions at the beginning of this chapter.

- How has your thinking changed?
- What new questions do you have?

Understanding Concepts

1. Describe the functions of the mouth, stomach, small intestine, and gall bladder in the digestion of a hamburger.

2. Use **Figure 1** to answer the following questions:
   (a) Identify the organ that initiates protein digestion.
   (b) Which areas of the digestive system secrete enzymes that break down starch?
   (c) In which area of the digestive system would you find bacteria that manufacture vitamin K?
   (d) Identify the gall bladder, esophagus, and pancreas.

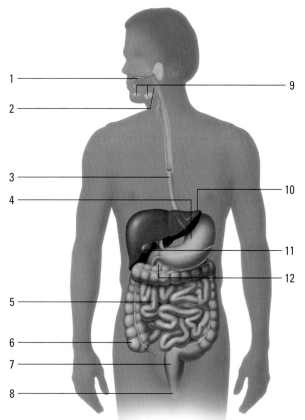

Figure 1

3. Why are you able to drink water while standing on your head?

4. Why don't the chemicals in the stomach digest the stomach itself?

5. Why are pepsin and trypsin stored in inactive forms? Why can erepsins be stored in active forms?

6. Under certain abnormal conditions, the stomach does not secrete hydrochloric acid. Identify two functions that hydrochloric acid has in the digestive process and describe how the failure to secrete hydrochloric acid will affect these processes.

7. Explain how the backwash of bile salts into the stomach can lead to stomach ulcers.

8. Why do individuals with gallstones experience problems digesting certain foods?

9. Why might individuals with an obstructed bile duct develop jaundice?

10. The total surface area of the small intestine is about 300 m^2, approximately the size of a tennis court. Explain the structural features of the small intestine that contribute to this tremendous surface area.

11. Trace the digestion of a spaghetti dinner from the mouth to the colon. Describe the enzymes and hormones involved. How would digestion differ depending on whether the sauce had meat or not?

12. Use the information in **Table 1** to calculate John's energy intake during one day.

Table 1: Food Consumed by John on a Normal Day

Food	Single serving size	Food energy (kJ/serving)	Amount consumed
cheddar cheese	7 g	710	21 g
2% milk	250 mL	540	1000 mL
scrambled eggs	1 egg	400	3
ground beef	1 patty	1080	2
bacon	1 strip	380	5
peas	250 mL	267	125 mL
apple	1	290	1
French fries	10 fries	650	30
lemonade	250 mL	420	250 mL
brownie	1 piece	400	4
cola	200 mL	320	400 mL

   (a) Use the following formula to calculate John's energy needs while cross-country skiing:
   Energy required for 1 day = energy factor × body mass × time
   energy factor for cross-country skiing = 42 kJ/kg/h
   body mass = 75 kg
   time = 2 h
   (b) Make suggestions for possible changes in John's diet. What foods would you substitute and why?
   (c) For how long would John have to cross-country ski to use all of the energy provided by a banana split (2436 kJ)? Show your calculations.

13. What conclusion might you draw from the fact that in wealthy nations the incidence of colon cancer is higher than in developing nations? Justify your answer.

14. Explain how it is possible to eat a balanced diet yet suffer from a nutritional deficiency.

Applying Inquiry Skills

15. The following experiment was designed to investigate factors that affect lipid digestion. A pH indicator, which turns pink when the pH is 7.0 or above and clear when the pH is below 7.0, is added to 3 test tubes as shown in **Figure 2**. The initial colour of each test tube is pink. Predict the colour changes for each of the test tubes shown in the diagram and provide your reasons.

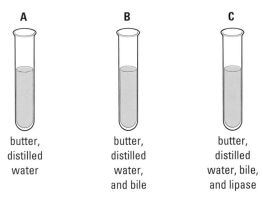

A	B	C
butter, distilled water	butter, distilled water, and bile	butter, distilled water, bile, and lipase

Figure 2

16. A mixture was made of water and egg white (albumin). Upon heating, the mixture coagulated. The following tests were then performed on the coagulated mixture:

    • test tube 1: The mixture was added to trypsin and a pH buffer of 9.0.
    • test tube 2: The mixture was added to trypsin and bile.
    • test tube 3: The mixture was added to trypsin and a pH buffer of 3.0.

    The 3 test tubes were placed in an incubator at 37°C for 1 h. The experimenter observed that the egg white was absent in test tubes 1 and 2, but still present in test tube 3.
    (a) Describe what happened to the egg white in test tubes 1 and 2 to explain why it disappeared.
    (b) Describe the role of bile in this experiment.
    (c) Identify and describe another test that would provide evidence for the process being investigated in the experiment.

17. Salivary amylase breaks down starch to small-chain polysaccharides and disaccharides. Benedict's reagent changes colour in the presence of these products. In the procedures shown in **Figure 3**, 5 mL of starch has been added to each of the test tubes. Which of the procedures would best determine the optimal pH for the digestion of starch by salivary amylase? Outline the problems with the procedures not chosen.

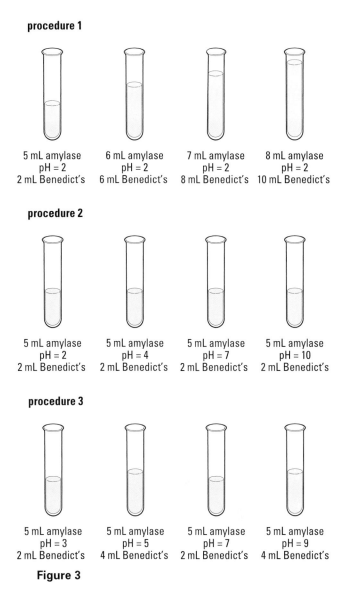

procedure 1

| 5 mL amylase pH = 2 2 mL Benedict's | 6 mL amylase pH = 2 6 mL Benedict's | 7 mL amylase pH = 2 8 mL Benedict's | 8 mL amylase pH = 2 10 mL Benedict's |

procedure 2

| 5 mL amylase pH = 2 2 mL Benedict's | 5 mL amylase pH = 4 2 mL Benedict's | 5 mL amylase pH = 7 2 mL Benedict's | 5 mL amylase pH = 10 2 mL Benedict's |

procedure 3

| 5 mL amylase pH = 3 2 mL Benedict's | 5 mL amylase pH = 5 4 mL Benedict's | 5 mL amylase pH = 7 2 mL Benedict's | 5 mL amylase pH = 9 4 mL Benedict's |

Figure 3

Making Connections

18. Coal-tar dyes are used to enhance the colour of various foods. Some dyes, however, have been found to be carcinogenic (cancer causing). Many of the artificial food colours used in Europe are banned in North America, and many used in North America are banned in Europe. Explain how two groups of scientists could test a food dye and come to different conclusions.

19. The incidence of colon cancer is highest in countries where people eat the greatest quantities of animal proteins and fat. Countries where cereal grains form the basic diet have a much lower incidence of colon cancer. What conclusions might you draw from these facts? Do you think colon cancer can be eliminated by a change in diet? Justify your answer.

7

In this chapter, you will be able to

- explain the role of the circulatory system in the transport of substances within the body;
- compare the circulatory systems of different organisms;
- describe how fitness is related to the efficiency of the cardiovascular system;
- carry out investigations that demonstrate how the body adjusts to exercise;
- identify lifestyle activities that could improve or impair health;
- identify contributions made by Canadian scientists in the study of blood and circulation.

Circulation and Blood

Table 1 outlines an exciting breakthrough in the field of medical biotechnology. Cardiovascular disease is the leading cause of death in North America. About 44 000 Canadians, 40% of them younger than 65, die each year from cardiac disease. Over 4000 patients in Canada and the United States are on the waiting list for a new heart. Only the sickest patients make the list and not all of them will receive a new heart—some will die waiting. Aggressive campaigns to educate people about the importance of organ donation have resulted in increased numbers of donors. However, it may not be enough. Over the past few years, the demand for organs has been rising by about 15% per year and this rate will likely increase. Fewer than 3000 patients worldwide receive heart transplants annually.

Table 1: Procedure for Making a Heart

1. Cells are placed along plastic scaffolding.	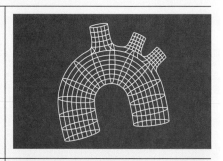
2. The scaffolding, seeded with cells, is placed in a bioreactor that provides nutrients and oxygen.	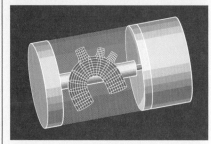
3. The cells divide and fill in the open spaces on the scaffolding.	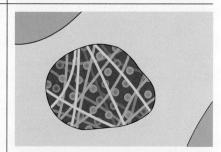
4. This technique can be used to grow parts of the heart, and perhaps eventually the entire organ.	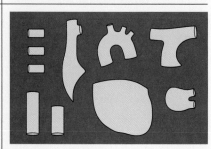

Dr. Michael Sefton (**Figure 1**), director of the Institute of Biomaterials and Biomedical Engineering at the University of Toronto, has a solution that would provide an almost unlimited number of hearts for transplant. What Sefton calls a "heart in a box" is a transplantable heart that can be grown in the laboratory.

First, researchers must create scaffolding that the cells will grow around. Typically, biodegradable plastics are used. The next step is to seed the scaffolding with living cells. The scaffolding and cells are then placed in a bioreactor—a sort of incubator that maintains constant body temperature and provides the nutrients and oxygen required to support cell division. Cells secrete proteins and growth factors that bind them together to form living tissues. Although researchers have not yet been able to grow a complete living heart, they have successfully grown components of the heart.

Figure 1
Dr. Michael Sefton

 Try This Activity

Monitoring Your Pulse

Walking or mild exercise will increase your heart rate by 20% to 30%. For those in good health, increased energy demands during extreme exercise can raise the heart rate to an incredible 200 beats per min. Although few individuals can sustain such a rapid heart rate, it indicates the capacity of the heart to adjust to changing situations.

 Do not perform this activity if you are not allowed to participate in physical education classes.

- While sitting still, place your index and middle finger near your wrist, as shown in **Figure 2**. The **pulse** you feel is blood rushing through the brachial **artery** in your arm. Count the number of heartbeats in 30 s. Record your pulse rate at rest and then calculate the heart rate as beats per min.
- Remain sitting quietly and place your index finger and middle finger on the side of your neck. You will feel blood pulse through the carotid artery, which is an artery that carries blood to the head. Take your pulse rate for 30 s and then calculate the heart rate for 1 min.
- Run on the spot for approximately 2 min.
- Take your pulse immediately after exercise using either the carotid artery or the brachial artery. Record your heart rate.
    (a) Compare the strength of the pulse in the carotid artery with that in your arm.
    (b) Compare your heart rate before and after exercise.
    (c) Do you think the difference between resting heart rate and the heart rate after exercise would be greater for athletes? Explain your answer.

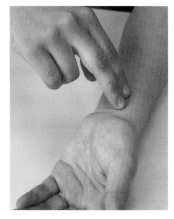

Figure 2
Arteries near the surface permit taking of the pulse.

Reflect on your Learning

1. People with heart problems often experience a racing and pounding heart even after mild exercise. Why does this occur?

2. Great athletes literally have "big hearts." How would the resting heart rate of someone with cardiovascular disease compare to that of an athlete? Suggest a reason for the difference.

3. If scientists wanted to grow a heart, would it be best to obtain cells from the individual who had the heart problem or another individual?

4. Although scientists have successfully grown cells in the shape of certain heart components, these tissue cultures lack the arteries and **veins** found in a normal heart. Why are blood vessels necessary?

pulse: change in the diameter of the arteries following heart contractions

artery: a blood vessel that carries blood from the heart to the body

veins: blood vessels that carry blood from the body to the heart

Figure 1
David, "the boy in the plastic bubble," had severe combined immunodeficiency syndrome.

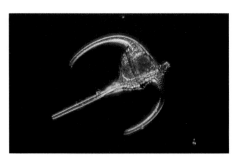

Figure 2
Diatom bathed in seawater

mesoderm: the middle tissue layer of an animal embryo. In more complex animals, this layer will become the muscles and other connective tissues, the blood vessels and blood cells, and various other organs, structures, and systems, depending on the organism.

ectoderm: the outermost tissue layer of an animal embryo. In more complex animals, this layer will become the skin, the outermost parts of the nervous system, and various other outer organs, structures, and systems, depending on the organism.

endoderm: the innermost tissue layer of an animal embryo. In more complex animals, this layer will become the digestive tract, the respiratory tract, and various other inner organs, structures, and systems, depending on the organism.

7.1 The Importance of a Circulatory System

Your circulatory system carries nutrients to cells, wastes away from cells, and chemical messages from cells in one part of the body to distant target tissues. It distributes heat throughout the body and, along with the kidneys, maintains acceptable levels of body fluid.

No cell is further than two cells away from a blood vessel that carries nutrients. Your circulatory system has 96 000 km of blood vessels to sustain your 100 trillion cells. No larger than the size of your fist and with a mass of about 300 g, the heart beats about 70 times/min from the beginning of life until death. During an average lifetime, the heart pumps enough blood to fill two large ocean tankers.

Every minute, 5 L of blood cycles from the heart to the lungs, picks up oxygen, and returns to the heart. Next, the heart pumps the oxygen-rich blood and nutrients to the tissues of the body. The oxygen aids in breaking down high-energy glucose into low-energy compounds and releases energy within the tissue cells. The cells use the energy to build new materials, repair existing structures, and for a variety of other energy-consuming reactions. Oxygen is necessary for these processes to occur, and the circulatory system plays a central role in providing that oxygen.

The circulatory system is also vital to human survival because it transports wastes and helps defend against invading organisms. It permits the transport of immune cells throughout the body. To appreciate the importance of the immune system, consider the story of David, "the boy in the plastic bubble"(**Figure 1**). David was born without an immune system, which meant that his body was unable to produce the cells necessary to protect him from disease. As a result, David had to live in a virtually germ-free environment. People who came in contact with him had to wear plastic gloves. Eventually, David received a bone marrow transplant from his sister. Unfortunately, a virus was hidden in the bone marrow. The sister, who had a functioning immune system, was able to protect herself from the virus, but David was not.

The Challenge of Transporting Oxygen and Nutrients

Single-celled organisms, such as the diatom in **Figure 2**, have no need for a circulatory system because oxygen can diffuse directly into them from the surrounding seawater and wastes can diffuse out.

Even simple multicellular organisms, such as the sponge in **Figure 3**, have no need for a circulatory system in an aquatic environment. Because the sponge has only two cell layers, all cells remain in contact with water. The flagella pull water into the sponge, bringing oxygen and nutrients into the body. Water and wastes are expelled through the large pore, or osculum, at the top of the body.

Relying on diffusion to deliver oxygen and nutrients is too limiting for more complex multicellular animals, which have three cell layers. The middle cell layer, or **mesoderm** layer, is sandwiched between the **ectoderm** and **endoderm** and does not come into direct contact with circulating fluids or water. Cells of the mesoderm need a circulatory system to bring them into contact with oxygen and nutrients.

Open and Closed Circulatory Systems

In an open circulatory system, blood carrying oxygen and nutrients is pumped into body cavities where it bathes the cells directly. This low-pressure system is commonly found in snails (**Figure 4**), insects, and crustaceans. There is no distinction

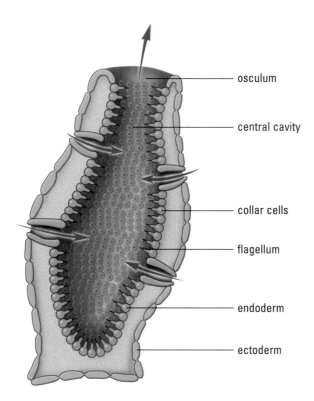

osculum

central cavity

collar cells

flagellum

endoderm

ectoderm

Figure 3
The sponge is composed of two cell layers: the ectoderm, on the outside of the sponge, and the endoderm, on the inside. Seawater acts as a transport system, carrying nutrients and removing wastes. Nutrients and wastes diffuse across cell membranes.

interstitial: refers to the space between cells

sinus: a body cavity or air space surrounding an internal organ

between the blood and the **interstitial** fluid in this system. Interstitial fluid is a fluid that occupies the spaces between cells. The contraction of one or more hearts pushes blood from one body cavity, or **sinus**, to another. Muscular movements by the animal during locomotion can assist blood movement, but diverting blood flow from one area to another is limited. When the heart relaxes, blood is drawn back toward the heart through open-ended pores.

In a closed circulatory system, the blood is always contained within blood vessels. This system, commonly found in earthworms (**Figure 5**), squids, octopuses, and vertebrates, separates blood from the interstitial fluid by enclosing blood inside tubes or vessels. The earthworm has five heartlike vessels that pump blood through three major blood vessels. Larger blood vessels branch into smaller vessels that supply blood to the various tissues.

Practice

Understanding Concepts

1. Describe the main functions of the circulatory system.
2. Differentiate between an open and a closed circulatory system.
3. Identify one advantage and one disadvantage of an open circulatory system.
4. Describe the similarities and differences between the circulatory systems of sponges, snails, and earthworms.
5. Why do multicellular animals need a circulatory system?

heart

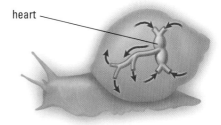

Figure 4
The snail has an open circulatory system.

capillaries

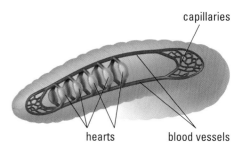

hearts   blood vessels

Figure 5
The worm has a closed circulatory system.

SUMMARY The Importance of a Circulatory System

1. The circulatory system
   • brings oxygen and nutrients to cells;
   • takes wastes away from cells;

- relays chemical messages throughout the body;
- helps maintain acceptable levels of fluid;
- is important in the body's immune system, permitting the transport of immune cells throughout the body.

2. Oxygen is needed to break down glucose, which creates the energy needed for various cellular processes.

3. In an aquatic environment, water acts as a transport system so a circulatory system is not needed by single-celled organisms and some multicellular organisms.

4. In open circulatory systems, blood bathes the cells of the body directly.

5. In closed circulatory systems, blood is always contained in blood vessels.

7.2 Components of Blood

The average 70-kg individual is nourished and protected by about 5 L of blood. Approximately 55% of the blood is fluid; the remaining 45% is composed of blood cells (**Figure 1**). The percentage of red blood cells in the blood is called the hematocrit. The fluid portion of the blood is referred to as the **plasma**, which is about 90% water, allowing it to be described as a fluid tissue. As in other **tissues**, the individual cells in the blood work together for a common purpose.

The plasma also contains blood proteins, glucose, vitamins, minerals, dissolved gases, and waste products of cell metabolism. The large plasma proteins help maintain homeostasis. One group of proteins is called the albumins; they, along with inorganic minerals, establish an osmotic pressure that draws water back into capillaries and helps maintain body fluid levels. A second group of proteins, the globulins, help provide protection against invading microbes. Fibrinogens, the third group of proteins, are important in blood clotting. **Table 1** summarizes the types of plasma proteins and their functions.

Erythrocytes

The primary function of **erythrocytes**, red blood cells, is the transport of oxygen. Although some oxygen diffuses into the plasma, the presence of hemoglobin increases the ability of the blood to carry oxygen by a factor of almost 70. Without hemoglobin, your red blood cells would supply only enough oxygen to maintain life for approximately 4.5 s. With hemoglobin, life can continue for 5 min. This is not much time, but remember that the blood returns to the heart and is pumped to the lungs, where oxygen supplies are continuously replenished. Cells deprived of oxygen for longer than 5 min start to die. This might indicate why people survive even when the heart stops for short periods of time. Children who have been immersed in cold water for longer than 5 min have survived with comparatively minor cell damage because colder temperatures slow body metabolism and decrease oxygen demand.

An estimated 280 million hemoglobin molecules are found in a single red blood cell. The hemoglobin is composed of heme, the iron-containing pigment, and globin, the protein structure. Four iron molecules attach to the folded protein structure and bind with oxygen molecules. The oxyhemoglobin complex gives blood its red colour. Once oxygen is given up to cells of the body, the shape of the hemoglobin molecule changes, causing the reflection of blue light. This explains why blood appears blue in the veins.

Red blood cells appear as biconcave (meaning concave on both sides) disks approximately 7 μm (micrometres) in diameter. This shape provides a greater

plasma: the fluid portion of the blood

tissues: groups of cells that work together to perform specialized tasks

hematocrit

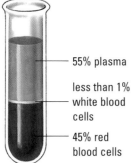

- 55% plasma
- less than 1% white blood cells
- 45% red blood cells

Figure 1

Table 1: Plasma Proteins

Type	Function
albumins	osmotic balance
globulins	antibodies, immunity
fibrinogens	blood clotting

erythrocytes: red blood cells that contain hemoglobin and carry oxygen

surface area for gas exchange—between 20% and 30% more surface area than a sphere. The outer membranes of red blood cells become brittle with age, causing them to rupture as they file through the narrow capillaries. Since red blood cells live only about 120 days, cell reproduction is essential. One estimate suggests that at least five million red blood cells are produced every minute of the day.

Red blood cells do not contain a nucleus when mature, which allows more room for the cell to carry hemoglobin. This enucleated condition raises two important questions. First, since cells, by definition, contain a nucleus or nuclear material, are red blood cells actually cells? The second question addresses cell reproduction: how do cells without a nucleus and chromosomes reproduce? The answer to both of the above questions can be found in bone marrow, where red blood cells are produced by nucleated stem cells. The young cells lose their nuclei as they are discharged into the blood stream.

The average male has about 5.5 billion red blood cells per millilitre of blood, while the average female has about 4.5 billion red blood cells per millilitre. Individuals living at high altitudes can have red blood counts as high as 8 billion red blood cells per millilitre. How does the body ensure that adequate numbers of red blood cells are maintained? Specialized white blood cells, located primarily in the spleen and liver, monitor the age of red blood cells and remove debris from the circulatory system. Following the breakdown of red blood cells, the haemoglobin is released. Iron is recovered and stored in the bone marrow for later use. The heme is transformed into bile pigments.

A deficiency in hemoglobin or red blood cells decreases oxygen delivery to the tissues. This condition, known as **anemia**, is characterized by low energy levels. The most common cause of a low red blood cell count is hemorrhage. Physical injury, bleeding due to ulcers, or hemorrhage in the lungs due to tuberculosis can cause anemia. If more than 40% of the blood is lost, the body is incapable of coping. Anemia may also be associated with a dietary deficiency of iron, which is an important component of hemoglobin. The red blood cells must be packed with sufficient numbers of hemoglobin molecules to ensure adequate oxygen delivery. Raisins and liver are two foods rich in iron.

Leukocytes

White blood cells, or **leukocytes**, are much less numerous than red blood cells. It has been estimated that red blood cells outnumber white blood cells by a ratio of 700 to 1. White blood cells have a nucleus, making them easily distinguishable from red blood cells. In fact, the shape and size of the nucleus, along with the granules in the cytoplasm, have been used to identify different types of leukocytes. **Figure 2**, page 246, shows the different types of leukocytes. The granulocytes are classified according to small cytoplasmic granules that become visible when stained. The agranulocytes are white blood cells that do not have a granular cytoplasm. Granulocytes and agranulocytes are both produced in the bone marrow, but agranulocytes are modified in the lymph nodes. Some leukocytes destroy invading microbes by phagocytosis; they squeeze out of capillaries and move toward the microbe like an amoeba. Once the microbe has been engulfed, the leukocyte releases enzymes that digest the microbe and the leukocyte itself. Fragments of remaining protein from the white blood cell and invader are called **pus**. Other white blood cells form special proteins, called antibodies, which interfere with invading microbes and toxins.

Platelets

Platelets, like red blood cells, do not contain a nucleus and are produced from large nucleated cells in the bone marrow. Small fragments of cytoplasm break

> **DID YOU KNOW?**
>
> The word erythrocyte comes from the Greek *erythros*, meaning "red." However, a single red blood cell does not appear red but pale orange—the composite of many red blood cells produces the red colour.

> **DID YOU KNOW?**
>
> Because red blood cells live only 120 days, they are continually breaking down and being replenished. The misconception that "young blood" is better than "old blood" persists even today. The blood of elderly people is virtually the same as the blood of young people.

anemia: the reduction in blood oxygen due to low levels of hemoglobin or poor red blood cell production

leukocytes: white blood cells

pus: protein fragments that remain when white blood cells engulf and destroy invading microbes

platelets: component of blood responsible for initiating blood clotting

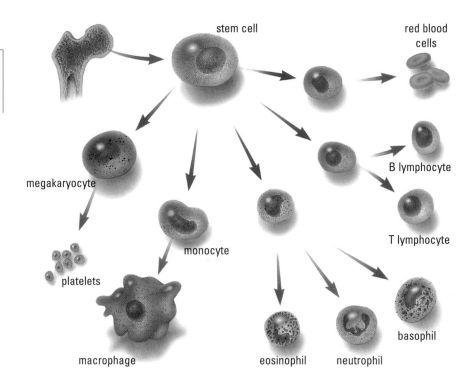

Figure 2
Stem cells of the bone marrow give rise to blood cells. Two classes of white blood cells are shown. The agranulocytes include the monocytes and lymphocytes. The granulocytes include the eosinophils, basophils, and neutrophils.

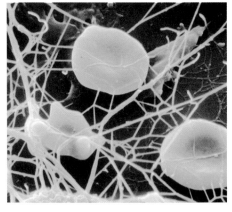

Figure 3
The formation of a blood clot where red blood cells are caught in a mesh of fibres

from the large megakaryocyte, a large cell in the bone marrow, to form platelets. The irregularly shaped platelets move through the smooth blood vessels of the body but rupture if they strike a sharp edge, such as that produced by a torn blood vessel. This is how the fragile platelets initiate blood-clotting reactions (**Figure 3**).

Practice

Understanding Concepts

1. Why is blood considered to be a tissue?
2. Name the two major components of blood.
3. List three plasma proteins and indicate the function of each.
4. What is the function of hemoglobin?
5. List factors that initiate red blood cell production.
6. What is anemia?
7. What causes the production of pus?
8. What is the role of platelets?

7.3 Blood Groups

Early attempts at blood transfusion, at times, were successful and, at other times, resulted in death. The variable results were explained when Karl Landsteiner discovered, at the turn of the 20th century, that different blood types exist. Successful transfusions require the correct matching of blood types. Special markers are located on the membrane of some red blood cells. Individuals with blood type A have the A marker attached to their cell membrane. Individuals with blood type B have the B marker attached to their cell membrane.

Individuals with blood type AB have both A and B markers on their red blood cell membranes, while those with blood type O have no special markers. If type A blood is transfused into a type O individual, special proteins, called **antibodies**, are produced in response to the A-type protein. The A-type protein acts as an **antigen**, a substance that stimulates the production of antibodies. The antibodies attach themselves to the A-type proteins and cause them to clump together. The clumped cells clog local capillaries and prevent oxygen and nutrient delivery to tissue cells. Without treatment, extensive tissue damage, and even death, may result.

Antigen A would not cause this same immune response if transfused into an individual with blood types A or AB because A-type antigens exist on the red blood cells of these individuals, and antibodies against A-type antigens are not normally produced.

Thus, people with type A blood will not produce antibodies against A-type antigens. People with type B blood will not produce antibodies against B-type antigens but will produce antibodies against A markers. Type AB individuals will not produce antibodies against A or B markers, so can tolerate transfusions of A, B, AB, or O blood. Such people are called universal acceptors. Type O blood normally contains antibodies against A and B antigens. These people cannot accept blood from A, B, or AB individuals. They can only accept type O blood. But since their blood cells possess neither A nor B markers, small amounts of this blood may be transfused into A, B, AB, and O individuals. They are universal donors. Table 1 summarizes the antigens and antibodies found in the **serum** for the various blood groups.

antibodies: proteins formed within the blood that react with antigens

antigen: a substance, usually protein in nature, that stimulates the formation of antibodies

serum: the liquid that remains after the solid and liquid components of blood have been separated

Table 1: Antigens and Antibodies Found in Blood Groups

Blood group	Antigen found on red blood cell	Antibody in serum
O	none	A and B
A	A	B
B	B	A
AB	A and B	none

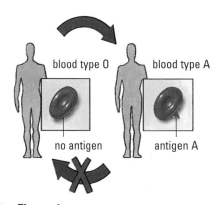
blood type O   blood type A
no antigen   antigen A

Figure 1
Individuals with blood type A can receive blood type O during a transfusion. However, individuals with blood type O cannot receive blood type A during a transfusion.

Rhesus Factor

During the 1940s, scientists discovered another antigen on the red blood cell: the rhesus factor. Like the ABO blood groups, the rhesus factor is inherited. Individuals who have this special antigen are said to be Rh-positive. Approximately 85% of Canadians have the antigen. The remaining 15% of individuals who do not have the antigen are said to be Rh-negative. Individuals who are Rh-negative may donate blood to Rh-positive individuals, but should not receive their blood. The human body has no natural antibodies against Rh factors, but antibodies can be produced in response to antigens following a transfusion. However, the immune reaction is subdued compared with that of the ABO group.

Frontiers of Technology: Artificial Blood

On March 1, 1982, a precedent-setting legal case brought attention to an emerging medical technology. A man and woman, trying to push their car, were critically injured when they were hit by another car. For personal reasons, the couple refused a blood transfusion. During the legal dispute that ensued, the wife died, and the courts ruled that action must be taken to save the husband's life. Five litres of Fluosol—artificial blood—were transfused into the man over a period of five days. Doctors believed that the artificial blood could maintain adequate oxygen levels until the man's bone marrow began replenishing red blood cells.

Fluosol, a nontoxic liquid that contains fluorine, was developed in Japan. Fluosol carries both oxygen and carbon dioxide. It requires no blood matching, and when frozen, can be stored for long periods of time. Artificial blood, unlike human blood, does not have to undergo expensive screening procedures before being used in transfusions. Artificial blood will not carry HIV, hepatitis, or any other virus. However, despite its advantages, artificial blood is not as good as the real thing. Although it carries oxygen, it is ill-suited for many of the other functions associated with blood, such as blood clotting and immunity. The real value of artificial blood is that it provides time until natural human blood can be administered. It could also serve as a supplement for patients with diseases like thalassemia (Cooley's anemia) or aplastic anemia, which require the patients to undergo multiple transfusions. Artificial blood might also help prevent an overload of iron.

SUMMARY Blood

1. Blood is composed mainly of water, plasma, and blood cells.
2. Plasma proteins play roles in maintaining homeostasis, in producing antibodies, and in blood clotting.
3. Erythrocytes function primarily to transport oxygen.
   - Erythrocytes contain hemoglobin, which increases the capacity of oxygen that can be carried in the blood.
   - Erythrocytes are produced in the bone marrow; once they leave, they have no nucleus and can no longer reproduce.
4. Leukocytes are an important part of the immune system.
5. Platelets initiate blood clotting.
6. The hematocrit is the ratio of red blood cell volume to the total blood volume. This value can also be expressed as a percent.
7. Blood type A has the A marker, type B has the B marker, type AB has both, and type O has neither.
8. Blood transfusion is successful if the recipient has a marker but the donor does not, and if the recipient and donor have the same marker.
   - The incompatible marker acts as an antigen in the recipient's body. The recipient will produce antibodies to fight the antigen, causing blood clotting.
   - AB is the universal recipient and O is the universal donor.
9. The rhesus (Rh) antigen is another potential source of incompatibility.

Practice

Understanding Concepts
1. Define antigen and antibody.
2. How does Rh-positive blood differ from Rh-negative blood?
3. What is artificial blood? How is it different from real blood?

Sections 7.2–7.3 Questions

Understanding Concepts

1. What are erythrocytes and what is their primary function?
2. Explain the mechanism by which hemoglobin increases the ability of blood to carry oxygen.
3. Are erythrocytes true cells? Why or why not?
4. State two situations that result in a deficiency of hemoglobin.
5. How do white blood cells differ from red blood cells?
6. State two major functions associated with leukocytes.
7. Explain why type O blood is considered to be the universal donor. Why is type AB considered to be the universal acceptor?
8. List the advantages and disadvantages associated with using artificial blood.

Applying Inquiry Skills

9. Cancer of the white blood cells is called leukemia. Like other cancers, leukemia is associated with rapid and uncontrolled cell production. Examine the hematocrits shown in **Figure 2** and predict which subject might be suffering from leukemia. Give your reasons.
10. Although hematocrits provide some information about blood disorders, most physicians would not diagnose leukemia on the basis of one test. What other conditions might explain the hematocrit reading you chose in question 9? Give your reasons.
11. Lead poisoning can cause bone marrow destruction. Which of the subjects in **Figure 2** might have lead poisoning? Give your reasons.
12. Which subject in **Figure 2** lives at a high altitude? Give your reasons.

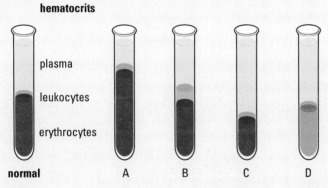

hematocrits

plasma

leukocytes

erythrocytes

normal   A   B   C   D

Figure 2
Hematocrits for four subjects: A, B, C, and D

Making Connections

13. Athletes can take unfair advantage of the benefits of extra red blood cells. Two weeks prior to a competition, a blood sample is taken and centrifuged, and the red blood cell component is stored. A few days before the event, the red blood cells are injected into the athlete. Why would athletes remove red blood cells only to return them to their body later? What problems could be created should the blood contain too many red blood cells? Give your reasons.

7.4 Blood Vessels

Arteries

Arteries are the blood vessels that carry blood away from the heart. They have thick walls composed of distinct layers. The outer and inner layers are primarily connective tissue, while the middle layers are made up of muscle fibres and elastic connective tissue, as shown in **Figure 1(a)**. Every time the heart contracts, blood surges from the heart and enters the arteries. The arteries stretch to accommodate the inrush of blood. The pulse you can feel near your wrist and on either side of your neck is created by changes in the diameter of the arteries following heart contractions. Heart contraction is followed by a relaxation phase. During this phase, pressure drops and elastic fibres in the walls of the artery recoil. It is interesting to note that the many cells of the artery are themselves supplied with blood vessels that provide nourishment.

A birth defect or injury can cause the inner wall of the artery to bulge. Known as an **aneurysm**, this condition is infrequent in young people, but can lead to serious problems for those who have the condition. In much the same way as the weakened wall of an inner tube begins to bulge, the weakened segment of the artery protrudes as blood pulses through. The problem escalates as the thinner wall offers less support and eventually ruptures. Less oxygen and nutrients are delivered to the tissues, resulting in cell death. A weakened artery in the brain is one of the conditions that can lead to a stroke.

Blood from the arteries passes into smaller arteries, called arterioles. The middle layer of arterioles is composed of elastic fibres and smooth muscle. The **autonomic nervous system**, which controls the motor nerves that maintain homeostasis, regulates the diameter of the arterioles. A nerve impulse causes smooth muscle in the arterioles to contract, reducing the diameter of the blood vessel. This process is called **vasoconstriction**. Vasoconstriction decreases blood flow to tissues. Relaxation of the smooth muscle causes dilation of the arterioles, and blood flow increases. This process, called **vasodilation**, increases the delivery of blood to tissues. This, in turn, permits the cells in that localized area to perform energy-consuming tasks.

Precapillary sphincter muscles regulate the movement of blood from the arterioles into capillaries. Blushing is caused by vasodilation of the arterioles leading to skin capillaries. Red blood cells close to the surface of the skin produce the pink colour. Vasodilation helps the body release some excess heat that is produced when you become nervous and blush. Have you ever noticed someone's face turn a paler shade when they are frightened? The constriction of the arteri-

aneurysm: a fluid-filled bulge found in the weakened wall of an artery

autonomic nervous system: the part of the nervous system that controls the motor nerves that regulate homeostasis. Autonomic nerves are not under conscious control.

vasoconstriction: the narrowing of blood vessels. Less blood goes to the tissues when the arterioles constrict.

vasodilation: the widening of blood vessels. More blood moves to tissues when arterioles dilate.

Figure 1
Simplified diagram and photo of an artery and a vein. Arteries have strong walls capable of withstanding great pressure. The middle layer of arteries contains both muscle tissue and elastic connective tissue. The low-pressure veins have a thinner middle layer. The photo shows a cross section of an artery and a vein.

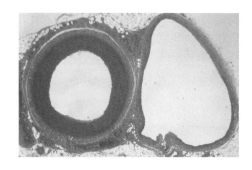

artery   vein

(a)   (b)

olar muscles diverts blood away from the outer capillaries of the skin toward the muscles. The increased blood flow to the muscles provides more oxygen and glucose for energy to meet the demands of a response to a threat or danger.

Arterioles leading to capillaries open only when cells in that area require blood. It has been estimated that the body would need 200 L of blood if all the arterioles were open at one time. Although the majority of brain capillaries remain open, as few as one fiftieth of the capillaries in resting muscle remain open.

Fat in the Arteries: Atherosclerosis

Anyone who has ever washed dishes is aware of how fat floats on water. You may have noticed that when one fat droplet meets another, they stick together and form a larger droplet. Unfortunately, the same thing can happen in your arteries. As fat droplets grow into larger and larger blockages, they slowly close off the opening of the blood vessel. Calcium and other minerals deposit on top of the lipid, forming a fibrous net of plaque. This condition is known as **atherosclerosis**, the most common form of a group of disorders called **arteriosclerosis,** or arterial disease. It can narrow the artery to one quarter of its original diameter and lead to high blood pressure (**Figure 2**). To make matters worse, blood clots, which are normally a life-saving property of blood, form around the fat deposits. As fat droplets accumulate, inadequate amounts of blood and oxygen are delivered to the heart muscle, resulting in chest pains.

Every year heart disease kills more Canadians than any other disease. Lifestyle changes must accompany any medical treatment. A low-fat diet, plus regular, controlled exercise are keys to prevention.

atherosclerosis: a degeneration of the blood vessel caused by the accumulation of fat deposits along the inner wall

arteriosclerosis: a group of disorders that cause the blood vessels to thicken, harden, become winding, and lose their elasticity

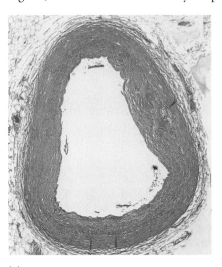

(a)

(b)

Figure 2
(a) Cross section of a normal artery
(b) Cross section of an artery from a person with atherosclerosis. Notice that fat deposits have narrowed the passageway.

Capillaries

Capillaries, composed of a single layer of cells, are the sites of fluid and gas exchange between blood and body cells. No cell is farther than two cells away from a capillary, and many active cells, such as muscle cells, may be supplied by more than one capillary. Most capillaries are between 0.4 and 1.0 mm long with a diameter of less than 0.005 mm. The diameter is so small that red blood cells must travel through capillaries in single file (**Figure 3**, page 252).

The single cell layer of capillaries, although ideal for diffusion, creates problems. Capillary beds are easily destroyed. High blood pressure or any impact, such as that caused by a punch, can rupture the thin-layered capillary. Bruising occurs when blood rushes into the spaces between tissues.

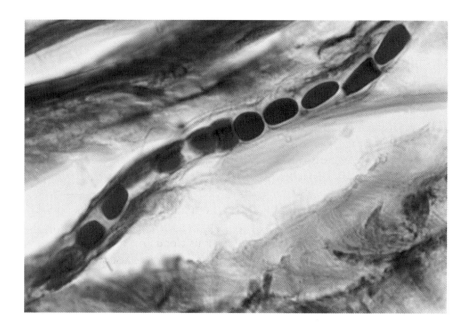

Figure 3
Red blood cells in a capillary. Notice that the capillary is only wide enough for cells to pass through one at a time.

Oxygen diffuses from the blood into the surrounding tissues through the thin walls of the capillaries into the body cells. Oxygenated blood, which appears red in colour, assumes a purple-blue colour as it leaves the capillary. The deoxygenated blood collects in small veins called venules and is carried back to the heart. Some protein is also exchanged, but the process is believed to involve endocytosis and exocytosis rather than diffusion. Water-soluble ions and vitamins are believed to pass through spaces in the walls of the capillary vessels. The fact that some spaces are wider than others may explain why some capillaries seem to be more permeable than others.

Veins

Capillaries merge and become progressively larger vessels, called venules. Unlike capillaries, the walls of venules contain smooth muscle. Venules merge into veins, which have greater diameter, as shown in **Figure 1(b)**. Gradually the diameter of the veins increases as blood is returned to the heart. However, the process of returning the blood to the heart is difficult. As blood flows from arteries to arterioles to capillaries, blood flow is greatly reduced. As blood passes through a greater number of narrower vessels with weaker walls, fluid pressure is reduced. By the time blood enters the venules, the pressure is between 15 mm Hg and 20 mm Hg. This pressure is not enough to drive the blood back to the heart, especially from the lower limbs.

How then does blood get back to the heart? William Harvey, an English physiologist, who lived from 1578 to 1657, conducted experiments to answer that question. In one experiment, he tied a band around the arm of one of his subjects, restricting venous blood flow. The veins soon became engorged with blood and swelled. Harvey then placed his finger on the vein and pushed blood toward the heart. The vein closed up or collapsed. Harvey repeated the procedure, but this time he pushed the blood back toward the hand. Bulges appeared in the vein at regular intervals. What caused the bulges? Dissection of the veins confirmed the existence of valves.

The valves open in one direction, steering blood toward the heart. They do not allow blood back in the other direction. When Harvey tried to push blood toward the hand, the valves closed, causing blood to pool in front of the valves and distend the vein. When he directed blood toward the heart, the valves opened and blood flowed from one compartment into the next. The vein collapsed because the band tied around the arm prevented the blood from passing.

Skeletal muscles also aid venous blood flow. Venous pressure increases when skeletal muscles contract and push against the vein. The muscles bulge when they contract, thereby reducing the vein's diameter. Pressure inside the vein increases and the valves open, allowing blood to flow toward the heart. Sequential contractions of skeletal muscle create a massaging action that moves blood back to the heart (**Figure 4**). This may explain why you feel like stretching first thing in the morning. It also provides a clue as to why some people faint after standing still for long periods of time. Blood begins to pool in the lower limbs and cannot move back to the heart without movement of the leg muscles.

The veins serve as more than just low-pressure transport canals; they are also important blood reservoirs. As much as 50% of your total blood volume can be found in the veins. During times of stress, venous blood flow can be increased to help you meet increased energy demands. Nerve impulses cause smooth muscle in the walls of the veins to contract, increasing fluid pressure. Increased pressure drives more blood to the heart.

Unfortunately, veins, like other blood vessels, are subject to problems. Large volumes of blood can distend the veins. In most cases, veins return to normal diameter, but if the pooling of blood occurs over a long period of time, the one-way valves are damaged. Without proper functioning of the valves, gravity carries blood toward the feet and greater pooling occurs. Surface veins gradually become larger and begin to bulge. This disorder is known as varicose veins. Although there is a genetic link to weakness in the vein walls, lifestyle can accelerate the damage. Prolonged standing, especially with restricted movement, increases pooling of blood. Prolonged compression of the superficial veins in the leg can also contribute to varicose veins.

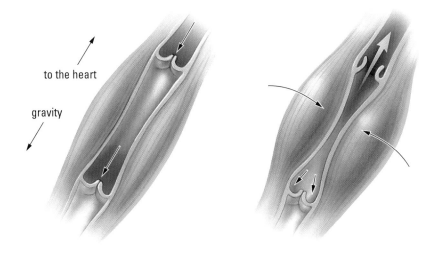

to the heart

gravity

Figure 4
Venous valves and skeletal muscle work together in a low-pressure system to move blood back to the heart.

Practice

Understanding Concepts

1. How do arteries differ from veins?
2. What causes a pulse?
3. Define vasodilation and vasoconstriction.
4. What are the functions of capillaries?

Figure 5 shows the major veins and arteries that make up the human circulatory system.

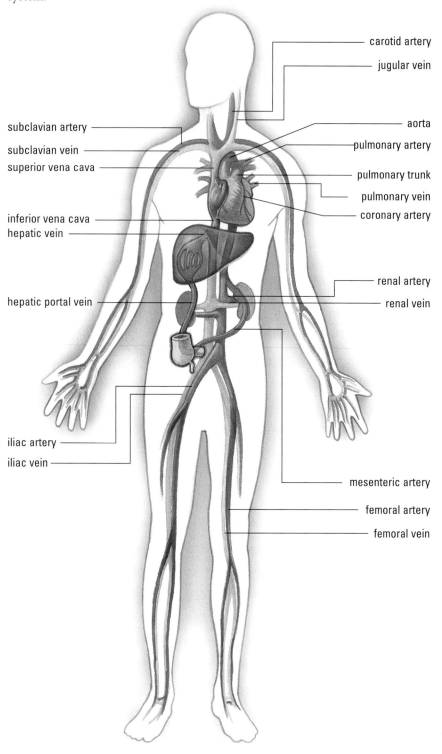

carotid artery

jugular vein

subclavian artery

subclavian vein

superior vena cava

aorta

pulmonary artery

pulmonary trunk

pulmonary vein

coronary artery

inferior vena cava

hepatic vein

renal artery

renal vein

hepatic portal vein

iliac artery

iliac vein

mesenteric artery

femoral artery

femoral vein

Figure 5
Circulatory system

SUMMARY Blood Vessels

1. Arteries carry blood away from the heart.
2. Vasoconstriction is a reduction in the diameter of the blood vessel, decreasing blood flow and the amount of oxygen to the tissues.

Vasodilation is an increase in the diameter of the blood vessel, increasing blood flow and the amount of oxygen to the tissues.

3. Atherosclerosis is a narrowing of the arteries due to a buildup of plaque and fat.

4. Capillaries are the site of fluid and gas exchange between the blood and body cells.

5. Veins and venules carry blood back to the heart.
   - Pressure in the venules is much lower than in the arteries.
   - One-way valves and skeletal muscles help venous blood flow.

Try This Activity

Mapping Veins

- Place a pressure cuff (sphygmomanometer) over a subject's upper arm and inflate it to 30 mm Hg.
- Locate one of the veins on the inside of the subject's arm and use your index finger to push blood in the vein toward the elbow.
- Describe the appearance of the vein. Draw a diagram to illustrate your description.
- Now push the blood in the vein toward the fingers.
- Describe the appearance of the vein. Draw a diagram to illustrate your description.
  - (a) How do you know that the blood vessel is a vein and not an artery?

CAUTION: Do not leave the sphygmomanometer inflated past 30 mm Hg and on longer than 5 min.

Section 7.4 Questions

Understanding Concepts

1. Explain what happens in the blood vessels when someone blushes. Why does this happen?

2. Are all the capillaries open all the time? Why or why not? What determines whether a capillary is open?

3. What are the advantages and disadvantages of capillaries being composed of a single cell layer?

4. Explain the importance of William Harvey's theory of blood circulation.

5. Why are aneurysms dangerous?

6. Why are fat deposits in arteries dangerous?

7. Fluid pressure is very low in the veins. Explain how blood gets back to the heart.

8. What causes varicose veins? What lifestyle changes could prevent the development of varicose veins?

9. Atherosclerosis is a disease caused by the buildup of fat and plaque on the inside of an artery.
   - (a) Explain how it occurs.
   - (b) What problems can be created by the buildup of fat and plaque?
   - (c) Suggest a treatment for the disorder.

7.5 The Mammalian Heart

septum: a wall of muscle that separates the right heart pump from the left

pulmonary circulatory system: the system of blood vessels that carries deoxygenated blood to the lungs and oxygenated blood back to the heart

systemic circulatory system: the system of blood vessels that carries oxygenated blood to the tissues of the body and deoxygenated blood back to the heart

atria: thin-walled chambers of the heart that receive blood from veins

ventricles: muscular, thick-walled chambers of the heart that deliver blood to the arteries

A fluid-filled membrane called the pericardium surrounds the heart. The fluid bathes the heart, preventing friction between its outer wall and the covering membrane.

The heart consists of two parallel pumps separated by the **septum**. The pumping action is synchronized; muscle contractions on the right side mirror those on the left. The pump on the right receives deoxygenated blood from the body tissues and pumps it to the lungs. The pump on the left receives oxygenated blood from the lungs and pumps it to the cells of the body. Vessels that carry blood to and from the lungs comprise the **pulmonary circulatory system**. Vessels that carry blood to and from the body comprise the **systemic circulatory system**. **Figure 1** illustrates the two systems.

The four-chambered human heart is composed of two thin-walled **atria** (singular: atrium) and two thick-walled **ventricles**. Blood from the systemic system enters the right atrium, and blood from the pulmonary system enters the left atrium. The stronger, more muscular ventricles pump the blood to distant tissues.

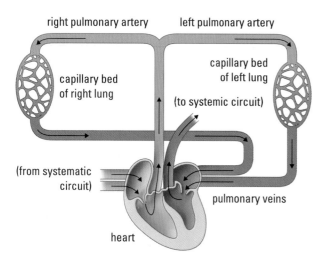

The Pulmonary Circuit

right pulmonary artery   left pulmonary artery

capillary bed of right lung

capillary bed of left lung

(to systemic circuit)

(from systematic circuit)

pulmonary veins

heart

The Systemic Circuit   capillary beds of head and upper extremities

(to pulmonary circuit)   aorta

(from pulmonary circuit)

heart

capillary beds of rest of the body

Figure 1
The pulmonary and systemic circuits of the circulatory system. The blood vessels carrying oxygenated blood are in red; the vessels carrying deoxygenated blood are in blue.

One-Way Blood Flow

Blood is carried to the heart by veins. The superior vena cava carries deoxygenated blood from the head and upper body to the right atrium. The inferior vena cava carries deoxygenated blood from all veins below the diaphragm to the same atrium. Oxygenated blood flowing from the lungs enters the left atrium by way of the pulmonary veins. Blood from both atria is eventually pumped into the larger ventricles (**Figure 2**).

Valves called **atrioventricular (AV) valves** separate the atria from the ventricles. In much the same way as the valves within veins ensure one-directional flow, the AV valves prevent blood from flowing from the ventricles back to the atria. The AV valves are supported by bands of connective tissue called chordae tendinae. A second set of valves, called **semilunar valves**, separate the ventricles from the

atrioventricular (AV) valves: heart valves that prevent the backflow of blood from the ventricles into the atria

semilunar valves: valves that prevent the backflow of blood from arteries into the ventricles

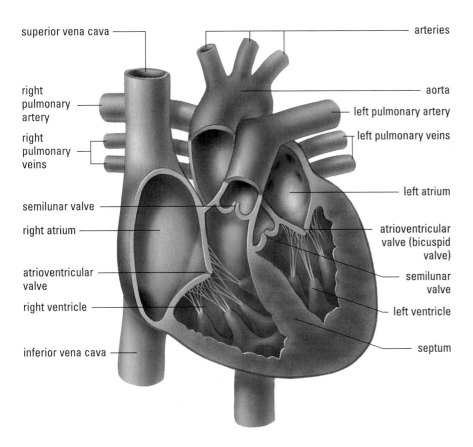

superior vena cava

arteries

right pulmonary artery

aorta

left pulmonary artery

right pulmonary veins

left pulmonary veins

left atrium

semilunar valve

right atrium

atrioventricular valve (bicuspid valve)

atrioventricular valve

semilunar valve

right ventricle

left ventricle

inferior vena cava

septum

Figure 2
Deoxygenated blood enters the right atrium. Blood from the right atrium is pumped to the right ventricle, which, in turn, pumps it to the lungs. Oxygenated blood moves from the lungs to the left atrium. Blood from the left atrium is pumped to the left ventricle, which, in turn, pumps it to the cells of the body.

arteries. These valves are half-moon shaped (hence, the name semilunar) and they prevent blood that has entered the arteries from flowing back into the ventricles.

Blood is carried away from the heart by arteries. The **aorta**, the largest artery in your body, carries oxygenated blood away from the heart. The **coronary arteries**, arteries that comprise an important branch of the aorta, supply the muscle cells of the heart with the oxygen and nutrients they require. A blocked artery illustrates the importance of proper coronary circulation. Chest pains, or angina, occur when too little oxygen reaches the heart. The heart, unlike other organs that slow down if they cannot receive enough nutrients, must continue beating no matter what demands are placed on it. It has been estimated that the heart may use 20% of the body's total blood oxygen during times of stress.

As in other arteries, fat deposits and plaque can collect inside coronary arteries. Drugs are often used to increase blood flow, but in severe situations blood flow must be rerouted. A coronary bypass operation involves removing a vein from another part of the patient's body and grafting it into position in the heart (**Figure 3**, page 258). However, in order to graft the vein, the heart must be temporarily stopped. During the operation, the patient's heart is cooled and a heart-lung machine is used to supply oxygen and push blood to the tissues of the body.

Frontiers of Technology: Cardiac Catheterization

At one time, doctors had to rely on external symptoms to detect coronary artery blockage. An inability to sustain physical activity, rapid breathing, and a general lack of energy are three of the symptoms of coronary distress. However, these same symptoms can also indicate a wide variety of other circulatory and respiratory diseases. One of the greatest challenges in medicine is matching symptoms with the correct disease. One way to determine whether or not a patient is suffering from coronary artery problems is to perform surgery. Unfortunately, the

aorta: the largest artery in the body. The aorta carries oxygenated blood to the tissues of the body.

coronary arteries: arteries that supply the cardiac muscle with oxygen and other nutrients

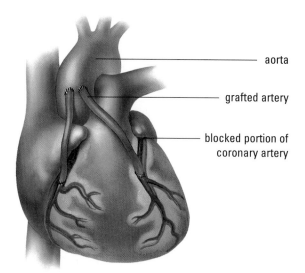

- aorta
- grafted artery
- blocked portion of coronary artery

Figure 3
Coronary bypass operation. Blood flow is rerouted around the blockage.

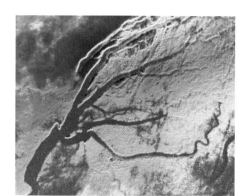

Figure 4
The image collected from an obstructed coronary artery

surgery is not without risks. Clearly, less invasive means for diagnosing the problem would be desirable.

One of the most useful techniques is that of cardiac catheterization. In this procedure, a small, thin plastic tube, called a catheter, is passed into a leg vein as the patient lies on an examination table. A dye visible on X-ray film is then injected into the catheter. The dye travels through the blood vessels and into the heart while its image is traced by means of a fluoroscope (a fluorescent screen). The image can also be projected on a television monitor (**Figure 4**).

A longer catheter can even reach the heart so that the dye can be injected into a coronary artery. As the blood and dye move through the artery, an X-ray movie can be made. An area of restricted blood flow pinpoints the region of blockage. The catheter helps direct the surgeon to the problem prior to surgery.

Blood samples can also be taken with the catheter to determine how much oxygen is in the blood in the different chambers. This tells the physician how well the blood is being oxygenated in the lungs. Low levels of oxygen in the left side of the heart can provide information about whether the circulatory and respiratory systems are working together efficiently. The catheter can even be used to monitor pressures in each of the heart chambers.

In a technique known as angioplasty, a catheter with a tiny balloon attached can be used to scrape and collect debris from a blockage.

Practice

Understanding Concepts

1. Differentiate between the systemic circulatory system and the pulmonary circulatory system.
2. What is the function of the AV valves and of the semilunar valves?
3. What is angina and what causes it?
4. What are coronary bypass operations and why are they performed?

SUMMARY   The Mammalian Heart

1. The pulmonary circulatory system is the system of blood vessels that carries blood to and from the lungs. The systemic circulatory system is the system of blood vessels that carries blood to and from the body.

2. The heart consists of two parallel pumps separated by the septum.
   - Blood enters the heart through the atria.
   - Ventricles pump the blood to the body tissues.
   - Atrioventricular valves prevent blood from flowing from ventricles back into the atria.
   - Semilunar valves prevent blood from flowing from arteries into ventricles.
   - Coronary arteries supply the heart with oxygen and nutrients.

3. A coronary bypass operation can be performed if a coronary artery is severely blocked.

4. In cardiac catheterization, dye is injected into a vein, and the pathway of blood can be traced.

7.6 Setting the Heart's Tempo

Heart, or cardiac, muscle differs from other types of muscle. Like skeletal muscle, cardiac muscle appears striated (grooved or ridged) when viewed under a microscope. But unlike skeletal muscle, cardiac muscle displays a branching pattern (**Figure 1**). The greatest difference stems from the ability of this muscle to contract without being stimulated by external nerves. Muscle with this ability, called **myogenic muscle**, explains why the heart will continue to beat, at least for a short time, when removed from the body.

The remarkable capacity of the heart to beat can be illustrated by a simple experiment. When a frog's heart is removed and sliced into small pieces while in a salt solution that simulates the minerals found within the body, each of the pieces continues to beat, although not at the same speed. Muscle tissue from the ventricles follows a slower rhythm than muscle tissue from the atria. Muscle tissue closest to where the venae cavae enter the heart has the fastest tempo. The unique nature of the heart becomes evident when two separated pieces are brought together. The united fragments assume a single beat. The slower muscle tissue assumes the tempo set by the muscle tissue that beats more rapidly.

The heart's tempo or beat rate is set by the **sinoatrial (SA) node** (**Figure 2**). This bundle of specialized nerves and muscles is located where the venae cavae enter the right atrium. The SA node acts as a pacemaker, setting a rhythm of about 70 beats per min for the heart. Nerve impulses are carried from the pacemaker to other muscle cells by modified muscle tissue. Originating in the atria, the contractions travel to a second node, the **atrioventricular (AV) node**. The AV node

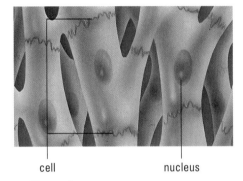

cell   nucleus

Figure 1
The heart is composed of cardiac muscle. The branching pattern is unique to cardiac muscle.

myogenic muscle: a muscle that contracts without external nerve stimulation

sinoatrial (SA) node: a small mass of tissue in the right atrium that originates the impulses stimulating the heartbeat

atrioventricular (AV) node: a small mass of tissue in the right atrioventricular region through which impulses from the sinoatrial node are passed to the ventricles

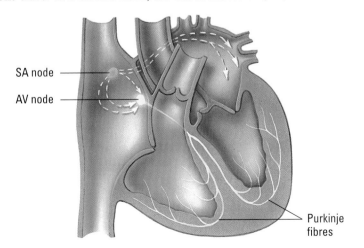

SA node

AV node

Purkinje fibres

Figure 2
The SA node initiates heart contractions. Modified muscle tissue passes a nerve impulse from the pacemaker down the dividing septum toward the ventricles.

serves as a conductor, passing nerve impulses via two large nerve fibres, called **Purkinje fibres**, through the septum toward the ventricles. The Purkinje fibres run along the septum that separates the right and left ventricles, carrying impulses from the AV node to the bottom tip of the heart. From here, these branching fibres carry impulses up along the outer walls of the ventricles back toward the atria. A wave of cardiac contraction follows the nerve pathway. Both right and left atria contract prior to the contraction of the right and left ventricles.

One of the greatest challenges for surgeons performing open-heart surgery is to make incisions at the appropriate location. A scalpel placed in the wrong spot could cut fibres that conduct nerve impulses.

Heart rate is influenced by autonomic nerves. Two regulatory nervous systems—the **sympathetic** and **parasympathetic nervous systems**—conduct impulses from the brain to the SA node. Stimulated during times of stress, the sympathetic nerves increase heart rate. This increases blood flow to tissues, enabling the body to meet increased energy demands. Conditions in which the heart rate exceeds 100 beats per min are referred to as tachycardia. Tachycardia can result from exercise or from the consumption of such drugs as caffeine or nicotine. During times of relaxation, the parasympathetic nerves are stimulated to slow the heart rate.

Diagnosing Heart Conditions

Doctors can use electrocardiographs, which map electrical fields within the heart, to make tracings to diagnose certain heart problems. Electrodes placed on the body surface are connected to a recording device. The electrical impulses are displayed on a graph called an electrocardiogram (ECG) (**Figure 3**). Changes in electrical current reveal normal or abnormal events of the cardiac cycle. The first wave, referred to as the P wave, monitors atrial contraction. The larger spike, referred to as the QRS wave, records ventricular contraction. A final T wave signals that the ventricles have recovered. A patch of dead heart tissue, for example, will not conduct impulses and produces abnormal line tracings. By comparing the tracings, doctors are able to locate the area of the heart that is damaged (**Figure 4**).

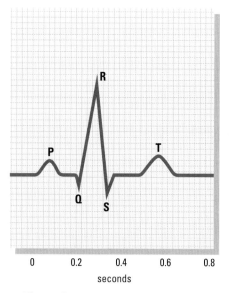

Figure 3
Electrocardiograph (ECG) tracing showing the duration of a single beat. The flat lines show the resting period between beats.

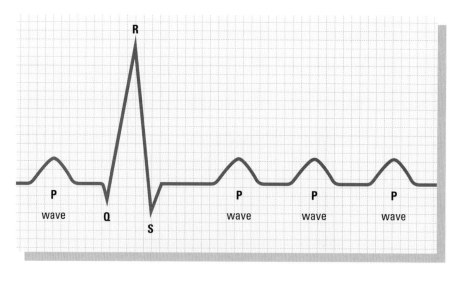

Figure 4
An abnormal electrocardiograph tracing. Can you determine what has gone wrong?

The electrocardiograph is especially useful for monitoring the body's response to exercise. Stress tests are performed by monitoring a subject who is riding a stationary bike or running on a treadmill. Some heart malfunctions remain hidden during rest, but can be detected during vigorous exercise.

Practice

Understanding Concepts

1. What is myogenic muscle?
2. What is the difference between the sympathetic and the parasympathetic nervous systems?

7.7 Heart Sounds

The familiar *lubb-dubb* heart sounds are caused by the closing of the heart valves. When the atria are relaxed, they begin to fill with blood. The relaxation is called **diastole**. As the atria fill, the muscular walls contract, increasing fluid pressure and forcing the AV valves open. Blood flows from the atria into the ventricles. In turn, as the ventricles fill with fluid, they contract. The increasing pressure forces the AV valves shut, producing a *lubb* sound and pushing blood through the semilunar valves and into the arteries. The contraction is called **systole**. The ventricles then begin to relax, and the volume in the chambers increases. With increased volume, pressure in the ventricles begins to decrease and blood tends to be drawn from the arteries toward the area of lower pressure. Blood is prevented from reentering the ventricles by the semilunar valves. The closing of the semilunar valves creates the *dubb* sound (**Figure 1**).

Occasionally, the valves do not close completely. This condition, referred to as a heart murmur, occurs when blood leaks past the closed heart valve because of an improper seal. The AV valves, especially the left AV valve (the bicuspid or mitral valve), must withstand increased pressure and are especially susceptible to defects. The rush of blood from the ventricle back into the atrium produces a gurgling sound that can be detected with the use of a stethoscope. Blood flowing back toward the atrium is inefficient because it is not directed to the systemic or pulmonary systems. The hearts of individuals who experience murmurs compensate for decreased oxygen delivery by beating faster.

diastole: relaxation (dilation) of the heart, during which the cavities of the heart fill with blood

systole: contraction of the heart, during which blood is pushed out of the heart

atria contracted,
ventricles relaxed

atria relaxed,
ventricles contracted

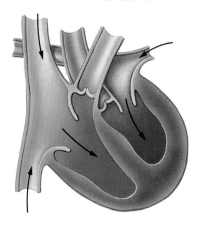 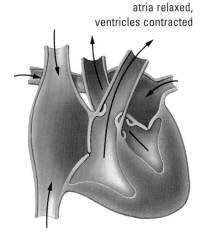

Figure 1
Right and left atria contract in unison, pushing blood into the right and left ventricles. Ventricular contractions close the AV valves and open the semilunar valves. The relaxation of the ventricles lowers pressure and draws blood back to the chamber. The closing of the semilunar valves prevents blood from reentering the ventricles.

Make certain to disinfect the earpieces of the stethoscope with rubbing alcohol before and after use.

A second mechanism helps compensate for a decreased blood flow in people with heart murmurs. Like an elastic band, the more cardiac muscle is stretched, the stronger is the force of contraction. When blood flows from the ventricle back into the atrium, blood volume in the atrium increases. The atrium accepts the normal volumes and the additional blood from the ventricle. The extra volume stretches the atrium and drives blood to the ventricle with greater force. The increased blood volume in the ventricle causes it to contract with greater force, driving more blood to the tissues.

Listening to Heart Sounds

- Place a stethoscope on your own chest and listen for a heart sound (**Figure 2**). Locate the area where the heart sounds are loudest and clearest.

Figure 2

- After 1 min of moderate exercise (e.g., walking on the spot), listen for your heart sounds again.

(a) Draw a diagram of a chest showing where you located the clearest sound.

(b) Did the sound of your heartbeat change after exercise? Describe what differences you heard.

Research in Canada: Dr. Tofy Mussivand

A team of Ottawa researchers, under the direction of Dr. Tofy Mussivand, is closing in on the production of an artificial heart. The electrohydraulic ventricular device (EVAD) is capable of taking over partial or complete functioning of the ventricles. Unlike other attempts at producing an artificial heart, EVAD is powered and controlled externally by a small battery-powered coil taped to the chest. This eliminates the need for a permanent opening in the skin and allows freedom of movement to patients, as they would no longer need to be attached to a machine.

While attending university, Dr. Mussivand learned that millions of people throughout the world were dying of heart disease, yet no one really had a solution. The best available treatment, heart transplants, did not provide the answer. The need for hearts far outstripped the supply of donor hearts and the cost of the operation was prohibitive. The more Dr. Mussivand looked at the options, the more convinced he became that the answer would be an artificial heart.

The challenge is largely one of engineering. The question is whether humans can produce a machine that moves fluid as well as a biological system that has evolved over millions of years. A device that assists or takes over the function of a faulty ventricle could save thousands of lives per year.

Practice

Understanding Concepts

1. Explain the terms diastole and systole.
2. What causes the characteristic heart sounds?
3. What causes heart murmurs?

SUMMARY Heartbeat

1. The heart's tempo is set by the sinoatrial (SA) node. Contractions in the SA node travel to the atrioventricular (AV) node and then travel along the Purkinje fibres to the rest of the heart.

2. Diastole is heart relaxation. Systole is heart contraction.

3. The *lubb-dubb* sound is caused by the AV valves and the semilunar valves closing in turn as blood is pushed from the atria through the ventricles and out of the heart. If the valves do not close completely, the heart compensates by beating faster and pumping blood with more force.

4. The electrohydraulic ventricular device (EVAD) is an artificial heart that can take over complete or partial functioning of the ventricles. Because it is powered externally, it frees the patient from being attached to a machine.

Sections 7.5–7.7 Questions

Understanding Concepts

1. What are the atria and the ventricles? How do they differ in structure and function?
2. In what sense is blood flow in the body one way?
3. Draw and label the major blood vessels and chambers of the mammalian heart. Trace the flow of deoxygenated and oxygenated blood through the heart.
4. Explain the importance of cardiac catheterization and describe the technique.
5. Explain differences in the strength of a pulse between the carotid artery (neck area) and the brachial artery (wrist area).
6. Explain changes in the pulse rate after exercise.
7. Describe the pathway of nerve impulses through the heart. Refer to the terms sinoatrial node, atrioventricular node, and Purkinje fibres.
8. How does the heart compensate for the improper function of the AV valves?
9. What is an electrocardiogram? Why is it useful? Explain what the different parts of an electrocardiogram indicate.

(continued)

Circulation and Blood **263**

10. Medical technologies are often patented, bringing in great profits to the owners. Using print or electronic media, find out about some of these technologies. Do you think that technology such as an artificial heart should be owned? What are the social and moral implications of such ownership?
Follow the links for Nelson Biology 11, 7.7.

GO TO www.science.nelson.com

11. Explain what effect drugs such as caffeine might have on the circulatory system and on the homeostatic range. How are these changes similar to or different from the changes brought on by exercise?

12. Predict some advantages and disadvantages of artificial hearts over donor hearts.

Reflecting

13. When researching the impact of scientific knowledge or technology on society, what kinds of sources do you consult? Do you think that medical or scientific sources will give an impartial point of view?

sphygmomanometer: a device used to measure blood pressure. Blood pressure is traditionally measured in millimetres of mercury (mm Hg).

cardiac output: the amount of blood pumped from the heart each minute. Cardiac output is determined by multiplying the heart rate by the stroke volume (the quantity of blood pumped with each heartbeat).

Figure 1
A sphygmomanometer

7.8 Blood Pressure

Blood surges through the arteries with every beat of the heart. Elastic connective tissue and smooth muscle in the walls of the arteries stretch to accommodate the increase in fluid pressure. The arterial walls recoil much like an elastic band as the heart begins the relaxation phase characterized by lower pressure. Even the recoil forces help push blood through arterioles toward the tissues.

Blood pressure is the force of the blood on the walls of the arteries. It can be measured indirectly with an instrument called a **sphygmomanometer** (Figure 1). A cuff with an air bladder is wrapped around the arm. A small pump is used to inflate the air bladder, thereby closing off blood flow through the brachial artery, one of the major arteries of the arm. A stethoscope is placed below the cuff and air is slowly released from the bladder until a low-pitched sound can be detected. The sound is caused by blood entering the previously closed artery.

Each time the heart contracts, the sound is heard. A gauge on the sphygmomanometer measures the pressure exerted by the blood during ventricular contraction. This pressure is called systolic blood pressure. Normal systolic blood pressure for young adults is about 120 mm Hg. (Blood pressure is measured in the non-SI units of millimetres of mercury, or mm Hg.) The cuff is then deflated even more, until the sound disappears. At this point, blood flows into the artery during ventricular relaxation or filling. This pressure is called diastolic blood pressure. Normal diastolic blood pressure for young adults is about 80 mm Hg. This normal blood pressure would be reported as 120/80 (120 over 80): a systolic pressure of 120 mm Hg and a diastolic pressure of 80 mm Hg. Reduced filling, such as that caused by an internal hemorrhage, will cause diastolic blood pressure to fall. **Figure 2** shows that fluid pressure decreases with distance from the ventricles; thus blood pressure readings are not the same in all arteries.

Blood pressure depends on two factors. The first is **cardiac output**. Any increase in cardiac output will increase blood pressure. Another factor is arteriolar resistance. The diameter of the arteriole is regulated by coiling, smooth muscles. Constriction of the smooth muscles surrounding the arteriole closes the

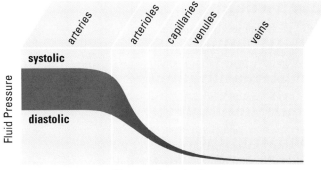

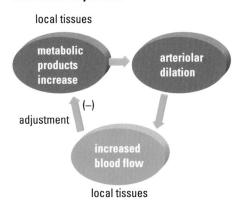

Figure 2
Fluid pressure decreases the further blood moves from the heart.

Homeostatic adjustment

Figure 3
Dilation of the arterioles increases blood flow to local tissues.

opening and reduces blood flow through the arteriole. With this reduced blood flow, more blood is left in the artery. The increased blood volume in the artery produces higher blood pressure. Conversely, factors that cause arteriolar dilation increase blood flow from the arterioles, thereby reducing blood pressure.

The smooth muscles in the walls of the arterioles respond to neural and hormonal controls that regulate blood pressure. The diameter of the arterioles also adjusts to metabolic products, such as those produced during the breakdown of sugar. When there is sufficient oxygen to break down sugar, carbon dioxide and water are produced. When there is insufficient oxygen, lactic acid is produced. Accumulation of carbon dioxide and lactic acid causes the relaxation of smooth muscles in the walls of the arterioles, dilating them. The dilated arterioles increase blood flow to local tissues, delivering more oxygen. Arteriolar dilation in response to increased metabolic products is a good example of homeostasis (**Figure 3**). Activities such as exercise cause an increase in metabolic products. Because these products accumulate in the most active tissues, the increased blood flow helps provide greater nutrient supply and carries the potentially toxic materials away. Tissues that are less active produce fewer metabolic products. These arterioles remain closed until the products accumulate.

Regulation of Blood Pressure

Regulation of blood pressure is essential since low blood pressure reduces your capacity to transport blood. The problem is particularly acute for tissues in the head where blood pressure works against the force of gravity. High blood pressure creates equally serious problems. High fluid pressure can weaken an artery and eventually result in its rupture.

Special blood pressure receptors are located in the walls of the aorta and the carotid arteries, which are major arteries found on either side of the neck. These receptors are sensitive to high pressures. When blood pressure exceeds acceptable levels, the receptors respond to the increased pressure on the wall of the artery. A nerve message travels to the medulla oblongata, the blood pressure regulator located at the stem of the brain. Sympathetic (stress) nerve impulses are decreased and parasympathetic (slow-down) nerve impulses are increased. By decreasing sympathetic nerve stimulation, arterioles dilate, increasing outflow of blood from the artery. Stimulation of the parasympathetic nerve causes heart rate and stroke volume to decrease. The decreased cardiac output slows the movement of blood into the arteries, lowering blood pressure.

Low blood pressure is adjusted by the sympathetic nerve. Without nerve information from the pressure receptors of the carotid artery and aorta, the sympathetic nerve will continue to be stimulated causing cardiac output to increase and arterioles to constrict. The increased flow of blood into the artery accompanied by decreased outflow raises blood pressure to acceptable levels.

DID YOU KNOW ?

Many people believe that a drink of alcohol will warm them up on a cold day. Alcohol causes dilation of the arterioles leading to the skin capillaries, causing the sensation of warmth. However, the sensation is misleading. The dilation of these arterioles actually speeds cooling.

Drugs and the Heart

Many traditional homeopathic medical treatments are now being supported by science. Foxglove, a popular garden plant in Britain, has long been used in tea as a tonic. Scientists have found that the active ingredient in the plant, digitalis, initiates strong, regular heart contractions and is now used to treat congestive heart failure. Nitroglycerine, an explosive, has also been used to prevent heart attacks.

Medical therapy for heart failure has improved greatly with the development of beta-blockers. These drugs are especially important for people with irregular heartbeats or who display the effects of high blood pressure.

Receptor sites located on the surface of cells receive molecules, such as hormones, that affect the way cells behave. Epinephrine, a stress hormone, attaches to receptors on the heart and blood vessels, producing increased heart rate and a narrowing of the blood vessels. Both effects lead to an increase in blood pressure. Beta-blockers tie up receptor sites designed to accept epinephrine.

There are two types of beta-receptors on cell surfaces, beta 1 and beta 2. Beta 1 receptors are found on the surface of the cardiac muscle. These affect the speed and strength of heart contractions and directly influence blood pressure. Beta 2 receptors are found primarily in the blood vessels and the bronchioles, small tubelike structures leading into the lungs. By blocking the harmful effects of the stress hormone, the heart rate slows and blood vessels relax, leading to a reduction in blood pressure.

Like most drugs, beta-blockers can have side effects. Since they reduce the effects of stress hormones by slowing the heart, patients may feel tired, be less able to exercise vigorously, and have lower blood pressure which may result in lightheadedness or dizziness.

Practice

Understanding Concepts

1. Define cardiac output, stroke volume, and heart rate.
2. Differentiate between systolic and diastolic blood pressure.
3. How does arteriolar resistance affect blood pressure?
4. How do blood pressure regulators detect high blood pressure?
5. Draw a flow chart or diagram to show how a beta-blocker works.

Making Connections

6. All drugs that block beta 2 receptors also block beta 1 receptors. Some drugs work selectively by blocking beta 1 receptors without affecting the beta 2 receptors. Indicate which drug, a beta 2 nonselective or beta 1 selective drug, would produce fewer side effects. Give your reasons.

7.9 Variation in Cardiovascular Output

Cardiovascular disease includes hypertension (sustained high blood pressure), arteriosclerosis, atherosclerosis, heart attack (destruction of the heart muscle), and stroke (interrupted blood flow to the brain). The following seven factors can affect the health of the heart and result in cardiovascular disorders:

1. high levels of cholesterol in the blood
2. smoking

3. diabetes mellitus

4. high blood pressure

5. lack of regular exercise

6. rapid weight gain or loss

7. genetic factors (congenital problems)

Various factors can be linked with cardiovascular disorders. It is important to remember that usually more than one factor contributes to the development of a cardiovascular problem.

Hypertension: The Silent Killer

Hypertension is caused by increased resistance to blood flow, which results in a sustained increase in blood pressure. If blood pressure remains high, blood vessels are often weakened and eventually rupture. The body may attempt to compensate for weakened vessels by increasing the support provided by connective tissues. Unfortunately, when the body increases the amount of connective tissue, arteries often become hard and less elastic. During systole, as blood pulses through these reinforced vessels, blood pressure increases more than usual, which in turn causes further stress leading to greater weakening.

Although hypertension is sometimes hereditary, diet is often a primary factor in the development of the disease. For example, in susceptible individuals, using too much salt can cause blood pressure to rise and the heart to work harder. Hypertension is often described as a silent killer because symptoms are usually not noticeable until the situation becomes very serious. A heart attack or stroke can be the first indication that something is wrong.

Arrhythmia

Physicians often refer to an irregular heartbeat as arrhythmia (**Figure 1**). When a coronary artery is blocked, it delivers less blood and causes the heart to beat in an irregular pattern. The buildup of toxic products associated with poor oxygen delivery can initiate contractions of the heart muscle. Rather than synchronized heartbeats, where muscle cells from within the ventricles pick up electrical signals from surrounding muscle fibres, each cell within the ventricle responds to the toxins surrounding it and begins to contract wildly. This is referred to as ventricular fibrillation.

As the heart fibrillates, blood is not pushed from one area to another in a coordinated fashion. The twitching heart pushes blood back and forth, reducing its ability to deliver needed oxygen. The heart responds by beating faster, but without a controlled pattern of muscle contraction, blood delivery to the tissues will not improve.

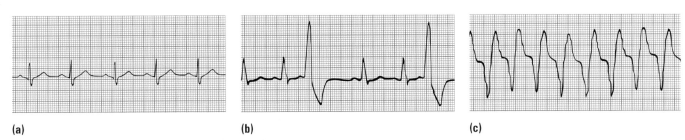

(a) (b) (c)

Figure 1
Not all arrhythmias are abnormal. The heart rate shown in **(a)** is a slowed heart rate associated with athletes. Tachycardia, in **(b)**, is a rapid heart rate which can be caused by exercise or by an inefficient heart. The ECG in **(c)** shows an uncontrolled heart rate, or ventricular fibrillation.

Adjustment of the Circulatory System to Exercise

Your body's response to exercise is an excellent example of a homeostatic mechanism. Exercise places considerable demands on the circulatory system, but this system does not act alone in monitoring the needs of tissues or in ensuring that adequate levels of oxygen and other nutrients are delivered to the active cells. The nervous and hormonal systems also play important roles in adjustment.

During times of stress, the sympathetic nerve stimulates the adrenal glands. The hormone epinephrine (adrenaline) is released from the adrenal gland and travels in the blood to other organs of the body. Epinephrine stimulates the release of red blood cells from the spleen. Although the significance of the response is not yet understood, it is clear that increased numbers of red blood cells aid oxygen delivery. Epinephrine and direct stimulation from the sympathetic nerve increase heart rate and breathing rate. The increased heart rate provides for faster oxygen transport, while the increased breathing rate ensures that the blood contains higher levels of oxygen. Both systems work together to improve oxygen delivery to active tissues. A secondary but important function is the increased efficiency of waste removal from the active tissues.

Blood cannot flow to all capillaries of the body simultaneously. The effect of dilating all arterioles at one time would be disastrous—blood pressure would plunge. Only the most active tissues receive priority in times of stress. As a result, epinephrine causes vasodilation of the arterioles leading to the heart, brain, and muscles, preparing the organism for the fight-or-flight reaction. At the same time, the blood vessels leading to the kidney, stomach, and intestines constrict, depriving these areas of the body of nutrients until the stress situation has been overcome.

Practice

Understanding Concepts

1. Name some factors that can result in cardiovascular disorders.
2. What is hypertension?
3. What is arrhythmia?
4. How does exercise affect your heart rate? Provide an explanation for any change.
5. How does exercise affect your blood pressure? Provide an explanation for any change.
6. How is it possible that two different people have different pulse rates after doing exactly the same exercise?

SUMMARY   Cardiac Output and Blood Pressure

1. Blood pressure is the force of blood on the walls of the arteries. It is measured as systolic and diastolic blood pressure in millimetres of mercury (mm Hg).

2. Blood pressure will be higher in vessels closer to the heart.

3. Blood pressure depends on the following:
   - cardiac output: Increased cardiac output increases blood pressure.
   - arteriolar resistance: If arteries are constricted, blood flow is slower and blood pressure is higher.

4. Some cardiovascular problems are related to blood pressure.
   - Hypertension is caused by increased resistance to blood flow, resulting in sustained increase in blood pressure.

- Arrhythmia is an irregular heartbeat, sometimes caused by a blocked coronary artery.
- Tachycardia is a rapid heart rate.

5. In times of stress, arterioles to more essential body areas dilate and those to less essential areas constrict.

Investigation 7.9.1

Effects of Posture on Blood Pressure and Pulse Rate

INQUIRY SKILLS

- ◉ Questioning
- ◉ Hypothesizing
- ◉ Predicting
- ◉ Planning
- ◉ Conducting
- ◉ Recording
- ◉ Analyzing
- ○ Evaluating
- ○ Communicating

Blood pressure is different in different areas of the body. It is affected by factors such as exercise, drugs, and even posture.

Question

What effect will posture have on your blood pressure and pulse rate?

Hypothesis/Prediction

Using what you know about blood pressure, predict what will happen. Formulate a hypothesis to explain your prediction.

Materials

sphygmomanometer
watch with second hand

Procedure

1. Ask your partner to sit quietly for 1 min.

2. Expose the arm of your partner and place the cuff of the electronic blood pressure gauge just above the elbow.

3. Close the valve on the rubber bulb and inflate it by squeezing the rubber ball until a pressure of 180 mm Hg registers.

4. Release the pressure by opening the valve on the blood pressure gauge and watch the digital readout.

5. Completely deflate the cuff and take your partner's pulse if the electronic blood pressure gauge does not provide it. Place your index and middle fingers on the arm near the wrist. Count the number of pulses in 1 min.

6. Record the systolic and diastolic blood pressures.

7. Record the pulse rate.

8. Repeat steps 2 to 7 while your partner is in a standing position and then in a lying position.

9. Record your results in a table similar to **Table 1**, on page 270.

Do not leave the pressure on for longer than 1 min. If you are unsuccessful, release the pressure and try again.

Analysis

(a) Compare the results with your predictions.

(b) Which varied more with the change in posture: systolic blood pressure or diastolic blood pressure? Explain.

(c) What factors other than posture might have contributed to the change in blood pressure?

Table 1

Position	Systolic blood pressure (mm Hg)	Diastolic blood pressure (mm Hg)	Pulse rate (beats/min)
standing	?	?	?
sitting	?	?	?
lying	?	?	?

 Do not perform this activity if you are not allowed to participate in physical education classes.

Investigation 7.9.2

Effects of Exercise on Blood Pressure and Pulse Rate

In this investigation, you will design ways to test the effects of exercise on blood pressure and pulse rate.

Question

How does exercise affect blood pressure and pulse rate?

Hypothesis/Prediction

With a partner, discuss how exercise might affect blood pressure and pulse rate. Make a prediction and formulate a hypothesis to explain your prediction.

Design

Design a controlled experiment to test your hypothesis and prediction. Include the following in your design:
- descriptions of the independent, dependent, and controlled variables
- a step-by-step description of the procedure
- a list of safety precautions
- a table to record observations

Materials

Make a list of the materials and apparatus needed to carry out the procedure.

Procedure

1. Submit the procedure, safety precautions, observation table, and list of materials and apparatus to your teacher for approval.
2. Carry out the procedure.
3. Write a report.

Analysis

(a) State how exercise affected blood pressure and pulse rate.

Evaluation

(b) Was your prediction correct? Was your hypothesis supported?
(c) Describe any problems or difficulties in carrying out the procedure.
(d) Include in your report ways to improve your current design.
(e) If you were to repeat this experiment, what new factors would you investigate? Write a brief description of the new procedure.

Explore an Issue

Debate: Access to Health Care

Heart disease is the number one killer of North Americans. Although congenital heart defects account for some of the problems, the incidence of heart disease can often be traced to lifestyle. The relationship between heart problems and smoking, excessive alcohol consumption, stress, high-cholesterol diets, and high blood pressure has long been established.

Statement

People who refuse to alter lifestyle choices that are dangerous to their health should not be permitted equal access to health care.

- In your group, research the issue.
- Search for information in newspapers, periodicals, CD-ROMs, and on the Internet.
  Follow the links for Nelson Biology 11, 7.9.

GO TO www.science.nelson.com

- Write a list of points and counterpoints that your group considered. You might look at the following: the roles of genetics and lifestyle in contributing to heart disease, the proportion of the population requiring heart transplants, the number that receive one, the criteria for being a qualified recipient, incidence of heart disease, user-pay concept, and cost of treatments.
- Decide whether your group agrees or disagrees with the statement.
- Prepare to defend your group's position in a class discussion.

Sections 7.8–7.9 Questions

Understanding Concepts

1. How does stroke volume affect cardiac output?

2. How do metabolic products affect blood flow through arterioles? What causes the accumulation of metabolic products and where is accumulation most likely to occur?

3. Outline homeostatic adjustment to high blood pressure. Refer to the sympathetic and parasympathetic nerves.

4. Would you expect blood pressure readings in all the major arteries to be the same? Explain your answer.

5. Why is systolic pressure lower when you are lying down than when you are standing up?

6. Why might diastolic blood pressure decrease as heart rate increases?

Making Connections

7. Arteriosclerosis is a group of disorders that cause high blood pressure. Name at least three other cardiovascular disorders that may be partially caused by lifestyle. How could lifestyle choices (e.g., related to nutrition or exercise) be changed to lessen a person's likelihood of getting the disorder?

7.10 Capillary Fluid Exchange

It is estimated that nearly every tissue of the body is within 0.1 mm of a capillary. Capillaries provide cells with oxygen, glucose, and amino acids and are associated with fluid exchange between the blood and surrounding **extracellular fluid (ECF)**. Most fluids simply diffuse through capillaries whose cell membranes are also permeable to oxygen and carbon dioxide. Water and certain ions are thought to pass through spaces between the cells of the capillary while larger molecules and a very small number of proteins are believed to be exchanged by endocytosis or exocytosis. This section will focus on the movement of water molecules.

Two forces regulate the movement of water between the blood and ECF: fluid pressure and osmotic pressure. The force that blood exerts on the wall of a capillary is about 35 mm Hg at the arteriole end of the capillary and approximately 15 mm Hg at the venous end. The reservoir of blood in the arteries creates pressure on the inner wall of the capillary. Much lower pressure is found in the ECF. Although fluids bathe cells, no force drives the extracellular fluids. Water moves from an area of higher pressure, the capillary, into an area of lower pressure, the ECF (**Figure 1**). The outward flow of water and small mineral ions is known as **filtration**. Because capillaries are selectively permeable, large materials such as proteins, red blood cells, and white blood cells remain in the capillary.

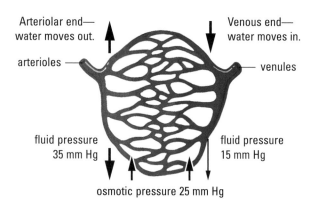

Arteriolar end	
osmotic pressure	25 mm Hg
fluid pressure	35 mm Hg
absorption	−10 mm Hg

Venous end	
osmotic pressure	25 mm Hg
fluid pressure	15 mm Hg
absorption	+10 mm Hg

Figure 1
Fluid movement into and out of the capillaries

The movement of fluids out of the capillary must be balanced with a force that moves fluid into the capillary. The fact that large proteins are found in the blood but not in the ECF may provide a hint as to the nature of the second force. Osmotic pressure draws water back into the capillary. The large protein molecules of the blood and dissolved minerals are primarily responsible for the movement of fluid into capillaries. The movement of fluid into capillaries is called absorption. Osmotic pressure in the capillaries is usually about 25 mm Hg, but it is important to note that the concentration of solutes can change with fluid intake or excess fluid loss caused by perspiration, vomiting, or diarrhea.

Application of the capillary exchange model provides a foundation for understanding homeostatic adjustments (**Figure 2**). The balance between osmotic pres-

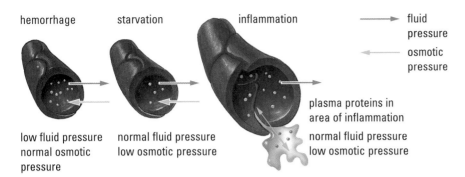

sure and fluid pressure is upset during hemorrhage (excessive discharge of blood from the blood vessels). The decrease in blood volume resulting from the hemorrhage lowers blood pressure. Although proteins are lost with the hemorrhage, so are fluids. Fewer proteins are present, but the concentration has not been changed. The force that drives fluid from the capillaries diminishes, but the osmotic pressure, which draws water into the capillaries, is not altered. The force drawing water from the tissues and ECF is greater than the force pushing water from the capillary. The net movement of water into the capillaries provides a homeostatic adjustment. As water moves into the capillaries, fluid volumes are restored.

Individuals who are suffering from starvation often display tissue swelling, or edema. Plasma proteins are often mobilized as one of the last sources of energy. The decrease in concentration of plasma proteins has a dramatic effect on osmotic pressure, which draws fluids from the tissues and ECF into the capillaries. The decreased number of proteins lowers osmotic pressure, thereby decreasing absorption. More water enters the tissue spaces than is pulled back into the capillaries, causing swelling.

Inflammation and Allergic Reactions

Why do tissues swell during inflammation or allergic reactions? When you eat a food to which you are allergic, cells react as though they are endangered and release a chemical messenger, which stimulates the release of another chemical stimulator, histamine. Histamine increases the permeability of the capillary membranes, enlarging the capillary and causing the area to redden. Proteins and white blood cells leave the capillary in search of the foreign invader while also altering the osmotic pressure. The proteins in the ECF create another osmotic force that opposes the osmotic force in the capillaries. Less water is absorbed into the capillary, and tissues swell.

Anaphylactic Reactions

Severe allergies are anaphylactic reactions (hypersensitivity) that can be life threatening. An anaphylactic reaction involves both the respiratory and circulatory systems. Often it is accompanied by the swelling of different body parts, welts, and itching. Anaphylactic reactions can be brought on by drugs, vaccines, bee stings, and some foods (peanuts, shellfish, eggs, berries, milk) in someone who is sensitive to these substances.

Anaphylactic shock can occur very quickly. It is characterized by weakness, sweating, and breathing difficulties. Nausea, diarrhea, and a drop in blood pressure may also occur. People who experience anaphylactic reactions often carry a kit containing adrenaline (epinephrine) or antihistamines. Physicians advise that anyone with a severe allergy should wear a medical alert bracelet or necklace and read all food and drug labels carefully.

Understanding Concepts

1. Is fluid pressure greater in the arterioles or in the veins? Give reasons.
2. Is fluid pressure inside the capillary greater or less than the pressure in the ECF? How does this affect the movement of water?
3. What process allows water to flow out of the capillary but keeps cells—such as proteins, red blood cells, and white blood cells—inside the capillary?
4. What is histamine? What is its function?

SUMMARY **Capillary Fluid Exchange**

1. Capillaries are associated with fluid exchange between blood and the extracellular fluid (ECF).
2. The movement of water between blood and the ECF is regulated by fluid pressure and by osmotic pressure.
   - Water moves from an area of high fluid pressure, the capillary, to an area of low fluid pressure, the ECF.
   - Proteins and dissolved minerals in the blood cause fluid from the ECF to move into the blood by osmosis.

7.11 The Lymphatic System

Normally, a small amount of protein leaks from capillaries to tissue spaces. Despite the fact that the leak is very slow, the accumulation of proteins in the ECF would create a major problem: osmotic pressure would decrease and tissues would swell.

The proteins are drained from the ECF and returned to the circulatory system by way of another network of vessels: the lymphatic system. **Lymph**, a fluid similar to blood plasma, is transported in open-ended lymph vessels that are similar to veins (**Figure 1**). The low-pressure return system operates by slow muscle contractions against the vessels, which are supplied with flaplike valves that prevent the backflow of fluids. Eventually, lymph is returned to the venous system.

Enlargements called **lymph nodes** are located at intervals along the lymph vessel. These house white blood cells that, by the process of phagocytosis, filter out any bacteria that might be present. The lymph nodes also filter damaged cells and debris from the lymph and they supply **lymphocytes** to the body. The lymph nodes sometimes swell when you have a sore throat.

lymph: the fluid found outside capillaries. Most often, the lymph contains some small proteins that have leaked through capillary walls.

lymph nodes: round masses of tissue that supply lymphocytes to the bloodstream and remove bacteria and foreign particles from the lymph

lymphocytes: white blood cells that produce antibodies

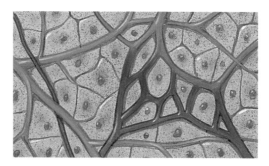

Figure 1
Lymph vessels (shown in green) are open-ended canals.

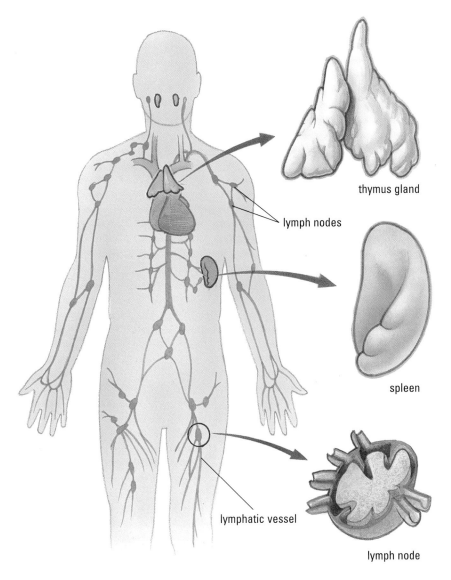

thymus gland

lymph nodes

spleen

lymphatic vessel

lymph node

Figure 2
The lymphatic system. Debris is filtered out from the lymph, and the lymph is returned to the circulatory system.

Lymphoid Organs

Red bone marrow is where all types of blood cells are produced. Stem cells, which are contained in the marrow, divide at incredible rates and differentiate into different types of white blood cells to meet the needs of the body. These specialized blood cells enter the circulatory system from a variety of sinuses. In children, red bone marrow is found in most bones; by adulthood, however, the cranium, sternum (breastbone), ribs, spinal column, and the long bones of the arms and legs have become the primary locations for blood cell production.

The **spleen** is one of the body's largest lymphoid organs (**Figure 2**). Located in the upper left side of the abdominal cavity, just below the diaphragm, the spleen is richly supplied with blood sinuses. The sinuses allow the spleen to hold approximately 150 mL of blood, making it an excellent blood reservoir. The spleen releases red blood cells in response to low blood pressure or when blood oxygen levels drop dramatically.

The **thymus gland** is one of the few glands that tends to get smaller as you age. Located in front of the trachea, just above the heart, the thymus gland is where T lymphocytes, or T cells, mature (**Figure 2**). The T cells within the

spleen: a lymphoid organ that acts as a reservoir for blood and a filtering site for lymph

thymus gland: a lymphoid organ in which lymphocytes mature, multiply, and differentiate

thymus gland are designed to ensure that the body does not initiate an immune response against its own proteins. T cells within the thymus do not produce antibodies in response to antigens. During fetal development and early postnatal life, a number of the body's own proteins enter the thymus gland and act as antigens, causing T cells within the thymus to become activated. However, once activated, the T cells are themselves destroyed. The thymic cells that remain cannot produce antibodies against your own proteins.

Practice

Understanding Concepts

1. What is lymph? How is lymph transported in the body? Where does lymph eventually go?
2. Why are lymphocytes important to the immune system?
3. What is the importance of the spleen to the body?

SUMMARY The Lymphatic System

1. Proteins in the ECF are returned to the circulatory system by the lymphatic system.
2. Lymph nodes house white blood cells that filter bacteria.
3. Red bone marrow is where all types of blood cells are produced.
4. The spleen stores and purifies blood. The spleen releases blood in response to low blood pressure or low oxygen levels in blood.

Sections 7.10–7.11 Questions

Understanding Concepts

1. What two factors regulate the exchange of fluids between capillaries and ECF?
2. Use the capillary exchange model to explain homeostatic readjustment following hemorrhage.
3. Why does a low concentration of plasma protein cause edema?
4. Why does an allergic reaction result in swelling of the tissues?
5. What are lymph vessels and how are they related to the circulatory system?
6. Antihistamines are useful drugs that relieve allergy symptoms by blocking the production or effects of histamines. Describe the effects of antihistamines on homeostasis.

Making Connections

7. Some schools have banned all nut products because of the fear that children who are allergic to peanuts might go into anaphylactic shock if exposed. Should schools ban food products that might cause problems for some people?

Key Expectations

Throughout this chapter, you have had opportunities to do the following:

- Compare the anatomy of different organisms—vertebrate and/or invertebrate (7.1).
- Explain the role of the circulatory system in the transport of substances in an organism (7.1, 7.2, 7.4, 7.10, 7.11).
- Identify examples of technologies that have enhanced scientific understanding of internal systems (7.3, 7.5, 7.6, 7.7).
- Analyze and explain how societal needs have led to scientific and technological developments related to internal systems (7.3, 7.5, 7.7).
- Present informed opinions about how scientific knowledge of internal systems influences personal choices concerning nutrition and lifestyle (7.4, 7.5, 7.8, 7.9, 7.10).
- Select appropriate instruments and use them effectively and accurately in collecting observations and data (7.4, 7.7, 7.9).
- Select and use appropriate modes of representation to communicate scientific ideas, plans, and experimental results (7.4, 7.7, 7.9).
- Demonstrate an understanding of how fitness level is related to the efficiency of the cardiovascular system (7.4, 7.7, 7.9).
- Describe how the use of prescription and nonprescription drugs can disrupt or help maintain homeostasis (7.5, 7.6, 7.7, 7.10).
- Select and integrate information about internal systems from various sources (7.7, 7.9).
- Design and carry out, in a safe and accurate manner, an experiment on feedback mechanisms, identifying specific variables (7.9).

Key Terms

anemia
aneurysm
antibodies
antigen
aorta
arteriosclerosis
artery
atherosclerosis
atrioventricular (AV) node
atrioventricular (AV) valves
atria
autonomic nervous system
cardiac output
coronary arteries
diastole
ectoderm
endoderm
erythrocytes
extracellular fluid (ECF)
filtration
interstitial
leukocytes
lymph
lymph nodes
lymphocytes
mesoderm
myogenic muscle
parasympathetic nervous system
plasma
platelets
pulmonary circulatory system
pulse
Purkinje fibres
pus
semilunar valves
septum
serum
sinoatrial (SA) node
sinus
sphygmomanometer
spleen
sympathetic nervous system
systemic circulatory system
systole
thymus gland
tissues
vasoconstriction
vasodilation
veins
ventricles

Make a Summary

In this chapter, you studied the circulatory system and its importance in transporting substances within the body. Create a concept map that shows how the circulatory system maintains homeostasis. Label the concept map with as many of the key terms as possible. Check other concept maps and use appropriate designs to make your sketch clear.

Reflect on your Learning

Revisit your answers to the Reflect on Your Learning questions at the beginning of this chapter.

- How has your thinking changed?
- What new questions do you have?

Understanding Concepts

1. Use the diagram of the heart in **Figure 1** to answer the following questions:

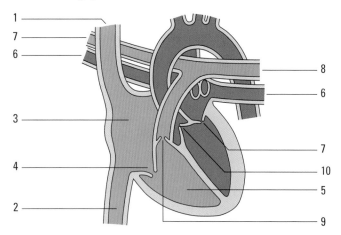

Figure 1

   (a) Identify the number(s) of the vein(s) that return blood to the heart from the body.
   (b) Identify the ventricle that contains deoxygenated blood.
   (c) Identify the heart valves that produce the *lubb* sound when closing.

2. Identify differences in the structures of veins, arteries, and capillaries and describe how they are related to the functions of each vessel.

3. "Oxygenated blood is found in all arteries of the body." Is this statement true or false? Give reasons to explain your answer.

4. Why does the left ventricle contain more muscle than the right ventricle?

5. Why does blood pressure fluctuate in an artery?

6. Arteriosclerosis is a condition referred to as "hardening of the arteries." It results from a reduction in the elasticity of the arteries. Identify two circulatory problems that might arise from this effect on the vessels.

7. The victim of an accident has had a large blood vessel severed and bleeding is severe. How does excessive bleeding endanger life? Outline two physiological responses that will help the victim to survive.

8. Why does an athlete have a lower resting heart rate than someone who is not fit?

9. Explain what happens to heart rate and blood pressure when a person moves quickly from a reclining position to a standing position.

10. High blood pressure is said to be a silent killer. Give one reason why high blood pressure is dangerous.

11. Using a capillary exchange model, explain why the intake of salt is regulated for patients who suffer from high blood pressure. (Hint: The salt is absorbed from the digestive system into the blood.)

12. Why do some people faint after standing still for a long time?

13. Explain why someone who suffers a severe cut might develop a rapid and weak pulse. Why might body temperature begin to fall?

14. Why does the blockage of a lymph vessel in the left leg cause swelling in that area?

15. A fetus has no need for pulmonary circulation. Oxygen diffuses from the mother's circulatory system into that of the fetus through the placenta. Therefore, the movement of blood through a baby's heart is highly modified: blood flows from the right atrium through an opening in the septum to the left atrium and then to the left ventricle. The opening between the right and left atria becomes sealed at birth. Explain why the failure of the opening to seal results in what has been termed a "blue baby."

16. What would happen if blood type A were transfused into people with blood types B, A, O, and AB? Provide an explanation for each case.

17. Explain why multicellular animals with a mesoderm cell layer must have a specialized transport system.

Applying Inquiry Skills

18. Which area of **Figure 2** represents blood in a capillary? Explain your answer.

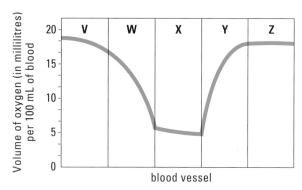

Figure 2

19. Compare open and closed circulatory systems by listing advantages and limitations of each.

20. **Table 1** is a record of a person's blood pressure taken in a sitting position before and after exercise.

(a) Why does systolic blood pressure increase after exercise?

(b) Why does diastolic blood pressure decrease after exercise?

Table 1

Condition	Systolic blood pressure (mm Hg)	Diastolic blood pressure (mm Hg)	Pulse rate (beats/min)
Resting	120	80	70
After exercise	180	45	160

21. A physician notes fewer red blood cells and prolonged blood clotting times in a patient. White blood cell numbers appear to have increased, but further examination reveals that only the granulocyte numbers have increased while the agranulocytes have decreased. In an attempt to identify the cause of the anomaly, the physician begins testing the bone marrow. Why did the physician test the bone marrow? Predict what might have caused the problem. Provide the reasons behind your prediction.

22. Use the data in **Table 2** to answer the following questions:
    (a) Which individual has recently moved from Calgary to Halifax? (Note: Halifax is at or near sea level, while Calgary is at a much higher altitude.) Give your reasons.
    (b) Which individual is suffering from dietary iron deficiency? Give your reasons.

Table 2

Individual	Breathing rate (breaths/min)	Hemoglobin (g/100 mL of blood)	O_2 content (mL/100 mL of blood)
A (normal)	15	15.1	19.5
B	21	8.0	13.7
C	12	17.9	22.1
D	22	16.0	14.1

23. How does the respiratory system depend on the circulatory system?

24. Nicotine causes the constriction of arterioles.
    (a) Explain why pregnant women are advised not to smoke.
    (b) Mothers who smoke give birth to babies who are, on average, 1 kg smaller than normal. Suggest a possible relationship between the effects of nicotine on the mother's circulatory system and the lower body mass of babies.

Making Connections

25. Heart disease is currently the number one killer of middle-aged males, accounting for billions of dollars every year in medical expenses and productivity loss. Should males be required by law to undergo heart examinations? Justify your answer. Consider the social and moral implications of such a law.

26. Caffeine causes heart rate to accelerate; however, a scientist who works for a coffee company has suggested that blood pressure will not increase due to coffee consumption. This scientist states that homeostatic adjustment mechanisms ensure that blood pressure readings will remain within an acceptable range. Design an experiment that will test the scientist's hypothesis. What other reasons might the scientist have for suggesting that caffeine does not increase blood pressure?

27. It has been estimated that for every extra kilogram of fat on a person, an additional kilometre of circulatory vessels is required to supply the tissues with nutrients. Indicate why obesity has often been associated with high blood pressure. Do only overweight people suffer from high blood pressure? Explain your answer.

28. Individuals who work in a certain chemical plant are found to have unusually high numbers of leukocytes. A physician calls for further testing. Hypothesize about the physician's reasons for concern. Why might the physician check both bone marrow and lymph nodes?

Exploring

29. Use the Internet to investigate the use of blood and blood components from animals, such as pigs, to provide temporary or long-term replacement for human blood. What are the social and ethical concerns about using animal tissues for humans?

    Follow the links for Nelson Biology 11, Chapter 7, Review

    GO TO www.science.nelson.com

30. Examine your own cardiovascular health. Consider such factors as diet, exercise, and heredity. Find out the extent to which each of these factors affect cardiovascular health, and identify ways in which your cardiovascular health can be improved.

    Follow the links for Nelson Biology 11, Chapter 7, Review

    GO TO www.science.nelson.com

In this chapter, you will be able to

- describe the process of ventilation and gas exchange in humans;
- compare how different organisms obtain oxygen from the atmosphere and compare their anatomical structures;
- carry out investigations that provide indicators of oxygen delivery to the cells and that can be used to determine fitness levels;
- identify lifestyle activities that could improve or impair health;
- describe how the use of prescription and nonprescription drugs can disrupt or help maintain homeostasis.

Respiratory System

Water polo players are superb athletes. The sport requires the strength of a rower, the endurance of a cross-country skier, and the scoring touch of a soccer player. Having the aggressiveness of most hockey players doesn't hurt either—water polo is a rough game.

What separates athletes, like Waneek Horn-Miller (**Figure 1**), from the majority of us? The exceptional physical fitness of an athlete depends largely on the superior ability to deliver oxygen and chemical fuels to the cells of the body. To sustain life-giving processes, all cells require nutrients and oxygen. Within the mitochondria, oxygen reacts with organic chemicals to provide energy. In muscle cells, this energy is used for movement. Athletes tend to have a superior

Figure 1
Waneek Horn-Miller led Canada's women's national water polo team in the 2000 Olympics in Sydney.

ventilation system that provides an excellent exchange of air, ensuring plentiful oxygen for the cells.

Training can increase your ability to take in oxygen (through the respiratory system) and deliver it (through the circulatory system) to cells of the body. However, not everyone who trains will become an elite athlete. Your inherited physiology may not allow for sufficient ventilation for a particular sport, or it may provide a structural support that makes you better suited for a different sport or activity. Athletes involved in endurance events, such as long-distance running, have more "slow-twitch" muscle cells, which use great amounts of oxygen. Athletes involved in events that rely more on quickness, such as weightlifting and sprinting, have more "fast-twitch" muscle cells, which require less oxygen. Your genes determine how much fast- and slow-twitch muscle you have in your body. Elite athletes are products of training and genes (**Figure 2**).

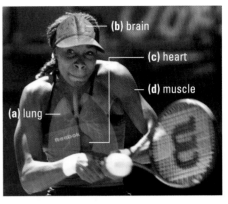

Figure 2
(a) The lungs deliver oxygen for the cells of the body via the blood.
(b) The brain regulates breathing and heart rates to meet demands for oxygen and produces chemicals that dull the pain produced by extreme exercise.
(c) The heart pumps blood, which carries oxygen to the cells of the body. Athletes tend to have larger hearts.
(d) "Slow-twitch" and "fast-twitch" muscle cells are determined by your genes.

Fitness Test ($VO_{2\,max}$)

"Aerobic fitness" refers to endurance, or the ability to sustain high levels of activity for prolonged periods. "Aerobic" refers to the consumption of oxygen by the body to meet the energy needs of exercise. One indicator of aerobic fitness is $VO_{2\,max}$, the maximum amount of oxygen that a person can use when exerting maximum effort. It is measured in millilitres of oxygen per kilogram of body mass per minute.

One method of calculating your $VO_{2\,max}$ is as follows:

- If you perform the test outdoors, choose a windless day.
- Run around a track or on a treadmill, if available, for 15 min. Otherwise, you will have to choose a road or trail where you can measure distances accurately.
- Record the distance you ran in metres.
- Calculate your speed using this formula:

$$speed = \frac{distance\ in\ metres}{time\ in\ minutes}$$

- Use the following formula to work out your predicted $VO_{2\,max}$:

$$VO_{2\,max} = (speed - 133) \times 0.172 + 33.3$$

- Compare the $VO_{2\,max}$ for students who are involved in different types of activities.
Follow the links for Nelson Biology 11, 8.0.

 www.science.nelson.com

- There are many different ways to calculate $VO_{2\,max}$. You might like to try several different methods for calculating your $VO_{2\,max}$ to compare the results.

Do not perform this activity if you are not allowed to participate in physical education classes.

Reflect on your Learning

1. What everyday experiences indicate the importance of providing oxygen to living cells?
2. Fitness can be measured by the body's ability to provide oxygen for the tissues of the body. Make a list of sports that you believe require great fitness. Make another list of sports that you believe require less fitness. Be prepared to justify your answers.

**Composition of
Earth's Atmosphere**

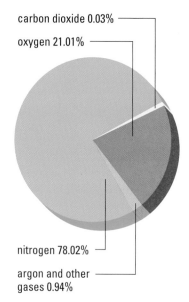

carbon dioxide 0.03%

oxygen 21.01%

nitrogen 78.02%

argon and other
gases 0.94%

Figure 1
Nitrogen and oxygen are the two most abundant components in atmospheric air.

respiration: all processes involved in the exchange of oxygen and carbon dioxide between cells and the environment. Respiration includes breathing, gas exchange, and cellular respiration.

The Importance of the Respiratory System

You live in a sea of air. Nitrogen, oxygen, carbon dioxide, and trace gases are taken into and expelled from your body with every breath. Earth's atmosphere is made up of approximately 78% nitrogen and 21% oxygen; the remaining gases, argon, carbon dioxide, and others, make up about 1% (**Figure 1**). The second most common component, oxygen, is vital to life. Cells obtain energy through a chemical reaction called oxidation, in which organic compounds are broken down using oxygen. Although a small amount of energy can be obtained in anaerobic conditions (in the absence of oxygen), life processes in humans cannot be maintained without an adequate supply of oxygen.

Oxygen is so essential to the survival of humans that just a few minutes without oxygen will result in death. By comparison, individuals can live for a number of days without water and several weeks without food. It has been estimated that an average adult utilizes 250 mL of oxygen every minute while resting. Oxygen consumption may increase up to 20 times with strenuous exercise.

Respiration and Breathing

The term **respiration** can be used to describe all processes that supply oxygen to the cells of the body for the breakdown of glucose and to describe the process by which wastes are transported to the lungs for exhalation. **Figure 2** shows the processes involved in respiration.

Animals use oxygen for cellular respiration. In section 1.3, you learned that small organelles called mitochondria are the centres of cellular respiration. During the process of cellular respiration, oxygen and sugar molecules react, resulting in the production of carbon dioxide and water. The energy released is used to maintain cell processes, such as growth, movement, and the creation of new molecules.

Breathing is the process by which air enters and leaves the lungs.

External respiration involves the exchange of gases (O_2 and CO_2) between the air and the blood.

Internal respiration involves the exchange of gases (O_2 and CO_2) between the blood and tissue fluids.

Cellular respiration involves the production of ATP in body cells.

Figure 2
The process of respiration

The concentration of oxygen in cells is much lower than in their environment because cells continuously use it for cellular respiration. Oxygen must be constantly replenished if a cell is to survive.

Breathing, or ventilation, involves the movement of gases between the external environment and the location where they can enter and leave the body. In aquatic organisms, such as fish, oxygen is extracted from the surrounding water. Land animals have approximately 20 times more oxygen available to them through the atmosphere. The intake of oxygen and the release of carbon dioxide by cells take place across a **respiratory membrane**.

The Challenge of Getting Oxygen

As animals increase in size, more oxygen is required to meet their energy needs. In order to deliver greater amounts of oxygen to cells, the respiratory membranes of the more complex animals must have an increased surface area. Some animals, such as earthworms, use their skin as a respiratory membrane (**Figure 3(a)**). The skin must be kept moist at all times to allow the proper diffusion of gases.

Fish, some salamanders, clams, starfish, and crayfish exchange gases through their gills. Gills are, essentially, extensions of the outer surface of the body. The extensive folding and branching of the gills provide increased surface area for the diffusion of gases, improving the efficiency of the respiratory organ (**Figure 3(b)**). Fish also use a countercurrent flow—the water moves over the gills in one direction while the blood, contained within the capillaries inside the gill, moves in the opposite direction (**Figure 4**). Countercurrent flow increases the efficiency of

breathing: the movement of gases between the respiratory membrane of living things and their external environment

respiratory membrane: the living membrane where the diffusion of oxygen and other gases occurs. Gases are exchanged between the living cells of the body and the external environment (the atmosphere or water).

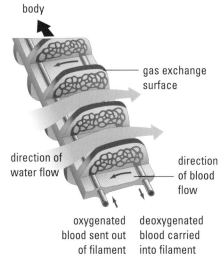

body

gas exchange surface

direction of water flow

direction of blood flow

oxygenated blood sent out of filament

deoxygenated blood carried into filament

Figure 4
A blood vessel carries deoxygenated blood into each filament, or fold, of the gills. Another carries oxygenated blood out to the body. Blood flowing from one vessel to the other runs counter to the direction of water flowing over gas exchange surfaces.

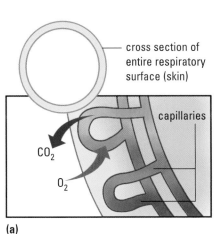

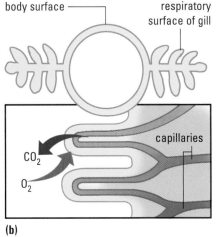

cross section of entire respiratory surface (skin)

capillaries

CO_2

O_2

(a)

body surface

respiratory surface of gill

capillaries

CO_2

O_2

(b)

Figure 3
(a) The outer surface of the earthworm's skin acts as the respiratory membrane for the diffusion of gases.
(b) Fish have an improved system with folded branching membranes called gills. With an increased membranous surface area, gills can provide more oxygen to the tissues.

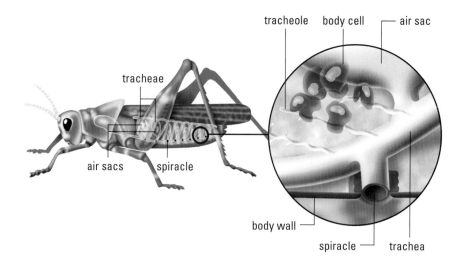

Figure 5
Tracheal system of insects. The branching tubes, or tracheae, often end in larger air sacs, which supply oxygen to organs that require it.

oxygen intake and ensures that the oxygen diffuses into the blood over the entire length of the blood vessel inside of the gill. Because the blood and water move in opposite directions, oxygen-poor blood is in contact with oxygen-rich water, while blood that is relatively rich in oxygen is in contact with oxygen-poor water. This ensures that oxygen always diffuses from the water into blood that has a low concentration of oxygen. If the blood and water flowed in the same direction, the amount of oxygen diffusion would be reduced because there would be only a small difference between the oxygen concentrations in the water and in the blood.

Although gills are ideal for aquatic environments, they are poorly adapted for land. Exposing the large surface area of the respiratory membrane to air causes too much evaporation. If the gills become dry, the membrane becomes impermeable to the diffusion of gases.

Insects do not have the problem of losing too much water by evaporation through their respiratory membrane because it is located inside their body. A tracheal system, consisting of branching respiratory tubes, connects cells directly to the atmosphere by openings in the exoskeleton called spiracles. Oxygen enters the body through the spiracles and is then delivered to the cells by the tracheae and blood (**Figure 5**). For land animals, the tracheal system is a vast improvement over gills, but it does not provide enough oxygen for larger animals. The limited oxygen delivery by a tracheal system is one factor that keeps insects small.

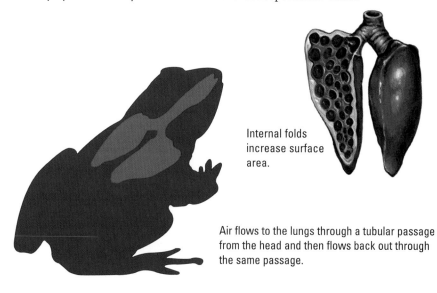

Internal folds increase surface area.

Air flows to the lungs through a tubular passage from the head and then flows back out through the same passage.

Figure 6
Respiratory membranes that are housed inside of the body lose less water to evaporation. The lung of the amphibian has many folds to increase surface area.

Although the atmosphere contains 20 times more oxygen than water, the size and activity of land animals are still limited by their ability to secure oxygen. Larger animals require an even more efficient system for the delivery of gases. The respiratory system consists of a windpipe or trachea that branches into the lungs. Frog lungs are balloonlike structures that allow the exchange of gases between the air and the blood. Although frogs, like earthworms, can absorb oxygen through their skin, the lungs provide substantially greater surface area (**Figure 6**). The internal location helps solve the problem of evaporation. Internal folds inside the lung increase the surface area for diffusion. The greater the surface area, the greater the amount of oxygen that can be absorbed.

SUMMARY **The Importance of the Respiratory System**

1. The cells of the body obtain energy through oxidation. Thus, oxygen is essential to survival.

2. Respiration includes all the processes involved in the exchange of oxygen and carbon dioxide between cells and the environment.

3. Some animals have an external respiratory membrane. They exchange gases through their skin or gills.

4. The larger the organism and the more active it is, the more oxygen is required to fulfill its energy needs. This means the respiratory membrane requires an increased surface area.

Practice

Understanding Concepts

1. Why is oxygen so essential for survival?
2. Differentiate between breathing and cellular respiration.
3. What advantage is gained from increasing the size of the respiratory surface?
4. Explain why the respiratory system of a fish is more efficient than that of an earthworm.
5. Why are gills not well suited for life on land?
6. Explain why the frog's respiratory system is more efficient than that of insects.

8.2 The Mammalian Respiratory System

In mammals, air enters the respiratory system either through the two nasal cavities or the mouth (**Figure 1**, page 286). Foreign particles are prevented from entering the nasal cavities by tiny hairs lining the passageways that act as a filtering system. The nasal cavities warm and moisten incoming air and contain mucus, which traps particles and keeps the cells lining the cavities moist.

The nasal cavities open into an air-filled channel in the mouth called the pharynx. Two openings branch from the pharynx: the **trachea**, or windpipe, and the esophagus, which carries food to the stomach. Mucus-producing cells, some of which are ciliated, line the trachea. The mucus traps debris that may have escaped

trachea: the windpipe

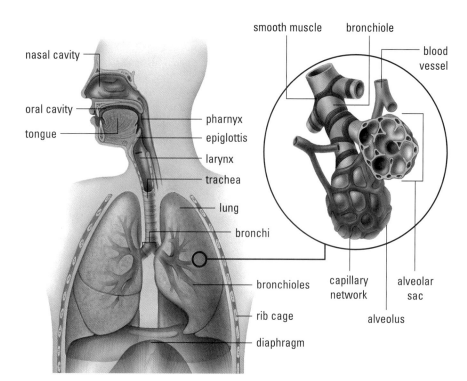

Figure 1
The human respiratory system

cilia: tiny hairlike protein structures found on some cells that sweep foreign debris from the respiratory tract

epiglottis: a structure that covers the glottis (opening of the trachea) during swallowing

larynx: the voice box

the filters in the nasal passage. This debris is swept by the **cilia** (singular: cilium) from the windpipe back into the pharynx. The wall of the trachea is supported by cartilage rings, which keep the trachea open. An enlarged segment of cartilage (the larynx) supports the **epiglottis**, a flaplike structure that covers the glottis, or opening of the trachea, when food is being swallowed. When food is chewed, it is forced to the roof of the mouth and pushed backward. This motion initiates a reflex action, which closes the epiglottis, allowing food to enter the esophagus rather than the trachea. If you have ever taken in food or liquids too quickly, you will know how it feels to bypass this reflex. Food or liquid entering the trachea stimulates the cilia, and particles too large to be swept out of the respiratory tract are usually expelled by a second more powerful reflex: a violent cough.

Air from the pharynx enters the **larynx**, or voice box, located at the upper end of the trachea. The larynx contains two thin sheets of elastic ligaments called the vocal cords (**Figure 2**). The vocal cords vibrate as air is forced from the lungs toward the pharynx. Different sounds are produced by a change in tension on the vocal cords. Your larynx is protected by a thick band of cartilage commonly known as the Adam's apple. Following puberty, the cartilage and larynx of males

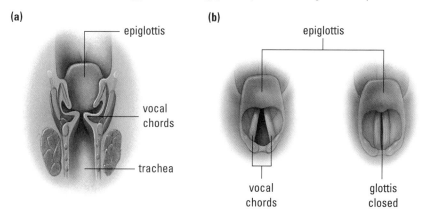

Figure 2
(a) Larynx, showing the vocal cords
(b) Position of the vocal cords when the glottis is open and closed during speech

increase in size and thickness. In the same way as a larger drum creates a lower-pitched sound, the larger voice box in males produces a deeper sound. Rapid growth of the larynx creates problems for adolescent boys who have difficulty controlling the pitch of their voices. Have you ever noticed how your voice lowers when you have a cold? Inflammation of the vocal cords causes swelling and produces lower-frequency vibrations. Should the infection become severe and result in a condition referred to as laryngitis, you may temporarily lose your voice.

Inhaled air moves from the trachea into two **bronchi** (singular: bronchus), which, like the trachea, contain cartilage rings. The bronchi carry air into the right and left lungs, where they branch into many smaller airways called **bronchioles**. Unlike the trachea and bronchi, the bronchioles do not contain cartilaginous rings. Smooth muscle in the walls of the bronchioles can decrease their diameter. Any closing of the bronchioles increases the resistance of air movement and can produce a wheezing sound. Air moves from the bronchioles into tiny sacs called **alveoli** (singular: alveolus). Measuring between 0.1 and 0.2 µm (micrometres) in diameter, each alveolus is surrounded by capillaries. In the alveoli, gases diffuse between the air and blood according to concentration gradients. Oxygen and carbon dioxide both move from areas of higher concentration to areas of lower concentration. Therefore, oxygen moves from the air within the lung to the alveoli, while carbon dioxide moves from the alveoli into the air inside the lung. The alveoli are composed of a single layer of cells, which permits more rapid gas exchange. Each lung contains about 150 million alveoli. That provides enough surface area to cover half a tennis court, or about 40 times the surface area of the human body.

Have you ever tried to pull the cover slip from a microscope slide, only to discover that it seems to be fused to the slide? This phenomenon is caused by water molecules adhering to the glass. A similar problem faces the alveoli. During inhalation the alveoli appear bulb shaped, but during exhalation the tiny sacs collapse. The two membranes touch but are prevented from sticking together by a film of fat and protein called lipoprotein. This film lines the alveoli, allowing them to pop open during inhalation. Some newborn babies, especially premature babies, do not produce enough of the lipoprotein. Extreme force is required to overcome the surface tension created, and the baby experiences tremendous difficulty inhaling. This condition, referred to as respiratory distress syndrome, often results in death.

The outer surface of the lungs is surrounded by a thin membrane called the **pleural membrane**, which also lines the inner wall of the chest cavity. The space between the pleural membranes is filled with fluids that reduce the friction between the lungs and the chest cavity during inhalation. Pleurisy, the inflammation of the pleural membranes and the buildup of fluids in the chest cavity, is most often caused when the two membranes rub together. This buildup of fluids puts great pressure on the lungs, making expiration (exhalation) easier, but inspiration (inhalation) much more difficult and painful.

Breathing Movements

Pressure differences between the atmosphere and the chest, or thoracic, cavity determine the movement of air into and out of the lungs. Atmospheric pressure remains relatively constant, but the pressure in the chest cavity may vary. An understanding of breathing hinges on an understanding of gas pressures.

Gases move from an area of high pressure to an area of low pressure. Inspiration occurs when pressure inside the lungs is less than that of the atmosphere, and expiration occurs when pressure inside the lungs is greater than that of the atmosphere.

bronchi: the passage from the trachea to either the left or right lung

bronchioles: the smallest passageways of the respiratory tract

alveoli: sacs of the lung in which the exchange of gases between the atmosphere and the blood occurs

DID YOU KNOW?

All of us have had the hiccups at one time or another. An irritation of the diaphragm causes air to become trapped in the respiratory tract and the diaphragm experiences a muscular spasm. The hiccup sound is produced when air is taken in as the glottis closes.

pleural membrane: a thin, fluid-filled membrane that surrounds the outer surface of the lungs and lines the inner wall of the chest cavity

diaphragm: a sheet of muscle that separates the organs of the chest cavity from those of the abdominal cavity

Expiration

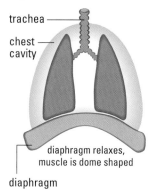

trachea

chest cavity

diaphragm relaxes, muscle is dome shaped

diaphragm

Inspiration

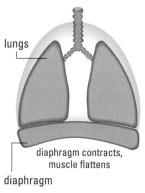

lungs

diaphragm contracts, muscle flattens

diaphragm

Figure 3
The diaphragm can be compared to a piston.

external intercostal muscles: muscles that raise the rib cage, decreasing pressure inside the chest cavity

internal intercostal muscles: muscles that pull the rib cage downward, increasing pressure inside the chest cavity

Figure 4
Changes in chest volume during inspiration and expiration.
(a) The external intercostal muscles contract and the rib cage pulls upward. Because pressure in the chest cavity is lower than the atmospheric pressure, air moves into the lungs.
(b) The external intercostal muscles relax and the rib cage falls. Because pressure in the chest cavity is higher than the atmospheric pressure, air moves out of the lungs.

The **diaphragm**, a dome-shaped sheet of muscle that separates the chest cavity from the abdominal cavity, regulates the pressure in the chest cavity. During inspiration, the diaphragm muscle contracts, or shortens, pulling downward. The chest volume increases and pressure in the lungs decreases. The atmospheric pressure is now greater than the pressure in the chest cavity, and air moves into the lungs. During expiration, the diaphragm relaxes and returns to its dome shape due to the force exerted by the organs in the abdomen. The chest volume decreases and pressure increases. The pressure in the chest cavity is now greater than the atmospheric pressure, and air moves out of the lungs. Have you ever found yourself gasping for air after receiving a blow to the solar plexus (the bottom of the rib cage)? The blow drives abdominal organs upward, causing the dome shape of the diaphragm to be exaggerated. Volume in the chest cavity is reduced, causing a large quantity of air to be expelled.

Think of the diaphragm as a piston. As the piston moves down, the volume in the chest cavity increases and pressure begins to fall—air is drawn into the lungs. As the piston moves up, the volume in the chest cavity decreases and pressure begins to increase—air is forced out of the lungs (**Figure 3**).

The diaphragm is assisted through the action of the intercostal muscles, which cause the ribs to move. Have you ever noticed how your ribs rise when you inhale? The ribs are hinged to the vertebral column, allowing them to move up and down. Bands of muscle, the **external intercostal muscles**, are found between the ribs. A nerve stimulus causes the external intercostal muscles to contract, pulling the ribs upward and outward. This increases the volume of the chest, lowers the pressure in the chest cavity, and air moves into the lungs. If the intercostals are not stimulated, the muscle relaxes and the rib cage falls. The fluids inside the pleural membrane push against the lungs with greater pressure, and air is forced out of the lungs. A second set of intercostals, the **internal intercostal muscles**, pulls the rib cage downward during times of extreme exercise or forced exhalation, such as blowing out a candle. The internal intercostal muscles are not employed during normal breathing.

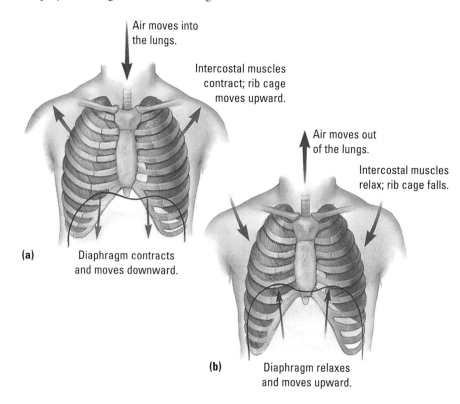

Air moves into the lungs.

Intercostal muscles contract; rib cage moves upward.

Air moves out of the lungs.

Intercostal muscles relax; rib cage falls.

(a) Diaphragm contracts and moves downward.

(b) Diaphragm relaxes and moves upward.

The importance of establishing pressure differences between the lungs and the atmosphere is best illustrated by a pneumothorax, or collapsed lung. A bullet wound or a stab wound to the ribs creates a hole in the pleural cavity, making it impossible for the chest cavity to establish pressure differences. When the diaphragm contracts and the rib cage is raised, the pressure inside the chest cavity is reduced; however, significantly less air is drawn into the lungs. Much of the air flows directly through the hole left by the wound. In most cases, the lungs remain collapsed. The hole in the chest cavity must be sealed and any excess air or fluid withdrawn before the lungs are capable of functioning normally.

Practice

Understanding Concepts

1. Describe the function of cilia in the respiratory tract.
2. Explain how the functions of the trachea, esophagus, and epiglottis are related.
3. Trace the pathway of a breath of air from its point of entry to its diffusion in the lungs. Refer to the structures that the breath passes by or through.

SUMMARY   The Mammalian Respiratory System

1. Air enters the respiratory system through the nose or the mouth; then it enters the pharynx, trachea, bronchi, and the bronchioles and alveoli in the lungs.

2. In the alveoli, gases diffuse between air and blood according to concentration gradients. Oxygen moves into the alveoli and carbon dioxide moves out of the alveoli.

3. The movement of gases into and out of the lungs is determined by the difference in pressure between the atmosphere and the chest, or thoracic, cavity. Pressure in the thoracic cavity is regulated by the diaphragm. The diaphragm is assisted by the movement of the intercostal muscles.
   • During inspiration (inhalation), the diaphragm flattens and pulls downward, the intercostal muscles contract, the rib cage pulls up and outward, chest volume increases, pressure in the lungs decreases, and air moves into the lungs.
   • During expiration (exhalation), the diaphragm is relaxed and becomes dome shaped, the intercostal muscles relax, the rib cage falls, chest volume decreases, pressure in the lungs increases, and air moves out of the lungs.

Sections 8.1–8.2 Questions

Understanding Concepts

1. Describe the similarities and differences between the bronchi, bronchioles, and alveoli. How is each type of structure well suited for its purpose in the lungs?

2. Explain how and why oxygen and carbon dioxide diffuse between the alveoli and the air in the lungs.

(continued)

3. Why does a throat infection cause your voice to produce lower-pitched sounds?

4. What is respiratory distress syndrome?

5. Why does the buildup of fluids in the chest cavity, as occurs with pleurisy, make exhalation easier but inhalation more difficult?

6. Describe the movements of the ribs and the diaphragm during inhalation and exhalation.

7. List five animals that would have a relatively high energy demand and five animals that would have a relatively low energy demand. Justify your selections.

8. Bronchitis is an inflammation of the bronchi or bronchioles that causes them to swell. What problems would be caused as the airways swell and decrease in diameter?

9. Nicotine inhaled with cigarette smoke causes blood vessels to narrow. What problems would this cause for the cells of the body?

Reflecting

10. What approach do you take when comparing two things? Do you use a table or any other method to organize your information?

8.3 Gas Exchange and Transport

An understanding of gas exchange in the human body is tied to an understanding of the physical nature of gases. As mentioned in the previous section, gases diffuse from an area of higher pressure to an area of lower pressure.

Dalton's law of partial pressure states that each gas in a mixture exerts its own pressure, or partial pressure. The pressure of each gas in the atmosphere can be calculated using the percentage by volume of each gas present. Atmospheric pressure at sea level is 101 kPa. The partial pressure exerted by oxygen can be calculated since we know that 21% of the air is oxygen. Therefore, $21\% \times 101$ kPa $\doteq 21.21$ kPa. By similar calculations, carbon dioxide exerts a partial pressure of 0.0303 kPa.

The graph in **Figure 1** shows the partial pressures of oxygen and carbon dioxide in the body. Gases diffuse from an area of high partial pressure to an area of lower partial pressure. The highest partial pressure of oxygen is found in atmospheric air. Oxygen diffuses from the air (21.2 kPa) into the lungs (13.3 kPa for the alveoli).

The partial pressure of oxygen in the blood differs depending on location. Arteries carry blood away from the heart while veins carry it back to the heart. Arteries are connected to veins by capillaries, where gas exchange takes place and oxygen diffuses into the tissues. (Remember that energy is continuously released from nutrients by reactions within the cells that require oxygen. Oxygen will never accumulate in the cells.) Therefore, the largest change in the partial pressure of oxygen is observed as oxygen travels from the arteries (12.6 kPa) into the capillaries (5.3 kPa).

Carbon dioxide, the product of cell respiration, follows an opposite pattern. Partial pressure of carbon dioxide is highest in the tissues and venous blood. The partial pressure of nitrogen, although not shown in the graph, remains relatively constant. Atmospheric nitrogen is not involved in cellular respiration.

Dalton's law of partial pressure: the total pressure of a mixture of nonreactive gases is equal to the sum of the partial pressures of the individual gases

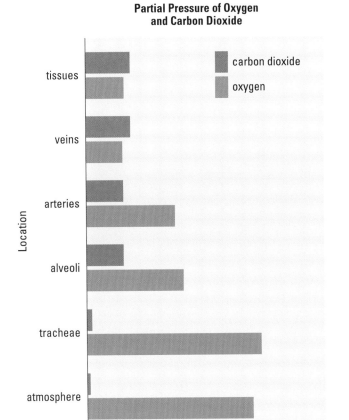

Partial Pressure of Oxygen and Carbon Dioxide

Legend: carbon dioxide, oxygen

Location (y-axis): tissues, veins, arteries, alveoli, tracheae, atmosphere

Partial Pressure (kPa) (x-axis): 0, 5, 10, 15, 20, 25, 30

Figure 1
The partial pressure of oxygen is lowest in the systemic veins and tissues, but highest in atmospheric air. Conversely, the partial pressure of carbon dioxide is highest in the tissues and systemic veins, but lowest in atmospheric air.

Oxygen Transport

Oxygen moves from the atmosphere, the area of highest partial pressure, to the alveoli. It then diffuses from the alveoli into the blood and dissolves in the plasma. Oxygen is not very soluble in blood—about 0.3 mL of oxygen per 100 mL of blood. However, even at rest, the body requires approximately 10 times that amount of oxygen. Hemoglobin greatly increases oxygen-carrying capacity. When oxygen dissolves into the plasma, hemoglobin forms a weak bond with the oxygen molecule to form oxyhemoglobin. Once oxyhemoglobin is formed, other oxygen molecules can dissolve in the plasma. With hemoglobin, the blood can carry 20 mL of oxygen per 100 mL of blood, almost a 70-fold increase.

The amount of oxygen that combines with hemoglobin is dependent on partial pressure. The partial pressure in the lungs is approximately 13.3 kPa. Thus, blood leaving the lungs is still nearly saturated with oxygen. As blood enters the capillaries, the partial pressure drops to about 5.3 kPa. This drop in partial pressure causes the dissociation, or split, of oxygen from the hemoglobin, and oxygen diffuses into the tissues. **Figure 2** on page 292 shows an oxygen–hemoglobin dissociation curve. You will notice that very little oxygen is released from the hemoglobin until the partial pressure of oxygen reaches 5.3 kPa. This ensures that most of the oxygen remains bound to the hemoglobin until it gets to the tissue capillaries. Also note that venous blood still carries a rich supply of oxygen. Approximately 70% of the hemoglobin is still saturated when blood returns to the heart.

DID YOU KNOW ?

Approximately 20% of the air taken into the lungs is not exchanged. Some air remains in the lung and respiratory tract even after you exhale. This is referred to as residual air.

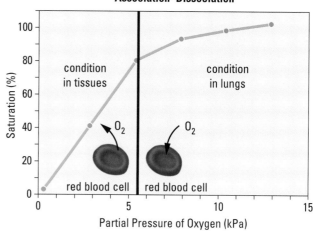

Oxygen–Hemoglobin Association–Dissociation

Figure 2
Hemoglobin gives up very little oxygen until the partial pressure of oxygen reaches 5.3 kPa in the surrounding tissues. As the partial pressure of oxygen rises, hemoglobin picks up more oxygen.

Carbon Dioxide Transport

Carbon dioxide is about 20 times more soluble than oxygen. About 9% of the carbon dioxide produced by the tissues of the body is carried in the plasma. Approximately 27% of the body's carbon dioxide combines with hemoglobin to form carbaminohemoglobin. The remaining 64% of the body's carbon dioxide combines with water from the plasma to form carbonic acid (H_2CO_3) as shown in **Figure 3(a)**. An enzyme called **carbonic anhydrase** increases the rate of this chemical reaction by about 250 times. The rapid conversion of free carbon dioxide into carbonic acid decreases the concentration of carbon dioxide in the plasma. This maintains a low partial pressure of carbon dioxide, ensuring that carbon dioxide continues to diffuse into the blood (**Figure 4**).

The formation of acids, such as carbonic acid, can create problems. Because acids can change the pH of the blood and eventually bring about death, they must be buffered. This is where the second function of hemoglobin comes into effect. Being unstable, the carbonic acid dissociates into bicarbonate ions (HCO_3^-) and hydrogen ions (H^+) as shown in **Figure 3(b)**. The hydrogen ions help dislodge oxygen from the hemoglobin, and then combine with it to form reduced hemoglobin. When hemoglobin combines with the hydrogen ions, it is removing H^+ from the solution; that is, the hemoglobin is acting as a **buffer**. Meanwhile the bicarbonate ions are transported into the plasma. Oxygen is released from its binding site and is now free to move into the body cells.

Once the venous blood reaches the lungs, oxygen dislodges the hydrogen ions from the hemoglobin binding sites. Free hydrogen and bicarbonate ions combine to form carbon dioxide and water as shown in **Figure 3(c)**. The highly concentrated carbon dioxide diffuses from the blood into the alveoli and is eventually eliminated during exhalation.

carbonic anhydrase: an enzyme found in red blood cells. The enzyme speeds the conversion of carbon dioxide and water to form carbonic acid.

buffer: a substance capable of neutralizing acids and bases, thus maintaining the original pH of the solution

(a) $CO_2 + H_2O \rightarrow H_2CO_3$

(b) $H_2CO_3 \rightarrow H^+ + HCO_3^-$

(c) $H^+ + HCO_3^- \rightarrow H_2O + CO_2$

Figure 3

Practice

Understanding Concepts

1. What does Dalton's law of partial pressure state?
2. (a) Where is the partial pressure of oxygen the highest? the lowest?
   (b) How is this related to the diffusion of oxygen into the tissues?
3. Where is the partial pressure of carbon dioxide the highest? the lowest?

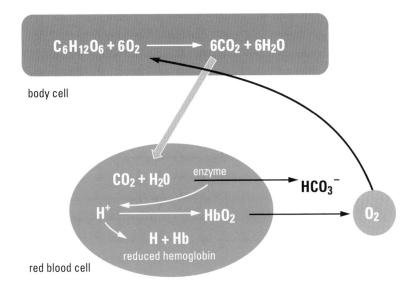

CO$_2$ transport

C$_6$H$_{12}$O$_6$ + 6O$_2$ $\longrightarrow$ 6CO$_2$ + 6H$_2$O

body cell

CO$_2$ + H$_2$O   enzyme   HCO$_3^-$

H$^+$ HbO$_2$ O$_2$

H + Hb

reduced hemoglobin

red blood cell

Figure 4
Under the influence of carbonic anhydrase, an enzyme found in red blood cells, carbon dioxide combines with water to form carbonic acid (H$_2$CO$_3$), which then dissociates into H$^+$ and HCO$_3^-$ ions.

Maintaining Gas Levels

A variety of homeostatic mechanisms exist to help maintain appropriate levels of oxygen and carbon dioxide. For example, a chemical receptor helps ensure that carbon dioxide, the waste product of cellular respiration, does not accumulate. During exercise, cellular respiration increases, causing carbon dioxide levels to increase. This stimulates chemical receptors in the brainstem. The activated nerve cells from the brain carry impulses to muscles that increase breathing movements. Increased breathing movements help flush excess carbon dioxide from the body. Other chemical receptors in the walls of the carotid artery are able to detect low levels of oxygen in your blood. A nerve is stimulated and a message is sent to the brain. The brain relays the information, by way of another nerve, to the muscles that control breathing. Thus, a system of "turning on" and "turning off" mechanisms is used to help maintain homeostasis.

SUMMARY   Gas Exchange and Transport

1. Diffusion of a gas occurs from an area of high pressure to an area of low pressure.

2. The partial pressure of oxygen is highest in the atmosphere and lowest in the veins and tissues.
   • Oxygen diffuses from the atmosphere into alveoli and then into the blood.
   • Hemoglobin increases the oxygen-carrying capacity of blood by bonding to oxygen molecules to form oxyhemoglobin. Hemoglobin and oxygen dissociate in the capillaries, and oxygen diffuses into the tissues.

3. The partial pressure of carbon dioxide is highest in the tissues and veins and lowest in the atmosphere.
   • Some carbon dioxide combines with water from plasma to form carbonic acid; this decreases carbon dioxide concentration, ensuring that carbon dioxide continues to diffuse into the blood.
   • Carbonic acid dissociates into HCO$_3^-$ and H$^+$. Hemoglobin combines with H$^+$, releasing oxygen and acting as a buffer.

- In the lungs, H^+ and HCO_3^- combine to form carbon dioxide and water. Carbon dioxide is highly concentrated; it diffuses from blood into alveoli and is eliminated through exhalation.

4. Homeostatic mechanisms help maintain gas levels. Chemical receptors detect a change in gas levels and send a message, via nerve cells, to increase or decrease breathing movements.

Section 8.3 Questions

Understanding Concepts

1. How does partial pressure affect the movement of oxygen from the alveoli to the blood?
2. How is carbon dioxide transported in the blood?
3. Describe the importance of hemoglobin as a buffer.
4. Trace the pathway of an oxygen molecule from the atmosphere to its combination with a hemoglobin molecule.
5. How do carbon dioxide levels regulate breathing movements?
6. Why would exposure to carbon monoxide (CO) increase breathing rates?

8.4 Disorders of the Respiratory System

All respiratory disorders share one common characteristic: they all decrease oxygen delivery to the tissues.

Bronchitis

bronchitis: an inflammation of the bronchial tubes

Bacterial or viral infections, as well as reactions to environmental chemicals, can cause a variety of respiratory problems. **Bronchitis** refers to a wide variety of ailments characterized by a narrowing of the air passages. It is characterized by an inflammation of the mucous lining in the bronchial tubes. The excess production of mucus leads to inflammation and tissue swelling, a narrowing of the air passages, and decreased air movement through the bronchi. The condition becomes even more serious in the bronchioles. Unlike the trachea and the bronchi, the bronchioles are not supported by rings of cartilage that help keep them open.

bronchial asthma: a respiratory disorder characterized by a reversible narrowing of the bronchial passage

emphysema: a respiratory disorder characterized by an overinflation of the alveoli

Two conditions, **bronchial asthma** and **emphysema**, are associated with the inflammation of the bronchioles. In both conditions, greater effort is required to exhale than to inhale. This occurs because lower pressure is produced in the lungs and bronchioles during inspiration. The expansion of the lungs not only opens the alveoli but also the small bronchioles, increasing the diameter of the bronchioles and decreasing resistance to airflow. During expiration, increased pressure in the chest cavity compresses the lungs and the bronchioles, decreasing the diameter of the bronchioles and increasing resistance to airflow. Therefore, less air leaves the lungs. The imbalance between the amount of air entering the lungs and the amount of air leaving the lungs must be met by increasing the exertion of expiration.

Although all the causes of asthma are not yet known, the condition is often associated with allergies. An allergic reaction initiates swelling of the tissues. In the case of asthma, the tissues that line the walls of the bronchioles swell. It is also suspected that an allergic reaction causes spasms of the muscles lining the bronchiole walls. Both responses greatly increase the resistance to air flowing out of the lungs. The discomfort experienced during an asthmatic attack is associated not with getting air into the lungs, but with getting it out.

Emphysema

Emphysema is associated with long-term bronchitis. Like bronchitis, the disorder involves an increased resistance to airflow through the bronchioles. Although air flows into the alveoli fairly easily, the decreased diameter of the bronchioles creates resistance to the movement of air out of the lungs. Air pressure builds up in the lungs (in fact, the word emphysema means overinflated). Unable to support the building pressure, the thin walls of the alveoli stretch and eventually rupture. The fact that there are fewer alveoli means there is less surface area for gas exchange which, in turn, leads to decreased oxygen levels. In the body's attempt to maintain homeostasis, the breathing rate increases and exhalation becomes more laboured. The circulatory system adjusts by increasing the heart rate.

Another problem is that adjoining pulmonary capillaries and alveoli are both destroyed. Fortunately, blood clotting prevents any major internal hemorrhaging, but the healing process creates a secondary problem. Because the elastic tissue of the lungs is replaced by scar tissue, the lungs are less able to expand and, therefore, they hold less air. The broken-down cells of the alveoli do not regenerate, leaving less surface area for diffusion.

Lung Cancer

More Canadian men and women die from lung cancer than from any other form of the disease (**Figure 1**). As in other cancers, there is uncontrolled growth of cells. The solid mass of cancer cells in the lungs greatly decreases the surface area for diffusion. Tumours may actually block bronchioles, thereby reducing airflow to the lungs and potentially causing the lung to collapse.

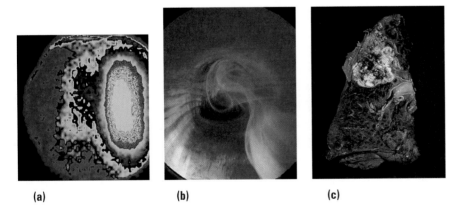

(a)   **(b)**   **(c)**

Figure 1
(a) A lung scan reveals cancer of the left lung. The colours in the healthy right lung indicate normal ventilation. On the left side, the absence of the normal colours and the presence of the purple colour instead indicate a nonfunctioning lung.
(b) Smoke descends toward the lungs.
(c) Postmortem specimen of a human lung showing a cancerous tumour of the upper lobe as a black and white area. The entire lung is permeated with black, tarry deposits, suggesting a history of heavy cigarette smoking.

SUMMARY   Disorders of the Respiratory System

1. All respiratory disorders decrease oxygen delivery to the tissues. Healthy lungs are much more efficient at gas exchange than unhealthy lungs are.

2. Bronchitis is an inflammation of the bronchioles, which results in narrowed air passages and decreased air movement.

3. Bronchial asthma is characterized by a narrowing of the bronchial passage.

4. Emphysema is overinflation of the alveoli. Air pressure builds up in the lungs, causing rupturing of the alveoli. Scar tissue replaces lung tissue, reducing the surface area available for diffusion.

5. Lung cancer is characterized by tumours, which reduce the surface area for diffusion.

Practice

Understanding Concepts

1. What is bronchitis? What are its effects on the respiratory system?
2. Describe the pressure changes that occur in the lungs during breathing for someone with emphysema.
3. Why does the uncontrolled growth of cells associated with lung cancer interfere with proper lung functioning?

Activity 8.4.1

Determining Lung Volume

Healthy lungs can take in more oxygen and expel wastes from the body with much greater efficiency than unhealthy lungs can. In this activity, you will examine indicators of general health by measuring lung capacity at rest.

Materials

respirometer with disposable mouthpieces

Procedure

1. Set the gauge to zero before you place a new, unused mouthpiece in the respirometer.

2. Be careful not to inhale at any time through the mouthpiece. Develop a regular, relaxed breathing pattern so you will obtain accurate results. After inhaling normally, place the mouthpiece attached to the respirometer in your mouth and exhale normally. Read the gauge on the respirometer. Record the volume exhaled as **tidal volume**.

3. Reset the respirometer to zero. Inhale normally, then place the mouthpiece attached to the respirometer in your mouth and exhale normally. Read the gauge on the respirometer and then exhale forcibly. Record the difference as **expiratory reserve volume**.

4. Reset the respirometer to zero. Inhale as much air as possible and then exhale for as long as you can into the respirometer. Read the gauge on the respirometer. Record the value as **vital capacity**.

5. Repeat steps 1 to 4 for two more trials, without changing the mouthpiece.

tidal volume: the amount of air inhaled and exhaled in a normal breath

expiratory reserve volume: the amount of air that can be forcibly exhaled after a normal exhalation

vital capacity: the maximum amount of air that can be exhaled

Analysis

(a) Determine your **inspiratory reserve volume** by using the following formula:

vital capacity = inspiratory reserve volume + expiratory reserve volume + tidal volume

inspiratory reserve volume: the amount of air that can be forcibly inhaled after a normal inhalation

Synthesis

(b) Predict how the tidal volume and vital capacity of a marathon runner might differ from that of the average Canadian.

(c) How might bronchitis affect your expiratory reserve volume? Provide your reasons.

(d) Predict how the respiratory volumes collected for a person with emphysema would differ from those you collected.

Practice

Applying Inquiry Skills

4. The following readings were taken for patients A and B on a respirometer (**Table 1**). Use the equation for vital capacity (page 296) to determine the unknown value in each case.

Table 1

	Patient A	Patient B
vital capacity (L)	?	5.1
inspiratory reserve volume (L)	3.0	2.8
expiratory reserve volume (L)	1.2	?
tidal volume (L)	0.6	0.8

5. The amount of air that remains in the lungs after a forced exhalation is called the residual volume. Why is it difficult to measure the residual volume experimentally?

6. Under conditions of heavy exercise, tidal volume is different than at rest. Would it be closer to one's expiratory reserve volume, inspiratory reserve voume, or vital capacity?

7. Why might respiratory volumes be measured during exercise? Provide a list of what could be investigated or diagnosed.

Reflecting

8. How has your ability to gather information from diagrams, graphs, and other figures improved since you started this course?

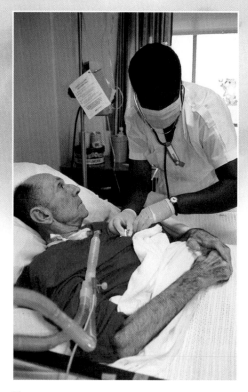

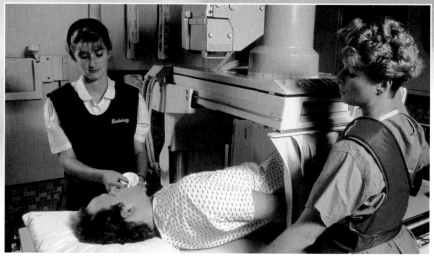

X-Ray Technicians

X-ray technicians perform various services such as advising patients on how to prepare for the X ray, as well as ensuring that the X ray itself is taken safely. X-ray technicians need to know about the fundamentals of anatomy, diseases, and traumas that affect the various body systems. They work in clinics and hospitals.

Nurse

Nurses require not only a thorough knowledge of human anatomy, but also a knowledge of drugs and how to administer them. They must also be sensitive to the psychological needs of their patients. Nurses work in a variety of places, including hospitals, doctors' offices, senior citizens' homes, and private homes.

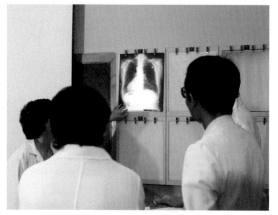

Oncologist

Oncology is the study of the causes, properties, disease progression, and treatment of tumours and cancers. Oncologists are doctors primarily involved in research. They work closely with physicians and other researchers, and conduct much of their studies through hospitals and universities.

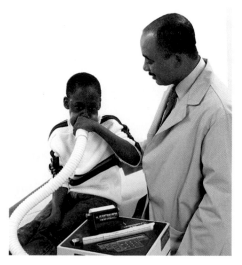

Respiratory Therapists

Respiratory therapists work in clinics and hospitals. Their duties include diagnosing respiratory problems, administering and monitoring oxygen and breathing devices, identifying treatments, and helping patients use the treatments.

Practice

Making Connections

9. Identify several careers that require knowledge about organ systems. Select a career you are interested in from the list you made. What university program do you need to take that will lead to that career? Which universities offer degree programs in this area? Which high-school subjects are required to enter the university program? Does the career require further training beyond university?

10. How are the concepts you have learned in this chapter used in the career?

11. Investigate and describe the responsibilities and duties involved in this career. What appeals to you about the career? What do you find less attractive?

12. Survey the newspaper or conduct a Web search to identify career opportunities in that area.

Investigation 8.4.1

The Effects of Exercise on Lung Volumes

The total lung capacity of fully grown, healthy lungs is about 5 L to 6 L of air. However, a person normally inhales only about 0.5 L and exhales the same volume. Various factors can affect the lung volume of a single breath. In this investigation, you will design ways to test the effects of exercise on lung volume during a single breath, that is, during one inhalation and exhalation.

Do not perform this activity if you are not allowed to participate in physical education classes.

Question

How does exercise affect the lung volume during a single breath?

Hypothesis/Prediction

Predict how exercise will affect lung volume. Create a hypothesis to explain your prediction.

Design

Design a controlled experiment to test your hypothesis. Include the following in your design:

- descriptions of the independent, dependent, and controlled variables
- a step-by-step description of the procedure
- a list of safety precautions
- a table to record observations

Materials

Make a list of the materials and apparatus needed to carry out the procedure.

Procedure

1. Submit the procedure, safety precautions, table, and list of materials and apparatus to your teacher for approval.

2. Carry out your investigation.

3. Write a report.

Analysis

(a) State how exercise affects lung volumes.

Evaluation

(b) Was your prediction correct? Was your hypothesis supported?
(c) Describe any problems or difficulties in carrying out the procedure.
(d) Include in your report ways to improve your current design.
(e) If you were to repeat this experiment, what new factors would you investigate? Write a brief description of the new procedure.

Case Study: Smoking and Lung Cancer

In contrast to skin cancers, lung cancers are almost always fatal—the five-year survival rate is not much better than 15%. Lung cancer is the second most common cancer, yet it is one of the most preventable. Prior to the use of tobacco, lung cancer was relatively rare. Smoking increased in popularity in the 1920s and it was usually men who smoked. In the 1940s, lung cancer began to increase at a

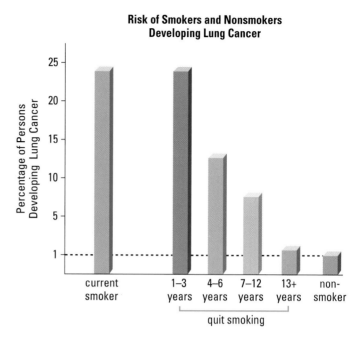

Risk of Smokers and Nonsmokers Developing Lung Cancer

Figure 3
When smokers stop smoking, their risk of lung cancer decreases with time.

dramatic rate, becoming the most common cancer in men. As more and more women began to smoke, lung cancer cases among women also rose significantly. In 1995, lung cancer surpassed breast cancer as the number one cancer killer of women. The World Health Organization estimates that, every year, 4 million people die as a result of smoking tobacco. This figure is expected to rise to 10 million by 2010.

When smokers quit the habit, their risk of developing lung cancer lessens over time (**Figure 3**). Also, as with most cancers, if lung cancer is detected at an early stage, there is a greater chance of surviving for five years. Some common symptoms include an unusual cough, sputum containing blood, hoarseness, and shortness of breath which is noticeable during physical activity. **Figure 4**

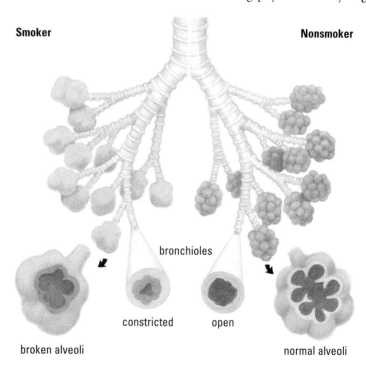

Figure 4
A comparison of the bronchioles and alveoli of a smoker and of a nonsmoker

shows how the bronchioles and alveoli of a smoker appear in comparison to those of a nonsmoker. In this case study, you will investigate how smoking can lead to lung cancer.

Evidence

1. Cancer usually begins in the bronchi or bronchioles. Components of cigarette smoke contribute to the development of cancerous tumours. The four diagrams in **Figure 5** show the development and progression of lung cancer.

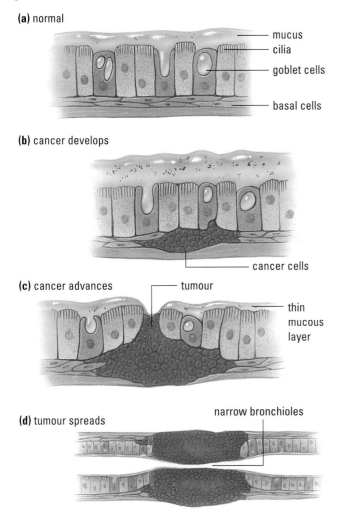

(a) normal
— mucus
— cilia
— goblet cells
— basal cells

(b) cancer develops
— cancer cells

(c) cancer advances
— tumour
— thin mucous layer

(d) tumour spreads
narrow bronchioles

Figure 5
Development of a tumour in the tissues of the bronchiole walls

2. Cigarette smoke travels through the bronchi and irritates the cells. Special cells produce mucus, which is designed to trap foreign particles. Compare the mucous layers in diagrams (a) and (b).

3. Ciliated cells line the bronchi. Cilia sweep away the debris trapped by the mucus. Unfortunately, the tar found in cigarette smoke slows the action of the cilia. The sludgelike tar becomes trapped in the mucus.

4. Diagram (b) shows the beginning of a cancerous tumour and diagram (c) shows how the tumour begins to grow. Note the location of the tumour and its growth.

5. While the cancerous tumour is still walled in by the basal membrane in diagram (c), the tumour has broken through the membrane in diagram (d).

Analysis

(a) How does cigarette smoke affect the mucous layer in the bronchi?

(b) Why does the buildup of tar in the bronchi limit airflow?

(c) In what area does a tumour begin to develop?

(d) Why has the mucous layer in diagram (c) decreased in size?

(e) At what stage might metastasis start to occur?

(f) Cancer cells often use lymph vessels as pathways to other parts of the body, where they continue to divide. Why does this characteristic make cancer especially dangerous?

DECISION-MAKING SKILLS

○ Define the Issue   ◉ Analyze the Issue
○ Identify   ◉ Defend a Decision
○ Alternatives   ◉ Evaluate
◉ Research

Explore an **Issue**

Debate: Smoking

The correlation between cigarette smoking and lung cancer has been established. Smoking has also been linked to high blood pressure, heart disease, and a host of other diseases. Individuals who smoke two or more packs of cigarettes a day are about 25 times more likely to develop lung cancer than nonsmokers. With increasing evidence of the dangers of secondhand smoke, nonsmokers' rights groups and health officials have successfully lobbied for a ban on smoking in the workplace and other public areas.

In 1996, the Canadian tobacco industry employed just under four thousand full-time and part-time workers and paid almost $250 million in wages. Thousands of seasonal workers were also employed, and thousands of wholesale and retail workers profit from the sale of tobacco products.

The government raises money through tobacco taxation. Federal and provincial taxes together account for 67% of the retail price of a pack of cigarettes. From 1996 to 1997, the federal government raised over $2 billion through tobacco taxes and Ontario earned over $350 million.

The tobacco industry's contributions to the Canadian economy are large. However, the costs associated with smoking are also large. It is estimated that, in 1992, the economic costs related to tobacco use in Canada (through, for example, the burden to the health-care system) were $9 billion. These costs are expected to increase.

Statement

The government should ban the sale of tobacco products.

- In your group, research the issue.
- Search for information in newspapers, periodicals, CD-ROMs, and on the Internet.
  Follow the links for Nelson Biology 11, 8.4.

 www.science.nelson.com

- Write a list of points and counterpoints that your group considered. You might consider the following: how smoking affects society, the influence of the tobacco industry on the economy, citizens' rights, the influence of peer pressure, self-image, and contentious lifestyle habits other than smoking.
- Decide whether your group agrees or disagrees with the statement.
- Prepare to defend your group's position in a class discussion.

Section 8.4 Questions

Understanding Concepts

1. Why is the slowing down of the cilia especially dangerous?

Applying Inquiry Skills

2. Nicotine, one of the components of cigarettes, slows the cilia lining the respiratory tract, causes blood vessels to constrict, and increases heart rate. Another component of cigarette smoke is carbon monoxide. Carbon monoxide competes with oxygen for binding sites on the hemoglobin molecule found in red blood cells. Analyze the data presented in this chapter, and describe the potential dangers associated with smoking.

3. Survey several people who smoke and calculate the amount of tar taken in each day. Most cigarettes contain about 15 mg of tar, with 75% of the tar being absorbed. Show your calculations.

Making Connections

4. On an X ray, a cancerous tumour shows up as a white spot (**Figure 6**). A healthy lung appears dark. Why would the tumour appear white?

Reflecting

5. How does what you have learned in this chapter relate to the information from Chapter 7? Has learning about the circulatory system helped you understand some of the processes of the respiratory system?

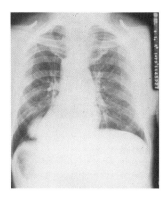

Figure 6
An X ray showing the presence of a tumour in the lower right lung

8.5 Organ Transplants in Canada

Canadian medical centres are among the world's leaders in transplanting donated human organs to save lives (**Figure 1**). **Table 1** lists some of the achievements that have occurred in Canada, including those that were the first in the world.

Table 1: First Successful Transplants Completed in Canada

Organs involved	Location	Date
heart valves*	Toronto	1956
kidney from living relative	Montreal	1958
kidney transplant using a deceased donor	Montreal	1963
heart	Montreal	1968
liver	Montreal	1970
bone marrow for a child	Toronto	1974
lung*	Toronto	1983
double lung*	Toronto	1986
liver and bowel*	London	1988
liver from living relative	London	1993
pig liver used to keep patient alive until a human liver became available	Montreal	1995
double lung (using two donors)	Winnipeg	1999

first in world

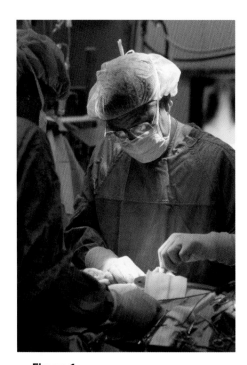

Figure 1
Heart transplant operation

The Multi-Organ Transplant Program at the London Health Sciences Centre, where about two hundred transplants are conducted annually, is one of the leading programs of its kind in the world. In 1998, Dr. David Grant, the lead surgeon in a nine-hour operation, successfully replaced the liver, bowel, stomach, and pancreas of a six-month-old girl who had been born with a rare disorder that prevented the normal functioning of her bowel and bladder. She also had a ruptured appendix and a twisted bowel, which meant she could be fed only through an intravenous tube. This method of feeding caused her liver to fail as well. The new organs were accepted by her body and she is expected to resume a normal lifestyle.

The success of the multi-organ program and the publicity surrounding it have resulted in an organ donation rate in the area surrounding London of twenty-six donors per million people. This compares with a national rate of about fourteen donors per million. If members of the public can be made aware of the possibility of saving lives, they may be motivated to sign the organ donor cards that are part of their driver's licence. Another organization helping to publicize the success rate of transplants is the Canadian Transplant Association, which sponsors the Canadian Transplant Games and other activities to show how recipients can excel in physical competition.

8.6 The Effect of Psychoactive Drugs on Homeostatic Adjustment

psychoactive drugs: drugs that affect the nervous system and often result in changes to behaviour

Psychoactive drugs are a group of legal and illegal drugs that exert their effect on the nervous system, disrupting its ability to receive information about the external or internal environment. Because the nervous system is the primary way in which your body receives information about changes in your internal and external environment, anything that distorts the nervous system's operation will create problems and prevent proper homeostatic adjustment.

stimulant: a drug that speeds the action of the central nervous system, often causing an increase in heart and breathing rate

While a person rests, heart rate and breathing rate are kept within a normal range by homeostatic adjustments, shown in **Figure 1**, graph (a). After taking a **stimulant**, such as caffeine, the homeostatic range for heart rate and breathing rate increase, shown in graph (b). After taking a **depressant**, such as alcohol, the ranges are lowered, shown in graph (c). Stimulants and depressants cause the homeostatic range to adjust to a different level than is normal.

depressant: a drug that slows down the action of the central nervous system, often causing a decrease in heart and breathing rate

Under normal circumstances, impulses are relayed between nerve cells in the brain by transmitter chemicals. A transmitter chemical released from one nerve

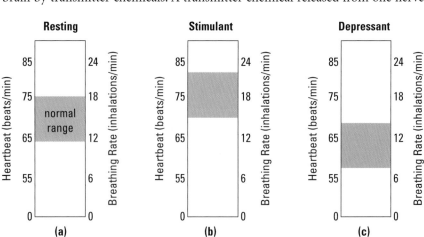

Figure 1
Stimulants and depressants can reset the operating range for heart rate and breathing rate.

cell attaches to receptor sites on another nerve cell. When enough receptor sites have been filled by the transmitter chemicals, the nerve cell membrane is disrupted and an impulse is initiated—the nerve cell fires. Psychoactive drugs interfere with either the movement of these transmitter molecules or their attachment to the receptor sites.

Depressants, such as tranquilizers, barbiturates, and alcohol, are a group of psychoactive drugs that slow down the action of the central nervous system (the brain and spinal cord). Some depressants delay the effect of transmitter chemicals by slowing the reaction of connecting nerves. Stimulants, such as cocaine, nicotine, and caffeine, are psychoactive drugs that speed up the action of the central nervous system. Some stimulants prevent the transmitting chemicals from breaking down or recycling. The transmitting chemical remains longer than it normally would and keeps the receptor sites on the nerve cell full, resulting in more frequent firing of the nerve cell.

Alcohol

Alcohol, a depressant, is one of the most widely used and abused of the psychoactive drugs. By interfering with the nervous system, alcohol affects other systems within the body. Nerves that conduct breathing movements from the brain stem are depressed. Alcohol consumption slows the heart, which in turn lowers oxygen delivery to the tissues of the body. Alcohol also affects nerve cells in the brain that control the release of a hormone that regulates water reabsorption by the kidneys. Under the influence of alcohol, the kidneys' ability to reabsorb and store water is impaired, and urine output increases. In turn, the loss of fluids from the body affects blood pressure (**Figure 2**).

One of the most pronounced effects of alcohol on homeostatic mechanisms occurs in the liver. Alcohol is readily broken down and used for energy, preventing the body from using other nutrients such as sugars, amino acids, and fatty acids. These nutrients are converted to fat and stored in the liver. The accumulation of fats in the liver is known as cirrhosis. In cirrhosis, normal liver cells

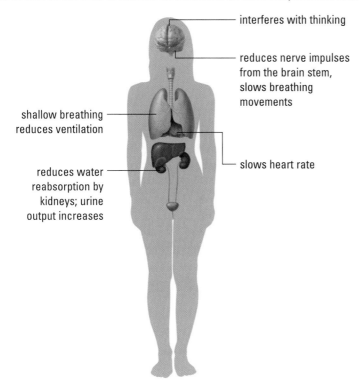

interferes with thinking

reduces nerve impulses from the brain stem, slows breathing movements

shallow breathing reduces ventilation

slows heart rate

reduces water reabsorption by kidneys; urine output increases

Figure 2
Effects of alcohol

are replaced by the fats, and the liver can no longer eliminate chemical wastes from the body, produce needed proteins, or carry out many of its other vital functions. A buildup of toxic wastes or a lack of needed proteins affects every organ system.

Nicotine

Nicotine is one of the most widely used, and most addictive, stimulants. A component of the tobacco plant, it is commonly taken in with cigarette smoke. When inhaled with cigarette smoke, nicotine reaches the brain in approximately 10 seconds. The binding of nicotine to certain receptor sites creates a feeling of pleasure in the brain. This is the same area affected by cocaine, another stimulant. Nicotine increases heart and breathing rates. It also increases the metabolic rate of cells by resetting the rate at which oxygen and sugars are broken down during cellular respiration.

Addiction is the body's attempt to cope with the chemical disruption caused by a drug. Nerve cells in the brain adjust to prolonged exposure to a drug by producing fewer receptor proteins. As a result, the neurons become less sensitive to the drug. As time passes, more of the drug is needed to maintain the same pleasurable feeling.

addiction: a compulsive need for a harmful substance. Addiction is characterized by tolerance to the substance and physiological withdrawal symptoms.

Practice

Understanding Concepts

1. Differentiate between stimulants and depressants.
2. What are the effects of stimulants and of depressants on heart rate and breathing rate?

SUMMARY The Effect of Psychoactive Drugs on Homeostatic Adjustment

1. While resting, homeostatic adjustments keep heart rate and breathing rate within a normal range.
2. Transmitter chemicals relay impulses between nerve cells in the brain. A transmitter chemical released from one nerve cell attaches to another. When enough receptor sites have been filled, the nerve cell fires.
3. Psychoactive drugs disrupt the ability of the nervous system to receive information about the external or internal environment. Psychoactive drugs work by affecting either the movement of the transmitter chemicals or their ability to attach to the receptor sites.
   - Depressants slow down the action of the central nervous system; some work by slowing the reaction of connecting nerves. Depressants lower the homeostatic range for heart rate and breathing rate.
   - Stimulants speed up the action of the central nervous system; some work by preventing transmitting chemicals from breaking down, causing frequent firing of the nerve cell. Stimulants raise the homeostatic range for heart rate and breathing rate.
4. Alcohol is a depressant.
   - Some of its effects are slowed heart rate and breathing rate, slowed oxygen delivery to the body, and decreased water reabsorption by the kidneys.

- The body uses alcohol for energy and converts other nutrients into fats that are stored in the liver. The fats replace normal liver cells, preventing the liver from performing its functions.

5. Nicotine is a highly addictive stimulant; it increases heart rate, breathing rate, and metabolic rate.

Sections 8.5–8.6 Questions

Understanding Concepts

1. Explain why a stimulant might increase the respiratory rate.

2. Describe how transmitter chemicals in the brain work, and the effects that depressants and stimulants have on their function.

3. Describe the short-term and potential long-term effects of alcohol on the body.

4. Explain addiction as a response to chemical disruption in the brain.

5. Using homeostatic mechanisms, explain why using a stimulant prior to exercise is dangerous.

Making Connections

6. Describe some reasons why people would and would not want to donate their organs. Which reasons do you find most compelling?

7. How might an understanding of the effect of depressants (such as alcohol) and stimulants (such as caffeine) on internal homeostasis affect a person's decisions about their use?

8.7 Digestive, Circulatory, and Respiratory Systems of the Fetal Pig

Like humans, the pig is a placental mammal, meaning that the fetus receives nourishment from the mother through the umbilical cord. In this investigation, you will study the circulatory, digestive, and respiratory systems of the fetal pig. Because the anatomy of the fetal pig resembles that of other placental mammals, this activity serves two important functions: it provides a representative overview of vertebrate anatomy, and it provides the framework for understanding functioning body systems. Pigs resemble humans in two other important ways: their skin lacks fur or feathers, and they are omnivores. Because of these similarities, the digestive tracts and layers underlying the skin in pigs and humans are more alike than those of some other mammals.

In Chapter 11 of this text, you will be examining the various systems of an invertebrate when you dissect an earthworm. Comparing the internal structures of the fetal pig and earthworm will allow you to see how the larger body of the fetal pig requires much more complex systems to allow nutrients and dissolved gases to enter and leave individual cells throughout the body.

Examining the Systems of a Fetal Pig

Read and follow the procedure carefully. Accompanying diagrams are included for reference only. Use the appropriate dissecting instruments. This activity has been designed to minimize the use of a scalpel.

Materials

safety goggles	string	dissecting pins
lab apron	scalpel	scissors
dissecting gloves	hand lens	ruler
preserved pig	dissecting tray	forceps and probe

Procedure

Part 1: External Anatomy

1. Place your pig in a dissecting tray. Using a ruler, measure the length of the pig from the snout to the tail. Use the graph in **Figure 1** to estimate the age of the fetal pig. Record your estimate.

Age/Size Ratio of Fetal Pig

Length (mm) vs *Age (days)*

Figure 1

2. Use **Figure 2** to help you identify the four regions of the pig's body: the head, the neck, the trunk, and the tail.

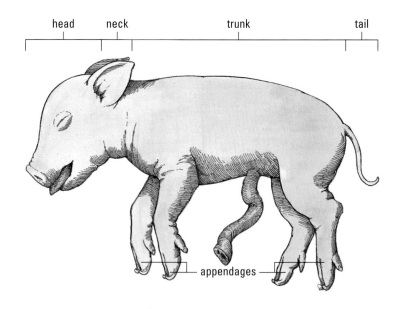

head   neck   trunk   tail

appendages

Figure 2
Lateral view of a fetal pig

- Wear safety goggles and an apron at all times.
- Wear plastic gloves when handling the preserved specimen and when performing a dissection to prevent any chemicals from coming in contact with your skin.
- Wash all splashes of preservative from your skin and clothing immediately. If you get any chemical in your eyes, rinse for at least 15 min.
- Work in a well-ventilated area. To reduce your exposure to any fumes from the preservative, make sure to avoid placing your face directly over the dissecting tray.
- Always cut away from yourself and others sitting near you, in case the scalpel slips.
- When you have finished the activity, clean your work area, wash your hands thoroughly, and dispose of all specimens, chemicals, and materials as instructed by your teacher.

3. Place the pig on its back (dorsal surface) and observe the umbilical cord. Locate the paired rows of nipples along the ventral surface of the pig. Both males and females have these nipples.

4. Examine the feet of the fetal pig. Indicate the position and number of toes.

Part 2: Abdominal Cavity

During the dissection, you will be directed to examine specific organs as they become visible. Remove only those organs indicated by the dissection procedure. Proceed cautiously to prevent damaging underlying structures.

5. With the pig still on its dorsal surface, attach one piece of string to one of the pig's hind legs, pull it under the dissecting pan, and tie it to the other hing leg. Repeat the procedure for the fore legs.

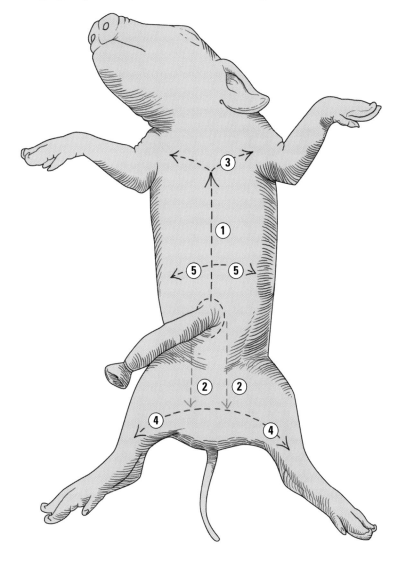

Figure 3
Ventral view of a fetal pig

6. Using scissors, make the incision indicated as #1 in **Figure 3**. Start by cutting around the umbilical cord, and then cut straight toward the anterior (head) of the pig.

7. Make incision #2 toward the posterior of the pig. Make incision #3 near the neck, and then incision #4. Make lateral incision #5; this incision runs

parallel to the diaphragm, which separates the thoracic cavity from the abdominal cavity.

8. Pull apart the flaps along incision #5, exposing the abdominal cavity. Use the probe to open the connective tissue (peritoneum) that holds the internal organs to the lining of the body cavity. Now pull apart the flaps of skin covering incision #4 to expose the posterior portion of the abdominal cavity. Use pins to hold back the flaps of skin (see **Figure 4**).

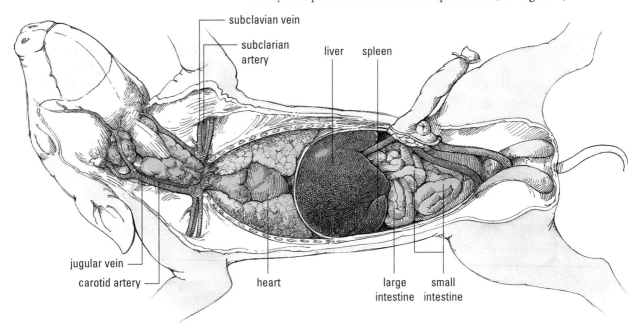

Figure 4
Abdominal cavity and thoracic cavity of the fetal pig. Organs of the digestive system and circulatory system are highlighted in the diagram.

9. Locate the liver near the anterior of the abdominal cavity. Record the number of lobes in the liver.

10. Using a probe, lift the lobes and locate the saclike gall bladder. Describe the location of the gall bladder.

11. Follow the thin duct from the gall bladder to the coiled small intestine. Bile salts, produced in the liver, are stored in the gall bladder. The bile duct conducts the fat-emulsifying bile salt to the small intestine.

12. Locate the J-shaped stomach beneath the liver. Using forceps and a probe, lift the stomach and locate the esophagus attached near its anterior end. Locate the small intestine at the posterior junction of the stomach. The coiled small intestine is held in place by mesentery (a thin, somewhat transparent, connective tissue). Note the blood vessels that transport digested nutrients from the intestine to the liver.

13. Using a probe and forceps, lift the junction between the stomach and small intestine, removing supporting tissue. Uncoil the junction and locate the creamy-white pancreas. The pancreas produces a number of digestive enzymes and a hormone called insulin, which helps regulate blood sugar. Describe the appearance of the pancreas.

14. Locate the spleen, the elongated organ found around the outer curvature of the stomach. The spleen stores red and white blood cells. The spleen also removes damaged red blood cells from the circulatory system.

15. Using a scalpel, remove the stomach from the pig by making transverse (crosswise) cuts near the junction of the stomach and the esophagus, and near the junction of the stomach and small intestine. Make a cut along the

midline of the stomach, and open the cavity. Rinse as instructed by your teacher. View the stomach under a hand lens. Describe the appearance of the inner lining of the stomach.

Part 3: Thoracic Cavity

16. Carefully fold back the flaps of skin that cover the thoracic cavity. You may use dissecting pins to attach the ribs to the dissecting tray. List the organs found in the thoracic cavity (**Figure 5**).

Figure 5
Thoracic cavity and urogenital system

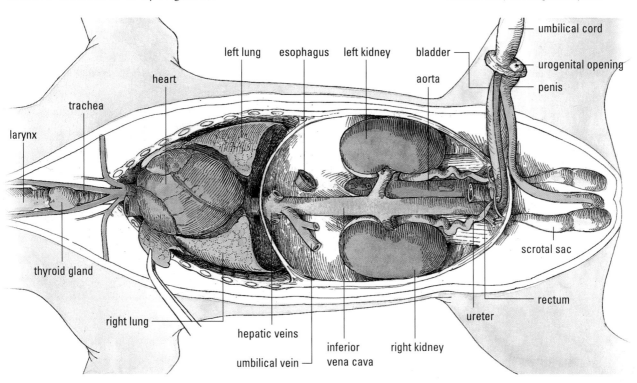

17. Locate the heart. Using forceps and a probe, remove the pericardium (a thin connective tissue covering the heart) from the outer surface of the heart. The large blood vessel that carries blood from the lower parts of the body to the right side of the heart is called the inferior vena cava (**Figure 6**). (The right side refers to the pig's right side.)

Figure 6
(a) Ventral view of the heart
(b) Dorsal view of the heart

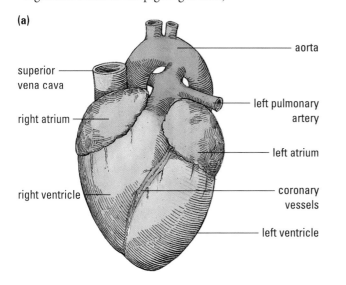

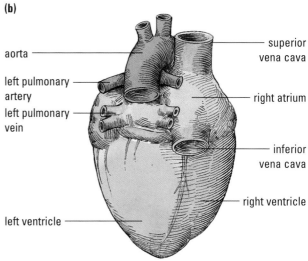

18. Blood from the head and upper body enters the right side of the heart through the superior vena cava. Both the superior and inferior venae cavae are considered to be veins because they bring blood to the heart.

19. Trace blood flow through the heart. Blood entering the right side of the heart collects in the right atrium. Blood from the right atrium is pumped into the right ventricle. Upon contraction of the right ventricle, blood flows to the lungs by way of the pulmonary artery. Arteries carry blood away from the heart. Blood, rich in oxygen, returns from the lungs by way of the pulmonary veins and enters the left atrium. Blood is pumped from the left atrium to the left ventricle and out the aorta.

20. Make a diagonal incision across the heart and expose the heart chambers. Compare the thickness of the wall of a ventricle with that of the wall of an atrium.

21. Locate the spongy lungs on either side of the heart and find the trachea leading into the lungs (**Figure 7**).

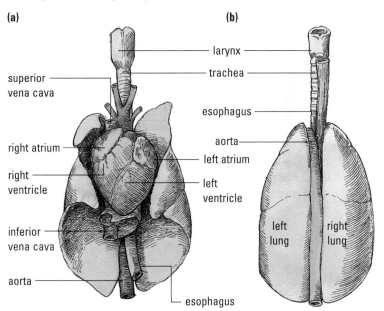

Figure 7
(a) Ventral view of heart and lungs
(b) Dorsal view of heart and lungs

22. Place your index finger on the trachea and push downward. Describe what happens.

Analysis

(a) What is the function of the umbilical cord?
(b) State the function of the following organs: stomach, liver, small intestine, gall bladder, pancreas, large intestine, and spleen.
(c) What is the function of the mesentery?
(d) Why does the left ventricle contain more muscle than the right ventricle?
(e) Why do the lungs feel spongy?
(f) What function do the cartilaginous rings of the trachea serve?
(g) Make labelled diagrams of the following:
   • digestive system
   • heart and blood vessels associated with it
   • the respiratory system
(h) Write a report in which you point out the similarities and differences between the anatomy of the pig and of the human.

Key Expectations

Throughout this chapter, you have had opportunities to do the following:

- Compare the anatomy of different organisms—vertebrate and/or invertebrate (8.1, 8.7).
- Describe the process of ventilation and gas exchange from the environment to the cell (8.1, 8.2, 8.3).
- Explain the role of the transport systems in the transport of substances in an organism (8.1, 8.2, 8.3).
- Select and integrate information about internal systems from various print and electronic sources, or from several parts of the same source (8.4).
- Locate, select, analyze, and integrate information on topics under study, working independently and as part of a team, and using appropriate library and electronic research tools, including Internet sites (8.4).
- Identify and describe science- and technology-based careers related to the subject area under study (8.4).
- Select and use appropriate modes of representation to communicate scientific ideas, plans, and experimental results (8.4, 8.7).
- Select appropriate instruments and use them effectively and accurately in collecting observations and data (8.4, 8.7).
- Design and carry out, in a safe and accurate manner, an experiment on feedback mechanisms, identifying specific variables (8.4).
- Demonstrate the skills required to plan and carry out investigations, using laboratory equipment safely, effectively, and accurately (8.4, 8.7).
- Demonstrate an understanding of how fitness level is related to the efficiency of the respiratory system (8.4).
- Present informed opinions about how scientific knowledge of internal systems influences personal choices concerning nutrition and lifestyle (8.4, 8.6).
- Describe how the use of prescription and nonprescription drugs can disrupt or help maintain homeostasis (8.6).

Key Terms

addiction
alveoli
breathing
bronchi
bronchial asthma
bronchioles
bronchitis
buffer
carbonic anhydrase
cilia
Dalton's law of partial pressure
depressant
diaphragm
emphysema
epiglottis
expiratory reserve volume

external intercostal muscles
inspiratory reserve volume
internal intercostal muscles
larynx
pleural membrane
psychoactive drugs
respiration
respiratory membrane
stimulant
tidal volume
trachea
vital capacity

Make a Summary

In this chapter, you studied how gases are exchanged between your body and the external environment. To summarize your learning, create a flow chart or diagram that shows how the respiratory system maintains homeostasis. Label the diagram with as many of the key terms as possible. Check other flow charts or diagrams and use appropriate designs to make your sketch clear.

Reflect on your Learning

Revisit your answers to the Reflect on Your Learning questions at the beginning of this chapter.

- How has your thinking changed?
- What new questions do you have?

Understanding Concepts

1. **Figure 1** shows the components of the human respiratory system.
   (a) Indicate the structure and functions of labels w, x, and y.
   (b) Which structure(s) have cartilaginous rings?
   (c) The inflammation or restriction of airflow in which structure is associated with asthma?

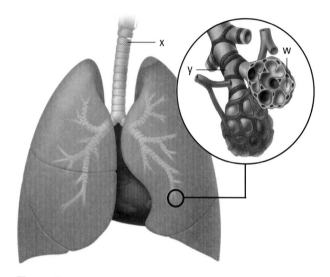

Figure 1

2. Explain the advantage of countercurrent flow in the gills of fish.

3. Compare the gills of fish and the lungs of frogs by identifying advantages and limitations of each.

4. Describe the pressure changes that occur during inhalation and exhalation.

5. Place the following structures in the order in which air passes through them during exhalation: bronchioles, trachea, pharynx, nasal cavities, alveolus, and bronchi.

6. A list of breathing actions is provided below. Identify the four that would be involved in expiration following strenuous exercise.
   (a) thoracic volume increases
   (b) thoracic volume decreases
   (c) diaphragm relaxes
   (d) diaphragm contracts
   (e) external intercostal muscles relax
   (f) external intercostal muscles contract
   (g) internal intercostal muscles relax
   (h) internal intercostal muscles contract

7. Why do breathing rates increase in crowded rooms?

Applying Inquiry Skills

8. A man has a chest wound. The attending physician notices that the man is breathing rapidly and gasping for air.
   (a) Why does the man's breathing rate increase? Why does he gasp?
   (b) What could be done to restore normal breathing?

9. The composition of air was analyzed from inhaled and exhaled air (**Table 1**).
   (a) Explain why more water is found in exhaled air.
   (b) Explain the difference in oxygen levels in inhaled and exhaled air.
   (c) If nitrogen is not used by the cells of the body, account for the different composition between inhaled and exhaled air.

Table 1

Gas component	Inhaled air (%)	Exhaled air (%)
oxygen	20.71	14.60
carbon dioxide	0.41	4.00
water	1.25	5.90
nitrogen	78.00	75.50

10. Changes in the partial pressure of gases in arterial blood were monitored over time as a subject began to perform light exercise (**Figure 2**).
   (a) At which time would the breathing rate likely be greatest? Provide reasons for your answer.
   (b) Predict when the subject began exercising. Give your reasons.
   (c) When would the breathing rate return to normal? Give your reasons.

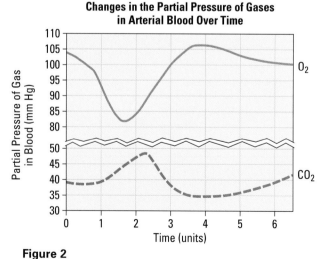

Figure 2
Changes in the partial pressure of gases in arterial blood over time

11. During mouth-to-mouth resuscitation, exhaled air is forced into the victim's trachea. Exhaled air contains a higher level of CO_2 than atmospheric air. Would the higher level of CO_2 create problems or would it be beneficial? Provide reasons for your answer.

12. Prior to swimming underwater, a diver breathes deeply and rapidly for a few seconds. How does hyperventilating help the diver hold her breath longer?

13. In **Figure 3**, which hemoglobin is more effective at absorbing oxygen? What adaptive advantage is provided by hemoglobin that allows it to combine readily with oxygen?

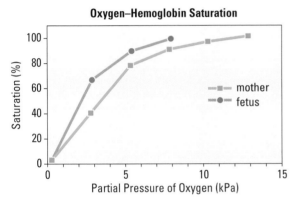

Figure 3
Fetal and adult hemoglobin

14. A scientist sets up the following experimental design (**Figure 4**). Note that sodium hydroxide absorbs carbon dioxide, and limewater turns cloudy when it absorbs carbon dioxide.
    (a) Indicate the purpose of the experiment.
    (b) Which flask acts as a control?
    (c) Why is sodium hydroxide used?

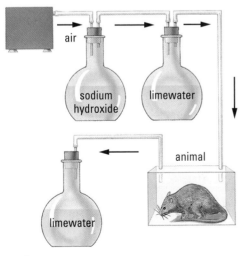

Figure 4

Making Connections

15. A patient is given a sedative that inhibits nerves leading to the pharynx, including those that control the epiglottis. What precautions would you take with this patient? Give your reasons.

16. Nicotine causes blood vessels, including those in the placenta, to constrict. Babies born to women who smoke are, on average, about 1 kg smaller than normal. This may be related to decreased oxygen delivery. Speculate about other problems that may face developing embryos due to the constriction of blood vessels in the placenta.

17. Cigarette smoke has been found to have the following effects on the respiratory system:
    • There is destruction of many of the cilia that line bronchi and bronchioles.
    • There is a buildup of mucus along the walls of the bronchioles. This reduces the interior diameter of the tubes.
    • There is an increase in blood pressure that causes the rupturing of the walls of some of the alveoli.

    Refer to each of the effects listed above. Indicate specific ways in which the normal functioning of the respiratory tract is altered by smoking tobacco.

Exploring

18. Allan Becker of the University of Manitoba studies dogs to learn more about how asthma works in people. Asthma is frequently associated with allergies, especially in children. Research how allergies have been linked with asthma. Indicate the benefits of using modelling experiments on dogs. Are there any disadvantages? Follow the links for Nelson Biology 11, Chapter 8 Review.

GO TO www.science.nelson.com

Determining Fitness Level

Figure 1
Khalid Khannouchi

Since the Amateur Athletics Federation began keeping records, there has been a steady increase in the performance level of elite athletes. Today, Olympic athletes run faster, jump higher, and throw farther than ever before. Since the turn of the century, records for power events have greatly improved, such as the 100-m sprint which improved by 10%, and the long jump which improved by 20%. In the endurance events, previous records have been surpassed by an even greater margin. At the 1908 Olympics in London, the winning time for the marathon was 2:55:18. In 1999, Khalid Khannouchi (**Figure 1**) covered that same distance in 2:05:42, almost 30% faster. Better nutrition and improvements in technology and training help to explain the progress.

Despite improvement in the performances of elite athletes over the past century, the fitness level of the general public has declined. A sedentary lifestyle and changing eating habits are factors in the increased mass of the average person. It has been estimated that one out of every three meals for most North Americans is consumed at a fast food outlet, where meals are high in fat content.

The ability to exercise depends upon your ability to deliver oxygen and nutrients to the muscles and brain. Oxygen demand increases as activity becomes more vigorous. An Olympic cross-country skier and an out-of-shape "couch potato" with the same size body have the same oxygen demands while resting or running. The skier's superior performance during exercise is due to the fact that she delivers more oxygen to her cells during vigorous exercise; her cardiac output and respiratory system make more oxygen available to the cells.

Investigation

For this task, you and a partner will design and carry out a fitness test that indicates, by indirect means, the amount of oxygen being delivered to the tissues. You will not actually be measuring the amount of oxygen delivered. However, your ability to perform certain physical tasks and the speed at which they can be done are indicators of the amount of oxygen being delivered by the respiratory and circulatory systems and of the amount used by the muscles.

To design a fitness test, you need to consider a number of factors, including the following:

- the indicators you will use to determine fitness level
- the equipment you will use (diagnostic and exercise)
- the availability of equipment (at school, at home, at a fitness centre)
- safety precautions (regarding subject's physical health and use of equipment)
- the number of subjects who will be tested

Present your design to your teacher for approval before beginning your fitness testing. After your design has been approved, carry out the test.

Once your investigation is completed, write a detailed report to communicate the Procedure, Observations, and Analysis of your investigation. Use appropriate scientific vocabulary, tables, and correct significant figures and SI units where appropriate.

Question

How can you determine the fitness level or oxygen uptake of an individual during exercise?

Materials/Equipment

respirometer (commercial or homemade model)

Complete the list of materials and equipment (diagnostic and exercise) required.

Procedure

1. Describe in detail each component of the test.
2. Be sure to provide instructions on the use of all equipment.
3. Include safety precautions where appropriate.

Observations

- Include base-line measurements of lung volumes, pulse rate, and other appropriate data.
- Display your data in a table, chart, and graph.

Analysis

(a) Determine the fitness level of the test subject. Provide the formula(s) necessary to determine this value and be sure to use the correct SI units for the measurements involved. Include the measurement of uncertainty, as a percentage, if appropriate.

(b) You may want to compare your results against a chart of standards, either standards determined by you and your group or from a source on the Internet.

Follow the links for Nelson Biology 11, Unit 3 Performance Task.

GO TO www.science.nelson.com

Evaluation

(c) Were there difficulties in obtaining the data? If so, suggest ways to improve the design.

(d) Evaluate the chart of standards. What are the limitations of using these standards to evaluate a person's fitness?

Synthesis

(e) Often people associate slimness with being physically fit. Provide arguments to the contrary.

(f) People who diet are often advised to incorporate exercise into their daily routine. Describe the relationship between diet and exercise in terms of the improvements in cardiovascular fitness and metabolism. Provide a case study example.

(g) Describe how you might customize your fitness test for specific sports (e.g., hockey, football, golf) or for specific positions (e.g., goalie, quarterback).

Understanding Concepts

1. Select the phrase to complete the following statement correctly:
   A heart attack will result from the lack of nutrients and oxygen to the heart muscles due to blockage by athero-sclerosis of the
   (a) pulmonary arteries;   (b) coronary arteries;
   (c) pulmonary veins;   (d) coronary veins.

2. Identify the blood vessel that is being referred to in each of the following statements:
   (a) This blood vessel is the site of diffusion of oxygen and nutrients.
   (b) This blood vessel has the highest blood pressure.

3. Select the phrase to correctly complete the following statement:
   A person with type A blood can
   (a) donate to blood types B and A, but only receive from type O;
   (b) donate to blood types AB and A, and receive from types O and A;
   (c) donate to blood types O and A, and receive from types O and B;
   (d) donate to blood types AB and A, but only receive from type O.

4. Identify the process by which gases move from the alveoli into the capillaries.
   (a) active transport   (b) osmosis
   (c) filtration   (d) diffusion

5. Select the phrase to correctly complete the following statement:
   Complete separation of the pulmonary and systemic circulation systems is necessary to provide
   (a) increased cardiac output;
   (b) more efficient operation of the lungs;
   (c) more efficient oxygenation of the blood;
   (d) increased ventricular contractions.

6. Select the phrase to correctly complete the following statement:
   Optimum gas exchange between the atmosphere and respiratory surface will occur if
   (a) the respiratory membrane is moist, the distance for diffusion is small, and the surface area for diffusion is large;
   (b) the respiratory membrane is dry, the distance for diffusion is large, and the surface area for diffusion is small;
   (c) the respiratory membrane is dry, the distance for diffusion is small, and the surface area for diffusion is small;

   (d) the respiratory membrane is moist, the distance for diffusion is large, and the surface area for diffusion is large.

7. **Figure 1** shows the components of the human respiratory system.
   Identify the structure by number and name that is described in each of the following statements:
   (a) This muscular structure relaxes during exhalation, causing the volume of the chest cavity to decrease.
   (b) This structure conducts air into the left lung.
   (c) This structure prevents food from entering the trachea.
   (d) Inhalation and exhalation are indicated by pressure changes within these structures.

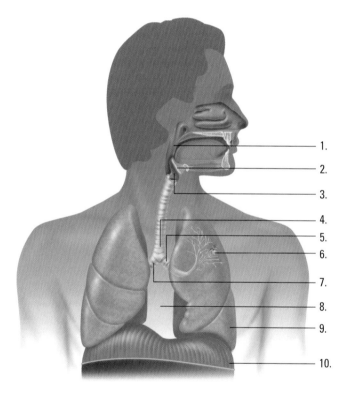

Figure 1

8. Which statement correctly describes how breathing rate is regulated?
   (a) The heart controls the breathing rate by monitoring oxygen levels.
   (b) The heart controls the breathing rate by monitoring carbon dioxide levels.
   (c) The brain controls the breathing rate by monitoring oxygen levels.
   (d) The brain controls the breathing rate by monitoring carbon dioxide levels.

9. Describe two ways the temperature regulating system in a building is similar to the feedback mechanism that regulates temperature in the human body.

Applying Inquiry Skills

10. Amylase digestion of starch is tested by the experiment shown in **Figure 2**. Each of the flasks is filled with 100 mL of 4% starch suspension. A 1% amylase solution is added to flask 1. The amylase solution is boiled for 2 min then added to flask 3.
    (a) Identify the control in this experiment.
        (i) flask 2   (ii) flask 3
        (iii) flasks 1 and 2   (iv) flasks 1 and 3
    (b) Iodine was added to the flasks after 10 min. The blue-black colour, indicating the presence of starch, would likely be observed in which flask(s)? Explain why.

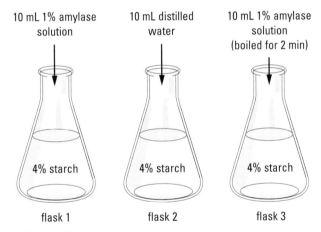

10 mL 1% amylase solution   10 mL distilled water   10 mL 1% amylase solution (boiled for 2 min)

4% starch   4% starch   4% starch

flask 1   flask 2   flask 3

Figure 2

11. **Table 1** shows the food energy in equal volumes of various dairy products.

    Different research groups proposed different explanations after examining the data.

    Group 1: All milk has the same energy value.

    Group 2: The greater the mass of the milk, the greater the energy value.

    Group 3: The greater the fat content, the greater the energy value.

    Group 4: You can't determine the energy value of different types of milk by looking at the volume used.

    (a) Which of the following hypotheses are reasonable?
        (i) 1 and 3   (ii) 2 and 4
        (iii) 3 and 4   (iv) 1 and 4
        (v) 2 and 3   (vi) none
    (b) Identify the independent and dependent variables.

(c) If someone were concerned about excessive energy intake, which dairy products would you recommend avoiding? Explain why.

Table 1: Dairy Products and Their Food Energy

Milk product	Mass (g) per 250 mL	Food energy (kJ) per 250 mL
whipping cream	252	3640
whole milk	257	660
2% milk	258	540
skim milk	258	380
buttermilk	258	430
evaporated milk	356	1490

12. A young child (2 to 3 years of age) requires approximately 6000 kJ of energy per day. An adult office worker can require 11 000 kJ of energy per day. The difference could be explained by which of the following? Choose as many answers as are applicable.
    (a) This adult is more active than all children.
    (b) This child is more active than this adult.
    (c) This adult is much larger than this child.
    (d) All adults lose heat faster than children do.

13. Pancreatin is a commercially prepared mixture of the components of the pancreas, including trypsin and lipase. An experiment was conducted to determine the effect of pancreatin and bile on the digestion of egg yolk. Egg yolk contains lipids and proteins. The scientist placed 10 g of egg yolk in each of 4 test tubes and incubated at 37°C for 24 h. As shown in **Table 2**, the pH of the solution was recorded at the beginning of the experiment and after 24 h. The amount of digestion is indicated by plus signs (+). The greater the number of plus signs, the greater the amount of digestion.

Table 2: Effects of Pancreatin on Lipids and Proteins

Test tube #	Initial pH of solution	Pancreatin	Bile	pH after 24 h	Amount of digestion
1	9	no	no	9	none
2	9	✓	no	7	+++
3	9	none	✓	9	+
4	9	✓	✓	6	++++

    (a) Which test tube acted as a control? What does the control indicate?
    (b) Explain why the pH of the solution changes after 24 h in test tubes 2 and 4.

(c) Explain why test tube 4 shows a greater amount of digestion than test tube 2.

(d) Draw a conclusion from the results of test tube 3.

14. Which of the following respiratory volumes cannot be measured directly using a respirometer?
    (a) tidal volume
    (b) expiratory reserve volume
    (c) inspiratory reserve volume
    (d) vital capacity

15. A respirometer was used to collect the following data from three different subjects (**Table 3**).
    (a) Calculate x, y, and z.
    (b) Predict which subject is likely the smallest. Justify your answer.

16. Hemoglobin and myoglobin are two proteins that carry oxygen. Myoglobin, found in muscle cells, has the ability to combine with one molecule of oxygen. Hemoglobin, found in red blood cells, has the ability to combine with four oxygen molecules. **Figure 3** shows the ability of hemoglobin and myoglobin to combine with oxygen in varying levels of partial pressure.

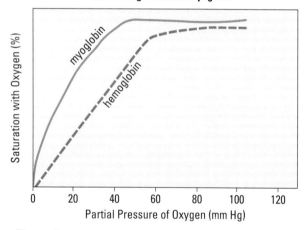

Oxygen Saturation Curves for Hemoglobin and Myoglobin

Figure 3

(a) Which protein accepts oxygen more readily?
(b) At what partial pressure does hemoglobin become saturated?
(c) At what partial pressure does myoglobin become saturated?
(d) Describe the adaptation for exercise by comparing the saturation curves for hemoglobin and myoglobin.

17. A student is given eight samples of red blood cell suspensions and the serum from each sample. The suspensions only contain red blood cells and the serum is blood plasma from which fibrinogen has been removed. She is asked to identify each of the four blood types present. To test these samples, the only materials she has at her disposal are a sample of type B red blood cell suspension and the serum from type B blood.
    (a) Describe the procedure, step by step, that she must use to identify the four blood types present. In each step, interpret what it means if the cells in the sample clump together, or if they do not clump together.
    (b) Explain why serum was used instead of plasma.

Making Connections

18. A glassful of milk contains lactose, proteins, butterfat (mostly triglycerides), vitamins, and minerals. Explain what happens to each component in your digestive tract.

19. As a person ages, the number of body cells steadily decreases and energy needs decline. If you were planning an older person's diet, what foods would you emphasize and why? Which ones would you de-emphasize?

20. Often, holiday meals are larger than regular meals and have a higher fat content. You may have noticed after eating one of these meals that you feel uncomfortably full for longer than normal. Based on what you have learned about digestion, suggest a biochemical explanation for the discomfort.

21. Lung cancer is the leading cause of cancer death in both men and women in Canada. It is also a disease that can be prevented. Controllable environmental factors seem

Table 3: Lung Volumes

Subject	Breaths per minute	Tidal volume (mL)	Exhaled air (mL/min)	Expiratory reserve volume (mL)	Inspiratory reserve volume (mL)	Vital capacity (mL)
1	10	400	4000	1500	2400	x
2	12	500	6000	y	2500	5000
3	15	z	9000	1700	2700	5000

to stimulate a number of cancer-causing genes over a period of time to become active, causing cells to develop into one or more of the various forms of lung cancer. Interpret the information presented in **Figure 4**.

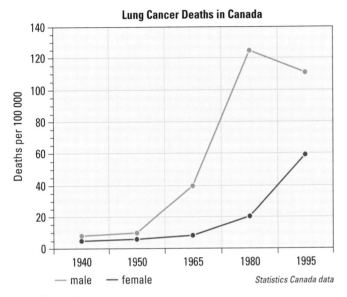

Figure 4

(a) In the early 1920s, shortly after the First World War, smoking became fashionable for men. Hypothesize why lung cancer rates did not increase until the 1950s.

(b) Suggest a reason why no comparable increase occurred in lung cancer in women during the same period. Justify your answer.

(c) Predict trends in lung cancer over the next 10 to 20 years.

(d) Compare the trends between males and females between 1980 and 1995.

(e) Copy **Table 4** into your notebook. Fill in the blank spaces in your table by calculating the survival rates for the various cancers.

(f) Based on the data, which type of cancer is most deadly?

Table 4: Survival of Cancer Patients After Five Years

Type of cancer	New cases	Deaths	Survival rate
lung	19 600	16 600	15%
breast	17 000	5 400	
colon	16 300	6 300	
prostate	14 300	4 100	
bladder	4 800	1 350	
kidney	3 700	1 350	
leukemia	3 200	1 110	

22. The money spent on cancer treatment continues to escalate every year. One politician has suggested that medical problems caused by inappropriate lifestyle choices should be given a lower priority for treatment.

(a) Identify two diseases that could be reduced by changing lifestyles.

(b) Evaluate the politician's statement. Should money be used first to treat people who have not contributed to their own health problem? Justify your answer.

23. In some forms of heart failure, the left side of the heart is the weaker and fails to perform properly while the right side continues to pump blood into the lungs with near normal vigour.

(a) Explain why fluid flows from the lung capillaries into the alveoli and bronchioles of the lungs, resulting in a condition called pulmonary edema.

(b) Describe the effect of pulmonary edema on the normal functioning of the lungs.

(c) Suggest a technological solution for this condition.

24. Prolonged starvation reduces the amount of protein in the blood. One consequence of this is an increased amount of tissue fluid, which tends to gather in the abdomen and lower limbs.

(a) How is this related to capillary fluid exchange?

(b) Indicate a possible reason why scientists have been unable to solve the problem.

Diversity of Living Things

Dr. Rudy Boonstra of the University of Toronto researches biodiversity in the boreal forests of Canada.

"About every 10 years, a dramatic natural phenomenon is played out in the boreal forests of Canada. Populations of snowshoe hares and their predators go through a cycle of boom and bust. Why do cycles occur?

"When predator populations are low, the hares reproduce at high rates. Predators then thrive because of their plentiful prey. However, the predators are so efficient at killing hares that soon the hare population is almost eliminated. As the predators' food source declines, so does their survival. With predator popula-

tions low again, hare populations increase—and on goes the cycle.

"During the period of heavy predation, hares exhibit stress symptoms, indicating they are aware of the danger. Reproduction declines,
they lose weight, and they cannot

Dr. Rudy Boonstra,
University of Toronto fight off disease as well."

Overall Expectations

In this unit, you will be able to

- explain how organisms have adapted to their specific environments;
- classify organisms using the rules of scientific classification (taxonomy);
- use techniques of sampling and classification to illustrate the fundamental principles of taxonomy;
- explain how the similarities and differences within the kingdoms of life are important in maintaining biodiversity within natural ecosystems;
- explain the use of microorganisms in biotechnology.

Are You Ready?

Knowledge and Understanding

1. Is a volcano a living organism? Using the list of criteria for a living organism below, create a table that contrasts the characteristics of a living organism with those of a nonliving object. Then make your conclusion.
   - grows and develops
   - respires
   - has cells
   - uses energy
   - responds to stimuli
   - has a heart
   - generates heat
   - reproduces

2. State whether you think each of the following statements is true or false. If false, rewrite the statement to make it true.
   (a) All bacteria are harmful to humans.
   (b) Viruses are larger than bacteria and human cells.
   (c) An antibiotic will continue to be equally effective against bacteria no matter how much it is used.
   (d) The amoeba feeds by surrounding its prey and engulfing it.
   (e) Algae can reproduce sexually and asexually.
   (f) Some trees, such as poplars, have separate sexes.
   (g) Plants move fluids from roots to stems through a series of tubes.
   (h) All animals have a heart and circulatory system.
   (i) Earthworms can often be seen swimming in small pools of water on sidewalks.
   (j) Parasites, such as tapeworms, have no need for a digestive system.

3. Match each structure in the images (**Figure 1**) to the functions listed.
   (i) movement or motility
   (ii) carries out photosynthesis
   (iii) movement and feeding
   (iv) stores eggs
   (v) food storage and embryo growth

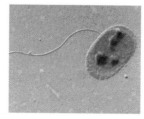

(a) Flagellum of bacterium

(b) Pseudopods of amoeba

(c) Green leaf

(d) Bean seed

(e) Egg pouch of daphnia

Figure 1

Inquiry and Communication

4. An experiment was performed to determine the effect of salt concentration on the development of brine shrimp. **Table 1** shows the number of eggs that hatched over six days in various concentrations of salt solution.
   (a) State a hypothesis for the experiment being investigated.
   (b) Identify the independent variable and the dependent variable.
   (c) Plot the results on a line graph. Use a different colour for each salt concentration.
   (d) State a conclusion for the experiment.
   (e) A student concludes that 8% salt is best for causing shrimp eggs to hatch. Would this have been your conclusion? Give reasons to support your answer.

Table 1

Day	2% salt	4% salt	6% salt	8% salt
1	0	0	0	1
2	0	10	2	3
3	0	14	8	6
4	2	22	16	8
5	4	38	27	17
6	6	49	38	29

5. Each of the birds shown has special adaptations (**Figures 2** to **6**).
   (a) Which bird or birds are adapted for eating meat?
   (b) List special adaptations for birds that eat meat.
   (c) The hummingbird uses a long tongue to get nectar from flowers. What other adaptations does the hummingbird have that make it successful at feeding?
   (d) Based on its physiology, speculate about special feeding adaptations of the pelican.
   (e) Swallows are insect eaters. Would you expect that the swallows catch insects in the air or scoop them from lakeshores? Give your reasons.

Figure 2
Hummingbird

Figure 3
Pelican

Figure 4
Owl

Figure 5
Swallow

Figure 6
Hawk

Taxonomy and the World of Microorganisms and Viruses

In this chapter, you will be able to

- classify organisms based on inferred relationships among them using appropriate terminology;
- show the usefulness of such a classification system by categorizing organisms found in a sample of pond water;
- describe the evolutionary history of a group of organisms;
- compare and contrast the structure and function of different types of cells;
- describe anatomical and physiological characteristics of microorganisms and viruses;
- compare and contrast the life cycles of microorganisms and viruses;
- explain the importance of viruses and bacteria in biotechnology.

People use various forms of classification in their everyday activities. They organize and group things to avoid confusion among objects, ideas, and events. What would life be like without classification systems? Imagine trying to find an article in your newspaper's classified section without the benefit of categories, or using a telephone directory without an alphabetical listing.

The science of classifying organisms, called **taxonomy**, has two main purposes: to identify organisms and to represent relationships among them. Most taxonomists use the hierarchical system of classification where organisms are arranged in a graded series or are ranked. However, like all systems created by people to organize and classify things, taxonomic systems are in some ways artificial, arbitrary, and limited.

The Greek philosopher Aristotle (384–322 B.C.) proposed that all creatures could be arranged in a hierarchy of complexity. He proposed that sponges and simple organisms occupy the lowest rung, while humans, nature's most advanced organisms, occupy the top rung. The dominance of humans over all living things was described as the *scala naturae*, or "ladder of nature."

Reflect on your Learning

1. List some of the classification systems you use in your life.

2. The original classification system used by Aristotle grouped organisms as either plants or animals. Explain why this initial grouping is limited.

3. Many years ago scientists debated whether or not sponges were plants or animals. Aristotle used an observation to settle the question. He poured ink next to a sponge and what he saw is shown in **Figure 1**. Examine (**b**). Does the experiment show that a sponge is a plant or an animal? Explain.

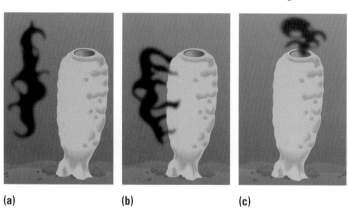

(a)   (b)   (c)

Figure 1
(a) Ink is placed next to the sponge.
(b) Ink is drawn into the sponge.
(c) Ink is expelled from the sponge.

Classifying Organisms

It is not always easy to classify an organism. To decide, a scientist works much like a detective. Both scientists and detectives make careful observations. Clues are compiled based on organizing similarities and differences. Eventually, the clues can be linked and a conclusion can be drawn. Scientists, like good detectives, look for ways of testing their conclusions.

- Examine the organisms in **Figures 2** to **7**.
  - (a) Speculate about which organisms are plants and which are animals. Give your reasons.
  - (b) What test would you conduct on each of these organisms to determine whether they are plants or animals?

taxonomy: the science of classifying organisms

DID YOU **KNOW** ?

Many scientists point out that humans may not be the most highly evolved creatures on the planet. The placement of humans at the top of the ladder is highly subjective.

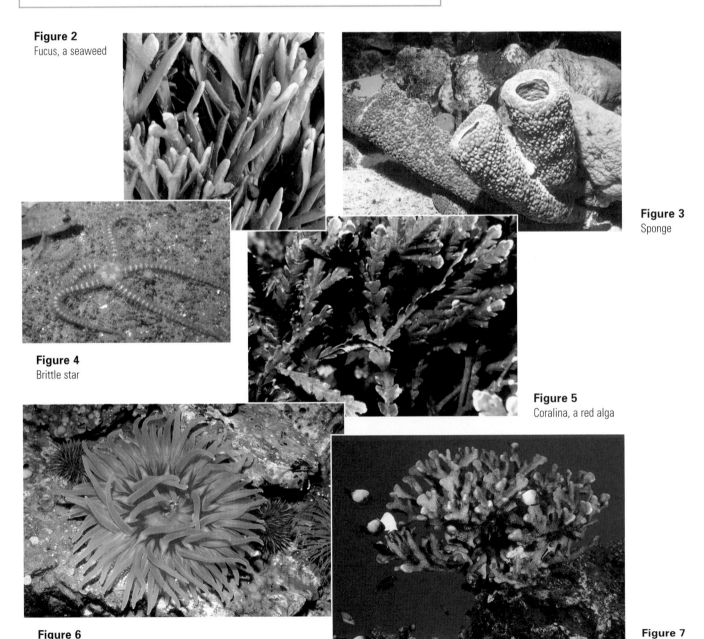

Figure 2
Fucus, a seaweed

Figure 3
Sponge

Figure 4
Brittle star

Figure 5
Coralina, a red alga

Figure 6
Sea anemone

Figure 7
Coral

9.1 Taxonomic Systems

Our present biological system of classification was developed from the system created by Swedish botanist Carl Linnaeus (1707–1778). His system was based on an organism's physical and structural features, and operated on the idea that the more features organisms have in common, the closer their relationship.

Linnaeus created rules for assigning names to plants and animals. He was the first to use **binomial nomenclature**, which assigns each organism a two-part scientific name using Latin words. Latin (and sometimes Greek) is still used today for naming organisms, and provides a common language for all scientists, regardless of their national origin. A scientific name is often based on some characteristic such as colour or habitat; an example is *Castor canadensis* (*Castor* meaning "beaver," and *canadensis* meaning "from Canada"). The first part of any scientific name is called the genus (plural: genera). Its first letter is always capitalized and can be written alone; for example, the *Acer* genus refers to maple trees. The second part is called the species and is never used alone; for example, *Acer rubrum* refers to the red maple. Living organisms within a **species** can only breed with members of their own species.

The two-name system provides an added advantage by indicating similarities in anatomy, embryology, and evolutionary ancestry. For example, binomial nomenclature suggests that the North American black bear (*Ursus americanus*) and the grizzly bear (*Ursus horribilis*) are closely related. Similar organisms are grouped into the same genus. The giant Alaskan brown bear (*Ursus arctos*) and polar bear (*Ursus maritimus*) are other relatives belonging to the same genus. By contrast, the koala bear and panda do not belong to the genus *Ursus* and are not considered true bears.

Levels of Classification

In our present classification system, there are seven main levels or **taxa** (singular: taxon), as shown in **Table 1**. Originally, the first level consisted of only two kingdoms: plants and animals. Later, single-celled organisms that displayed both plant and animal traits were discovered. To recognize this unique group, scientists created a third kingdom: **Protista**. However, shortly after the introduction of the protist kingdom, it was noted that certain microorganisms within this group shared an additional distinct feature. Bacteria and cyanobacteria, unlike protists, lack a true nucleus. This distinction resulted in the establishment of a fourth kingdom: **Monera**. The monerans are referred to as prokaryotes since they lack a true nucleus. All other groups of living organisms are know as eukaryotes. Later, taxonomists acknowledged that mushrooms and moulds are sufficiently different

binomial nomenclature: a method of naming organisms by using two names—the genus name and the species name. Scientific names are italicized.

species: a group of organisms that look alike and can interbreed under natural conditions to produce fertile offspring

taxa: categories used to classify organisms

Protista: a kingdom originally proposed for all unicellular organisms such as the amoeba. More recently, multicellular algae have been added to the kingdom.

Monera: in a five-kingdom system, a kingdom that includes organisms that lack a true nucleus

Table 1: Levels of Classification

Levels of classification	Dandelion	Housefly	Human
kingdom	Plantae	Animalia	Animalia
phylum	Tracheophyta	Arthropoda	Chordata
class	Angiospermae	Insecta	Mammalia
order	Asterates	Diptera	Primates
family	Compositae	Muscidae	Hominidae
genus	*Taraxacum*	*Musca*	*Homo*
species	*officinale*	*domestica*	*sapiens*

from plants and thus were placed in a separate kingdom called Fungi. This five-kingdom classification system, which includes animalia, plantae, fungi, protista, and monera, was originally proposed by Robert Whittaker in 1969. It enjoyed wide acceptance until recently. In the 1970s, microbiologist Carl Woese and other researchers at the University of Illinois conducted studies indicating that a group of prokaryotic microorganisms called **archaebacteria** are sufficiently distinct from bacteria and other monerans that they, in fact, constitute their own kingdom.

Archaebacteria have been known for a long time. They thrive in harsh habitats such as salt lakes, hot springs, and the stomach chambers of cattle and other ruminants. Archaebacteria possess cell walls and ribosome components that are very different from those in **eubacteria**. Eubacteria possess a rigid cell wall composed of peptidoglycan, a three-dimensional polymer containing carbohydrate and protein subunits. Thus, Woese and his colleagues proposed that the kingdom Monera be divided into two kingdoms, Archaebacteria and Eubacteria (true bacteria). The resulting six-kingdom system includes animalia, plantae, fungi, protista, eubacteria, and archaebacteria. **Table 2** summarizes a six-kingdom system of classification.

Today, most scientists believe that organisms have changed over time. The history of the evolution of organisms is called **phylogeny**. Relationships are often shown in a type of diagram called a phylogenetic tree, where the tree starts from

Archaebacteria: in a six-kingdom system, a group of prokaryotic microorganisms distinct from eubacteria that possess a cell wall not containing peptidoglycan and that live in harsh environments such as salt lakes and thermal vents

Eubacteria: in a six-kingdom system, a group of prokaryotic microorganisms that possess a peptidoglycan cell wall and reproduce by binary fission

phylogeny: the history of the evolution of a species or a group of organisms

Table 2: A Six-Kingdom System of Classification

	Kingdom	General characteristics	Cell wall	Representative organisms
	1. Eubacteria	• simple organisms lacking nuclei (prokaryotic) • either heterotrophs or autotrophs • all can reproduce asexually • live nearly everywhere	often present (contains peptidoglycan)	bacteria, cyanobacteria
	2. Archaebacteria	• prokaryotic • heterotrophs • live in salt lakes, hot springs, animal guts	present (does not contain peptidoglycan)	methanogens, extreme thermophiles, extreme halophiles
	3. Protista	• most are single celled; some are multicellular organisms; some are eukaryotic • some are autotrophs, some heterotrophs, some both • reproduce sexually and asexually • live in aquatic or moist habitats	absent	algae, protozoa
	4. Fungi	• most are multicellular • all are heterotrophs • reproduce sexually and asexually • most are terrestrial	present	mushrooms, yeasts, bread moulds
	5. Plantae	• all are multicellular • all are autotrophs • reproduce sexually and asexually • most are terrestrial	present	mosses, ferns, conifers, flowering plants
	6. Animalia	• all are multicellular • all are heterotrophs • most reproduce sexually • live in terrestrial and aquatic habitats	absent	sponges, worms, lobsters, starfish, humans

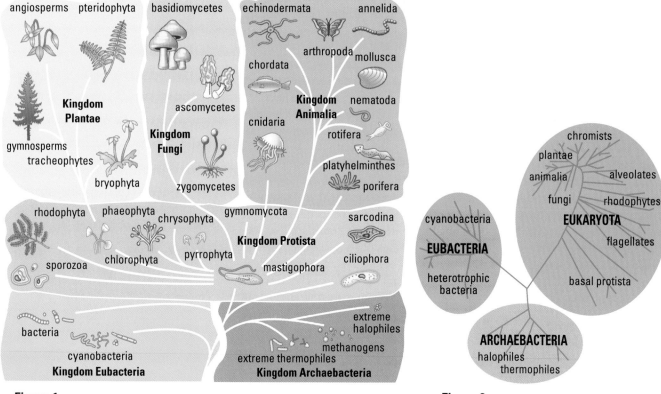

Figure 1
This phylogenetic tree shows relationships within the six kingdoms.

Figure 2
A three-domain system of classification

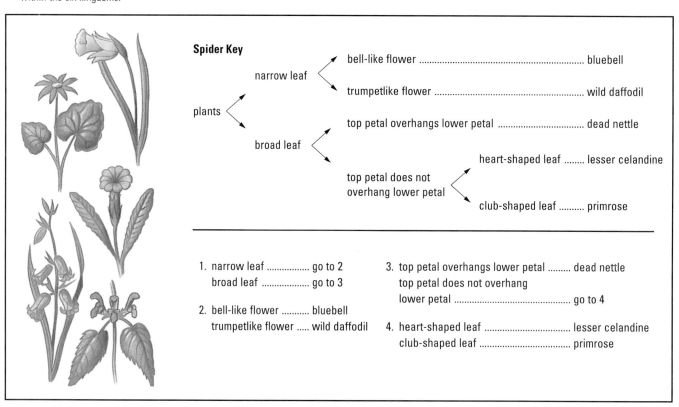

Figure 3
Sample dichotomous keys. The top key is sometimes called a spider key because of its shape.

the most ancestral form and includes branchings that lead to all of its descendants. **Figure 1** (page 330) shows an overall picture of the relationships, but more specific diagrams are possible.

DNA sequencing studies conducted by Carol Bult and Carl Woese in 1996 revealed that some of the genes in archaebacteria are more closely related to the genes of humans and other eukaryotes than to those of eubacteria. Woese proposed that archaebacteria are so different from other prokaryotes that their name should not contain the term bacteria. He suggested the name *Archaea* for this group. While the six-kingdom system grows in popularity among biologists in general, many microbiologists feel that all traditional kingdom systems should be replaced with a system that better reflects the evolutionary history of life. This has led to a three-domain classification system (**Figure 2**). Scientists continue to update evolutionary history and classification schemes as more DNA evidence is collected and analyzed.

Many scientists regularly use classification manuals to conduct their identification work. Usually it involves the use of a **dichotomous key** (**Figure 3**). The key is constructed so that a series of choices must be made, and each choice leads to a new branch of the key. If choices are made accurately, the end result is the name of the organism being identified.

dichotomous key: a two-part key used to identify living things. Di menas two.

Organisms in Pond Water

Work with a partner and follow your teacher's instructions.

- Collect a sample of water from a pond or from an aquarium filter.
- Prepare a wet mount slide of the water sample.
- Observe the organisms (**Figures 4** and **5** provide examples) under a microscope and use a key provided by your teacher to classify the organisms.

  (a) Draw at least three different organisms you viewed. Label and indicate the size of each of the organisms.

  (b) Classify the organisms you viewed as either single celled or multicellular.

  (c) How would you determine if the multicellular organisms you viewed should be grouped with plants or animals?

  (d) Describe at least two different ways that the organisms you observed moved.

  (e) Suggest a procedure for determining the number of organisms in a water droplet.

  (f) Speculate on how two of the organisms obtain their nutrients.

Figure 4
Common single-celled pond organisms

Practice

Understanding Concepts

1. Describe a situation in which classification affects your life.
2. What is taxonomy? What is meant by hierarchical classification?
3. What is meant by the term binomial nomenclature?
4. Indicate the advantage of using a Latin name over a common (e.g., English) name. Provide at least one example.
5. List, in order, the major levels of classification, starting with kingdom.
6. What is a phylogenetic tree?

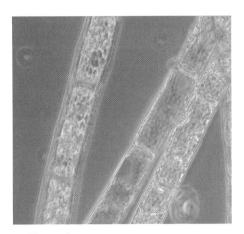

Figure 5
Filamentous algae, plantlike organisms

Figure 6

anterior

dorsal fin

mouth

dorsal surface

posterior

pectoral fin

ventral surface

tail fin

Using a Classification Key

Whales are often grouped as either toothed or baleen whales. Baleen whales have a series of vertical plates that branch and crisscross at the opening of the mouth. Each plate acts as a filter, straining small marine life from the seawater. In this activity, you will identify various species of whales using a dichotomous key.

Procedure

1. Use **Figure 6** to help you identify the whale's body structures referred to in the key.

2. To identify each whale in **Figures 7** to **13**, start by reading part 1(a) and (b) of the key. Then follow the "Go to" direction at the end of the appropriate sentence until the whale has been properly classified.

Figure 7
- teeth
- adult length: 6.0 m (females), 6.7 m (males)
- adult mass: 7.4 tonnes (t) (females), 10.5 t (males)

Figure 8
- baleen plates
- adult length: 13.7 m (females), 12.9 m (males)
- adult mass: 25–30 t

Figure 9
- teeth
- adult length: 3.5 m (females), 4.5 m (males)
- adult mass: 1.0 t (females), 1.2 t (males)

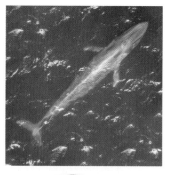

Figure 10
- teeth
- adult length: 4.2 m (females), 4.7 m (males)
- adult mass: 900 kg (females), 1.6 t (males)

Figure 11
- teeth
- adult length: 11 m (females), 15 m (males)
- adult mass: 20 t (females), 45 t (males)

Figure 12
- baleen plates
- adult length: 26.5 m (females), 25 m (males)
- adult mass: 200 t (females), 100 t (males)

Figure 13
- baleen plates
- adult length: 14–15 m
- adult mass: 50–60 t

The Key

1. (a) baleen plates	Go to 2.
(b) teeth	Go to 4.
2. (a) dorsal fin	Go to 3.
(b) no dorsal fin	bowhead whale *(Balaena mysticetus)*
3. (a) long pectoral fin	humpback whale *(Megaptera novaeangline)*
(b) short pectoral fin	blue whale *(Balaenoptera musculus)*
4. (a) no dorsal fin	Go to 5.
(b) large dorsal fin	killer whale *(Orincus orca)*
5. (a) small nose	Go to 6.
(b) large projection from nose	narwhal *(Mondon monoceros)*
6. (a) mouth on ventral surface (underside) of head	sperm whale *(Physeter macrocephalus)*
(b) mouth at the front of head	beluga *(Delphinapterus leucas)*

Analysis

(a) What are four characteristics used to classify whales?

(b) Why might biologists use a key?

(c) Provide an example of when a biologist might use a key to classify whales.

(d) Make a list of other characteristics that could be used to classify whales.

Evaluation and Synthesis

(e) Research to find out more about whales, for example, their distribution ranges and whether a species is threatened or endangered.

(f) As a group, give each of the insects (**Figure 14**) a genus and species name, and record the names in a notebook. On a separate piece of paper, make a dichotomous key that allows others in your class to identify each of the insects.

   (i) Hand in the dichotomous key that your group constructed.

   (ii) Comment on how successfully another group was at using the key. What changes would you make to your key?

(g) Identify five different trees or shrubs native to your locale and make a dichotomous key that allows others to identify them.

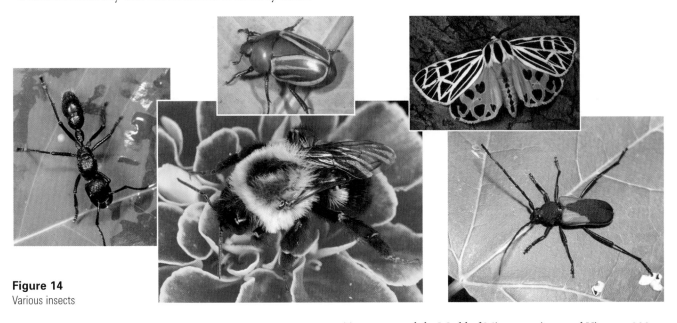

Figure 14
Various insects

SUMMARY Taxonomic Systems

1. Taxonomy is used to help biologists identify organisms and recognize natural groupings of living things.
2. Different organisms have different scientific names, which consist of a genus name and a species name.
3. In taxonomy, there are seven main taxa arranged in order: kingdom, phylum, class, order, family, genus, and species. The taxa are used to group organisms by their similarities according to structure and/or evolutionary history.

Section 9.1 Questions

Understanding Concepts

1. Why is the classification of organisms important?
2. Discuss the importance of the use of scientific names in the study of biology.
3. Why is phylogeny sometimes called the foundation of taxonomy?
4. Which of the kingdoms is at the bottom of a phylogenetic tree? Why is it placed there?

Applying Inquiry Skills

5. Use the information in **Table 3** to answer the following questions.
   (a) Which of the species are the most closely related? Explain.
   (b) Is the river otter more closely related to the muskrat or the weasel? Why?
   (c) Is the groundhog more closely related to the chipmunk or the ferret? Why?
   (d) Which of the species is (are) the closest relative(s) of the squirrel? Explain.

6. The following is a list of some Latin (Lat.) and Greek (Gr.) words and their English definitions:
   - alopekos (Gr.): fox
   - felis (Lat.): cat
   - articus (Lat.): arctic
   - pous (Gr.): foot
   - alpinous (Lat.): mountain
   - canis (Lat.): dog
   - lagos (Lat.) or lepus (Gr.): rabbit
   - aquaticus (Lat.): found in water
   - mephitis (Lat.): bad odour
   - rufus (Lat.): reddish

   Match each scientific name with the correct common name.
(a) *Felis concolor*	(i)	arctic shrew
(b) *Sorex arcticus*	(ii)	swamp rabbit
(c) *Canis rufus*	(iii)	skunk
(d) *Mephitis mephitis*	(iv)	red wolf
(e) *Alopex lagopus*	(v)	alpine chipmunk
(f) *Eutamias alpinus*	(vi)	arctic fox
(g) *Sciurus arizonensis*	(vii)	mountain lion
(h) *Sylvilagus aquaticus*	(viii)	Arizona gray squirrel

Table 3

Common name	Scientific name	Family
red squirrel	*Tamiasciurus hudsonicus*	Sciuridae
shorttail weasel	*Mustela erminea*	Mustelidae
groundhog	*Marmota monax*	Sciuridae
mink	*Mustela vison*	Mustelidae
eastern chipmunk	*Tamias striatus*	Sciuridae
river otter	*Lutra canadensis*	Mustelidae
fisher	*Martes pennanti*	Mustelidae
muskrat	*Ondatra zibethica*	Cricetidae
black-footed ferret	*Mustela nigripes*	Mustelidae

9.2 Viruses

Classification systems provide a framework for examining existing living organisms and a method for comparing modern organisms with extinct forms. However, **viruses** (*virus* is the Latin word for poison) do not fit the six-kingdom system because they do not display most of the characteristics of living cells.

viruses: microscopic particles capable of reproducing only within living cells

What makes a virus unique? Outside a living cell, a virus is a lifeless chemical and carries out no life function on its own. However, once it invades a living cell, the virus displays an important trait it shares with all living things: it reproduces. On this basis, viruses occupy a position between nonliving and living matter.

Viruses (**Figure 1**), such as those that cause the common cold, smallpox, cold sores, influenza, measles, polio, and the mumps, were a mystery for many years because of their microscopic size. It was not until 1934 that an early electron microscope enabled scientists to view the tiny particles. A virus is so small that it must be measured in units called nanometres (nm). One nanometre equals one billionth of a metre (10^{-9} m). Viruses range in size from about 20 to 400 nm in diameter. Because it is difficult to visualize such small measurements, imagine over 5000 of the influenza viruses fitting on the head of a pin.

A virus is much less complex than the simplest living organism. It consists of an inner nucleic acid core surrounded by an outer protective protein coat called a **capsid**. The capsid accounts for 95% of the total virus and gives the virus its particular shape (**Figure 2**). **Bacteriophages**, also referred to as phages, are a category of viruses known as "eaters of bacteria" that have a unique tadpole shape with a distinct head and tail region.

Sometimes a lipid membrane encloses the capsid, as happens with the human immunodeficiency virus (HIV), otherwise known as the AIDS virus. The lipid layer is thought to come from the host cell's membrane when the virus leaves the host. Unlike living organisms, whose nucleic acid core possesses both DNA and RNA, viruses contain only one or the other. This genetic material may be a single or double strand, depending on the type of virus.

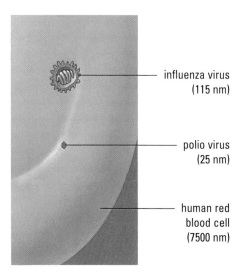

Figure 1
Comparative sizes of viruses and the human red blood cell. Note that viruses are not smaller than molecules. Each virus is composed of hundreds of thousands of large protein molecules, which surround a core of nucleic acids.

capsid: the protective protein coat of viruses

bacteriophages: a category of viruses that infect and destroy bacterial cells

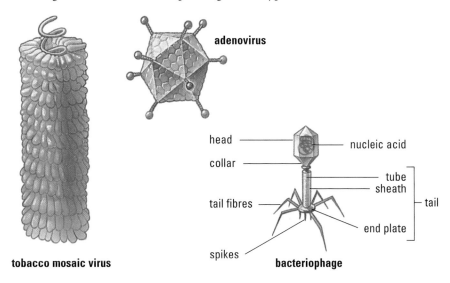

tobacco mosaic virus

adenovirus

head — nucleic acid
collar
tube
sheath
tail fibres
tail
end plate
spikes
bacteriophage

Figure 2
The protein coat, or capsid, of a virus may display various geometric shapes.

Viral Diversity

While viruses must enter cells to carry out life processes, not every virus is considered to be a disease-causing agent or pathogen. Some, such as the tobacco mosaic virus, which infects the leaves of the tobacco plant, do not appear to destroy plant tissue on any large scale. Generally, viruses are selective and, in most cases, specific viruses enter only specific host cells.

The first viruses to be described were disease agents of higher plants and animals. Around 1915, it was discovered that certain viruses can infect bacteria. Since 1940, bacteriophages have served as the main experimental objects for examining the fundamental biological properties of viruses.

Viral Specificity

Generally, viruses have a highly specialized relationship with their hosts; that is, they infect only bacteria, only animals, or only plants. Most bacteriophages have a very restricted **host range**. In fact, if two different bacteria are susceptible to infection by a single phage, this usually indicates that these bacteria are closely related to one another. In contrast to bacteriophages, most plant viruses are capable of infecting a wide range of plants, at least under laboratory conditions.

Some animal viruses have a broad host range; for example, the swine flu virus can infect hogs and humans, and the rabies virus can infect many mammalian species, including rodents, dogs, and humans. Other animal viruses have a very narrow host range; for example, the human cold virus usually infects only the cells of the human upper respiratory tract, and the AIDS virus affects the immune system because it attaches only to a specific site on the surface of certain types of white blood cells.

Viral Replication

Replication, the process by which genetic material is duplicated before a cell divides, occurs in viruses in a variety of ways, but there are generally four basic steps.

1. **Attachment** and entrance: The virus chemically recognizes a host cell (e.g., a bacterium) and attaches to it. Either the whole virus or only its DNA or RNA material enters the cell's cytoplasm.

2. **Synthesis** of protein and nucleic acid units: Molecular information contained in the viral DNA or RNA directs the host cell in replicating viral components (nucleic acids, enzymes, capsid proteins, and other viral proteins).

3. **Assembly** of the units: The viral nucleic acids, enzymes, and proteins are brought together and assembled into new virus particles.

4. **Release** of new virus particles: The newly formed virus particles are released from the infected cell, and the host cell dies.

The entire process, known as the lytic cycle, may be completed in as little as 25 to 45 min and produce as many as 300 new virus particles (under laboratory conditions). A bacteriophage that causes **lysis** of the host cell is said to be a virulent phage.

Certain types of viruses, such as cancer-causing viruses, have a lysogenic cycle. In a lysogenic cycle, the virus does not kill the host cell outright. It may coexist with the cell and be carried through many generations without apparent harm to the host. A bacteriophage that does not cause lysis of the host is called a temperate phage.

The temperate phage injects its nucleic acid into the host bacterium, similar to the way in which the virulent phage acts, but it does not take control of the cell. Instead, its nucleic acid becomes integrated into the bacterium's DNA and acts as another set of genes on the host chromosome. Then it is replicated along with the host DNA and passed along to all the daughter cells. During this period of normal replication, the virus appears to be in a dormant state, called **lysogeny**. Sometimes the dormant virus may be activated by a stimulus, such as damage to the DNA, separation of its nucleic acid from the host chromosome, or some other event such as changes in temperature or available nutrients. This triggers the lytic cycle and, once again, the virus becomes virulent. **Figure 3** shows both the lytic and lysogenic cycles.

host range: the limited number of host species, tissues, or cells that a virus or other parasite can infect

lysis: the destruction or bursting open of a cell, e.g., when an invading virus replicates in a bacterium and many viruses are released

lysogeny: the dormant state of a virus

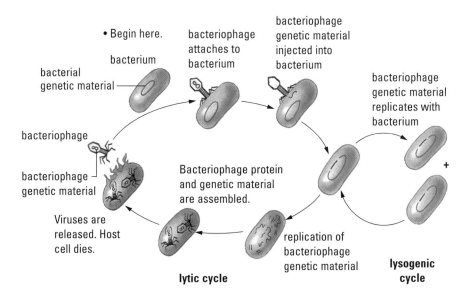

Figure 3
Viral replication

Viruses and Human Health

In many diseases caused by viruses (**Table 1**), the virus attacks cells as it reproduces. The destruction of these cells causes symptoms of the disease. Unfortunately, most viral infections are difficult to treat and are not destroyed by the sulfa drugs or antibiotics that have been so effective against diseases caused by bacteria. As well, some viruses remain dormant in the body for years before disease symptoms appear. Certain cancer-causing viruses, known as oncogenic viruses, cause disease by adding specific genes to an infected cell, thereby turning it into a cancer cell. Other viral diseases can be prevented with **vaccines** (e.g., polio, smallpox, hepatitis B). When people are vaccinated, the body reacts to the vaccine as if it were a real virus and produces antibodies. These antibodies stay with us and, as a result, the body is immune to that disease.

vaccines: solutions that are prepared from viral components or inactivated viruses

Table 1: Examples of Human Viral Diseases

Pathogen	Disease(s)	Transmission
DNA viruses		
Epstein-Barr	infectious mononucleosis	direct contact, air-transmitted droplets
poxviruses	smallpox	direct contact, air-transmitted droplets
varicella-zoster	chicken pox	direct contact, air-transmitted droplets
RNA viruses		
enteroviruses	polio, infectious hepatitis	direct contact, fecal contamination
rhinoviruses	common cold	direct contact, air-transmitted droplets
paramyxoviruses	measles, mumps	direct contact
rhabdoviruses	rabies	bite by infected animal
orthomyxoviruses	influenza	direct contact, air-transmitted droplets
retroviruses (HIV)	AIDS, associated with cancer	direct contact

The hepatitis B virus comes in different forms, the polyhedral form being infectious in humans (**Figure 4**).

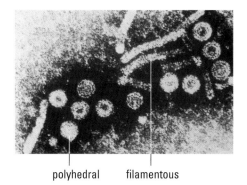

polyhedral   filamentous

Figure 4
An electron micrograph of two forms of the hepatitis B virus, isolated from the blood of an infected patient. Only the polyhedral forms are infectious.

Phylogeny

There is no fossil evidence that indicates the evolutionary history of viruses; however, a number of theories have been proposed. The following three theories are the most widely considered and have not been disproved. The third explanation appears to be the most commonly accepted.

1. At one time viral ancestors were cellular organisms that lived as parasites on other cells. Due to their parasitic existence, they depended less and less on their own cellular components and eventually evolved to their present form.

2. Viral ancestors were once free-living, precellular forms that later became parasites of cellular organisms.

3. Viruses arose from detached fragments of the genetic material of cellular organisms.

Practice

Understanding Concepts

1. Why aren't viruses placed on a phylogenetic tree?
2. How might a bacteriophage prove to be useful?
3. List the characteristics of a virus that make it different from living things.
4. Compare the size of a virus with that of a human red blood cell.
5. Name two human diseases caused by DNA viruses and two caused by RNA viruses. How are they transmitted?
6. Retroviruses, which are RNA viruses that infect animal cells, have a lysogenic cycle. What does this mean?

Making Connections

7. Contact a local veterinary hospital and inquire about the vaccinations they provide for cats, dogs, cows, and horses. What viruses are the vaccinations for?
8. What are simple precautions you can take to reduce the risk of infection by a virus when a family member has an infection?

Frontiers of Technology: Viral Vectors and Gene Therapy

Gene therapy has been established as the new frontier in medical research, and viruses are being used as vectors (carriers) to carry specifically altered DNA into cells. On September 14, 1990, the long-awaited first use of gene therapy on a human patient began when an attempt was made to correct a rare genetic disease called adenosine deaminase (ADA) deficiency in a four-year-old girl. Children develop ADA deficiency when they inherit one defective ADA gene from each parent. These children have little immunity and do not produce antibodies following vaccination against diseases such as tetanus and diphtheria. If they contract a serious disease at any time in their lives, their chances for recovery are slim.

Gene therapy to correct this genetic disease has taken almost 10 years to develop. One step required isolating the normal ADA gene from human T-cell lymphocytes (the type of white blood cells responsible for fighting off infections and cancer). The next step required the development of an efficient method to transfer the normal gene into human immune cells.

In the early 1980s, scientists successfully transferred genes into mammalian cells by using viral vectors that were genetically altered mouse viruses (**Figure 5**). To create a viral vector, scientists remove the genes in a virus and replace them with the gene to be transferred. Then, the vector is mixed with growing cells in the laboratory. The vector enters a cell and deposits the new gene in the chromosome of that cell. The gene remains in the cell as long as the cell survives and is passed onto daughter cells as the cell divides. Extensive studies with mice, monkeys, and, more recently, humans have resulted in the transfer of a new gene into as many as 90% of the cells in the laboratory cultures. These studies have shown that the gene technology has a very low risk of causing problems.

When the ADA-afflicted child was injected with her own gene-corrected T-cell lymphocytes, her body developed the ability to make antibodies of the appropriate type in normal amounts. (This is the normal immune response following infection or immunization.) Repeated injections will enable the child to live a relatively normal life.

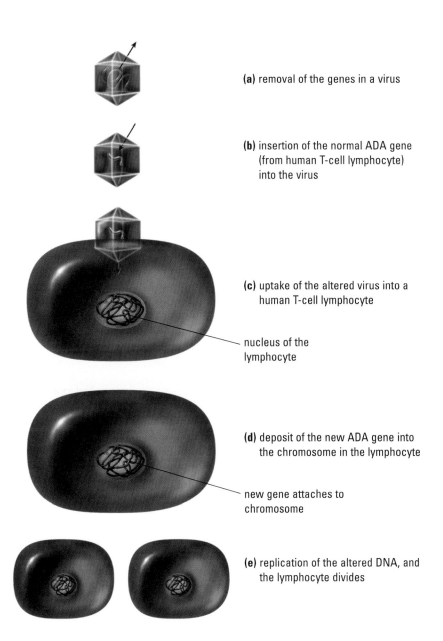

(a) removal of the genes in a virus

(b) insertion of the normal ADA gene (from human T-cell lymphocyte) into the virus

(c) uptake of the altered virus into a human T-cell lymphocyte

nucleus of the lymphocyte

(d) deposit of the new ADA gene into the chromosome in the lymphocyte

new gene attaches to chromosome

(e) replication of the altered DNA, and the lymphocyte divides

Figure 5

Genetic alteration of T-cell lymphocytes has potential in the treatment of many diseases, including AIDS and cancer. However, although viral vectors work reasonably well with T-cell lymphocytes, liver cells, and skin, they do not work well with cells that are not multiplying, such as those in the spinal cord and brain. Thus, for a wide variety of neurological disorders, such as Parkinson's and Alzheimer's diseases, other types of gene-transfer methods must be developed.

SUMMARY Viruses

1. Viruses are nonliving microscopic particles capable of reproducing only in living cells.

2. A virus consists of a nucleic acid core and a protein capsid.

3. The nucleic core of a virus contains either DNA or RNA genetic material.

4. Bacteriophages act in one of two ways:
   - Virulent phages replicate actively and cause lysis.
   - Temperate phages lie dormant for varying periods of time, and their genetic material can be carried through many generations attached to a host chromosome.

5. Viruses replicate and produce disease by the lytic cycle.

Section 9.2 Questions

Understanding Concepts

1. Why are viruses, although considered to be nonliving, frequently referred to as parasites?

2. Speculate on how viruses may have evolved.

3. How does a virulent phage differ from a temperate phage?

4. What is the significance of the movement of genes from one organism to another?

Making Connections

5. Research which viral diseases now have vaccines. Find out for how long each vaccine is effective. Suggest possible reasons why some vaccines are longer lasting than others.
Follow the links for Nelson Biology 11, 9.2
GO TO www.science.nelson.com

6. Many parents do not have their children immunized against polio because "nobody gets polio anymore." What do you think of this reasoning?

7. Why should you be concerned with the use of genetically engineered viruses?

8. What future applications do you see for viruses as vectors of foreign genes?

9.3 Kingdoms Eubacteria (Bacteria) and Archaebacteria

Archaebacteria, while appearing to be similar to eubacteria in size and shape, are in fact, very different. Archaebacteria (*archae* meaning early or primitive) often live in environments where most other organisms cannot survive. They are found in swamps and in habitats with high salt concentrations, high temperatures, or high acidity. Most live in environments without oxygen (anaerobic conditions). These harsh environments are believed to closely resemble conditions that existed when life first evolved on Earth. Three major groups of archaebacteria include the methanogens (methane-producing organisms), extreme thermophiles (organisms that thrive in temperatures up to 110°C), and extreme halophiles (organisms that live in very salty water).

Unlike viruses and cells of multicellular organisms, archaebacteria and eubacteria are capable of independent life. In this respect, they share a number of common characteristics, listed in **Table 1**.

Table 1: Common Characteristics of Archaebacteria and Eubacteria

1. Cells are prokaryotic. All are single celled.
2. Cells contain no membrane-bound organelles (like nuclei or mitochondria).
3. Cells have a single chromosome.
4. Cells reproduce asexually by binary fission.

Identification and Classification of Eubacteria

Eubacteria can be classified according to appearance since most bacteria display one of three basic shapes:

1. spherical (plural: cocci; singular: coccus)

2. rod-shaped (plural: bacilli; singular: bacillus)

3. spiral (plural: spirilla; singular: spirillum)

Figure 1 shows a typical bacterium and its feature. After division, many bacteria stay together in groups, or clusters, or individual cells. Cocci and bacilli, and sometimes spirilla, form pairs, cluster colonies, or chains (filaments) of cells. For example, *Streptococcus mutans*, which is the main cause of tooth decay, forms chains. *Staphylococcus aureus*, a common bacterium found on the skin, forms clumps (**Figure 2**). Myxobacteria form colonies (**Figure 3**).

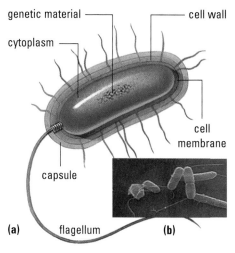

Figure 1

(a) Not all eubacteria possess every feature shown. Some species may have an outer sticky capsule (slime layer), and/or spikelike projections called pili.

(b) An electron micrograph of the rod-shaped bacterium *Escherichia coli*, or *E. coli*.

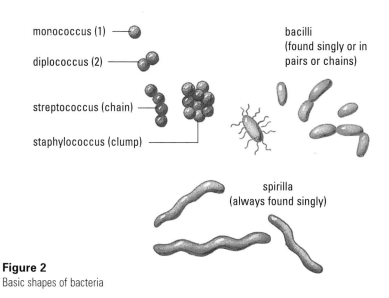

Figure 2
Basic shapes of bacteria

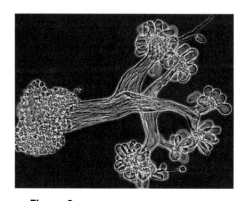

Figure 3
Although bacteria are unicellular, one group, the myxobacteria, provides a good imitation of multicellularity. The millions of cells that form the colony behave like a predatory animal by trapping other microorganisms.

Respiration and Nutrition

Eubacteria can also be grouped according to respiration and modes of nutrition. Classification by respiration reveals that some bacteria, such as those that cause tuberculosis, must have oxygen to survive. These organisms are called **obligate aerobes**. Other bacteria, such as those found in deep soils and marine and freshwater sediments, can only grow in the absence of oxygen, and hence are called **obligate anaerobes**. However, most bacteria are referred to as **facultative anaerobes**, meaning they can survive and grow with or without oxygen.

When oxygen is absent, both obligate and facultative anaerobes obtain energy by fermentation. The products of fermentation are carbon dioxide and alcohol. During fermentation, different groups of bacteria produce a wide array of organic compounds, such as ethyl alcohol, lactic acid, acetone, and acetic acid.

Classification by modes of nutrition reveals that eubacteria show diversity in how they obtain energy. Most bacteria are heterotrophs, obtaining their nutrients from other organisms. Some bacteria are parasites, obtaining nutrients without contributing to the health of the host. This is the case with many disease-causing bacteria. On the other hand, bacteria that obtain their nutrients from dead and decomposing organic matter are called saprophytes. This type of bacteria is used in treatment of sewage.

Some bacteria are autotrophs, making the food they require from inorganic substances by the processes of photosynthesis or chemosynthesis. Photosynthetic bacteria convert carbon dioxide and water into carbohydrates by using the energy from sunlight. Chemosynthetic bacteria use chemical reactions rather than sunlight as their energy source for producing carbohydrates.

Reproduction and Growth

Eubacteria and archaebacteria reproduce asexually by binary fission. Although this bears some resemblance to mitosis, binary fission is much simpler. The single strand of bacterial DNA replicates, resulting in identical genetic material being transferred to each new cell. Following replication of the genetic material, the bacterium produces a cross wall and divides into two identical cells, which may separate or remain attached (**Figure 4**).

Sexual reproduction is not common in bacteria. However, conjugation does occur among some intestinal bacteria, like *E. coli* and *Salmonella*. In conjugation, two conjugal bacteria, referred to as donor and recipient, make cell-to-cell contact

obligate aerobes: bacteria that require oxygen for respiration

obligate anaerobes: bacteria that conduct respiration processes in the absence of oxygen

facultative anaerobes: bacteria that prefer environments with oxygen, but can live without oxygen

DID YOU **KNOW ?**

Bacteria show great biodiversity and are found virtually everywhere, including the upper atmosphere, the Arctic, the deepest parts of the ocean, and most areas of any animal's body. It is evident that bacteria can adjust to life anywhere on Earth. A modern species of bacteria has adapted to living in the fuel tanks of jet aircraft! Another species has adapted to living in highly radioactive sites.

As more species are being discovered, information about bacteria is continually being updated on the Internet.

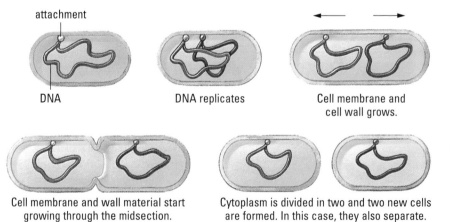

Figure 4
Binary fission. Features normally associated with mitotic cell division, such as centrioles, spindle fibres, and visible chromosomes, are not involved in this process.

attachment

DNA | DNA replicates | Cell membrane and cell wall grows.

Cell membrane and wall material start growing through the midsection. | Cytoplasm is divided in two and two new cells are formed. In this case, they also separate.

by means of a cytoplasmic bridge. Plasmids (pieces of DNA) are transferred from the donor to the recipient. Recombination of the DNA occurs in the recipient, so that it now has an altered set of characteristics. Following the transfer, the two bacteria separate.

Some bacteria of the bacillus type have adapted to survival during unfavourable environmental conditions by forming dormant or resting cells known as **endospores**. An endospore is formed inside a bacterial cell (**Figure 5**). During endospore formation, a thickened wall forms around the genetic material and cytoplasm. The remainder of the original cell eventually disintegrates. Endospores are resistant to heat and cannot easily be destroyed. When suitable growing conditions return, the wall breaks down and an active bacterium emerges.

endospores: dormant cells of bacilli bacteria that contain genetic material encapsulated by a thick, resistant cell wall. These forms of cells develop when environmental conditions become unfavourable.

Beneficial Effects of Bacteria

Contrary to the popular belief that most microorganisms are harmful, the usefulness of bacteria far outweigh the damage they do. Bacteria constitute most of the decomposers of dead plants and animals and are essential for converting and recycling nature's raw materials into nutrients for living plants and animals. Table 2 provides some other examples of the usefulness of bacteria.

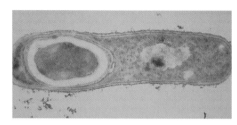

Figure 5
An electron micrograph showing an endospore within a bacterium.

Table 2: Examples of Beneficial Effects of Bacteria

Type of bacteria	Beneficial effects
clostridia	production of butanol and acetone from molasses
acetobacter	production of vinegar from alcohol
intestinal bacteria	food digestion; synthesizing of vitamins in humans (e.g., to regulate blood clotting)
lactobacilli	production of lactic acid from sugar
aztobacter, nitrobacter	fixation of nitrogen in soils
streptococci, lactobacilli	production of dairy products (e.g., cheese, buttermilk, and yogurt)
streptomyces	source of antibiotics (e.g., streptomycin, terramycin, neomycin, and erythromycin)

Frontiers of Technology: Poison-Eating Microorganisms

A toxic wood preservative, called penta-chlorophenol, has been known to seep from storage containers, contaminating nearby soil and underground water. The cost of cleaning the soil by traditional means of excavation and incineration usually ranges between two hundred and three hundred dollars per cubic metre. In some cases, hundreds of tonnes of soil must be excavated. However, a bacterium from the genus *Flavobacterium* has proved to be a dramatically economical and efficient alternative to these traditional methods. The bacteria dismantle complex toxic penta molecules, leaving nontoxic carbon dioxide, water, and harmless chlorides; and they work cheaply, requiring only oxygen and nutrients normally found in the soil. One biotechnology company quotes between thirty and fifty dollars per cubic metre of soil for cleanup using poison-eating microorganisms. Besides cost, another advantage is that the bacterial population grows as long as penta molecules remain; once the toxic chemical has been removed, the bacterium dies.

Gene-splicing techniques can greatly enhance the ability of bacteria to destroy pollutants. Combining genes from different organisms creates a multitude of possibilities. Super poison-eating bacteria can be produced by selecting genetic information that promotes destruction of the poison and that accelerates the rate of growth and reproduction of the bacteria.

Some environmental spills require different kinds of microorganisms used together. For example, three types of bacteria are employed to eliminate three components of gasoline: benzene, xylene, and toluene. The combination approach has been used successfully to break down harmful PCBs (polychlorinated biphenyls), which were once used in hydraulic and electrical systems.

Practice

Understanding Concepts

1. What features are shared by most bacteria?
2. How are bacteria similar to and different from viruses?
3. What feature(s) might cause cyanobacteria to be classified as plants by some biologists?
4. Draw three bacteria of different shapes and classify them according to shape.
5. How has endospore formation guaranteed the survival of bacteria?
6. Describe the binary fission process in bacteria.
7. What is conjugation in bacteria? Why is it important?

Harmful Effects of Bacteria

Bacteria are probably best known for causing disease. *Bacillus anthracis* was the first bacterium proven to cause a disease (**Figure 6**). The 1976 Legionnaire's disease outbreak in Philadelphia took scientists months to identify and was caused by a bacterium named *Legionella pneumophilia*. Some of the best-known bacterial scourges over the years have been tuberculosis, diphtheria, typhoid fever, and bubonic plague (Black Death). More recently, the contamination of the water supply in Walkerton, Ontario, by a deadly strain of *E. coli* has prompted new government legislation on the testing of municipal water.

Bacteria cause disease symptoms in a variety of ways. In some cases, their sheer numbers place such a tremendous material burden on the host's tissues that they interfere with normal function. In other cases, they actually destroy cells and tissues. In still other cases, bacteria produce poisons called toxins. **Table 3** provides some examples of other diseases and/or destruction caused by bacteria.

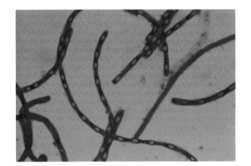

Figure 6
Bacillus anthracis was the first bacterium proven to cause a disease. It results in the deadly disease anthrax, which affects cattle, sheep, and people.

Table 3: Examples of Harmful Effects of Bacteria

Type of bacteria	Disease/destruction produced
clostridia	botulism, tetanus, and gangrene
streptococci	strep throat, scarlet fever, and pneumonia
staphylococci	boils, food poisoning, and skin infections
lactobacilli	souring of milk
pseudomonads	gasoline spoilage
bacilli	destruction of silkworms
staphylococci and pseudomonads	food spoilage
coliform bacteria	pollution of water sources, soft rot in plants, gastroenteritis and dysentery in humans
spirilla	cholera and syphilis

Infectious diseases spread from one individual to another in a variety of ways: moisture droplets in the air, dust, direct contact, fecal contamination, animal bites, and wounds (cuts and scratches). Fortunately for most people, protection from disease is provided by the body's own defence mechanisms. Also, other methods provide protection: sterilization/disinfection and use of antiseptics; extermination of animals that carry disease; immunization; and administration of antibiotics.

Bacterial Resistance to Antibiotics

Since the beginning of the "age of antibiotics" in the mid-1940s, researchers have identified more than twenty-five hundred naturally occurring **antibiotics**, which are used to treat infections caused by some of the most dangerous strains of bacteria. However, over the past 50 years, many disease-causing bacteria have slowly developed resistance to antibiotics.

Antibiotic resistance appears to develop from variations within a bacterial population. When the bacteria are first exposed to an antibiotic, the weaker strains of the bacteria are killed. Other members may have slight variations in their genetic material that allow them to survive the antibiotic. These individuals then reproduce and pass their resistance to the next generation. Scientific research clearly indicates that the most common type of bacterial resistance to antibiotics is caused by bacteria that contain R (resistance) factors. These various R factors are plasmids with special genes that code for enzymes that inactivate specific drugs. The plasmids can be transferred and recombined in conjugation.

antibiotics: chemicals produced synthetically or by microorganisms that inhibit the growth of or destroy certain other microorganisms

DID YOU KNOW ?

The threats of antibiotic-resistant bacteria and emerging infectious agents remain a major concern to both biomedical researchers and the public. Two recent reports have created real fear about antibiotic resistance—one from Britain, warning about a new strain of "flesh-eating" bacteria that killed three people, and another report from Canada, where the leader of a federal political party lost a limb to the same disease. While the bacterium responsible (a virulent subtype of group A streptococcus) is not new and does not actually eat flesh, it does appear to be more prevalent than in the past.

DECISION-MAKING SKILLS

- ● Define the Issue
- ○ Identify Alternatives
- ● Research
- ● Analyze the Issue
- ● Defend a Decision
- ○ Evaluate

Explore an Issue

Debate: New Strains of Drug-Resistant Bacteria

The use of antibiotics in the 1960s increased dramatically and, in effect, changed the nature of bacterial populations by favouring the resistant forms. Misuse of the drugs likely occurred when antibiotics were taken inappropriately at any sign of sickness, or were not taken for the full term of the prescription.

Today, we are witnessing strains of previously controllable infectious bacteria that have become resistant to prescribed drugs. These resistant forms are appearing not only in humans but also in cattle, poultry, and hogs. New drugs are being sought and expensive research is being conducted to combat these new strains.

Statement

The misuse of antibiotics could cause serious problems for society.

- In your group, research the issue. Learn more about the misuse of antibiotics, new strains of drug-resistant bacteria, the resurgence of diseases like cholera and tuberculosis, and the development of a new type of antibiotic.
- Search for information in newspapers, periodicals, CD-ROMs, and on the Internet.
  Follow the links for Nelson Biology 11, 9.3.

GO TO www.science.nelson.com

- Write a list of points and counterpoints that your group considered.
- Decide whether your group agrees or disagrees with the statement.
- Prepare to defend your group's position in a class discussion.

The question then arises whether such plasmids existed prior to the "age of antibiotics." British scientists, studying pre–1945 bacterial collections, have determined that, indeed, the plasmids were widespread but that genes encoding for antibiotic resistance were uncommon. This study suggests that the resistant strains become the dominant or common type.

Practice

Understanding Concepts

8. What are three of the main modes of disease transmission?

9. What is an antibiotic?

10. How does antibiotic resistance appear to develop within a bacterial population?

Making Connections

11. Research to find out what causes Toxic Shock Syndrome.

INQUIRY SKILLS

○ Questioning   ● Recording
○ Hypothesizing   ● Analyzing
● Predicting   ● Evaluating
○ Planning   ○ Synthesizing
● Conducting   ● Communicating

Investigation 9.3.1

Effects of Antiseptics

Antiseptics, as well as disinfectants, are germicidal substances that are used to destroy microorganisms or to prevent their development. For example, alcohol is used to disinfect surfaces and instruments before surgery, and areas in which infectious bacteria are believed to be present, such as floors, walls, and linens. Antiseptics such as mouthwashes and antibacterial soaps are applied primarily to living tissue. A number of different antiseptics prevent reproduction of bacteria by reducing the bacteria's ability to produce new proteins, which are the structural components of cells.

In this investigation, you will grow bacteria on an agar culture medium, and then place disks containing various antiseptics on the culture medium. After two days, you will measure the zones of inhibition, which are areas of no bacterial growth.

Question

Which of the antiseptics tested is most effective at limiting bacterial growth?

Prediction

Antiseptics such as Lysol or alcohol are stronger than mouthwash or antibacterial soap. Predict the relative size of the zones of inhibition when treated with the different antiseptics.

Materials

Follow these precautions when using a Bunsen burner:
- Tie back long hair.
- Secure the burner to a stand using a metal clamp (Ontario Fire Code).
- Check that the rubber hose is properly connected to the gas valve.
- Close the air vents on the burner and using a sparker —not matches—light the burner.
- Open the air vents just enough to get a blue flame.

lab apron
petri dish with nutrient agar
Bunsen burner
three 50-mL beakers
hole punch
forceps
incubator
bleach (or other sterilizing solution)

wax marker
inoculating loop
alcohol swab
filter paper
soft pencil
masking tape
ruler (in millimetres)

bacteriological culture
  (e.g., *Bacillus subtilis*)

antiseptics (e.g., Lysol, alcohol,
  mouthwash, antibacterial soap)

Procedure

1. Put on your lab apron.

2. Using a wax marker, draw on the outside of one petri dish, dividing the bottom into four sections and numbering each section as shown in **Figure 7**.

3. Sterilize an inoculating loop in the flame of a Bunsen burner for 30 s and then allow it to cool for 30 s. Remove the cap from the bacterial culture. Place the loop inside the liquid culture, remove the loop, and sterilize the opening of the vial. If using a glass vial, sterilize by quickly passing the opening of the vial through a Bunsen burner flame 2 or 3 times. (This is also referred to as "flaming" the opening.) If using a plastic vial, sterilize by wiping the opening with an alcohol swab. Then replace the cap.

4. Remove the upper lid of the petri dish and streak the entire plate with the inoculating loop by first running it along the surface of the agar in horizontal strokes and then in vertical strokes, as shown in **Figure 8**. Make sure that you streak the entire plate. Close the petri dish.

5. Add approximately 10 mL of different antiseptics to 3 different 50 mL beakers. Label each beaker with the antiseptic used.

6. Using a hole punch, make 3 disks of filter paper and then, with a soft pencil, label the disks to signify the 3 types of antiseptics. Using forceps, place each disk into one of the 3 beakers and allow the disks to soak for 5 min (**Figure 9**).

7. Using the forceps, place the 3 different antiseptic disks in the centres of the sections of the petri dish marked 1, 2, and 3, and close the lid (**Figure 10**). Seal the lid by running tape along the side of the petri dish. Write your initials, the date, class period, and your teacher's name on the petri dish.

8. Record which antiseptic was placed in each numbered section of the petri dish.

9. Invert the petri dish and place it in an incubator, set at 32°C (if possible).

10. Wash the workstation and your hands thoroughly.

11. Check the petri dish after 48 h. Using the ruler, measure the growth ring around each disk. Four separate diameter measurements (a, b, c, and d) should be taken around each disk as shown in **Figure 11**.

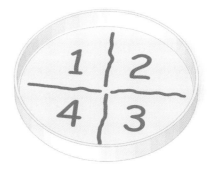

Figure 7

Figure 8

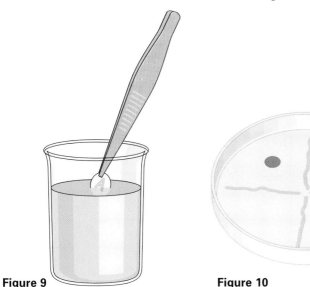

Figure 9

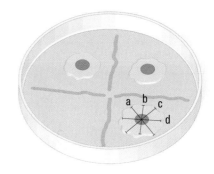

Figure 10   **Figure 11**

12. Record your data in a table similar to **Table 4**.

Table 4

Petri dish section	Measurement (mm)				
	(a)	**(b)**	**(c)**	**(d)**	**Average**
1.	?	?	?	?	?
2.	?	?	?	?	?
3.	?	?	?	?	?
4.	Description:				

13. Return all petri dishes to your teacher for proper disposal.
14. Using a sterilizing agent (e.g., bleach), wipe down all the surfaces used. Also sterilize all containers and tools that were used.
15. Wash your hands thoroughly.

Analysis

(a) On the basis of the experimental evidence, indicate which antiseptic was the most effective.
(b) What was the purpose of area 4 on the petri dish? What did this show?

Evaluation and Synthesis

(c) Look up the terms antiseptic and disinfectant in a dictionary. What is the main difference is between disinfectants and antiseptics?
(d) How do you think bleach would affect bacterial growth?
(e) Contact several manufacturers of antibacterial soaps or creams. (There may be a Web site or toll-free phone number on the package.) Try to find out what kinds of bacteria the product is effective against. Share your results with the class and compare findings.

SUMMARY Kingdoms Eubacteria and Archaebacteria

1. Eubacteria and archaebacteria are called prokaryotes and are different from all other organisms, which are referred to as eukaryotes. The DNA of prokaryotes is not contained in a membrane-bound nucleus.
2. Eubacterial and archaebacterial reproduction is mainly asexual by binary fission. Their abundance is largely a result of their rapid rate of reproduction.
3. Eubacteria and archaebacteria display a variety of modes of nutrition.
4. Eubacteria take part in many ecological processes, e.g., decomposition, nitrogen fixation, and oxygen production. They are also pathogens.

Section 9.3 Questions

Understanding Concepts

1. Explain the different methods of bacterial reproduction.
2. Where would you expect to find anaerobic bacteria in nature?
3. List as many products as you can that rely on a bacterium for part of their production.

4. How do you explain the increase of infections in hospitals caused by antibiotic-resistant bacteria?

Applying Inquiry Skills

5. Describe how you would attempt to find a new antibiotic.

6. What would you do to find out if your new antibiotic from question 5 works in humans?

Making Connections

7. (a) Comment on the dangers associated with the following common practice: Raw hamburger patties are carried on a plate to the barbecue and then the burgers are placed on the grill, where they are cooked thoroughly. Then the cooked hamburgers are taken from the barbecue and are placed back on the same plate (which has not been washed) and carried into the house for serving.
   (b) Why do you think eating a medium-rare steak does not pose the same risk as eating a medium-rare hamburger?

8. Humans have various bacteria that are normally present in most of their body systems. These "good" bacteria are essential for maintaining health. Comment on the use of antibacterial creams to cure vaginal infections, which wipe out virtually all bacteria that live there, including both good and bad types.

9. What potential problems might be created by placing small dosages of antiseptics in skin-care products?

Reflecting

10. Analyze your hygiene habits. Do you wash before and after every meal? after using the washroom? How conscientious should you be about being "germfree"?

9.4 Kingdom Protista

Protists first appeared in the fossil record about 1.5 billion years ago. Although an ancient group, they are more recent than bacteria and demonstrate an important evolutionary advancement: a discrete, membrane-bound nucleus. For that reason, protists are called eukaryotic cells. Also, they contain organelles, such as ribosomes, mitochondria, and lysosomes. These structures provide a more efficient method of using available nutrients and carrying out metabolic activities.

Most protists are microscopic and unicellular and are found in fresh or salt water. Plankton, tiny floating organisms that include protists, is an important source of biological energy for nearly all food webs in aquatic environments and, in turn, is connected to terrestrial food webs. Zooplankton, animal-like protists, is heterotrophic. Phytoplankton, plantlike protists, is photosynthetic.

Protists were once considered to be "first animals" and were placed in a phylum called Protozoa. Recently, many biologists have argued that all single-celled eukaryotes (animal-like or not) should be placed in a separate kingdom called Protista. Now, the kingdom also includes the simplest multicellular organisms—those that do not have true tissues. Protista is made up of three distinct groups: plantlike protists, animal-like protists, and fungilike protists.

Table 1: Phyla in Kingdom Protista

Group	Phyla
animal-like protists (Protozoa)	Sarcodina Mastigiphora Ciliophora Sporozoa
plantlike protists	Euglenophyta Chrysophyta Pyrrophyta Chlorophyta Phaeophyta Rhodophyta
fungilike protists	Gymnomycota

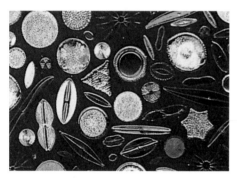

Figure 1
Phytoplankton found in salt water

While variations are still used, this trilevel division has gained widespread acceptance. **Table 1** shows eleven phyla in the kingdom.

In section 9.1, you studied a sample of pond water. You may have observed some of the types of organisms shown in **Figure 1** in your sample.

Plantlike Protists

This group of protists is plantlike because the organisms contain chlorophyll, the type of pigment that traps sunlight energy for photosynthesis. Traditionally, plantlike protists have been called algae. However, now the term algae has no formal taxonomic significance. Today, algae refers to any photosynthetic organism that is either unicellular or multicellular without tissues. In some modern classification schemes, algae are grouped in two different kingdoms: Eubacteria (cyanobacteria, once commonly known as blue-green algae), and Protista (red, brown, and green algae). The plantlike protists consist of six phyla, as shown in **Table 2**.

Table 2: Plantlike Protists

Phylum	Description	
1. Euglenophyta (e.g., *Euglena*)	Euglenophytes live mainly in fresh water and are particularly abundant in stagnant waters. The photosynthetic euglenids can also take in solid food, if necessary. Euglenids, like many flagellated protists, reproduce asexually.	
2. Chrysophyta (e.g., golden-brown algae and diatoms)	This group is found in both fresh and salt water. They contain chlorophyll and are autotrophs. Many are flagellated and encased in shells or skeletons. Diatoms, the most abundant plantlike protist, are not flagellated but are encased in two thin silica valves or shells joined together.	
3. Pyrrophyta (e.g., dinoflagellates)	Dinoflagellates are autotrophs and contain chlorophyll and red pigments. They are important primary producers and a major component of the oceanic phytoplankton.	
4. Chlorophyta (green algae, e.g., *Spirogyra*)	Green algae live mainly in fresh water, but also are found in moist soils and coastal tropical seas. They are autotrophs and contain the pigments chlorophyll and carotene. Some organisms are unicellular, such as *Chlamydomonas*, while others are multicellular, such as *Ulva* (sea lettuce).	
5. Phaeophyta (brown algae, e.g., *Fucus*)	Brown algae live mainly in colder seawater and include kelp and rockweed. Members of this phylum are multicellular, autotrophic, and contain the pigments chlorophyll and fucoxanthin.	
6. Rhodophyta (red algae, e.g., *Porphyra*)	Some red algae live in fresh water, but most inhabit warmer seawater. These multicellular organisms include dulce and Irish moss. Red algae are autotrophic and include various pigments of chlorophyll, carotene, and phycobilin.	

The euglenids (belonging to the Euglenophyta phylum) have several features unique to protists. In most cases, they obtain nourishment by means of photosynthesis, but during periods of darkness, euglenids become heterotrophic and take in solid food, a trait commonly associated with animals. Over eight hundred species inhabit ponds and lakes around the world.

One of the most widely studied plantlike protists, *Euglena*, represents a typical euglenid and illustrates many of the group's characteristics (see **Figure 2**). Two striking features of *Euglena* are the eyespot and flagellum. The eyespot is believed to be part of the organism's sensory-motor system, used to detect light. The flagellum is used to propel the organism through aquatic environments in a whiplike fashion. Most species have two flagella. The entire outer boundary of *Euglena* beneath the plasma membrane is surrounded by a firm yet flexible covering called a pellicle. There is no cell wall.

Other conspicuous structures are the central nucleus, the large green chloroplasts, and vacuoles. Interestingly, the chloroplast pigments are identical to those in green algae and land plants. The vacuoles are used to collect and remove excess water. Food is stored in the form of starch granules, a common practice in plants.

Euglena, like many flagellated protists, reproduce asexually. Following nuclear division, the rest of the cell divides lengthwise. This process, called longitudinal fission, involves a growth in cell circumference while the organelles are being duplicated. During unfavourable conditions, *Euglena* may form a thickly coated resting cell.

Green, Brown, and Red Algae

The plantlike protists belonging to the three phyla Chlorophyta, Phaeophyta, and Rhodophyta are mostly multicellular but without tissues. This distinguishes them from higher plants, which are multicellular and have tissues.

Algae are extremely well adapted to wet or moist environments (**Figure 3**). Normally considered to be aquatic organisms, they can also be found in soils, on the lower trunks of trees, and on rocks. Both brown and red algae, commonly called seaweeds, are generally large multicellular oceanic plants. They contain chlorophyll and other coloured pigments that permit them to carry out photosynthesis with the particular wavelengths of light that occur with changes in water depth. Unicellular green algae, usually referred to as phytoplankton, are found in both marine and freshwater environments, floating on or near the water surface.

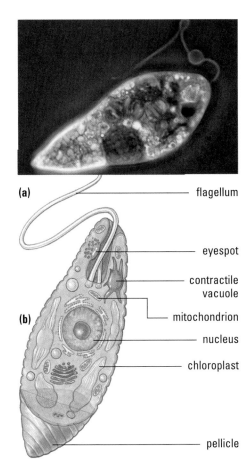

(a) flagellum

(b) eyespot · contractile vacuole · mitochondrion · nucleus · chloroplast · pellicle

Figure 2
(a) A photomicrograph of a living *Euglena*
(b) This longitudinal section shows the internal organelles of *Euglena*.

(a)

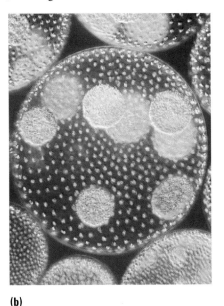

(b)

Figure 3
(a) *Chlamydomonas*, a common single-celled green alga that lives in freshwater habitats
(b) *Volvox*, a colonial green alga, also lives in fresh water. It is composed of thousands of interdependent flagellated cells that closely resemble the free-living cell *Chlamydomonas*.

Algae, especially green algae (**Figure 4**), play an important role in the overall global environment. They are the primary food producers in aquatic food chains. More significantly, they supply about 67% of the global supply of oxygen, through photosynthesis. It has become a concern that human wastes and industrial contaminants may be reducing some algal populations, particularly in the oceans. Excessive growth of algae can also be a serious problem (**Figure 5**).

(a)   **(b)**   **(c)**

Figure 4
Green algae may have been ancestors to land plants more than 400 million years ago.
(a) A species that lives in a shallow stream
(b) A marine species called the mermaid's wineglass
(c) A green alga that lives on snow. Red accessory pigments protect the alga's chlorophyll.

Humans utilize algae in a variety of ways. Some algae are consumed directly as food, providing an excellent source of vitamins and trace minerals, while others are used as fertilizers. Agar is a mucilaginous material extracted from the cell walls of certain red algae. It is used in the production of drug capsules, gels, and cosmetics. A similar substance, carrageenan, is used in cosmetics, paints, ice cream, and pie filling.

Besides being directly used for consumption, algae are instrumental in creating petroleum resources. Brown algae store food as oils. Once filled with oil, the algae die and sink to the bottom of the ocean or sea where they may be buried under mud and sand. Over millions of years and under extreme heat and pressure within the earth, the oil from the algae can be transformed into crude-oil deposits. Examples of brown algae are shown in **Figure 6**.

Figure 5
Excessive growth of algae due to nutrient pollution can be devastating for other life forms in this pond.

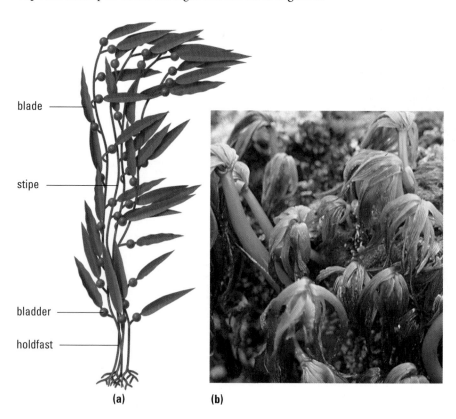

blade

stipe

bladder

holdfast

(a)   **(b)**

Figure 6
Examples of brown algae.
(a) Diagram of a kelp, showing gas-filled bladders that function as floats. Kelp stay submerged even at low tide.
(b) A population of sea palms, which grow in the intertidal zone and are alternately submerged and exposed to air

Plants: Anatomy, Growth, and Functions

Dr. Vojislava Grbic studies plant growth and diversity. "The secret to a plant's diversity and growth patterns lies in the cellular activities occurring in areas called meristems. Meristems have the potential to form many plant cell types. My research is focused on understanding how branches develop and grow in certain patterns, giving each species its own distinct shape.

"We wonder why some cells stay unspecialized while others differentiate into various cell types. Our studies will provide answers and allow us to control branching patterns of various plants, which will lead to improved production in agriculture, forestry, and horticulture."

**Dr. Vojislava Grbic,
University of
Western Ontario**

Overall Expectations

In this unit, you will be able to

- describe how plants grow, develop, and supply various products needed by other organisms;
- demonstrate an understanding of how certain factors affect plant growth and of the adaptations made by plants to their environments;
- demonstrate and describe human uses of plants;
- evaluate how the needs of humans influence the development and use of plant science and technology.

Cellular Functions

A whole new field of science has opened up in the area of nanotechnology. Nanotechnology involves creating structures and machines of molecular dimensions. Nano means one billionth (10^{-9}). Researchers have already produced tweezers made of DNA and a motor containing just 56 atoms.

The scanning tunnelling microscope (STM) is a miniscule electrical probe that creates topographical images of objects at an atomic level. The STM has revealed that the tremendous strength of an abalone shell is due to the folded nature of an organic adhesive in the shell. The idea of additional folds increasing the strength of an adhesive is expected to be applied to synthetic polymers.

**Dr. Bob Wolkow,
National Research
Council of Canada**

Dr. Bob Wolkow speculates that "this is only the start of combining organic and inorganic materials, and understanding them at an atomic level. From here, I expect an emergence of hybrid biology and nanotechnology devices that could be used as new medical diagnostic tools." Such tools could be used to view our genes.

Overall Expectations

In this unit, you will be able to

- understand cell structure and function, including metabolism and transport of substances across membranes;
- investigate the energy-converting activities of living organisms;
- understand the relationship between cellular functions and their applications in technology and the environment.

Are You Ready?

Knowledge and Understanding

1. Using the periodic table of elements, in Appendix C, identify the elements carbon, hydrogen, oxygen, and nitrogen. List everything you know about these elements.

2. Use the terms molecule, ion, element, atom, or compound to describe each substance: Mg^{2+}, Na^+, SO_4^{2-}, CH_4.

3. Match each component of the compound light microscope (**Figure 1**) to its correct name.

condenser lens   stage   revolving nosepiece
fine adjustment knob   diaphragm   objective lens
tube   ocular lens   coarse adjustment knob

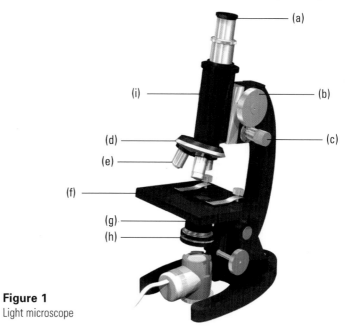

Figure 1
Light microscope

4. Copy **Table 1** into your notebook. Match each part of the microscope in **Figure 1** with the description of its function in **Table 1**.

Table 1: Parts of a Microscope

Part	Function
?	directs light to the object or specimen
?	moves tube up and down to focus on specimen used with low-power objective lens
?	moves tube up and down for sharp focus on specimen
?	magnifies object, usually by 10 times
?	supports the ocular lens
?	changes objective lenses
?	regulates amount of light reaching specimen
?	enlarges image of specimen under three different magnifications
?	supports the microscope slide

5. **Figure 2** shows a typical animal cell and plant cell. Prepare a table with the headings Animal Cell and Plant Cell and list the structures that allow you to distinguish between the two types of cells.

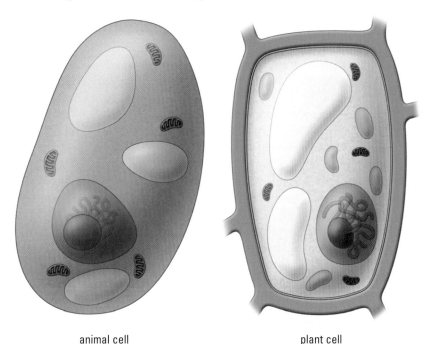

animal cell   plant cell

Figure 2
Typical animal and plant cells

6. (a) Name the chemical process that occurs within the mitochondria of all plant and animal cells.
   (b) Write a chemical (word) equation for this process.

Technical Skills and Safety

7. (a) Write the following steps for microscope use in the correct order.
   - Using the coarse adjustment knob, focus the specimen.
   - Using the fine adjustment knob, focus the specimen.
   - Place the slide onto the stage and secure it with the clips.
   - While looking through the ocular lens, centre the specimen within the field of view.
   - Ensure that the low-power objective lens is in place.
   - Once the specimen is centred and in focus, switch to a higher-power objective lens.
   (b) List the safety precautions that you should follow when using a microscope.

8. **Figure 3** shows plant cells as viewed under a microscope. Draw a diagram of a cell. Label as many parts of the cell as you can.

Math Skills

9. Refer to question 8. Using an ocular lens with a magnification of 15×, and an objective lens magnification of 50×, what is the total magnification?

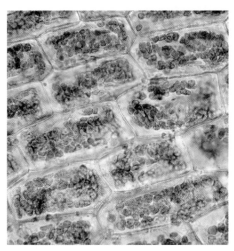

Figure 3
Plant cells

- describe how organelles and other cell components carry out various cell processes, and explain how these processes are related to organ functions;
- describe the structure of the cell membrane, and use this understanding to explain how materials move into and out of cells;
- explain the relationships between important cellular processes in living cells;
- investigate factors affecting cellular functions;
- discuss research advances in biotechnology and evaluate their applications.

Cell Biology

A Window on the Invisible World

Inventions in one field of science often contribute to advancements in others. The two greatest inventions of the 1600s, the telescope and the microscope (**Figure 1**), changed the way people understood and explained the universe. No one scientist was responsible for the development of the microscope. Like many other inventions, the development of the microscope was an ongoing process that involved technological advances in glass making and lens polishing, along with refinements to existing models.

The light microscope allowed scientists to view the contents of cells, and led to the realization that plants and animals share many common cellular features. Increasing the magnification eventually reached a point where resolution (clarity of the image) became a major problem.

In 1938, James Hillier and Albert Prebus, graduate students at the University of Toronto, produced the first functional electron microscope (**Figure 2**). The main advantage of the electron microscope is in its greater resolution. This new microscope allows us to view viruses and subcellular parts, and together with biochemical evidence, assign functions to these structures. Since then, further developments in the field of microscopy have led to three types of microscopes:

1. Transmission Electron Microscope (TEM): The image is produced by a beam of electrons passing through a very thin slice of the specimen. The image appears on a screen and is a flat, two-dimensional image (**Figure 3**).
2. Scanning Electron Microscope (SEM): The image is produced by a beam of electrons which scans across the surface of the sample. As secondary electrons are released by the sample, they form an image which is three-dimensional in quality (**Figure 4**).
3. Scanning Tunnelling Microscope (STM): A miniscule electrical probe is placed near the surface of the sample, and an image is created which can be converted by computer to a topographical image. The images produced are used for atomic-level imaging, and for manipulating molecules and atoms. It is used in the field of nanotechnology (**Figure 5**).

Reflect on your Learning

1. How have advances in technology allowed a greater understanding of cellular structures and functions?
2. How do the cells of animals differ from the cells of plants?
3. (a) In your group, brainstorm a list of substances you think must enter and exit a cell.
   (b) Make a separate list of substances you think a cell would need to keep out.
4. Write a short description of how you think substances enter and exit cells.

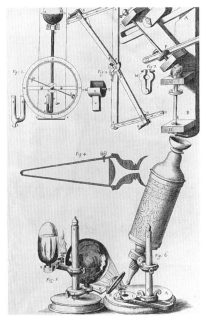

Figure 1
The impact of the microscope was so great that most scholars of the 17th century proudly displayed their microscopes in full view.

Figure 2
The 1938 version of an electron microscope, made at the University of Toronto by James Hillier and Albert Prebus

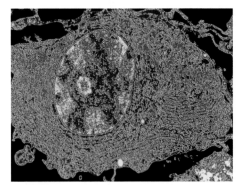

Figure 3
Animal cell as seen through a transmission electron microscope (10 000×)

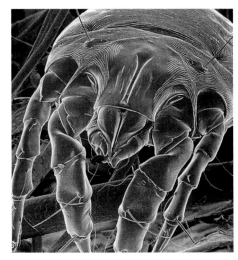

Figure 4
Scanning electron micrograph of a dust mite (400×)

Try This
Activity

Comparing Plant Cells

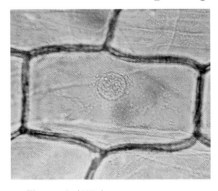

Figure 6 (400×)

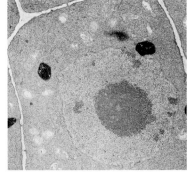

Figure 7 (12 000×)

- Refer to Appendix B to review the use of a microscope.
- Place a slide of a plant tissue (*Elodea, Spirogyra,* or other) under your light microscope.
- Examine under low power, then medium power, and finally high power.

(a) Compare what you see to the plant cells shown in **Figures 6** and **7**.
(b) Name any cell parts that you recognize.
(c) Which cell parts can be seen in all images?
(d) Describe any structures that you can see only in the electron micrograph (**Figure 7**).

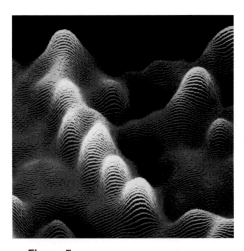

Figure 5
Scanning tunnelling micrograph of double-stranded DNA. The peaks represent the ridges of the helixes in the DNA molecule (2 000 000×).

1.1 The Cell Theory

The cell is the smallest living unit and is the basic building block of organisms. It has been estimated that the human body is made up of about 100 trillion cells. Different cells in the body are specialized to perform various tasks. Muscle cells, for example, are capable of rapid contraction. Nerve cells transmit electrochemical messages to the body about our environment and enable us to respond. Vision, hearing, taste, smell, and touch all depend on nerves. The transport of oxygen, defence against disease, and communication between different parts of the body all depend on specialized body cells.

Most of the cells in your body are invisible to the naked eye. A colony of small, rod-shaped bacteria, called *Escherichia coli* (**Figure 1**), lives in your gut. These microbes supply you with important vitamins and will likely be with you until the end of your life. Plant spores and pollen may find their way into your lungs, and a host of microbes and microscopic animal eggs may enter your body with the food you consume. Your body is a living ecosystem.

Cells outside your body also play an important role in your life. For example, a survey of your skin may reveal different types of fungi. Athlete's foot is an example of a fungus. Your skin also harbours a sphere-shaped bacterium called *Staphylococcus epidermis*.

Other types of cells have a variety of functions that are useful to us. Yogurt is a living community of bacteria. A bacterium called *Lactobacillus bulgaricus* causes the milk in yogurt to sour, but a second bacterium, *Streptococcus lactis*, enhances the taste. The processing of cheese, beer, and wine depends on cells. Even the tanning of leather could not be accomplished if it were not for cells.

In spite of their varied size, shape, appearance, and function, cells have some things in common. All cells digest nutrients, excrete wastes, synthesize needed chemicals, and reproduce.

Cells are the basic unit of all living things. An understanding of cell function provides the basis for determining how tissues and organs work. This understanding arose from the observations and research of many scientists. In 1665, Robert Hooke used the word cell to describe the repeated honeycomb structures he observed while viewing a thin slice of cork with his primitive microscope.

Figure 1
Electron micrograph of *Escherichia coli*. Large numbers of bacteria live in our intestines where they help us digest food.

Cork, the spongy tissue from cork oak trees, has few living cells; what Hooke observed was the rigid cell walls that surround the once-living plant cells. In 1674, Anton Van Leeuenhoek described living blood cells, bacteria, and single-celled organisms in a drop of water. Around 1820, Robert Brown described the appearance of a tiny sphere in plant cells, which he called the nucleus. Nuclei were soon discovered in animal cells. In 1838, a zoologist, Theodor Schwann, and a botanist, Mathias Schleiden, concluded that plant and animal tissues are composed of units called cells. It was the work of Louis Pasteur with microorganisms that led Rudolph Virchow, in 1858, to theorize that all cells come from preexisting cells.

The cell theory states the following:

- All living things are composed of one or more cells.
- The cell is the smallest entity that retains the properties of life.
- New cells arise only from cells that already exist.

Making a Model of Primitive Cells

Scientists believe that life began somewhere between 3.9 and 3.5 billion years ago. One of the important steps in the process was the formation of a cell membrane. One hypothesis for how the first cell membranes formed points to tiny structures called microspheres (**Figure 2**), made of protein and fats.

Materials: lab apron, safety goggles, latex gloves, medicine dropper, 10-mL graduated cylinder, water, large test tube, test-tube stopper, 10 mL of liquid vegetable fat or corn oil, Sudan IV indicator

  Sudan IV is dissolved in alcohol and is flammable. Keep covered before and after use. Keep away from a flame. Avoid breathing the fumes. Keep away from your skin. If it splashes in your eyes, wash with water for 15 minutes. You may have to seek medical attention.

Figure 2
Microspheres, tiny droplets of fat, water, and protein, might have given rise to cell membranes.

- Put on your lab apron, safety goggles, and gloves.
- Measure 6 mL of water. Pour the water into a large test tube.
- Using a medicine dropper, add 10 drops of vegetable fat and then a single drop of Sudan IV indicator.
- Stopper the test tube and shake it vigorously with your index finger firmly on the stopper, away from you and other students.
  (a) Describe the microspheres.
  (b) What happens when two microspheres touch?
  (c) How is the barrier created by the microsphere similar to a cell membrane?

1.2 Overview of Cell Structure

Cells vary in size, shape, and function, but all plant cells and animal cells have many common features (**Figure 1**). The **cell membrane** (or plasma membrane) forms the outer boundary of the cell. All material inside the cell is called the **protoplasm**. Inside the protoplasm, generally near the centre, is the **nucleus**. The fluid and other materials between the cell membrane and the nuclear membrane or envelope, collectively, are called the **cytoplasm**.

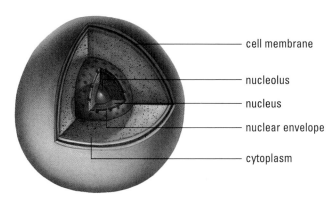

Figure 1
Basic structures common to plant and animal cells

Nucleus

The nucleus is the control centre of the cell, just like your brain is the control centre of your body. A cell is classified as **eukaryotic** or **prokaryotic**, depending on whether or not it has a membrane-bound or true nucleus (see **Figure 2**). It directs all of the cell's activities. Inside the nucleus, hereditary or genetic information is organized into threadlike structures called **chromosomes**. Each chromosome contains a number of different characteristic-determining units, or **genes**. Each gene contains the instructions to produce protein. All chromosomes are composed of nucleic acids and proteins. The nucleic acid in chromosomes is **deoxyribonucleic acid (DNA)**, the genetic material of life.

The nucleus of every cell in your body contains DNA. The fascination that scientists have for DNA arises from the fact that it is one of the few molecules

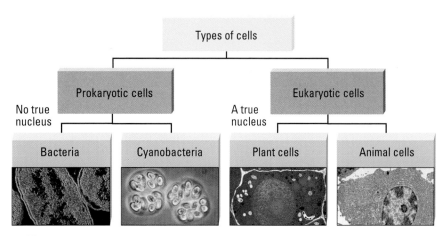

Figure 2
Classification of cells

known that is capable of replicating itself, thereby permitting cell division. Before a cell divides, each strand of genetic information is duplicated. During cell division, the duplicate strands separate and each daughter cell receives a complete set of DNA molecules.

The nucleus directs other cell activities. Some cells contain a dark-stained, spherical structure inside the nucleus. This is the **nucleolus**. The nucleolus may be associated with a second nucleic acid, **ribonucleic acid (RNA)**. There is evidence that the nucleolus is involved in making a genetic material called ribosomal RNA (rRNA). The formation of rRNA is directed by DNA. It is believed that rRNA directs the formation of specialized structures called **ribosomes** that are important in the synthesis of proteins.

Pores in the nuclear envelope (the outermost portion of the nucleus) allow RNA and other chemicals to pass through, while DNA remains inside. Messenger RNA (mRNA) is a copy of the genetic information contained in DNA. This mRNA is a molecule that can leave the nucleus and attach to the ribosomes, where its information can be used to form a protein.

nucleolus: a small, spherical structure located inside the nucleus

ribonucleic acid (RNA): a single-stranded genetic messenger that carries genetic information from the nucleus to the cytoplasm

ribosomes: structures within the cell, where protein synthesis occurs

Cell Membrane

The cell membrane is the outermost boundary of the cell. Composed of proteins and a double layer, or bilayer, of **lipid** (fat) molecules, the cell membrane holds the contents of the cell in place and regulates the movement of materials into and out of the cell. The cell membrane also contains receptor sites, which allow the entry of materials that affect cell activity. This is a **selectively permeable membrane**, that is, it allows some materials in while keeping others out. This function is particularly important in nervous, intestinal, and kidney tissues. Each of these tissues requires that materials be moved in or out in a carefully controlled manner (**Figure 3**).

lipids: a chemical group which includes fats and oils

selectively permeable membrane: a barrier that allows some molecules to pass through, but prevents other molecules from penetrating

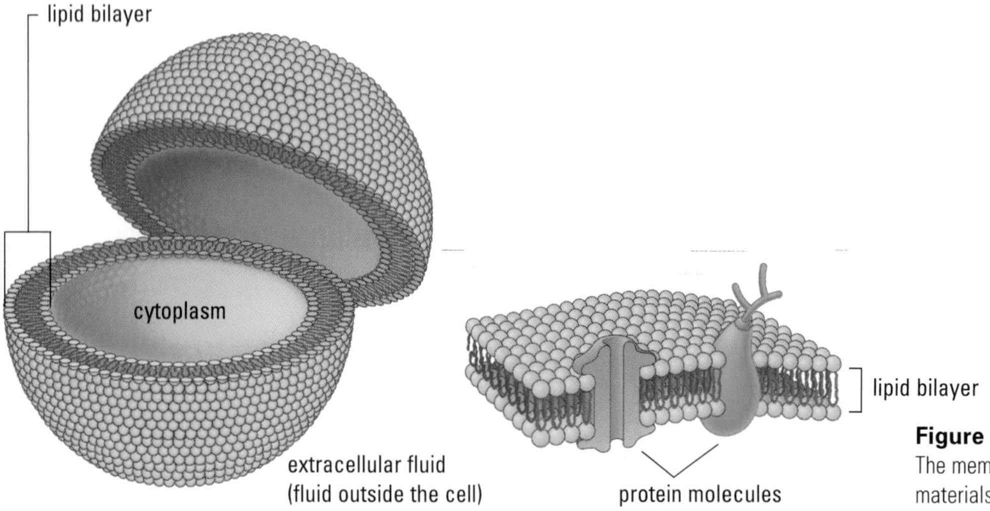

lipid bilayer

cytoplasm

extracellular fluid (fluid outside the cell)

lipid bilayer

protein molecules

Figure 3
The membrane controls the exchange of materials into and out of the cell.

Cytoplasm

The cytoplasm is a fluid that contains all the parts of a cell inside the cell membrane and outside the nucleus. Many of the cell's chemical activities take place in the cytoplasm. Absorbed nutrients are transported and processed there. During the processing of nutrients, waste products build up. The cytoplasm stores the wastes until proper disposal can be carried out. The cytoplasm can be compared to your blood in terms of its transporting abilities.

Flagella and Cilia

Some animal cells have flagella (singular: flagellum), or whiplike tails. A flagellum helps a cell move by using contractile proteins to spin it in a corkscrew motion, much like the propeller of a boat. Many cells have a single whiplike tail; however, some have multiple flagella. The human sperm cell is one example of a cell that uses a flagellum for propulsion (**Figure 4**).

The cell membranes of other specialized cells are equipped with a number of shorter, hairlike structures called cilia (singular: cilium). Like flagella, cilia have contractile proteins that cause them to move. Most often, cilia appear to move in a coordinated fashion. In some cells, the main function of cilia is locomotion; in others, it is to create fluid currents to move materials. The cells that line your windpipe are equipped with hundreds of thousands of cilia that move debris trapped in mucus away from your lungs.

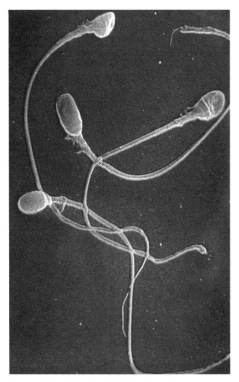

Figure 4
Sperm cells have flagella to help them swim to the egg.

SUMMARY Overview of Cell Structure

1. The cell theory states the following:
   - All living things are composed of one or more cells.
   - The cell is the smallest entity that retains the properties of life.
   - New cells arise only from cells that already exist.
2. Cells are composed of a nucleus and cytoplasm bound by a cell membrane.
3. The nucleus is the control centre of the cell.
   - It contains chromosomes, genes, DNA, and RNA.
   - It directs cell division and the formation of cell structures.
4. The cell membrane forms the boundary of the cell.
   - It is composed of proteins and lipids. It is selectively permeable.
5. The cytoplasm is where the work of the cell is done.
6. Flagella and cilia help cells move; cilia sometimes help move materials.

Practice

Understanding Concepts

1. Copy **Table 1** into your notebook. Complete the third column.

Table 1: Functions of the Nucleus in Eukaryotic Cells

Area of cell	Cell part	Function
nucleus	chromosomes	
	nucleolus	
	nuclear envelope	

1.3 Looking at Cells and the Cytoplasmic Organelles

A cell can be compared to a factory. Like a factory, the cell erects new cell structures and repairs damaged structures. Nutrient molecules are taken in and fashioned

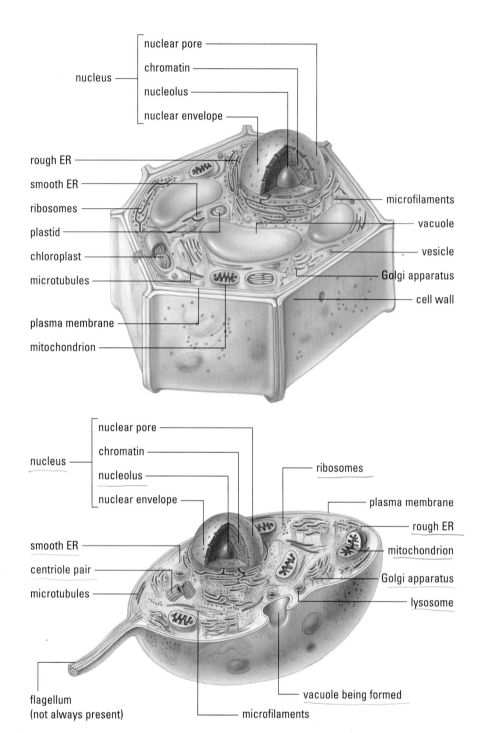

Figure 1
Plant cell and organelles

Figure 2
Animal cell and organelles

into molecules essential for life. The cytoplasm is the region of the cell in which work is accomplished, while the nucleus provides the directions. The cytoplasm has specialized structures called organelles (small organs), which are visible when viewed under an electron microscope (**Figures 1** and **2**).

The development of the transmission electron microscope opened a new window into the cell, allowing scientists to observe things they had previously only imagined. However, the electron microscope has its limits. Because cell structures must be embedded in plastic to be viewed, scientists are limited to observing dead cells. A clearer understanding of a living cell requires different technologies.

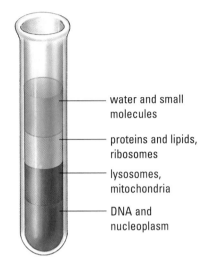

water and small molecules

proteins and lipids, ribosomes

lysosomes, mitochondria

DNA and nucleoplasm

Figure 3

Figure 4
An electron micrograph and a diagram of the mitochondrion. In animal cells, mitochondria are the largest of the cytoplasmic organelles. The one in the drawing is about a hundred thousand times larger than the real size.

Cell fractionation provides information about the thousands of chemical reactions that occur simultaneously in a cell. In this technique, cells are broken open to release the organelles and other parts of the cytoplasm and then are placed in a test tube that is spun in a centrifuge. The centrifuge is a machine that rotates at high speeds to produce a force hundreds of thousands of times greater than normal gravity. Denser structures are driven toward the bottom of the test tube while less dense objects remain nearer the surface. Because each layer contains different cell parts, the layers can be separated and the chemical reactions in each layer studied (**Figure 3**).

In another technique, living cells are treated with **radioisotopes**. The radioactivity emitted from the radioisotopes can be traced with special equipment. The radioactive chemicals can be tracked through chemical reactions.

SUMMARY **Looking at Cells**

1. The transmission electron microscope limits observation to dead and non-living things.

2. Cell fractionation provides information about chemical reactions that occur in cells.
   • Cells are ground up, placed in a test tube, and spun in a centrifuge.
   • Denser structures fall to the bottom and less dense ones stay on top.
   • Layers with components of different density are separated and each layer is studied.

3. Living cells can be treated with radioisotopes.
   • Radioactivity emitted can be traced.
   • Radioactive chemicals can be followed to determine the chemical reactions in which they participate.

The Cytoplasmic Organelles
Mitochondria
Tiny oval-shaped organelles called **mitochondria** are often referred to as the power plant of the cell (**Figure 4**). Each mitochondrion provides the cell with needed energy in a series of chemical processes called **cellular respiration**. During these processes, sugar molecules combine with oxygen to form carbon dioxide and water. Energy is also released and is temporarily stored in a compound called **adenosine triphosphate (ATP)**.

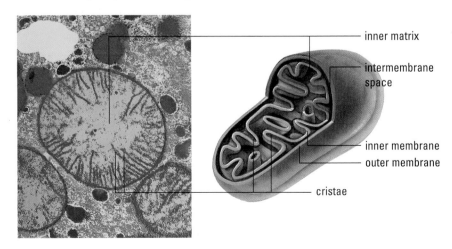

inner matrix

intermembrane space

inner membrane

outer membrane

cristae

It is important to note that energy is not created in the mitochondria. Nutrient molecules such as glucose (a sugar) are transported into the cell. It is the process of breaking down chemical bonds in sugar molecules that releases energy. This energy is stored in ATP.

Mitochondria have two separate membranes: a smooth outer membrane and a folded inner membrane. The inner membrane consists of fingerlike projections called cristae. Proteins called **enzymes** are located on the cristae. These enzymes speed up the reactions of cellular respiration.

The more active a particular cell is, the more mitochondria it will contain. As a comparison, muscle cells have thousands of mitochondria, while adipose (fat-storing) cells have a much lower number.

The evolutionary origin of mitochondria presents one of the most baffling, but intriguing, questions for biologists. Mitochondria contain their own DNA unlike that found in the nucleus. Could this mean that the mitochondria were once separate organisms that invaded eukaryotic cells? The answer to this question is one of the many remaining mysteries of science.

Research in Canada: Killing Off Cancer Tumours

A difficulty with many methods of attacking tumour cells is that the tumours learn to resist the agents we use against them. But as cancer researchers like Dr. Gurmit Singh (**Figure 5**) learn more about tumour biology, we can find new methods of attack. Working at the Hamilton Regional Cancer Centre (and McMaster University), Dr. Singh studies the way the mitochondria of a tumour's cells differ from those of normal cells. In healthy cells, any disruption of the mitochondria causes cellular respiration to stop, and the cell ultimately dies. In tumour cells, the abnormal mitochondria allow the cells to continue to live and grow. Dr. Singh is trying to activate the normal cell death signals in tumour cells so that they will die as normal cells do.

Ribosomes

Ribosomes are the organelles on which proteins are synthesized (**Figure 6**). Cell growth and reproduction require the constant synthesis of many different protein molecules. Proteins are composed of chains of smaller molecules called **amino acids**. There are 20 different amino acids. The properties of a protein are determined by the number and sequence of amino acids in the chain. Amino acids are chemically bonded together by enzymes at the ribosomes. The specific sequence of amino acids is determined by instructions encoded in the DNA. A change in position of a single amino acid can create a different protein.

Ribosomes are made of rRNA and proteins. Measuring just 20 nm in length, ribosomes are among the smallest organelles found in the cytoplasm. (There are

enzymes: protein molecules that increase the rate at which biochemical reactions proceed

Figure 5
Dr. Gurmit Singh

amino acids: organic chemicals that can be linked together to form proteins

Figure 6
A simpified illustration of a ribosome

1 000 000 nm in 1 mm.) Yet, despite their minute size, ribosomes make up a great portion of the cytoplasm. For example, it is estimated that in *Escherichia coli*, ribosomes account for one quarter of the cell mass. The large number of ribosomes permits the simultaneous construction of many proteins within a single cell.

Endoplasmic Reticulum

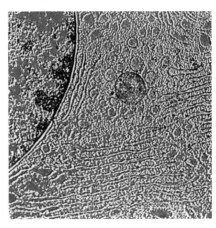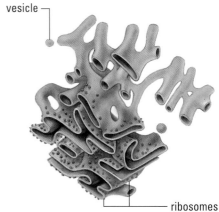

vesicle

ribosomes

Figure 7
Electron micrograph and diagram of rough endoplasmic reticulum. The micrograph shows the endoplasmic reticulum as parallel yellow-and-green linear structures.

endoplasmic reticulum: a series of canals that transport materials throughout the cytoplasm

A network of interconnected canals carries materials throughout the cytoplasm. The canals, composed of parallel membranes, are referred to as **endoplasmic reticulum** (ER). The membranes can appear either rough or smooth when viewed under the electron microscope. The rough endoplasmic reticulum (RER) has many ribosomes attached to it (**Figure 7**). It is especially prevalent in cells that specialize in secreting proteins. For example, RER is highly developed in cells of the pancreas that secrete digestive enzymes. The smooth endoplasmic reticulum (SER) has no ribosomes attached and is the structure in which fats or lipids are synthesized. SER is prevalent in cells of developing seeds and in animal cells that secrete steroid hormones.

Golgi Apparatus

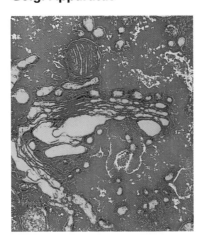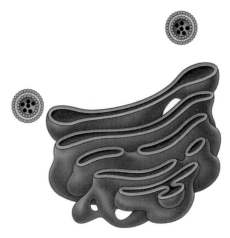

Figure 8
Electron micrograph and diagram of the Golgi apparatus. The Golgi apparatus appears as red-and-green structures in the micrograph.

Golgi apparatus: a protein-packaging organelle composed of membranous sacs

The **Golgi apparatus** (Figure 8) was first described by the Italian physician Camillo Golgi in 1898. Golgi had stained cells from a barn owl and found a new cytoplasmic structure. Half a century later, electron microscopy confirmed Golgi's observations. The Golgi apparatus stores, modifies, and packages proteins from the rough endoplasmic reticulum. The Golgi apparatus looks like as a stack of flattened balloons, which are actually membranous sacs piled on top of each other.

These membrane-bound structures pinch off at the ends to produce smaller protein-filled sacs called **vesicles**. The vesicles move toward the plasma membrane, fuse with it, and empty their contents outside the cell in a process called **exocytosis**. Through exocytosis, large molecules, such as hormones and enzymes, are released from cells.

Vesicles can also be formed when the plasma membrane brings materials into the cell by a process called **endocytosis** (**Figure 9**). Many different vesicles are produced within the cell and their main function is transport.

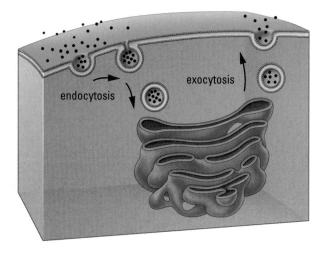

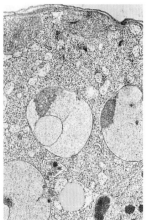

vesicles: small sacs or packets that are released by the Golgi apparatus. Vesicles are important in the processes of exocytosis and endocytosis.

exocytosis: a process by which particles are released from a cell by fusing a particle-filled vesicle with the cell membrane

endocytosis: a process by which the cell membrane wraps around a particle and pinches off a vesicle inside the cell

Figure 9
Exocytosis and endocytosis

Lysosomes

Lysosomes are organelles bound by a single membrane and formed by the Golgi apparatus. They contain a variety of enzymes that break down large molecules and cell parts within the cytoplasm. Food particles that are brought into the cell are broken down into smaller molecules, which can then be used by the cell. Lysosomes are found only in animal cells.

Lysosomes also play an important role in the human body's defence mechanism by destroying harmful substances that find their way into the cell. When white blood cells encounter and engulf invading bacteria, the lysosomes release their digestive enzymes, destroying the bacteria and the white blood cell. The fluid and protein fragments that remain after the cells have been destroyed make up a substance called pus. The enzymes released from the lysosomes also destroy damaged or worn-out cells (**Figure 10**).

More than 30 different hereditary diseases have been linked to defective digestive enzymes in the lysosomes. Tay-Sachs disease, for example, is caused by an

lysosomes: vesicles that contain a variety of enzymes able to break down large molecules

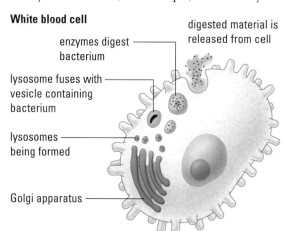

Figure 10
Acting as "suicide sacs," lysosomes release enzymes that destroy damaged or worn-out cells.

enzyme deficiency that results in the accumulation of waste materials. A buildup of waste products in the cells of the body can cause brain damage.

Microfilaments and Microtubules

Microfilaments are pipelike structures found in the cytoplasm that help provide shape and movement for the cells (**Figure 11**). Muscle cells have many microfilaments. Microtubules are tiny tubelike fibres that transport materials throughout the cytoplasm. Composed of proteins, the microtubules are found in cilia and flagella.

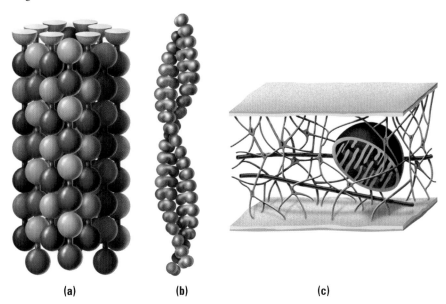

Figure 11
The structural organization of **(a)** microfilaments and **(b)** microtubules. Diagram **(c)** shows how both structures form a network throughout a cell, attaching to the cell membrane and organelles.

(a)   **(b)**   **(c)**

Research in Canada: Artificial Cells

Figure 12
Dr. Thomas Chang

Dr. Thomas Chang (**Figure 12**), a scientist at McGill University, invented artificial cells in 1957 when he was still an undergraduate student. Because artificial cells function in much the same way as natural cells, they have the capacity to replace biological cells.

Continuing research by Dr. Chang and other scientists around the world has resulted in artificial blood cells and artificial cells being used in the treatment of diabetes and liver failure. Artificial blood cells are not true blood cells but rather membrane-bound sacs containing modified hemoglobin molecules that can be used for short-term oxygen transport.

For patients suffering from blood poisoning, charcoal-filled artificial cells can be used to filter toxins from the blood. In a procedure called hemoperfusion, a patient's contaminated blood is circulated through a column filled with artificial cells containing charcoal. The charcoal absorbs the toxins and remains in the artificial cells due to the artificial cell's membrane, which acts as a barrier. Other types of artificial cells are being tested that would treat hereditary diseases and metabolic disorders.

Practice

Understanding Concepts

1. Copy **Table 1** into your notebook. Complete column 3.

Table 1: Parts of Eukaryotic Cells

Area of cell	Cell part	Function
nucleus	chromosome	• contains the genetic information that regulates cell function
		• composed of DNA and protein
		• makes a chemical messenger, called mRNA, which carries the genetic information from the nucleus to the ribosomes
	nucleolus	• makes rRNA that goes to the cytoplasm to form ribosomes
	nuclear envelope	• functions principally as a barrier
cytoplasm	ribosome	?
	endoplasmic reticulum	?
	mitochondrion	?
	Golgi apparatus	?
	lysosome	?

2. What is an artificial cell?

Making Connections

3. Other than the types that have already been discussed, what other artificial cells have been created? How are they used? To learn more about artificial cells, search the Internet.
Follow the links for Nelson Biology 11, 1.3.

 GO TO www.science.nelson.com

4. Find out how Dr. Gurmit Singh uses radioactive labelling in his research. Write a report outlining some of his work.
Follow the links for Nelson Biology 11, 1.3.

GO TO www.science.nelson.com

1.4 Special Structures of Plant Cells

The organelles discussed up to this point are found in both plant and animal cells (except for lysosomes, which are present in animal cells only). However, plant cells differ from animal cells in that they can produce and store their own food in specialized organelles called **plastids**. In addition, a large part of the cytoplasm of plant cells is composed of a fluid-filled space called a **vacuole**. The vacuole serves as a storage space for sugars, minerals, proteins, and water. The vacuole also increases the size, and hence the surface area, of the cell, allowing for the high rate of absorption of minerals that is necessary for plant nutrition. This vacuole also serves to provide physical support through **turgor pressure**.

Plastids

Plastids are chemical factories and storehouses for food and colour pigments. **Chloroplasts** are plastids that contain the green pigment chlorophyll (**Figure 1**, page 20). Chloroplasts specialize in photosynthesis, a process by which plants combine carbon dioxide from the air, with water from the roots, in the presence of

plastids: organelles that function as factories for the production of sugars or as storehouses for starch and some pigments

vacuole: a large, fluid-filled compartment in the cytoplasm of a plant cell that stores sugars, minerals, proteins, and water and is important in maintaining turgor pressure

turgor pressure: the pressure exerted by water against the cell membrane and the cell walls of plant cells

chloroplasts: plastids that contain the green pigment chlorophyll and specialize in photosynthesis

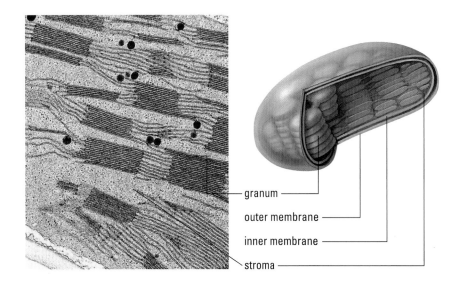

Figure 1
Electron micrograph and diagram showing the general internal structure of a chloroplast

granum —
outer membrane —
inner membrane —
stroma —

chromoplasts: plastids that store orange and yellow pigments

amyloplasts: colourless plastids that store starch

light, producing sugar and releasing oxygen. Like mitochondria, chloroplasts contain their own DNA and ribosomes, distinct from the cell's DNA. As well, during cell division, the chloroplasts and mitochondria are self-replicating.

Chromoplasts, another type of plastid, store the orange and yellow pigments found in numerous plant parts, including fruits and flowers. Colourless plastids, called **amyloplasts**, are storehouses for starch. Seeds and potato tubers (roots) contain many amyloplasts.

Plant Cell Walls

Most plant cells are surrounded by a cell wall composed of cellulose. Its main function is to protect and support plant cells. Some plants have a single cell wall, referred to as the primary cell wall, but others also have a secondary cell wall, which provides the cell with extra strength and support. The petals of a flower are composed of cells that have thin primary cell walls, as do cells in cherries, strawberries, and lettuce. Particularly rigid secondary cell walls can be found in cells of trees. The secondary wall can remain even after the plant cells have died. In fact, most of a tree trunk is made up of hollow cells with only the cell walls remaining (**Figure 2**).

In plant tissues, cells are arranged in regular patterns. The layer between the cell walls is referred to as the middle lamella (plural: lamellae). The middle lamella contains a sticky fluid, called pectin, which helps hold the cells together.

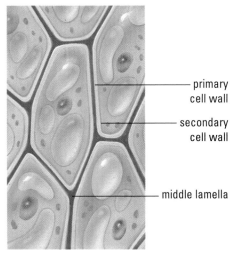

— primary cell wall
— secondary cell wall
— middle lamella

Figure 2
Plant cell wall arrangement

Practice

Understanding Concepts

1. Outline the major concepts of the cell theory.
2. Describe the function of the nucleus.
3. What are chromosomes and genes?
4. How is DNA different from RNA?
5. Describe the function of the cell membrane.
6. Describe the activities that take place within the cytoplasm.
7. List three cytoplasmic organelles found in animal cells and state the function of each.
8. Why would you expect to find more amyloplasts in a potato tuber than in a potato leaf?

Activity 1.4.1

Plant and Animal Cells

In this activity, you will prepare slides for viewing under a light microscope. You will identify the components and compare the structures of plant cells and animal cells.

Materials

lab apron	safety goggles
latex gloves	compound microscope
toothpicks (3)	lens paper
cover slip	glass slides (3)
scalpel	forceps
medicine dropper	*Elodea* sprig or onion peel
prepared slide of human cheek epithelium	tomato
prepared slide of muscle tissue	banana
paper towels	methylene blue solution
Lugol's solution (iodine solution)	warm water

Procedure

1. Use lens paper to clean the ocular and objective lenses of the microscope. Put on your lab apron, safety goggles, and gloves. Refer to Appendix B to review the use of a microscope.

2. Using the forceps, pull off one leaflet from the *Elodea* sprig. Prepare the leaflet as a wet mount. Place the leaflet on a slide and add two drops of warm water. Holding the cover slip with your thumb and forefinger, touch it to the surface of the slide at a 45° angle. Gently lower the cover slip, allowing the air to escape. If air bubbles are present, gently tap the slide with the eraser end of a pencil to remove them. (If you are using an onion, carefully peel away the thin skin on the inside of a layer of the onion. Use this thin skin to make the wet mount slide.)

3. Focus the cells under low-power magnification. Identify a group of cells that you wish to study, and move the slide so that the cells are in the centre of the field of view.

4. Rotate the nosepiece of the microscope to the medium-power objective and view the cells. Using your fine-adjustment focus, bring the cells into clear view. Draw and describe what you see.

5. Slowly decrease the light intensity by adjusting the diaphragm of the microscope. Record which light intensity reveals the greatest detail.

6. Rotate the revolving nosepiece to high-power magnification. Using the fine adjustment, focus on a group of cells. Draw a four-cell grouping and label as many cell structures as you can see. Estimate the diameter of one cell.

7. Rotate the nosepiece back to low-power magnification and remove the *Elodea* slide.

8. Using a toothpick, scrape a tiny amount of tomato pulp (immediately under the skin) onto a clean slide. Add one drop of methylene blue and then complete the wet-mount preparation. Repeat steps 3 to 6.

 Methylene blue solution and Lugol's solution are toxic and are irritants. They can stain your skin and clothing. Any spills on the skin, in the eyes, or on clothing should be washed away immediately. If they splash in your eyes, wash with water for 15 min. You may have to seek medical attention.

9. Rotate the nosepiece back to low-power magnification and remove the slide containing the tomato cells.

10. Using a toothpick, scrape a tiny amount of banana pulp onto a clean slide. Add one drop of Lugol's solution and then complete the wet-mount preparation. Repeat steps 3 to 6 but, this time, draw 3 different cells.

11. Rotate the nosepiece back to low-power magnification and remove the slide containing the banana cells. Place the prepared slide of human cheek epithelium on the stage of the microscope. Repeat steps 3 to 6 but draw 3 different cells.

12. Rotate the nosepiece back to low-power magnification and remove the slide containing the human cheek epithelium. Place the prepared slide of muscle tissue on the stage of the microscope. Repeat steps 3 to 6 but draw 3 different cells.

13. Wrap the plant tissue in paper towel and dispose of it in the garbage. Wash all equipment and return it to its storage area. Wash your hands thoroughly.

Analysis

(a) How does the arrangement of plant cells and animal cells differ?

(b) Create a table that compares your estimates of the size of each tissue cell you studied. Which cells were larger, animal or plant?

(c) Compare the *Elodea* (or onion) cells to the tomato cells. Explain why the cells of the tomato do not appear to have any chloroplasts.

(d) Compare the muscle cells to the cheek epithelium cells. Why is their shape so different?

(e) What structures did methylene blue stain in the plant tissue?

(f) What structures did Lugol's solution stain in the plant tissue?

(g) What careers would make use of the techniques learned in this activity?

SUMMARY Cytoplasmic Organelles

1. Mitochondria are the organelles where cellular respiration occurs.
   - Cellular respiration changes energy into the form of adenosine triphosphate (ATP).
   - More active cells have more mitochondria.

2. Ribosomes are the sites where proteins are synthesized.
   - Proteins are a component of several cell structures.
   - Ribosomes receive messages from the mRNA about what type of protein to produce.
   - Amino acids bond together to make protein.

3. The endoplasmic reticulum carries messages through cytoplasm.
   - The rough endoplasmic reticulum (RER) has ribosomes attached. RER is prevalent in cells that secrete proteins.
   - The smooth endoplasmic reticulum (SER) has no ribosomes. SER is the organelle in which fats or lipids are synthesized. It is prevalent in cells that secrete hormones.

4. Golgi apparatus is a protein-packaging structure, important in endocytosis and exocytosis.
   - Vesicles move to the outer membrane and are released, removing large molecules from the cell.

5. Lysosomes are vesicles.
   - They contain a variety of enzymes that break down large molecules and cell parts.
   - They destroy harmful substances.

6. Plants have plastids, which are associated with the production and storage of food. Types of plastids:
   - Chloroplasts have chlorophyll (green pigment) and specialize in photosynthesis.
   - Chromoplasts have orange and yellow pigment.
   - Amyloplasts have colorless pigment and store starch.

7. Plants also have vacuoles which are fluid-filled spaces that store sugars, minerals, and proteins.

8. Cell walls of plants are composed of cellulose.

Sections 1.1–1.4 Questions

Understanding Concepts

1. List the ways in which animal cells and plant cells differ. Offer an explanation for each difference.
2. What kind of endoplasmic reticulum are you likely to find in cells that line the stomach? in a plant seed?
3. Why would a mammal's cells require more mitochondria than a lizard's? (*Hint:* A lizard is a cold-blooded animal.)
4. Why would muscle tissue cells require more mitochondria than fat cells?
5. What role does the Golgi apparatus play in the transport of materials into and out of the cell? What kinds of materials does it package?

Applying Inquiry Skills

6. An electron micrograph shows a cell that contains very large numbers of mitochondria and Golgi apparatuses. This cell also has a great deal of rough endoplasmic reticulum. What can you conclude about the activities that take place in this cell?

1.5 Cell Research in Medicine

Researchers throughout the world are looking for ways to put cells to work in beneficial ways. In Canada, for example, scientists at Queen's University in Kingston have modified cells so they produce antibodies to fight diseases such as cancer, leprosy, and tetanus. Antibodies are one of the body's defences against foreign invaders. They attack bacteria, viruses, and other substances that the body identifies as harmful.

Scientists from McGill University in Montreal have fused normal white blood cells to cancer cells. Referred to as **hybridomas**, these cells produce huge

hybridomas: cells that result from the fusion of two different cells

amounts of antibodies for defence against infectious diseases (**Figure 1**). Imagine large vats of these cells, producing antibodies on an industrial scale.

Other scientists study lysosomes. During the metamorphosis of insects, the membranes of the lysosomes become permeable, releasing digestive enzymes. The discovery of the reason behind this process may give scientists a better understanding of arthritis. It is believed that inflammation and pain from arthritis may be caused by the leakage of enzymes from lysosomes, in the same way that the lysosome membranes of insects leak enzymes. This hypothesis is supported by the fact that cortisone, a drug known to reduce swelling, strengthens the lysosome membrane. Lysosomes may hold the key to controlling aging and disease.

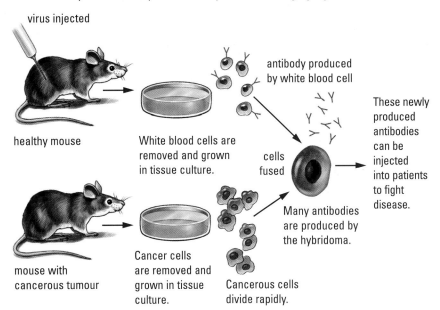

Figure 1
A hybridoma is a fusion of a cancer cell with a white blood cell. Like a cancer cell, it divides rapidly. Like a white blood cell, it produces antibodies.

Tissue Cultures

Tissue cultures, groups of cells grown out of the body in petri dishes in laboratories, have several fascinating applications. In 1989, when 21-month-old Alyssa Smith was admitted to hospital with a failing liver, doctors used cells from her mother to save her life. Doctors removed a section of her mother's liver, grew the cells in a tissue culture, and then placed small groups of these cells in Alyssa's liver. These cells continued to divide and grow until they formed a normal liver.

Similar methods have been developed for skin cells. As a result, burn victims are able to have new artificial skin from cells grown in tissue cultures (**Figure 2**). A bad burn destroys all of the skin tissue in the affected area. Previously, surgeons replaced the burned skin by taking pieces of skin, called grafts, from unburned parts of the patient's body. The grafts grow in the new location, but the area the skin was taken from must also grow new skin cells. This process can be painful.

The cornea (the clear surface of the eye) is a tissue that is commonly transplanted. In most cases, the replacement cornea is obtained following death of the donor. In a new technique, corneal cells are harvested from a patient's living relative. These cells are cultured into a coin-sized tissue and then transplanted into the damaged eyes of the patient. This process has resulted in a reduced reliance on corneal donation as a source of tissue.

A Halifax hospital has begun transplanting brain tissue from fetuses into patients with Parkinson's disease. This hereditary disease affects the nervous system; motor activity is impaired by uncontrollable tremors. Although Parkinson's disease sufferers would be helped immensely by the procedure,

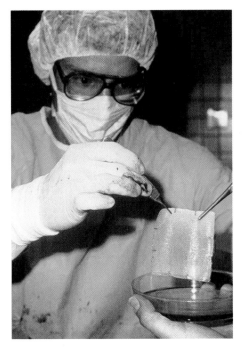

Figure 2
The first use of artificial skin to treat a third-degree burn occurred in 1981.

Dr. B.W.D. Bailey, president of Victoria General Hospital in Halifax, cautions that the repercussions of this controversial procedure must be given much consideration.

Practice

Understanding Concepts

1. What is a tissue culture?
2. What is a hybridoma?
3. Compare a hybridoma and a tissue culture. What are the similarities and differences, if any?

Explore an
Issue

Debate: Cell Research

Statement

Cell research is progressing so quickly that many of the long-range effects of the research have not been considered. A public body representing a wide range of social, moral, and ethical viewpoints should be established to oversee the technological applications of cell biology.

- In your group, research the issue. Learn more about the techniques discussed and find out about other techniques.
- Search for information in newspapers, periodicals, CD-ROMs, and on the Internet.
  Follow the links for Nelson Biology 11, 1.5.

GO TO www.science.nelson.com

- Write a list of points and counterpoints that your group considered.
- Decide whether your group agrees or disagrees with the statement.
- Prepare to defend your group's position in a class discussion.

DECISION-MAKING SKILLS

○ Define the Issue | ● Analyze the Issue
○ Identify | ● Defend a Decision
  Alternatives | ○ Evaluate
● Research

SUMMARY Cell Research in Medicine

1. Hybridomas are cells created by fusing white blood cells and cancer cells. They produce large amounts of antibodies for defence against disease.

2. The secretion of digestive enzymes by lysosomes may explain arthritis and other aging processes.
   - This hypothesis is supported by the fact that cortisone reduces swelling. Cortisone is known to strengthen the lysosome membrane.

3. Cells grown in tissue cultures can be used for skin, liver, and other cells.
   - In 1989, Alyssa Smith's life was saved by a tissue culture of her mother's liver cells. The cells were placed in Alyssa's liver, and they divided and grew until a normal liver was formed.
   - On burn victims, pieces of skin from unburned parts of the body are grafted to the damaged area. Artifical skin would relieve the pain caused by the grafting process.
   - In a new technique, cornea cells from a living relative can be cultured and then transplanted into the eyes of the patient. This reduces reliance on donors.

4. A controversial procedure involving transplanting fetal tissue into patients with Parkinson's disease raises social, moral, and ethical concerns.

Section 1.5 Questions

Making Connections

1. (a) Identify three areas of cell research other than those discussed. Select one or two representative examples and research them in depth. What are the potential medical applications of each type of research? Describe the research techniques and the medical procedure in administering the treatment.
   (b) How does each technique work? Explain the components, processes, and mechanisms involved in each.
   (c) How long has each technique been studied? Does this affect your confidence in its potential for success?
   (d) Who or what might be harmed during either the research process or the medical procedure itself? What are side effects of the treatment? How common are these complications likely to be? What are the social, moral, and ethical implications of the research or medical procedure?
   (e) How many people's lives could be improved or saved by each technique? What is the likelihood of success for each technique?
   Follow the links for Nelson Biology 11, 1.5.

   GO TO www.science.nelson.com

2. The tissue-engineering technology that created artificial skin may lead to a new field of science: regenerative medicine. What other fields of science were created in the wake of a new technology? Follow the links for Nelson Biology 11, 1.5.

   GO TO www.science.nelson.com

Reflecting

3. What personal, social, moral, and ethical considerations might influence your stance on the value of new medical technologies?

Key Expectations

Throughout this chapter, you have had opportunities to do the following:

- Describe how organelles and other cell components carry out various cell processes and explain how these processes are related to the function of organs (1.1, 1.2, 1.3, 1.4).
- Explain how scientific knowledge of cellular processes is used in technological applications (1.1, 1.3, 1.5).
- Illustrate and explain important cellular processes, including their function in the cell, the ways in which they are interrelated, and the fact that they occur in all living cells (1.1, 1.3).
- Present informed opinions on advances in cellular biology and possible applications through related technology (1.3, 1.5).
- Analyze ways in which societal needs have led to technological advances related to cellular processes (1.3, 1.5).
- Locate, select, analyze, and integrate information on topics under study, working independently and as part of a team, and using appropriate library and electronic research tools, including Internet sites (1.3, 1.5).
- Select appropriate instruments and use them effectively and accurately in collecting observations and data (1.4).
- Select and use appropriate modes of representation to communicate scientific ideas, plans, and experimental results (1.4).

Key Terms

adenosine triphosphate (ATP)
amino acids
amyloplasts
cell fractionation
cell membrane
cellular respiration
chloroplasts
chromoplasts
chromosomes
cytoplasm
deoxyribonucleic acid (DNA)
endocytosis
endoplasmic reticulum
enzymes
eukaryotic
exocytosis
genes
Golgi apparatus
hybridomas
lipids
lysosomes
mitochondria
nucleolus
nucleus
plastids
prokaryotic
protoplasm
radioisotopes
ribonucleic acid (RNA)
ribosomes
selectively permeable membrane
turgor pressure
vacuole
vesicles

Make a Summary

In this chapter, you have examined cells and cellular functions in detail. Prepare a table summarizing the parts of cells. Use headings such as Cell Areas (nucleus/cytoplasm, etc.), Cell Parts/Organelles (names), Function.

Reflect on your Learning

Revisit your answers to the Reflect on Your Learning questions at the beginning of this chapter.

- How has your thinking changed?
- What new questions do you have?

Understanding Concepts

1. The main advantage of an electron microscope over a light microscope is
   (a) its larger size
   (b) its greater magnification
   (c) its greater resolution
   (d) its ability to view dead cells

2. Which of the following is not composed of cells?
   (a) viruses
   (b) bacteria
   (c) animals
   (d) plants

3. Which nucleic acid stays confined within the nucleus of eukaryotes?
   (a) rRNA
   (b) mRNA
   (c) DNA
   (d) none of the above

4. Identify those organelles that are surrounded by a membrane and those that are not.

5. Compare the structure and function of
   (a) mitochondria and chloroplasts;
   (b) rough endoplasmic reticulum and smooth endoplasmic reticulum.

6. Name the organelles involved in each of the following processes and state the role of each organelle in the process.
   (a) protein synthesis
   (b) cellular respiration
   (c) lysosomal digestion

7. Explain why stomach cells have a large quantity of ribosomes and Golgi apparatuses.

8. Many organelles have purposes that relate to the function of organs. List as many links as possible that you read about in the chapter and then add your own to this list.

9. Why is turgor pressure not a feature of animal cells?

10. If the total magnification of a microscope is 500×, and the ocular lens is 10×, what is the magnification of the objective lens?
    (a) 50×
    (b) 60×
    (c) 510×
    (d) 5000×

11. What happens to the size of the field of view and the brightness of illumination as you move from low power to medium power?
    (a) the field of view becomes larger, while the view becomes darker
    (b) the field of view becomes larger, while the view becomes lighter
    (c) the field of view becomes smaller, while the view becomes darker
    (d) the field of view becomes smaller, while the view becomes lighter

12. Which of the following samples would have a cell wall visible?
    (a) human cheek epithelium cell
    (b) tomato cells
    (c) muscle cell
    (d) sperm cells

13. Give three possible reasons for seeing only darkness when you first look in a light microscope.

14. Determine the actual size of the image in **Figure 1**. The view is shown at medium power, with a field of view measuring 1500 μm in diameter (1 μm = 1.0×10^{-6} m).

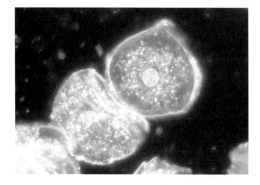

Figure 1

Applying Inquiry Skills

15. A student tries to view a specimen under a light microscope. He notices a large, dark object in view. The object does not move when the slide is moved. What steps should be taken to improve his ability to view the image?

16. A student looks through a microscope and views a cell (**Figure 2**). There do not appear to be any chloroplasts. She concludes that the cell must be an animal cell. Is she correct? Explain your answer.

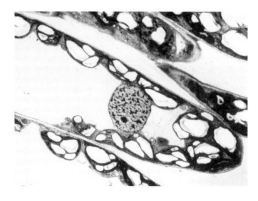

Figure 2

17. Images of algae were viewed through a light microscope, transmission electron microscope, and scanning electron microsope (**Figures 3, 4, 5**). Indicate which image is seen through which microscope. Justify your answers.

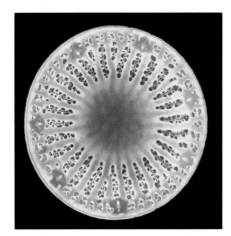

Figure 3

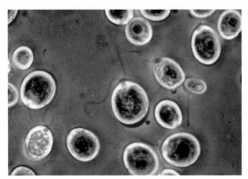

Figure 4

Figure 5

Making Connections

18. Which of the following could be used to determine what happens to oxygen during cellular respiration?
    (a) radioisotopes
    (b) hybridomas
    (c) scanning tunnelling microscope
    (d) cell fractionation

19. Other than for costs, why is the light microscope the viewing instrument most often used in school laboratories?

20. Outline some of the factors to consider if you wished to view different types of cells and compare their mitochondria numbers. Consider microscope type and slide preparation.

21. Why are white blood cells fused with cancer cells to produce hybridomas? (Hint: What is special about each of these types of cells?)

22. What is the advantage of using artificial cells to replace blood cells?

23. Where do scientists obtain the cells that are grown in tissue cultures?

24. Some unrelated diseases have an interesting link: the mitochondria. Diseases like viral hepatitis, obstructive jaundice, and some forms of muscle disease are characterized by abnormally shaped mitochondria. Large amounts of nutrients appear to accumulate in the mitochondria. Indicate why this link might be important. Does it provide any clues to the cause of the diseases? How might scientists use this information to help develop a cure?

25. Throughout this chapter, you have encountered topics in cell research that are still not understood and require more research. In the past, new discoveries have sometimes led to major upheavals in scientific thinking about a subject. What does this reveal about the ability of science to explain biological and other scientific phenomena definitively?

Exploring

26. Research the history of scientific inquiry into the cell and cellular processes. Pay special attention to the theories and hypotheses that were made and how they were supported or refuted.
    Follow the links for Nelson Biology 11, Chapter 1 Review.
    GO TO www.science.nelson.com

27. Many new instruments have been developed to aid a surgeon during an operation. In most cases, the result has been a smaller incision, resulting in a faster recovery for the patient. One such example is an arthroscope. Find out more about how an arthroscope works.
    Follow the links for Nelson Biology 11, Chapter 1 Review.
    GO TO www.science.nelson.com

Chemistry of Life

"Low fat," "lite," "cholesterol-free": these are the promises made by food companies to entice weight-conscious consumers to buy their products. "No-fat" potato chips, "just-one-calorie" diet pop—how can these be possible? Both products, along with the drinks in **Figure 1**, have something in common: they contain artificially produced ingredients that replace the usual nutrients.

Olestra, an artificial fat, was produced in a laboratory and is designed to replace fats. It is constructed of a carbohydrate and a particular kind of fat. Our bodies do not have the enzymes necessary to break down this material, so, like fibre, it passes through our systems without being digested and absorbed. However, this product can have some unpleasant side effects and is currently not approved for use in Canada. In the United States, it is sold under the trade name Olean®.

Artificial sweeteners are sweet but are not sugars. One artificial sweetener, cyclamate, was discovered by accident in 1937. A graduate student at the University of Illinois was working with a group of chemical compounds called sulfamates, of which cyclamate is a member. Sulfamates were expected to have important medicinal properties. By chance, the student touched his hand, which was covered with tiny cyclamate crystals, to his mouth. To his surprise, it tasted sweet. Cyclamate, about 30 times sweeter than sucrose, became a popular artificial sweetener. In the 1970s, however, cyclamates were banned in the United States (but not in Canada) because they were found to cause cancer in laboratory animals. Saccharin, another artificial sweetener, is about three hundred times sweeter than sucrose. It is currently banned in Canada but not in the United States. In fact, many artificial substances that are supposed to mimic nutrients are thought to be potentially harmful. Long-term studies on large populations need to be conducted before the effects are fully known.

The most common sweetener used today is aspartame, which is sold under the name NutraSweet®. Aspartame is not a sugar; it is closely associated with proteins. Like cyclamate, this chemical compound was also discovered accidentally. A research project on anti-ulcer drugs needed a pill for the control group. One group would receive the ulcer pill, and the other would receive a pill that looked like the ulcer medicine but had no medicinal value (called a placebo). During the preparation of the control pill, one of the chemists tasted it. To his astonishment, the chemical was sweet. Aspartame is about two hundred times sweeter than sucrose.

Both sugar and fats have a bad reputation. But sugar is your body's first choice for fuel and fats are your body's storage materials of choice. They are necessary parts of a healthy diet.

In this chapter, you will be able to

- describe the structure and function of biochemical compounds such as carbohydrates, proteins, lipids, and nucleic acids;
- conduct and interpret tests to identify the major biochemical compounds;
- carry out investigations on cellular function and the identification of major biochemical compounds;
- describe the structure of the cell membrane;
- describe passive and active transport into and out of cells;
- conduct and interpret tests to identify and measure rates of diffusion;
- compare how energy flows between photosynthesis and respiration and between plants and animals;
- compare the two types of respiration (aerobic vs. anaerobic);
- identify how scientific knowledge of cellular biology is used in technological applications, and present opinions about their societal impact.

Reflect on your Learning

1. List the five nutrients required to maintain good health.

2. List two ways the body uses lipids, proteins, carbohydrates, vitamins, and minerals.

3. List some foods that are mainly carbohydrates.

4. List some foods that are mainly fats.

5. Are fats good or bad for you? Justify your answer.

Figure 1
Which artificial sweetener do you think is found in the diet soda? the sports drink?

Try This
Activity

Guidelines for Good Health

The following list outlines the dietary guidelines for healthy eating:

- One gram of fat yields 37 kJ of energy. Total fat intake should provide no more than 30% of the total energy requirement per day.
- Cholesterol intake should not exceed 300 mg per day.
- One gram of protein yields 17 kJ of energy. Protein should provide approximately 15% of total energy intake.
- One gram of carbohydrate yields 17 kJ of energy. Carbohydrates should provide between 50% to 55% of total energy intake. Foods containing complex carbohydrates are recommended.
- Sodium should not exceed 3 g per day.
- Total energy consumed should be sufficient to maintain a person at an appropriate body mass.
- A healthy diet contains a variety of foods.

Examine the food label from a food item. Analyze the nutritional information using the guidelines given. Present your information in a table.
- (a) How many kilojoules of energy does the food item provide?
- (b) If an average adult needs a total of about 10 000 kJ per day, what percentage of the total does the energy provided by the food item represent?

Chemistry of Life **31**

2.1 Nutrients

Living things are composed of nonliving chemicals. Proteins, carbohydrates, lipids (fats), nucleic acids, vitamins, and minerals are chemicals of living things, as is illustrated in **Figure 1**. All of the **macromolecules** (proteins, carbohydrates, lipids, and nucleic acids) are primarily made up of carbon, hydrogen, oxygen, and sometimes nitrogen. Scientific investigations have shown that the principles of chemistry apply to biology. An understanding of the chemistry of life comes in part from an understanding of how chemical reactions are regulated within cells. Chemicals from your surroundings are altered within your cells. Bonds are broken and new bonds are formed in a continuous cycle of matter. The uptake of nutrients, the synthesis of chemicals, and the excretion of wastes are vital to life.

Consider the oxygen you breathe. Mitochondria, the "power plant" organelle in cells, use oxygen to break the bonds in nutrient sugar molecules. The breakdown of the sugar yields usable energy, carbon dioxide, and water. The water and carbon dioxide released from all cells are utilized by plants and some bacteria to make sugars and to release oxygen (**Figure 2**). An oxygen molecule that you have absorbed may once have been part of a potato, a giant maple tree, or Julius Caesar! Your next glass of water may have been shared by a million people before you!

macromolecules: large molecules that are made by joining several separate units, such as joining several sugar units to form a starch molecule

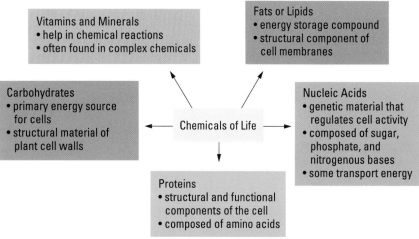

Figure 1
The chemicals of living things include carbohydrates, lipids (fats), nucleic acids, proteins, and vitamins and minerals.

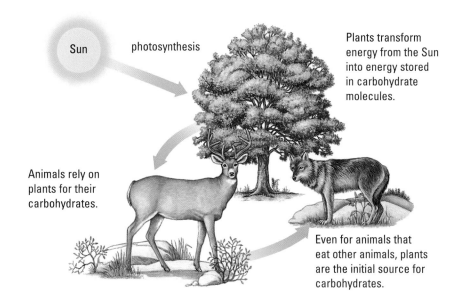

Figure 2
Only plants (and some bacteria) can manufacture carbohydrates. Animals rely on plants for this energy nutrient.

The foods you eat can be classified into three major groups of **nutrients**: carbohydrates, proteins, and fats, which are also called lipids. These nutrients make up the bulk of what you eat. Vitamins and minerals are also required, but in much smaller amounts. Most of the food you eat is a combination of nutrients. For example, the cereal you eat for breakfast or the bowl of vegetable soup you have for lunch is a combination of carbohydrates, proteins, and fats, as well as some vitamins and minerals.

nutrients: the raw materials needed for cell metabolism

Activity 2.1.1

Investigating 3-D Molecules

In this activity, you will view and manipulate computer-generated, three-dimensional molecular models of various biochemicals, the chemicals of living matter. You will also compare the different nutrient molecules.

1. Log on to your computer.
   Follow the links for Nelson Biology 11, 2.1.

GO TO www.science.nelson.com

2. Select the carbohydrates category. View glucose (**Figure 3**) and fructose. Name the elements present in these two molecules. Count the number of atoms of each element. Record the information as a ratio.

3. View lactose and sucrose. For each molecule, record the number of atoms of each element.

4. View cellulose and amylose. Describe the differences between cellulose and amylose. Provide an illustration of each molecule.

5. Select the lipids category. View glycerol, a fatty acid, and a triglyceride. A triglyceride is made up of glycerol and fatty acids. Record the number of glycerol and fatty acids molecules that can be found in one triglyceride.

6. View a saturated fatty acid and an unsaturated fatty acid. Compare these two by the number of hydrogens per carbon.

7. View a phospholipid. Compare the phospholipid to a triglyceride by the types of elements that each contains and by their structure.

8. Select the proteins category and view haemoglobin (**Figure 4**). Compare hemoglobin to carbohydrates and lipids by size and complexity. Record the new elements that are present.

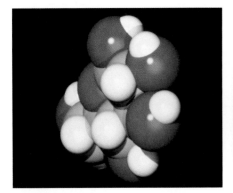

Figure 3
Computer-graphics model of a glucose molecule

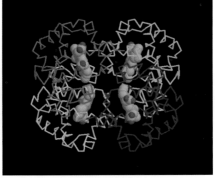

Figure 4
Computer-graphics model of a hemoglobin molecule

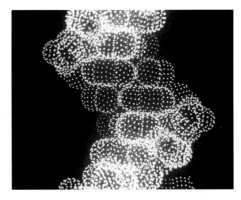

Figure 5
Computer-graphics model of DNA

9. View the following amino acids: glycine, alanine, and cystine. Indicate which parts of these molecules are the same.

10. Select the nucleic acids category. View DNA (**Figure 5**, page 33). Compare the size and complexity of this molecule to the other three categories viewed so far.

11. Highlight the sugar-phosphate region. Describe its overall structure and location in the double helix. The "rungs" of the helix "ladder" are made up of four different nitrogenous bases: adenine, thymine, guanine, and cytosine. View each of these. Record which bases always pair together.

12. View an RNA molecule. Compare this molecule to the DNA molecule in terms of its bases and overall structure.

Analysis

(a) What is the ratio of elements in the simplest carbohydrates? How does this relate to their group name, carbohydrates?

(b) How does the total number of atoms in sucrose compare to the total number of atoms in one glucose plus one fructose? What atoms are missing? What common material do these atoms form?

(c) Construct a table to summarize your comparisons of the four major categories of biochemicals: carbohydrates, lipids, proteins, and nucleic acids. Compare these four in terms of elements present; relative size (the number of atoms will determine whether the molecules are small, medium, or large); and relative complexity (they could be simple chains or groups or branching structures).

(d) List the basic subunits for each of the four major categories of biochemicals.

(e) Make up a sequence of DNA nitrogenous bases that is the complementary match to ATT GCG TCG AAA. How would the match appear if it were an RNA molecule?

2.2 Carbohydrates

carbohydrates: nutrients made up of a single sugar molecule or many sugar molecules. Carbohydrates contain only carbon, hydrogen, and oxygen.

Carbohydrates are the body's most important source of energy and make up the largest component in most diets. As an energy source, they can be quickly utilized.

Figure 1
Some sources of carbohydrates. How many of these foods do you regularly include in your diet?

However, the human body is not able to make these vital chemicals by itself. You rely on plants as your source of carbohydrates. **Figure 1** shows some sources of carbohydrates. Potatoes, bread, corn, rice, and fruit contain large amounts of carbohydrates. Using the energy provided by the Sun, plants combine carbon dioxide and water to synthesize carbohydrates during photosynthesis.

Simple Sugars: Monosaccharides

Carbohydrates are made up of either single sugar molecules or chains containing many sugar molecules. The term saccharide is from the Greek word for sugar. A sugar made up of a single sugar molecule is called a monosaccharide, "mono" meaning one. A monosaccharide can exist in the form of a straight chain or a ring, as illustrated in **Figure 2**. It usually contains carbon, hydrogen, and oxygen in the ratio 1:2:1. For example, glucose, the most common single sugar, has the formula $C_6H_{12}O_6$. Glucose is found in all the cells of your body and is your primary source of energy. Fructose, another single sugar, is found in fruits. Fructose makes an ideal sugar for many diet foods because it tastes much sweeter than glucose. A smaller quantity can be used to achieve the same sweet taste that many people enjoy. Galactose, yet another single sugar, is one of the sugars found in milk. Deoxyribose is a single sugar component of the large DNA molecule. You can recognize sugars on food labels because the names of most sugars end with -*ose*.

Combining Single Sugars: Disaccharides

More complex sugars are combinations of single sugars. Sugars made up of two sugar molecules are called disaccharides, "di" meaning two. All disaccharides are formed by a process called **dehydration synthesis** (**Figure 3**). A hydroxyl group (-OH) is removed from one monosaccharide and a hydrogen (-H) atom is removed from a second monosaccharide. These form the water molecule that is extracted from the two monosaccharide sugar molecules, enabling these molecules to bond together.

Maltose, a malt sugar used in making beer, is a dissaccharide composed of two glucose molecules. Lactose, the most common sugar in milk, is composed of glucose and galactose. Sucrose, or table sugar, is composed of glucose and fructose. This common sugar is harvested from sugar cane, sugar beets, or the sugar maple tree (**Figure 4**).

Sucrose is manufactured in several forms, the most common being brown sugar and white sugar. Brown sugar is white sugar combined with molasses (a thick, dark brown syrup made by refining raw sugar), which gives it the brown colour, rich taste, and sticky feel.

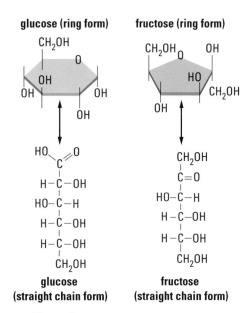

Figure 2
Many single sugars can exist in either the straight-chain or ring form.

dehydration synthesis: a series of chemical reactions that allow two molecules to bond by the formation of a water molecule

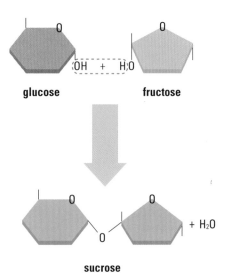

Figure 3
The formation of a disaccharide by dehydration synthesis (or condensation reaction)

(a)   (b)   (c)

Figure 4
Some sources of sucrose: **(a)** sugar beet, **(b)** sugar maple, and **(c)** sugar cane

Complex Carbohydrates: Polysaccharides

starch: a large carbohydrate molecule used by plants to store energy

Starch (a combination of two types of polysaccharides, amylose and amylopectin) is a large carbohydrate composed of many sugar molecules linked into long, branching chains. Thus, it is referred to as a polysaccharide, "poly" meaning many. Some starches contain between two thousand and six thousand glucose molecules. Plants generally store excess sugar molecules as starch in the roots and stems. When a plant needs energy, the starches are broken down into single sugar molecules, which are used as the cells need them. Wheat products (such as bread and pasta) and potatoes are rich sources of starches.

cellulose: the carbohydrate that forms the cell walls of plant cells

Cellulose, a component of plant cell walls, is also a polysaccharide composed of a great many glucose units. However, because of the way that the glucose molecules are linked together, cellulose has properties quite different from those of starch. Unlike starch, which serves as an energy storage nutrient in plants, cellulose is a structural component of plant cells. Cellulose cannot be digested by humans; that is, it cannot be broken down into simpler molecules and used as a source of energy. However, it still makes up an important component of your diet. Cellulose is often called fibre or roughage, because it functions to hold water in the large intestine, thereby aiding in the elimination of wastes.

glycogen: the form of carbohydrate storage in animals

Animals store carbohydrates in the form of a polysaccharide called **glycogen**. The structure of glycogen resembles that of the starch molecule, except for slight differences in its chain branching. The excess sugars carried by the blood are linked together to form glycogen, which is then stored in the liver and muscles. As the concentration of glucose in the blood begins to drop, glycogen is converted back to individual glucose units.

Chitin is a polysaccharide that forms the hard external skeleton of animals such as insects and crustaceans.

Figure 5 shows three different polysaccharides.

cellulose   amylose (starch)   glycogen

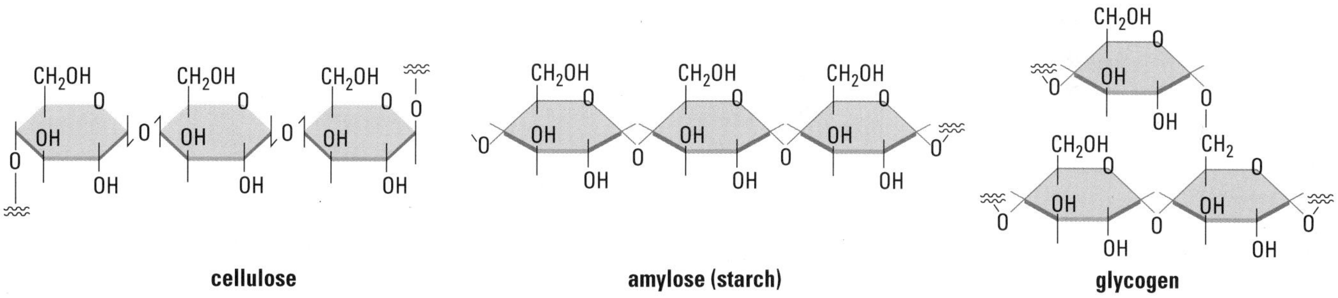

Figure 5
Three different polysaccharides, each composed of chains of simple sugars. The squiggles indicate that the chain continues repetitively.

Practice

Understanding Concepts

1. Describe the relationship between simple carbohydrates, such as monosaccharides and disaccharides, and complex carbohydrates.
2. All organic compounds contain carbon and hydrogen together and, thus, carbohydrates are organic compounds. Compounds that do not contain carbon and hydrogen together are called inorganic. Water contains hydrogen and oxygen. Is water organic or inorganic? Why or why not? Carbon dioxide contains carbon and oxygen. Is carbon dioxide organic or inorganic? Why or why not?
3. What makes water special with regard to living things?
4. What are the functions of carbohydrates?

5. Name three single sugars and indicate where you would expect to find these sugars.
6. Copy and complete **Table 1**.
7. How are starch and cellulose alike? How do they differ?
8. How can you recognize ingredients on food labels that are sugars?
9. (a) Name two substances that are sweet but are not sugars.
   (b) What is each substance made of?

Table 1

Name of dissacharide	Monosaccharide components
maltose	?
sucrose	?
lactose	?

2.3 Lipids

Humans have a love-hate relationship with lipids. They taste good when used as butter on toast, but they are also the cause of serious illnesses such as coronary heart disease and obesity.

Lipids are divided into three groups: (a) fats, oils, and waxes; (b) phospholipids; and (c) steroids.

Like carbohydrates, lipids supply energy to the cells of the body. But, unlike carbohydrates, they are difficult for your body to break down. That is why you usually feel satisfied longer after eating lipids than after eating simple carbohydrates and proteins.

Lipids are an excellent energy storage compound. One gram of lipid contains approximately twice as much energy as one gram of carbohydrate or protein. Lipids aid in the absorption of vitamins, serve as insulation for the body, are key components in cell membranes, protect some of the delicate organs of the body, and act as the raw materials for the synthesis of hormones and other important chemicals.

Lipids are compounds that contain carbon, hydrogen, and oxygen but in different proportions from carbohydrates. Unlike carbohydrates, lipids are insoluble in water; they are only soluble in substances such as acetone (nail polish remover), alcohol, ether, and chloroform. You may have noticed how fat floats on the surface of water while you are doing dishes. The most common type of lipid is composed of two structural units: one **glycerol** molecule and three **fatty acids**, which when combined are referred to as a **triglyceride**. **Figure 1**, page 38, shows how triglycerides are formed by a dehydration synthesis reaction.

An important function of lipids is the storage of energy. Glycogen supplies are limited in most animals. If glycogen stores are full, excess carbohydrates are converted into fat. This helps explain why eating carbohydrates can cause an increase in fat storage.

glycerol: a three-carbon molecule containing three hydroxyl (-OH) groups

fatty acids: long chains of carbon and hydrogen joined together. The end of the chain has an acid group (–COOH).

triglyceride: a lipid composed of glycerol and three fatty acids which are bonded together.

Saturated and Unsaturated Fats

If a fatty acid contains only single bonds between carbon atoms, then the fatty acid is described as saturated. Saturated fatty acids contain the maximum number of hydrogen atoms possible. In the fatty acid, if some hydrogen atoms are missing, that is, if there are double bonds between some of the carbon atoms, the fatty acid is said to be unsaturated. This concept is illustrated in **Figure 2**, page 38.

Liquid fats are often referred to as oils. Plants contain oils, examples being sunflower seed oil and corn oil. Plant fats are often described as polyunsaturated since their fatty acids contain many double bonds between carbon atoms and could hold more hydrogen atoms. Animal fats have the maximum number of hydrogen atoms and are saturated. In most cases, the greater the number of hydrogen atoms, the firmer the fat. Lard and butter are examples of saturated

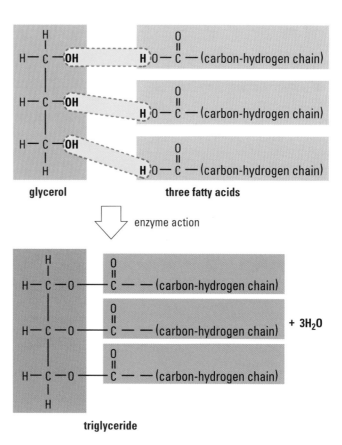

Figure 1
Fats, or triglycerides, are formed by the union of glycerol and three fatty acids. The structural diagram shows an example of a dehydration synthesis reaction.

Figure 2
Saturated fats, found in animal fats, do not have double bonds between carbon atoms in the fatty acids, making them more stable. Unsaturated fats, found in plant oils, contain double bonds. The –COOH group at the end of each chain of carbon atoms is an acid group, thereby giving the molecule the name fatty acid.

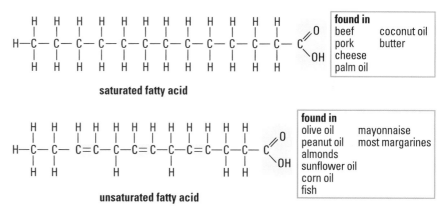

animal fats. Sometimes hydrogen atoms are added to the unsaturated plant fats to make them firmer through a process called hydrogenation. Margarine is an example of a plant fat or oil that is hydrogenated. Hard margarine has most of its double bonds filled with hydrogen atoms.

Double bonds react more easily, which means that the double bonds of the unsaturated fats are more easily broken down by the cells of your body. Saturated animal fats, by comparison, tend to be very difficult to break down. Which type of fat would you prefer in your diet?

Animal fats provide one important advantage over the less stable plant fats. Because animal fats do not have weak chemical bonds, they permit cooking at higher temperatures than the less stable plant fats do. That is why many fast-food restaurants use animal fats (such as lard) for cooking. By cooking foods at higher temperatures, cooking time can be reduced.

Other Lipids

A second group of lipids is called **phospholipids**. These differ from triglycerides in that one of the fatty acids attached to the glycerol backbone of the molecule is replaced by a phosphate group. This creates a polar end that is soluble in water, while the two fatty acids form a nonpolar end that is insoluble. Thus, they are similar to soaps in their chemical properties. Soaps are made from fats treated with lye (NaOH), which attaches a sodium ion onto the glycerol instead of a fatty acid. This molecule has one end that is water soluble; the other end is soluble in oils and greases. This allows soap to dissolve oil and grease in water. These similar special properties, as shown in **Figure 3**, make phospholipids well suited for cell membranes, which will be discussed later in this chapter.

Waxes are another type of lipid. Waxes are insoluble in water, making them highly suitable as a waterproof coating for plant leaves or animal feathers and fur. Steroids are also lipids. Although chemically different from other lipids, they are also made up of carbon, hydrogen, and oxygen, and are insoluble in water. One common feature of all steroids, including cholesterol, testosterone, and estrogen, is their carbon-based, multiple-ring structure.

phospholipids: the main components of cell membranes; composed of a phosphate group and two fatty acids attached to the glycerol backbone

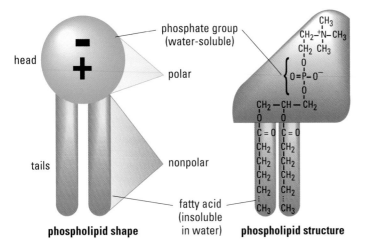

phospholipid shape

phospholipid structure

Figure 3
Phospholipid shape and structure. The phosphate group makes these lipids soluble in water. The fatty acid chains make it soluble in lipids.

Cholesterol

Cholesterol serves many vital functions. Cholesterol is one of the chemicals used by your body to make certain hormones. The male and female sex hormones are made from this chemical. Without cholesterol, the differences between the sexes would be less obvious. Cholesterol is also an important part of the cell membrane.

Certain forms of cholesterol have been associated with heart disease and circulatory problems. If cholesterol combines with other fats, it can form a plaque that may block blood vessels (**Figure 4**). The reduced blood flow to that tissue can cause it to die from lack of oxygen and nutrients. The buildup of cholesterol in blood vessels that serve the heart muscle is especially dangerous.

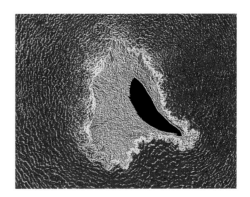

Figure 4
Plaque formation causes restricted blood flow in blood vessels and can lead to heart attack or stroke.

Practice

Understanding Concepts

1. What are fats?
2. Describe some groups of lipids that are not triglycerides.
3. What are the two structural components of triglycerides?
4. What is cholesterol?
5. How do saturated fats differ from unsaturated fats?

Carbohydrates and Lipids

Carbohydrates

1. Carbohydrates are our most important source of energy. We get carbohydrates from plants.

2. A monosaccharide is a sugar made up of a single sugar molecule.

3. Disaccharides are formed by dehydration synthesis (condensation reaction).

4. Polysaccharides (starch, cellulose, glycogen) are formed of many sugar molecules linked together.

5. Starch is an energy storage unit; when energy is needed, starch is broken into single sugar molecules for use by cells.

6. Cellulose is a polysaccharide that cannot be digested by humans.

7. In animals, excess sugars are linked together to form glycogen; when energy is needed, glycogen converts back to individual glucose units.

Lipids

8. Lipids are excellent energy storage compounds but are difficult for your body to break down.

9. Lipids are insoluble in water.

10. Triglycerides are composed of one glycerol and three fatty acids.

11. Plant fats or oils are unsaturated because they do not have all of the hydrogen atoms that they could; animal fats are saturated because they do have the maximum number of hydrogen atoms.

12. Plant fats are easier to break down because they have double bonds; animal fats do not have double bonds and are harder to break down.

13. Cholesterol is an important part of cell membranes. It is also associated with heart disease and circulatory problems.

14. Phospholipids have a water soluble end and a water insoluble end.

Sections 2.1–2.3 Questions

Understanding Concepts

1. What happens to carbohydrates that are not immediately used by your body? Why might you want to limit your carbohydrate intake?

2. Are fats essential to your diet? Explain your answer.

3. How can cholesterol be both beneficial and harmful to you? Explain.

Making Connections

4. Research the dangers associated with artificial fats and sweeteners. Often products like these are initially approved and only later is it discovered that they are harmful. Why do you think products are approved before they are proven safe? What would you suggest be done to ensure the safety of a new food product before approval?
Follow the links for Nelson Biology 11, 2.3.

GO TO www.science.nelson.com

2.4 Proteins

Proteins are used by cells to build structures and are used in chemical activities. Enzymes are proteins that control the rates of many reactions, including the chemical reactions involved with processes such as digestion and cellular respiration. Unlike carbohydrates and fats, proteins are not primarily used for energy. Like lipids and carbohydrates, proteins are composed of carbon, hydrogen, and oxygen. However, proteins contain nitrogen and, sometimes, sulfur atoms as well.

Proteins are associated with building cell structures. Whenever cells are damaged and require repair, proteins are manufactured. Your cells also make proteins to build structures for new cells. Consider the amount of protein that your body must be making this very second. The rate at which proteins used to build red blood cells are created is equal to the rate at which your red blood cells die. The fact that you are not experiencing any discomfort is because red blood cells are replaced at a rate of one million per second. This means that proteins must be constructed for one million cell membranes and many million cell organelles each and every second, and this is just for red blood cells!

Proteins are composed of 20 different amino acid building blocks (**Figure 1**). The order and number of amino acids determine the type of protein. The sequencing of amino acids is regulated by the genes located on the chromosomes. The genes are patterns of **nucleotides** in DNA. This pattern is used to build messenger RNA (mRNA) during a process called **transcription**. The mRNA leaves the nucleus of the cell and attaches to ribosomes in the cytoplasm. There is another type of RNA in the cytoplasm called transfer RNA (tRNA). Each tRNA is attached to a specific amino acid. The tRNA is brought into position to join with mRNA on the ribosome. As this takes place, each amino acid is brought into the correct position in the chain to build the protein. This process is called **translation**. A small protein may contain as few as 8 amino acids (insulin has 51 amino acids), while some of the longer chains can have more than 4000 amino acids. A chain of several amino acids can also be referred to as a **polypeptide** (**Figure 2**). Refer to **Figure 3** for a diagram depicting protein synthesis.

The importance of proteins in the diet cannot be underestimated. Although the body is capable of making many amino acids, there are eight amino acids that the body cannot synthesize. These eight amino acids, called essential amino acids, must be obtained from your foods. The lack of any one of the essential amino acids will lead to specific protein deficiencies and disease.

nucleotides: the basic structural units of nucleic acid. Each unit is composed of a five-carbon sugar, a phosphate, and a nitrogenous base.

transcription: the process by which an mRNA molecule is built using the sequence of nucleotides in DNA as a template

translation: the process by which polypeptides (proteins) are produced at the ribosomes. Transfer RNA (tRNA) positions amino acids according to the sequence of nucleotides in mRNA.

polypeptide: a chain of amino acids held together by peptide bonds

Amino group   Acid group

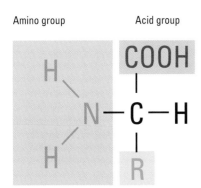

Figure 1
A generalized plan for an amino acid. The R group can represent a number of different chemical groups.

Figure 2
A polypeptide composed of four amino acids

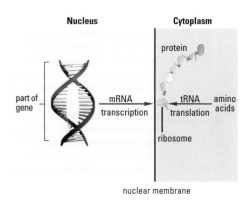

Figure 3
Protein synthesis

Figure 4
Dr. Charles Deber

denature: to disrupt amino acid bonds (that hold a protein molecule together) by physical or chemical means, changing the protein's shape, which may or may not be temporary

coagulation: the permanent change in a protein's shape due to amino acid bond disruption

Research in Canada: Proteins

The membranes of all living cells are the links between the inside and the outside worlds. They receive and transport information, nutrition, and medication for the cells. Key to the correct functioning of cell membranes are transmembrane proteins, which are a group of modified proteins that are embedded in the cell membrane and which act as channels for certain ions moving into and out of the cell. A shortage in the number of these membrane proteins, or mutation of them, is the cause of many human diseases such as diabetes, multiple sclerosis, cystic fibrosis, muscular dystrophy, and several forms of cancer.

Current studies tell us about the shape of these proteins. The next step is finding out about the specific molecular structure of the proteins, which would help scientists deduce the molecular mechanism behind some of the diseases mentioned. Such knowledge could be vital to the development of new drugs and therapy to help people with these diseases. Dr. Charles Deber (**Figure 4**) at the Hospital for Sick Children in Toronto has developed a new technique to obtain more information about this characteristic of the proteins.

Dr. Deber has isolated and cloned key segments of the defective membrane protein associated with cystic fibrosis. Through the use of computer modelling, he hopes to produce and study normal and cystic fibrosis mutants to gain insight into the molecular defect(s) which lead to the disease.

Denaturation

Exposing a protein to heat, radiation, or a change in pH will alter its shape. Physical or chemical factors that disrupt the bonds between the amino acids, changing the configuration of the protein, are said to **denature** the protein. The protein may uncoil or assume a new shape. The result is a change in the protein's physical properties as well as its biological activities. Once the physical or chemical factor is removed, the protein may assume its original shape. When the change in protein shape is permanent, **coagulation** is said to have occurred.

Practice

Understanding Concepts

1. What are proteins?
2. What are amino acids?
3. List foods that are considered to be a good source of proteins.
4. How can only 20 amino acids account for the very large number of different proteins?
5. Define protein denaturation and coagulation.

INQUIRY SKILLS

- ○ Questioning
- ○ Hypothesizing
- ○ Predicting
- ○ Planning
- ● Conducting
- ● Recording
- ● Analyzing
- ● Evaluating
- ● Communicating

Investigation 2.4.1

Identifying Nutrients

In this activity, you will use laboratory tests to identify single sugars (monosaccharides), starches, proteins, and lipids. You will then use these tests to identify the nutrients present in an unknown sample.

Question

What nutrients are present in the unknown sample?

Materials

goggles	lab apron	8–16 test tubes
test tube racks	test tube brushes	hot plate
thermometer	400-mL beaker	10-mL graduated cylinder
distilled water	wax pencil	test-tube holder
medicine droppers	depression spot plate	rubber stoppers
test tube brush	detergent solution	ring clamp
utility stand		

For monosaccharide test:

Benedict's reagent	5% glucose (dextrose) solution	5% fructose solution
5% sucrose solution	5% starch suspension	unknown solution

 Benedict's reagent, Lugol's solution, Sudan IV indicator, and Biuret reagent are toxic and can cause an itchy rash. Avoid skin and eye contact. Wash all splashes off your skin and clothing thoroughly. If you get any chemical in your eyes, rinse for at least 15 min and inform your teacher.

For starch test:

Lugol's solution	5% glucose solution	5% sucrose solution
5% starch suspension	unknown solution	

For lipid tests:

Sudan IV indicator	unglazed brown paper	unknown solution
vegetable oil	(2 letter-sized sheets)	
skim milk	whipping cream	

 Sudan IV indicator is flammable. Keep away from the hot plate.

For protein test:

Biuret reagent	gelatin	egg albumin
skim milk	unknown solution	

Procedure

Before you begin

- make sure that all the glassware is clean and well rinsed;
- note the location of the eyewash station;
- put on your apron and goggles.

Part 1: Monosaccharide Test

Benedict's reagent is an indicator for monosaccharides and some disaccharides, such as maltose. **Table 1** summarizes the quantitative results obtained when a simple sugar reacts with Benedict's reagent.

1. Prepare a water bath by heating 300 mL of tap water in a 400-mL beaker on a hot plate. Heat the water until it reaches approximately 80°C. (Use the thermometer to monitor the temperature.) See **Figure 5**.

Table 1

Colour of Benedict's reagent		Approximate sugar concentration
	blue	nil
	light green	0.5%–1.0%
	green to yellow	1.0%–1.5%
	orange	1.5%–2.0%
	red to red brown	>2.0%

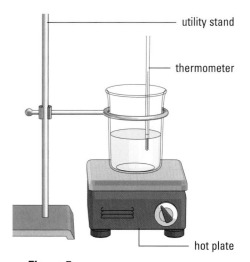

Figure 5
A water bath

2. Using a 10-mL graduated cylinder, measure 3 mL each of distilled water, fructose, glucose, sucrose, starch, and the unknown solution. Pour each solution into a separate test tube. Clean and rinse the graduated cylinder after pouring each solution. Label each test tube using the wax pencil. Record the code for your unknown sample. Add 1 mL of Benedict's reagent to each of the test tubes.

3. Using a test tube holder, place each of the test tubes in the hot water bath. Observe for 6 min. Record any colour changes in a chart.

Part 2: Starch Test

Lugol's solution contains iodine and is an indicator for starch. Iodine turns blue-black in the presence of starch.

4. Using a medicine dropper, place a drop of water on a depression spot plate and add a drop of Lugol's solution. Record the colour of the solution.

5. Repeat the procedure, this time using drops of starch, glucose, sucrose, and your unknown solution. Record the colour of the solutions in a chart. Which solutions indicate a positive test?

Part 3: Sudan IV Lipid Test

Sudan IV solution is an indicator of lipids, which are soluble in certain solvents. Lipids turn from a pink to a red colour. Polar compounds will not assume the pink colour of the Sudan IV indicator.

6. Using a 10-mL graduated cylinder, measure 3 mL each of distilled water, vegetable oil, skim milk, whipping cream, and the unknown solution. Pour each solution into a separate labelled test tube. Clean and rinse the graduated cylinder after each solution. Add 6 drops of Sudan IV indicator to each test tube. Place stoppers on the test tubes and shake them vigorously for 2 min. Record the colour of the mixtures in a chart.

Part 4: Translucence Lipid Test

Lipids can be detected using unglazed brown paper. Because lipids allow the transmission of light through the brown paper, the test is often called the translucence test.

7. Draw one circle (10-cm diameter) on a piece of unglazed brown paper. Place 1 drop of water in the circle and label the circle accordingly. Using more sheets, draw a total of 7 more circles (10-cm diameter). Place 1 drop of vegetable oil, skim milk, whipping cream, and unknown solution, each inside its own circle, labelling the circles as you do. When the water has evaporated, hold both papers to the light and observe. In a chart, record whether or not the papers appear translucent.

Part 5: Protein Test

Proteins can be detected by means of the Biuret reagent test. Biuret reagent reacts with the peptide bonds that join amino acids together, producing colour changes from blue, indicating no protein, to pink (+), violet (++), and purple (+++). The + sign indicates the relative amounts of peptide bonds.

8. Measure 2 mL of water, gelatin, albumin, skim milk, and the unknown solution into separate labelled test tubes. Add 2 mL of Biuret reagent to each of the test tubes, then tap the test tubes with your fingers to mix the contents. Record any colour changes in a chart.

Analysis

(a) What laboratory evidence suggests that not all sugars are monosaccharides?
(b) Explain the advantage of using two separate tests for lipids.

 To mix the contents of a stoppered test tube thoroughly, make sure the stopper is on tight. Place your index finger on top of the stopper, gripping the test tube firmly with your thumb and other fingers. Then shake away from others.

(c) Which test tube or depression plate served as a control in the test for monosaccharides, starches, lipids, and proteins?

(d) Summarize your group's findings about the nutrients present in your unknown solution. Be sure to include the identifying code.

Evaluation

(e) Why should the graduated cylinder be cleaned and rinsed after measuring out each solution?

(f) List possible sources of error, and indicate how you could improve your method.

Synthesis

(g) A student places a sugar cube into one of the unknown solutions in the experiment. Explain the effect this would have on the results.

(h) A drop of iodine accidently falls on a piece of paper. Predict the colour change, if any, and provide an explanation for your prediction.

(i) A student heats a test tube containing a large amount of protein and Biuret reagent. She notices a colour change from violet to blue. Explain why heating causes a colour change.

(j) Several artificial sweeteners are to be tested. Which reagents would you use on a solution of each sweetener? Why?

(k) Predict the results of a lipid test on samples of butter and margarine.

(l) If you were to test the nutrient content of a sports drink or a diet drink that acts as a meal substitute, which nutrients would you expect to find in each? Make your predictions, then compare your list with the list of nutrients on each product.

(m) Identify careers where knowledge of nutrient testing would be useful.

2.5 Nucleic Acids

The two nucleic acids, DNA and RNA, are not nutrients but they are essential for all living things.

Figure 1 (page 46) shows the basic structure of DNA. DNA consists of sugars, phosphate groups, and four nitrogenous bases: adenine (A), guanine (G), cytosine (C), and thymine (T). (Note: RNA does not contain thymine but another base called uracil (U), and the sugar is different.) A sugar, a phosphate group, and a nitrogenous base are bonded together to form a subunit called a nucleotide. The nitrogenous bases of adjacent DNA nucleotides always pair as follows: A-T (or T-A) and G-C (or C-G) (**Figure 1**, page 46).

Figure 2 (page 46) illustrates some of the great variety of life forms on Earth. Despite the diversity, the chromosomes of all living things are composed of the same chemical: DNA. So what causes the differences? Since proteins are largely responsible for the structural and functional characteristics of living things, organisms of one kind must contain a different set of proteins than organisms of another type. Proteins are long chains of amino acids. The number and sequence of amino acids in the chain determines the unique characteristics of a particular protein. The sequence of amino acids in a protein molecule is determined by the sequence of nucleotide bases in a segment of DNA called a gene. This means that each of the genes in a DNA molecule (a particular sequence of nucleotide bases) codes for the production of one protein (a corresponding sequence of amino acids). We call this the one gene, one protein hypothesis.

DID YOU KNOW ?

Evolutionary biologists have theorized that the RNA molecule existed before the DNA molecule and may even have given rise to the DNA molecule.

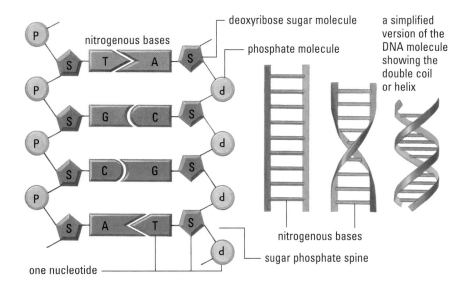

Figure 1
DNA is organized in a double coil or helix.

What a protein does in a cell, or can do, is almost entirely determined by its structure. Remember that there are 20 different amino acids available to a cell for making proteins. A protein molecule composed of 254 amino acids will have different chemical properties than one with 102 amino acids. Two protein molecules with the same 77 amino acids in each, but with a different sequence of those amino acids, will behave differently. Since the sequence of amino acids for all the protein molecules in an organism is coded by the sequence of nucleotides in the DNA molecules, organisms differ from one another on account of the unique set of DNA molecules that are passed on from one generation to the next.

As it turns out, each of the amino acids in a protein is coded not by one nucleotide base in DNA, but by a triplet of bases. With 4 different bases (A, T, C, and G) available in DNA, and 3 bases needed to code for a particular amino acid, $4^3 = 64$ different base triplets are possible as codes for the 20 available amino acids. This is more than enough to code for all 20 amino acids.

(a) German shepherd

(b) rabbit

(c) mouse

(d) black molly

Figure 2
How does your genetic information differ from that of a dog, rabbit, mouse, or black molly? A look into one of your cells reveals 46 chromosomes, forming 23 pairs. The chromosomes contain hereditary or genetic information in the form of genes. One chromosome in each pair contains information that was passed on to you by your mother, the other by your father. A dog **(a)** has 39 pairs of chromosomes, a rabbit **(b)** has 22 pairs, and a mouse **(c)** has 20 pairs. The black molly **(d)** has 23 pairs, the same number as humans. Since a black molly contains the same number of chromosomes as a human, it is obvious that it's not the number of chromosomes that distinguishes one type of organism from another. What then is responsible for the great variety of life forms that exist on Earth today? A closer look at the chromosomes reveals something surprising. All of the chromosomes are composed of the same chemical: DNA.

In fact, each amino acid has more than one DNA base triplet code, very much like synonyms in the English language. Since three DNA bases are used as a code for each amino acid in a protein, a gene on DNA has at least three times as many bases as there are amino acids in the protein it codes for.

A DNA molecule contains thousands of base pairs along its length. This allows it to code for a great variety of proteins. Much in the same way that the 26 letters of the alphabet may be arranged for an almost endless number of words, so too can the 4 base pairs. The almost endless combinations of the 4 base pairs, that is, the base-pair sequence and the number of base pairs in the sequence, provide for great diversity in the genetic code.

Practice

Understanding Concepts

1. What is the chemical of heredity?
2. What are the three major groups of chemicals in DNA?
3. If DNA is composed of only three types of chemicals, why are living things so different from each other?
4. What are the base pairs in DNA and RNA?
5. Describe the mechanism that is responsible for placing the specific amino acids of a protein in the proper order.
6. If a nucleotide triplet contains A, T, and C, how many different arrangements can you obtain?

SUMMARY   Nucleic Acids and Proteins

Nucleic Acids

1. There are two nucleic acids, DNA and RNA.
2. DNA provides the instructions for building proteins.
3. DNA is composed of sugar, phosphate, and nitrogenous bases.
4. Nitrogenous bases pair up: adenine-thymine (adenine-uracil in RNA), and guanine-cytosine.

Proteins

5. Proteins are structural and functional components of the cell.
6. Proteins are made of amino acids; the number and sequence is determined by DNA.
7. Enzymes, hormones, antibodies, and hemoglobin are examples of chemicals in cells and the body that are made from protein.

2.6 The Living Cell Membrane

The cell membrane separates a cell's protoplasm from its external environment. However, the cell membrane is much more than a flexible envelope that holds the cytoplasmic organelles in place. It regulates what enters and leaves the cell. If the cell membrane is pierced, some cytoplasm will ooze out but the puncture will soon be sealed. Cell membranes are found only in living cells.

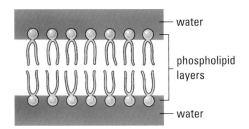

Figure 1
The cell membrane is composed of two layers of phospholipids.

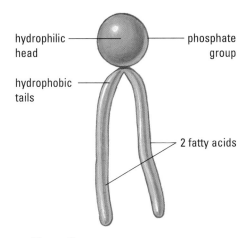

Figure 2
Phospholipid molecule

In order to survive, cells must extract nutrients from the environment. As nutrients are processed, wastes build up inside the cell. The environment acts as a waste disposal site. Without it, the accumulation of poisonous wastes within a cell would soon cause cell death.

The fact that the cell membrane is involved in the transportation of materials into and out of the cell raises some questions: How does the cell determine which materials to absorb? How does the cell select molecules?

To understand the movement of molecules across the cell membrane, the chemical composition of the membrane must be examined. **Figure 1** illustrates that the membrane appears as two layers of phospholipids (lipids with phosphate groups). As shown in **Figure 2**, each phospholipid molecule can be represented by a head and two tails. The head is hydrophilic, or water attracting, and is soluble in water. The tails are hydrophobic, or water repelling, and are not soluble in water. The water-soluble ends of one layer face the outer aqueous environment and the water-soluble ends of the second layer face the aqueous cytoplasm. The hydrophobic tails are forced to face inward, away from the watery environment and toward each other (**Figure 3**).

A variety of different protein molecules is embedded within the phospholipid bilayer. Many of these proteins carry special sugar molecules and are thus referred to as glycoproteins. The glycoproteins provide the cell with a unique identity. Glycoproteins vary between different organisms, and even between individuals within the same species. They help distinguish a type A red blood cell from a type B red blood cell. Your immune system identifies foreign invaders by recognizing the unique glycoprotein structure on their cell membrane. This helps explain why transplanted organs are often rejected by recipients.

The protein molecules in the phospholipid bilayer serve different functions. Some act as gatekeepers, opening and closing paths through the cell membrane. For example, specialized protein gatekeepers, located in nerve cells, allow potassium ions to move through the pores when the membrane is at rest. However, when the nerve is excited, a sodium ion gateway is opened. This is what causes the nerve impulse to fire.

Other proteins act as receptor sites for hormones. Hormones are chemical messengers that allow cells to communicate with one another. For example, insulin, a hormone secreted by the pancreas, attaches itself to specific receptor

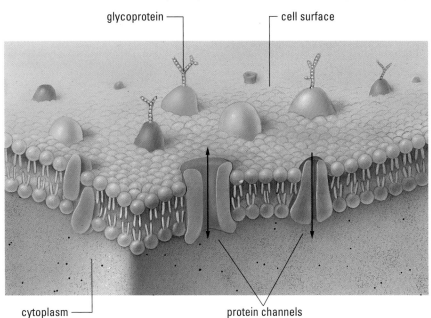

Figure 3
A cell membrane is formed by a double layer of phospholipid molecules in which protein molecules are embedded.

proteins that are embedded in the membrane, causing it to open a pore and allow in sugar from the blood. Another class of protein is involved in transport, using cell energy to pick up needed materials and move them into or out of the cell. You will learn more about transport proteins later in this chapter.

Liposomes and Aquaculture

In 1965, Alec Bangham discovered that lipids could initiate their own assembly into double-layered spheres approximately the size of a cell. The bubbles are now known as **liposomes**. Liposomes function like cell membranes because they can fuse with a cell membrane and deliver their contents to the cell's interior. Today, liposomes are used together with cancer-fighting drugs to help the drug target the tumours. This helps to reduce unwanted side effects from drug interaction with healthy tissues. Using liposomes as a drug delivery system also enables patients to accept higher doses of anti-cancer drugs. Only two liposomal drugs have been approved for use so far. One of these was discovered by Dr. Theresa Allen and her research group at the University of Alberta. Research is in the clinical trial stage for liposome-delivered drugs to be used against breast cancer, as well as lung cancer, melanoma, leukemia, and prostate cancer.

Using liposomes to deliver materials is showing promise as a means of increasing the efficiency of gene therapy. Gene therapy is the process of introducing new genes into the DNA of a person's cells to correct a genetic disease or flaw. In a process similar to endocytosis, researchers have successfully inserted DNA into liposomes which have fused with target cells.

DID YOU KNOW ?

Some scientists have speculated that many hormone disorders are due not to inadequate hormone production but to low numbers of receptor proteins on the membranes of cells that hormones act on. An excess number of receptors could also cause severe problems. Hypertension, or high blood pressure, may be caused by an excess number of receptor sites for stress-related hormones.

liposomes: artificial lipid vesicles

Practice

Understanding Concepts

1. Describe the molecular structure of the cell membrane.
2. The cell membrane is often referred to as a fluid mosaic. What makes the membrane fluid? What makes it a mosaic?
3. What are liposomes?
4. What structural feature of a cell membrane is missing from liposomes?
5. What are some functions of proteins in the cell membrane?

SUMMARY The Cell Membrane

1. The cell membrane appears as two layers of phospholipids; each phospholipid has a hydrophilic, water-soluble end and a hydrophobic, non-water-soluble end. One water-soluble end faces the outer environment and the second water-soluble end faces the cytoplasm.

2. Glycoproteins and other proteins are embedded in the phospholipid bilayer; they impart a unique identity to the cell.
   - Some proteins open and close the paths through the cell membrane, while others act as receptor sites for hormones.
   - Some proteins are involved in transport.

3. Liposomes are artificial lipid vesicles that can be used to improve delivery of materials to cells.

Understanding Concepts

1. In what ways does a cell membrane differ from a cell wall?
2. What is the relationship between
   (a) nucleic acids and proteins?
   (b) proteins and a cell membrane?

2.7 Passive Transport

passive transport: the movement of materials across a cell membrane without the use of energy from the cell

Cells must take in food and eliminate wastes to maintain a constant internal environment. The movement of materials across a cell membrane without the expenditure of cell energy is called **passive transport**. The following section discusses three types of passive transport: diffusion, osmosis, and facilitated diffusion.

Diffusion

Brownian motion: the random movement of molecules

diffusion: the movement of molecules from an area of higher concentration to an area of lower concentration

Molecules in a solution or gas move about randomly and collide. This random movement is referred to as **Brownian motion**. Molecules move in all directions with equal frequency, bouncing off each other when they collide. This causes molecules to move from an area of high concentration to an area of lower concentration. (Concentration can be described as the number of molecules per unit volume.) This process is called **diffusion**. You can experience diffusion simply by opening a perfume bottle. The molecules inside the perfume bottle move into the surrounding air. Diffusion is not confined to gases, as you can see if you drop ink into a glass of water and watch the colour of the water gradually change (**Figure 1**). Water molecules collide with the molecules of ink, spreading them apart. The water molecules alone are not responsible for diffusion, however. The ink molecules are also colliding with water molecules and other ink molecules.

Molecular collisions cause diffusion. Can you predict how an increase in water temperature would affect the rate of diffusion? The faster the molecules move, the more often they collide and the faster they will move apart. In other words, diffusion rates increase with temperature. In gas phase, pressure increases the frequency of particle collisions. Molecules are bunched close together if the pressure is high. Molecules in high-pressure areas will collide more frequently. Molecules in areas of lower pressure are spread out and collide less frequently. Gas molecules move by diffusion from areas of high pressure to areas of low pressure.

Figure 1
Ink diffusing in water over time

To summarize, diffusion is affected by concentration, temperature, and pressure and will continue until molecules are equally distributed within a space.

Oxygen and carbon dioxide move across cell membranes by diffusion. Oxygen diffuses from the blood, an area of high concentration, into the cell, where the oxygen concentration is lower. Because oxygen is continuously consumed within a cell, the concentration of oxygen does not build up. Carbon dioxide, by contrast, accumulates within a cell and diffuses from the cell to the blood.

Osmosis

Osmosis is the diffusion of water across a selectively permeable membrane. Consider the system shown in **Figure 2(a)**. Membrane X is permeable to water, but impermeable to the larger protein molecules. Protein cannot diffuse from side A into side B.

The concentration of protein in side A is greater than in side B. Did you notice that the volume of fluid in side A is equal to that in side B? Which side do you think has the greater concentration of water? There are fewer protein molecules in side B, but many more water molecules. The spaces between protein molecules are filled with water molecules. Therefore, the fewer the number of protein molecules, the greater the amount of space for water molecules.

osmosis: the diffusion of water molecules across a selectively permeable membrane

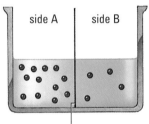

(a) Membrane X is permeable to water but not to protein.

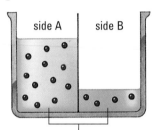

(b) There are equal concentrations of protein and water in side A and side B.

Figure 2

Try This
Activity

The Effect of Salt Water on Potato Strips

- Take a potato and slice it into 5 thin strips, each about 2 cm wide and 5 cm long. Try to make all of the strips the same width and length.
- Obtain salt solutions with the following concentrations: 0% salt, 1% salt, 2% salt, 5% salt, and 10% salt.
- Place a potato strip in each of 5 test tubes. Label the test tubes with the above salt concentrations.
- Pour equal volumes of each salt solution into their respective test tubes, so that each strip is completely covered.
- Allow the strips to remain in the solutions for 20 min and then remove and measure the length of each strip.
- Prepare an observation chart to record the original and final measurements.
- Calculate the change in length. Remember to use negative numbers to show a strip that became shorter.

(a) Describe the changes in the length of the potato strips as salt concentration increases. Suggest an explanation for the observed changes.

Remember that osmosis is just a diffusion of water. Water will diffuse from side B, the area of higher water concentration, to side A, the area of lower water concentration. Like other molecules, water follows the **concentration gradient**. **Figure 2(b)**, page 51, shows what happens when water moves from side B into side A by osmosis.

As water leaves side B, the protein in side B becomes more concentrated. As water enters A, the protein in side A becomes less concentrated. Eventually, the concentrations of protein and water in sides A and B will become equal. Once the system reaches **equilibrium**, water molecules will continue to move between the two sides; however, the number of molecules gained from side A equals the number of molecules gained from side B. There is no net movement of water at equilibrium.

The movement of water into and out of living cells is vital to life processes. Ideally, cells are bathed in **isotonic solutions**—solutions in which the solute concentration outside the cell is equal to that inside the cell, as shown in **Figure 3(c)**. Blood's major function in your body is to keep your internal environment in an isotonic balance called **homeostasis**. Maintaining an isotonic balance is only part of homeostasis. In isotonic solutions, the water movement into a cell is balanced by the water movement out of the cell.

A **hypotonic solution**, shown in **Figure 3(a)**, is a solution that has a lower concentration of solute and, therefore, a higher concentration of water than

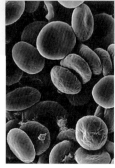 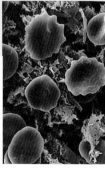 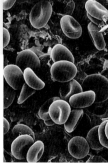

(a) hypotonic
water diffuses inward

(b) hypertonic
water diffuses outward

(c) isotonic
no net change in water movement

Figure 3
(a) Red blood cells in hypotonic solution
(b) Red blood cells in hypertonic solution
(c) Red blood cells in isotonic solution

inside the cell. Cells that are placed in hypotonic solutions are not at equilibrium. Water molecules move from the area of high water concentration, the solution outside the cell, into the region of lower water concentration, the inside of the cell, by osmosis. The cell expands as water moves in.

In a freshwater environment, the low solute concentration can result in excess water moving into the cells. Organisms have developed ways to expel the incoming water; for example the paramecium uses a contractile vacuole to expel excess water from the cytoplasm, as illustrated in **Figure 4**. Any failure of the contractile vacuole would be disastrous for the cell; water would continue to diffuse into the cell, eventually causing it to explode.

On the other hand, cells placed in a **hypertonic solution** tend to shrink. In hypertonic solutions, the concentration of solutes is higher than in the cell and the concentration of water is lower. Water moves out of the cell by osmosis into the external solution causing the cell to shrink in size. For an example, see **Figure 3(b)**.

Have you ever noticed that the salt sprinkled on sidewalks during winter kills the surrounding grass in spring? Salt creates a hypertonic solution that draws water out of the cells of the grass causing the plants to wilt. Water pressure, referred to as turgor pressure, pushes the cytoplasm of the plant cell against the cell wall. Turgor pressure is the reason plants are rigid (**Figure 5**). The process of osmosis explains why plants die if they are exposed to too much fertilizer and why vegetables are sprayed with water in your local grocery store.

Facilitated Diffusion

Protein carrier molecules, located in the cell membrane, can aid in passive transport. Although the precise action of these carriers is not well understood, scientists believe that the protein carriers speed up the movement of molecules already moving across the cell membrane. Glucose diffuses into red blood cells hundreds of times faster than other sugar molecules that have similar properties. Why would one sugar molecule diffuse faster than another? Scientists have proposed that the diffusion is facilitated. The carrier proteins must be specialized to aid the diffusion of glucose molecules, but not other sugars.

Activity 2.7.1

Observing Diffusion and Osmosis

In this activity, you will construct a model of a selectively permeable membrane using dialysis tubing. You will observe osmosis and diffusion through this membrane.

Materials

lab apron	goggles	utility stand
ring clamp	scissors	string
medicine dropper	2 test tubes	25-mL graduated cylinder
100-mL graduated cylinder	250-mL beaker	500-mL beaker
dialysis tubing	funnel	paper towels
distilled water	4% starch suspension	glucose solution
Lugol's solution	glucose indicator such as Tes-tape®	

 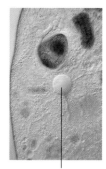

contractile vacuole

Figure 4
A contractile vacuole expels water from a paramecium, permitting it to exist in a hypotonic environment.

hypertonic solution: a solution where the concentration of solutes outside a cell is higher than that found inside the cell

Figure 5
As the plant cells lose turgor pressure, the plant begins to wilt.

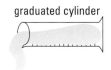

graduated cylinder

funnel ——

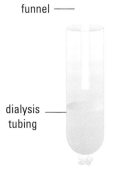

dialysis —— tubing

Figure 6

Procedure

1. Put on your lab apron and goggles.

2. Cut two strips of dialysis tubing (about 15 cm long) and soak them in a beaker of tap water for approximately 2 min.

3. Rub the dialysis tubing between your fingers to find an opening. (This is similar to the way you would open a flattened plastic bag.) Tie a knot near one of the ends of the dialysis tubing. Repeat with the other piece of tubing.

4. Using a graduated cylinder, measure 15 mL of the 4% starch suspension. Pour the suspension into the open end of one of the dialysis tubes through the funnel.

5. Fill the second dialysis tube with 15 mL of the glucose solution.

6. Twist the open end of each dialysis tube and tie in a knot. Try to have as little air as possible. Trim off the excess tubing.

7. Rinse the outside of both dialysis tubes with distilled water to remove any fluids that may have leaked out during the tying process.

8. Cut two equal lengths of string. Suspend each tube with string from the ring clamp as shown in **Figure 7**.

9. Pour 250 mL of water into the small beaker. Add 5 drops of Lugol's solution.

10. Pour 245 mL of the dilute Lugol's solution into the large beaker, and test the solution for the presence of glucose using a glucose indicator.

11. Suspend the dialysis tubes in the dilute Lugol's solution so that they are submerged but do not touch the bottom of the beaker (**Figure 7**).

12. Pour the remaining 5 mL of dilute Lugol's solution into a test tube and place in the beaker with the dialysis tubes.

13. Pour 5 mL of the starch suspension into a second test tube and place in the beaker. Wait for 20 min.

14. While waiting, construct a table similar to **Table 1** in your notebook. Record your observations for the second column.

Table 1: Colour of the Solution or Suspension

	Initial observations	After 20 min
dilute Lugol's solution in beaker	?	?
starch suspension in dialysis tube	?	?
glucose solution in dialysis tube	?	?
dilute Lugol's solution in test tube	?	?
starch suspension in test tube	?	?

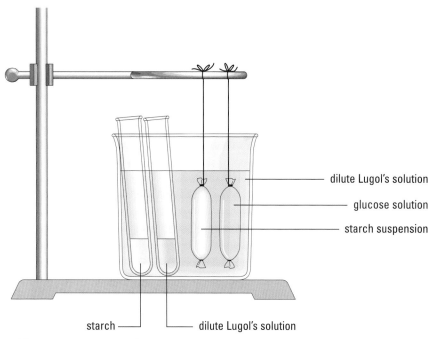

dilute Lugol's solution

glucose solution

starch suspension

starch —— —— dilute Lugol's solution

Figure 7

15. After the 20 min, complete the third column of your table.

16. Test the dilute Lugol's solution in the beaker and test tube for the presence of glucose. Indicate the colour the glucose indicator changed. Add this information to your table.

17. When you have finished, empty the contents of the dialysis tubes down the sink drain, flushing with lots of water. The empty tubes may be placed in the garbage. Be sure to wash the equipment and your hands thoroughly.

Analysis

(a) Which substance diffused from the dilute Lugol's solution in the beaker into the dialysis tube? Provide laboratory evidence and suggest a reason to explain this.

(b) Which substance diffused from the dialysis tube into the dilute Lugol's solution in the beaker? Provide laboratory evidence and suggest a reason to explain this.

(c) Explain the function of the test tube with Lugol's solution. What evidence does it provide?

(d) Explain the function of the test tube with the starch suspension. What evidence does it provide?

(e) What properties does this experiment demonstrate about the dialysis tube as a model for a cell membrane?

(f) Which substance tested has the largest molecules? Explain your answer.

Evaluation

(g) Suggest possible sources of error in the procedure.

(h) Prepare a report.

Investigation 2.7.1

Factors Affecting the Rate of Osmosis

You will design a controlled experiment to examine the factors which affect the rate of diffusion. Possible variables might include different concentrations of syrup and/or starch; temperature; ratios of syrup and starch in a mixture.

INQUIRY SKILLS

○ Questioning	○ Recording
○ Hypothesizing	○ Analyzing
○ Predicting	○ Evaluating
○ Planning	○ Communicating
○ Conducting	

Prediction

Based on what you have learned, indicate which factor(s) will affect the rate of osmosis and how.

Design

Choose one variable to examine. Write a detailed plan for carrying out your investigation.
Your design should include the following:
- control(s)
- how you will record and display your observations

Materials

(a) List all of the materials and apparatus that you will need.

Procedure

When your teacher has approved your procedure and the materials that you will use, carry out the procedure with your lab group. Be sure to include diagrams of the apparatus setup and safety precautions.

Analysis

(b) Summarize your results in a report. Include a table to display your observations.

Evaluation

(c) In your report, comment on your predictions. Consider any possible sources of error. Prepare your conclusions.

(d) Describe how you could improve your method, as well as how you could further investigate your variable.

(e) Indicate another variable(s) you would like to investigate. Write a brief outline of an experiment to test the new variable(s).

2.8 Active Transport

Molecular Active Transport

When a cell uses its own energy to move materials from an area of low concentration to an area of higher concentration, this is referred to as **active transport**. It has been estimated that, while you sleep, between 30% and 40% of your total energy budget is used for active transport. Without active transport, your kidneys would fail to reabsorb needed materials, your muscles would not contract, and your nerves would not carry impulses.

active transport: involves the use of cell energy to move materials across a cell membrane against the concentration gradient

Endocytosis and Exocytosis

Cells take in smaller solutes by means of transport carrier molecules. However, some molecules, many of which are essential to life, will not pass through the pores of the cell membrane. How do cells absorb the larger molecules? The cell must expend energy to transport these larger substances.

Endocytosis is the process by which cells ingest materials. The membrane folds around materials outside the cell. The ingested particle is trapped within a pouch, or vacuole, inside the cytoplasm. Often enzymes from the lysosomes are then used to digest the molecules absorbed by endocytosis.

There are two types of endocytosis. In **pinocytosis**, cells take up dissolved molecules by engulfing small volumes of the external solution. Cells of your small intestine engulf fat droplets by pinocytosis. **Phagocytosis** is the process by which cells engulf solid particles (**Figure 1**). Some white blood cells are often referred to as phagocytes (eater cells) because they consume invading microbes by engulfing them. The microbe, trapped in the vacuole, is digested when the vacuole fuses with lysosomes. Some membrane proteins have a unique shape to match specific molecules, such as insulin or cholesterol, allowing these materials to enter the cell by endocytosis.

pinocytosis: a form of endocytosis in which liquid droplets are engulfed by cells

phagocytosis: a form of endocytosis in which solid particles are engulfed by cells

Exocytosis is the process by which large molecules held within the cell are transported to the external environment. Waste materials are often released by exocytosis. Useful materials, like transmitter chemicals emitted from nerve cells, are also released by exocytosis. The Golgi apparatus holds the secretions inside fluid-filled membranes. Small vesicles break off and move toward the

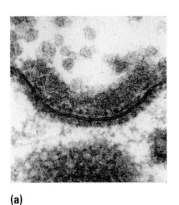

(a)

(b)

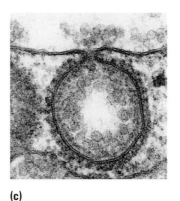

(c)

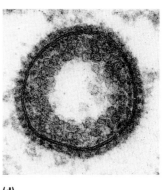

(d)

cell membrane. The vesicles fuse with the cell membrane and the material is released into the external environment.

Figure 1
Electron micrographs illustrate phagocytosis, a form of endocytosis.
(a) Membrane begins to fold around molecules.
(b) Membrane traps molecules.
(c) Cell membrane comes together and vacuole is formed.
(d) Large molecules are digested within the vacuole.

Practice

Understanding Concepts

1. List three factors that affect diffusion rates.
2. (a) Define isotonic, hypertonic, and hypotonic solutions.
   (b) Does it make sense to use these terms to describe inherent properties of a solution, or do they only apply when comparing concentrations?
3. How does facilitated diffusion differ from simple diffusion and osmosis?
4. How does passive transport differ from active transport?
5. Describe the two types of endocytosis.
6. Contrast endocytosis with exocytosis.

SUMMARY Passive Transport and Active Transport

Passive Transport

1. Passive transport is the movement of materials across a cell membrane without expenditure of cell energy.
2. Diffusion is movement of molecules from a region of higher concentration to a region of lower concentration.
3. Osmosis is the diffusion of water molecules across a selectively permeable membrane.
4. In an isotonic solution, the concentration of solute molecules outside the cell is equal to the concentration inside.
5. In a hypotonic solution, the concentration of solutes outside the cell is lower than inside.
6. In a hypertonic solution, the concentration of solutes outside the cell is higher than inside.
7. In facilitated diffusion, protein carrier molecules speed up the movement of molecules already moving across the cell membrane. Cell energy is not used.

Active Transport

8. Active transport uses the cell's own energy to move materials into or out of a cell.

9. In endocytosis, cells engulf large particles by extending their cytoplasm around the particle.

   • In pinocytosis, cells engulf liquid droplets.

   • In phagocytosis, cells engulf solid particles.

10. In exocytosis, large molecules held within the cell are transported to the external environment.

Sections 2.7–2.8 Questions

Applying Inquiry Skills

1. Two cells are filled with the same concentration of sugar solution and placed in separate beakers of distilled water. Both cells expand as water diffuses into the cells, but one expands faster. What are possible reasons for the different rates of expansion?

Making Connections

2. The disease cholera is caused by a bacterium which enters the body through drinking water. Once inside the body, it adheres to the intestinal lining and secretes a toxic product. In response, the lining secretes chloride ions. Sodium ions follow the chloride ions out of the cells. How do these events explain the fluid loss associated with cholera?

2.9 Energy Flow in Photosynthesis and Cellular Respiration

autotrophs: organisms that use energy, usually light, to synthesize their own food from inorganic compounds

photosynthesis: the process by which plants use chlorophyll to trap sunlight energy and use it to produce carbohydrates

For life to continue on our planet, two conditions are necessary. First, matter must be continuously recycled. With the exception of fragments of rock that have struck the earth over time from outer space, the number of atoms on earth is unchanging. All things that are currently living have recycled these atoms from other organisms. Although the atoms may be restructured into new and different molecules, matter is continuously exchanged between living and non-living things, between plants and animals, between different animals, and between animals and microorganisms. The carbon and oxygen atoms exhaled from your lungs once may have been part of a sugar molecule found in corn, which in turn may have been made using a carbon dioxide molecule exhaled by Sir John A. Macdonald!

The second condition for life on Earth is that energy must be continuously added to the planet by sunlight (**Figure 1**). Because energy is lost from a system each time it changes form, energy must be added to the system to ensure that it keeps working. Plants are the key to keeping the energy flowing. Plants are **autotrophs** that use energy from sunlight to convert the low-energy compounds carbon dioxide and water into carbohydrates, through a process called **photosynthesis**, as shown in **Figure 2**. The word photosynthesis is a composite of photo, meaning "light," and synthesis, meaning "to make or build." Photosynthesis, then, is the process of using energy in light to build carbohydrates (**Figure 3**, page 60). The bonds in these

Figure 1
The Sun is the source of life on Earth. Sunlight provides the energy that powers the process of photosynthesis.

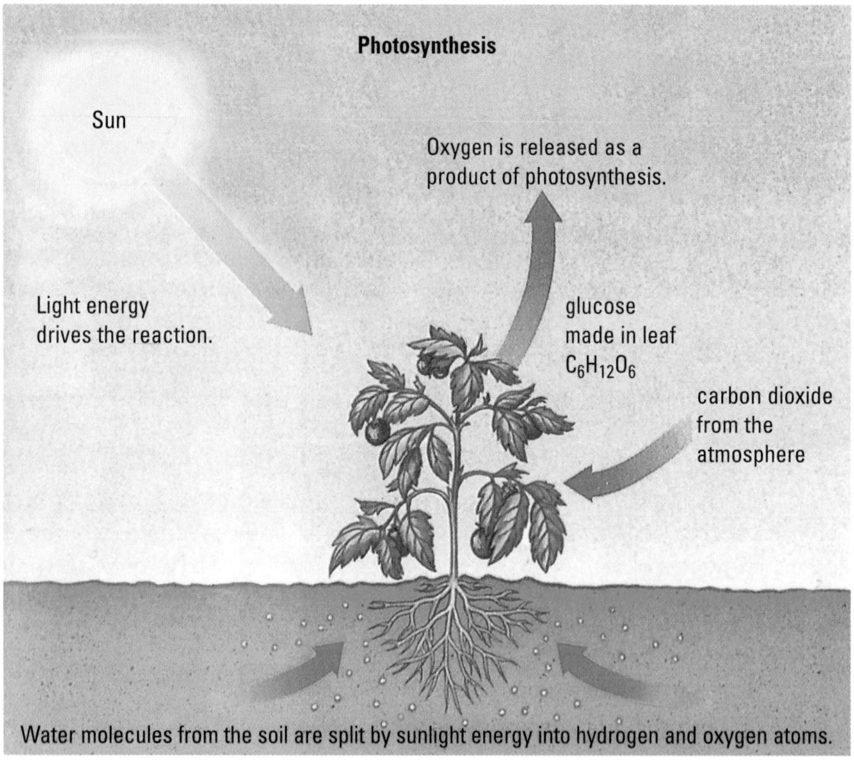

Photosynthesis

Sun

Oxygen is released as a product of photosynthesis.

Light energy drives the reaction.

glucose made in leaf $C_6H_{12}O_6$

carbon dioxide from the atmosphere

Water molecules from the soil are split by sunlight energy into hydrogen and oxygen atoms.

Figure 2
Plants use light energy to form glucose from water and carbon dioxide.

carbohydrates are a source of chemical energy. The following is a simplified chemical equation to show the overall reaction of photosynthesis:

$$6CO_2 + 12H_2O + \text{energy from light} \rightarrow C_6H_{12}O_6 + 6O_2 + 6H_2O$$

carbon dioxide + water + energy from light → glucose + oxygen gas + water

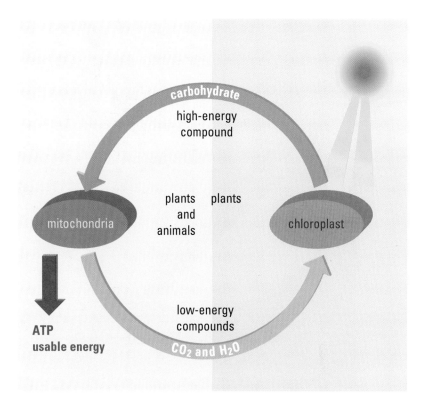

Figure 3
Two important organelles for energy transformation. The chloroplasts are the sites for photosynthesis and the mitochondria are the sites for cellular respiration.

Unlike autotrophs, heterotrophic organisms are not capable of making their own food and, therefore, they must take in food that is already made. Animals are heterotrophs. Your food comes either directly from plants or from animals that have eaten plants. Heterotrophs break chemical bonds in food molecules, thereby returning carbon dioxide and water to the environment. As the chemical bonds break, the energy contained in them is released. This process, which takes place in the mitochondria (**Figure 3**), is cellular respiration, and it is what provides usable energy to your body. Plants also carry out cellular respiration to provide cell energy. In the sections that follow, you will investigate the transformation and exchange of energy within our living world. The following reaction summarizes one type of cellular respiration called aerobic respiration:

$$C_6H_{12}O_6 + 6O_2 \rightarrow 6CO_2 + 6H_2O + energy$$

glucose + oxygen $\rightarrow$ carbon dioxide + water + energy

Practice

Understanding Concepts
1. Explain the meaning of the word photosynthesis.
2. Explain the role of chlorophyll in photosynthesis.
3. Describe how all our energy can be traced back to the Sun.
4. Explain how photosynthesis results in a gain of energy.
5. What is the difference between how autotrophs and heterotrophs obtain energy?

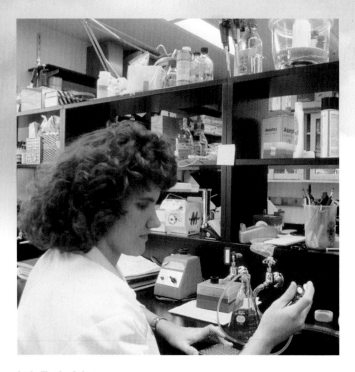

Lab Technician

Often people in this career work in clinics, private laboratories, or hospitals. They use microscopes and other equipment to complete lab tests for doctors and their patients. Programs to become certified as a lab technician are available at community colleges as well as private colleges. They range from two to three years in duration, with numerous upgrading courses available after graduation.

Dietician or Nutritionist

Dieticians and nutritionists work in hospitals, clinics, diet centres, and health clubs. They help people plan meals that contain the required nutrients in the proper proportions. Some dieticians plan diets for athletes in training or for people with medical problems that limit the foods they can consume. Dieticians and nutritionists need an undergraduate degree as well as additional training specifically in the field of nutrition and dietetics.

Practice

Making Connections

6. Identify several careers that require knowledge about cellular functions. Select a career you are interested in from the list you made or from those identified above. Imagine that you have been employed in your chosen career for five years and that you are applying to work on a new project that you are interested in.
   (a) Describe the project. It should be related to some of your new learning in this unit. Explain how the concepts from this unit are applied in the project.
   (b) Create a résumé listing your credentials and explaining why you are qualified to work on the project. Include in your résumé
     • your education, what university program you graduated from, which university you attended, any graduate training;
     • your duties and skills in previous jobs;
     • your salary expectation.

Follow the links for Nelson Biology 11, Chapter 2 Career.

GO TO www.science.nelson.com

Botanist

Botanists are scientists who study many aspects of plant life, including photosynthesis and cellular respiration. They might work as a researcher for a university, or as a consultant to advise other industries about which kinds of plants are best suited for a given purpose. For example, they could test to see which type of wheat grows the fastest, or which plant is capable of filtering impurities out of contaminated land. As well as having, at a minimum, an undergraduate degree, botanists need to have experience both in field work and in the lab.

Cellular Respiration in Plants and Animals

Cellular respiration includes all the chemical reactions that provide energy for life. Carbohydrates, most notably in the form of glucose, are the most usable source of energy. Only after glucose supplies have been depleted do cells turn to other fuels. For animals, glycogen, a storage carbohydrate composed of many glucose units, breaks down, releasing single glucose units into the blood in an attempt to maintain blood glucose levels. Once glycogen supplies from the liver and muscles are depleted, lipids become the preferred energy source. Proteins, the organic compounds of cell structure, are used as a final resort once lipid supplies have been exhausted. The utilization of proteins for energy means that the cell begins breaking down its own structures in order to obtain energy.

Plants use starch as an energy storage compound in much the same manner as animals use glycogen. When needed, starch can be broken down to maltose, a disaccharide. Unlike starch, maltose is soluble. Eventually, the disaccharide can be broken down into monosaccharide units.

The Role of ATP

Cellular respiration in plants and animals involves chemical reactions that take place in the cells' cytoplasm and mitochondria. The overall process is a series of chemical reactions that involves the reaction of glucose with oxygen to produce carbon dioxide, water, and heat. At certain steps in the process, several adenosine triphosphate (ATP) molecules are formed by the addition of inorganic phosphate, Pi, to adenosine diphosphate, ADP, in a process called **phosphorylation**. Carbon dioxide is released as a waste product. The water molecules add to the water content of the cell, while the thermal energy helps maintain a constant body temperature in mammals and birds. ATP is an important compound used throughout the cell to help drive cellular processes that require a net input of energy, including active transport, cell division, movement, and endocytosis. It has been estimated that, at most, only 36% of the energy released in the overall process is used to make ATP; the remaining 64% is released as heat.

All chemical reactions involve breaking chemical bonds between the atoms of reactant molecules and the creation of new bonds between the separated atoms, which form product molecules. Breaking bonds always requires energy; forming bonds always releases energy (**Figure 1**).

breaking bonds: molecules + energy → separate atoms
forming bonds: separate atoms → molecules + energy

Figure 1

The difference between the energy absorbed when reactant bonds break, and the energy released when product bonds form, determines whether energy will be released in the overall process. If energy is absorbed, the reaction is **endergonic** (**Figure 2**). If energy is released in the process, the reaction is **exergonic** (**Figure 3**).

ATP is a source of chemical potential energy for cellular processes that require energy. ATP is composed of the nitrogenous base adenine (also found in RNA and DNA), ribose (the sugar in RNA), and three phosphate groups (**Figure 4**, page 63).

In the cell, an enzyme called ATPase facilitates the extraction of a phosphate group from the ATP molecule. This reaction also involves water molecules. The overall reaction is exergonic and results in the production of separate ADP and Pi components (**Figure 5**, page 63).

phosphorylation: the addition of one or more phosphate groups to a molecule

endergonic: any process that requires (consumes) energy

exergonic: any process that gives off (releases) energy

electron transport system: a series of progressively stronger electron acceptors, with energy release at each step

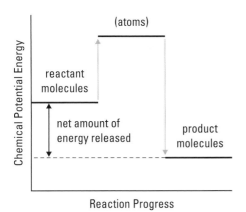

Figure 2
Energy is released in an exergonic reaction

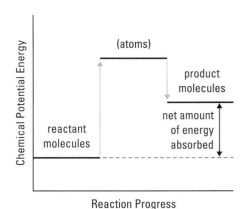

Figure 3
Energy is absorbed in an endergonic reaction

When this reaction is coupled to energy requiring processes such as active transport or the separation of sister chromatids in anaphase of mitosis, energy is transferred to the endergonic steps in the process, allowing them to occur. Since there is a limited number of ATP molecules in a cell, how does a cell provide a continuous supply of energy to the many energy-requiring processes that keep the cell alive?

Production of ATP

In the overall equation for cellular respiration, the hydrogen atoms of the glucose molecule eventually combine with oxygen atoms (from O_2) to produce water molecules. Carbon dioxide is produced from the carbon and oxygen atoms in the glucose molecule. Outside of a living cell, an electrical spark can be used to combine hydrogen and oxygen. As shown in **Figure 6(a)**, such a combination produces water. Because water contains much less chemical potential energy than the reactants (hydrogen and oxygen), energy is released. Unfortunately, this rapid release of energy is not ideally suited for cells. In many situations, cells could be damaged by the uncontrolled release of heat. The heat may actually distort proteins and damage the cell. However, a step-by-step release of energy, using the **electron transport system**, allows cells time to transform and store much of the energy that is stored in food molecules such as glucose.

In many ways, the electron transport system resembles the set of stairs shown in **Figure 6(b)**. As a glucose molecule is broken down, energy is released. Electrons move from one electron acceptor (a step on the stairs) to progressively stronger acceptors, releasing energy at each step. The final electron acceptor, oxygen, combines with two hydrogen ions to form a low-energy compound: water. Electrons found on the top step possess high energy levels and thus are able to provide the greatest amount of energy for ATP formation. Electrons found on the lower steps possess low potential energy levels and are less able to provide energy for ATP formation.

Thus, ATP, ADP, and Pi are continuously recycled within the cell. The electron transport system transfers the chemical potential energy of food molecules to ADP and Pi, forming ATP. In turn, ATP transfers chemical potential energy to the various energy-requiring processes in the cell.

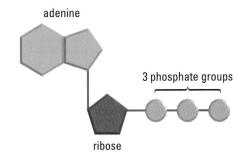

Figure 4
Adenosine triphosphate (ATP)

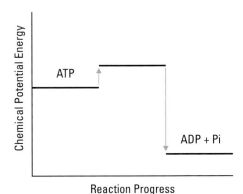

Figure 5
The breakdown of ATP into ADP and Pi is an exergonic process.

Figure 6
(a) Outside of living cells, the combination of oxygen and hydrogen releases energy as a burst of heat.
(b) The step-by-step release of energy by electron transport systems enables cells to convert much of the energy into ATP.

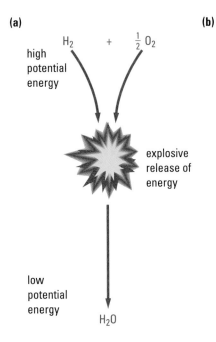

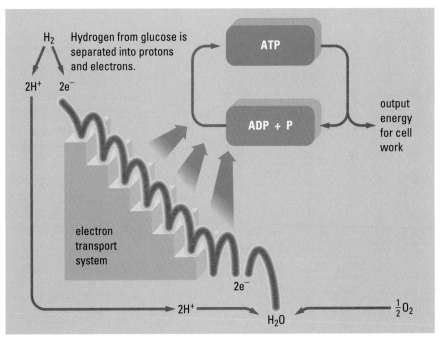

Understanding Concepts

1. What is ATP? What is ADP?
2. Why does energy need to be transferred from glucose to ATP?
3. Describe the electron transport system and explain why it is important in living things.
4. What is produced from the energy that is released by the electron transport system?
5. How could you determine if the products of cellular respiration have less chemical potential energy than the reactants? Where did the energy that was released during cellular respiration come from?

2.11 Types of Respiration

The major difference in the type of respiration that takes place in a cell is determined by the presence or absence of oxygen. Although we usually think of oxygen as necessary for all living things, some organisms do not need oxygen. Even some of our own cells will continue their activities with little oxygen present.

Aerobic Respiration

Aerobic respiration takes place in the presence of oxygen and involves the complete oxidation of glucose. Carbon dioxide and water are the products of aerobic respiration. For a single glucose molecule, 36 ATP molecules are also formed. The following reaction summarizes aerobic respiration:

$$C_6H_{12}O_6 + 6O_2 + 36 \text{ ADP} + 36 \text{ Pi} \rightarrow 6CO_2 + 6H_2O + 36 \text{ ATP}$$

During aerobic respiration more hydrogen atoms are removed from the glucose molecule than in anaerobic respiration. Each time an additional hydrogen is extracted, another electron is released to travel along the electron transport chain. The final products—carbon dioxide and water—are lower-energy compounds than lactic acid or alcohol.

The key to understanding why aerobic respiration provides greater amounts of energy than anaerobic respiration is found within the electron transport system. As the electron travels along the electron transport chain, ATP molecules are formed. Oxygen acts as the final electron acceptor for aerobic respiration. Oxygen accepts the electron and combines with the hydrogen proton to form water. Without oxygen as the final electron acceptor, the electrons could not be passed along the electron transport chain, and aerobic respiration would soon stop.

In aerobic respiration, the hydrogen atoms move through the electron transport chain within the mitochondria and eventually combine with oxygen atoms to form water. The final products of aerobic respiration are carbon dioxide, water, and ATP.

Anaerobic Respiration

anaerobic respiration: respiration that takes place without oxygen

Anaerobic respiration takes place in the absence of oxygen. There are two major types of anaerobic respiration: lactic acid fermentation and alcoholic fermentation.

Lactic Acid Fermentation

lactic acid: an organic molecule that is half of a glucose molecule with the molecular formula $C_3H_6O_3$

In animals, especially in muscle cells that are working hard, **lactic acid** fermentation can partially break glucose down to lactic acid, releasing energy in the

aerobic respiration: glucose + oxygen → carbon + water dioxide

typical energy yield = 36 ATP

When oxygen is present, vast energy supplies come from ATP, which is synthesized during aerobic respiration; however, when energy demands exceed the oxygen supply, some ATP is formed by anaerobic respiration. One product of anaerobic respiration is lactic acid. It is now known that some athletes can tolerate higher than average levels of lactic acid, enabling them to continue energy production longer.

process. However, many hydrogen atoms remain attached to the carbon atoms, which means there is still energy that has not been released. Only a limited number of chemical bonds can be broken during anaerobic respiration. A total of two ATP molecules is produced. If oxygen were available, the lactic acid would **oxidize** and more energy could be released.

Lactic acid accumulates in muscles during strenuous exercise if sufficient amounts of oxygen are not delivered to the tissues (**Figure 1**). Have you ever felt a sharp pain in your side while running? This pain may have been caused by a buildup of lactic acid. If you continue to run, the pain often intensifies. Anaerobic respiration does not supply enough energy to meet the demands that you are placing on your body, and the depletion of ATP reserves causes fatigue. The pain is merely providing a warning message.

oxidize: the loss of electrons from an atom or molecule

Alcoholic Fermentation

Another type of anaerobic respiration, **alcoholic fermentation**, occurs in the cytoplasm of yeast cells. Like anaerobic lactic acid fermentation, alcoholic fermentation yields only two ATP molecules. In alcoholic fermentation, special enzymes break down glucose so that rather than yielding lactic acid as an end product, carbon dioxide and alcohol are produced (**Figure 2**).

alcoholic fermentation: chemical decomposition of a carbohydrate in the absence of oxygen. It usually involves the conversion of carbohydrate into alcohol and carbon dioxide gas.

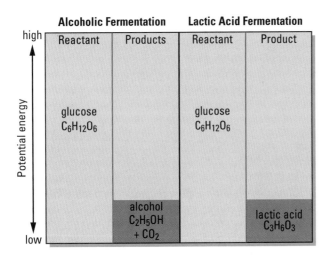

Figure 2
Glucose is broken down to lower-energy products. Alcohol and lactic acid have less potential energy than glucose.

Bread dough rises when carbon dioxide gases are released during fermentation. The bubbles released in champagne are caused by the same chemical process. Both champagne and bread dough also contain alcohol as a product of fermentation, although most of the alcohol that would be found in bread evaporates during the baking.

This alcohol still contains about 93% of the energy from glucose in its bonds and, therefore, could be a valuable fuel for industry or tranportation. This stored energy can be released by burning when oxygen is present. The result is a very clean burning fuel, as the only products are carbon dioxide and water.

Investigation 2.11.1

Yeast Fermentation

Fermentation is a process in which microorganisms convert sugar to alcohol in the absence of oxygen. Carbon dioxide gas is produced during this process. Collecting this gas and allowing it to displace water makes it possible to measure the quantity of gas produced.

In this activity, you will examine the products of yeast fermentation and measure the quantity of gas produced by yeast fermentation.

Question

How much gas is produced during yeast fermentation?

Design

Yeast is added to a sugar solution and to distilled water. Each solution is heated and the effects are observed. The volume of gas produced is measured by determining the volume of water displaced by the gas. Data will be recorded in a table.

Materials

2 large test tubes	apron
2 rubber test-tube stoppers (1-holed and with a short glass tube already inserted)	rubber tubing to fit over the glass tubing
water bath incubator set at 35°C–40°C	2% sucrose solution
5-mL graduated pipette with suction device	distilled water
basting bulb or small beaker	dry yeast suspension
25-mL or 50-mL graduated cylinder	cooking oil in dropper bottles
1-L beaker	plastic basin

Procedure

1. Put on your apron.
2. Prepare a water bath for the yeast. Heat 300–400 mL of water in a 1-L beaker until it reaches 35°C–40°C.
3. Obtain 2 test tubes and label test tube 1 as "sugar" and test tube 2 as "control."
4. Your team will test the sucrose solution and one control solution (distilled water). Using a pipette, measure 2.5 mL of your sugar solution and place into the sugar test tube . Using a clean pipette, measure 2.5 mL of distilled water and place into the control test tube.

5. Set the test tubes into the water bath.

6. Obtain the yeast suspension. Gently swirl the yeast suspension to mix the yeast that settled to the bottom of the container. Put 2.5 mL of the yeast suspension into the sugar test tube. Mix the yeast with the sugar solution. Put several drops of oil into the test tube. The oil should completely cover the surface of the mixture.

7. Incubate the test tube for 10 min in the water bath. Be sure to keep the temperature of the water bath constant. If you need to add more hot or cold water, first remove about as much water as you will be adding, otherwise the beaker may overflow. Use a basting bulb or small beaker to remove excess water.

8. Firmly place a rubber stopper with a glass tube into the sugar test tube. Fit a length of rubber tubing to the end of the glass tube.

9. Fill the graduated cylinder completely with water. Add water to the basin to a depth about the length of your hand.

10. Cover the top of the graduated cylinder with your hand and carefully place the cylinder upside down in the basin. If water escapes during this process, refill the cylinder and try again.

11. One partner will hold the cylinder slightly at an angle to allow the rubber tubing from the test tube to fit into the cylinder.

12. As gas from the test tube enters the graduated cylinder, water is pushed out of the cylinder. Record the water levels in the cylinder at 5 min intervals for 20 min.

13. Repeat steps 6 to 12, using the control test tube.

Analysis

(a) What is the source of the measured gas?
(b) Why was the apparatus incubated?
(c) What else was produced by the yeast in the sugar test tube?

Evaluation

(d) Identify possible sources of error, and indicate how you could improve the procedure.
(e) Identify careers for which knowledge of these procedures would be useful.

Practice

Understanding Concepts

1. Contrast ATP production in anaerobic and aerobic respiration. What accounts for this difference?

2. What are the advantages and disadvantages of anaerobic and aerobic respiration?

3. Which has higher potential energy: lactic acid or water? How do you know?

SUMMARY   Photosynthesis and Respiration

1. Photosynthesis occurs in chloroplasts. Light energy is converted to chemical potential energy.

Table 1: Comparison of Aerobic Respiration and Photosynthesis

Aerobic respiration	Photosynthesis
energy released from food	energy added in the production of food
high-energy reactants	low-energy reactants
low-energy products	high-energy products
oxygen required	oxygen released
glucose required	CO_2 and H_2O required
CO_2 and H_2O produced	glucose produced

2. Cellular respiration occurs in mitochondria. Energy in carbohydrates is converted into ATP, a usable form of energy.

3. Energy is needed to create ATP. When phosphate is added to ADP in phosphorylation, energy is transferred into ATP.

4. There are two types of anaerobic respiration:
   - In lactic acid fermentation, glucose is partially broken down to lactic acid.
   - In alcoholic fermentation, which occurs in yeast, the products are carbon dioxide and alcohol.

5. Anaerobic respiration does not provide much energy for ATP; the end products are not low energy so there is still energy that was not released.

6. Aerobic respiration involves the complete oxidation of glucose. More hydrogen atoms are removed and the final products are low-energy compounds. (See **Table 1**.)

7. During aerobic cellular respiration, electrons move by the electron transport system from one electron acceptor to progressively stronger acceptors, releasing energy at each step. This process provides energy for ATP production.

DECISION-MAKING SKILLS

- ○ Define the Issue
- ○ Identify Alternatives
- ● Research
- ● Analyze the Issue
- ● Defend a Decision
- ○ Evaluate

Figure 3

Explore an Issue

Take a Stand: Global Warming and the Interdependence of Cellular Respiration and Photosynthesis

Satellite images, such as **Figure 3**, show that tropical rain forests are disappearing at a rate of 171 000 km^2 each year. Plants are important to the global balance of gases because of the role they play in photosynthesis and respiration. Rain forests, in particular, because of their great number of plants, are vast reservoirs of carbon dioxide. By burning the forests, the carbon dioxide is released. Plant energy that could have been converted into a form useful to humans is simply burned off. Any substantial changes in the global plant population are likely to affect animals and humans as well because of our dependence on plants as the primary producers of energy.

Another factor that affects the global balance of gases is the burning of fossil fuels. When fossil fuels are burned, carbon dioxide that was contained in inorganic matter (e.g., coal and oil) is released. Each year, about 40% of the energy fixed by plants is used or destroyed by humans. One estimate suggests that we consume 2 T (tonnes) of coal and produce 150 kg of steel annually for every person on our planet. Carbon dioxide is produced by each of these processes. Many scientists speculate that the increases in the carbon dioxide levels are causing global warming.

Statement

Legislation should be introduced to reduce fossil fuel emissions and the cutting of rain forests.

- In your group, research the issue. Search for information in newspapers, periodicals, CD-ROMs, and on the Internet.
  Follow the links for Nelson Biology 11, 2.11.

GO TO www.science.nelson.com

- Identify current and past conferences that have addressed this issue.
- Identify the perspectives of each of the opposing positions.
- Write a position paper summarizing your opinion.

Key Expectations

Throughout this chapter, you have had opportunities to do the following:

- View and manipulate computer-generated, three-dimensional molecular models of important biochemical compounds (2.1).
- Identify and describe the structure and function of important biochemical compounds (2.1, 2.2, 2.3, 2.4, 2.5).
- Explain how scientific knowledge of cellular processes is used in technological applications (2.3, 2.4, 2.6, 2.11).
- Locate, select, analyze, and integrate information on topics under study, working independently and as part of a team, and using appropriate library and electronic research tools, including Internet sites (2.3, 2.4, 2.11).
- Carry out, in a safe and accurate manner, biological tests for macromolecules found in living organisms (2.4).
- Select and use appropriate modes of representation to communicate scientific ideas, plans, and experimental results (2.4, 2.7, 2.11).
- Select appropriate instruments and use them effectively and accurately in collecting observations and data (2.4, 2.7, 2.11).
- Illustrate and explain important cellular processes, including their function in the cell, and the ways in which they are interrelated (2.4, 2.9, 2.10, 2.11).
- Identify and describe science- and technology-based careers related to the subject area under study (2.4, 2.9, 2.11).
- Express the result of any calculation involving experimental data to the appropriate number of decimal places or significant figures (2.7).
- Design and carry out an investigation on cellular function, controlling the major variables (2.7, 2.11).
- Identify new questions and problems stemming from the study of metabolism in plant cells and animal cells (2.9).
- Explain the flow of energy between photosynthesis and respiration (2.9, 2.10).
- Compare anaerobic and aerobic respiration and state the advantages and disadvantages for an organism or tissue of using either process (2.11).

Key Terms

active transport
alcoholic fermentation
anaerobic respiration
autotrophs
Brownian motion
carbohydrates
cellulose
coagulation
concentration gradient
dehydration synthesis
denature
diffusion
electron transport system
equilibrium
fatty acids
glycerol
glycogen
homeostasis
hypertonic solution
hypotonic solution
isotonic solution
lactic acid
liposomes
macromolecules
nucleotides
nutrients
osmosis
oxidize
passive transport
phagocytosis
phospholipids
phosphorylation
photosynthesis
pinocytosis
polypeptide
starch
transcription
translation
triglyceride

Make a Summary

In this chapter, you examined biochemicals and biochemical processes.

1. Prepare a table summarizing the chemicals of life. Use headings such as name, structure, function, and formation.
2. Prepare a table comparing respiration and photosynthesis.

Reflect on your Learning

Revisit your answers to the Reflect on your Learning questions at the beginning of this chapter.

- How has your thinking changed?
- What new questions do you have?

Understanding Concepts

1. Compare the structures of
   (a) starch and protein;
   (b) phospholipids and triglycerides;
   (c) proteins and nucleic acids.

2. Compare the functions of carbohydrates and lipids.

3. Provide an example of dehydration synthesis (condensation reaction) by showing how two monosaccharides form a disaccharide. Show the reactants and end products of the reaction.

4. Why is cellulose, or fibre, considered to be an important part of the diet, even though it cannot be digested?

5. Why are phospholipids well suited for cell membranes?

6. In what ways is DNA unique among molecules?

7. Where in the cell are the different types of RNA found?

8. Compare the following pairs of processes
   (a) facilitated diffusion and active transport;
   (b) diffusion and active transport;
   (c) exocytosis and endocytosis.

9. Why does grass wilt if it is over-fertilized?

10. Will a plant cell burst if placed in a hypotonic solution? Explain.

11. A red blood cell is placed in distilled water and examined under a microscope. Describe and explain the changes in the shape of the red blood cell.

12. If a tiny hole is made in a cell membrane, it usually closes over immediately. What features of cell membranes allow this to happen so quickly?

13. Why is turgor pressure not a feature of animal cells?

14. Why could a salt solution be used as an antiseptic for a cut?

15. How would a marine fish react in a tank of freshwater?

16. Explain **Figure 1**. Discuss how the energy flows. What is the relationship between mitochondria and chloroplasts in plant cells?

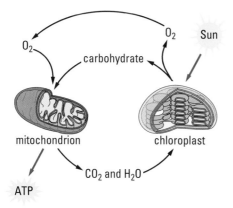

Figure 1

17. Explain how the breakdown of ATP provides energy for cells.

18. Copy **Table 1** and fill in the blanks with the correct answers.

Table 1

Characteristic	Anaerobic respiration	Aerobic respiration
amount of ATP produced		
site of activity in cell		
final products		

19. Use **Figure 2** to answer the following questions.
    (a) Identify the energy source K
    (b) Identify processes L, M, and N.
    (c) Identify substance O.

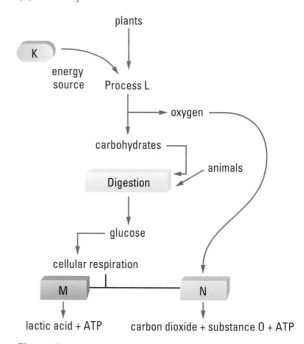

Figure 2

20. Predict what would happen to a cell if its mitochondria failed to work.

21. In reindeer, the membrane lipids of cells in the legs was compared to the membrane lipids in the hooves. The hoof cell membranes have a higher degree of unsaturation than the leg cell membranes. What survival value would this have for reindeer?

22. What is the link between membrane proteins and diseases?

Applying Inquiry Skills

23. Identify the following as water soluble or fat soluble. Describe how you could test your answer.
    - (a) glucose
    - (b) starch
    - (c) amino acids
    - (d) proteins
    - (e) phospholipids
    - (f) triglycerides
    - (g) waxes
    - (h) sucrose
    - (i) butter

24. A student began heating several test tubes of unknown solutions with Benedict's reagent. He recorded his observations at 5 min, and planned to make a second observation 1 min later. After 20 min had elapsed, he examined his test tubes and noticed that all the solutions were a red-brick colour. Comment on whether his results are reliable. Explain.

25. Describe a method to examine the effects of osmosis on onion cells at the cellular level.

26. **Figure 3** shows a carrot used as an osmometer, a device used to measure osmosis. Investigators set up two different experiments by placing carrots in beakers containing different solutions—distilled water (**a**) and 10% salt (**b**). Procedures (**a**) and (**b**) were conducted at the same time of day and in almost identical conditions. Unfortunately, the investigators forgot to record which beaker contained the salt solution.
    - (a) What problem was being investigated in this experiment?
    - (b) Identify the independent and dependent variables in the experiment.
    - (c) On the basis of your observations, predict which beaker contained the distilled water.
    - (d) Suggest a method for determining the rate of osmosis for condition (**a**).
    - (e) Based on the observations after 10 min, draw a conclusion from the experiment.

27. Athletes lose salt and water as they perspire. The hotter the conditions are, the greater the amount of fluid lost. If only water is replaced following extreme exercise, the concentration of salt in the blood will decrease and the body cells will take up more water. Fragile cells, such as red blood cells, can burst if too much water moves into the cell.
    - (a) Suggest why the uptake of water by human body cells occurs.
    - (b) Propose an experimental design that will allow you to determine how much salt should be added to the water that an athlete drinks following exercise.

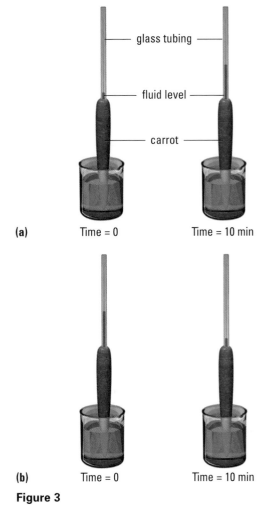

Figure 3

Making Connections

28. Indicate some of the symptoms of individuals who are deficient in carbohydrates, proteins, and lipids, respectively. Indicate why problems might occur with individuals who take in too much of each type of nutrient.

29. Margarine is processed by attaching hydrogen atoms to unsaturated double bonds of plant oils. The oil becomes solid or semi-solid. Have you ever noticed that some margarines are stored in plastic tubs, while others are stored in wax paper? Compare the two types of margarine. Which would you recommend?

30. What are liposomes useful for?

31. Construct a table, summarizing the importance of photosynthesis and respiration to life on Earth. Which important points are increasing or decreasing as a result of human activity?

Factors Affecting Cell Membrane Permeability

Red beet cells are surrounded by a selectively permeable membrane. This membrane can control the movement of materials (such as water) in and out of the cells. Red beets contain a strong red pigment called anthocyanin. Anthocyanins are largely responsible for the red colouring of buds and young shoots and for the purple and purple-red colours of autumn leaves. Anthocyanins are also present in blossoms, fruits, and other roots. Occasionally, they are found in larval and adult flies and "true" bugs. These compounds are thought to have antioxidant properties; that is, the ability to absorb harmful chemicals called free radicals. It is thought that free radicals are the causes of cardiovascular disease and cancer. When the membrane of the beet is disrupted, the red pigment leaks out.

Your task is to use your knowledge about cell membrane structure and osmosis to design an experiment to determine which conditions affect the rate of osmosis or amount of pigment leakage through an intact cell membrane.

Investigation

For this task, you will design and perform an experiment to determine how one of three variables—temperature, salinity of the surrounding environment, or surface area of the tissue—affects the rate of osmosis or pigment leakage in red beets.

Before you begin your design, select one variable to test. In your design, list the materials and equipment you have selected. Describe your method and criteria for determining the amount of pigment leakage. Provide any safety precautions needed, and include a step in your procedure to appropriately dispose of materials. Indicate what you think you will observe. After your design has been approved by your teacher, perform the experiment, under supervision.

Write a detailed report to communicate the procedure you used, the observations made, your analysis of the observations, and your evaluation of the investigation. Use appropriate scientific vocabulary, tables, correct significant figures, and SI units, where appropriate.

Question

How does the variable you have chosen affect the rate of osmosis or amount of pigment leakage?

Materials

List the materials you require to carry out the investigation.

Procedure

Students should follow the procedure based on their own design. Special considerations must be taken into account for the testing of each variable. Include this information in your procedure. Be sure to include a control.

The following are specific instructions for investigating each of the three variables:

Cut pieces of beet must be rinsed thoroughly first to ensure that all pigment that has leaked out due to cutting is reduced.

Temperature—Use different temperatures. Do not exceed 40°C or the membranes will be destroyed completely.

Salinity—Describe the salinity of the surrounding environment in terms of grams of salt per unit volume of water.

Surface Area—Use cubes or slices of beet.

Observations

(a) Record the rate of osmosis or amount of leakage for several (but equal) time intervals or at various temperatures, depending on the variable you have chosen. Display your observations in a table or chart.

Analysis

(b) Describe how the variable you have chosen to investigate affects the rate of osmosis or pigment leakage.

(c) Provide explanations for what you observed.

Evaluation

(d) Suggest ways to improve the design of your investigation.

(e) What were some sources of error in this experiment?

Synthesis

(f) From what you have learned, describe a procedure that would extract the pigment from a beet completely. Suggest reasons for why this would be desirable.

(g) If you were preparing a meal with beets, how could you cook them but still retain as much of their colour as possible?

(h) Beet roots are the site for nutrient storage in beets. What tests could you perform to determine which nutrients are present? Name the process by which these nutrients have originally been produced.

(i) Beet root cells are surrounded by cell membranes. Considering the components of cell membranes, explain the effects of the following and why.
  • beet root pieces in an organic solvent
  • placing beet root pieces in an acidic solution

(j) What other investigations could you complete using beet roots?

Understanding Concepts

1. Which of the following lists shows the biochemical compounds in order from highest potential energy to lowest potential energy?
   (a) carbon dioxide, glucose, lactic acid, alcohol
   (b) alcohol, carbon dioxide, glucose, lactic acid
   (c) glucose, alcohol, lactic acid, carbon dioxide
   (d) glucose, lactic acid, alcohol, carbon dioxide

2. If a red blood cell (**Figure 1**) is placed in a hypotonic solution, it may
   (a) shrink   (b) stay the same size
   (c) grow and maybe burst   (d) begin cell division

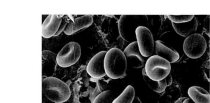

Figure 1

3. Which of the following organelles are NOT surrounded by a membrane?
   (a) ribosomes   (b) mitochondria
   (c) chloroplasts   (d) nuclei

4. Which of the following groups of biochemical compounds contain nitrogen?
   (a) carbohydrates   (b) lipids
   (c) proteins   (d) nucleic acids

5. Which process is a form of passive transport?
   (a) diffusion   (b) active transport
   (c) endocytosis   (d) exocytosis

6. Which of the following are the building blocks of carbohydrates (**Figure 2**)?
   (a) monosaccharides (b) phosphates
   (c) amino acids   (d) fatty acids

Figure 2

7. What is formed when glycerol combines with fatty acids (**Figure 3**)?
   (a) carbohydrates   (b) lipids
   (c) proteins   (d) nucleic acids

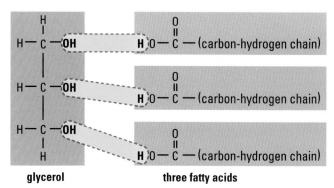

glycerol   three fatty acids

Figure 3

8. Which process requires the use of cellular energy?
   (a) simple diffusion   (b) osmosis
   (c) facilitated diffusion (d) active transport

9. Both facilitated diffusion and active transport involve
   (a) passive transport (b) the utilization of energy
   (c) carrier proteins   (d) vesicle formation

10. Which nutrients contain nitrogen?
    (a) glucose   (b) sucrose
    (c) proteins   (d) lipids

11. Distinguish between the following:
    (a) cilia and flagella
    (b) amyloplasts and chromoplasts
    (c) simple diffusion and facilitated diffusion
    (d) hypotonic solutions and hypertonic solutions
    (e) pinocytosis and phagocytosis
    (f) mRNA and DNA

12. Identify the organelles involved in the following processes. How do the organelles accomplish each function?
    (a) digestion   (b) transport
    (c) gas exchange   (d) excretion

13. Compare protein synthesis in cells with an automobile assembly plant. Match the parts of the auto assembly plant with the correct part of the cell. Provide reasons for each of the matches.

14. How have cell fractionation and the use of radioisotopes helped advance the knowledge of cell biology?

15. Why did these new technologies need to be developed?

16. Compare lysosomal digestion with a recycling operation. Match the stages and give reasons for your choices.

17. Healthy cells have selectively permeable membranes. What would happen to a cell with a fully permeable membrane?

18. Hormones are the body's chemical messengers. Protein hormones, such as insulin, must attach themselves to a receptor site on the cell membrane; fat-soluble hormones, such as sex hormones, pass directly into the cell. Explain why steroid hormones pass directly into the cytoplasm of the cell.

19. While preparing a meal, you notice that your celery is wilted and soft. Why did this happen? How could you freshen it up?

20. In preparing a roast beef, you notice that your cookbook specifies that salt should only be added at the end of roasting. Why? What could happen to your roast if you add salt before roasting?

21. Nerve cells use active transport to move ions (charged particles) to create nerve signals. During prolonged, intense activity, some athletes experience a severe drop in blood sugar levels. They become faint, are unable to think clearly, and lose their co-ordination. Use your knowledge of active transport to explain the reactions.

22. High cholesterol levels and high risk of atherosclerosis are in part related to diet and in part determined by genetics. Low-density lipoproteins (LDLs) are considered to be "bad" cholesterol, and have been associated with the clogging of arteries through plaque formation (**Figure 4**). High-density lipoproteins (HDLs), the "good" cholesterol, carry LDLs back to the liver for breakdown. The LDL cholesterol receptor sites located on cell membranes are controlled by genetics. Explain how the number of receptor sites may cause a predisposition to atherosclerosis.

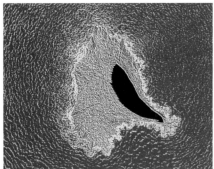
Figure 4
Plaque formation in a blood vessel

23. Explain why the following processes are essential for life.
    (a) protein synthesis   (b) cellular respiration
    (c) lysosomal digestion   (d) active transport
    (e) photosynthesis

24. Distinguish between the following:
    (a) glucose and sucrose
    (b) animal fats and plant fats or oils
    (c) starch and glycogen
    (d) protein, polypeptide, and amino acid
    (e) lactic acid and alcohol

25. What do proteins have in common with sugars?

26. Explain why runners consume large quantities of carbohydrates a few days prior to a big race.

27. What makes nucleic acids different from the other three categories of important biological molecules (carbohydrates, proteins, lipids)?

28. If plant cells can make their own food, why do they need mitochondria?

29. Why do plants produce more starch than they can use for food?

30. Why is photosynthesis viewed to be the most important chemical process on the earth?

31. List the possible energy-releasing paths for
    (a) proteins   (b) fats

32. Explain **Figure 5**.

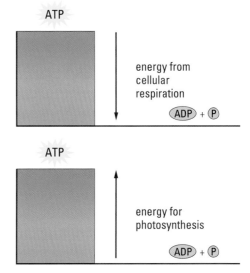
Figure 5

33. What kind of fermentation do yeast cells perform?

34. Compare alcoholic fermentation, lactic acid fermentation, and aerobic respiration. Consider starting materials, end products, and energy produced.

35. Consider the process of alcoholic fermentation. Why is alcohol a good fuel to use?

Applying Inquiry Skills

36. A cell is viewed under low power magnification. When the revolving nosepiece is turned to the high power magnification, the object disappears from view. Despite repeated attempts to refocus the view, the cell cannot be seen. Provide a possible explanation.

37. A student examines a prepared slide of human cheek epithelium tissue under a light microscope. She states that she does not see any mitochondria. List possible reasons for not seeing mitochondria.

38. Investigators set up the following experimental design in **Figure 6** to examine diffusion in living cells. A student is able to see how far the dye has moved up the celery stalk by examining each of the cuts.
    (a) What question did the investigators attempt to answer with this experiment?
    (b) State a hypothesis for the experiment.
    (c) Identify the independent and dependent variables.
    (d) How would you measure the rates of diffusion?
    (e) Predict which celery stalk would have the greatest movement of dye. Explain why.
    (f) Suggest improvements to the experimental design.

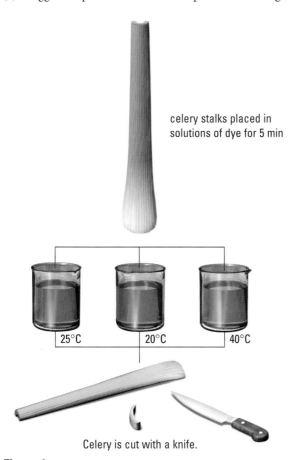

celery stalks placed in solutions of dye for 5 min

25°C   20°C   40°C

Celery is cut with a knife.

Figure 6

39. Interpret **Figure 7**. Why does the sugar move into the cell? Explain why more sugar is found inside the cell in diagram B. Why has the concentration of sugar decreased in diagram C?

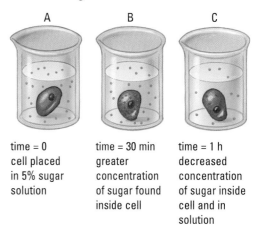

A   B   C

time = 0	time = 30 min	time = 1 h
cell placed	greater	decreased
in 5% sugar	concentration	concentration
solution	of sugar found	of sugar inside
	inside cell	cell and in
		solution

Figure 7

40. A scientist determines that the concentration of glucose (sugar) in a cell is constantly greater than that of its surroundings. She decides to look at the cell under an electron microscope and discovers a greater than normal number of mitochondria.
    (a) Why might she have counted mitochondria?
    (b) Provide a possible explanation for the greater number of mitochondria.
    (c) Predict what would happen to the glucose concentration if the cell's mitochondria were destroyed.

41. A marathon runner collapses after running on a hot day. Although the runner consumed water along the route, analysis shows that many of the runner's red blood cells had burst. Why did the red blood cells burst? (Hint: On hot days many runners consume drinks that contain sugar, salt, and water.)

42. Draw structural diagrams to show the dehydration synthesis reactions of
    (a) glucose + glucose;   (b) glucose + fructose.
    What is the name for each new type of nutrient which you have drawn?

43. After chewing a cracker for several minutes, you may notice that it begins to taste sweet. What has happened in your mouth? How could you test your hypothesis?

44. A student was eating a diet chocolate bar, and decided to test for the presence of monosaccharides. What test results would you expect? Are there any other tests you would perform?

45. **Table 1** shows the results submitted by students in their nutrient analysis lab report. Help them to interpret their results by stating what nutrients are present in each unknown sample.

Table 1

Unknown sample	Benedict's reagent	Lugol's solution	Sudan IV indicator	Biuret reagent
A	blue	black	clear	violet
B	orange	black	red	blue
C	blue	yellow	clear	blue

46. **Figure 8** shows an experiment designed to demonstrate the relationship between plants and animals. Bromothymol blue indicator was placed in each of the test tubes. High levels of carbon dioxide will combine with water to form carbonic acid. Acids will cause the bromothymol blue indicator to turn yellow. The initial colour of the test tubes is blue. Predict the colour change, if any, in each of the test tubes. Explain your predictions.

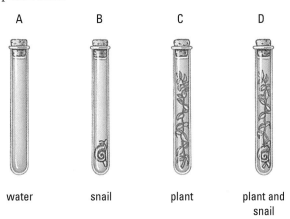

A   B   C   D

water   snail   plant   plant and snail

Figure 8

47. While completing an investigation involving yeast fermentation, you realize that your partner has allowed air to get into the fermentation flask. Although the yeast seem to still be active, you notice that the odour is very different: it smells like vinegar. What should this flask smell like? What can you infer about making vinegar with yeast?

Making Connections

48. What is the Canadian contribution to microscopy?

49. What are two functions of the scanning tunneling microscope?

50. Summarize the various links between abnormal number of mitochondria and diseases.

51. Why are more advanced techniques other than better microscopes required for future cell research?

52. Through photosynthesis, algae fix carbon dioxide in the atmosphere and release oxygen. A number of scientists have suggested that increasing algae growth in the oceans may reverse escalating carbon dioxide levels and reduce atmospheric warming. One suggestion is to seed the oceans with iron and phosphates—nutrients that will increase plankton blooms. Comment on this strategy.

Exploring

53. In the early 1900s, malnutrition was the leading cause of infant death in Canada. Children were being fed but not nourished. Scientists at the Hospital for Sick Children in Toronto worked on improving infant nutrition. Their result was Pablum™. Research the development of Pablum.
Follow the links for Nelson Biology 11, Unit 1 Review.
GO TO www.science.nelson.com

54. A research team from the University of South Florida built a robot that is powered entirely by food. Known as The Eating Machine, the robot has a metallic stomach which contains bacteria capable of producing the necessary enzymes to break down carbohydrate molecules. Electrons released by this process are used to charge a battery which powers the robot. Research the efforts made by the team.
Follow the links for Nelson Biology 11, Unit 1 Review.
GO TO www.science.nelson.com

55. A new biodegradable polymer spray has been produced, specifically designed for skin burn treatment. The spray provides a lattice upon which the body can grow new collagen, thus closely mimicking the healing process of real skin. The result is new skin with little scarring. When collagen is grown haphazardly, scarring results. Research the current status of this new treatment.
Follow the links for Nelson Biology 11, Unit 1 Review.
GO TO www.science.nelson.com

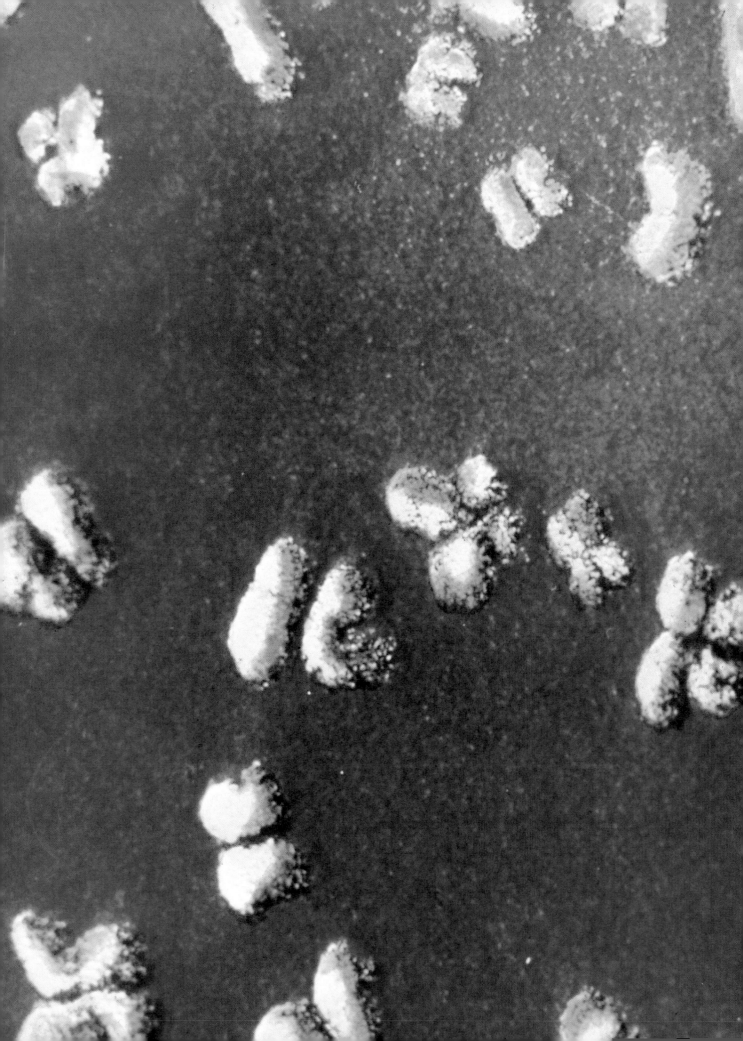

Genetic Continuity

Cystic fibrosis is caused by a defective gene. The gene causes a person's airways to release a sticky mucus that can cause blockages in the lungs, along with digestive problems. It affects about 1 in 2000 children. A group of Toronto doctors—Lap-Chee Tsui, Frank Collins, and Jack Riordan—working with a group from Michigan, identified and located the gene that causes the disorder.

Although finding the gene is only the first step of many in creating a cure, it does pinpoint the source of the problem. As new techniques become available, the defective gene may one day be deactivated, removed, or replaced.

Lap-Chee Tsui made this statement about his work: "I think the most wonderful part of doing research in biology and medicine is its unpredictability, at least at the present time. There is so much to learn about the life process, and it continues to amaze us."

Dr. Lap-Chee Tsui

Overall Expectations

In this unit, you will be able to

- demonstrate an understanding of how cell division, genes, and heredity are related;
- perform and analyze laboratory experiments that explore heredity;
- outline advances in scientific understanding of genes and in genetic technology, and demonstrate awareness of the social and political issues raised by genetic research and reproductive technology.

Are You Ready?

Knowledge and Understanding

1. Identify the cell structures shown in **Figure 1** and explain the importance or function of each.

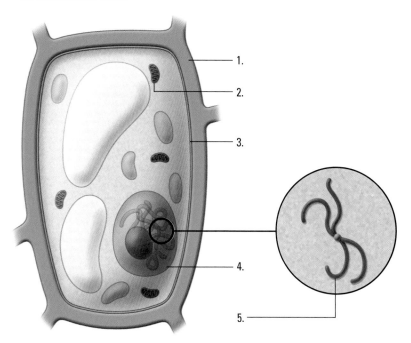

Figure 1

2. (a) Organize the following structures from largest to smallest.

   organ, chromosome, organism, nucleus, tissue, DNA molecule, cell, gene

   (b) Copy **Figure 2.** Use the listed structures in (a) as labels for your diagram.

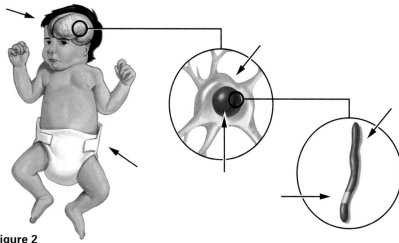

Figure 2

3. If a muscle cell for humans contains 46 chromosomes, indicate the number of chromosomes that you would expect to find in each of the following figures:

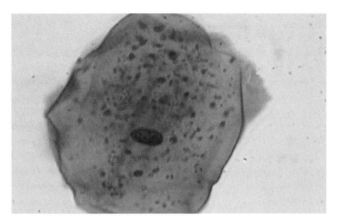

Figure 3
Skin cell, 450×

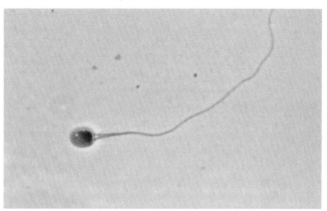

Figure 4
Sperm cell, 1000×

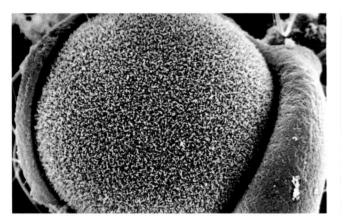

Figure 5
Unfertilized egg cell, 2000×

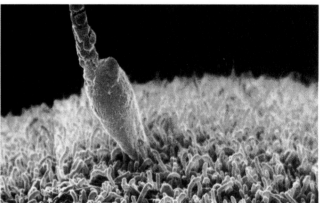

Figure 6
Fertilized egg cell, 5500×

4. Provide examples of hereditary traits that are
   (a) determined by genes;
   (b) influenced by the environment.

5. Many single-cell organisms divide by a process called binary fission. One cell divides into two cells identical to each other and identical to the original cell. More complex organisms form specialized sex cells. When sex cells combine from two different organisms, they form a fertilized egg or zygote.
   (a) Identify one advantage of binary fission as a means of reproduction.
   (b) Identify and explain an advantage of reproduction by the union of sex cells from different individuals.

6. Explain why the duplication of genetic material is essential for cell division.

Math Skills

7. **Table 1** shows the stages of a typical cell cycle. Draw and label a circle graph to represent the data.

8. A couple are expecting their third child. After the birth of two boys, they reason that the next child will be a girl.
   (a) Determine the probability of having three boys in a row.
   (b) Determine the probability that the next child will be a girl.

Table 1

Stage	Time (h)
rapid growth	15
growth and DNA replication	20
preparation for division	10
mitosis	5

Cell Division

Daphnia (**Figure 1**), close relatives of lobsters, crabs, shrimp, and barnacles, are smaller than grains of sand. They are usually found in ponds and marshes. Despite their small size, daphnia are easy to see because of their jerky swimming movements. A closer examination of this microscopic organism reveals a surprisingly sophisticated organ system for a creature so small. Under low-power magnification, dark eyes are visible beneath the antennae. A digestive system is also visible and a tiny transparent heart can be seen beating along the animal's back. Immediately behind the heart is the structure that makes the daphnia a truly remarkable animal. The brood pouch is the place where two different kinds of eggs can be stored.

Females can produce offspring without a mate since they can produce eggs that require no **fertilization**. Reproduction without sexual intercourse is called **asexual reproduction**. Upon development, these eggs become females, which in turn produce females, all of which are identical to each other and to the parent. Then, in response to some environmental cue, daphnia begin producing eggs that develop as either males or females. The males and females produce sex cells. **Sexual reproduction** occurs when the sperm cells fertilize the egg cells, producing many offspring with a variety of traits. Asexual reproduction occurs when food is plentiful, while sexual reproduction is triggered during times of environmental stress.

This animal is often used in scientific experiments on the effects of drugs because, with the aid of a microscope, its heart and other organs can clearly be seen.

Reflect on your Learning

1. Make a list of the advantages of being multicellular.

2. Suggest possible advantages of reproducing
   (a) asexually
   (b) sexually

3. If 22 chromosomes are found in the muscle cell of a mouse, predict the number of chromosomes found in each cell of the following types:
   (a) brain cell
   (b) sperm cell
   (c) fertilized egg cell

   Explain your predictions.

In this chapter, you will be able to

- describe the process of cell division (mitosis and cytokinesis) and the formation of sex cells (meiosis);
- use simulations and your own observations to describe the process of meiosis and to explain how abnormal meiosis can produce genetic disorders in offspring;
- find and describe examples of recent research in cell biology and evaluate ethical issues associated with the application of the research.

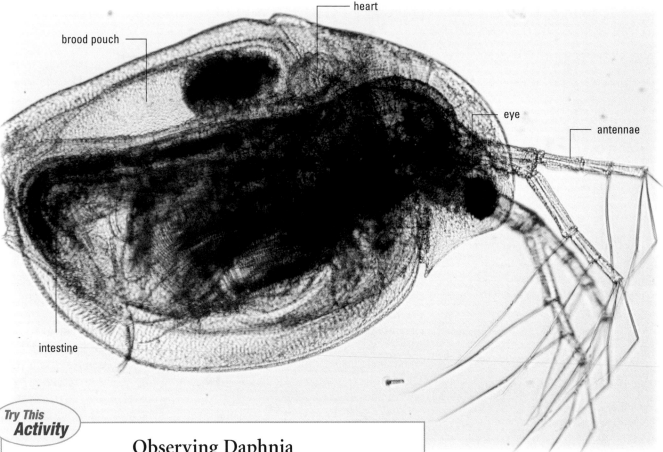

brood pouch

heart

eye

antennae

intestine

Try This
Activity

Observing Daphnia

Materials: prepared slide of daphnia, concave depression slide, glycerin, cover slip, daphnia culture, medicine dropper, microscope, ice cubes, cotton swab

- If available, look at a prepared slide of daphnia. Take note of the daphnia's general appearance and the location of certain features (e.g., eyes, antennae, heart) so that you will be able to identify them more easily in the daphnia culture.
- Remove the prepared slide. Obtain the other materials. Using a cotton swab, smear some glycerin into the depression on the slide. Then, using a medicine dropper, place a small drop from a daphnia culture onto the glycerin. Prepare a wet mount by adding a cover slip. Examine the slide for daphnia under low-power magnification. Pay attention to the movement and heart rate.
- Place the slide on an ice cube for 3 min, then dry the bottom of the slide with a paper towel and observe once again under low-power magnification.

(a) Why did you smear glycerin on the slide?
(b) Why did you put the slide on an ice cube?
(c) Make a drawing of a daphnia.
(d) Compare the movement and heart rate at room temperature and under cold conditions.
(e) Describe the appearance of the brood pouch.
(f) Do you think that daphnia are composed of many cells or one cell? Explain your reasoning.
(g) Try viewing the daphnia under medium power. (Hint: You may have to adjust the diaphragm.) Draw what you see.

Figure 1
Daphnia is also known as a water flea, but it is a crustacean, not an insect.

fertilization: the union of male and female sex cells

asexual reproduction: the production of offspring from a single parent; offspring inherit the genes of that parent only

sexual reproduction: the production of offspring from the union of two sex cells, one from each different parent. The genetic makeup of the offspring is different from that of either parent.

3.1 Principles of Cell Division

Cell division is one of the most studied, yet least understood, areas of biology. We know that the estimated 100 trillion cells that make up your body began from a single fertilized egg. Tissue growth requires cell division (**Figure 1**). Cell division also maintains a fully grown individual. For example, red blood cells die and are replaced at a rate of one million every second (**Figure 2**). Despite all that has been observed, many questions remain unanswered. Under what circumstances do cells divide? How does the rate of cell division change (in the formation of calluses, for example)? Why do bone marrow cells divide at enormous rates, while adult brain cells do not? Why do some cells specialize and others do not? Why do certain cells divide while other cells do not?

The first part of the cell theory states that all living things are made up of one or more cells and that all cells are formed from pre-existing cells by cell division.

Early studies showed that a single cell gives rise to two cells by an equitable distribution of the nuclear contents in a process called **mitosis**, and a roughly equal division of the cytoplasm and its constituent organelles in a process called **cytokinesis**. The cell that divides is referred to as the parent cell and the two cells that result from cell division are referred to as daughter cells. Note that, although terminology used to describe cell division includes references to "daughters" or "sisters," the components are genderless.

A cell's genetic information is contained in the DNA molecules in its nucleus. These molecules are found within structures called chromosomes. A human body cell contains 46 chromosomes.

To prepare for mitosis, the cell makes a replica of each DNA molecule. This causes each chromosome to become doubled. The replication and then separation of duplicate chromosomes ensure that the daughter cells are genetically identical to each other and to the parent cell. The replication of genetic information

mitosis: a type of cell division in which a daughter cell receives the same number of chromosomes as the parent cell

cytokinesis: the division of cytoplasm

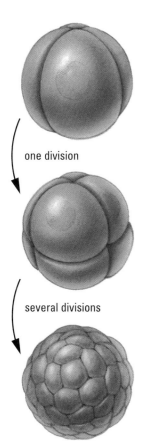

one division

several divisions

Figure 1
Early stages of cell division of a fertilized egg in a frog

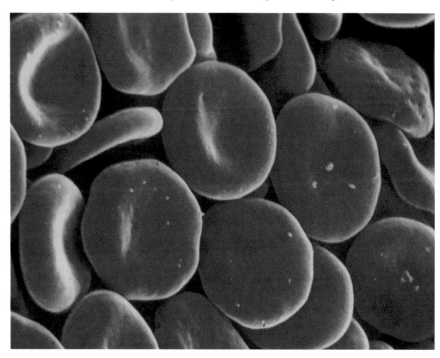

Figure 2
Human erythrocytes (red blood cells). A normal person contains about 25 trillion red blood cells, yet the life span of a red blood cell is only 20 to 120 days.

Try This Activity

Comparing Surface Areas of Small and Large Cells

Why is it that cells divide rather than continue to increase in size? It is because the relationship of the surface area of a cell to the volume of its cytoplasm affects the exchange of materials and wastes through the cell. Nutrients enter and wastes exit through the cell membrane. Cell division results in an increase in the surface area-to-volume ratio, thereby permitting the exchange of more materials through the cell membrane.

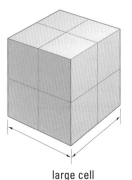 

large cell   small cell

Figure 3
This cell's dimensions have doubled, which increases its surface area by a factor of 4 and its volume by a factor of 8.

- Using a ruler, measure the dimensions of a sugar cube in millimetres. Calculate the surface area of one side. Then calculate the surface area of the cube.
- Arrange eight cubes as shown in the diagram and calculate the surface area of the larger cube.

  (a) Compare the surface area of the large cube to that of eight smaller cubes.
  (b) If these cubes represented cells, how would the area of the cell membrane of the large cell compare with that of the many smaller cells?
  (c) For a small and a large cell, compare the ratios of cell surface area to cell volume.
  (d) Explain why metabolically active cells, such as nerve cells, tend to be somewhat smaller in size, while inactive cells, such as fat cells, are often larger.

also ensures future cell divisions. Each daughter cell is a potential parent cell for the next generation.

The number of chromosomes that result from cell division is the same as the number of chromosomes before the division. Each cell in your body has 46 chromosomes; all succeeding generations of cells will also contain 46 chromosomes.

All body cells have the same genetic information because all cells in the human body are derived from the same fertilized egg. A muscle cell, for example, has all of the chromosomes of a cell found in your brain, or of a cell found in your heart. However, not all cells in the human body have the same shape or carry out the same functions. One of the most puzzling questions that confront cell biologists is why different cells do different jobs. What makes a brain cell conduct nerve impulses, and what makes a muscle cell contract? How do specialized cells know which genes to use? These questions are currently under investigation.

3.2 The Cell Cycle

The sequence of events from one cell division to another is called the cell cycle and is often described as taking place in phases. However, the cycle does not pause after each phase; it is a continuous process. The phases help scientists describe the events of mitosis and cytokinesis. For most cells, the nuclear division that occurs during mitosis marks only a small part of their cycle. The stage between nuclear divisions, called **interphase**, is marked by a period of rapid growth, the replication of chromosomes, another period of growth, and preparation for further divisions (**Figure 1**).

Interphase

During interphase, cells grow, make structural proteins that repair damaged parts, transport nutrients to where they are needed, eliminate wastes, and prepare themselves for mitosis by building proteins. These proteins also function in the construction of enzymes that aid chemical reactions. The most important of these reactions are those that control the synthesis of DNA and the replication of the genetic information in the chromosomes.

During interphase, the genetic material is called **chromatin**. Chromatin is all the DNA molecules and associated proteins in the nucleus. (Keep in mind that an individual chromosome is composed of a single DNA molecule and its proteins.) When referred to as chromatin, the chromosomes are long, thin strands dispersed throughout the nucleus in a tangled, fibrous mass.

Each chromosome duplicates itself during interphase. The original chromosome and its duplicate are attached to each other by a structure called the **centromere**. While attached to one another, the original chromosome and its duplicate are each referred to as **sister chromatids**. Since sister chromatids contain identical genetic information, the pair, attached at the centromere, is considered to be one chromosome (**Figures 2 and 3**).

interphase: the time interval between nuclear divisions. During this phase, a cell increases in mass, roughly doubles the cytoplasmic components, and duplicates its chromosomes.

chromatin: the tangled fibrous complex of DNA and protein within a eukaryotic nucleus

centromere: the structure that holds chromatids together

sister chromatids: a chromosome and its duplicate, attached to one another by a centromere. The pair remains attached until separated during mitosis.

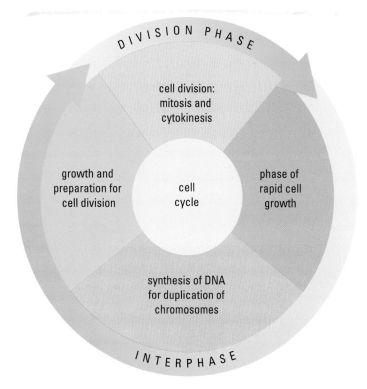

Figure 1
The cell cycle. The circle represents the entire life cycle of the cell, which can be divided into two major phases: interphase and the cell division phase. Most cells spend the majority of their time in interphase.

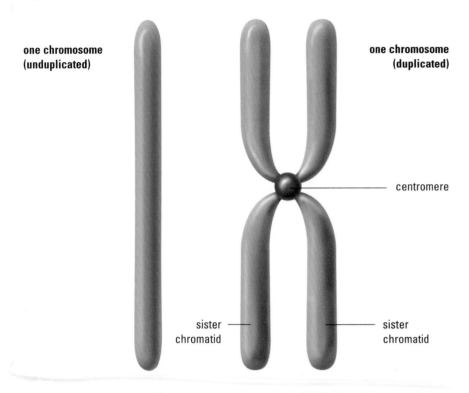

one chromosome (unduplicated)

one chromosome (duplicated)

centromere

sister chromatid

sister chromatid

Figure 2
Diagram of an unduplicated and a duplicated chromosome

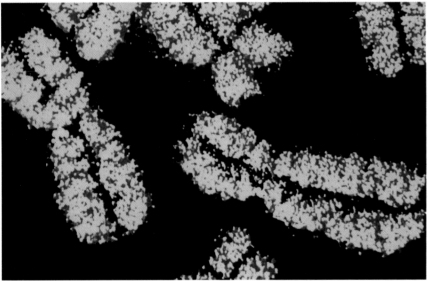

Figure 3
Chromosomes as sister chromatids

Prophase

Prophase is the first phase of mitosis. The chromosomes in the nucleus become visible under a microscope as they shorten and thicken (**Figure 4**, page 88). In animal cells, a small body in the cytoplasm separates and its parts move to opposite poles of the cell as the chromosomes become visible. These tiny structures, called **centrioles**, provide attachment for the **spindle fibres**, which serve as guide wires for the attachment and movement of the chromosomes during cell division. Collectively, the centrioles and spindle fibres make up the spindle apparatus. Most plant cells do not have centrioles, but spindle fibres still form and serve a similar purpose. The centromere joining the two chromatids helps anchor the chromosomes to the spindle fibres. When viewed under a microscope during prophase,

centrioles: small protein bodies that are found in the cytoplasm of animal cells

spindle fibres: protein structures that guide chromosomes during cell division

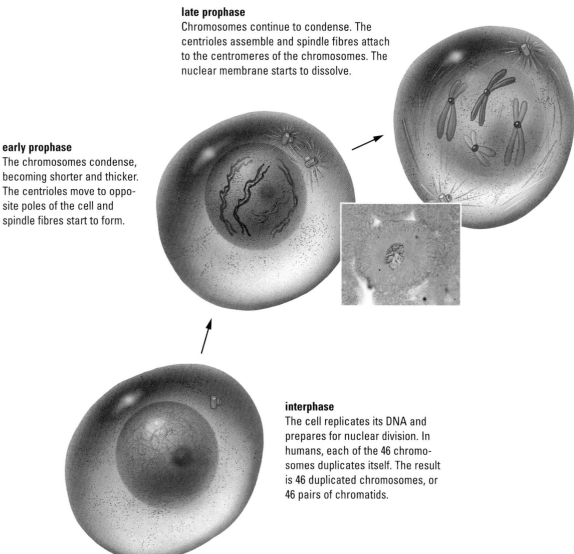

late prophase
Chromosomes continue to condense. The centrioles assemble and spindle fibres attach to the centromeres of the chromosomes. The nuclear membrane starts to dissolve.

early prophase
The chromosomes condense, becoming shorter and thicker. The centrioles move to opposite poles of the cell and spindle fibres start to form.

interphase
The cell replicates its DNA and prepares for nuclear division. In humans, each of the 46 chromosomes duplicates itself. The result is 46 duplicated chromosomes, or 46 pairs of chromatids.

Figure 4
Mitosis in an animal cell

the nuclear membrane appears to fade; in effect, it is dissolving to allow the separation of chromosomes and cell organelles.

Metaphase

The second phase of mitosis is metaphase. Chromosomes composed of sister chromatids move toward the centre of the cell. This centre area is called the equatorial plate, because, like the equator of Earth, it is midway between the poles of the cell. The chromosomes appear as dark, thick filamentous structures that are attached to the spindle fibres. Even though they are most visible at this stage, it is still very difficult to count the number of chromosomes in most cells because the chromosomes are entangled. Chromatids can become intertwined during metaphase.

Anaphase

Anaphase is the third phase of mitosis. The centromeres divide and the sister chromatids, now referred to as chromosomes, move to opposite poles of the cell. If mitosis proceeds correctly, the same number and type of chromosomes will

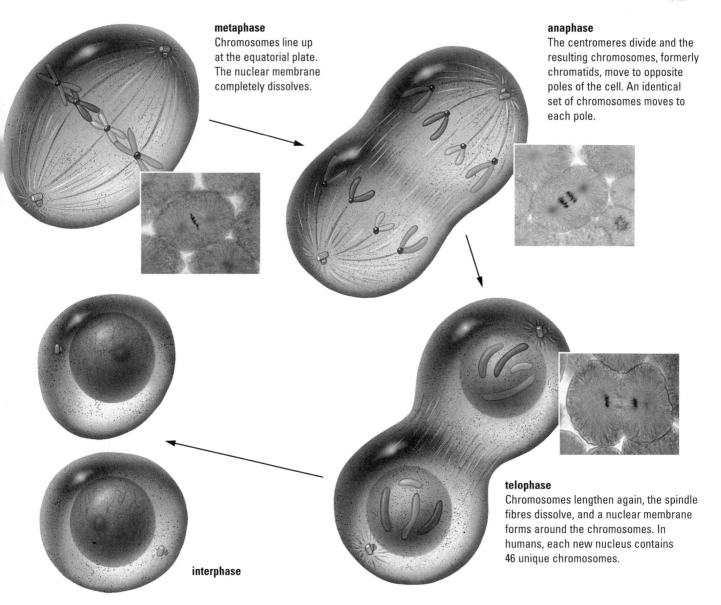

metaphase
Chromosomes line up at the equatorial plate. The nuclear membrane completely dissolves.

anaphase
The centromeres divide and the resulting chromosomes, formerly chromatids, move to opposite poles of the cell. An identical set of chromosomes moves to each pole.

telophase
Chromosomes lengthen again, the spindle fibres dissolve, and a nuclear membrane forms around the chromosomes. In humans, each new nucleus contains 46 unique chromosomes.

interphase

be found at each pole. Occasionally, segments of the chromatids will break apart, and may reattach, in anaphase.

Telophase

The last phase of mitosis is telophase. The chromosomes reach the opposite poles of the cell and begin to lengthen. The spindle fibres dissolve and a nuclear membrane forms around each mass of chromatin. Telophase is followed by cytokinesis, the division of the cytoplasm.

Cytokinesis

Once the chromosomes have moved to opposite poles, the cytoplasm begins to divide. Cytokinesis appears to be quite distinct from nuclear division. In an animal cell, a furrow develops, pinching off the cell into two parts. This is the end of cell division. In plant cells, the separation is accomplished by a cell plate that forms between the two chromatin masses. The cell plate will develop into a new cell wall, eventually sealing off the contents of the new cells from each other.

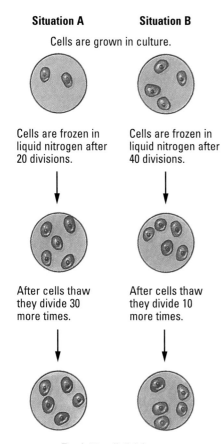

Situation A   Situation B

Cells are grown in culture.

Cells are frozen in liquid nitrogen after 20 divisions.

Cells are frozen in liquid nitrogen after 40 divisions.

After cells thaw they divide 30 more times.

After cells thaw they divide 10 more times.

Total: 50 cell divisions

Figure 1
Cell division appears to be controlled by a biological clock.

spermatocytes: sperm-producing sex cells

Practice

Understanding Concepts

1. List the stages of mitosis. Briefly describe what occurs in each stage. To help in your description, sketch the sequence of events that occurs in an animal cell. Include labels for different structures.

2. A cell with 10 chromosomes undergoes mitosis. Indicate how many chromosomes would be expected in each of the daughter cells.

3. During interphase, what event must occur for the cell to be capable of undergoing future divisions?

4. Using a dictionary, look up the meaning of the prefixes used in the stages of mitosis: pro-, meta-, ana-, and telo-. Why would they be used in the naming of the phases of mitosis?

5. Compare and contrast the structure of the daughter cells with that of the original parent cell.

6. Describe the structure and explain the function of the spindle fibres.

7. What is the significance of cytokinesis? Speculate what would happen if cytokinesis did not occur.

8. When a cell has reached its maximum size, what two alternatives does it have? When does the cell carry out one alternative over the other?

9. What would happen if you ingested a drug that prevented mitosis? What if it only prevented spindle fibre formation?

10. A cell from a tissue culture has 38 chromosomes. After mitosis and cytokinesis, one daughter cell has 39 chromosomes and the other has 37. What might have occurred to cause the abnormal chromosome numbers?

11. Suppose that during mitosis, both sister chromatids moved to the same pole, resulting in daughter cells with a different number of chromosomes than the parent cell. How might this abnormality affect cell structure, cell function, or both?

3.3 A Cell Clock

How old can cells become? If cells continue to divide, why can't an organism stay eternally young and live forever? Research on cultured cells (cells grown in a nutrient medium) seems to indicate that a biological clock regulates the number of cell divisions available to cells. When immature heart cells maintained in tissue culture were frozen, they revealed an internal memory that counted the number of cell divisions they had undergone. If a cell had undergone twenty divisions before freezing, the cell completed another thirty divisions once it thawed. Then the cell died. When a cell was frozen after ten divisions, it completed another forty divisions when thawed. The magic number of fifty divisions was maintained no matter how long the freezing or at what stage the division was suspended (**Figure 1**).

Not all cells of the body have the same ability to reproduce. Age is one reason cells stop dividing. However, usually division is stopped by cell specialization. Skin cells and the cells that line the digestive tract reproduce more often than do the more specialized muscle cells, nerve cells, and secretory cells. Only two cell types in the human body seem able to divide continuously: the sperm-producing cells, called **spermatocytes**, and the cells of a cancerous tumour. Males are

capable of producing as many as one billion sperm cells a day from the onset of puberty well into old age. However, once the spermatocyte specializes and becomes a sperm cell, it loses its ability to divide further. Cancer cells divide at such an accelerated rate that the genes cannot regulate the proliferation and cannot direct the cells toward specialization. For example, people who have leukemia—white blood cell cancer—have too many white blood cells that divide too rapidly to specialize and carry out their normal functions. A person with leukemia has a reduced ability to fight infections. This happens because cells need time after division in order to specialize.

It would appear that the more specialized a cell is, the less able it is to undergo mitosis. The fertilized egg cell is not a specialized cell; differentiation begins to occur only after many divisions. Interestingly, it is at the point where differentiation begins that the biological clock within the cell is turned on.

Once again, new information raises even more questions. Do all cells age in a similar fashion? Does specialization cause aging? Will an understanding of cancer be linked to the study of aging?

SUMMARY Mitosis

Cell division produces new cells for cell growth and for the replacement of worn-out cells in the body.

1. Cell division involves a series of steps that produce two genetically identical daughter cells. Two divisions occur during cell division: nuclear division (mitosis) and cytoplasmic division (cytokinesis).

2. During interphase, genetic material is replicated.

3. Cells seem able to divide only a finite number of times.

4. Cells lose the ability to divide as they specialize.

Activity 3.3.1

Frequency of Cell Division

One of the goals of the biotechnology industry is to create cells that divide rapidly. The agriculture industry devotes large amounts of money to the development of chemicals that increase plant growth. One such chemical, gibberellin, a plant hormone, can be used to accelerate growth (**Figure 2**).

In medicine, rapidly dividing cells can be used to generate tissues quickly for such uses as skin grafts or growing new organ parts. To determine if efforts yield useful results, the frequency of cell division (the number of cells dividing in proportion to the total number of cells) is calculated.

In this activity, you will view and compare plant and animal cells during mitosis. You will examine prepared slides of an onion tip root and a whitefish **blastula** to identify cells that are dividing. Because slides are used, the cell divisions you are viewing are frozen in time. Therefore, it will not be possible for you to watch a single cell progress through the stages of mitosis. Based on your observations, you will determine the frequency of cell division and construct a clock representing the division cycle, given the time taken to complete one cycle of mitosis. In a table, you will record the number of cells in each stage of mitosis.

Figure 2
Two untreated cabbage plants (controls) and three cabbage plants treated with gibberellin (tests)

blastula: an embryonic stage consisting of a ball of cells produced by cell division following the fertilization of an egg

Materials

microscope

prepared slides of onion root tip

lens paper

prepared slides of whitefish blastula

Procedure

Part 1: Observing Dividing Cells

1. Obtain an onion root tip slide and place it on the stage of your microscope. View the slide under low-power magnification. Focus using the coarse-adjustment knob.

2. Centre the root tip in the field of view and then rotate the nosepiece to the medium-power objective lens. Focus the image using the fine-adjustment knob. Observe the cells near the root cap. This area is referred to as the meristematic region of the root.

3. Move the slide to view the cells away from the root tip. These are the mature cells of the root. Record the differences between the cells of the meristematic area and the mature cells of the root. Draw a diagram to help you (**Figure 3**).

4. Return the slide to the meristematic area and centre the root tip. Rotate the nosepiece to the high-power objective lens. Use the fine adjustment to focus the image.

5. Locate and observe cells in each of the phases of mitosis. It will be necessary to move the slide to find each of the four phases. Use **Figure 3** as a guide. Draw, label, and title each of the phases of mitosis. It is important to draw only the structures that you can actually see under the microscope.

6. Return your microscope to the low-power objective lens and remove the slide of the onion. Place the slide of the whitefish blastula on the stage. Focus

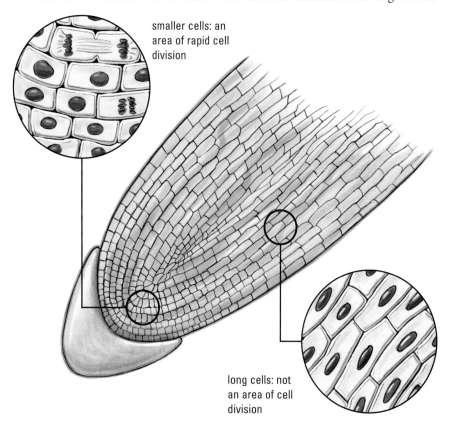

smaller cells: an area of rapid cell division

long cells: not an area of cell division

Figure 3

Meristematic region of the onion root tip where the cells are actively growing and dividing

with the coarse-adjustment knob. Repeat the procedure that you followed for the onion cells and, in the whitefish blastula, locate dividing cells under high-power magnification.

Part 2: Determining the Frequency of Cell Division

7. Count 20 adjacent whitefish blastula cells and record whether the cells are in interphase or division phase. Record the number of cells in interphase and the number of cells that are actively dividing.

8. Repeat the same procedure for the meristematic region of the plant root.

Part 3: Creating a Cell-Division Clock

9. Under high-power magnification, locate 50 onion root cells that are dividing. Do not include cells that are between divisions. Identify the phase of mitosis each cell is in. Record the number of cells in each phase.

10. Repeat the procedure for the cells of the whitefish blastula.

Analysis

Part 1: Observing Dividing Cells

(a) How do the cells of the meristematic area differ from the mature cells of the root?

(b) Why were plant root tip cells and animal blastula cells used for viewing cell division?

(c) Explain why the cells that you viewed under the microscope do not continue to divide.

(d) Compare and contrast cell division in plant and animal cells. Use a Venn diagram to organize your ideas.

Part 2: Determining the Frequency of Cell Division

(e) For both the plant and animal cells, calculate the percentage of cells that are dividing. Use the following formula:

$$\frac{\text{Number of cells dividing} \times 100}{\text{Total number of cells counted}} = \_\_ \text{ \% dividing}$$

(f) For both plant and animal cells, create a circle graph showing the percentage of cells in division phase and the percentage of cells in interphase. Label the diagrams appropriately. Compare the graphs. How are they different? How are they the same?

Part 3: Creating a Cell-Division Clock

(g) For both plant and animal cells, calculate the percentage of cells that are in each of these four phases: prophase, metaphase, anaphase, and telophase.

(h) For each cell type, construct a circle graph showing the percentage of cells in each phase of mitosis. Include labels and titles.

(i) If it takes 16 h to complete one cycle of mitosis for whitefish and 12 h for onions, determine the time spent in each phase. Include this information in your circle graphs.

Synthesis

(j) The number of animal cells in each phase of mitosis was recorded in **Table 1**. If the time taken to complete one cycle of mitosis was 15 h, create a cell-division clock to represent the data.

Table 1

Mitotic phase	Number of cells in phase
prophase	15
metaphase	20
anaphase	10
telophase	5

Cell Cycle for Cell A: 36 h

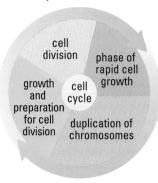

cell division
phase of rapid cell growth
cell cycle
growth and preparation for cell division
duplication of chromosomes

Cell Cycle for Cell B: 25 h

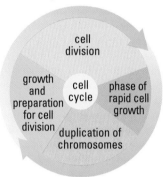

cell division
cell cycle
phase of rapid cell growth
growth and preparation for cell division
duplication of chromosomes

Figure 4

Understanding Concepts

1. Explain the concept of the "cell clock."

2. Suggest reasons why skin cells, blood cells, and the cells that line the digestive tract reproduce more often than other types of cells such as muscle cells.

3. (a) Describe the differences between the two cell cycles in **Figure 4**.
   (b) Which cell cycle do you believe would represent a cell of an embryo and which would represent an unspecialized cell in an adult? Give your reasons.

4. List areas of the body where you think cell division is most rapid. Also, indicate the comparative level of specialization of the cells in each area. Explain your predictions.

Making Connections

5. It is believed that herbicides like 2,4-D and 2,4,5-T may work by stimulating cell division. Why would the stimulation of cell division make these chemicals effective herbicides?

6. At one time, blood was transfused only from younger individuals to the elderly. It was believed that younger blood would provide the elderly with more energy. Do older people actually have older blood cells? Support your answer.

7. X rays and other forms of radiation break chromosomes apart. Physicians and dentists will not X-ray pregnant women. Even women who are not pregnant wear a lead apron when being X-rayed near the uterus. The apron bars the passage of X rays. Why is it undesirable to X-ray the reproductive organs? Why is it especially undesirable to X-ray pregnant women?

8. Scientists have developed techniques aimed at getting highly specialized cells to act as if they are immature cells that have not yet become specialized. Why would scientists want to be able to get a mature nerve cell to respond like a cell that hasn't undergone specialization?

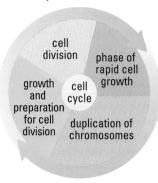

Figure 1

Hydra reproduce asexually by budding. Note the bud forming on the side of the body. Eventually the bud will break off from the body of the parent and form a separate, but identical, organism. This is an example of natural propagation.

3.4 Cloning

Cloning is the process of forming identical offspring from a single cell or tissue. Because the clone originates from a single parent cell, it is identical, or nearly identical, to the parent. Although some clones show accidental changes in genetic information, cloning does not result in the variation of traits that would occur with the combination of male and female sex cells. It is for this reason that cloning is referred to as asexual reproduction (see **Figure 1** and **Figure 2**, page 95).

Plant Clones

Fredrick Stewart created great excitement in the scientific world in 1958 when he revealed that he'd created a plant from a single carrot cell (**Figure 4**, page 95). Today, the technique he used is commonly referred to as cloning. Most orchids are now produced from clones. Unlike plants that reproduce sexually, cloned plants are identical to their parents. This allows for the production of strains of plants with predictable characteristics.

Plant tissue culture and cloning techniques have laid the groundwork for **genetic engineering**. However, many questions have yet to be answered (**Figure 5**). Carrots, ferns, tobacco, petunias, and lettuce respond well to cloning, but the grass and legume families do not. No one really knows why.

Each regenerated cell contains the complete complement of chromosomes and genes from the parent. Yet, some cells specialize and become roots, stems, or leaves. The cells within the leaf use only certain parts of their DNA, while the cells of the root use other segments of the DNA. Huge sections of the genetic information remain dormant in specialized plant cells. The trick in cloning plant cells appears to be in delaying specialization, or differentiation.

genetic engineering: intentional production of new genetic material by substituting or altering existing material

Figure 2
The strawberry plant reproduces by sending out runners from its main stem. These new plants have the advantage of relying on the parent for nutrients. This is an example of natural propagation.

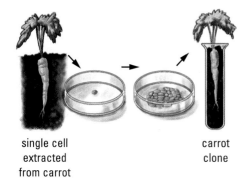

single cell extracted from carrot

carrot clone

Figure 4
It is possible to create a carrot from a single cell or small group of cells. This is an example of artificial propagation.

**Try This
Activity**

Cloning from a Plant Cutting

In some plants, asexual reproduction is accomplished naturally when a portion of the plant, such as a stem or leaf, breaks off and develops roots at the base of the broken portion. It is possible for the broken part to become a new plant. This activity is an example of artifical propagation.

Materials: coleus plant, scissors, goggles, gloves, tongs, fungicide, flower pot, potting soil, apron

 The fungicide is poisonous. Any spills on the skin, in the eyes, or on clothing should be washed immediately with cold water. Report any spills to your teacher.

• Follow the steps shown in **Figure 3**.

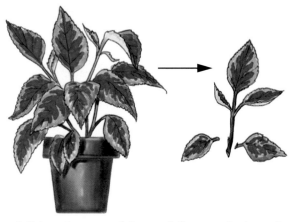

1. Using scissors, carefully cut off the tips of three coleus stems. Cut on an angle. Include several leaves on each stem.

2. Remove a few leaves from the bottom. Put on splash goggles, and wear gloves and/or use tongs to immerse the stem in fungicide.

3. Plant the stem in moist soil.

Figure 3

• Record the plant's initial height and number of leaves. Take these measurements every week for two months.
• Describe the new coleus each time.

(a) What evidence suggests that coleus can regenerate parts of the plant that were lost?
(b) Without removing the plant from the pot, how can you demonstrate that the roots from the cutting are growing?

Figure 5
A scientist conducts plant research.

enucleated: the condition where a cell does not contain a nucleus

totipotent: having the ability to support the development of an egg to an adult

Animal Clones

While plant cloning experiments were being conducted, Robert Briggs and Thomas King were busy investigating nuclear transplants in frogs. Working with the common grass frog, the scientists extracted the nucleus from an unfertilized egg cell by inserting a fine glass tube, or micropipette, into the cytoplasm and sucking out the nucleus (**Figure 6**). A cell without a nucleus is referred to as **enucleated**.

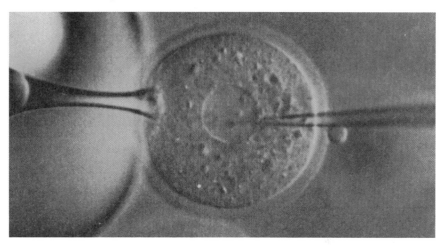

Figure 6
A small glass tube, called a micropipette, is used to remove the nucleus from a cell and later introduce a new nucleus.

Next, the nucleus of a cell from a frog embryo in the blastula stage of development was removed and inserted into the enucleated cell. The egg cell with the transplanted nucleus began to divide much like any normal fertilized egg cell. In later trials, the cell with the transplanted nucleus occasionally grew into an adult frog. Not surprisingly, the adult frogs displayed the characteristics from the transplanted nucleus. Careful analysis proved that the adults were clones of the frog that donated the nucleus (**Figure 7**).

However, different results were obtained when the nucleus was taken from cells at later stages of development. A nucleus that can bring a cell from egg to adult is referred to as **totipotent**, but not all nuclei are totipotent. For example, the nucleus from cells in a later stage, called the gastrula stage, did not bring the enucleated egg from the single-cell stage to the adult. If cell division occurred at all, it did not progress as far as for eggs that received a blastula nucleus. The difference is that the nucleus of a cell in the gastrula stage of development, unlike a cell in the earlier blastula stage, has specialized. As cells begin to specialize, a regulatory mechanism must turn off some of the genes that allow cell division.

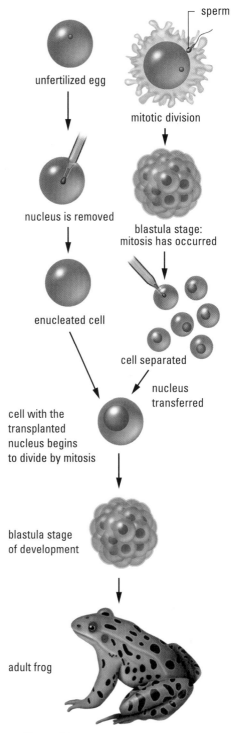

sperm

unfertilized egg

mitotic division

nucleus is removed

blastula stage: mitosis has occurred

enucleated cell

cell separated

cell with the transplanted nucleus begins to divide by mitosis

nucleus transferred

blastula stage of development

adult frog

Figure 7
The nucleus is transferred from one embryo cell into an egg cell.

Figure 8
Nature's clones. The left photograph shows identical twins. They originate from a single fertilized egg that splits into two.

The right photograph shows fraternal twins. They originate from two different eggs and each is fertilized by a separate sperm cell.

In another experiment, a scientist took the nuclei from gut cells, which were fully specialized, of African clawed toad tadpoles and inserted them into egg cells whose nuclei had been destroyed by ultraviolet radiation. Many of these cells failed to develop, but some did grow into adults. Analysis of the adults confirmed that they were clones.

Cloning from Adult Cells

Adult mammalian cells tend to be highly specialized. Transferring nuclei from the highly specialized cells of an adult mammal into an enucleated cell will not stimulate cell division. The genes controlling cell division have been switched off. Until recently, the only way to get clones was by splitting off cells from a developing embryo. The fertilized egg divides again and again to attain the structure of an adult, but, in dividing, the cells aggregate and specialize into different tissues and organs. It would appear that cells must be taken before the eight-cell stage of development to ensure that their nuclei are totipotent. After the eight-cell stage, the cells specialize and their ability to stimulate cell division is gone (**Figure 9**).

The long-held scientific belief that adult cells can't be used to clone animals was disproved with the appearance of a sheep named Dolly. Dr. Ian Wilmut, of the Rosalind Institute in Scotland, extracted the nucleus from an udder cell of an adult Finn Dorsett sheep and placed the nucleus into the enucleated egg cell from a Poll Dorsett sheep. The egg was allowed to develop in a petri dish until an early embryo stage was reached. Then this embryo was placed into the womb of a third

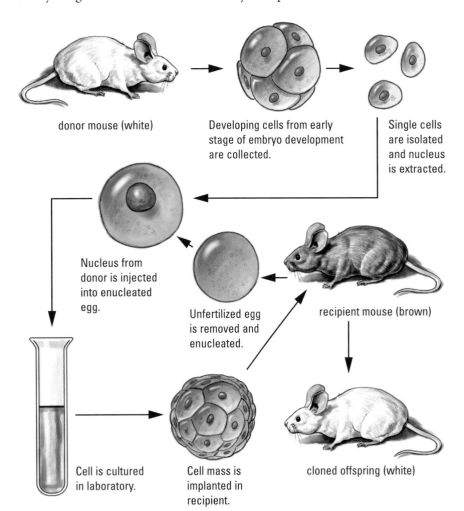

donor mouse (white)

Developing cells from early stage of embryo development are collected.

Single cells are isolated and nucleus is extracted.

Nucleus from donor is injected into enucleated egg.

Unfertilized egg is removed and enucleated.

recipient mouse (brown)

Cell is cultured in laboratory.

Cell mass is implanted in recipient.

cloned offspring (white)

Figure 9
Mammal cells have been cloned by embryo splitting.

Figure 10
Dolly could claim three different sheep as mothers. The genetic mother died before Dolly was even born.

sheep, a Scottish Blackface. The offspring, named Dolly, looked nothing like the birth mother or the Poll Dorsett. Her genetic information was identical to that of the Finn Dorsett adult; Dolly was a clone (**Figure 10**).

Not only did Dolly's genetic information come from an adult sheep, it came from a sheep that was no longer living. Dr. Wilmut extracted the nucleus from frozen mammary tissue. Dr. Wilmut's team concluded that previous attempts to clone mammals from adult cells had failed for one of three reasons: The cells were too actively replicating DNA and dividing, they had been in the wrong stage of the cell cycle, or the wrong set of genes had been turned off as the cell specialized. The Scottish research team starved the cells for a few days before extracting the nucleus. Soon after starvation, the adult cells began to act like unspecialized cells within an embryo. The nucleus could then be transferred into the cytoplasm of the enucleated egg cell. It is possible that starving a cell slows its metabolism and may turn genes back on that are shut off during cell specialization.

The potential benefits of animal cloning are numerous. Animals that have organs suitable for transplant to humans could be cloned, providing many more organs to patients who need them. Medical experimentation and research would also benefit from the availability of cloned animals. For example, experiments on the effectiveness of a drug are often difficult to measure because of the genetic variation among the individuals tested. Did the drug work because it is effective or because the individual is sensitive to it? If all the test subjects were genetically identical, clearer results could be obtained. In agriculture, the strongest livestock could be cloned, decreasing farmers' losses due to disease, and thereby increasing yield. Despite the great medical potential for cloning, many people have moral and ethical problems with the idea and worry about the impact on society and on individuals of cloning humans.

Practice

Understanding Concepts

1. What is cloning?
2. Why are identical twins often called "nature's clones"?
3. Do all the cells of your body divide at the same rate? Explain.
4. What is an enucleated cell?

Liver Tissue Transplants

Alyssa Smith (**Figure 11**) has biliary atresia, a common and fatal childhood liver disease where the ducts that drain bile from the liver to the intestine are absent or closed. Because bile cannot be supplied to the intestine, the proper digestion of food cannot occur. As bile builds in the liver, damage occurs. As a result, some of the liver's functions—i.e., maintenance of blood sugar levels, detoxification of poisons, extraction of wastes from the blood, production of bile salts for fat digestion, and storage of minerals and vitamins—can be impaired.

A revolutionary medical procedure performed on the 21-month-old girl may provide a new direction in transplant surgery. In this procedure, doctors removed the left lobe of her mother's liver and implanted fragments into Alyssa. Liver tissue is regenerative, which means that cells from the transplanted liver fragments will continue to divide and grow into a normal liver. Doctors hope that the transplanted liver will prevent illness as the diseased liver begins to fail. The idea of using a living donor has many potential benefits, the most obvious being a much wider availability of donors.

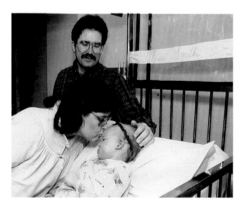

Figure 11
Alyssa Smith, shown with her mother and father, received transplanted cells from her mother's liver in a transplant operation involving a living liver donor.

Biotechnology in Agriculture

Biotechnology in agriculture includes a variety of techniques designed to increase food production or protect existing strains of crops and trees. Cross-pollination, cell fusion, cloning, and other techniques produce plants that have the best characteristics of their parents.

In one of the most controversial areas of genetic engineering research, scientists are able to combine genes from unrelated species. Using a technique called recombinant DNA, genes can be extracted from the nucleus of one organism and spliced into the chromosomes of a new organism. Remarkably, the donated genes affect the traits of the new organism. One of the worries about this type of research focuses on the potential consequences if genetically engineered organisms were accidentally released into the environment. Biotechnology is used to produce food crops that grow quickly, are resistant to diseases and pests, are efficient at absorbing nutrients from the soil, etc.—in short, to produce a "superplant" that excels at survival and growth. However, naturally growing species are not as well-equipped and would not be able to compete with this superplant. Natural populations of plants could be eliminated. The implications of the accidental release of a genetically engineered germ cell or virus are even greater.

biotechnology: the use of living things in industrial or manufacturing applications

DID YOU **KNOW ?**

Agriculture Canada preserves seeds and cell cultures for future use in breeding new crop varieties. Currently, some 82 000 seed samples, mainly barley and wheat, are stored at the research centre in Ottawa to ensure that their genes are preserved after the plants are no longer grown commercially.

Practice

Understanding Concepts

5. What is recombinant DNA?
6. What is the purpose of seed banks?
7. How does biotechnology change the normal "rules" of plant reproduction?

Explore an Issue

Debate: The Use of Biotechnology in Agriculture

The crops that have been modified by genetic engineering techniques have been part of a global controversy. The "superplant" has great agricultural value but has implications with regard to the diversity of plants.

Statement

Biotechnology will provide cheaper and higher quality food for future generations.

- In your group, research the issue. Learn more about a technique (tissue cloning, recombinant DNA, cell fusion, etc.) used in agriculture. Describe the technique. What impact has the technique had on food production? On the environment, if any?
- Search for information in newspapers, periodicals, CD-ROMs, and on the Internet.
Follow the links for Nelson Biology 11, 3.4.

GO TO www.science.nelson.com

- Write a list of points and counterpoints which your group has considered.
- Decide whether your group agrees or disagrees with the statement.
- Prepare to defend your group's position in a class discussion.

DECISION-MAKING SKILLS

- Define the Issue
- Identify Alternatives
- Research
- Analyze the Issue
- Defend a Decision
- Evaluate

SUMMARY **Cloning**

1. Cloning is the process of forming identical offspring from a single cell or tissue.

2. Cloning permits the production of offspring with characteristics identical to those of the parent.

3. Some plants and animals naturally clone themselves (reproduce asexually).

4. Artificial animal cloning involves injecting the nucleus from a fertilized egg in the blastula stage into an enucleated cell.

5. A nucleus that can bring a cell from egg to adult is called totipotent.

Section 3.4 Questions

Understanding Concepts

1. Describe how nuclear transplants are used to clone frogs.

2. Dolly was not the first cloned animal, nor was she the first mammal clone. What made her cloning so special?

3. Explain why male animals would no longer be needed if cloning became the accepted method of reproduction.

4. If the nucleus is extracted from an adult animal cell and placed into an enucleated egg, how would it be possible to distinguish the cloned individual from the original?

Making Connections

5. Make a list of benefits and potential problems associated with cloning farm animals.
   Follow the links for Nelson Biology 11, 3.4.

   GO TO www.science.nelson.com

6. Speculate on the potential benefits and problems associated with cloning humans.

7. Research the nature versus nurture debate and the evidence provided by studies of twins. Find out about some psychological conditions that have both a genetic and an environmental component. What are the advantages and disadvantages of each approach? Think about the social, moral, and ethical implications of each viewpoint.
   Follow the links for Nelson Biology 11, 3.4.

   GO TO www.science.nelson.com

3.5 Cancer

Cancer is a broad group of diseases associated with the uncontrolled, unregulated growth of cells. Much more active than normal cells, cancer cells divide at rates that far exceed those of their ancestors (**Figure 1**).

After the cells in your body have specialized, they divide only to replace damaged cells. A balance between cell destruction and cell replacement maintains a healthy organism. Cells communicate information about the body's needs from one cell to another. With the growth of a callus, accelerated cell divisions not only replace damaged cells, but also increase the cell numbers of the protective outer layers of skin to shield the delicate nerve and blood vessels that occupy the inner layers. This occurs through cell-to-cell communication.

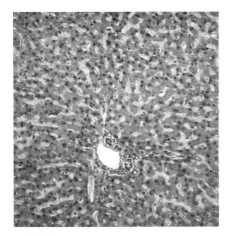

(a) Normal cells

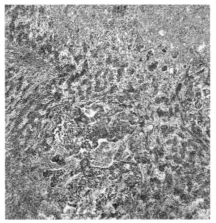

(b) Some cancerous cells can be identified by an enlarged nucleus.

Figure 1

Figure 2
Terry Fox

Normal cells usually cannot divide when isolated from one another. Cancer cells, by comparison, are capable of reproducing in isolation. A cancer cell growing in an artificial culture is capable of dividing once every twenty-four hours. At this rate of division, a single cancer cell would generate over one billion descendants in a month. Fortunately, cancer cells do not reproduce that quickly in the body of an organism. Such rapid growth would create a tumour containing trillions of cells with an estimated mass of ten kilograms within six weeks. Even the rapidly dividing cells of the embryo are not capable of such proliferation.

The cancer cell does not totally ignore messages from adjacent cells that are meant to regulate its rate of reproduction. Scientists have discovered that some cancers progress at a very slow rate, often stopping for many years, only to become active at a later date. However, all cancer cells can reproduce without directions from adjacent cells. The isolated cancer cell creates yet another problem. Most normal cells adhere to one another which helps communication between cells in a given tissue. Kidney, liver, and heart cells, even when isolated from each other in the laboratory, show a remarkable quality of attraction when placed close together. Cancer cells do not adhere well to other cancer cells, nor do they stick to normal cells. This ability to separate from other cell masses makes cancer growth dangerous. Cancer cells can dislodge from a tumour and move to another area, which means **metastasis** is occurring. This ability to metastasize makes the source of cancer difficult to locate and to control. The cancer that began in Terry Fox's leg was not removed completely by the amputation of the limb (**Figure 2**). During his run across Canada, cancer was diagnosed in his lung; it had metastasized.

metastasis: the event where a cancer cell breaks free from the tumour and moves into another tissue

Another important difference between normal cells and cancer cells is that cancer cells lack the ability to differentiate. It has been estimated that there are about 100 different types of cells in the human body. Each cell type has a unique shape that enables it to carry out a specialized function. Cancer cells do not mature and specialize like normal cells. Therefore, another threat to health arises because the cancer cells cannot carry out some of the functions of the normal cells.

Practice

Understanding Concepts

1. In what ways does cellular communication regulate the division of normal cells?

2. How rapidly can a cancer cell grown in tissue culture divide? Is it likely that the cancer cell could divide that quickly in the human body? Explain.

3. How would the cell cycle of a cancer cell differ from that of a normal cell? Give reasons.

Causes of Cancer

The causes of cancer are still not fully understood. The incidence of some cancers has been linked with a family history of that cancer. For example, statistics show that many women with breast cancer had a mother or grandmother with breast cancer. And yet some women who come from families with a long history of breast cancer will not develop it, and others who have no breast cancer in their families will. Genetics plays a relatively small part in explaining the incidence of most cancers.

Certain lifestyle habits, such as smoking or excessive exposure to Sun, are known to greatly increase the risk of cancer—lung cancer and skin cancer, respectively. Other factors are exposure to radiation and to toxic chemicals. The importance of each factor in determining a person's risk is uncertain, and it is possible that there are other cancer-causing factors that we do not yet know about.

SUMMARY   Cancer

1. Cancer is associated with uncontrolled growth of cells.

2. Cancer cells, unlike normal cells, are capable of reproducing in isolation. They do not adhere to other cancer cells or to normal cells. This means that cancer cells can metastasize.

3. Cancer cells do not specialize, so they cannot carry out some of the functions of normal cells, making cancer cells inefficient in terms of carrying out normal cell functions.

Section 3.5 Questions

Understanding Concepts

1. Provide a hypothesis that explains why the skin is so susceptible to cancer.

Applying Inquiry Skills

2. A scientist finds a group of irregularly shaped cells in an organism. The cells demonstrate little differentiation, and the nuclei in some of the cells stain darker than others. Based on these findings, would it be logical to conclude that the organism has cancer?

3. It is thought that two plant extracts, A and B, may fight cancer. A pharmaceutical company wants to test A and B on samples of cancerous tissue.
   (a) Design an experiment to test the effects of A and B.
   (b) Hypothesize what you would expect to see if an extract has positive effects.
   (c) Include in your design a way to test effectiveness.

Making Connections

4. Find out about the treatments for different kinds of cancer. How do they work?

5. Taxol, a chemical extract from the bark of the Pacific yew tree, may be useful in the treatment of breast and ovarian cancer. As a result of past intense efforts to harvest this extract, many yew trees were being killed, resulting in the endangerment of the Pacific yew. Research the Web to learn how taxol may work as an anti-cancer drug. List at least two ways that taxol can now be obtained while saving the Pacific yew tree.
Follow the links for Nelson Biology 11, 3.5.

GO TO www.science.nelson.com

6. Many pharmaceuticals are derived from natural sources, and it is thought that other exotic plants might contain cancer-fighting extracts. Survey medical journals, the newspaper, and other resources to learn about another plant that is currently being researched. Describe how this plant's extracts might possibly work in fighting the disease.
Follow the links for Nelson Biology 11, 3.5.

GO TO www.science.nelson.com

7. What opportunities might humans be losing by permitting the destruction of forests and other ecosystems?

8. High-level exposure to Xrays and gamma rays emitted from radioisotopes can result in radiation poisoning. Radiation poisoning results in damage to DNA, especially in cells engaged in DNA replication. Hair loss and damage to the lining of the stomach and intestines are two early symptoms. Explain why.

9. High levels of radiation may have some serious side effects, as noted earlier. Why is highly focused radiation therapy used against some cancers?

Reflecting

10. Science is often incorrectly presented as a list of facts, rather than a list of questions. What questions remain to be answered about cell division?

3.6 Meiosis

Meiosis is the process by which sex cells, or **gametes**, are formed. (In humans, this takes place in the testes and ovaries.) It involves two stages of cell division that have some similarities to the phases in mitosis. In mitosis, the chromosome number of the daughter cells is the same as in the parent cell. In meiosis, the chromosome number of the daughter cells is half that of the parent cell. A human cell containing 46 chromosomes will undergo meiosis and produce gametes that have 23 chromosomes. Each gamete will contain both the same number and the same kind of chromosomes. The number of chromosomes in a gamete is called the **haploid** chromosome number, or *n*; the number of chromosomes in all other cells is twice the haploid number and is called the **diploid** number, or *2n*. In humans, the haploid chromosome number is 23 and the diploid chromosome number is 46.

Offspring carry genetic information from each of the parents. This explains why you might have your father's eyes but your mother's hair. Although you may look more like one parent than another, you receive genetic information from each parent. For example, your father gives you a chromosome with genes that code for eye colour, but so does your mother. Each of the 23 chromosomes

meiosis: two-stage cell division in which the chromosome number of the parental cell is reduced by half. Meiosis is the process by which gametes are formed.

gametes: sex cells that have a haploid chromosome number

haploid: refers to the number of chromosomes in a gamete

diploid: refers to twice the number of chromosomes in a gamete. Every cell of the body, with the exception of sex cells, contains a diploid chromosome number.

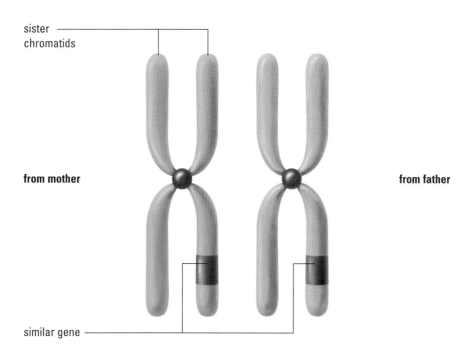

sister
chromatids

from mother

from father

similar gene

Figure 1
Homologous chromosomes

homologous chromosomes: paired chromosomes similar in shape, size, gene arrangement, and gene information

zygote: a cell resulting from the union of a male and female sex cell, until it divides and then is called an embryo

that you receive from your father is matched by 23 chromosomes from your mother. The paired chromosomes are called **homologous chromosomes** because they are similar in shape, size, and gene arrangement. The genes in homologous chromosomes deal with the same traits. Each cell in your body, except the sex cells, contains 23 pairs of homologous chromosomes, or 46 chromosomes in total. Homologous chromosomes interact during meiosis. Your characteristics are determined by the manner in which the genes from homologous chromosomes interact (**Figure 1**).

During fertilization, a haploid ($n = 23$) sperm cell unites with a haploid ($n = 23$) egg cell to produce a diploid ($2n = 46$) **zygote**. The fusion of human male and female gametes restores the diploid chromosome number in the zygote. The zygote will begin dividing by mitosis and will produce a multicellular human baby. Thus, mitosis and meiosis are linked in the cycle of life.

Stages of Meiosis

Meiosis involves two nuclear divisions that produce four haploid cells. Meiosis I is often called reduction division because the diploid, or $2n$, chromosome number is reduced to the haploid, or n, chromosome number (**Figure 2**). The second phase, meiosis II, is marked by a separation of the two chromatids.

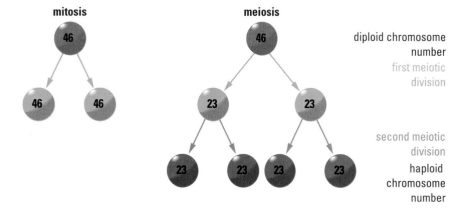

Figure 2
Comparison of mitosis and meiosis in humans. Mitosis produces two diploid cells from one diploid cell. Meiosis produces four haploid cells from one diploid cell.

homologous
chromosome
pair

As the chromosomes
move closer together,
synapsis occurs.

Chromatids break,
and genetic information
is exchanged.

Figure 3
Crossing over occurs between homologous pairs of chromosomes during prophase I of meiosis.

The phases used to describe the events of mitosis can also be used to describe meiosis. As with mitosis, assume that DNA replication occurs prior to the cell division phase.

Meiosis I

During prophase I, the nuclear membrane begins to dissolve, the centriole splits and its parts move to opposite poles within the cell, and spindle fibres are formed. The chromosomes come together in homologous pairs. Each chromosome of the pair is a homologue and is composed of a pair of sister chromatids. The whole structure is then referred to as a **tetrad** because each pair is composed of four chromatids.

This process is referred to as **synapsis**. As the chromosomes synapse, the chromatids often intertwine. Sometimes the intertwined chromatids from different homologues break and exchange segments or undergo **crossing over** (see **Figure 3**). Crossing over permits the exchange of genetic material between homologous pairs of chromosomes.

Metaphase I follows prophase I (see **Figure 4**). The homologous chromosomes attach themselves to the spindle fibres and line up along the equatorial plate.

tetrad: a pair of homologous chromosomes, each with two chromatids

synapsis: the pairing of homologous chromosomes

crossing over: the exchange of genetic material between two homologous chromosomes

prophase I   **metaphase I**   **anaphase I**   **telophase I**

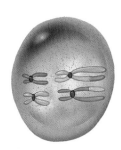

The replicated chromosomes condense. Homologous chromosomes come together in synapsis and crossing-over occurs. Chromosomes attach to the spindle.

Chromosomes line up at the equatorial plate.

Each chromosome separates from its homologue. They move to opposite poles of the cell.

The nucleus completes its division. The chromosomes are still composed of sister chromatids. The cytoplasm divides after telophase.

Figure 4
During meiosis I, homologous chromosomes are segregated.

During anaphase I, the homologous chromosomes move toward opposite poles. The process is known as segregation. At this point of meiosis, reduction division occurs. One member of each homologous pair will be found in each of the new cells. Each chromosome consists of two sister chromatids.

During telophase I, a membrane begins to form around each nucleus. However, unlike in mitosis, the chromosomes in the two nuclei are not identical. Each of the daughter nuclei contains one member of the chromosome pair. Although homologous chromosomes are similar, they are not identical. They do not carry exactly the same information. The cells are now ready to begin the second stage of meiosis.

Meiosis II

Meiosis II occurs at approximately the same time in each of the haploid daughter cells. However, for simplicity, consider the events in only one of the cells. (In **Figure 5**, both cells from meiosis I are shown). During meiosis II, pairs of chromatids will separate and move to opposite poles. Note that, unlike with mitosis and meiosis I, there is no replication of chromosomes prior to meiosis II.

Prophase II signals the beginning of the second division. During this stage, the nuclear membrane dissolves and the spindle fibres begin to form.

Metaphase II follows prophase II. It is signalled by the arrangement of the chromosomes, each with two chromatids, along the equatorial plate. The chromatids remain pinned together by the centromere.

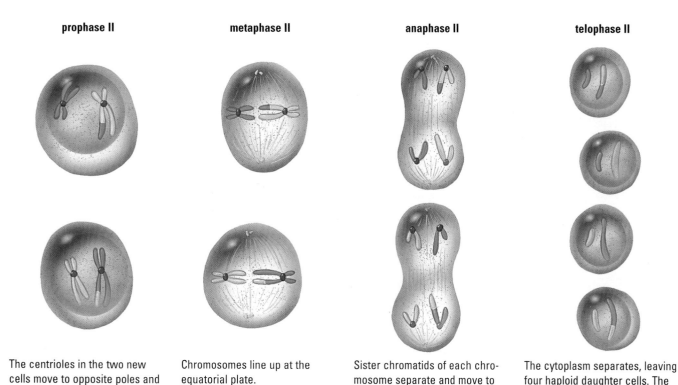

prophase II

metaphase II

anaphase II

telophase II

The centrioles in the two new cells move to opposite poles and new spindle fibres form. The chromosomes become attached to the spindle.

Chromosomes line up at the equatorial plate.

Sister chromatids of each chromosome separate and move to opposite poles.

The cytoplasm separates, leaving four haploid daughter cells. The chromosome number has been reduced by half. These cells may become gametes.

Figure 5
During meiosis II, sister chromatids separate.

Anaphase II can be identified by the breaking of the attachment between the two chromatids and by their movement to the opposite poles. This stage ends when the nuclear membrane begins to form around the chromatids, now referred to as chromosomes.

The cell then enters its final stage of meiosis: telophase II. During this stage, the second nuclear division is completed and then the second division of cytoplasm occurs. Four haploid daughter cells are produced from each meiotic division.

Practice

Understanding Concepts

1. Define meiosis. Describe the main stages in the process. Sketch the sequence of stages to help you in your description. Label your diagrams appropriately.
2. How are haploid cells different from diploid cells in humans?
3. What is a tetrad?
4. What are homologous chromosomes?
5. Do homologous chromosomes have the same number of genes? Explain.
6. Do homologous chromosomes have identical genes? Explain.

Gamete Formation in Grasshoppers

Figure 6

Obtain prepared slides of grasshopper (**Figure 6**) testes and identify cells undergoing meiosis. Make a few sample diagrams of cells at various stages of cell division.

(a) Label the chromosomes.
(b) Are you able to count the chromosome number? Explain why or why not.
(c) Explain and compare what happens in prophase, metaphase, and anaphase of meiosis I and II.
(d) How do cells undergoing meiosis II differ from cells undergoing meiosis I?

3.7 Comparing Mitosis and Meiosis

All multicelled, eukaryotic species grow and repair tissue by mitosis, followed by cytokinesis. Single-celled eukaryotic species follow the same process to reproduce asexually. Mitosis occurs in nonreproductive cells. In interphase, the cells have a diploid chromosome number. Before mitotic division occurs, their DNA is replicated. Reproductive cells also have a diploid chromosome number in interphase. Before meiosis I, their DNA is replicated. However, there is no DNA replication between meiosis I and meiosis II. **Figures 1** and **2** summarize the similarities and differences between mitosis and meiosis. As you examine **Figures 1** and **2**, make note of the chromosome number of the cell or cells, whether the chromosome number is haploid or diploid, and during which stage the chromosome number changes.

The most significant difference between mitosis and meiosis is the end results. Mitosis results only in clones of the original; all daughter cells are genetically identical to each other and to the parent cell. Meiosis results in four cells that are different from each other and from the parent. Meiosis, combined with fertilization, explains the variation in traits that is observed in species that reproduce sexually. The variation occurs through three mechanisms. First, crossing over during prophase I exchanges genes on the chromosomes. Second, during metaphase I, the paternal and maternal chromosomes are randomly assorted. Although homologues always go to opposite poles, a pole could receive all the maternal chromosomes, all the paternal ones, or some combination. Lastly, during fertilization, different combinations of chromosomes and genes occur when two gametes unite.

In later chapters, you will learn about the ways in which meiosis and fertilization contribute to the enormous diversity exhibited by sexually reproducing organisms. You will learn about how sexual reproduction promotes genetic variability, and how this contributes to the survival of individuals, populations, and entire species.

Meiosis I

| prophase I | metaphase I | anaphase I | telophase I |

 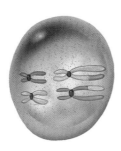  

The replicated chromosomes condense. Homologous chromosomes come together in synapsis and crossing-over occurs. Chromosomes attach to the spindle.

Homologous chromosomes line up at the equatorial plate.

Each chromosome separates from its homologue. They move to opposite poles of the cell.

The nucleus completes its division. The chromosomes are still composed of sister chromatids. The cytoplasm divides after telophase.

Figure 1
Stages of meiosis I.

Mitosis

prophase

The chromosomes condense, becoming shorter and thicker. The centrioles assemble and spindle fibres attach to the centromeres of the chromosomes. The nuclear membrane starts to dissolve.

metaphase

Chromosomes line up at the equatorial plate. The nuclear membrane completely dissolves.

anaphase

The centromeres divide and the resulting chromosomes, formerly chromatids, move to opposite poles of the cell. An identical set of chromosomes moves to each pole.

telophase

Chromosomes lengthen again, the spindle fibres dissolve, and a nuclear membrane forms around the chromosomes.

Meiosis II

prophase II

The centrioles in the two new cells move to opposite poles and new spindle fibres form. The chromosomes become attached to the spindle.

metaphase II

Chromosomes line up at the equatorial plate.

anaphase II

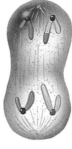

Sister chromatids of each chromosome separate and move to opposite poles.

telophase II

The cytoplasm separates, leaving four haploid daughter cells. The chromosome number has been reduced by half. These cells may become gametes.

Figure 2
Comparsion of the stages in mitosis and meiosis II

Comparing Mitosis and Meiosis

In this investigation, you will model and compare the events of mitosis and meiosis.

Materials

blue modelling clay   plastic knife
red modelling clay   sheets of paper
green modelling clay   pencil

Procedure

For each step, make a coloured sketch of your model with appropriate labels. Include brief descriptions of your steps and make sure to use the same step numbers as given.

Part 1: Mitosis

1. Take some red clay and roll it between your hands to create a piece 10 cm long and about as thick as your finger. Make another piece about 5 cm long.

2. Repeat step 1 with the blue clay.

3. Make an identical copy of each piece of clay. Then attach the identical pieces with a green ball of clay (**Figure 3**).

4. Draw a line down the length of a sheet of paper. Line up the four chromosomes along the line (**Figure 4**).

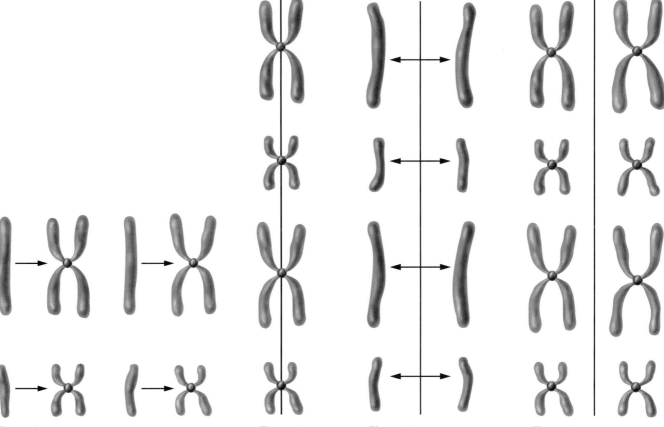

Figure 3   **Figure 4**   **Figure 5**   **Figure 6**

Figure 7

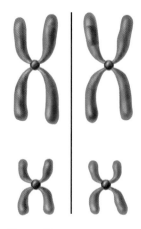

Figure 8

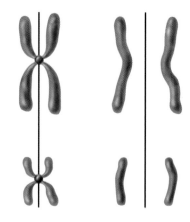

Figure 9

5. Remove the green balls and move each of the single pieces of clay to opposite ends of the paper (**Figure 5**).

6. If the cell is going to divide again, each single chromosome must synthesize a duplicate during interphase. Make an identical copy of each piece of clay as before (**Figure 6**).

Part 2: Meiosis

7. Follow steps 1 to 3 from part 1.

8. Demonstrate crossing-over. Break off a piece of clay from one chromosome and attach it to the other chromosome (**Figure 7**). Repeat a few times if you like.

9. To simulate metaphase I, place the chromosomes on either side of the equatorial plate, represented by a line drawn on a piece of paper (**Figure 8**).

10. Choose one of the haploid daughter cells and line the chromosomes up along the equatorial plate. Remove the centromere and move chromosomes to opposite poles (**Figure 9**).

Analysis

Part 1: Mitosis
(a) In step 3, what process did you model?
(b) What do the red and blue pieces of clay represent? What do the green balls of clay represent?
(c) In step 4, what is the diploid chromosome number of the cell?
(d) What phase of mitosis does the model represent?
(e) In step 5, what structure do the single pieces of clay represent after separation?
(f) What phase of mitosis does the model represent?
(g) In step 6, how many chromosomes are in each of the daughter cells?
(h) Compare the daughter cells with the parent cell.

Part 2: Meiosis
(i) In steps 1 to 3, on what basis are chromosomes considered to be homologous?
(j) What is the diploid chromosome number?
(k) In step 8, what must happen before the homologous chromosomes can cross over?
(l) In which phase does crossing over occur?
(m) What happens during crossing over?

(n) In step 9, how does metaphase I of meiosis differ from metaphase of mitosis?

(o) What is the haploid chromosome number?

(p) In step 10, compare the resulting daughter cells of mitosis and meiosis.

Sections 3.6–3.7 Questions

Understanding Concepts

1. How does the first meiotic division differ from the second meiotic division?

2. Explain why synapsis may lead to the exchange of genetic information.

3. Compare meiosis with mitosis. How does meiosis differ from mitosis?

Applying Inquiry Skills

4. Copy and complete **Table 1**. Compare the chromosome number in the organisms before, during, and as a result of meiosis. Indicate whether the chromosome number is haploid or diploid.

Table 1

	Human	Cat	Shrimp	Bean
Before meiosis				
chromosome number (haploid or diploid?)	46	?	?	?
number of pairs of homologous chromosomes	23	?	127	?
After meiosis I				
chromosome number (haploid or diploid?)	23	19	?	?
After meiosis II				
chromosome number (haploid or diploid?)	23	?	?	11
number of pairs of homologous chromosomes	0	?	?	?

5. Suggest an approach for completing Activity 3.7.1 using computer technology.

3.8 Reproduction and Cell Division

Both plants and animals grow by mitotic cell division and form gametes by meiosis. In flowering plants, pollen contains the male sex cells. The egg cells of plants are stored in a variety of structures. Plant gametes contain a haploid chromosome number. A zygote forms from the fusion of a male and female gamete, and, following mitosis, either spores or seeds will result, which will produce new generations of plants.

In many familiar plants, such as pine trees and roses, spore-producing bodies (sporophytes) form from the zygote after mitosis. Spores are haploid cells that often contain a protective wall that is resistant to drought and other

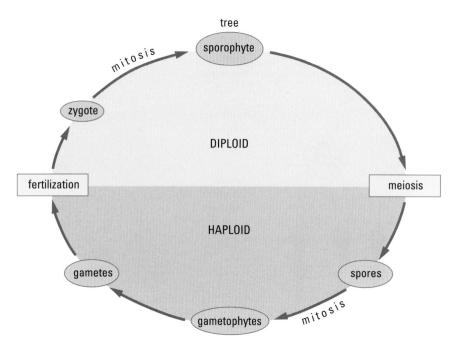

Figure 1
Plant life cycle. For a pine tree, the tree itself is the sporophyte. The gametophytes develop in the cones.

unfavourable environmental conditions. Spores germinate and give rise to the gamete-producing bodies, or gametophytes. Alternation between the spore-producing bodies and the gamete-producing bodies is shown in the generalized plant life cycle (**Figure 1**).

In animals (**Figure 2**), meiosis takes place in the testes and ovaries, which produce sperm cells and egg cells, respectively. Two haploid gametes (an egg and a sperm) unite to form a diploid zygote, which in turn undergoes mitosis. During cell division, diploid cells are formed from other diploid cells. During maturation, some of the cells specialize and become gamete-producing cells. These gamete-producing cells undergo meiosis and form haploid cells, bringing the life cycle full circle.

The formation of sex cells during meiosis is referred to as **gametogenesis**. Although human male and female gametes both follow the general process of

gametogenesis: the formation of gametes (sex cells) in animals

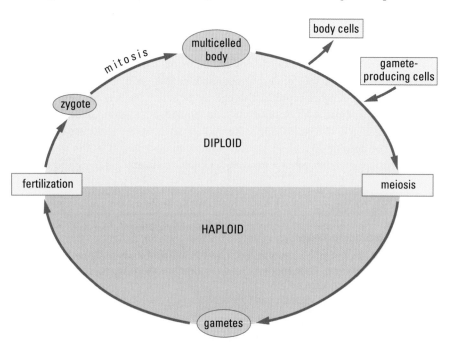

Figure 2
Animal life cycle. Animals alternate between meiosis (forming haploid cells) and mitosis (forming diploid cells).

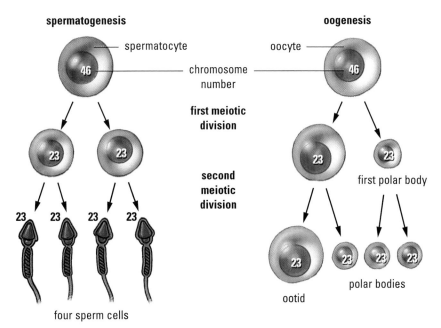

spermatogenesis

spermatocyte

chromosome number

first meiotic division

second meiotic division

23 23 23 23

four sperm cells

oogenesis

oocyte

first polar body

polar bodies

ootid

Figure 3
Generalized diagram of sperm and egg cell formation in humans

ootid: an unfertilized ovum

polar bodies: cells that contain all the genetic information of a haploid ovum but lack sufficient cytoplasm to survive; formed during meiosis in females

oocytes: immature eggs

meiosis, some differences do exist. The cytoplasm of the female gametes does not divide equally after each nuclear division. As shown in **Figure 3**, one of the daughter cells, called the **ootid**, receives most of the cytoplasm. The other cells, called the **polar bodies**, die, and the nutrients are absorbed by the body of the organism. Only one ovum (egg cell) is produced from meiosis. In contrast, with sperm cells, there is an equal division of cytoplasm. Because of their function, sperm cells have much less cytoplasm than egg cells. Sperm cells are specially designed for movement: they are streamlined and cannot carry excess weight. For the egg, movement is unimportant. Egg cells use the nutrients and organelles carried within the cytoplasm to fuel future cell divisions in the event that the egg cell becomes fertilized.

Males make many more sex cells than females. The diploid spermatocytes—the cells that give rise to sperm cells—are capable of many mitotic divisions before meiosis ever begins. Males can produce one billion sperm cells every day. Baby females have about two million primary **oocytes** in their ovaries. Most of these are absorbed into the body, with about 300 000 oocytes remaining, 400 to 500 of which will be released during the reproductive years. Primary oocytes have already entered meiosis I, but they will remain suspended in prophase I until the female reaches reproductive age, or puberty. Starting at the first menstrual cycle, meiosis will resume in one oocyte at a time, once a month. Menopause marks the end of the menstrual cycle in women and usually occurs between the ages of 40 and 55. At this point, the oocytes left in the ovaries are nonfunctional or unresponsive and are no longer released.

Practice

Understanding Concepts

1. A muscle cell of a mouse contains 22 chromosomes. Based on this information, how many chromosomes are there in the following types of mouse cells?
   (a) daughter muscle cell formed from mitosis
   (b) egg cell
   (c) fertilized egg cell

2. Compare the mechanisms of gametogenesis in males and females.

3. When meiosis occurs in females, the cytoplasm is not divided equally among the resulting four cells. Explain why.

4. Draw a diagram of meiosis to show how a female child is produced from the union of a sperm and an egg.

5. Compare the life cycles of plants and animals.

Sex Chromosomes

In 1906, scientists, while observing meiosis in the testes cell of a fruit fly, noted that one of the chromosomes appeared hook shaped. One chromosome pair did not appear to have a homologous member. Did this mean that the fruit fly carried a genetic mutation? Another male was examined and found to have a small hook-shaped chromosome. Within the female cell, all the chromosomes were rod shaped. After observing different male and female cells during meiosis, the scientists concluded that one pair of chromosomes differed in males and females. In females, this pair was always two rod-shaped chromosomes of identical length, while the male had one rod-shaped chromosome and a smaller hook-shaped chromosome (**Figure 4**). These chromosomes are called **sex chromosomes**. Chromosomes that are not sex chromosomes are referred to as **autosomes**.

The rod-shaped chromosomes are called X chromosomes, while the hook-shaped chromosome is called a Y chromosome. Females have two homologous X chromosomes and males have one X and one Y chromosome. Although they are physically different, the X and the Y chromosomes are able to synapse for part of their length and function as homologous chromosomes during meiosis. Many male animals, including humans, have a Y chromosome that identifies the individual as a male.

sex chromosomes: the pair of chromosomes that have a role in the sex of an individual

autosomes: the chromosomes not involved in sex determination

female sex chromosomes

male sex chromosomes

double X chromosomes

an X and Y chromosome

Figure 4
Sex chromosomes

SUMMARY Meiosis

1. Meiosis involves the formation of sex cells or gametes.

2. Cells undergoing meiosis pass through two divisions.

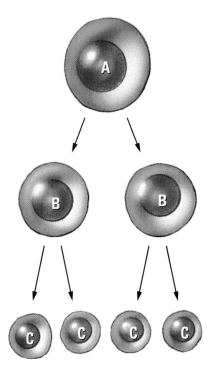

Figure 5
Sperm cell production in humans

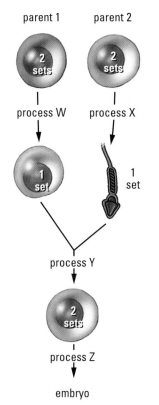

Figure 6
The processes and number of sets of chromosomes involved in the production of an embryo in humans

3. All gametes produced by meiosis have haploid chromosome numbers.

4. Homologous chromosomes are similar in shape, size, gene arrangement, and gene information.

5. Crossing over is the exchange of genetic material between homologous chromosomes that occurs during meiosis.

6. Sex chromosomes are pairs of chromosomes that determine the sex of an organism. All other chromosomes are autosomes.

Sections 3.8 Questions

Understanding Concepts

1. **Figure 5** shows sperm cell production following meiosis.
   (a) Which cells do not contain homologous pairs?
   (b) If the chromosome number for cell A is 12, indicate the chromosome number for cell C.

2. Use **Figure 6** to answer the questions below.
   (a) Which process(es) identify mitosis? Explain your answer.
   (b) Which process(es) identify meiosis? Explain your answer.

Making Connections

3. King Henry VIII of England had some of his wives executed for not producing sons. Indicate why a little knowledge of meiosis might have been important for Henry's wives.

4. A microscopic water animal called daphnia can be reproduced from an unfertilized egg. This form of reproduction is asexual because male gametes are not required. Indicate the sex of the offspring produced. Explain your answer.

3.9 Abnormal Meiosis: Nondisjunction

Meiosis, like most processes of the body, is not immune to mistakes. Nondisjunction occurs when two homologous chromosomes move to the same pole during meiosis. The result is that one of the daughter cells will be missing one chromosome while the other will retain an extra chromosome. Cells that lack genetic information, or have too much information, will not function properly. Nondisjunction can also occur in any cell during mitosis where separation of chromatids is involved, but the effects are most devastating during the formation of sex cells in meiosis.

In humans, nondisjunction produces gametes with 22 and 24 chromosomes (**Figure 4**, page 118). The gamete with 24 chromosomes has both chromosomes from one of the homologous pairs. If that gamete joins with a normal gamete of 23 chromosomes from the opposite sex, a zygote containing 47, rather than 46, chromosomes will be produced. The zygote will then have three chromosomes in place of the normal pair. This condition is referred to as **trisomy**. However, if the sex cell containing 22 chromosomes joins with a normal gamete, the resulting zygote will have 45 chromosomes. The zygote will have only one of the chromosomes rather than the homologous pair. This condition is called **monosomy**. Once the cells of the trisomic or monosomic zygotes begin to divide, each cell of the body will contain greater or less than 46 chromosomes.

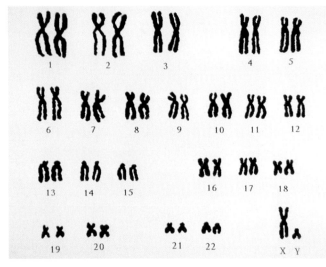

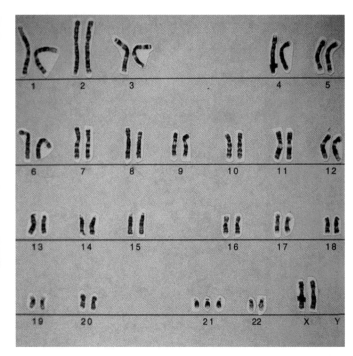

(a) Karyotype chart of a male with 46 chromosomes. Notice that the chromosome pair number 23 is not homologous. Males contain an X and a Y chromosome. They act as a homologous pair in meisois, but they are not similar in size and shape as are the other chromosome pairs.

Figure 1

(b) Down syndrome female. Note the trisomy of number 21. Down syndrome affects both males and females.

Nondisjunction Disorders

Nondisjunction is associated with many different human genetic disorders. Compare the chromosomes of a male shown in **Figure 1(a)**, with the chromosomes of a female who has **Down syndrome**, shown in **Figure 1(b)**. Notice how the chromosomes are arranged in pairs. Such a picture of the chromosomes is known as a **karyotype chart**. In about 95% of cases, a child with Down syndrome has an extra chromosome in chromosome number 21. This trisomic disorder is produced by nondisjunction; the person has too much genetic information.

The trisomic condition is referred to as a syndrome because it involves a group of disorders that occur together. People with Down syndrome (**Figure 2**) can be identified by several common traits, regardless of race: a round, full face; enlarged and creased tongue; short height; and a large forehead. It has been estimated that 1 in 600 babies is born with Down syndrome. Down syndrome is

trisomy: the condition where there are three homologous chromosomes in place of a homologous pair

monosomy: the condition where there is a single chromosome in place of a homologous pair

Down syndrome: a trisomic disorder in which a zygote receives three homologous chromosomes for chromosome pair number 21

karyotype chart: a picture of chromosomes arranged in homologous pairs

Figure 2
People with Down syndrome have a wide range of abilities.

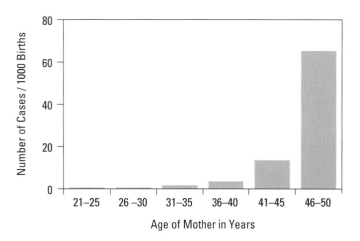

Figure 3
Incidence of Down syndrome

generally associated with mental retardation, although people with this condition retain a wide range of mental abilities. Statistics indicate that the risk of having a baby with Down syndrome increases with the age of the mother. A woman in her forties has a 1 in 40 chance of having a child with Down syndrome. This rate is 25 times greater than for a woman in her twenties (**Figure 3**).

Turner syndrome: a monosomic disorder in which a female has a single X chromosome

Turner syndrome occurs when sex chromosomes undergo nondisjunction (**Figure 4**). This monosomic disorder produces a female with a single X chromosome. In the egg cell, both homologous X chromosomes move to the same pole during meiosis I. When the egg with no X chromosome is fertilized by a normal sperm cell with an X chromosome, a zygote with 45 chromosomes is produced. Individuals with Turner syndrome appear female, but do not usually develop sexually, and tend to be short and have thick, widened necks. About 1 in every 3000 female babies is a Turner syndrome baby. Most Turner syndrome fetuses are miscarried before the 20th week of pregnancy.

Klinefelter syndrome: a trisomic disorder in which a male carries an XXY condition

Klinefelter syndrome is caused by nondisjunction in either the sperm or egg. The child inherits two X chromosomes—characteristic of females—and a single

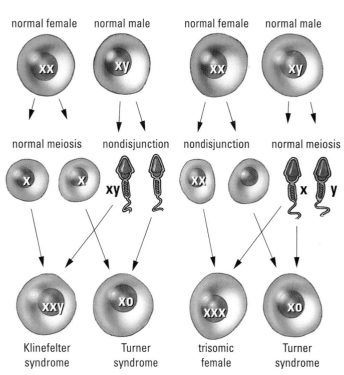

Figure 4
Nondisjunction disorders in humans

Y chromosome—characteristic of males. The child appears to be a male at birth; however, as he enters sexual maturity, he begins producing high levels of female sex hormones. Males with Klinefelter syndrome are sterile. It has been estimated that Klinefelter syndrome occurs, on average, in 1 of every 500 male babies.

Karyotype Charts

Technicians obtain a karyotype chart by mixing a small sample of tissue with a solution that stimulates mitotic division. A different solution is added which stops division at metaphase. Since chromosomes are in their most condensed form during metaphase—their size, length, and centromere location are most discernible—it is the best phase in which to obtain a karyotype. The metaphase cells are placed onto a slide and then stained, so that distinctive bands appear. A photograph of the chromosomes is taken. The image is enlarged, and each chromosome is cut out and paired up with its homologue. Homologous chromosomes are similar in size, length, centromere location, and banding pattern. Finally, all the pairs are aligned at their centromeres in decreasing size order. The sex chromosomes are always placed last.

Sample Problem

Figure 5 shows the incomplete karyotype of a human.

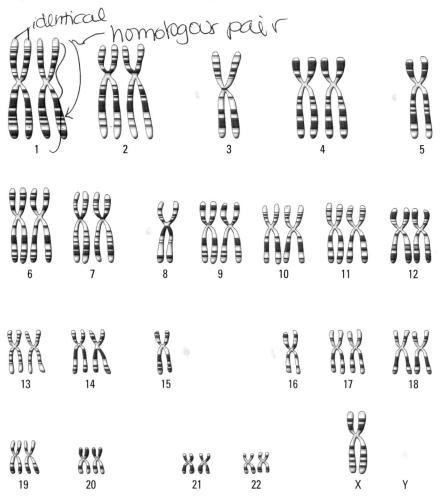

Figure 5
The karyotype chart of a human. Notice several chromosomes are missing.

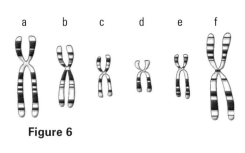

Figure 6

Figure 7 g

1. Identify where chromosomes a to f (**Figure 6**) should be in the karyotype chart.
2. This person has either Down syndrome or Klinefelter syndrome. Identify the placement of chromosome g (**Figure 7**) to identify which disorder the patient has.

Solution

- Start by scanning the karyotype chart to see which pairs are missing a chromosome. Pairs 3, 5, 8, 15, and 16 need a partner.
- Match the most obvious chromosomes first: the longest, shortest, or most distinctively banded chromosomes.
- For chromosome matches that are not as obvious, look carefully at the banding pattern and location of the centromere.
- Always pay attention to the X and Y chromosomes. In **Figure 5**, the missing chromosome might be X or Y. If it is Y, it will have to be found through elimination since it will not match X.

1. a, 5   b, 8   c, 16   d, Y   e, 15   f, 3
2. Klinefelter syndrome is the XXY condition. Chromosome g is too short to be an X chromosome, so the patient must have Down syndrome. Chromosome g belongs at number 21.

Practice

Understanding Concepts

1. What is nondisjunction?
2. Differentiate between monosomy and trisomy.
3. What is Down syndrome?
4. What is a karyotype?
5. What is Turner syndrome?

Research in Canada: Dr. Irene Uchida

Down syndrome was first described by the English doctor John Down in 1866. By the early 1930s, geneticist L.S. Penrose had shown a link between the incidence of Down syndrome and the mother's age. The Canadian researcher Dr. Irene Uchida (**Figure 8**) was intrigued by the higher incidence of nondisjunction disorders, specifically Down syndrome, in the babies of older women.

Dr. Uchida found a 1930s study on fruit flies, which showed that exposures to high dosages of radiation increased the frequency of nondisjunction disorders. Although meiosis in fruit flies is not identical to that in humans, Dr. Uchida decided to pursue the radiation link for clues in her research on Down syndrome. During her first set of experiments, Dr. Uchida verified the original fruit fly experiments. She found that the greater the exposure to radiation, the higher was the incidence of chromosome abnormalities. Assuming that the results for fruit flies would hold true for other chromosomes, Dr. Uchida hypothesized that the oocytes of older women have been exposed to radiation for a longer period of time and, therefore, are more likely to suffer chromosome damage.

A 1960 survey, conducted by Dr. Uchida at the Children's Hospital in Winnipeg, indicated that women who had been exposed to radiation prior to conception were more likely to have children with Down syndrome. Similar studies carried out in other hospitals supported Dr. Uchida's findings. In 1968,

Figure 8
Dr. Irene Uchida

Dr. Uchida carried out a larger study with women who had been exposed to radiation. Using a list of patients who had had abdominal X-ray examinations, or fluoroscopies, from the Winnipeg General Hospital between 1956 and 1959, Dr. Uchida compiled a second list of women who had given birth to a child with Down syndrome. The larger study supported her hypothesis that X rays may be one of the causes of Down syndrome.

How do you think the results of Dr. Uchida's research influenced the way X rays are now used and administered?

Frontiers of Technology: Prenatal Testing

Developments in medical technology now make it possible for doctors to detect genetic disorders like Down syndrome even before the baby is born. Tests for chromosomal abnormalities are generally offered to women over the age of 35.

The most widely performed technique, called amniocentesis, involves drawing fluid from the sac surrounding the developing fetus with a syringe, using ultrasound for guidance (**Figure 9**). The fluid, called amniotic fluid, also contains cells from the developing fetus. When these cells are treated with special stains, the chromosomes can be made visible for microscopic examination. A camera mounted to the microscope is often used to take a picture of the chromosomes. Amniocentesis can be performed in the 11th week of pregnancy, although it is safer to wait until about 14 weeks.

A complementary technique, called karyotyping, discussed earlier in the chapter, compares the number, size, and shape of homologous chromosomes. A chromosome count of 47, for example, would indicate the existence of a nondisjunction disorder. By comparing homologous chromosomes, physicians can identify the specific disorder or syndrome.

Another technique, chorionic villus sampling (CVS), draws cells from the outer membrane (chorion) surrounding the embryo. This procedure can be performed as early as eight weeks into pregnancy.

For women under 35, a more recent test called the multiple marker screen (MMS) identifies women who may be at higher risk of having a baby with a genetic disorder. The MMS is performed at 15 to 18 weeks of pregnancy, using the mother's blood. One such marker is a substance called alphafetoprotein, or AFP. Elevated levels of AFP in the mother's blood may indicate neural tube defects in the baby that result in malformations of the baby's spine or skull. Amniocentesis is recommended to verify the AFP levels. An ultrasound is performed to confirm the age of the baby and physical formation.

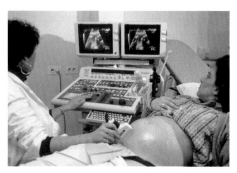

Figure 9
Ultrasound can be used to locate the position of the fetus within the uterus. It uses the energy from sound waves bouncing off the fetus to make its image.

SUMMARY Nondisjunction

1. Nondisjunction occurs when two homologous chromosomes move to the same pole during meiosis. In humans, this produces gametes with 22 and 24 chromosomes.
   - trisomy—a zygote contains 47 chromosomes—Down syndrome or Klinefelter syndrome
   - monosomy—a zygote contains 45 chromosomes—Turner syndrome

2. A karyotype chart is a picture of chromosomes arranged in homologous pairs in descending order by size, with the sex chromosomes placed last.

3. Amniocentesis involves withdrawing amniotic fluid, which can then be karyotyped to look for nondisjunction disorders.

Section 3.9 Questions

Understanding Concepts

1. Use a diagram to illustrate how nondisjunction in meiosis I differs from nondisjunction in meiosis II.

Making Connections

2. State advantages and disadvantages of ultrasound and amniocentesis.

3. Amniocentesis, like many other advances in reproductive technology, raises many moral and ethical questions. The knowledge of a disorder can help the parents and medical staff prepare for the birth of the child, and can help the parents prepare for the needs of a child with a genetic disorder. Do you think the technique should be used? Would you place any limits or restrictions on its use?

4. Ultrasound, amniocentesis, chorionic villus sampling, and the multiple marker screen test all give information about the fetus. Research to find out how improved scientific understanding about fetuses and amniotic fluid contributed to the development of each technique.
Follow the links for Nelson Biology 11, 3.9.

GO TO www.science.nelson.com

5. Do some research about Dr. Uchida's work, or find out about other Canadian contributions to research and technology in genetics.
Follow the links for Nelson Biology 11, 3.9.

GO TO www.science.nelson.com

Key Expectations

Throughout this chapter, you have had opportunities to do the following:

- Demonstrate an understanding of the process and importance of mitosis (3.1, 3.2, 3.3).
- Select appropriate instruments and use them effectively and accurately in collecting observations and data (3.3).
- Express the result of any calculation involving experimental data to the appropriate number of decimal places or significant figures (3.3).
- Select and use appropriate modes of representation to communicate scientific ideas, plans, and experimental results (3.3, 3.7).
- Locate, select, analyze, and integrate information on topics under study, working independently and as part of a team, and using appropriate library and electronic research tools, including Internet sites (3.4, 3.5, 3.9).
- Explain the process of meiosis in terms of the replication and movement of chromosomes (3.6, 3.7, 3.8).
- Describe genetic disorders in terms of the chromosomes affected, physical effects, and treatment (3.9).
- Explain the process of meiosis through their own investigations (3.6, 3.7).
- Organize data that illustrate the number of chromosomes in haploid cells and diploid cells, and the number of pairs of chromosomes in diploid cells, that occur in various organisms before, during, and as a result of meiosis (3.7).
- Describe and analyze examples of genetic technologies that were developed on the basis of scientific understanding (3.4, 3.9).
- Identify and describe Canadian contributions to knowledge about genetic processes (3.9).

Key Terms

asexual reproduction
autosomes
biotechnology
blastula
centrioles
centromere
chromatid
chromatin
crossing over
cytokinesis
diploid
Down syndrome
enucleated
fertilization
gametes
gametogenesis
genetic engineering
haploid
homologous
   chromosomes

interphase
karyotype chart
Klinefelter syndrome
meiosis
metastasis
mitosis
monosomy
oocytes
ootid
polar bodies
sex chromosomes
sexual reproduction
spermatocytes
spindle fibres
synapsis
tetrad
totipotent
trisomy
Turner syndrome
zygote

Make a Summary

Sketch the processes of meiosis and mitosis and show the differences between them. Label the sketch with as many of the key terms as possible. Check other sketches and use appropriate designs to make your sketch clear.

Reflect on your Learning

Revisit your answers to the Reflect on Your Learning questions at the beginning of this chapter.

- How has your thinking changed?
- What new questions do you have?

Understanding Concepts

1. The four stages of cell division are shown in **Figure 1**. Label each stage correctly:

   prophase, metaphase, anaphase, telophase

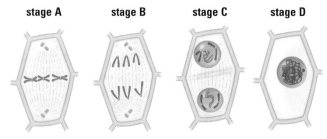

stage A   **stage B**   **stage C**   **stage D**

Figure 1

2. **Figure 2** shows plant and animal cells during cell division.
   (a) Identify each cell as either a plant or an animal cell. Justify your answer.
   (b) Identify the phases of cell division.

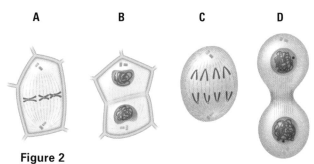

   A   B   C   D

Figure 2

3. Consider what would happen if you remained a single cell for your entire life. How would your life as a single-cell organism differ from that of a multicellular organism?

4. What evidence suggests that one of your nerve cells carries the same number of chromosomes and the same genetic information as one of your muscle cells?

5. Explain why orchids cloned from tissue cultures are so similar.

6. Explain why a better understanding of the mechanism of cell division may enable scientists to regenerate limbs.

7. Explain why the formation of calluses on the hands provides evidence that cell division can be stimulated by cell damage.

8. Explain why sexual reproduction promotes variation.

9. Indicate which of the following body cells would be capable of meiosis:
   (a) brain cells
   (b) fat cells
   (c) cells of a zygote
   (d) sperm-producing cells of the testes

10. **Figure 3** shows a cell with 36 chromosomes undergoing meiosis.
    (a) How many chromosomes would be in each cell during stage B?
    (b) How many chromosomes would be in each cell during stage C?
    (c) In which stage(s) would you find a cell with a diploid chromosome number?

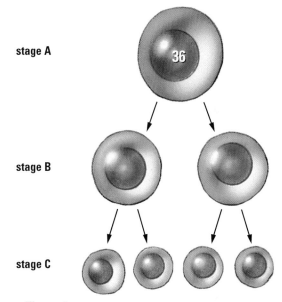

stage A

stage B

stage C

Figure 3

11. Use **Figure 4** to answer the following questions:
    (a) How many chromosomes were in the sperm cell?
    (b) Explain how it is possible for the sperm cell to be produced from a cell that has 46 chromosomes.
    (c) How many homologous pairs of chromosomes would be in the zygote if it were a female?
    (d) How many chromosomes would be in each cell following mitosis?

Fertilization in Humans

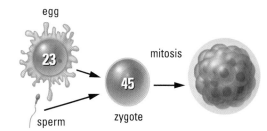

egg

sperm

zygote

mitosis

Figure 4

12. Use **Figure 5** to answer the following questions:
    (a) Identify the gender of the person.
    (b) Could this karyotype be taken from a human sperm cell? Give reasons for your answer.
    (c) Identify any disorder shown in the karyotype chart.

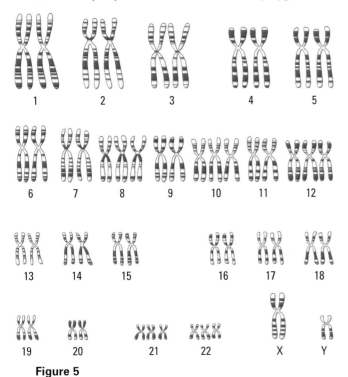

Figure 5

13. Explain how it is possible to produce a trisomic XXX female.

14. Diagram the kind of nondisjunction that would cause a male and female each with an abnormal number of chromosomes to produce an XYY offspring.

Applying Inquiry Skills

15. Some herbicides kill weeds by increasing the rate at which they grow.
    (a) Design an experiment that would test this hypothesis.
    (b) What is the control used in your experiment?
    (c) Identify the manipulated (independent) variable and responding (dependent) variable for your experiment.
    (d) How will you measure the rate of cell division?

16. **Table 1** shows data collected from two different fields of view while examining hamster embryo cells. The number of cells found in each of the cell phases was recorded. It took 660 min to complete one cycle from interphase to interphase.

Table 1

Cell phase	Area #1	Area #2	Total cell count	Time spent in phase
interphase	91	70	?	?
prophase	10	14	?	?
metaphase	2	1	?	?
anaphase	2	1	?	?
telophase	4	4	?	?

(a) Copy the table into your notebook and complete the calculations. To calculate the time spent in, for example, interphase:

$$\frac{\text{Number of cells in interphase}}{\text{Total number of cells counted}} = \frac{\text{Time spent in phase}}{\text{Total time of cycle (660 min)}}$$

(b) Using the data provided, draw a circle graph showing the amount of time spent in each phase of the cell cycle.

17. Use **Figure 6** to answer the following questions:
    (a) Which area of the body shows the most growth after birth?
    (b) What might account for the rapid change in the development of the heart beginning at about 12 years of age?

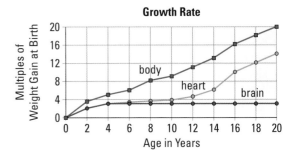

Figure 6

18. Identify one advantage of using a cutting instead of using seeds to grow a new plant.

19. A researcher compared the growth rate of malignant cells and normal cells in mice. The number of cells present in an area of 1 μm² was counted every two days over a period of two months.
    (a) Which of the graphs in **Figure 7** (page 126) represents the data collected? Explain your answer.
    (b) Why did the researcher need both malignant cells and normal cells?
    (c) Suggest three environmental conditions that the researcher needed to control while culturing these cells.

Graph A

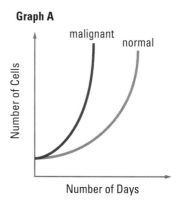

Graph B

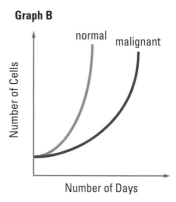

Graph C

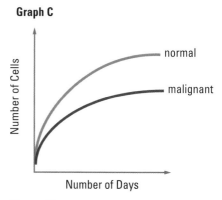

Figure 7

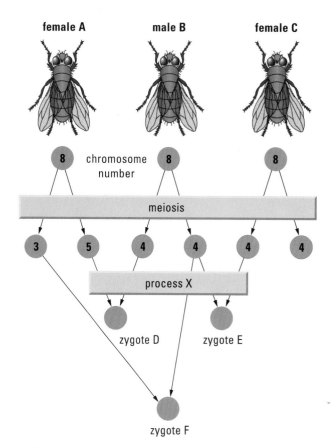

Figure 8

Making Connections

21. Copy **Table 2**. Place a check mark (✓) beside the statements that you believe are always, or almost always, true for fraternal and identical twins. Give reasons for each of your predictions.

Table 2

Descriptor	Fraternal twins	Identical twins
They have the same blood type.	?	?
They are the same sex.	?	?
They like the same hockey team.	?	?
They are the same weight.	?	?
They have the same hair colour.	?	?
They know what the other one is thinking.	?	?

20. Fruit flies normally have eight chromosomes. Flies with fewer chromosomes die before maturity. **Figure 8** shows the process of meiosis in three fruit flies.
    (a) In which parent did nondisjunction take place?
    (b) How many chromosomes would be in zygotes D, E, and F?
    (c) What is happening during process X?
    (d) Which zygote would most likely be healthy?
    (e) Name the conditions that the other zygotes have.

22. The cells in the outermost layer of your skin have no nucleus. A moisturizer claims to restore and rejuvenate these cells, meaning that new cells are grown.
    (a) Would these skin cells be capable of producing other skin cells?
    (b) How would you go about testing the claim?

23. Ionizing radiation, such as X rays, breaks apart chromosomes. **Figure 9** shows the effects of irradiation on a cell in metaphase.

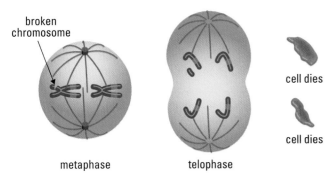

Figure 9

Food scientists will sometimes irradiate fruit and vegetables to improve shelf life. How does irradiation help preserve food?

24. A girl with malignant melanoma, a cancer of the pigment cells of the skin, had one of her cancer cells grafted onto the skin of her mother. Scientists believed that the invading cancer cell from the dying child would stimulate the production of antibodies in the mother. The invading cancer cells would be destroyed by the antibodies, which the scientists would then harvest and transfuse into the blood of the dying girl. The experiment did not work. The young girl died a few days after the procedure. However, the true failure of the experiment was not known until some months later. The mother developed skin cancer, which killed her. This situation raises moral and ethical questions. How might the following factors affect your decision about treatment?
    • risks versus benefits
    • quality of life during and after treatment

25. According to one report, the incidence of infertility appears to be increasing in countries like Canada. The sperm count in males has fallen more than 30% in the last half century and is continuing to fall. Although there is no explanation for this phenomenon at present, environmental pollution is suspected. Suggest other reasons for decreased fertility in males and females.

26. If nondisjunction disorders could be eliminated by screening sperm and egg cells, sperm and egg banks could all but eliminate many genetic disorders. Comment on the social, moral, and ethical implications to society of the systematic elimination of genetic disorders in humans.

27. A technique called egg fusion involves the union of one haploid egg cell with another. The zygote contains the full $2n$ chromosome number and is always a female. Discuss the implications to society if this technique were applied to humans.

Exploring

28. Progeria is a rare condition that causes premature aging and early death. A 7 year old with this genetic disease can resemble an individual of 70 or 80 years of age. Affected individuals usually die by age 14. Prepare a report outlining the symptoms and treatments for this condition, or for another condition mentioned in this chapter (for example, cystic fibrosis or cancer).
    Follow the links for Nelson Biology, Chapter 3 Review.
    GO TO www.science.nelson.com

29. On the Internet, browse to find Web sites that provide online karyotyping activities.
    Follow the links for Nelson Biology 11, Chapter 3 Review.
    GO TO www.science.nelson.com

Genes and Heredity

Have you ever been able to identify a stranger as a member of a particular family? Red hair, high cheekbones, or a prominent nose can often be traced through a family's lineage. The observation that a young child resembles her grandmother suggests that physical characteristics are inherited. Similar observations can be made in the world of plants and animals. Flowers with white petals most often produce flowers with white petals. Palomino horses most often produce other palomino horses. Characteristics appear to be repeated from generation to generation. The study of genetics examines the inheritance of biological traits. The passing of traits from parents to offspring is called **heredity**.

Your biological traits are controlled by genes located on the chromosomes that are found in every cell of your body. You inherited half of your chromosomes from your mother and the other half from your father. Your traits are a result of the interactions of the genes from both parents; however, your genes and traits are uniquely your own. It has been estimated that more than 8 million combinations are possible from the 23 chromosomes inherited from each parent. The more than 8 million combinations from each parent will produce more than $(8 \times 10^6)(8 \times 10^6)$ or 6.4×10^{13} (64 trillion) possible combinations for offspring. Your genetic makeup is one of those combinations.

Reflect on your Learning

There are many examples of diversity that do not fit the expected continuity pattern of biological traits.

1. How is it possible for two parents with dark brown hair to have a child with red hair?

2. How is it possible for parents who both have type A blood to have a child who has type O blood?

3. Why might a genetic trait, such as baldness, skip a generation?

4. How much of Brett Hull's ability to score goals comes from heredity? Are goal scorers born or can they be taught?

5. Keifer and Donald Sutherland are successful actors. Are there genes that might influence aptitude for a particular profession?

6. How much of Dawn Langstroth's singing ability can be attributed to genetics?

Throughout this chapter, note any changes in your ideas as you learn new concepts and develop your skills.

In this chapter, you will be able to

- describe and explain how Mendel formulated his laws of heredity;
- explain the concepts of dominance, codominance, incomplete dominance, and recessiveness;
- predict the outcome of various genetic crosses and solve genetic problems;
- compare data from laboratory investigations on genetic crosses and present the results;
- examine and describe contributions made by Canadians to the advancement of scientific understanding of heredity.

Identifying Similarities and Differences

- Observe the people in **Figures 1** to **4** and identify physical traits, such as eye colour, eye shape, freckles, and nose length and width, in which there are family resemblances.
  (a) Present the traits on a chart.
  (b) Indicate which traits you believe are inherited.

heredity: the passing of traits from parents to offspring

Figure 1
Brett Hull scoring a goal

Figure 2
Bobby and Brett Hull starred in the National Hockey League, becoming the first father and son to win the Hart Trophy.

Figure 3
Keifer Sutherland and his father, Donald, are successful actors.

Figure 4
Anne Murray performs with her daughter, Dawn Langstroth.

4.1 Early Beliefs and Mendel

The idea that biological traits are inherited existed long before the mechanisms of inheritance and gene interaction were understood. Stone tablets crafted by the Babylonians 6000 years ago show the pedigrees of successive generations of champion horses. Other carvings from the same period show the artificial cross-pollination of date palms. Early records kept by Chinese farmers provide evidence of methods used for improving different varieties of rice. The selection of desired traits was based on keen observation and, to a great extent, trial and error.

Early naturalists made assumptions about some incredible cross-species **hybrids**. The giraffe, for example, was thought to have been a cross between a leopard and a camel. (The fact that camels and leopards are not compatible did not seem to deter the theory.) The banana was thought to have been a hybrid of the acacia and the palm.

Humans also mated selectively to produce desirable traits. The ancient Egyptians encouraged the intermarriage of royalty to preserve bloodlines. For example, Cleopatra married her younger brother. Plato, a Greek philosopher of the early 4th century B.C., called for the segregated mating of the elite. To maintain a line of strong warriors, the ancient Spartans practised infanticide, killing babies with undesirable characteristics.

Pioneer of Genetics: Gregor Mendel

One of the classic scientific experiments on inheritance was performed by an Austrian monk named Gregor Mendel (1822–1884) during the mid-19th century (**Figure 1**). Mendel's work with garden peas not only explained the mechanism of gene inheritance in plants, but provided a basis for understanding heredity in general. When Mendel's work was rediscovered many years later, it provided the missing piece in the theory of how organisms survive and reproduce.

Why did Mendel choose the garden pea on which to perform his work? First, he observed that garden peas have a number of characteristics that are expressed in one of two ways (see **Figure 2**). For example, some garden peas produce green

hybrids: offspring that differ from their parents in one or more traits. Interspecific hybrids result from the union of two different species.

Figure 1
Gregor Mendel was an Austrian monk whose experiments with garden peas laid the foundation for the science of genetics.

Figure 2
The seven characteristics Mendel studied in his experiments with garden peas. Flower colour and seed colour are correlated. Plants with white flowers produce seeds that are yellow, and plants with violet-purple flowers produce seeds that are green.

characteristics	dominant trait	recessive trait		dominant trait	recessive trait
seed shape	round	wrinkled	flower position	side of stem	end of stem
seed colour	yellow	green			
pod shape	inflated	constricted	stem length	tall	short
pod colour	green	yellow			
flower colour	purple	white			

seeds (peas), while others produce yellow seeds. Some plants are tall, while others are short. Mendel also noticed different flower positions on the stem and different flower colours. The fact that there were only two ways for each trait to be expressed would make it easy to see which traits had been inherited from generation to generation.

A second reason for using garden peas is the way the plant reproduces. Garden peas are both self-fertilizing and cross-fertilizing. Fertilization occurs when pollen produced by the stamen, the male part of the plant, attaches to the pistil, the female part (**Figure 3**). The pistil consists of the stigma, style, and ovary. The pollen grains fertilize the egg cells in the ovary. This process is called pollination. In self-fertilization, pollination occurs within one flower and the traits of the offspring are easily predicted. Mendel cross-pollinated the pea plants rather than allowing them to self-pollinate. He made sure to use pure-breeding plants, that is, plants that always produce identical offspring. For example, tall plants produce only tall plants. If any offspring in any generation was not tall, then Mendel did not consider the parent plant to be pure and did not use it in his experiment. He transferred the pollen from one plant to the pistil of another plant, thus combining the male and female sex cells of different plants. To ensure that the recipient plant didn't pollinate itself, he first removed its anthers. The pollen present then had to originate from the donor plant, resulting in seeds produced from cross-pollinated plants (**Figure 4**).

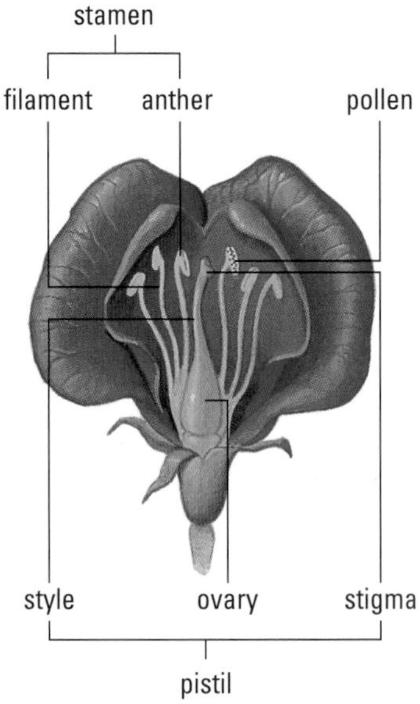

Mendel's Experiments

Mendel's predecessors had hypothesized that the crossing of different traits would create a blend. According to this theory, crossing a plant that produced round seeds with one that produced wrinkled seeds would result in slightly wrinkled seeds. However, Mendel proved that this was not the case. When he crossed the pollen from a plant that produced round seeds with the eggs of one that produced wrinkled seeds, the offspring were always round. Did this mean that the pollen determines the seed coat? To test this idea, Mendel crossed the pollen from a wrinkled seed plant with the eggs from a round seed plant. Once again, all the offspring were round. In fact, the round trait dominated, regardless of whether the trait came from the male (pollen) parent or the female (seed) parent.

Mendel repeated the procedure for other characteristics. He discovered that one trait always dominated another, whether the sex cell came from the male or female part of the plant. Tall plants produced tall offspring when cross-pollinated with short plants; likewise, plants that had yellow seeds produced offspring with yellow seeds when cross-pollinated with plants that had green seeds. Mendel reasoned that things called factors control the traits of a plant. The factors were later

Figure 3
The structure of a flower

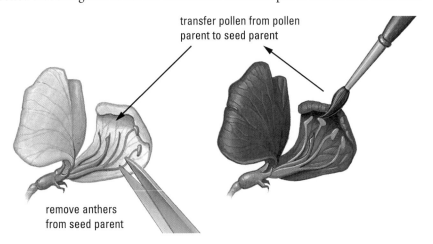

remove anthers from seed parent

transfer pollen from pollen parent to seed parent

Figure 4
The donor plant is also known as the pollen parent and the recipient is also known as the seed parent.

called genes. He assumed that the genes control the inheritance of particular traits, such as seed colour and plant stem height. He also realized that there are alternate forms of a gene. Today, the alternate forms of a gene are called **alleles**. Green and yellow are expressions of the different alleles for seed colour. Tall stems and short stems are expressions of the different alleles for stem height. In garden peas, the traits that were expressed most often were considered to be **dominant** and those expressed less frequently were **recessive**. The allele for a yellow seed is dominant over the allele for a green seed; the allele for tall stems is dominant over the allele for short stems.

Mendel cross-pollinated many plants and kept track of all the results. For each type of cross, he recorded the number of offspring that exhibited the dominant trait versus the recessive trait. He created a system of symbols to show what traits were passed to offspring. In this system, letters are used to represent traits. Uppercase letters stand for dominant traits, and lowercase letters stand for recessive traits. For the dominant trait of yellow seeds, Y represents the allele for yellow seeds; y represents the allele for green seeds, the recessive trait. Today, Mendel's system is still in use.

Mendel continued his experimentation by crossing two hybrid plants with round seeds from the first generation. He referred to the first generation as filial generation one, or F_1 generation. The word *filial* comes from the Latin for son. Both of these F_1 plants contain R and r alleles, one from each of their parents. This makes them hybrids. Remember, the R represents the round allele, while the r represents the wrinkled allele.

You might predict an equal number of round and wrinkled offspring in the second, or F_2, generation. However, this is not the ratio Mendel discovered when the two hybrids were crossed. He was astonished to find that 75% of the offspring expressed the dominant round trait, while only 25% expressed the wrinkled trait. How can these results be explained?

Figure 5 shows what happens when the sex cells, or gametes, from the F_1 generation recombine to form an F_2 generation. All members of the F_1 generation are round, but wrinkled offspring appear during the F_2 generation. Any members of the F_2 generation with an R allele will be round because the round allele is dominant over the wrinkled allele. To be wrinkled, the offspring must

Meiosis occurs. Each gamete has one of the homologous chromosomes.

F_2 generation inherits alleles from the gametes of the F_1 generation.

Figure 5
The result of crossing two hybrid pea plants with round seeds from the first generation

Creating a Personal Profile

Table 1 lists human traits controlled by dominant and recessive alleles.

Table 1

Trait	Dominant	Recessive
eye colour	brown or black or green	blue or grey
hair colour	brown or black	blonde or red
hairline	pointed on forehead	straight across forehead
freckles	present	absent
earlobe	suspended	attached to head
hair texture	curly	straight
eyesight	near or farsighted	normal vision
eyelashes	long	short
nose line	convex tip	concave or straight
fingers	6 fingers	5 fingers
Rh blood factor	positive *Rh* factor	negative *Rh* factor
ear rim	curled rim	not curled rim
thumb joint	last joint bends out	last joint is straight
finger hair	present	absent
folded hands	left thumb over right	right thumb over left
tongue rolling	can be rolled into U shape	cannot be rolled
clenched fist	two wrist cords	three wrist cords
chin dimple	dimple in middle	no dimple
blood type	type A, B, AB	type O
eyes	astigmatism	no astigmatism

Copy **Table 2**. Use information in **Table 1** to complete your personal profile. Which additional traits can you include?

Table 2

Trait (use the letter indicated)		Appearance or physical condition	Dominant or recessive	Possible genetic makeup
eye colour	*E/e*	?	?	?
hairline	*L/l*	?	?	?
earlobe	*T/t*	?	?	?
ear rim	*R/r*	?	?	?
freckles	*F/f*	?	?	?
thumb joint	*J/j*	?	?	?
finger hair	*P/p*	?	?	?
tongue rolling	*Y/y*	?	?	?
folded hands	*D/d*	?	?	?
nose line	*N/n*	?	?	?
hair colour	*H/h*	?	?	?
chin dimple	*G/g*	?	?	?
clenched fist	*K/k*	?	?	?

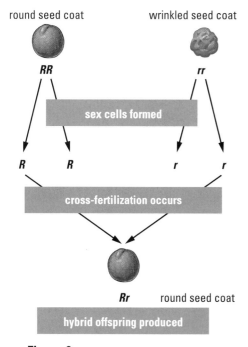

round seed coat   wrinkled seed coat

RR   **rr**

sex cells formed

R   **R**   **r**   **r**

cross-fertilization occurs

Rr   round seed coat

hybrid offspring produced

Figure 6
Cross-fertilization results from crossing a pea plant that is pure breeding for round seeds and a pea plant with wrinkled seeds.

segregation: the separation of paired alleles during meiosis

genotype: the alleles an organism contains

have two recessive *r* alleles. **Figure 5** illustrates why the expected ratio in the F_2 generation is three round to one wrinkled.

The following is a summary of Mendel's experiments:

1. Inherited characteristics are controlled by factors, now known as genes, that occur in pairs. Each member of a pair of genes is presently referred to as an allele. A self-fertilizing pea plant with round seeds receives an allele for round seeds from the pollen and a complementary allele for round seeds from the egg. Similarly, a self-fertilizing pea plant with wrinkled seeds has two alleles for wrinkled seeds. During cross-fertilization each parent contributes one of its alleles (see **Figure 6**).

2. One factor, or allele, masks the effect or expression of another. This process is known as the principle of dominance. The allele for round seeds masks the effect of the allele for wrinkled seeds. The dominant allele is indicated by an uppercase letter, in this case, *R*. The recessive allele is designated with the lowercase letter *r*.

3. A pair of factors, or alleles, separate or segregate during the formation of sex cells. This process is often referred to as the law of **segregation**. Mendel concluded that sex cells must contain only one member of the pair of factors, or alleles. Mendel came to the correct conclusion, despite the fact that he did not know the mechanism of meiosis. The principles of sex cell formation were not discovered until 25 years after Mendel had completed his experiments.

Significance of Mendel's Work

You may be surprised to learn that you already know more about genetics than Gregor Mendel did. Mendel did not know that what he called factors were located on chromosomes. He had not observed cells during cell division. However, Mendel was one of the first biologists to perform careful experiments and to record and interpret quantitative data. He often repeated procedures many times to support his conclusions. Prior to Mendel, the application of mathematical concepts to biology was not common. Gregor Mendel is considered to be the father of genetics. Even today, biologists refer to the study of a particular type of inheritance as Mendelian genetics.

SUMMARY   **Mendel's Laws of Heredity**

1. Each parent contributes one allele during cross-fertilization. If a pure-breeding plant self-fertilizes, each offspring receives two copies of the same allele.

2. The dominant allele is always expressed when the recessive allele is present.

3. Each pair of alleles segregates during the formation of sex cells.

4.2   Single-Trait Inheritance

Terms Used in Genetics

The following terms will help you read about and describe heredity:

Genotype refers to the alleles that an organism contains for a particular trait. One allele for each trait is inherited from each parent. Since the alleles occur in pairs, various combinations are possible. A tall-stem pea plant could have two different genotypes, *TT* and *Tt*. A short-stem pea plant can have only

one genotype, *tt*, since the allele for tall stems (*T*) is dominant over the allele for short stems (*t*).

Phenotype refers to the observable traits of an individual. Since a pea plant can be tall or short, there are only two possible phenotypes for this characteristic. The tall phenotype may have two different genotypes, *TT* or *Tt*.

Homozygous is a term used to describe the genotype of an organism that contains two alleles that are the same. The homozygous condition for a tall-stem plant would be *TT*. The homozygous condition for a short-stem plant would be *tt*. These are the pure-breeding individuals.

Heterozygous is a term used to describe the genotype of an organism that contains two different alleles. The only heterozygous condition possible for stem length in a garden pea is *Tt*. The phenotype for this heterozygous genotype is a tall-stem plant, which is a hybrid.

The cross between a *TT* tall pea plant and a *tt* short pea plant is a **monohybrid cross** because only one trait, plant height, is being tested. Crossing members of the F$_1$ generation, *Tt* tall pea plants, to find an F$_2$ generation is another example of a monohybrid cross.

A special chart, referred to as a **Punnett square**, helps us organize the results of a cross between the sex cells of two individuals. The Punnett square resembles the chart for an *X*s and *O*s game. The chart can be used to predict the genotypes and the phenotypes of the offspring.

Sample Problem 1

Consider a cross between a pea plant that is heterozygous for round seeds and a pea plant that has wrinkled seeds. The allele for round seeds is dominant over that for wrinkled seeds. *R* can be used to indicate the round dominant allele and *r* can be used to represent the wrinkled recessive allele. Determine the genotypes of the offspring.

Solution

Because the plant with the round seeds is heterozygous, the genotype for the plant is *Rr*. Since the gene for wrinkled seeds is recessive, the plant with wrinkled seeds must contain two wrinkled alleles. The genotype of this plant is *rr* (**Figure 1**).

The symbols for the gametes are written across the top and along the left side of the square. By drawing a line from each gamete at the top and another line from each gamete along the side, the possible allele combinations of the offspring can be determined (see **Figure 2**, page 136).

phenotype: the observable traits of an organism that arise because of the interaction between genes and the environment

homozygous: a genotype in which both alleles of a pair are the same

heterozygous: a genotype in which the alleles of a pair are different

monohybrid cross: a cross that involves one allele pair of contrasting traits

Punnett square: a chart used by geneticists to show the possible combinations of alleles in offspring

---

DID YOU **KNOW ?**

In 1910, Reginald Punnett became the first professor of genetics at Cambridge University. As a result of further investigations of Mendel's work, he discovered some basic principles of genetics, including sex linkage and sex determination. He also discovered some exceptions to Mendel's rules. He invented a table, now known as the Punnett square, for displaying the results of a genetic cross.

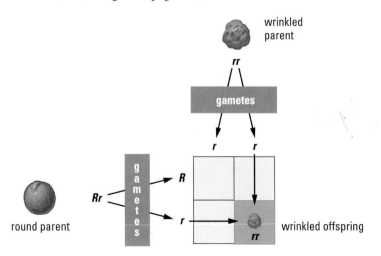

Figure 1
A partially completed Punnett square for a cross between a pea plant that is heterozygous for round seeds and a pea plant that has wrinkled seeds. The *rr* genotype shown in the chart is only one of the four possible combinations. When the chart is completed, all the possible genotypes are shown.

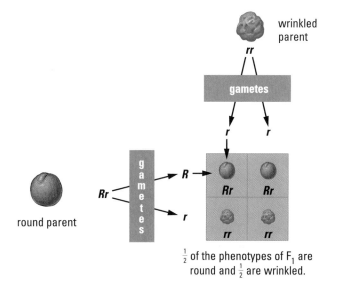

Figure 2
A completed Punnett square for a cross between a heterozygous round-seed pea plant and a wrinkled-seed pea plant

round parent

$\frac{1}{2}$ of the phenotypes of F$_1$ are round and $\frac{1}{2}$ are wrinkled.

Sample Problem 2

From the Punnett square, it is also possible to determine the probability of each phenotype occurring in the offspring.

Solution

For the cross shown in **Figure 3**, the probability that an offspring will be round is 75%, and the probability it will be wrinkled is 25%.

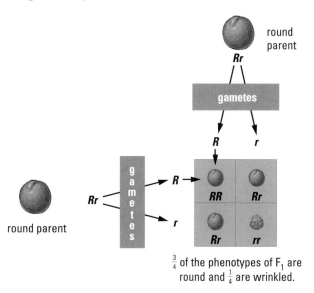

Figure 3
A Punnett square showing the results of a cross between two heterozygous plants with round seeds.

round parent

$\frac{3}{4}$ of the phenotypes of F$_1$ are round and $\frac{1}{4}$ are wrinkled.

Sample Problem 3

Consider a cross where only the offspring are observable. It is still possible to determine the genotypes of the parents in many of the cases even if the parents are unknown.

Offspring phenotype	Numbers
round-seed peas	5472
wrinkled-seed peas	1850

Solution

Determine the ratio of round plants to wrinkled plants. You will probably need to round out the ratio.

$$\frac{\text{round}}{\text{wrinkled}} = \frac{5472}{1850} \doteq \frac{3}{1}$$

Now list the possible genotypes for each phenotype.

Phenotype	Genotype
round-seed peas	*RR* or *Rr*
wrinkled-seed peas	*rr*

The only known genotype of the offspring is that of the wrinkled peas. Place *rr* in one of the boxes for the Punnett square. (Any box will work.) This means that each parent contributed a wrinkled allele (**Figure 4**). You also know that the other offspring are round (**Figure 5**). This indicates that the other alleles from the parents must be round. Both parents had round seeds (**Figure 6**).

Sample Problem 4

A plant that is homozygous for purple flowers is crossed with a plant that has white flowers. If the purple condition is dominant over the white condition, what are the genotypes and phenotypes of the F$_1$ generation?

Solution

parent phenotypes: homozygous purple flowers × white flowers
parent genotypes: *PP*   ×   *pp*
parent gametes: *P* or *P*   ×   *p* or *p*

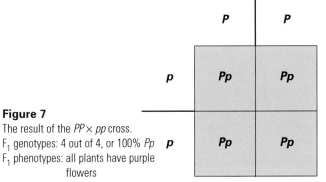

Figure 7
The result of the *PP* × *pp* cross.
F$_1$ genotypes: 4 out of 4, or 100% *Pp*
F$_1$ phenotypes: all plants have purple flowers

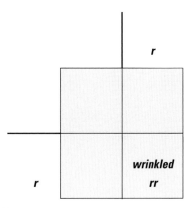

Figure 4

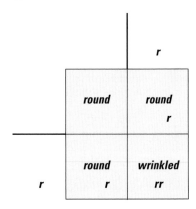

Figure 5

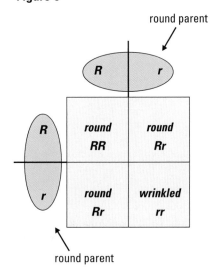

Figure 6

Practice

Understanding Concepts

1. In your own words, define the following terms: genotype, phenotype, homozygous, heterozygous, alleles, monohybrid cross, and Punnett square.

2. What is a purebred stock? Use the terms in question 1 in your explanation. Why was it important that Mendel use purebred plants in his experiments?

3. **Table 1** shows the results of Mendel's experiments.

Table 1

Trait	Alleles (DOMINANT/recessive)	Possible genotypes
plant height	TALL dwarf	*TT* or *Tt* *tt*
flower colour	PURPLE white	*PP* or *Pp* *pp*
flower position on stem	AXIAL (at branches) top	*AA* or *Aa* *aa*
pod colour	GREEN yellow	*GG* or *Gg* *gg*
pod shape	INFLATED constricted	*Il* or *Ii* *ii*
seed colour	YELLOW green	*YY* or *Yy* *yy*
seed shape	ROUND wrinkled	*RR* or *Rr* *rr*

For each of the crosses listed (a to e), create a Punnett square and determine the following information: parent phenotypes, parent genotypes, parent gametes, F_1 genotypes, and F_1 phenotypes.
(a) Two heterozygous tall parents are crossed.
(b) A heterozygous tall plant is crossed with a dwarf plant.
(c) Two plants that are heterozygous for purple flowers are crossed.
(d) A plant that is homozygous for green pods is crossed with a plant that has yellow pods.
(e) A plant that is homozygous for round seeds is crossed with a plant that is heterozygous for round seeds.

4. In guinea pigs, the allele for a black coat is dominant over the allele for a white coat. A black guinea pig was crossed with a white guinea pig. All F_1 offspring have black coats.
(a) Describe how you can determine whether or not the black parent is homozygous or heterozygous for the condition. Indicate the letter you will use to represent an allele.
(b) If 10 offspring were produced, indicate how many you would expect to have black coat colour, if the black parent were heterozygous.

Test Cross

You have probably heard someone referred to as the black sheep of the family. Do you know why black sheep are not preferred? Black wool tends to be brittle and is very difficult to dye. How can a sheep rancher avoid getting black sheep? Since the white condition is dominant, using a homozygous white ram (the male of the species) will ensure that all of the flock will have white hair. However, if the white ram is heterozygous, a number of black wool offspring may result. How can you know for sure if a ram is homozygous for the white phenotype?

A test cross is often performed to determine the genotype of a dominant phenotype. The test cross is always performed between the unknown genotype and a homozygous recessive genotype. In this case, the homozygous recessive individual would be a black ewe (the female of the species). So, in the cross

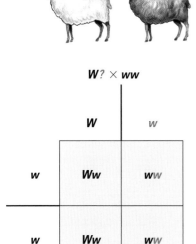

$W? \times ww$

	W	w
w	Ww	ww
w	Ww	ww

Half of the offspring are
black and half are white.

$W? \times ww$

	W	W
w	Ww	Ww
w	Ww	Ww

All of the offspring
are white.

Figure 8
A test cross involves crossing an individual of unknown genotype, which shows a dominant trait, with an individual that has a recessive genotype. If any offspring show the recessive trait, then the individual must be heterozygous. If all offspring show the dominant trait, then the individual must be homozygous.

between a white ram whose genotype is unknown (either *WW* or *Ww*) and a black ewe (whose genotype must be *ww*), if any offspring show the recessive trait, then the individual must be heterozygous. If all offspring show the dominant trait, then the individual must be homozygous (**Figure 8**).

Sections 4.1–4.2 Questions

Understanding Concepts

1. For Labrador retrievers, black fur colour is dominant over yellow.
   (a) What would be the genotype of a homozygous black dog? a heterozygous black dog?
   (b) Could the heterozygous black dog have the same genotype as a dog with yellow fur? Explain.

Applying Inquiry Skills

2. A pea plant with round seeds is cross-pollinated with a pea plant that has wrinkled seeds. For the cross, indicate each of the following:
   (a) the genotypes of the parents if the round-seed plant is heterozygous
   (b) the gametes produced by the round- and wrinkled-seed parents
   (c) the genotypes and the phenotypes of the F_1 generation
   (d) the genotypes and phenotypes of the F_2 generation, if two round-seed plants from the F_1 generation are cross-pollinated

3. For Dalmatian dogs, the spotted condition is dominant over non-spotted.
   (a) Using a Punnett square, show the results of a cross between two heterozygous parents.
   (b) A spotted female Dalmatian dog mates with an unknown father. From the appearance of the pups, the owner concludes that the male was a Dalmatian. The owner notes that the female had six pups, three spotted and three non-spotted. What are the genotype and phenotype of the unknown male?

(continued)

4. For Mexican hairless dogs, the hairless condition is dominant over the condition of being hairy. A litter of eight pups is found; six are hairless and two are hairy. What are the probable genotypes of their parents?

Making Connections

5. A human neurological disorder referred to as Huntington's chorea is caused by a dominant allele. Because the allele doesn't express itself until a person reaches about 50 years of age, early detection has been difficult. In one family, a woman begins to show symptoms. Her father had Huntington's chorea, but her mother never developed the disorder. The woman's husband shows no symptoms, nor does anyone in his immediate family.
   (a) What is the genotype of the woman who has developed Huntington's chorea?
   (b) What is the probable genotype of the woman's husband?
   (c) If the man and woman have 6 children together, how many are likely to develop Huntington's chorea?

4.3 Selective Breeding

selective breeding: the crossing of desired traits from plants or animals to produce offspring that have one or several of the favoured characteristics

inbreeding: the process by which breeding is limited to a number of desirable phenotypes

hybridization: the mating of two different parents to produce offspring with desirable characteristics of both parents

Farmers and ranchers have long been using **selective breeding** processes to improve varieties of plants and animals. For example, early farmers would identify plants with desirable characteristics and select those plants to form the seed crop for the following year. Plants are selected based on traits such as flavour, yield, and hardiness to environmental conditions. Using this technique over many generations has resulted in rust-resistant wheat; sweet, full-kernel corn; and canola which germinates and grows rapidly in colder climates. Selective breeding of wild cabbage has produced green cabbage, red cabbage, broccoli, and cauliflower.

Have you ever heard the term *purebred*? Many dogs and horses are considered to be purebreds or thoroughbreds, which means they have a genetic line that has been closely regulated by **inbreeding**. Inbreeding is the mating of closely related individuals for the purpose of maintaining or perpetuating certain characteristics. This often means that similar phenotypes are selected for breeding. The desirable traits vary from breed to breed. For example, Irish setters are chosen for their long, narrow facial structure and long, wispy hair (**Figure 1**). Another dog, the pit bull, was inbred for fighting; quick reflexes and strong jaws were chosen as desirable phenotypes. Some geneticists have complained that inbreeding has caused problems for the general public as well as for the breed itself.

New varieties of plants and animals can be developed by **hybridization**. This process is the opposite of inbreeding. Rather than involving plants or animals with similar traits, the hybridization technique involves combining desirable but different traits from different parents to have all of these favourable traits present in the offspring. Corn has been hybridized extensively. The hybrids tend to be more vigorous than either parent. **Figure 2** shows the most common method used.

Figure 1
Irish setter

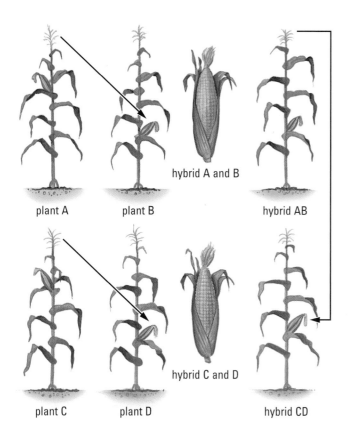

plant A plant B hybrid AB

hybrid A and B

hybrid ABCD

plant C plant D hybrid CD

hybrid C and D

Figure 2
Pollen from plant A fertilizes plant B, and pollen from plant C fertilizes plant D. The hybrids formed are then crossed to produce a final hybrid that contains genetic information from all four parent strains.

Research in Canada: The Plant Breeders

Canada boasts a long history of contributions to plant genetics. One of our first great plant breeders was Sir Charles Saunders. Born in Toronto in 1837, Saunders is noted for developing and introducing Marquis wheat, a high-quality wheat used in bread making. Marquis is a hybrid derived from a cross between Red Calcutta and Red Fife varieties of wheat. Prior to the Marquis strain, Red Fife wheat was used in most of Canada because it is resistant to diseases such as rust and smut, and yields more than other wheats, but is slow to mature. Red Calcutta, an import from India, matures two to three weeks earlier than Red Fife. The Marquis hybrid matures at least a week earlier than Red Fife and provides good yields.

Richard Downey, formerly of the University of Saskatchewan, and Baldur Steffasson, from the University of Manitoba, are recognized as world leaders in plant genetics. Downey was born in Saskatoon in 1927 and Steffasson was born in Manitoba in 1917. Their research has helped bring wealth to their respective provinces. Downey and Steffasson are known for developing a type of rapeseed with low levels of erucic acid and glucosinolate. The result was a high-quality oil-seed crop, known worldwide as canola (**Figure 3**).

Figure 3
Canola field

Pedigree Charts

Pedigree analysis is another way to solve genetic puzzles. This approach is useful when the traits of many generations of offspring have been recorded. A **pedigree chart** can be used to trace the passing of an allele from parents to offspring.

A pedigree chart contains a number of symbols that identify gender, relationships between individuals, and whether an individual expresses a trait or carries the allele as part of a heterozygous genotype. Some commonly used symbols are shown in **Figure 4**, page 142.

pedigree chart: a graphic presentation of a family tree that permits patterns of inheritance to be followed for a single gene

Pedigree Symbols

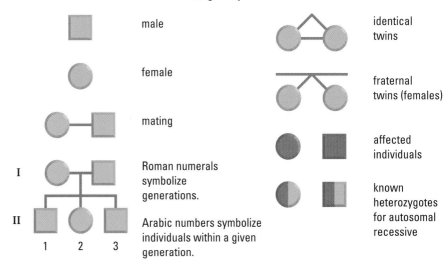

male

female

mating

I Roman numerals symbolize generations.

II Arabic numbers symbolize individuals within a given generation.

1 2 3

identical twins

fraternal twins (females)

affected individuals

known heterozygotes for autosomal recessive

Birth order within each group of offspring is drawn left to right, oldest to youngest.

Sample Pedigree

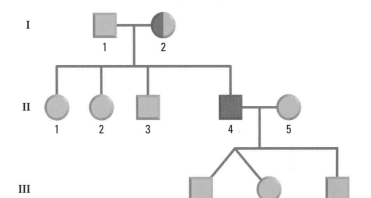

Figure 4
Squares represent males and circles represent females. Individuals that express the trait are shown in a coloured circle or square. If it is known, individuals that carry the allele as part of a heterozygous genotype are shown by partial colour or shading. A vertical line connects parents to offspring.

From the point of view of individual III - 1, the symbols represent the following relationships:

I - 1 = grandfather   I - 2 = grandmother
II - 1 and II - 2 = aunts   II - 3 = uncle   II - 4 = father   II - 5 = mother
III - 2 = fraternal twin sister   III - 3 = brother

Practice

Understanding Concepts

1. Copy **Figure 5** into your notebook. Indicate whether each family member is homozygous or heterozygous for shortsightedness, or homozygous for normal vision.

2. If couple 4 and 5 in row II had another child, what genotype might the child have? (Hint: What genotype is possible but not shown in the chart?) Would the child have normal vision or be shortsighted?

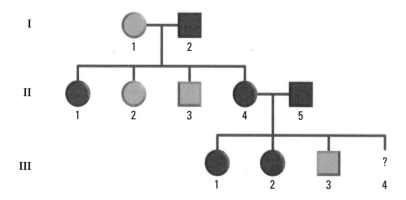

Figure 5
Pedigree chart that shows a family with the trait of shortsightedness. The allele for short-sightedness (S) is dominant over the allele for normal vision (s).

4.4 Multiple Alleles

For each of the traits studied by Mendel, there were only two possible alleles. The dominant allele controlled the trait. It is possible, however, to have more than two different alleles for one gene. In fact, there are many genes with multiple alleles.

Geneticists who study the tiny fruit fly called *Drosophila melanogaster* (**Figure 1**) have noted that many different eye colours are possible. The red, or wild type, is the most common, but apricot, honey, and white colours also exist. Although a fruit fly can have only two different alleles for eye colour at any one time, more than two alleles are possible. A fruit fly may have an allele for wild-type eyes and another for white. Its prospective mate may have an allele for apricot-coloured eyes and another for honey-coloured eyes. The dominance hierarchy is as follows: wild type is dominant over apricot, which is dominant over honey, which is dominant over white. In the case of multiple alleles, it is no longer appropriate to use uppercase and lowercase letters. Capital letters and superscript letters or numbers are used to express the different alleles and their combinations. Blood types are often denoted using superscripts. For example, blood type A, if homozygous, contains two A alleles, $I^a I^a$.

The dominance hierarchy and symbols for eye colour in *Drosophila* are shown in **Table 1**. For simplicity, the capital letter E is used for the eye colour gene and superscript numbers used to indicate the dominance hierarchy for the allele.

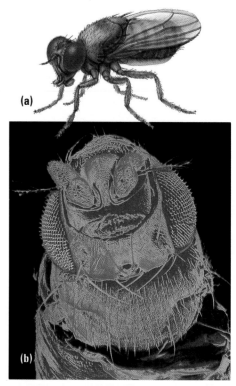

Figure 1
(a) *Drosophila melanogaster*, the fruit fly, is widely used for genetics studies.
(b) Wild type, or red, is the most common eye colour. The wild-type allele is the dominant allele.

Table 1: Dominance Hierarchy and Symbols for Eye Colour in *Drosophila*

Phenotype	Genotypes	Dominant over
wild type	E^1E^1, E^1E^2, E^1E^3, E^1E^4	apricot, honey, white
apricot	E^2E^2, E^2E^3, E^2E^4	honey, white
honey	E^3E^3, E^3E^4	white
white	E^4E^4	

Sample Problem

What is the phenotypic ratio of the offspring from the mating of the following *Drosophila*?

E^1E^4 (wild-type eye colour) $\times$ E^2E^3 (apricot eye colour)

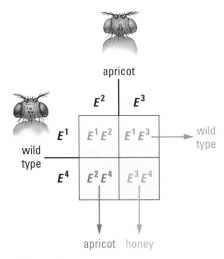

Figure 2
A cross between a fruit fly with wild-type eye colour and one with apricot-coloured eyes

apricot

	E^2	E^3
E^1	$E^1 E^2$	$E^1 E^3$
E^4	$E^2 E^4$	$E^3 E^4$

wild type → wild type

apricot   honey

F₁ parent

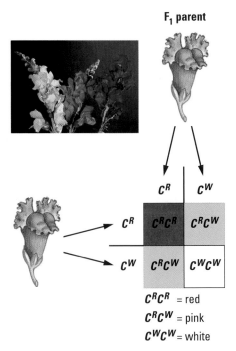

	C^R	C^W
C^R	$C^R C^R$	$C^R C^W$
C^W	$C^R C^W$	$C^W C^W$

$C^R C^R$ = red
$C^R C^W$ = pink
$C^W C^W$ = white

Figure 1
Colour in snapdragons is an example of incomplete dominance. Red-flowering and white-flowering snapdragons combine to produce pink-flowering plants in F₁. The F₂ generation produces one red to two pink to one white.

Solution

The phenotypic ratio of the F₁ offspring is two wild-type eye colour to one apricot eye colour to one honey eye colour. The Punnett square for this cross is shown in **Figure 2**.

SUMMARY **Single-Trait Inheritance**

1. Since one allele is inherited from each parent, various genotypes are possible.

2. The dominant phenotype is expressed if the offspring is either heterozygous or homozygous for the dominant allele.

3. The recessive phenotype is expressed only if the offspring is homozygous for the recessive allele.

4. When there are multiple alleles for a given characteristic, the alleles have a dominance hierarchy.

Practice

Understanding Concepts

1. Use the information in **Table 1** (page 143) to answer the following questions:
   (a) Of the genotypes listed in the table, which would you say represent the homozygous condition under simple single-trait inheritance? Explain.
   (b) Of the genotypes listed, which would you say represent the homozygous recessive condition? Explain.

Applying Inquiry Skills

2. Use the information in **Table 1** and the method shown in **Figure 2** to find the F₁ phenotypes if a white-eyed fly is crossed with one that has honey-coloured eyes.

3. Find the F₁ phenotypes if a fly with apricot-coloured eyes ($E^2 E^4$) is crossed with one that has honey-coloured eyes ($E^3 E^4$).

4.5 Incomplete Dominance and Codominance

Prior to Mendel's studies, many scientists believed that hybrids would have a blending of traits. Although Mendel never found any examples of new traits or blended traits produced by the combinations of different alleles, many do exist in nature. When two alleles are equally dominant, they interact to produce a new phenotype. This kind of interaction is known as incomplete dominance.

For example, if red snapdragons are crossed with white snapdragons, all of the F₁ offspring are pink. The pink colour is produced by the interaction of red and white alleles. This type of incomplete dominance is often called intermediate inheritance. If the F₁ generation is allowed to self-fertilize, the F₂ generation produces a ratio of one red to two pink to one white. The Punnett square in **Figure 1** helps to explain this result.

Another type of incomplete dominance is referred to as codominance. In this type of interaction, both alleles are expressed at the same time. Shorthorn cattle

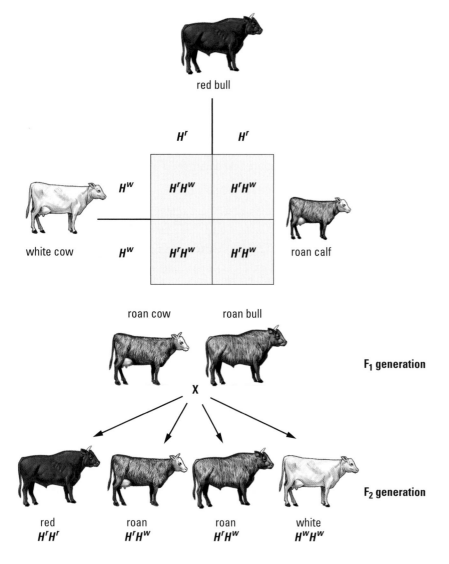

provide an excellent example of codominance (**Figure 2**). A red bull crossed with a white cow produces a roan calf. The roan calf has intermingled white and red hair. A roan calf would also be produced if a white bull were crossed with a red cow.

Practice

Understanding Concepts

1. Explain in your own words the meaning of dominance, codominance, and incomplete dominance.

Applying Inquiry Skills

2. Determine the F_1 phenotypes of a cross between a pink and a white snapdragon.

3. Find the F_1 phenotypes of a cross between a red cow and a roan bull.

4. A geneticist notes that crossing a round radish with a long radish produces oval radishes. If oval radishes are crossed with oval radishes, the following phenotypes are noted in the F_2 generation: 100 long, 200 oval, and 100 round radishes. Use symbols to explain the results obtained for the F_1 and F_2 generations.

5. Diabetes is a recessive genetic disorder. A defective gene reduces insulin production by the pancreas. Insulin is released into the circulatory system and allows the cells of the body to absorb glucose from the blood. Individuals who lack insulin have high blood sugar. In an attempt to trace the inheritance of the defective allele in one family, the data in **Table 1** was gathered.
   (a) Construct a pedigree chart showing the passage of the diabetes allele.
   (b) Indicate the probable genotypes of Jennifer and Ryan.
   (c) Indicate the probable genotypes of Susan and Walter.
   (d) Whose genotype cannot be determined with 100% certainty? Explain.

Table 1

Name	Relationship	Phenotype
Jennifer	mother	normal
Ryan	father	normal
Walter	son of Ryan and Jennifer	diabetic
Susan	wife of Walter	normal
Helen	daughter of Ryan and Jennifer	normal
James	son of Ryan and Jennifer	normal
Colin	son of Susan and Walter	diabetic

Figure 3
A genetics counsellor helps a couple understand the genetic factors involved in diseases and disorders.

Genetic Screening

Before the development of a process that permitted the extraction of insulin from animals, many people who had the recessive allele for diabetes in the homozygous condition died before passing on their genes to offspring. Genetic screening attempts to identify genetic conditions prior to birth or attempts to predict these conditions prior to conception (**Figure 3**). Genetic information is obtained through a variety of methods including detailed pedigrees and biochemical testing for known disorders. Methods of prenatal diagnosis can indicate the sex of the child as well as the presence of many genetic conditions. Amniocentesis and chorionic villi sampling (CVS) are the most widely used techniques.

Huntington's chorea is a neurological disorder caused by a dominant allele that only begins to express itself later in life. The disease is characterized by the rapid deterioration of nerve control, eventually leading to death. Early detection of this disease by genetic screening is possible.

Practice

Understanding Concepts

1. Define genetic screening. Describe some technologies used in genetic screening.
2. What are some advantages of genetic screening? Provide an example.
3. What are some physical dangers associated with genetic screening methods? Provide an example.

Making Connections

4. What are some social, moral, and ethical objections to genetic screening? Provide an example.
   Follow the links for Nelson Biology 11, 4.5.

 **GO TO** www.science.nelson.com

5. What laws, if any, do you think will arise regarding the use of genetic screening?

Explore an
Issue

Debate: Genetic Screening

Genetic screening techniques are coming of age, and the controversy that surrounds them is growing by the minute. You have researched the benefits and problems associated with genetic screening. Do the benefits outweigh the problems?

Statement

Genetic screening should be compulsory.

- Write a list of points and counterpoints which you and your group considered.
- Decide whether your group agrees or disagrees with the statement.
- Prepare to defend your group's position in a class discussion.

DECISION-MAKING SKILLS

○ Define the Issue	● Analyze the Issue
● Identify Alternatives	● Defend a Decision
● Research	● Evaluate

Case Study: A Mystery

There are four different ABO blood types as shown in **Table 2**. The alleles for blood types A and B are codominant but are dominant over the allele for type O. The rhesus factor is a blood factor that is regulated by a gene. The *Rh*-positive allele is dominant over the *Rh*-negative allele. In this activity, you will solve a murder mystery using genetics.

Evidence

As a bolt of lightning flashed above Black Mourning Castle, a scream echoed from the den of Lord Hooke. When the upstairs maid peered through the door, a freckled arm reached for her neck. Quickly, the maid bolted from the doorway, locked herself in the library, and telephoned the police. Inspector Holmes arrived to find a frightened maid and the dead body of Lord Hooke. Apparently, the lord had been strangled. The inspector quickly gathered evidence. He noted blood on a letter opener, even though Lord Hooke did not have any cuts or abrasions. The blood sample proved to be type O, *Rh*-negative. The quick-thinking inspector phoned the family doctor for each family member's medical history. **Figure 4** shows the relatives who were in the castle at the time of Lord Hooke's murder.

Table 2

Phenotypes	Genotypes
type A	$I^A I^A$, $I^A I^O$
type B	$I^B I^B$, $I^B I^O$
type AB	$I^A I^B$
type O	$I^O I^O$

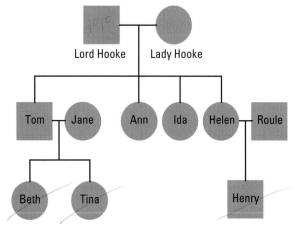

Figure 4
Pedigree chart of Lord and Lady Hooke

The inspector gathered the information in **Table 3**. Some of the family members were deeply tanned, so the inspector found it difficult to determine whether or not freckles were present on their arms. Note that having freckles is an inherited trait and the allele for freckles is dominant over the allele for no freckles.

Table 3

Family member	Blood type	*Rh* factor	Freckles
Lord Hooke	AB	+	no
Lady Hooke	A	+	no
Helen	A	+	no
Roule	O	+	no
Henry	Refused blood test		?
Ida	A	–	?
Ann	B	+	?
Tom	O	–	no
Jane	A	+	?
Beth	O	–	?
Tina	A	+	yes

The crafty inspector drew the family close together and, while puffing on his pipe, indicated that he had found the murderer. He explained that one of the heirs to the fortune was not Lord Hooke's biological child. The inspector believed that the child committed the murder.

Analysis

(a) Who was the murderer? State the reasons for your answer.
(b) Describe the procedure you followed to obtain your answer.
(c) How did the inspector eliminate the other family members?

Sections 4.3–4.5 Questions

Understanding Concepts

1. Explain the difference between a dominant and a recessive condition. Provide an example.

2. Guinea pigs with yellow coat colour have the genotype C^YC^Y. Guinea pigs with cream coat colour (cream-coloured hairs) have the genotype C^YC^W, and those with white coat colour have the genotype C^WC^W. Is the condition for coat colour one of complete dominance, incomplete dominance, or codominance? Explain.

Applying Inquiry Skills

3. Phenylketonuria (PKU) is a genetic disorder caused by a dominant allele. People with PKU are unable to metabolize a naturally occurring amino acid, phenylalanine. If phenylalanine accumulates, it inhibits the development of the nervous system, leading to mental retardation. The symptoms of PKU are not usually evident at birth but can develop quickly if the child is not placed on a special diet. **Figure 5** is a pedigree chart that shows the inheritance of the defective PKU allele in one family.

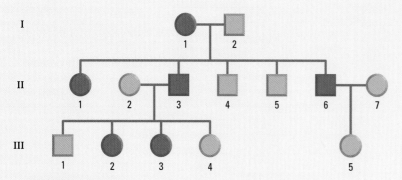

Figure 5

(a) How many generations are shown in the pedigree chart?

(b) How many children were born to the parents of the first generation?

(c) What are the genotypes of individuals 1 and 2 in generation I?

(d) How is it possible that in generation II, some of the children showed symptoms of PKU while others did not? (Hint: Use a Punnett square to help with your explanation.)

(e) Individuals 6 and 7 in generation II had a child without PKU. Does this mean that they can never have a child with PKU? Explain your answer.

4. Multiple alleles control the intensity of pigment in mice. The gene D^1 designates full colour, D^2 designates dilute colour, and D^3 is deadly when homozygous. The order of dominance is $D^1 > D^2 > D^3$. When a full-coloured male is mated to a dilute-coloured female, the offspring are produced in the following ratio: two full colour to one dilute to one dead. Indicate the genotypes of the parents.

5. Multiple alleles control the coat colour of rabbits. A grey colour is produced by the dominant allele C. The C^{ch} allele produces a silver-grey colour, called chinchilla, when present in the homozygous condition, $C^{ch}C^{ch}$. When C^{ch} is present with a recessive allele, a light silver-grey colour is produced. The allele C^h is recessive to both the full-colour allele and the chinchilla allele. The C^h allele produces a white colour with black extremities. This coloration is called Himalayan. An allele C^a is recessive to all alleles. The C^a allele results in a lack of pigment, called albino. The dominance hierarchy is $C > C^{ch} > C^h > C^a$. **Table 4** below provides the possible genotypes and phenotypes for coat colour in rabbits. Notice that four genotypes are possible for full colour but only one for albino and chinchilla.

Table 4

Phenotypes	Genotypes
full colour	CC, CC^{ch}, CC^h, CC^a
chinchilla	$C^{ch}C^{ch}$
light grey	$C^{ch}C^h$, $C^{ch}C^a$
Himalayan	C^hC^h, C^hC^a
albino	C^aC^a

(continued)

(a) Indicate the genotypes and phenotypes of the F_1 generation from the mating of a heterozygous Himalayan rabbit with an albino rabbit.

(b) The mating of a full-coloured rabbit with a light-grey rabbit produces two full-coloured offspring, one light-grey offspring, and one albino offspring. Indicate the genotypes of the parents.

(c) A chinchilla rabbit is mated with a light-grey rabbit. The breeder knows that the light-grey rabbit had an albino mother. Indicate the genotypes and phenotypes of the F_1 generation from this mating.

(d) A test cross is performed with a light-grey rabbit, and the following offspring are noted: five Himalayan rabbits and five light-grey rabbits. Indicate the genotype of the light-grey rabbit.

6. Palomino horses are known to result from the interaction of two different alleles. The allele C^r in the homozygous condition produces a chestnut, or reddish, coat. The allele C^m in the homozygous condition produces a very pale cream coat, called cremello. The palomino colour is caused by the interaction of both the chestnut and cremello alleles. Indicate the expected ratios in the F_1 generation from mating a palomino with a cremello.

7. There are four different ABO blood types (**Table 5**). The alleles for blood types A and B are codominant but are dominant over the allele for blood type O.

Table 5

Phenotypes	Genotypes
type A	$I^A I^A$, $I^A I^O$
type B	$I^B I^B$, $I^B I^O$
type AB	$I^A I^B$
type O	$I^O I^O$

Determine the possible phenotypes and genotypes of the F_1 generation offspring of a parent with type A blood and a parent with type B blood. Use Punnett squares to show your work.

4.6 Dihybrid Crosses

Mendel also studied the inheritance of two separate traits in crossbreeding, following the same procedure he had used for studying single traits. He cross-pollinated pure-breeding plants that produced yellow, round seeds with pure-breeding plants that produced green, wrinkled seeds to study the inheritance of two traits. The laws of genetics that apply for a single-trait inheritance (monohybrid cross) also apply for a two-trait inheritance (**dihybrid cross**).

Figure 1 shows a dihybrid cross. The pure-breeding round seed is indicated by the symbol *RR*, and the pure-breeding wrinkled seed by the recessive alleles *rr*. The pure-breeding yellow seed is indicated by the alleles *YY* and the green by *yy*. The genotype for the yellow, round parent is *YYRR*, while the genotype for the green, wrinkled parent is *yyrr*. The F_1 offspring produced from such a cross are heterozygous for both the yellow and round genotypes.

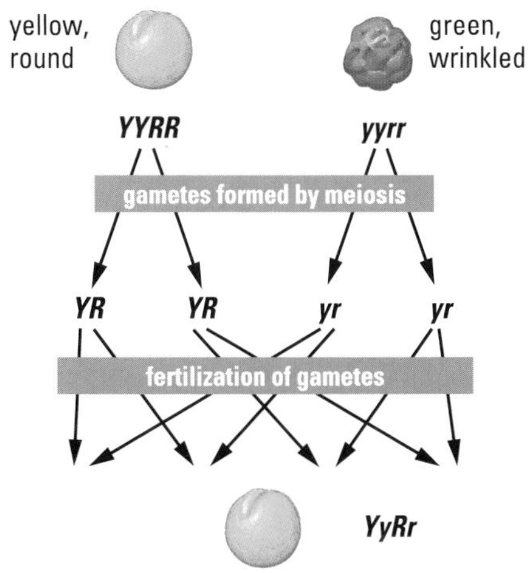

All members of the F_1 generation have the same genotype and phenotype.

Figure 1
A dihybrid cross between a pure-breeding pea plant with yellow, round seeds and a pure-breeding pea plant with green, wrinkled seeds

dihybrid cross: a type of cross that involves two genes, each consisting of nonidentical alleles

Now consider a cross between a pure-breeding green, round pea plant and a pure-breeding yellow, wrinkled pea plant. **Figure 2** shows the resulting offspring. Inheritance of the gene for colour is not affected by either the wrinkled or round alleles. By doing other crosses, Mendel soon discovered that the alleles assort independently, even though he did not know about the existence of chromosomes or the process of meiosis. Today, this phenomenon is referred to as the **law of independent assortment**. The genes that govern pea shape are inherited independently of the ones that control pea colour.

Mendel allowed plants of the F_1 generation to self-fertilize in order to produce an F_2 generation. Each heterozygous, yellow, round plant can produce four different phenotypes. As the homologous chromosomes move to opposite poles during meiosis, the yellow allele will segregate with the round and wrinkled alleles in equal frequency (see **Figure 3**). This means that the sex cells containing *YR* will equal the number of sex cells containing *yR*. Similarly, the green allele will segregate with round and wrinkled alleles in equal frequency. The number of gametes containing *Yr* will be equal to the number of gametes containing *yr*.

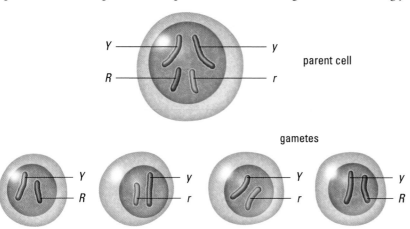

parent cell

gametes

A Punnett square can be used to predict the genotypes and phenotypes of the F_2 generation. The phenotype ratios of the F_2 generation shown in the Punnett square in **Figure 4** (page 152) are

yellow, round $\dfrac{9}{16}$ green, round $\dfrac{3}{16}$

yellow, wrinkled $\dfrac{3}{16}$ green, wrinkled $\dfrac{1}{16}$

Probability

Genotypic and phenotypic ratios are determined by the probability of inheriting a certain trait. The probability of an event is the likelihood that the event will occur. For example, you can calculate the probability of getting heads when you toss a coin. Probability can be expressed by the following formula:

$$\text{Probability} = \frac{\text{number of ways that a given event could occur}}{\text{total number of possible events}}$$

In the coin-toss example, there is only one way of tossing heads, so the numerator is 1. The denominator is 2 because there are two possible events in total, heads or tails. Therefore, the probability of tossing heads is $\frac{1}{2}$.

Two important rules will help you understand probability:

- Chance has no memory. For example, if you tossed two heads in a row, the probability of tossing heads once again would still be $\frac{1}{2}$. The probability of

law of independent assortment: If genes are located on separate chromosomes, they are inherited independently of each other.

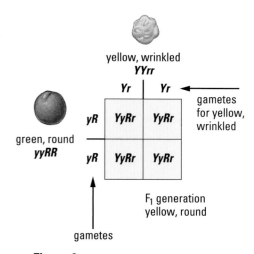

yellow, wrinkled
YYrr

green, round
yyRR

gametes for yellow, wrinkled

F_1 generation
yellow, round

gametes

Figure 2
The inheritance of the gene for pea colour is not affected by the gene for pea shape.

Figure 3
Four possible combinations for the gametes of the genotype *YyRr*

gametes	**YR**	**yR**	**Yr**	**yr**
YR	YYRR	YyRR	YYRr	YyRr
yR	YyRR	yyRR	YyRr	yyRr
Yr	YYRr	YyRr	YYrr	Yyrr
yr	YyRr	yyRr	Yyrr	yyrr

Figure 4
A Punnett square used to find the F₂ generation of a dihybrid cross. Each heterozygous, yellow, round plant can produce four different phenotypes.

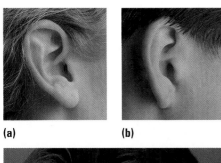

(a)   (b)

(c)

(d)

Figure 5
The shape of both earlobes and hairline are inherited characteristics in humans. The free earlobe **(a)** is dominant over the attached earlobe **(b)**, and the widow's peak **(c)** is dominant over the straight hairline **(d)**.

each event is not affected by the results of the others, so the events are said to be independent.

• The probability of independent events occurring together is equal to the product of those events occurring separately. For example, the chances of tossing heads once is $\frac{1}{2}$; the probability of tossing heads twice is $\frac{1}{2} \times \frac{1}{2} = \frac{1}{4}$, and the probability of tossing heads three times in a row is $\frac{1}{2} \times \frac{1}{2} \times \frac{1}{2} = \frac{1}{8}$.

Sample Problem 1

In humans, free earlobes are controlled by the dominant allele E, and attached earlobes by the recessive allele e (**Figure 5** (**a**) and (**b**)). The widow's peak hairline is regulated by the dominant allele H, while the straight hairline is controlled by the recessive allele h (**Figure 5** (**c**) and (**d**)). Consider the mating of the following genotypes:

$$EeHh \times EeHh$$

What are the probabilities of obtaining F₁ offspring with the following characteristics?

• widow's peak and free earlobes
• straight hairline and free earlobes
• widow's peak and attached earlobes
• straight hairline and attached earlobes

Solution

Dihybrids can be treated as two monohybrids, as shown in **Figure 6**. Isolate the gene for earlobes and for hairline and work with each as a monohybrid. The F₁ generation resulting from a cross between the heterozygous parents can be determined.

The following are the phenotype probabilities for earlobes:

free earlobes $\frac{3}{4}$ attached earlobes $\frac{1}{4}$

EeHh × *EeHh*

	$\frac{1}{2}$ *E*	$\frac{1}{2}$ *e*
$\frac{1}{2}$ *E*	$\frac{1}{4}$ *EE*	$\frac{1}{4}$ *Ee*
$\frac{1}{2}$ *e*	$\frac{1}{4}$ *Ee*	$\frac{1}{4}$ *ee*

	$\frac{1}{2}$ *H*	$\frac{1}{2}$ *h*
$\frac{1}{2}$ *H*	$\frac{1}{4}$ *HH*	$\frac{1}{4}$ *Hh*
$\frac{1}{2}$ *h*	$\frac{1}{4}$ *Hh*	$\frac{1}{4}$ *hh*

Figure 6
Punnett squares showing monohybrid crosses between heterozygous parents for free earlobes and for a widow's peak.

The following are the phenotype probabilities for hairlines:

widow's peak $\frac{3}{4}$ straight hairline $\frac{1}{4}$

The monohybrids can now be combined to calculate the probabilities of the dihybrid crosses. For example, the chances of producing an F_1 offspring from the mating of *EeHh* × *EeHh* who has

- a widow's peak and free earlobes is $\frac{3}{4} \times \frac{3}{4}$, or $\frac{9}{16}$;
- a straight hairline and free earlobes is $\frac{1}{4} \times \frac{3}{4}$, or $\frac{3}{16}$;
- a widow's peak and attached earlobes is $\frac{3}{4} \times \frac{1}{4}$, or $\frac{3}{16}$;
- a straight hairline and attached earlobes is $\frac{1}{4} \times \frac{1}{4}$, or $\frac{1}{16}$.

Figure 7 shows the results of the dihybrid cross.

gametes	*EH*	*eH*	*Eh*	*eh*
EH	*EEHH*	*EeHH*	*EEHh*	*EeHh*
eH	*EeHH*	*eeHH*	*EeHh*	*eeHh*
Eh	*EEHh*	*EeHh*	*EEhh*	*Eehh*
eh	*EeHh*	*eeHh*	*Eehh*	*eehh*

Figure 7
Punnett square showing the results of the dihybrid cross of *EeHh* × *EeHh*.

If 16 offspring were generated, it is expected that 9 would have a widow's peak and free earlobes (orange squares), 3 would have a straight hairline and free earlobes (yellow squares), 3 would have a widow's peak and attached earlobes (green squares), and 1 would have a straight hairline and attached earlobes.

Sample Problem 2

What is the probability that a child from the mating of the *EeHh* × *EeHh* parents would be a male with a widow's peak and have attached earlobes?

Solution

The calculation indicates that the dihybrid cross is equivalent to two separate monohybrid crosses. Consider each of the separate probabilities:

- The probability of producing a male is $\frac{1}{2}$.
- The probability of having a widow's peak is $\frac{3}{4}$.
- The probability of having attached earlobes is $\frac{1}{4}$.
- Therefore, the probability of producing a male with a widow's peak and attached earlobes is $\frac{1}{2} \times \frac{3}{4} \times \frac{1}{4}$, or $\frac{3}{32}$.

Practice

Applying Inquiry Skills

1. In guinea pigs, black coat colour (*B*) is dominant over white (*b*), and short hair length (*H*) is dominant over long (*h*). Indicate the genotypes and phenotypes from the following crosses:
   (a) A guinea pig that is homozygous for black and heterozygous for short hair is crossed with a white, long-haired guinea pig.
   (b) A guinea pig that is heterozygous for black and for short hair is crossed with a white, long-haired guinea pig.
   (c) A guinea pig that is homozygous for black and for long hair is crossed with a guinea pig that is heterozygous for black and for short hair.

2. Black coat colour (*B*) in cocker spaniels is dominant over white coat colour (*b*). Solid coat pattern (*H*) is dominant over spotted pattern (*h*). The gene for pattern arrangement is located on a different chromosome than the one for colour, and the pattern gene segregates independently of the colour gene. A male that is black with a solid pattern mates with three females. The mating with female A, which is white and solid, produces four pups: two black, solid; and two white, solid. The mating with female B, which is black and solid, produces a single pup, which is white, spotted. The mating with female C, which is white and spotted, produces four pups: one white, solid; one white, spotted; one black, solid; one black, spotted. Indicate the genotypes of the parents.

3. The alleles for human blood types A and B are codominant, but both are dominant over the type O allele. The *Rh* factor is separate from the ABO blood group and is located on a separate chromosome. The *Rh*-positive allele is dominant over *Rh*-negative. Indicate the possible phenotypes from the mating of a woman with type O, *Rh*-negative, with a man with type A, *Rh*-positive.

Making Connections

4. Two pea plants are crossbred. Using a Punnett square and probability analysis, you predict that $\frac{3}{4}$ of the offspring will be tall. However, less than $\frac{1}{4}$ grow to be tall. What other factors can affect phenotype? How much trust should be put on probability calculations?

Reflecting

5. When solving genetic problems, do you check over your work before giving your final answer? How do you check—by redoing the questions, working backwards, or some other method?

Activity 4.6.1

Genetics of Corn

Corn is one of the world's most important food crops. It has been subject to selective breeding techniques and hybridization for many years, which have resulted in vigorous, high-yielding varieties. Nearly all corn grown today is hybrid corn. Some varieties of corn are chosen for their sweet flavour while the mixed coloration of the Indian corn varieties makes them popular decorations during the autumn months.

In this activity, you will determine the probable genotypes of parents by examining the phenotypes of corn for two different and independently assorted traits.

Materials

dihybrid corn ears (sample A, sample B)

Procedure

1. Obtain a sample A corn ear from your teacher (**Figure 8**). The kernels display two different traits whose genes are located on different chromosomes.

2. Describe the two different traits: colour and shape. Predict which phenotypes are dominant and which are recessive.

3. Assume that the ear of corn is from the F_2 generation. The parents of the F_1 corn were pure-breeding homozygous for each of the characteristics. Assign the letters P and p to the alleles for colour, and A and a to the alleles for shape. Use the symbols $PPaa$ and $ppAA$ for the parents of the F_1 generation. Describe the phenotype of the $PPaa$ parent and $ppAA$ parent. The cross for the F_1 generation is $PPaa \times ppAA$.

4. Count 100 of the kernels in sequence; describe the phenotypes and record the number of each in a table similar to **Table 1**.

Figure 8
Sample A

Table 1

Phenotype	Number	Ratio
dominant alleles for colour and shape		
dominant allele for colour, but recessive allele for shape		
recessive allele for colour, but dominant allele for shape		
recessive alleles for colour and shape		

5. Obtain sample B. Assume that this ear was produced from a test cross. Count 100 kernels in sequence and record your results as in step 4.

Analysis and Evaluation

(a) What are the expected genotypes and phenotypes of the F_1 generation resulting from a cross between the parents $PPaa$ and $ppAA$?

(b) Use a Punnett square to show the expected genotypes and the phenotypic ratio of the F_2 generation. Compare your results with what you obtained in step 4. What factors might account for discrepancies? Would your results be any different if you took larger samples or took multiple samples and averaged the results?

(c) Assuming that sample B was produced from a test cross (i.e., B is from the F_1 generation), what is the phenotypic ratio of the F_1 generation?

(d) What is the phenotype of the unknown parent?

Virtual Fruit Fly Simulation

Follow the links for Nelson Biology 11, 4.6 to begin the activity.

GO TO www.science.nelson.com

Test your understanding of Mendelian genetics by performing crosses in the Virtual Fruit Fly lab. You will be able to select traits that demonstrate the concepts of dominant and recessive alleles, segregation, and independent assortment.

- Create a classroom list of all possible crosses.
- Working in groups of 2 or 3, select at least 2 crosses to examine.
- Write a report of the crosses you tested and the results.
- State the genotypes and phenotypes of the parents and F_1 and F_2 offspring.
- Indicate the number of each type of offspring produced in each generation.
- Record the information on the classroom list.
- As a class, analyze the data. Compare the number of flies for eye colour and for the other traits that were examined.

Section 4.6 Questions

Understanding Concepts

1. Why are test crosses important to plant breeders?

2. A dihybrid cross can produce 16 combinations of alleles, 9 of which are different. Explain why 100 seeds were counted rather than only 16 or 9.

Applying Inquiry Skills

3. A dominant allele Su, called starchy, produces smooth kernels of corn. The recessive allele su, called sweet, produces wrinkled kernels of corn. The dominant allele P produces purple kernels, while the recessive allele p produces yellow kernels. A corn plant with starchy, yellow kernels is cross-pollinated with a corn plant with sweet, purple kernels. One hundred kernels from the hybrid are counted, and the following results are obtained: 52 starchy, yellow kernels and 48 starchy, purple kernels. What are the genotypes of the parents and of the F_1 generation?

Making Connections

4. Thousands of years ago, the ancestor of corn grew only in Mexico. Scientists have used technology and selective breeding methods to develop varieties of corn that can grow in a wide range of environmental conditions. As a result, corn is now grown in many places where it would not occur by nature. What are some risks associated with growing a species in a foreign environment? Start your research on the Internet. Follow the links for Nelson Biology 11, 4.6.

GO TO www.science.nelson.com

Key Expectations

Throughout this chapter, you have had opportunities to do the following:

- Describe and explain the process of discovery that led Mendel to formulate his laws of heredity (4.1).
- Explain, using Mendelian genetics, the concepts of inherited traits, dominance, codominance, incomplete dominance, and recessiveness. (4.1, 4.2, 4.4, 4.5).
- Explain how the concepts of DNA, genes, chromosomes, and meiosis account for the transmission of hereditary characteristics from generation to generation (4.1, 4.6).
- Express the result of any calculation involving experimental data to the appropriate number of decimal places or significant figures (4.2).
- Solve basic genetic problems involving monohybrid crosses, incomplete dominance, codominance, and dihybrid crosses using the Punnett method (4.2, 4.5, 4.6).
- Predict the outcomes of various genetic crosses (4.2, 4.5, 4.6).
- Identify and describe examples of Canadian contributions to knowledge about genetic processes and to technologies and techniques related to genetic processes (4.3, 4.5).
- Describe and analyze examples of genetic technologies that were developed on the basis of scientific understanding (4.5).
- Research genetic technologies using sources from print and electronic media, and synthesize the information gained (4.5, 4.6).
- Locate, select, analyze, and integrate information on topics under study, working independently and as part of a team, and using appropriate library and electronic research tools, including the Internet (4.5, 4.6).
- Compile qualitative and quantitative data from a laboratory investigation on monohybrid and dihybrid crosses, and present the results, either by hand or by computer (4.6).
- Select and use appropriate modes of representation to communicate scientific ideas, plans, and experimental results (4.6).

Key Terms

alleles
dihybrid cross
dominant
genotype
heredity
heterozygous
homozygous
hybridization
hybrids
inbreeding

law of independent assortment
monohybrid cross
pedigree chart
phenotype
Punnett square
recessive
segregation
selective breeding

Make a Summary

In this chapter, you studied genes and heredity. To summarize your learning, create a concept map that shows the principles of genetics. Label the sketch with as many of the key terms as possible. Check other concept maps and use an appropriate design to make your concept map clear.

Reflect on your Learning

Revisit your answers to the Reflect on Your Learning questions at the beginning of this chapter.

- How has your thinking changed?
- What new questions do you have?

Understanding Concepts

1. Explain why Mendel's choice of the garden pea for his experiments was especially appropriate.

2. Explain why, under normal circumstances, an individual can carry only two alleles of a gene.

3. Differentiate between codominance and incomplete dominance.

4. Test crosses are valuable tools for plant and animal breeders.
    (a) Provide two practical examples of why a cattle rancher might use a test cross.
    (b) Why are test crosses most often done on bulls rather than on cows?

5. Cystic fibrosis is regulated by a recessive allele *c*. Explain how two normal parents can produce a child that has the disorder.

6. Cats with 6 toes carry a dominant allele. Draw a pedigree showing the mating of a male cat with 6 toes to a normal female. (Assume that the male cat had a normal mother.) Include this information in your pedigree chart:
    • The cats produce 6 offspring (4 females and 2 males).
    • Only one of the female offspring mates and produces a litter (3 males and 2 females).

7. Two different genes control the expression of kernel colour in Mexican black corn. Gene *B* produces black coloration. Gene *B* influences the expression of gene *D*, which produces a dotted pigmentation. The dotted variation appears only when gene *B* is homozygous. A colourless variation arises when both genes are homozygous recessive.
    (a) What is (are) the possible genotype(s) for corn with dotted pigmentation?
    (b) What would kernels with a genotype of *BBdd* look like?

8. For shorthorn cattle, the mating of a red bull and a white cow produces a roan calf that has intermingled red and white hair. Many matings between roan bulls and roan cows produce cattle in the following ratio: 1 red, 2 roan, 1 white. Is this an example of codominance or multiple alleles? Explain your answer.

Applying Inquiry Skills

9. For pea plants, long stems are dominant over short stems. Determine the phenotype and genotype ratios of the F_1 offspring from the cross-pollination of a heterozygous long-stem plant with a short-stem plant.

10. For horses, the trotter characteristic is dominant over the pacer characteristic. A male, who is described as a trotter, mates with three different females and each female produces a foal. The first female, who is a pacer, gives birth to a foal that is a pacer. The second female, also a pacer, gives birth to a foal that is a trotter. The third female, a trotter, gives birth to a foal that is a pacer. Determine the genotypes of the male, all three females, and the three foals sired.

11. For ABO blood groups, the A and B alleles are codominant, but both A and B are dominant over type O. Indicate the blood types possible from the mating of a male with blood type O to a woman with blood type AB. Could a female with blood type AB ever produce a child with blood type AB? Could she ever have a child with blood type O?

12. The following information was gathered on blood types in one family (**Figure 1**).

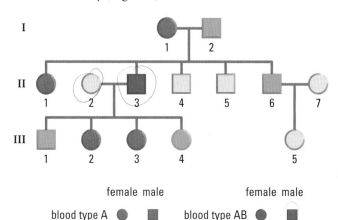

Figure 1

    (a) Indicate the genotypes for individuals 1 and 2 in generation I.
    (b) Would it ever be possible for individuals 2 and 3 in generation II to have a child with blood type O? Explain why or why not.
    (c) If individuals 6 and 7 had another child, what would be the probability of the child having blood type O?

13. Thalassemia is a serious human genetic disorder which causes severe anemia in the homozygous condition ($T^m T^m$). People with thalessemia die before sexual maturity. The heterozygous condition ($T^m T^n$) causes a less serious form of anemia. The genotype $T^n T^n$ causes no symptoms of the disease. Indicate the

possible genotypes and phenotypes of the offspring if a male with the genotype T^mT^n married and had children with a woman of the same genotype.

14. In guinea pigs, black coat colour is dominant over white. Short hair is dominant over long hair. A guinea pig that is homozygous for white and for short hair is mated with a guinea pig that is homozygous for black and for long hair. Indicate the phenotype(s) of the F_1 generation. If two hybrids from the F_1 generation are mated, determine the phenotype ratio of the F_2 generation.

15. The diabetes allele is recessive. Use the phenotype chart (**Figure 2**) to answer the following questions.
    (a) How many children do parents A and B have?
    (b) Indicate the genotypes of parents A and B.
    (c) Give the genotypes of M and N.

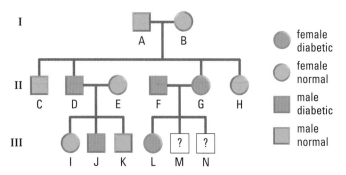

Figure 2

Making Connections

16. In Canada, it is illegal for individuals to marry their own immediate relatives. Using the principles of genetics, explain why inbreeding is discouraged.

17. Amniocentesis is a common prenatal procedure that is used to obtain cells to test for genetic abnormalities, such as cystic fibrosis. The test is usually done in the 15th to 18th week of pregnancy on a woman who has an increased risk of having children with genetic abnormalities. Cystic fibrosis is caused by a recessive allele found on chromosome 7.
    (a) See **Figure 3**. Woman (O), who has cystic fibrosis in her family history through marriage, is carrying a child. The lineage of her husband (K) is also linked with cystic fibrosis. On the basis of the information provided, would you recommend amniocentesis? Keep in mind that, like all invasive procedures, some risk, although small, is associated with amniocentesis. Provide reasons for your response.
    (b) Would you recommend the procedure if man K married woman O's cousin, woman J? Give your reasons.

(c) Should amniocentesis be performed even if there is no strong evidence suggesting genetic problems? Give your reasons.
(d) Should pedigrees be made public? Identify pros and cons before coming to a conclusion.

Father K's Family Tree

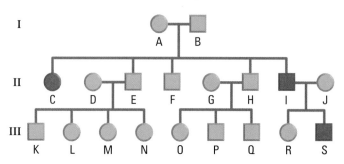

Mother O's Family Tree

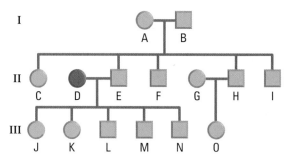

Figure 3

18. During the 19th century, individuals with genetic disorders were often shunned. As the 20th century emerged and society began to understand that many of these conditions were genetic, a movement emerged that aimed to eliminate defective genes or less desirable genes from the human population.
    (a) How might defective genes be eliminated from a population?
    (b) Identify moral and ethical issues associated with a policy that attempts to eliminate genes considered less advantageous.

Exploring

19. Reread "The Plant Breeders" in section 4.3. Find out about other Canadian contributions to plant genetics. What technologies were used or created in the discovery process?
    Follow the links for Nelson Biology 11, Chapter 4 Review.
    GO TO   www.science.nelson.com

The Source of Heredity

In this chapter, you will be able to

- explain how the concepts of DNA, genes, chromosomes, and meiosis are related to heredity;
- explain the concept of sex-linked genes and solve genetic problems related to sex linkage using a Punnett square;
- summarize how scientific discoveries in cell biology along with those in genetics have provided a clearer picture of inheritance;
- examine science-related social issues arising from new genetic research and technology.

In the Warner Brothers cartoon "Pinky and the Brain," scientists at Acme labs create a genius mouse named Brain. With the help of his friend Pinky, the Brain conspires to take over the world. Although no real mouse has the intelligence to challenge humans, scientists have created a strain of mice with superior intelligence. The genetically modified strain, dubbed Doogie, has greater memory.

The modification and insertion of a single gene, NR2B, into a chromosome of the mice improves the functioning of nerve receptors that play a key role in memory and learning. The laboratory-bred Doogie mice learn faster and remember more than normal mice. For example, scientists found that when a new and an old object were introduced into the cage with the Doogie mice, they spent most of their time exploring the new object (**Figure 1**). This indicated that they recognized and remembered the old object. Normal mice spent equal time with the new and old objects.

The Doogie mice generated great excitement because humans possess a corresponding gene embedded in their genetic material. Learning that a gene could affect how information is received by nerve cells may provide an important clue in understanding how memory works.

The Effect of Environment on Phenotype

All genes interact with the environment. At times, it is difficult to identify how much of the phenotype is determined by the genes (nature) and how much is determined by the environment (nurture). Fish of the same species show variable numbers of vertebrae if they develop in water of different temperatures. Primrose plants are red if they are raised at room temperature, but become white when raised at temperatures above 30°C. Himalayan rabbits are partially black when raised at low temperatures, but white when raised at high temperatures.

The water buttercup, *Ranunculus aquatilis*, provides another example of how genes can be modified by the environment. The buttercup grows in shallow ponds, with some of its leaves above and some below the water surface. Despite identical genetic information in the leaves above and beneath the water, the phenotypes differ. Leaves found above the water are broad, lobed, and flat, while those found below the water are thin and finely divided.

Reflect on your Learning

1. In what part of the cell would you find genes?
2. Can you distinguish males from females by looking at their genetic material?
3. Explain how a better understanding of chromosome structure could lead to a more complete understanding of gene function.
4. Why might some people be opposed to making mice smarter?
5. Why might the research with mice prove important for people with Alzheimer's disease?

Figure 1
Mice with a modified gene spent more time learning about a new object introduced into their environment.

Try This
Activity

Inherited Traits

For this activity, assume that the ability to roll your tongue is a dominant trait and is represented by the allele R.

You are given the following information:

- Al could roll his tongue.
- Betty can roll her tongue.
- Christina, daughter of Al and Betty, cannot roll her tongue.
- David can roll his tongue.
- Eileen could not roll her tongue.
- Fred, son of Eileen and David, can roll his tongue.
- Eileen and Al died.
- Betty and David married and had a daughter, Gina, who cannot roll her tongue.

(a) What is the genotype of each person?

Figure 1
The artist Leonardo da Vinci became interested in anatomy and dissection because of his desire to paint the human form better.

5.1 Early Developments in Genetics

During the Middle Ages (A.D. 500–1300), curious individuals would sneak into caves to dissect corpses. Despite strict laws prohibiting such behaviour, the inquiring minds of early physicians and scientists compelled them to conduct their investigations. Generations of artists sketched different parts of the body (**Figure 1**), creating a guide to anatomy in the process. As a composite structure of organs began to appear, theories about function arose.

The principle that structure gives clues about function also applies to genetics. However, the early geneticists had to wait for the emergence of the light microscope before investigations into genetic structure could seriously progress. The study of genes is closely connected with technology. In this chapter, you will discover how science and technology are intertwined. The light microscope, the electron microscope, X-ray diffraction, and gel electrophoresis have provided a more complete picture of the mechanisms of gene action.

Cytology and Genetics

Over 2000 years ago, the Greek philosopher Aristotle suggested that heredity could be traced to the "power" of the male's semen. He believed that hereditary factors from the male outweighed those from the female. Other scientists speculated that the female determined the characteristics of the offspring and that the male gamete merely set events in motion. Another popular theory was that hereditary traits were located in the blood. The term "bloodline" is still used today and is a reminder of this misconception. Now we know that genes, located along threadlike chromosomes inside the nucleus of each cell, are responsible for heredity.

The discovery of the nucleus in 1831 was an important step toward understanding the structure and function of cells and the genes they contain. By 1865, the year in which Mendel published his papers, many of the old misconceptions had been cleared up. Nineteenth-century biologists knew that the egg and sperm unite to form a zygote, and it was generally accepted that factors from the egg and sperm were blended in developing the characteristics of the offspring. Even though Mendel knew nothing about meiosis, or the structure or location of the hereditary material, or how the genetic code worked, he was able to develop many theories about inheritance that adequately explain how traits are passed on from generation to generation. However, the lack of complementary information and a comprehensive theory of gene action meant that Mendel's work was interpreted as a mere experiment with garden peas.

At about the same time that Mendel was conducting his experiments with garden peas, new techniques in lens grinding were providing better microscopes. The improved technology helped a new branch of biology, called cytology, to flourish. Cytology is the study of cell formation, structure, and function.

Aided by these technological innovations, in 1882, Walter Fleming described the separation of threads within the nucleus during cell division. He called the process mitosis. In the same year, Edouard van Benden noticed that the sperm and egg cells of roundworms had two chromosomes, but the fertilized eggs had four chromosomes. By 1887, August Weisman offered the theory that a special division took place in sex cells. By explaining the reduction division now known as meiosis, Weisman added an important piece to the puzzle of heredity and provided a framework in which Mendel's work could be understood. When scientists rediscovered Mendel's experiments in 1900, the true significance of his work became apparent.

5.2 Development of the Chromosomal Theory

In 1902, American biologist Walter S. Sutton and German biologist Theodor Boveri independently observed that chromosomes came in pairs that segregated during meiosis. The chromosomes then formed new pairs when the egg and sperm united. The concept of paired, or homologous, chromosomes supported Mendel's explanation of inheritance based on paired factors. Today, these factors are referred to as the alleles of a gene. One factor, or allele, for each gene comes from each sex cell.

The union of two different alleles in offspring and the formation of new combinations of alleles in succeeding generations could be explained and supported by cellular evidence. The behaviour of chromosomes during gamete formation could help explain Mendel's law of segregation and law of independent assortment.

Sutton and Boveri knew that the expression of a trait, such as eye colour, was not tied to only the male or only the female sex cell. Some structures in both the sperm cell and the egg cell must determine heredity. Sutton and Boveri deduced that Mendel's factors (alleles) must be located on the chromosomes. The fact that humans have 46 chromosomes, but thousands of different traits, led Sutton to hypothesize that each chromosome carries genes. Genes that are on the same chromosome are said to be linked genes.

Chromosomal Theory

The development and refinement of the microscope led to advances in cytology and the union of two previously unrelated fields of study: cell biology and genetics. As you continue exploring genetics, you will learn about ways in which other branches of science, such as biochemistry and nuclear physics, have integrated with genetics.

The chromosomal theory of inheritance can be summarized as follows:
- Chromosomes carry genes, the units of heredity.
- Paired chromosomes segregate during meiosis. Each sex cell or gamete has half the number of chromosomes found in the **somatic cells**. This explains why each gamete has only one of each of the paired alleles.

Chromosomes assort independently during meiosis. This means that each gamete receives one member from each pair of chromosomes, and that each chromosome pair has no influence on the movement of any other chromosome pair. This explains why in a dihybrid cross an F$_1$ parent, *AaBb*, produces four types of gametes: *AB, aB, Ab, ab*. Each gamete appears with equal frequency due to segregation and independent assortment. Each chromosome contains many different alleles and each gene occupies a specific locus or position on a particular chromosome.

Figure 1
While studying grasshoppers in 1902, Walter Sutton noted that chromosomes were paired and separated during meiosis.

somatic cells: all the cells of an organism other than the sex cells

5.3 Morgan's Experiments and Sex Linkage

Few people have difficulty distinguishing gender in humans. Mature females look very different from mature males. Even immature females and males can be differentiated on the basis of anatomy. But can you distinguish whether a skin cell or muscle cell comes from a female or a male? The work of the American geneticist Thomas Hunt Morgan (1866–1945) provided a deeper understanding of gender and inheritance.

Morgan was among the first of many geneticists who used the tiny fruit fly, *Drosophila melanogaster*, to study the principles of inheritance. There are several reasons why the fruit fly is an ideal subject for study. First, the fruit fly reproduces rapidly. Offspring are capable of mating shortly after leaving the egg, and females produce over 100 eggs after each mating. Female *Drosophila* can reproduce for the first time when they are only 10 to 15 days old, so it is possible to study many generations in a short period of time. Since genetics is based on probability, the large number of offspring is ideal. A second benefit arises from *Drosophila*'s small size. Many can be housed in a single culture tube. A small, solid nutrient at the bottom of the test tube can maintain an entire community. The third and most important quality of *Drosophila* is that males can easily be distinguished from females. Males are smaller and have a rounded abdomen with a dark-coloured posterior segment while the larger females have a pointed abdomen with a pattern of dark bands.

Morgan discovered a number of obvious **mutations** in *Drosophila*, some of which are summarized in **Table 1**. He noted that some of the mutations seemed to be linked to other traits. Morgan's observations added support to the theory that the genes responsible for the traits were located on the chromosomes.

mutations: a heritable change in the molecular structure of DNA. Many mutations change the appearance of the organism.

Table 1: Genes Identified by Morgan's Research on *Drosophila*

Trait	Dominant/recessive	Location
wingless *(wg)*	recessive lethal (all wingless offspring are born dead)	chromosome 2
congested *(cgd)*	recessive lethal (prevents first-stage larvae from leaving the egg case)	chromosome 1
curly wings *(Cy)*	dominant	chromosome 2
stubble bristles *(Sb)*	dominant	chromosome 3
purple eyes *(pr)*	recessive nonlethal	chromosome 2
ebony body *(e)*	recessive nonlethal	chromosome 3
miniature wings *(m)*	sex-linked recessive	chromosome 4
cut wings *(ct)*	sex-linked recessive	chromosome 4
white eyes *(w)*	sex-linked recessive	chromosome 4
vermillion eyes *(v)*	sex-linked recessive	chromosome 4

Figure 1

In *Drosophila,* the allele that codes for white eyes is recessive to the allele that codes for red eyes.

While examining the eye colour of a large number of *Drosophila*, Morgan noted the appearance of a white-eyed male among many red-eyed offspring (**Figure 1**). He concluded that the white-eyed trait must be a mutation. Morgan was interested in tracing the inheritance of the allele coding for white eyes, so he mated the white-eyed male with a red-eyed female. All members of the F_1 generation had red eyes. Normal Mendelian genetics indicated that the allele for red eyes was dominant. Most researchers might have stopped at that point, but Morgan did not. Pursuing further crosses and possibilities, he decided to mate two hybrids from the F_1 generation. An F_2 generation produced $\frac{3}{4}$ red and $\frac{1}{4}$ white, a ratio that could again be explained by Mendelian genetics. But further examination revealed that all the females had red eyes. Only the males had white eyes. Half of the males had red eyes and half had white eyes. Did this mean that the white-eyed phenotype only appears in males? Why could males express the white-eyed trait but not females? How did the pattern of inheritance differ between males and females? To find an answer, Morgan turned to cytology (**Figure 2**).

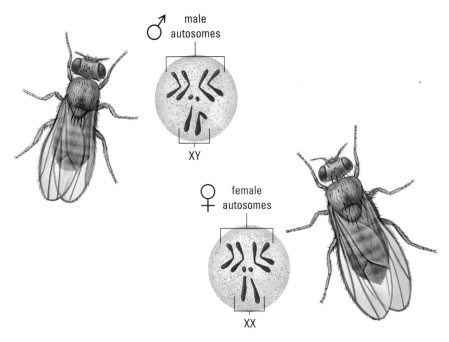

Previous researchers had stained and microscopically examined the eight chromosomes from the cells of the salivary glands of *Drosophila*. They found that females have four homologous pairs and males have only three homologous pairs. The fourth pair, which determines sex, is only partially homologous. Males were found to have one X chromosome paired with a small, hook-shaped Y chromosome. Females have two paired X chromosomes. Since the X and Y chromosomes are not completely homologous (although they act as homologous pairs during meiosis), it was concluded that they contain different genes.

Morgan explained the results of his experiments by concluding that the Y chromosome does not carry the gene to determine eye colour. We now know that the differences in the X and Y chromosomes in *Drosophila* explain that the genes are located on the part of the X chromosome that does not match the Y chromosome. Therefore, Morgan's conclusion was correct. The Y chromosome does not carry an allele for the eye-colour gene. Traits located on sex chromosomes are called **sex-linked traits**.

The initial problem can now be reexamined. The pure-breeding, red-eyed female can be indicated by the genotype $X^R X^R$ and the white-eyed male by the genotype $X^r Y$. The symbol X^R indicates that the allele for red eye is dominant and is located on the X chromosome. There is no symbol for eye colour on the Y chromosome because it does not contain an allele for the trait. A Punnett square, as shown in **Figure 3** (page 166) can be used to determine the genotypes and the phenotypes of the offspring. All members of the F_1 generation have red eyes. The females have the genotype $X^R X^r$, and the males have the genotype $X^R Y$.

The F_2 generation is determined by a cross between a male and female from the F_1 generation. Upon examination of the F_1 and F_2 generations, the question arises whether the males inherit the trait for eye colour from the mother or father. The male offspring always inherit a sex-linked trait from the mother. The father supplies the Y chromosome, which makes the offspring male.

The F_2 male *Drosophila* are $X^R Y$ and $X^r Y$. The females are either homozygous red for eye colour, $X^R X^R$, or heterozygous red for eye colour, $X^R X^r$ (**Table 2**). Although Morgan did not find any white-eyed females from his initial cross, some white-eyed females do occur in nature. For this to happen, a female with at least one allele for white eyes must be crossed with a white-eyed male. Notice that

sex-linked traits: traits that are controlled by genes located on the sex chromosomes

Table 2: Possible Genotypes for *Drosophila*

Females	Males
$X^R X^R$	$X^r Y$
$X^R X^r$	$X^R Y$
$X^r X^r$	

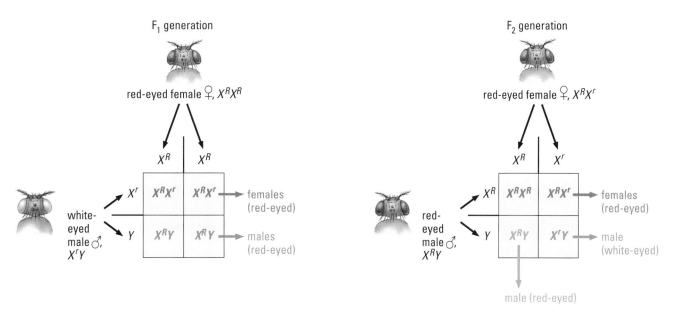

Figure 3
Punnett squares showing F$_1$ and F$_2$ generations for a cross between a homozygous red-eyed female and a white-eyed male.

recessive lethal: a trait that, when both recessive alleles are present, results in death or severe malformation of the offspring. Usually, recessive traits occur more frequently in males.

Barr body: a small, dark spot of chromatin located in the nucleus of a female mammalian cell

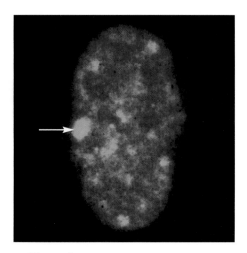

Figure 4
Micrograph of an inactivated X chromosome, called a Barr body, as it appears in a human female's somatic cell during interphase. The X chromosome is not condensed this way in a human male's cells.

females have three possible genotypes, but males have only two. Males cannot be homozygous for an X-linked gene because they have only one X chromosome.

Sex-linked genes are also found in humans. For example, a recessive allele located on the X chromosome determines red–green colourblindness. More males are colourblind than females because females require two recessive alleles to exhibit colourblindness. Since males have only one X chromosome, they require only one recessive allele to be colourblind.

This explains why **recessive lethal** X-linked disorders in humans occur more frequently in males. This could also explain why the number of females reaching the age of 10 and beyond is greater than the number of males. Males die at birth or before the age of 10 from recessive lethal X-linked disorders.

Research in Canada: Barr Bodies

The difference between male and female autosomal (non-sex) cells lies within the X and Y chromosomes. Dr. Murray Barr, working at the University of Western Ontario in London, recognized a dark spot in some of the somatic cells of female mammals during the interphase of meiosis (see **Figure 4**). This spot proved to be the sex chromatin, which results when one of the X chromosomes in females randomly becomes inactive in each cell. This dark spot is now called a **Barr body** in honour of its discoverer. This discovery revealed that not all female cells are identical; some cells have one X chromosome inactive, while some have the other. This means that some cells may express a certain trait while others express its alternate form, even though all cells are genetically identical. For example, if a human female is heterozygous for *anhidrotic ecto-dermal dysplasia*, she will have patches of skin that contain sweat glands and patches that do not. This mosaic of expression is typical of X chromosome activation and inactivation. In normal skin, the X chromosome with the recessive allele is inactivated. In the afflicted skin patches, the X chromosome with the recessive allele is activated.

Humans have 46 chromosomes. Females have 23 pairs of homologous chromosomes: 22 autosomes, and two X sex chromosomes. Males have 22 pairs of homologous chromosomes, and one X sex chromosome and one Y sex chromosome (see **Figure 5**). It has been estimated that the human X chromosome carries between 100 and 200 different genes. Therefore, there are numerous sex-

linked traits, including hemophilia, hereditary nearsightedness (myopia), and night blindness, that affect males much more often than females. The Y chromosome, considerably smaller than the X chromosome, carries the information that determines gender.

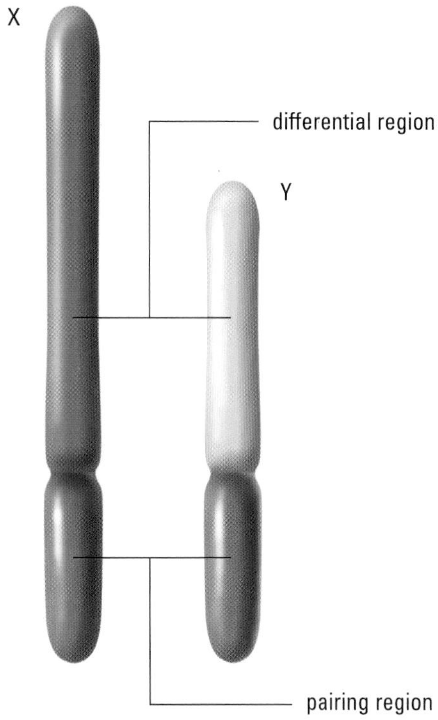

Figure 5
Sex chromosomes. Sections of the X and Y chromosomes are homologous; however, few genes are common to both chromosomes.

SUMMARY Chromosomes and Sex-Linked Traits

1. The chromosomal theory of inheritance:
   (a) Chromosomes carry genes, the units of heredity.
   (b) Each chromosome contains many different genes.
   (c) Paired chromosomes segregate during meiosis. Each sex cell or gamete has half the number of chromosomes found in a somatic cell.
   (d) Chromosomes assort independently during meiosis. This means that each gamete receives one member from each pair of chromosomes, and that each chromosome pair has no influence on the movement of any other chromosome pair.

2. Females have two X chromosomes, while males have one X and one Y chromosome.

3. Sex-linked traits are controlled by genes located on the sex chromosomes. A recessive trait located on the X chromosome is more likely to express itself in males than in females since males need only one copy of the recessive allele while females need two.

4. Female somatic cells can be identified by Barr bodies, which are actually dormant X chromosomes.

Practice

Understanding Concepts

1. Describe how the work of Walter Sutton and Theodor Boveri advanced our understanding of genetics.

2. How do sex cells differ from somatic cells?

3. Describe how Thomas Morgan's work with *Drosophila* advanced the study of genetics.

4. Identify two different sex-linked traits in humans.

5. What are Barr bodies?

6. A recessive sex-linked allele *(h)* located on the X chromosome increases blood-clotting time, causing hemophilia.
   (a) With the aid of a Punnett square, explain how a hemophilic offspring can be born to two normal parents.
   (b) Can any of the female offspring develop hemophilia? Explain.

Applying Inquiry Skills

7. In humans, the recessive allele that causes a form of red–green colourblindness *(c)* is found on the X chromosome.
   (a) Identify the F_1 generation from a colourblind father and a mother who is homozygous for colour vision.
   (b) Identify the F_1 generation from a father who has colour vision and a mother who is heterozygous for colour vision.
   (c) Use a Punnett square to identify parents that could produce a daughter who is colourblind.

Activity 5.3.1

Sex-Linked Traits in Fruit Flies

On the Internet, follow the links for Nelson Biology 11, 5.3.

GO TO www.science.nelson.com

Familiarize yourself with the software before starting this activity. Start with the tutorial. Note that the labelling of traits in the software is different from the conventions used in this textbook. Be sure you understand what each label in the software correlates to in the textbook.

In this activity, you will cross *Drosophila* (**Figure 6**) that carry genes for sex-linked traits using virtual fruit fly software. To determine if a trait is sex-linked, you will perform two sets of crosses: (a) and (b).

For cross (a), these conditions must be met if the trait is sex-linked:

- In the F_1 generation, female offspring inherit the trait of the male parent and male offspring inherit the trait of the female parent.
- In the F_2 generation, there is a 1:1 phenotypic ratio for the traits in both males and females.

For cross (b), you will confirm that the trait is sex-linked. You will cross parents with traits that are opposite to the traits of the parents in cross (a). By examining the phenotypic ratios in offspring of the F_1 generation, you can observe the greater frequency of one trait in either the male or the female offspring.

Questions

(a) If white eye colour in *Drosophila* is a sex-linked recessive trait, what are the phenotypic ratios of the F_1 generation when a homozygous red-eyed female and a white-eyed male are crossed?

(b) What other traits are sex-linked in *Drosophila*? Are they recessive or dominant?

Materials

virtual fruit fly simulation software
computers

Procedure

1. Log onto the software.
   Remember that each parent is homozygous for the trait chosen.

2. Select 1000 offspring.

3. For crosses (a) and (b), follow these algorithms:
   (a) P: white-eyed female × red-eyed male
       F_1: red-eyed female × red-eyed male
       F_2: red-eyed female × white-eyed male
   (b) P: homozygous red-eyed female × white-eyed male
       F_1: red-eyed female × red-eyed male

4. For each cross, create a Punnett square to show the expected phenotypic ratio of offspring in each generation. Also, be sure to indicate the genotype of each phenotype.

5. After each cross, analyze the results to show the number of offspring (out of 1000) of each sex and with each trait. Record the information in a table beside the corresponding Punnett square.

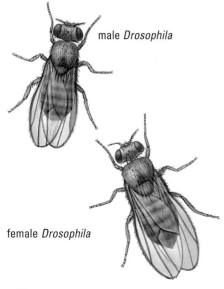

male *Drosophila*

female *Drosophila*

Figure 6
Drosophila males are smaller and have a rounded abdomen while the larger females have a pointed abdomen.

6. When you have finished cross (a), create a new parental generation.

7. Carry out cross (b). Follow steps 4–6.

8. Determine if other traits are sex-linked. Follow the same procedure as in step 3, using new traits. Indicate which traits you are examining.

Analysis

(a) In one or two paragraphs, describe the results of crosses (a) and (b). Is white eye colour in *Drosophila* sex-linked? If so, which sex does this trait appear in more frequently? Explain.

(b) In one or two paragraphs, describe the results with the other traits you examined. Is the trait sex-linked? If so, which sex does this trait appear in more frequently? Is the trait recessive or dominant? Explain.

Evaluation

(c) List and briefly explain any technical difficulties you had using the software.

(d) What improvements would you suggest to enhance the usefulness of the software?

Case Study: Following the Hemophilia Gene

In the 19th century, the allele for hemophilia A appeared in Queen Victoria's children (**Figure 7**) and spread to other European royal families through her descendants. This blood-clotting disorder occurs in about 1 in 700 males.

In this activity, you will trace the gene for hemophilia in the Royal British family and demonstrate your knowledge of pedigree charts.

Figure 7
Queen Victoria, Prince Albert, and family, 1845

Evidence

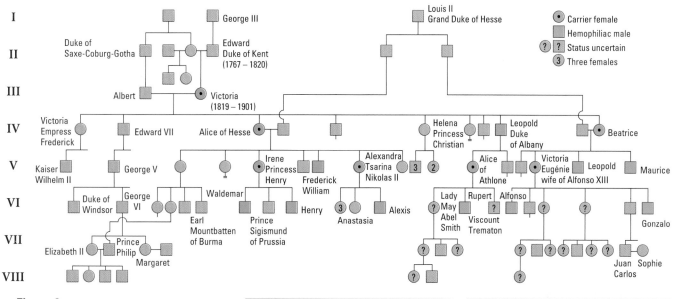

Figure 8
Pedigree chart of Queen Victoria and Prince Albert

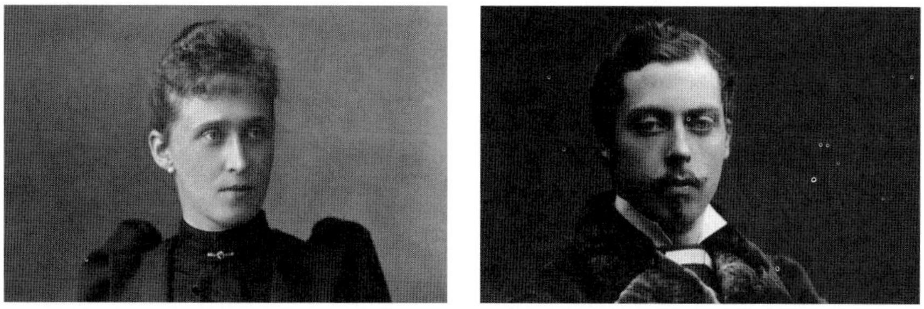

Figure 9
(a) Alice of Hesse
(b) Leopold, Duke of Albany

Figure 10
(a) Czar Nikolas II, and Alexandra
(b) The children of Nicholas and Alexandra: Tatiana, Marie, Anastasia, Olga, and Alexis

Procedure

1. Study the pedigree chart (**Figure 8**) of Queen Victoria and Prince Albert.

Analysis

(a) Who was Queen Victoria's father?

(b) How many children did Queen Victoria and Prince Albert have?

(c) What are the genotypes of Alice of Hesse and Leopold, Duke of Albany (**Figure 9**)?

(d) Alexandra, a descendant of Queen Victoria, married Nikolas II of Russia (**Figure 10(a)**). Nikolas and Alexandra had four girls and one son, Alexis. Explain why Alexis was the only child with hemophilia.

Synthesis

(e) Is it possible for a female to be hemophiliac? If not, explain why not. If so, identify a male and female from the pedigree chart who would be capable of producing a hemophiliac female offspring.

(f) On the basis of probability, calculate the number of Victoria and Albert's children who would be carriers of the gene for hemophilia.

Sections 5.1–5.3 Questions

Understanding Concepts

1. Explain how the chromosome theory, when combined with what you have already learned about genes and meiosis, gives a more complete understanding of heredity.

2. How might knowledge of meiosis have helped Mendel prove his laws of heredity?

3. A form of diabetes is caused by a recessive allele located on an autosomal chromosome. You already know that red–green colourblindness is caused by a recessive sex-linked allele. Explain why the ratio of women to men with diabetes is much closer than the ratio of women to men with red–green colourblindness.

4. Use a Punnett square to explain how a woman who is not colour-blind, but whose father is colourblind, can give birth to a son who is colourblind.

5. Red–green colourblindness in females is much more common than hemophilia A, another sex-linked disorder. Give a possible explanation.

Applying Inquiry Skills

6. In *Drosophila*, miniature wings are produced by a recessive sex-linked allele on chromosome 4 (the X chromosome). Wingless flies are produced by a recessive autosomal allele found on chromosome 2.
   (a) List the genotypes of
      (i) a female with one allele for miniature wings;
      (ii) a female with one allele for winglessness.
   (b) Use a Punnett square to compare the results of crossing a normal male with female (i) and then female (ii) above.
   (c) List the phenotypes of the members of the F_1 generation in each cross from (b).
   (d) Identify the two parent *Drosophila* that could produce an offspring that would be homozygous for winglessness.

7. A mutant sex-linked trait called notched (X^N) is deadly in female *Drosophila* when homozygous. Males who have a single allele (X^N) will also die. The heterozygous condition ($X^N X^n$) causes small notches on the wing. The normal condition in both males and females is represented by the allele X^n.
   (a) Indicate the phenotypes of the F_1 generation from the following cross: $X^n X^N \times X^n Y$.
   (b) Explain why dead females are never found in the F_1 generation, no matter which parents are crossed. Use a Punnett square to help you.
   (c) Explain why the mating of a female $X^n X^N$ and a male $X^N Y$ is unlikely. Use a Punnett square to help you.

(continued)

8. When you first learned about the Punnett square, it was in the context of single-trait inheritance. What other types of inheritance have you investigated using this method? Have you found it difficult to adapt the Punnett square to fit different models of inheritance? Have you discovered any shortcuts to using the Punnett square?

Figure 1
DNA contains the information that ensures that pea plants produce seeds that grow into other pea plants.

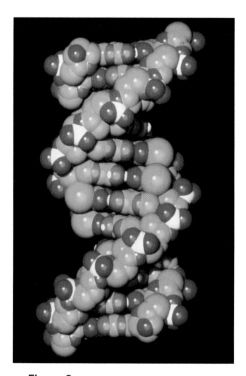

Figure 2
Computer-graphic model of DNA

5.4 Looking Inside the Chromosome

The nucleus of every cell in your body contains deoxyribonucleic acid, or DNA. This molecule is found in the cells of all organisms, from mushrooms to trees, from archaea to mammals. Scientists' fascination with DNA arises from the fact that it is the only molecule known to be capable of replication. Sugar molecules, protein molecules, and fat molecules cannot replicate. DNA can replicate, thereby permitting cellular reproduction.

DNA provides the means by which hereditary information is transmitted from generation to generation as new organisms are produced. DNA contains instructions that ensure the continuity of life. Pea plants produce seeds that grow into other pea plants because the DNA holds the chemical messages for producing the roots, stems, leaves, and seedpods of a pea (**Figure 1**). In a similar way, guinea pigs give birth to other guinea pigs, and humans conceive and give birth to other humans. However, you have learned that not all offspring are identical to their parents. The uniqueness of descendants can often be explained by new combinations of genes and by mutations.

When organisms grow, the new cells develop into specialized cells according to the instructions they receive from the DNA that has been passed down. In mature organisms, DNA carries the directions that guide the replacement and repair of worn cell parts. This information is carried by way of chemical messages between the nucleus, which is the control centre of the cell, and the cytoplasm, which contains the functioning organelles. Characteristics such as your hair colour, skin colour, and nose length are all coded within the chemical messages of DNA. Packed within the DNA are all the instructions that make you unique. Unless you are an identical twin, your DNA code is one of a kind.

In order to understand how genes affect the expression of an organism's traits, you will have to learn how DNA regulates the production of cell protein. This is a concept that is explored in detail in the next biology course. Proteins are the major structural and functional components of all cells. DNA not only provides for the continuity of life, but also accounts for the diversity of life forms.

Searching for the Chemical of Heredity

By the early 1940s, biologists began to accept the hypothesis that hereditary material resides within chromosomes. Chemical analysis indicated that a chromosome is made up of roughly equal amounts of proteins composed of amino acids and nucleic acids. By altering the sequence of amino acids in the protein chain, new proteins can be formed. Nucleic acids, like proteins, are very large molecules. However, nucleic acids share many similarities. The basic unit of nucleic acids is the nucleotide. DNA is a polymer, or long chain, of nucleotides (**Figure 2**). Each nucleotide is made of a phosphate group, a five-carbon sugar molecule, and one of four different nitrogenous bases.

At first, scientists speculated that the protein component of the chromosome was the genetic material that contained the hereditary message. Protein within the nucleus was thought to act as a master molecule, directing the arrangement of amino acids in the cytoplasm. Nucleic acids were thought to be monotonous repetitions of a single sugar molecule, identical phosphates, and four nitrogenous bases. The same sugar, phosphate, and nitrogenous bases were found in all living things. But how could a language be constructed upon such a limited alphabet? The hypothesis that the key to the genetic code lies in the proteins seemed logical, but it was actually incorrect.

Case Study: Evidence of Heredity Material

In 1928, an army officer named Frederick Griffiths attempted to create a vaccine against *Streptoccocus pneumoniae*. This bacterium, which causes pneumonia, exists in two forms. The first form is surrounded by a sugar-like compound called a capsule. The other variety has no capsule. When cells with a capsule divide, they form other cells with capsules. Cells without capsules form new cells without capsules.

The following is a summary of Griffiths's experimental procedure and results.

Procedure

1. Mouse A was injected with capsuled cells. Mouse B was injected with non-capsuled cells.
2. The capsuled pneumonia cells were heated and killed, and then injected into mouse C.
3. The heated capsuled cells were mixed with noncapsuled cells. The mixture was grown on a special growth medium. Cells from this culture medium were injected into mouse D.

Evidence

Mouse A contracted pneumonia and died, while mouse B continued to live.
Mouse C continued to live.
Mouse D died. An autopsy indicated that the mouse had died of pneumonia.

Analysis

A microscopic examination of the dead and live cell mixture revealed cells with and without capsules. Griffiths hypothesized that a chemical in the dead, heat-treated, capsuled cells must have altered the genetic material of the living non-capsuled cells. In similar previous experiments, scientists had dubbed this chemical the transforming principle.

Griffiths's unexpected results intrigued scientists. Oswald Avery, Colin MacLeod, and MacLyn McCarty at the Rockefeller Institute in New York City conducted a series of experiments that transformed harmless cells into disease-causing cells by mixing them with extracts of a capsuled or disease-causing strain (**pathogen**) that had been killed.

pathogen: a disease-causing organism

When Avery and his colleagues added protein-digesting enzymes to the extracts, the cells continued to be transformed, indicating that proteins were not the carriers of genetic information. However, when they added an enzyme that digests DNA but not proteins, the transformation of harmless cells stopped. In 1944, they reported that the hereditary material in the extracts was DNA and not the proteins found in the chromosome.

5.5 Discovering the Structure of DNA

After scientists confirmed that DNA was the material of heredity, their focus shifted to understanding how it works. Part of that understanding would come from knowing its structure, because, as mentioned earlier, structure provides many clues about function. In the race to be the first to discover the structure of DNA, scientists around the world employed emerging technologies to help them gain new insights into this mysterious "molecule of life." In the end, the honour would go to the scientists James Watson and Francis Crick.

James Watson (**Figure 1**) was considered a child prodigy when he entered the University of Chicago at the age of 15. He began studying ornithology (the study of birds), but eventually turned his attention to genetics and molecular biology. In 1951, he began studies at England's Cambridge University, where he met Francis Crick, a physicist who had served with the British army during World War II. Each would use their expertise in a different area of science to interpret and synthesize the experimental data that were being accumulated.

One important source of data came from the Cambridge laboratory of Maurice Wilkins, where researcher Rosalind Franklin used X-ray diffraction to help determine the structure of the DNA molecule. Franklin invented a technique in which she photographed the DNA molecule and obtained clear images showing its helical structure (**Figure 2**). She also identified where the phosphate sugars are located in the molecule.

Another clue came from comparing the chemical structure of DNA molecules in different organisms. Scientists already knew that molecules of DNA were made up of sugars (deoxyribose), phosphate, and four different nitrogenous bases: adenine, guanine, cytosine, and thymine. What they did not know was the way the components were arranged. New research revealed that, in a given species, the number of adenine molecules is the same as the number of thymine molecules and the number of guanine molecules is the same as those of cytosine. This suggested that the nitrogenous bases were arranged in pairs.

Figure 1
James Watson, Francis Crick, and Maurice Wilkins were awarded the Nobel Prize for physiology/medicine in 1962 for what many people believe was the most significant discovery of the 20th century. Unfortunately, Rosalind Franklin, whose work was a key part of the foundation on which the DNA model was built, had died of cancer by this time.

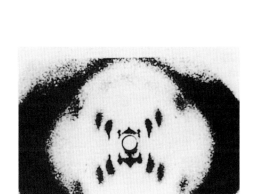

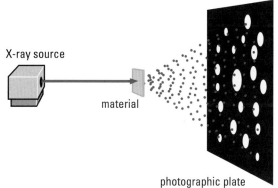

Figure 2
X rays are short, high-frequency electromagnetic waves. An X ray passing through a material scatters in a certain way. The degree and pattern of scattering depends on the organization of molecules that make up the material being scanned. When X rays are passed through a fragment of DNA and the diffraction patterns are captured on film, the resulting picture looks like this photo. Just as a shadow of your hand gives some information about its shape, the X-ray diffraction shadow of the DNA molecule provides a glimpse of the molecule's shape and the arrangement of its atoms.

Watson's background enabled him to understand the significance of the emerging chemical data, while Crick was better able to appreciate the significance of the X-ray diffraction results. With such data, Watson and Crick developed a three-dimensional model of the DNA molecule, which they presented to the scientific community in 1953. This model, which was visually confirmed in 1969, is still used by scientists today. The current version of the model incorporates some additional information gathered since Watson and Crick's time.

Models are very useful tools for scientists. A model airplane looks like the real thing, except that it is much smaller. By studying the replica, one can learn more about how actual planes work. The model of a molecule, however, cannot be scaled down for detailed study; molecules are already too small to see. Instead, molecules are made larger to show how the different atoms interact. X-ray diffraction, another type of modelling, provides a picture that indicates how different chemical bonds interact with each other. Scientists use models as visual devices that help them understand the relationship and interactions of different parts of the molecule.

Politics and Science

Watson and Crick might not have been the first to publish their findings on the structure of DNA were it not for politics. Franklin had adapted the X-ray diffraction technique to view an image of the DNA molecule, and Watson and Crick, also working in England, used her data to develop their model. The American scientist Linus Pauling (**Figure 3**), a leading investigator in the field, wanted to study the X-ray photographs, but was denied a visa to go to England. Pauling, along with others, had been identified by Senator Joseph McCarthy as a communist sympathizer for his support of the antinuclear movement (**Figure 4**). Many scientists believe that the United States passport office may have unknowingly determined the winners in the race for the discovery of the structure of DNA.

Figure 3
Linus Pauling

Figure 4
Senator McCarthy at hearings for communist activities

5.6 | The Structure of DNA

DNA is most often described as a double helix, which looks like a spiral ladder. The sugar and phosphate molecules form the backbone of the ladder, while the nitrogenous bases form the rungs. Nitrogenous bases from one spine of the ladder are paired with nitrogenous bases from the other spine by means of hydrogen bonds. A hydrogen bond is a weak bond that forms between the positive charge of the hydrogen atom at the end of one molecule and the negative charge on an electronegative nitrogen or oxygen atom on the end of another molecule. The backbone of the DNA molecule becomes twisted, which makes the molecule look like a winding, spiral ladder. The DNA molecule is made of individual units composed of deoxyribose sugars, phosphates, and nitrogenous bases. Each unit is referred to as a nucleotide. **Figure 1** shows two complementary nucleotides. Notice how the hydrogen bonds form between the **complementary base pair**. One estimate indicates that there are approximately 3.5 billion base pairs of DNA and 30 000 genes located on the 46 chromosomes of humans. Nitrogenous bases with double rings, called purines (adenine and guanine), always combine with nitrogenous bases with single rings, called pyrimidines (cytosine and thymine)(**Figure 2**, page 176). An adenine molecule always pairs with a thymine (with 2 hydrogen bonds), while a guanine molecule always pairs with a cytosine (with 3 hydrogen bonds)(**Figure 3**, page 176).

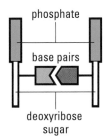

Figure 1

complementary base pair: a pair that forms between nitrogenous bases in the DNA molecule. The pairings are adenine and thymine (A–T) and guanine and cytosine (G–C).

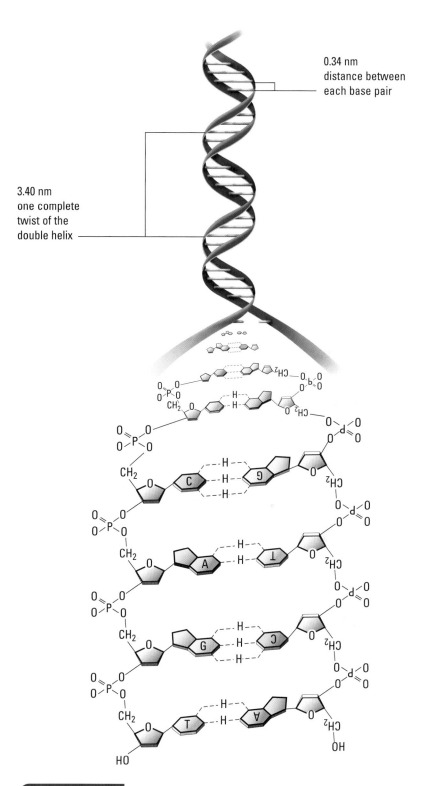

Figure 2
Structural formulas for the nitrogenous bases of DNA

0.34 nm distance between each base pair

3.40 nm one complete twist of the double helix

Figure 3
Representation of a DNA molecule. Note the complementary hydrogen bonds.

SUMMARY DNA

1. DNA occurs in every cell of all organisms and is the carrier of genetic information. DNA can replicate, permitting cell division.

2. Chromosomes are made of roughly equal parts protein and nucleic acids.

3. The amino acids that make up proteins can be organized in many ways, which means that a great number of different proteins are possible.

4. Nucleic acids are all quite similar and are made of nucleotides arranged in long chains. Nucleotides are composed of phosphates, sugar molecules, and one of four different nitrogenous bases: adenine, guanine, cytosine, and thymine.

5. Watson and Crick were first to publish a model of the structure of DNA. It is often described as a double helix. The helix is composed of specific nucleotide pairs: adenine and thymine, and guanine and cytosine. These pairs are called complementary base pairs.

Practice

Understanding Concepts

1. Name two ways in which the DNA molecule is important in the life of a cell.
2. What chemicals make up chromosomes?
3. What are nucleotides?
4. On what basis did some scientists believe that proteins provided the key to the genetic code?
5. Who were some of the scientists whose work led to the double-helix model of DNA?
6. How did the X-ray diffraction technique provide a vital part of the information needed for understanding the structure of DNA?
7. Which nitrogenous base pairs with guanine? Which base pairs with adenine?

5.7 Genes That Change Position

Until the 1980s, most scientists believed that genes occurred in fixed positions along the chromosome. With the exception of a few new combinations that might occur because of crossing over, chromosome structure was thought to be fixed. Barbara McClintock (**Figure 1**), an American biologist, shattered this traditional view with a new theory that genes can move to new positions. Her theory was dubbed the Jumping Gene Theory.

McClintock came to this conclusion after observing and interpreting the results of extensive experiments she had done on Indian corn. Colour variation in the kernels of Indian corn led her to hypothesize that there exist elements called **transposons**. The insertion of some genes into a new position along a chromosome inactivates the genes that affect the production of pigment. Barbara McClintock's well-documented experiments were largely ignored or dismissed by many in the scientific community. The idea of jumping genes opposed the widely accepted theory that genes are fixed in position along the chromosomes. Despite criticism, McClintock continued to gather data on jumping genes. Her contributions to science were recognized in 1983 when she was awarded the Nobel Prize. Even though she first announced her findings in the late 1940s, they were not well accepted for decades. The scientific community demands a high degree of support for new and conflicting ideas. This support involves the testing and retesting of the new ideas that appear to cast doubt on existing theories before they will be accepted. Without such careful scrutiny, the foundations of science might become too unstable.

Figure 1
In the late 1940s, Barbara McClintock's theory of jumping genes was greeted with skepticism by the scientific community.

transposons: specific segments of DNA that can move along the chromosome

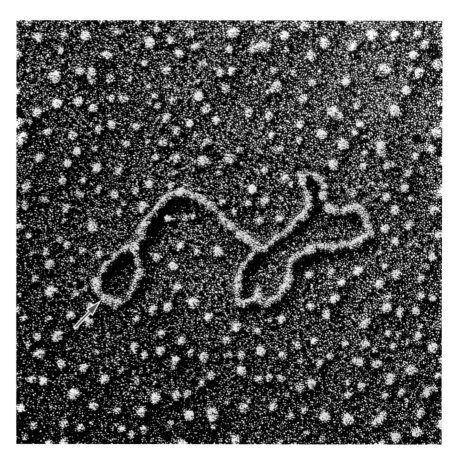

Figure 2
A bacterium which is found in the human gut has been treated so that the cell wall ruptures. The large circular chromosome is visible, as are the smaller plasmids, which are identified by a blue arrow.

plasmid: a small ring of genetic material

conjugation: a form of sexual reproduction in which genetic material is exchanged between two cells

recombinant DNA: DNA that is created when fragments of DNA from two or more different organisms are spliced together

restriction enzyme: an enzyme that attacks a specific sequence of nitrogenous bases on a DNA molecule to create fragments of DNA with unpaired bases

Recombinant DNA

Transposable genes have since been discovered in other organisms. Bacteria can insert genes apparently randomly along a circular chromosome. The transposon can even move from the chromosome in one bacterium to a chromosome in another bacterium. The inserted genes can be integrated with a secondary structure of DNA, called a **plasmid**, present in a bacterium. The plasmid is a small ring of genetic material that can be considered extra DNA (see **Figure 2**).

Some bacteria are capable of a form of sexual reproduction called **conjugation**. During bacterial conjugation, two or more cells fuse and plasmids are passed from one cell to another. This special property that makes it possible to insert DNA from one organism into the DNA of another was exploited by geneticists in a revolutionary new technology called genetic engineering. One technique led to the creation of **recombinant DNA**, DNA that is formed when DNA fragments from two or more different organisms are spliced together in a laboratory. In this technique, DNA from one organism is cut at specific sites by a **restriction enzyme**, leaving DNA fragments with unpaired nitrogenous bases or "sticky ends." Plasmid DNA is treated with the same restriction enzyme, resulting in plasmid DNA with "sticky ends." One fragment of DNA from one organism is inserted into the plasmid DNA where bonding occurs between complementary nitrogenous bases, creating recombinant DNA.

But bacterial conjugation (**Figure 3**) has also created some serious consequences for humans. Some disease-causing bacteria have evolved genes that are resistant to certain forms of antibiotics, like penicillin. Under normal conditions, antibiotics interfere with chemical reactions that occur within these harmful microbes. Genes that provide resistance to an antibiotic permit the disease-causing bacteria to continue living and damaging the cells of the body.

Cleaving and Recombining DNA

To create recombinant DNA, an enzyme called a restriction enzyme is added to DNA to cut it at specific sites. The result is fragments of DNA with unpaired nitrogenous bases.

Materials: a 20-cm strip of red Velcro (both the rough and the soft pieces), a 10-cm strip of blue Velcro, scissors, and a stapler or masking tape

Cleaving DNA

To simulate the effect of a restriction enzyme on DNA, do the following:

- Stick the 2 strips of blue Velcro together to represent 2 strands in a DNA molecule. Do not match the ends.
- To demonstrate the effect of the restriction enzyme, use scissors to cut the Velcro (DNA) into smaller strips (about 3 cm long) according to **Figure 4(b)**. This represents an organism's DNA that has been cut.
- Keep these strips for the next procedure.

Recombining DNA

- Stick the 2 strips of red Velco together, and staple or tape the ends to form a circle, as in **Figure 4(c)**. This circle represents plasmid DNA.
- Cut open the circle using the same procedure you used previously to demonstrate the effect of the restriction enzyme. Remove about a 3-cm long strip. See diagram **Figure 4(d)**.
- Now take one blue Velcro with the sticky ends (the organism's DNA) and combine it with the sticky ends of the red Velcro (the plasmid DNA) to form a circle. See **Figure 4(e)**.
- You have created a model of recombinant DNA. See **Figure 4(f)**.

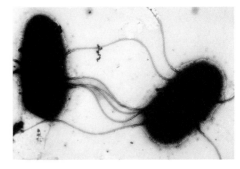

Figure 3
Bacteria exchange genetic material during conjugation. An appendage forms a bridge for the exchange of genetic material.

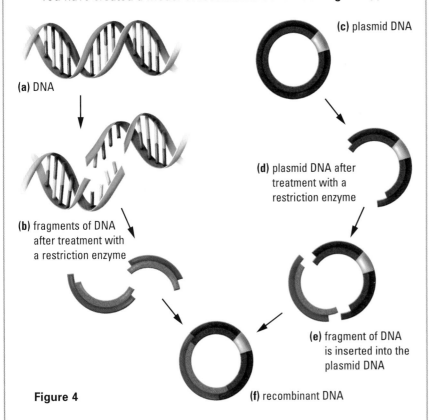

(a) DNA

(b) fragments of DNA after treatment with a restriction enzyme

(c) plasmid DNA

(d) plasmid DNA after treatment with a restriction enzyme

(e) fragment of DNA is inserted into the plasmid DNA

(f) recombinant DNA

Figure 4

5.8 Genetic Research and Technologies

Human Genome Project

In a series of meetings held in the mid-1980s, plans were developed to begin the process of producing maps of the entire genetic makeup of a human being (**Figure 1**). The international project began in the United States in October 1990 with James Watson, of Nobel Prize fame, as one of the first directors. The human **genome** consists of approximately 30 000 genes with the 23 pairs of chromosomes containing an estimated 3 billion pairs of nucleotides that make up the DNA. When the project began only about 4500 genes had been identified and the sequence of nucleotides that made up those genes determined. The collaborative efforts of many scientists from numerous countries and rapid improvements in sequencing techniques helped complete the gene map by May 2000 (see **Table 1**).

A DNA sequencing technique developed by the British biochemist Frederick Sanger was the most common method employed in the project. In this technique, pieces of DNA are replicated and changed so that the fragments, each ending with one of the four nucleotides, can be detected by a laser. In this way, automated equipment can determine the exact number of nucleotides in the chain. A computer is used to combine the huge amount of data and reconstruct the original sequence of nucleotides in the DNA.

The enormous quantity of DNA required for the project was initially produced through the cloning of human DNA in single-celled organisms, such as bacteria or yeast. Now a process developed by American biochemist Kary Mullins is used to make millions of copies of a single molecule of DNA in only a few hours.

Prior to the Human Genome Project, the genes for hereditary disorders such as cystic fibrosis, muscular dystrophy, and Huntington's chorea had been identified. The aim of the project is to add to this list so that new drugs and genetic therapies can be developed to combat genetic disorders. On the other hand, the project also has the potential of opening a Pandora's box full of controversial ethical questions, legal dilemmas, and societal problems. Who will own or control the information obtained and how will we prevent the misuse of the data?

genome: the complete set of instructions contained within the DNA of an organism

Table 1: Milestones in Genome Mapping

human genome completely mapped	June 2000
human chromosome 21 completely mapped	May 2000
human chromosomes 5, 16, 19 mapped	April 2000
Drosophila genome completed	March 2000
human chromosome 22 completed (the first chromosome to be mapped)	December 1999

DID YOU KNOW ?

On February 15, 2001, scientists from the Human Genome Project and Celera Genomics confirmed that there were approximately 30 000 genes in the human genome; a number far less than the original estimate of 120 000. This was determined using two different DNA sequencing techniques.

Other Facts

(1) 99.9% of the nitrogenous base sequences is the same in all humans.
(2) Only 5% of the genes contains the instructions for producing functional proteins; the remaining 95% does not have any real function.
(3) A worm has approximately 18 000 genes; a yeast cell has about 6000.

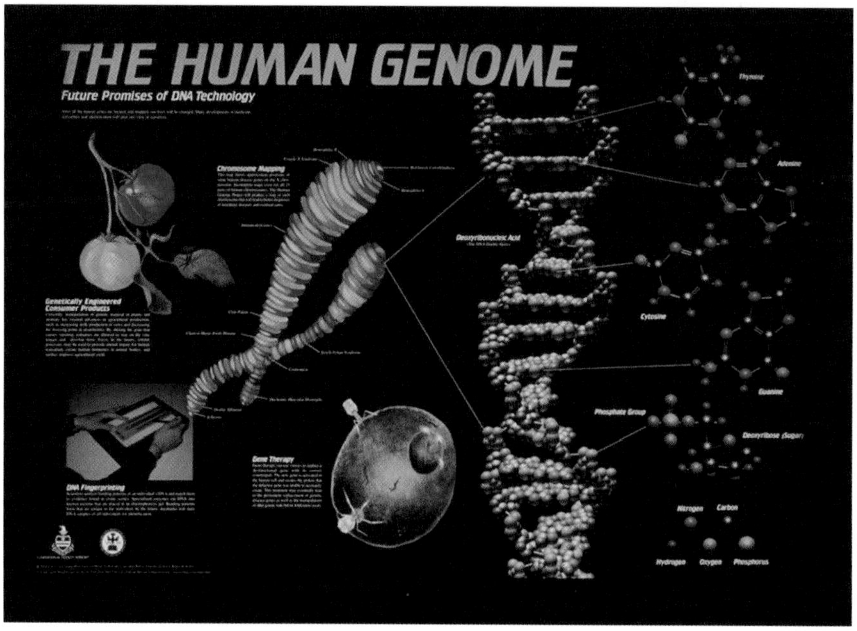

Figure 1
A poster of the Human Genome Project

Huntington's Chorea

Huntington's chorea, an incurable brain disorder, typically strikes people in the prime of life, setting them on an irreversible course of debilitating mental breakdown and eventual death. Victims have been described as being trapped inside their own bodies, unable to communicate. In 1993, Dr. Michael Hayden of the University of British Columbia led a team that isolated the gene responsible for Huntington's chorea. As a result, doctors can now use a simple DNA test to screen people with a family history of the disease and, with 98% accuracy, warn potential sufferers, or parents who might pass on the disorder.

Using another technique, known as animal modelling, laboratory mice are genetically manipulated to behave like people with Huntington's. They are then studied to find out how and why the disorder causes brain cells to die prematurely. The ultimate goal is to develop treatments to alleviate the effects of this fatal condition.

Cystic Fibrosis

Cystic fibrosis is an inherited disorder associated with a single gene, which produces a protein known as CFTR (cystic fibrosis transmembrane conductance regulator). Those who suffer from cystic fibrosis must inherit two defective alleles, one from each parent. Since approximately 1 in every 25 people of European ancestry carries the defective gene, cystic fibrosis is the most common recessive genetic disorder within this population. Families in Ontario who had more than one child suffering from cystic fibrosis provided most of the samples required for the laboratory research into locating the gene. The Hospital for Sick Children in Toronto is a leading centre for research into this disorder. Led by Dr. Lap-Chee Tsui, a team working at this hospital identified the gene in 1989. This group has mapped two modifier genes in animals that alter the severity of cystic fibrosis. The team continues to investigate the impact that these modifier genes can have on the treatment of this disorder.

Dr. Christine Bear of the University of Toronto, using a cystic fibrosis mouse model, has demonstrated that it is possible to correct the basic defect by delivering a normal version of the CFTR protein. Her team is working to develop a method of providing the protein therapy that would be able to correct the defect.

The field of genetics is also extending beyond the research laboratory and into daily medical practice. Dr. Judith Hall (**Figure 2**), head of pediatrics at the University of British Columbia (1999–2000) is a clinical geneticist with considerable experience dealing with families affected by genetic disorders such as cystic fibrosis. Communicating the genetic reasons for a disease to children and their families is both challenging and rewarding for Dr. Hall. Genetic research is advancing every day in finding ways to treat or prevent disorders.

Muscular Dystrophy

Muscular dystrophy is the name given to a group of genetic disorders that cause the weakening and deterioration of muscles. Some forms of muscular dystrophy result from defects on autosomal chromosomes and may occur in both males and females. Other forms are sex-linked and affect males. The rare form known as Emery-Dreifuss muscular dystrophy occurs in males as a result of a recessive X-linked gene. However, even female carriers, who have the normal dominant allele in addition to the recessive defective allele, can exhibit some mild symptoms.

Much of the research on muscular dystrophies has occurred in Canada. Dr. Ron Worton and his colleagues at the Ontario Hospital for Sick Children opened a new era in research and treatment for this group of disorders when they located

Figure 2
Dr. Judith Hall

the gene responsible for Duchenne muscular dystrophy in 1987. This is a sex-linked form that begins to affect boys between ages two and six. They suffer progressive damage beginning in the muscles of the pelvis, upper arms, and legs. The calf muscles enlarge because the enzyme creatinine kinase leaks out from the muscles causing them to swell. Most boys with Duchenne are confined to a wheelchair by age 12.

There are other forms of muscular dystrophy that affect males and females equally, since the defective genes are carried on the autosomes. Some forms are due to recessive alleles and others are caused by dominant alleles.

Canadian researchers who have identified genes for muscular dystrophy and related disorders are summarized in **Table 2**.

Table 2: Researchers of Some Genetic Disorders

Disorder	Year	Institution	Researchers
Duchenne muscular dystrophy (sex-linked, affecting males)	1987	Ontario Hospital for Sick Children	Dr. Ron Worton
amyotrophic lateral sclerosis (Lou Gehrig's disease)	1993	Montreal General Hospital	Dr. Denise Figlewicz Dr. Guy Rouleau
spinal muscular atrophy (SMA)	1995	Ontario Research Institute, Ottawa	Dr. Alex MacKenzie
oculopharyngeal muscular dystrophy (autosomal dominant)	1997	McGill University, Montreal	Dr. Guy Rouleau
autosomal spastic ataxia	2000	Montreal General Hospital/ Saint-Justine Hospital	Andrea Richter Dr. Serge Melancon Dr. Thomas Hudson

5.9 DNA Fingerprinting

In 1986, DNA matching was used to identify a rapist–murderer in Leicester, England. More than 1000 men were tested in three villages near Leicester. The technique, called DNA fingerprinting (**Figure 1**), was used to free one suspect, and led to the arrest and subsequent confession of another. Later in the same year, a rapist in Orange County, Florida, was convicted on the basis of genetic evidence.

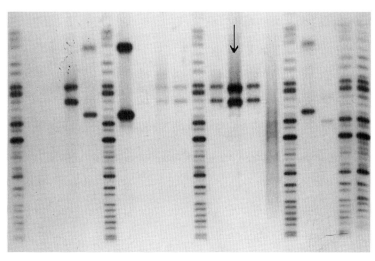

Figure 1
DNA profiling can be used to identify criminals. The width of the bands and the separation between them create a pattern that is unique to each individual.

The DNA fingerprinting test used in Britain was developed by Alec Jeffreys, a geneticist from the University of Leicester. Jeffreys found that long stretches of the DNA molecule are similar from one person to another. In fact, geneticists estimate that far less than one percent of DNA is unique to an individual. However, particular segments of DNA do contain unique sequences of nitrogenous bases, with only identical twins sharing the same nitrogenous base sequences in these sections. These short segments of DNA appear to have no function. Geneticists believe that the codes are nonsense codes, which seem to repeat as though they are a chemical type of stuttering.

In the DNA fingerprinting test, DNA may be isolated from samples of skin, hair, semen, or blood at the crime scene or from the victim and compared with DNA obtained from a blood sample from the suspect. The DNA samples are cut with restriction enzymes. These restriction enzymes make it possible to cut DNA at specific points. The result is reproducible DNA fragments. Differences in the molecular structure of each person's DNA mean that the location and number of cuts that can be made are distinct for each person, so that the profile of each person's DNA is unique. The DNA fragments are then transferred to a nylon sheet, where a radioactive marker identifies the unique sequence of the DNA chain. The nylon sheet is then placed against an X-ray film. Black bands appear where the markers have attached to the segments used to establish identity. A print is made from the film and used to compare samples.

DNA Fingerprinting and the Law

Assume that you are carrying out a criminal investigation. A rare and very expensive painting has been taken from a gallery. For forensics evidence, you have collected a blood sample from a broken window, the likely point of entry. Five known suspects have been identified. A court order allows the collection of a blood sample from each of the suspects. Gel electrophoresis is performed on the suspects' blood samples and on the sample obtained from the broken window. Gel electrophoresis is a process used to separate the fragments of DNA cleaved by restriction enzymes. An electric field is used to move the negatively charged DNA molecules to produce a pattern of bars, formed when the molecules migrate along a gel surface. The location of the bars produced by an unknown sample can be compared to the location of bars produced by a known sample. Matching bars serve to identify the unknown (**Figure 2**).

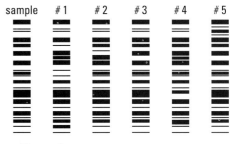

Figure 2

• Analyze the evidence provided.
  (a) Which suspect's gel electrophoresis pattern matches the sample's pattern?
  (b) Would you convict this individual on the basis of the DNA evidence? Explain why or why not.

Practice

Understanding Concepts

1. When you eat fish, you take in fish protein and fish DNA; however, you do not assume the characteristics of a fish. Explain why not.

2. What is DNA fingerprinting and how does it work?

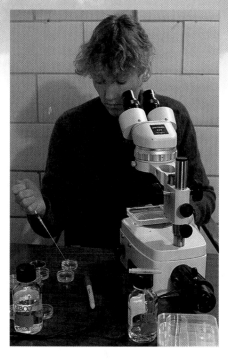

Forensic Scientist

Police are using more sophisticated technologies to determine the identity of criminals. DNA fingerprinting, for example, is a powerful tool in the hands of a trained technician.

Horticulturalist

A horticulturalist may work in a flower shop, in the plant import/export trade, or in a greenhouse or garden. It is essential for this person to have detailed knowledge about plant reproductive cycles, growing habits, and resistance to diseases.

Animal Breeder

Animal breeders use techniques such as artificial insemination and fertility drugs to obtain high-quality livestock. Such techniques require the identification of the best breeding period and of the most desirable traits.

Practice

Making Connections

3. Identify several careers that require knowledge about genetic continuity.
   Follow the links for Nelson Biology 11, Chapter 5 Career.

   GO TO www.science.nelson.com

4. Select a career you are interested in from the list you made.
   (a) What university program do you need to take that will lead to that career?
   (b) Which universities offer degree programs in this area?
   (c) Which high-school subjects are required to enter the university program?

5. How are the concepts you have learned in this unit used in the career?

6. Investigate and describe the responsibilities and duties involved in this career. What appeals to you about the career? What do you find less attractive?

7. Survey the newspaper or conduct a web search to identify career opportunities in that area. What is the average income of someone with your chosen career?

Genetics Counsellor

A genetics counsellor evaluates individuals or families who may be at risk of having a genetic condition that could be passed on to their children. A family history is gathered and analyzed. If necessary, the counsellor recommends clinical examinations and tests to establish diagnosis and determine the risk to family members and future children.

5.10 Gene Therapy

Currently over 3500 different genetic diseases have been linked to defective genes. Cystic fibrosis, diabetes, hemophilia, Huntington's chorea, and sickle cell anemia are but a few of the hereditary disorders that affect Canadians. Although there are treatments, like insulin injections for diabetes, that can control the affliction, no true cures have been found. Medical advances coupled with modified diet, restricted behaviour, and environmental adjustments have enabled people, who just a century ago would have been doomed by such disorders, to continue living a reasonably productive life.

Imagine the potential of transforming the defective gene. Although **gene therapy** is in the early stages of development, the prospects of correcting defective genes present an exciting possibility in the quest to conquer genetic disorders. There are three possible strategies for gene therapy. The first strategy involves gene insertion. The normal gene is inserted into position on the chromosome of an affected cell. This can be accomplished by employing a virus, or some other agent, to carry the gene into position where it begins to function in the intended manner. Because not every cell uses a particular gene, the insertion can be restricted to those cells in which the gene is active. For example, diabetes occurs because cells of the pancreas do not produce sufficient amounts of insulin. The normal gene for insulin production need only be inserted into specialized cells within the pancreas. Muscle cells do not produce insulin and, therefore, the gene would not have to be inserted into a muscle cell. A second method involves gene modification. The defective gene is modified chemically in an effort to recode the genetic message. This method is much more delicate and requires greater knowledge of the chemical composition of the normal and the defective genes. The third technique, gene surgery, is the most ambitious. The defective gene is actually removed and replaced with a normal gene.

The first clinical use of gene therapy occurred in 1990 when a four-year-old girl was given treatment for an inherited disorder of the immune system that resulted in an enzyme deficiency. In this usually fatal condition, insufficient amounts of the enzyme adenosine deaminase (ADA) were produced. A genetically modified virus was used to carry a normal ADA gene into her immune cells. The newly inserted gene caused the cells to begin producing the required enzyme.

Certain forms of viruses normally reproduce themselves by inserting their own DNA into normal body cells and reprogramming them to produce new viruses. As a result, they can be modified so that they will carry beneficial genetic material as well. The trick is to cause them to insert the material in the appropriate place so that it will work as intended rather than cause some unexpected result.

Canadian scientists have been in the forefront of research into gene therapy. In 1999, McGill professor Michel Tremblay and Dr. Brian Kennedy of the Merck Frosst Center for Therapeutic Research identified the gene responsible for controlling the production of the enzyme tyrosine phosphatase. Since this enzyme is involved in regulating blood glucose levels, it plays a significant role in diabetes mellitus (Type II diabetes) and obesity. This knowledge can lead to gene therapy techniques to control these problems. In 2000, a joint team from South Korea and Canada announced the successful treatment of juvenile (Type I) diabetes in mice through the use of gene therapy. Numerous projects underway offer great hope for those who suffer the effects of a wide variety of genetic disorders.

gene therapy: a procedure by which defective genes are replaced with normal genes to cure genetic disorders

○ Define the Issue   ○ Analyze the Issue
○ Identify   ○ Defend a Decision
   Alternatives   ○ Evaluate
○ Research

Explore an
Issue

Take a Stand: Gene Therapy

The successes of gene therapy are modest, but its boundaries are extended almost daily. The attempt to cure disease and reduce suffering has found few opponents. However, there are worries that gene therapy will be abused. In addition, some people believe that tampering with DNA is socially, morally, and ethically wrong.

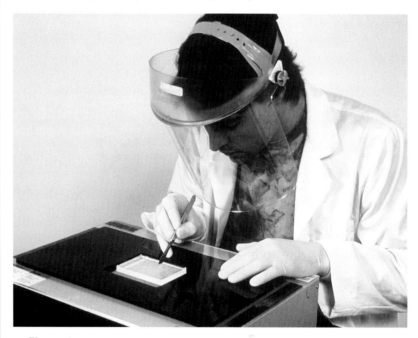

Figure 1

Statement

Gene therapy should be supported as one of the most promising medical technologies for the future.

- In your group, research the issue.
- Search for information in newspapers, periodicals, CD-ROMs, and on the Internet.
  Follow the links for Nelson Biology 11, 5.10.

  GO TO   www.science.nelson.com

- Identify the perspectives of each of the opposing positions.
- Develop and reflect on your opinion.
- Write a position paper summarizing your views.

DID YOU KNOW ?

It has been estimated that 8% of the human population will show some signs of a genetic disorder by age 25.

SUMMARY   DNA Fingerprinting and Gene Therapy

1. Some segments of DNA are different in each person, so each individual has a distinctive array of short segments. In DNA fingerprinting, these segments are used to identify people.

2. Gene therapy involves three possible techniques: inserting a normal gene into the chromosomes, chemically modifying the defective gene to make it behave normally, or extracting the defective gene and replacing it with a normal one.

Key Expectations

Throughout this chapter, you have had opportunities to do the following:

- Summarize main scientific discoveries of the 19th and 20th centuries that led to the modern concept of the gene (5.1–5.10).
- Explain how the concepts of DNA, genes, chromosomes, and meiosis account for the transmission of hereditary characteristics from generation to generation, and demonstrate an understanding that a genetic disorder linked to the sex chromosomes is more likely to be expressed in males than in females (5.2, 5.3, 5.4, 5.5, 5.8, 5.9).
- Demonstrate the skills required to plan and carry out investigations (5.3).
- Select and use appropriate modes of representation to communicate scientific ideas (5.3).
- Compile qualitative and quantitative data from a laboratory investigation on monohybrid and dihybrid crosses, and present the results by hand or computer (5.3).
- Predict the outcome of various genetic crosses (5.3).
- Solve basic genetic problems involving sex-linked genes using the Punnett method (5.3).
- Explain, using Mendelian genetics, the concept of sex linkage (5.3, 5.8).
- Identify and describe examples of Canadian contributions to knowledge about genetic processes, and to technologies and techniques related to genetic processes (5.3, 5.8, 5.10).
- Research genetic technologies using sources from print and electronic media, and synthesize the information gained (5.7, 5.8, 5.9, 5.10).
- Identify and describe science- and technology-based careers related to the subject area being studied (5.9).
- Locate, select, analyze, and integrate information on topics being studied (5.9, 5.10).

Key Terms

Barr body
complementary
  base pair
conjugation
gene therapy
genome
mutations
pathogen

plasmid
recessive lethal
recombinant DNA
restriction enzyme
sex-linked traits
somatic cells
transposons

Reflect on your Learning

Revisit your answers to the Reflect on Your Learning questions at the beginning of this chapter.

- How has your thinking changed?
- What new questions do you have?

Make a Summary

In this chapter, you studied cells and chromosome structure to better understand how genes work. To summarize your learning, create a poster of a human genome that shows the principles of sex-linked genes and helps show the relationship between genes and chromosomes. Label the sketch with as many of the key terms as possible. Check other posters and use appropriate ideas to make your poster clear.

Understanding Concepts

1. In what ways was the development of the chromosome theory linked with the development of the light microscope?

2. Discuss the contributions made by Walter Sutton, Theodor Boveri, Thomas Morgan, and Barbara McClintock in the development of the modern-day chromosome theory of genetics.

3. What led scientists to speculate that proteins were the hereditary material?

4. Science and technology share a relationship of mutual reciprocity. This means that not only do scientific breakthroughs provide information for technological applications, but technological advances also spur scientific progress. Explain the role X-ray diffraction techniques played in the discovery of the structure of DNA.

5. Using DNA as an example, explain why scientists use models.

6. Compare the amount of DNA found inside one of your muscle cells with the amount of DNA found in one of your brain cells.

7. Why are organ transplants more successful between identical twins than between other individuals?

8. Aristotle suggested that heredity could be linked to male semen. Other scientists suggested that only the female determined the traits of offspring. Based on the knowledge that you have gathered about genetics, cite physical evidence that would refute both of these theories.

Applying Inquiry Skills

9. In *Drosophila*, the allele for wild-type eye colour is dominant and sex-linked. The allele for white eye colour is recessive. The mating of a male with wild-type eye colour with a female of the same phenotype produces offspring that are $\frac{3}{4}$ wild type and $\frac{1}{4}$ white eyed. Indicate the genotypes of the parents, and of the offspring in the F_1 generation.

10. The autosomal recessive allele (*tra*) transforms a female *Drosophila* into a phenotypic male when it occurs in the homozygous condition. The females that are transformed into males are sterile. The *tra* allele has no effect in XY males. Determine the F_1 and F_2 generations from the following cross: XX, +/*tra* crossed with XY, *tra/tra*. (Note that the + indicates the normal dominant gene.)

11. Use the information from the pedigree chart (**Figure 1**) to answer the following questions:
    (a) Indicate the phenotypes of the parents.
    (b) If parents #1 and #2 were to have a fourth child, determine the probability that the child would have hemophilia.
    (c) If parents #1 and #2 were to have a second male, determine the probability that the boy would have hemophilia.
    (d) Indicate the genotypes of #4 and #5.

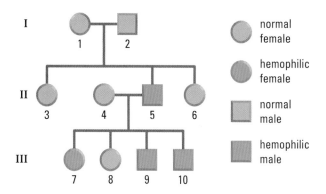

Figure 1

12. Edward Lambert, an Englishman, was born in 1717. Lambert had a skin disorder that was characterized by very thick skin, which was shed periodically. The hairs on his skin were very coarse and quill-like, giving him the name "porcupine man." Lambert had six sons, all of whom exhibited the same traits. The trait never appeared in his daughters. In fact, the trait has never been recorded in females. Provide an explanation for the inheritance of the "porcupine" trait.

Making Connections

13. Suppose you read the following headline in a supermarket tabloid: "Genes that produce chlorophyll in plants are inserted into the chromosomes of cattle." Indicate some of the possible advantages of this unlikely procedure.

14. People with Huntington's chorea, a dominant neurological disorder, often do not know they have the disorder until they are between 35 and 45 years of age, and the symptoms do not express themselves until the people are well into their 60s. Suggest why the later development of the disorder might be related to its greater frequency in the population.

15. Explain the significance of locating the cystic fibrosis gene.

16. The ability to use gene therapy to fix the genes of people who have genetic disorders lies on the horizons of genetic research. Gene therapy has been employed for some blood disorders and may one day be considered a cure-all. One genetic blood disorder, sickle cell anemia, results from a single defective gene that causes hemoglobin to form abnormally. Gene therapy could, in theory, turn off the defective gene and turn on the normal gene. Although inconclusive, the technique shows promise. Should scientists attempt to develop treatments that turn genes on and off? Do scientists have the right to alter human genes? Support your opinion.

17. Muscular dystrophies are a group of hereditary disorders characterized by the deterioration and weakening of the skeletal muscles. Duchenne muscular dystrophy (DMD) is the most common and serious of this group of disorders. Prior to DNA analysis that permitted screening for the defective gene, early diagnosis was difficult because symptoms are mild. By the time the child reaches approximately three years of age, the symptoms become more pronounced. **Figure 2** shows a chart for a family in which the gene is present.

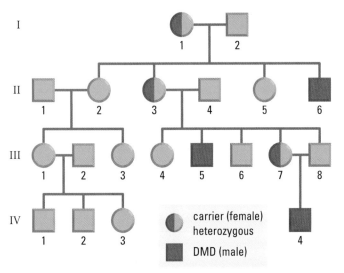

Figure 2

(a) Is DMD dominant or recessive? Explain your answer.
(b) Is the defective gene located on a sex chromosome? Explain your answer.
(c) For generation I, provide the probable genotypes of parents 1 and 2.
(d) Inheriting a defective gene is often described as a matter of chance. For parents 3 and 4 in generation II, compare the theoretical probability of passing on the defective gene to the actual result.

(e) Prenatal diagnosis for DMD allows parents to know if the defective gene is present before the birth of their child. People with the disorder rarely live beyond 20 years of age. Should prenatal testing be encouraged?
(f) Prior to genetic screening technology, women who had a family history of DMD would rely on ultrasound and karyotype analysis for sex determination. Why would these women want to know the sex of the child? Is it possible that some of the early screening techniques could have led to the termination of a genotypically normal child? Explain why or why not.

18. Despite meeting all of the criteria required by good scientific research, Barbara McClintock's work on Indian corn was ignored for almost 20 years. The significance of her work was only acknowledged once other scientists, working with bacteria, confirmed that genes move along chromosomes. Why do you think her work was not accepted?

Exploring

19. DNA fingerprinting has had a tremendous impact on law enforcement. Find some Canadian examples of criminal trials in which DNA fingerprinting has been submitted as evidence. How did the fingerprint evidence influence the trial? Should DNA fingerprinting ever be used to convict a criminal? Why or why not?
Follow the links for Nelson Biology 11, Chapter 5 Review.
GO TO www.science.nelson.com

20. Use the library and the Internet to research genetic technologies described in this chapter. Which technology did you choose to research and why? Prepare a short report about your topic, synthesizing the information you learned. Include in your report an analysis of the advantages and disadvantages of the technology, both scientific and social.
Follow the links for Nelson Biology 11, Chapter 5 Review.
GO TO www.science.nelson.com

Investigating Human Traits

Human traits are autosomal or sex-linked. Autosomal traits appear in equal frequency in female and male populations. Sex-linked traits appear with greater frequency in one sex than in the other.

Your task is to use your background knowledge in Mendelian genetics and inheritance to determine if certain traits are autosomal or sex-linked.

Investigation

For this task, you will design and carry out a correlational study to determine if certain traits are autosomal or sex-linked by examining the frequency of each trait for each sex. You will investigate the traits listed in **Table 1**. Refer to Appendix A to learn more about a correlational study.

Table 1: Traits That May Be Investigated

Trait	Dominant	Recessive
hairline	pointed on forehead	straight across forehead
ear lobe	suspended	attached to head
eyesight	nearsighted	normal vision
thumb joint	last joint bends out	last joint is straight
folded hands	left thumb over right	right thumb over left
tongue rolling	can be rolled into U-shape	cannot be rolled
clenched fist	two wrist cords	three wrist cords
chin dimple	dimple in middle	no dimple
colour vision	normal colour vision	red–green colourblind

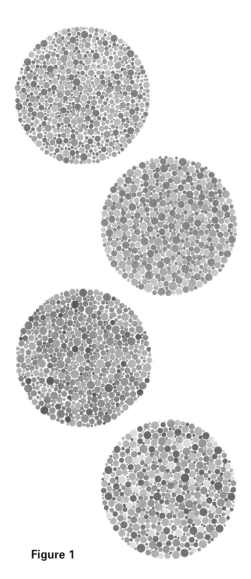

Figure 1

To determine whether or not an individual is red-green colourblind, use **Figure 1**. Those who can read the numbers embedded within the diagrams have normal colour vision. (One diagram does not have a number embedded in it.)

An individual who has myopia (or near-sightedness) has trouble seeing clearly at a distance. Objects that are at a distance will appear out of focus while close objects will be very clearly seen. **Figure 2** is an example of what a person with myopia will see.

For this task, you and your group will be assigned to survey a group of twenty people, with an equal number of males and females, either within the school population or outside. Data from each group will be recorded in one table. The larger table will represent the data for the population.

In your Design, list the materials and survey that you will use. Write a brief plan for carrying out your study. Your Design should include the following information:

- division of labour—who will do what
- the tools (e.g., survey) that you will use to gather your data
- how you will gather your data
- how will you display the data
- a description of your sample

After your teacher has approved of the Design, a general design will be provided for all groups to standardize the Procedure. Then you and your group will carry out the study.

Figure 2
People with myopia will see close objects very clearly, while those in the background will appear out of focus.

Write a detailed report to communicate the Procedure, Observations, and Analysis of your study. Use appropriate scientific vocabulary, tables, and correct significant figures, where appropriate.

Question

Which traits in **Table 1** are autosomal and which are sex-linked?

Materials

survey or questionnaire

(a) Complete the materials list.

Procedure

Equal numbers of males and females must be surveyed. Indicate where your sample was obtained.

1. Complete the procedure.

Observations

(b) Display the data for your sample in a chart, table, or graph.

(c) Display the data for the population in a chart, table, or graph.

Analysis

(d) Determine the frequencies and percentages of each trait by sex for your sample and for the population.

(e) Indicate which traits are autosomal and which are sex-linked.

(f) Compare your results for your sample with the entire school population. Explain any differences in the data, if any. You might consider sample size, population, chance, etc. For more information, on the Internet follow the links for Nelson Biology 11, Unit 2 Task.

GO TO www.science.nelson.com

Evaluation

(g) Evaluate your evidence.

(h) Critique your experimental design, the materials, the tools used to obtain the data, the sample size, and the population examined.

(i) Suggest how you could improve your study if you were to repeat it.

Synthesis

(j) Alfred, who suffers from disorder X, marries Betty, who does not suffer from the disorder.

Alfred and Betty have a son, Charles, and two daughters, Debbie and Emily. None of the children has disorder X.

Charles marries and has two sons, both of whom do not have the disorder.

Debbie marries and has one son and two daughters. None of her children has the disorder.

Emily marries and has one son who suffers from the disorder.

(i) Draw a pedigree chart to display the information given.

(ii) Give two reasons for deciding that the pedigree chart is for an X-linked trait.

(iii) Indicate Emily's genotype.

Understanding Concepts

1. Which event occurs before and which occurs at the end of the division phase of the cell cycle?
   (a) synapsis
   (b) crossing over
   (c) cytokinesis
   (d) DNA replication

2. The cell cycle of a whitefish embryo is shown in **Figure 1**.
   (a) Identify the numbers that represent anaphase and telophase, respectively.
   (b) Describe the events that take place during interphase.
   (c) Explain what happens during cytokinesis.

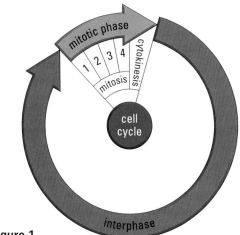

Figure 1

3. Refer to **Figure 2**. In which process does a reduction division occur?
   (a) process 1   (b) process 2   (c) process 3   (d) process 4

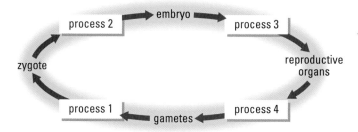

Figure 2

4. In **Figure 2**, which process represents a return to a diploid chromosome number?
   (a) process 1   (b) process 2   (c) process 3   (d) process 4

5. In **Figure 2**, which processes represent mitosis?
   (a) processes 1 and 2   (b) processes 2 and 3
   (c) processes 3 and 4   (d) processes 4 and 1

6. Use **Table 1** to identify which of the following statements are correct.
   (a) Animals with greater intelligence have more chromosomes.
   (b) Sperm cells from a cat and a horse have the same chromosome number.
   (c) The larger the animal, the greater the chromosome number.
   (d) An unfertilized egg from a dog contains 39 chromosomes.
   (e) A zygote from a fruit fly contains 16 chromosomes.
   (f) A brain cell of a human contains 92 chromosomes.
   (g) The chromosome number of a sperm cell from a cat is 19.
   (h) The somatic cell of a dog contains 39 chromosomes.

Table 1

Organism	Chromosome number (2n)
dog	78
human	46
fruit fly	8
cat	38
horse	66

7. Choose two organisms from **Table 1** and indicate the haploid chromosome number and the number of pairs of chromosomes in diploid cells before, during, and after meiosis.

8. Gregor Mendel did not have an understanding of mitosis or meiosis when he formulated his law of segregation. If he had viewed dividing cells under a microscope, he might have described the law of segregation in this way:
   (a) Dominant alleles are expressed following mitosis.
   (b) Dominant alleles are expressed following meiosis and the formation of gametes.
   (c) Paired alleles separate during mitosis and each of the paired alleles is found in a different gamete.
   (d) Paired alleles separate during meiosis and each of the paired alleles is found in a different gamete.

9. Examine **Figure 3**. Identify the processes that represent each of the following activities:
   (a) the process of mitosis
   (b) the division of diploid cells to produce haploid cells
   (c) the division of diploid cells to produce identical daughter cells
   (d) the union of haploid gametes to produce a diploid cell
   (e) the process of meiosis

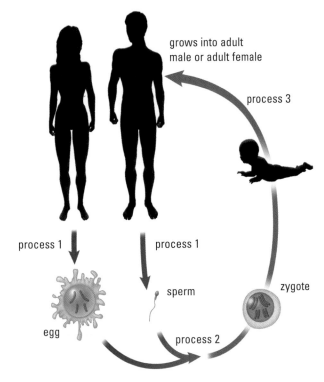

grows into adult
male or adult female

process 3

process 1

process 1

sperm

zygote

egg

process 2

Figure 3

10. The typical life cycle for a plant and for an animal are shown in **Figure 4**.
    (a) For both life cycles, identify structures that have a haploid chromosome number.
    (b) On the diagram, indicate where haploid cells become diploid cells.

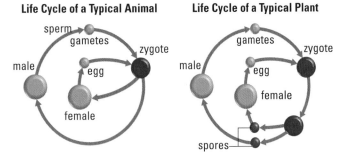

Life Cycle of a Typical Animal

sperm
gametes
zygote
male
egg
female

Life Cycle of a Typical Plant

gametes
zygote
male
egg
female
spores

Figure 4

11. How does Mendel's work apply to the concepts of dominance and recessiveness?

12. In cows, the polled trait (hornless) is dominant over the horned trait. The gene responsible for the trait is not located on a sex chromosome. A single bull mates with three different cows and produces offspring, as shown in **Figure 5**. What are the genotypes for the bull, cow A, and cow B?

(a) bull–*Pp*, cow A–*pp*, cow B–*Pp*
(b) bull–*PP*, cow A–*pp*, cow B–*Pp*
(c) bull–*Pp*, cow A–*pp*, cow B–*pp*
(d) bull–*PP*, cow A–*Pp*, cow B–*Pp*

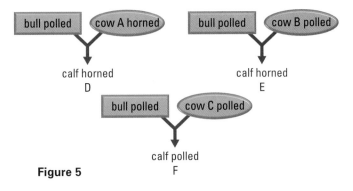

| bull polled | cow A horned | | bull polled | cow B polled |

calf horned
D

calf horned
E

| bull polled | cow C polled |

calf polled
F

Figure 5

13. In question 12, which of the cattle could have two possible genotypes?
    (a) cow C and calf F
    (b) cow B and calf E
    (c) cow A and calf D
    (d) bull and calf D

14. Describe one of the genetic disorders you learned about in this unit. Include in your description the chromosomes responsible for the trait, physical effect, and possible treatments.

15. Refer to **Figure 6**. The phosphate and sugar backbone of the DNA molecule can be identified as which of the following structures, respectively?
    (a) 1 and 2   (b) 2 and 3
    (c) 3 and 4   (d) 4 and 5

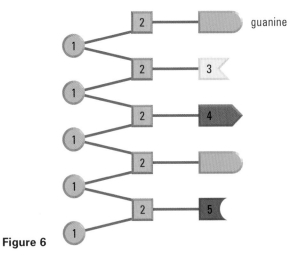

guanine

Figure 6

16. In **Figure 6**, the nitrogenous base cytosine would be represented by which structure?
    (a) 2   (b) 3   (c) 4   (d) 5

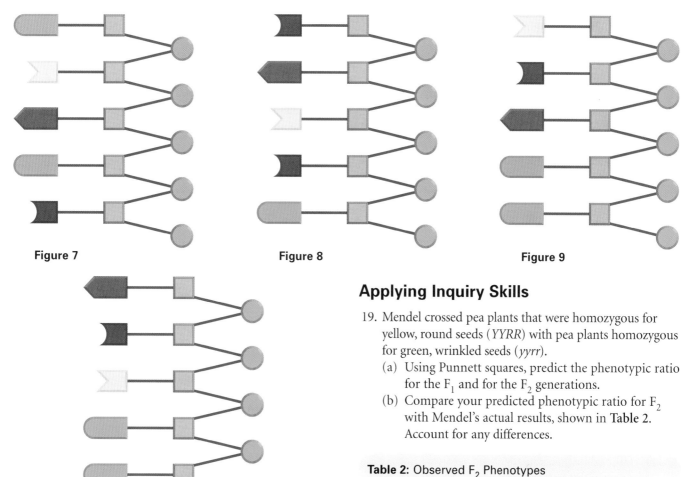

Figure 7

Figure 8

Figure 9

Figure 10

17. Which of **Figures 7–10** would represent the strand of DNA complementary to the one in **Figure 6**?

18. Scientists believe that an autosomal gene, called BRCA$_1$, can be linked to 5–10% of all cases of breast cancer. About 80% of the women who inherit the gene are likely to develop breast tumors. Men who carry the mutated gene rarely develop breast cancer, but they can pass on the gene to their offspring.
    (a) Is the mutant gene dominant or recessive? Explain.
    (b) A woman who carries the gene marries a man with no history of breast cancer in his family. If they have four daughters, what is the probability that a daughter could inherit the mutant gene for BRCA$_1$?
    (c) What evidence suggests that environmental factors affect the expression of the gene?
    (d) Explain how detection of the mutant BRCA$_1$ gene could help reduce the incidence of breast cancer.
    (e) Why would it be important not to release someone's genetic blueprint?

Applying Inquiry Skills

19. Mendel crossed pea plants that were homozygous for yellow, round seeds (*YYRR*) with pea plants homozygous for green, wrinkled seeds (*yyrr*).
    (a) Using Punnett squares, predict the phenotypic ratio for the F$_1$ and for the F$_2$ generations.
    (b) Compare your predicted phenotypic ratio for F$_2$ with Mendel's actual results, shown in **Table 2**. Account for any differences.

Table 2: Observed F$_2$ Phenotypes

Number of plants	Phenotype
315	round, yellow seeds
108	round, green seeds
101	wrinkled, yellow seeds
32	wrinkled, green seeds
Total: 556 F$_2$ plants observed	

20. Eye colour in *Drosophila* is controlled by different genes. Refer to **Table 1** in section 4.4.
    (a) Determine the genotypes and phenotypes of the F1 generation from the following cross: $E^1E^4 \times E^2E^3$.
    (b) Which parents would produce the following offspring: 50 wild-type eye colour, 26 honey eye colour, and 24 white eye colour?

21. Feather colour in parakeet birds is controlled by two genes, which are located on different chromosomes. One of the pigment genes is regulated by allele *B* that produces a blue colour. The recessive allele *b* does not produce any colour. The other pigment gene is controlled by allele *Y* that produces a yellow colour. The recessive allele *y* does not produce any colour. The presence of both the *Y* and *B* alleles produces parakeets with green feathers.

(a) Indicate the genotypes and phenotypes of the F$_1$ generation produced from the following cross: *BbYY* × *bbYy*.

(b) What is the probability of producing a yellow bird from two parents that carry heterozygous alleles for green feathers?

22. **Table 3** lists a number of inherited traits found in house cats.

Table 3

Dominant phenotype	Dominant allele	Recessive phenotype	Recessive allele
pointed ears	*E*	folded ears	*e*
smooth hair	*S*	curly hair	*s*
polydactyl (more than 5 digits)	*P*	5 digits	*p*

A female with pointed ears, smooth hair, and 5 digits mates with an unknown male and produces 8 kittens. The following phenotypes were noted in the offspring:

- All of the kittens have 5 digits.
- Six have smooth hair and 2 have curly hair.
- Four have pointed ears and 4 have folded ears.

(a) Indicate the genotype of the female for all 3 traits.

(b) Indicate the genotype and phenotype of the unknown male.

(c) If this female mated with a male (genotype *EESsPp*), indicate the probability of producing a cat with 6 or more digits.

(d) For the mating in part (c), indicate the probability of producing a cat with pointed ears.

23. Four babies born in the same hospital on the same day have been placed in the wrong beds. Use the blood types of the parents (**Table 4**) and of the babies to match the infants with their correct mothers and fathers. Explain your predictions.

- Alleles *A* and *B* are both dominant over the *O* allele. Alleles *A* and *B* are codominant, producing blood type AB.

Table 4

Couple	Mother's blood type	Father's blood type
1	A, *Rh*-positive	B, *Rh*-negative
2	O, *Rh*-negative	B, *Rh*-positive
3	AB, *Rh*-negative	O, *Rh*-negative
4	O, *Rh*-negative	O, *Rh*-negative

- The *Rh* gene (rhesus factor) is located on a different chromosome and segregates independently of the ABO genes. The *Rh*+ allele is dominant over the *Rh*− allele.

The following are the babies' blood types:
baby A   O, *Rh*-negative
baby B   AB, *Rh*-negative
baby C   O, *Rh*-positive
baby D   A, *Rh*-negative

Making Connections

24. List benefits and risks associated with cloning. Conduct a survey to determine public opinion.

25. Many people who were exposed to nuclear radiation from the Chernobyl nuclear disaster in 1986 developed tumours. Some of these cancers are the result of chromosomal damage. High-intensity radiation causes chromosomes to break apart, and small fragments become scattered throughout the nucleus.

(a) Why would the fragmentation of chromosomes affect cell division?

(b) Suggest a method that can be used to detect these changes in chromosomes.

Exploring

26. Find out about either Hugo de Vries or H. J. Muller and summarize his contribution to the modern concept of the gene.
Follow the links for Nelson Biology 11, Unit 2 Review.
GO TO www.science.nelson.com

27. Find some examples of Canadians who have contributed to knowledge about genetic processes and describe their work.
Follow the links for Nelson Biology 11, Unit 2 Review.
GO TO www.science.nelson.com

28. Use your library and the Internet to research genetic technologies. How did scientific understanding guide or help in the development of the technology?
Follow the links for Nelson Biology 11, Unit 2 Review.
GO TO www.science.nelson.com

Are You Ready?

Knowledge and Understanding

1. Plants and animals rely on each other. Analyze **Figure 1** and answer the following questions:
   (a) Name the chemical process that occurs within the leaves of green plants. Write a chemical equation for this process.
   (b) Name the chemical process that occurs within the mitochondria of living cells. Write a chemical equation for this process.
   (c) The process in part (a) provides two important substances that the animal obtains from the plant. Name the two substances.
   (d) The process in part (b) provides two important substances that the plant obtains from the animal. Name the two substances.

Figure 1

2. Create two columns and label them with the headings Asexual Reproduction and Sexual Reproduction. Place the following terms under the correct heading: vegetative, flower, grafting, pollen, ovary, stem cuttings, runners, meiosis, mitosis, seeds, cloning, hermaphroditic, eggs, fruit, budding, fragmentation, and genetic variation.

3. Draw a diagram similar to **Figure 2** in your notebook and add arrows indicating the direction of movement (the pathway) of water as it enters, moves within, and exits the plant. In addition, label as many plant parts as you can.

4. Referring to **Table 1**, match each plant tissue or organ in the first column with its function in the second column.

Table 1

Tissue or organ	Function
1. fern sporangium	(a) absorbs water and anchors moss plants
2. xylem	(b) contains the female sex cells
3. phloem	(c) contains nutrients for an embryo plant and young seedling
4. rhizoid	(d) manufactures carbohydrates
5. ovary	(e) contains seeds
6. pollen	(f) carries water and dissolved materials from roots to leaves
7. fruit	(g) where meiosis produces spores
8. cotyledon	(h) attracts pollinators
9. green algae in lichens	(i) carries carbohydrates within the plant
10. flower	(j) transfers male gametes in seed plants

Figure 2

5. Use specific, detailed examples to justify the following statement: Mitosis plays an important role in both growth and reproduction.

6. Choose two specific plants that normally grow in very different locations.
   (a) Describe the general soil and climatic conditions in those locations.
   (b) Explain how each is able to survive these different conditions.
   (c) Also suggest a type of plant which would not likely be able to survive in each location.

Inquiry and Communication

7. (a) Describe a diagnostic test for carbon dioxide gas which uses the indicator bromthymol blue.
   (b) Design an experiment that would employ this test to determine whether a living aquatic plant absorbs carbon dioxide from its environment. Describe the experiment as a series of numbered steps.

8. Choose any kind of plant (e.g., cactus, moss plant, pine tree, dandelion) and describe how its particular features enable it to survive in a particular environment.

9. What climatic and soil variables strongly influence what types of plants are able to grow in a given location?

Plants: Form and Function

In this chapter, you will be able to

- use photographs, diagrams, microscopes, and models to identify and describe the major types of cells and tissues, as well as the structures and functions of plant roots, stems, and leaves, which allow for efficient reproduction and transport of water, minerals, and nutrients;
- conduct your own investigations to understand the plant transport system and how plants are influenced by, and have adapted to, their environments;
- compare structures and functions of various vascular plants;
- investigate the differences between monocots and dicots;
- describe the various specific nutrients required for plant growth and development;
- describe various external and internal plant growth regulators;
- design and carry out an investigation of an external factor affecting plant growth.

Ontario's Space Age Tomatoes

Food and oxygen supplies are essential for human life support in space. As long space flights become more common, plants will be needed to provide a continuous supply of oxygen and fresh, nutritious food. The plants will consume carbon dioxide and require a source of water and nutrients. Scientists need to know what plants are best suited for space travel and how their germination and growth are affected by the conditions in space. Tomato plants may be one good choice for space travel. The plants are easy to grow and tomatoes are highly nutritious.

On November 30, 2000, Dr. Marc Garneau (**Figure 1**), Canada's most experienced astronaut, carried 200 000 tomato seeds with him on Space Shuttle Mission STS-97. These seeds were subjected to radiation and conditions of microgravity during the 11-day shuttle mission. After returning to Earth, the seeds were distributed to 3000 classrooms across Canada. Students grew the seeds and compared their germination and growth with similar seeds that did not go into space. The data were sent to scientists at the University of Guelph for careful analysis.

Plants are equally important on Earth. They supply us with building materials, fabrics, foods, and medicines. They cover our soccer fields and decorate our yards, homes, and offices. The paper in this book is made from plant materials. Forestry is Ontario's single largest industry. Wildlife requires balanced ecosystems with a diversity of plant species. Plant biologists study a wide range of topics, including plant cell chemistry, anatomy, growth, functions, and potential new uses of plants and plant parts.

Reflect on your Learning

1. What is a plant?
2. What is the purpose of photosynthesis? Why are both animals and plants dependent on this process?
3. Unlike most animals, plants are unable to move from place to place. What specific problems or limitations does immobility create for plants as they attempt to reproduce and to protect themselves from consumers? Describe any methods you can think of that plants use to overcome these difficulties.

Figure 1
Dr. Marc Garneau carried two hundred thousand tomato seeds into space.

Plant Structures and Functions

Like all highly evolved organisms, plants have developed specialized tissues, organs, and systems. Although not many people are familiar with plant physiology and microscopic structures, everyone is familiar with the terms leaves, stems, roots, cones, flowers, fruit, and seeds.

- Collect and examine one sample of a leaf, stem, root, flower, and fruit. If possible, choose these parts from a variety of plants.
- Carefully study the structure of each plant part, taking note of its size, colour, texture, and shape.
- In your notebook, write the name of one of these plant parts and, if possible, name the plant it came from. Then answer the following questions. Repeat this procedure for the other four plant parts.

(a) Describe the colour and shape of the structure. Include a drawing or traced outline of the structure.
(b) What is the function of the plant part?
(c) What characteristics does it have that might discourage insects and other animals from eating it? What characteristics might encourage insects and animals to eat it?
(d) Where does this plant grow naturally?
(e) Suggest why so many different shapes and sizes of this plant part exist among plants.
(f) Describe any features that you think are unusual.

13.1 Vascular Plant Structure and Function

Recall that plants can be placed into two major divisions: vascular plants and nonvascular plants. **Table 1** summarizes the key features of four plant taxa.

Table 1: Major Features of Plant Taxa

Taxon	Waxy cuticle	Vascular tissue	Pollen grains	Seeds	Flowers and fruits
mosses	X	-	-	-	-
ferns	X	X	-	-	-
gymnosperms	X	X	X	X	-
angiosperms	X	X	X	X	X

In Chapter 10, you learned about some of the features that allowed some early nonvascular plants to survive on land. You also learned about the specific features of vascular plants that permitted them to be extremely successful and form the basis of terrestrial ecosystems. This chapter will concentrate on the biology of vascular plants.

Photosynthesis, the process by which plants synthesize their own carbohydrates, is the most important distinguishing feature of the plant kingdom. It is these carbohydrates which supply both stored chemical energy and building blocks for cell growth and reproduction. They also provide the food for all other organisms through food webs. Photosynthesis is a cellular process. How do plants, like angiosperms, ensure that their green cells receive the sunlight, carbon dioxide, and water needed for photosynthesis? How do they distribute the manufactured carbohydrates to all their cells? How do plants store these products?

Cells of all multicellular organisms are organized into tissues, tissue systems, organs, and organ systems. These structures enable plants to successfully photosynthesize, grow, and reproduce (**Figures 1** and **2**).

Figure 1
Cells of all complex plants are organized into tissues, tissue systems, organs, and organ systems that enable the plants to carry out the basic processes of life, such as photosynthesis, distribution and storage of carbohydrates, growth, and reproduction.

meristems: regions of the plant where some cells retain the ability to divide repeatedly by mitosis

apical meristems: regions at the tips of all roots and shoots. They are responsible for the primary growth, which lengthens shoots and roots throughout the life of the plant.

Meristems

Unlike animal cells, which have many kinds of cells that undergo division, plant cells divide by mitosis only in specific regions called **meristems**. The meristems at the root tips and shoot tips are called **apical meristems**. In spermatophytes, root and shoot systems begin to form in the tiny embryo within the seed. As the seed germinates, cells at the root and shoot tips divide by mitosis and elongate. Meristems at locations other than the root and shoot tips form complete or incomplete cylinders of tissue within roots and stems. Cell division in these cylinders, called the **lateral meristems**, causes an increase in the diameters of roots and

lateral meristems: cylindrical regions in roots and stems. They are responsible for all increases in diameters of roots and stems.

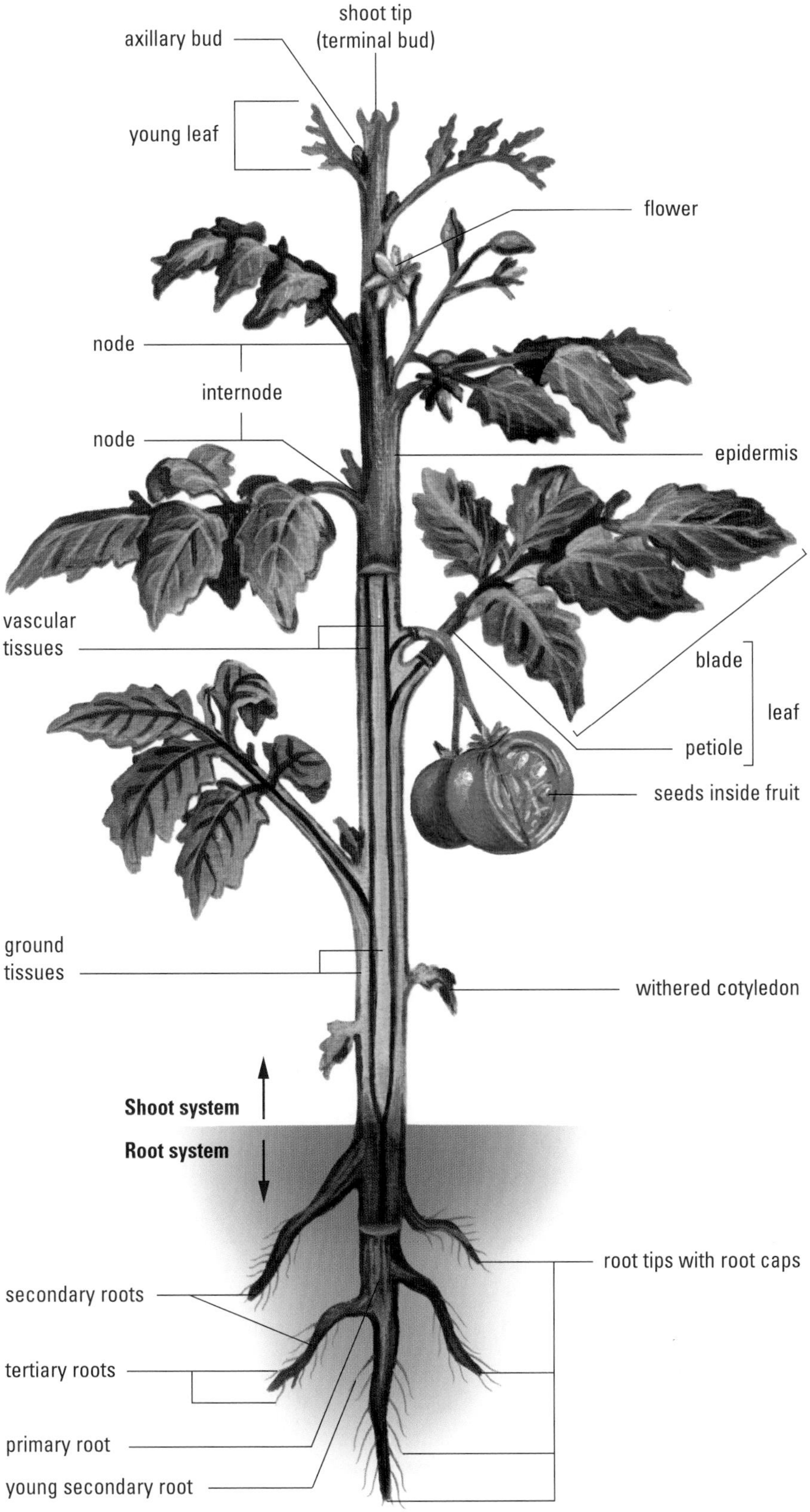

Labels (top to bottom, clockwise):
- axillary bud
- shoot tip (terminal bud)
- young leaf
- flower
- node
- internode
- node
- epidermis
- vascular tissues
- blade
- leaf
- petiole
- seeds inside fruit
- ground tissues
- withered cotyledon
- Shoot system
- Root system
- root tips with root caps
- secondary roots
- tertiary roots
- primary root
- young secondary root

Figure 2
The two main organ systems of plants are the root system and the shoot system. Also shown are the major tissues, organs, and organ systems of a typical angiosperm. Note that not all plants will have all the parts shown, nor will parts with similar functions look the same from species to species.

Plants: Form and Function **495**

vascular cambium: a lateral meristem which is responsible for creating new xylem and phloem tissue

primary growth: all plant growth originating at apical meristems resulting in increases in length, as well as growth originating at the lateral meristems in the first year of a plant's life

secondary growth: plant growth originating at lateral meristems which results in increased diameters of roots and stems in the second and all subsequent years of a plant's life

stems. Cambium (plural: cambia) is another name for a lateral meristem. Cell division within the **vascular cambium** produces cells which will become new xylem and phloem tissue.

Primary and Secondary Growth

Primary growth includes all growth in the length of roots and stems throughout a plant's entire life and all growth in the diameters of roots and stems which occurs in the first year of a plant's life. **Secondary growth** is the result of lateral meristem activity throughout the rest of a plant's life (**Figure 3**). Cells produced by meristematic tissue eventually differentiate into all other plant tissues. These tissues are arranged into the familiar specialized organs of the plant: roots, stems, leaves, and cones or flowers.

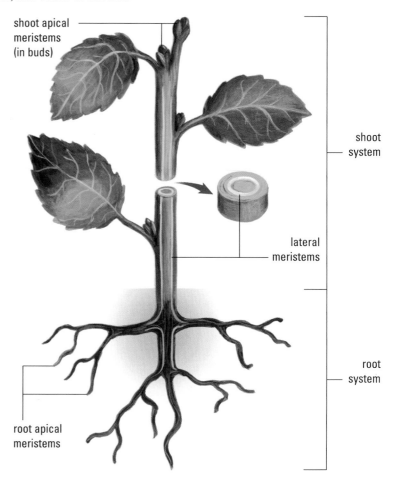

shoot apical meristems (in buds)

shoot system

lateral meristems

root system

root apical meristems

Figure 3
All seed plants have regions of primary (apical and lateral) growth in their first year. If a plant lives more than one year, it continues to have primary apical growth, but after the first year, it also has secondary growth, which is always lateral.

SUMMARY Vascular Plant Structure and Function

1. The cells of complex plants are organized into tissues, tissue systems, organs, and organ systems.

2. Primary growth is growth in length and occurs at apical meristems, plus at the lateral meristems during a plant's first year.

3. Secondary growth is growth in diameter and occurs at lateral meristems of plants which live more than one year.

4. Cells produced by meristematic tissue differentiate into all other plant tissues.

Understanding Concepts

1. What four things do plants require for photosynthesis?
2. Carbohydrates are produced during photosynthesis. What is supplied to plants by these carbohydrates and why are they important?
3. What are the two basic systems of most plants?
4. Where are apical meristems? What results from their activity?
5. What is the difference between primary and secondary growth?
6. How would plants that exhibit secondary growth differ in size and structure from plants that exhibit only primary growth?

13.2 Plant Tissues

Plant tissues are specialized for functions such as absorption, transport, storage, photosynthesis, and reproduction. There are three major types of plant tissue—dermal tissue, ground tissue, and vascular tissue—which are distributed throughout the plant body. Dermal tissue cells are found in the **epidermis** and the **periderm**. Ground tissue cells make up all of the internal nonvascular regions in the plant. Vascular tissue cells are specialized for conducting materials throughout the plant body.

Dermal Tissue System

The outermost cell layer of the main plant body is the epidermis. On parts of the shoot system, the epidermis produces on its exterior surface a waxy, noncellular layer called the cuticle. The cuticle protects against excessive water loss and infection by microorganisms. When you polish an apple, you are polishing this waxy cuticle. The cuticle also restricts gaseous exchange through the surfaces of the epidermal cells.

Epidermal tissue often contains highly specialized cells such as root hair cells and leaf guard cells, which you will learn about later. During secondary growth in roots and stems, the epidermis is replaced by the periderm, another dermal tissue. Some peridermal cells form many layers of special cells that soon die but leave behind a material that waterproofs the roots and stems and protects the inner tissues from structural damage (**Figure 1**).

epidermis: the outermost cell layer of a multicellular plant experiencing primary growth

periderm: a protective covering that replaces the epidermis in plants that show extensive secondary growth

(a)

(b)

Figure 1
(a) Epidermal cells cover the outer surfaces of these *Hibiscus* leaves. A layer of cuticle makes these leaves shine.
(b) Epidermal cells of this tree trunk have long ago been replaced by periderm cells. As trees grow laterally through secondary growth, the outer material splits, giving different trees their characteristic bark patterns. In addition, the bark here is covered by two kinds of crustose lichen.

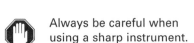

Always be careful when using a sharp instrument.

Apple Skin

Materials: two apples, knife, and ascorbic acid or lemon juice, 50–mL cylinder, two 400–mL beakers, balance

- Leave one apple unpeeled and uncut.
- Carefully peel the protective epidermis (the skin) from a second apple. Cut the peeled apple in half and cut one of these halves into two quarters.
- Submerge one peeled apple quarter in water.
- Submerge the second peeled apple quarter in a solution of ascorbic acid (15–20 g of powder in 250 mL of water) or lemon juice (30 mL of lemon juice in 250 mL of water).
- Leave the unpeeled apple and the peeled apple half exposed to open air circulation.
- After three days, cut open the unpeeled apple and compare it to the three peeled pieces.
  - (a) Describe the three peeled pieces compared to the unpeeled apple and compared to each other.
  - (b) What happened to the peeled apple left open to air circulation that did not happen to the apple in water, ascorbic acid, or lemon juice?
  - (c) Predict what you might find growing on the three peeled pieces if you waited long enough.
  - (d) How effective is the apple's epidermis in preventing these potential problems?

Ground Tissue System

There are three types of ground tissues: parenchyma, collenchyma, and sclerenchyma. **Parenchyma** consists of living cells that make up the bulk of the primary plant body. It is involved in photosynthesis and the storage of nutrients, carbohydrates, and water. It also functions in healing wounds and regenerating plant parts. Plants such as cacti have large amounts of parenchyma tissue for water storage; these plants are called **succulents**. The general term for ground tissue in the centre of roots and stems is **pith**. Pith consists of spongy parenchyma cells, and it functions in the storage of nutrients, carbohydrates, and water (**Figure 2(a)**). The ground tissue that surrounds the pith is called the **cortex** and is made up of more rigid cells.

Collenchyma, also consisting of living cells, helps strengthen the plant and is specialized for supporting the plant's primary growth regions. Collenchyma cells have thickened cell walls that provide a measure of flexibility to plant parts that must be able to bend to withstand windy conditions (**Figure 2(b)**). Celery stalks contain lots of collenchyma tissue.

Sclerenchyma cells have a secondary cell wall composed of cellulose and lignin, a material that provides added strength and rigidity to the cell (**Figure 2 (c)**). These cells may occur as a continuous mass, or in small clusters, or may be individually scattered throughout the plant. The gritty texture of pear fruit is due to individual sclerenchyma cells. The thickened walls of sclerenchyma cells strengthen and support various plant parts. They are particularly evident in parts where hardness is an advantage, such as in the shells of nuts and in cactus spines. Sclerenchyma cells are often dead at maturity but they continue to perform their functions.

parenchyma: a living ground tissue that makes up the bulk of the plant body. Parenchyma tissues take part in several tasks, including photosynthesis, storage, and regeneration.

succulents: plants which have thick, fleshy parts due to the presence of large amounts of parenchyma for water storage

pith: the parenchyma tissue at the very centre of roots and stems

cortex: the parenchyma tissue, usually with slightly thicker cell walls, surrounding the vascular tissue in roots and stems

collenchyma: a living ground tissue that offers flexible support for primary growth

sclerenchyma: a ground tissue whose mature cells are dead. These cells have thick walls composed of cellulose and lignin. Sclerenchyma supports mature plants and often protects seeds.

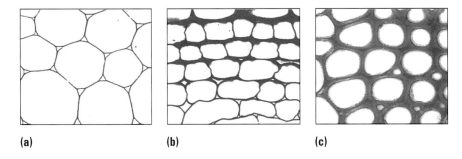

Figure 2
Photomicrographs showing cross-sectional views of (a) parenchyma with large spaces for water storage; (b) collenchyma with moderately thick cell walls for flexibility; and (c) sclerenchyma with very thick cell walls with lignin for strength and support

Vascular Tissue System

Vascular tissue includes xylem and phloem along with some collenchyma and parenchyma cells. As you learned in Chapter 10, xylem is the main tissue in plants for conducting water and minerals. Xylem tissue contains fibres and water-conducting cells called **tracheids** and **vessels** (Figure 3). Tracheids are longer than vessel elements and have tapered, overlapping ends. The cell walls have pits which are unthickened areas for easier transfer of materials between neighbouring cells. Vessels are long, continuous tubes of individual vessel elements that are joined end to end. The vessel elements have thickened walls and large perforations in their end walls. Both types of cells are found in angiosperms, but gymnosperms have only tracheids. Tracheids and vessels are dead at maturity; all that remains are the lignified cell walls. They continue to transport water and dissolved substances until they get filled with various deposits. Wood, no matter what kind, is comprised almost entirely of xylem tissue.

Phloem tissue transports sugars and other solutes throughout the plant body. Unlike mature xylem, which is dead, mature phloem is a living tissue. **Sieve tubes** provide an obstruction-free pathway for the movement of materials from one cell to the next. Sieve tubes are composed of sieve elements, which are long and thin phloem cells with sieve plates at the end walls. The sieve plates have large pores that allow easy passage of water and dissolved materials. Sieve elements also have pits on their side walls. These cells lack a nucleus, ribosomes, Golgi apparatus, cytoskeleton, and vacuoles.

Sieve elements are usually associated with **companion cells**, which serve to direct the activities of the sieve tubes and supply them with needed substances (**Figure 4**, page 500).

tracheids: xylem cells with tapered, overlapping ends and pits in their cell walls for conducting water and dissolved materials in plants

vessels: long tubes of vessel elements for conducting water and dissolved materials in plants

sieve tubes: long tubes formed by many sieve elements to allow easy passage of water and dissolved materials

companion cells: small cells lying next to the sieve elements and directing their activities

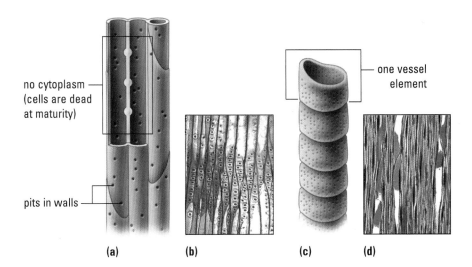

no cytoplasm (cells are dead at maturity)

pits in walls

one vessel element

(a) (b) (c) (d)

Figure 3
Longitudinal views of xylem cells
(a) Drawing of tracheids
(b) Photomicrograph of tracheids
(c) Drawing of a vessel
(d) Photomicrograph of vessels and fibres

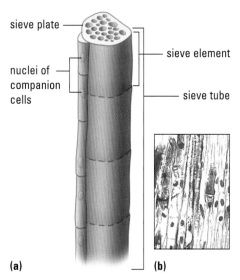

Figure 4
Longitudinal views of phloem cells
(a) Drawing of a sieve tube composed of sieve elements and their companion cells
(b) Photomicrograph of sieve tubes and companion cells

sieve plate
nuclei of companion cells
sieve element
sieve tube
(a)
(b)

Understanding Concepts

1. Name the three plant tissue systems.
2. Name the specific type of tissue or cell that performs each of the following functions:
   (a) provides a waterproof surface layer
   (b) rapidly divides to form new cells
   (c) transports sugars throughout the plant
   (d) stores carbohydrates
   (e) forms hard shells and provides support
   (f) conducts water and minerals upward in stems
   (g) directs or regulates the activities of phloem sieve elements

SUMMARY **Plant Tissues**

1. The dermal tissue system includes the epidermis (the outermost layer) and the periderm, which replaces the epidermis in plants that show secondary growth.
2. On the shoot system, the epidermis often forms a cuticle to prevent excessive water loss.
3. Epidermal tissue often has specialized cells such as root hair cells and leaf guard cells.
4. During secondary growth, the epidermis is often replaced by periderm.
5. The ground tissue system makes up all the nonvascular internal cells.
   • Parenchyma makes up the bulk of the plant body and participates in many tasks, including photosynthesis, storage, and regeneration.
   • Collenchyma offers flexible support for primary growth.
   • Schlerenchyma has thicker walls of cellulose and lignin to offer stronger support.
6. The vascular tissue system consists of xylem, phloem, and some collenchyma and parenchyma cells and conducts materials through the plant body.
   • Xylem conducts water and dissolved materials via tracheids and vessels.
   • Phloem transports dissolved carbohydrates throughout the plant through sieve tubes.

Section 13.2 Questions

Understanding Concepts

1. In what ways does the epidermis protect plants? Explain why the following statement is incorrect: A tree trunk's bark is its epidermis.
2. What is the difference between pits and pores in vascular tissue? What purpose is served by the pits and pores in xylem and phloem cells?
3. Woody tree trunks are primarily xylem tissue. What is another major function of xylem in addition to the transport of water and dissolved minerals?
4. Explain how it is possible for dead cells to perform functions in plants.

5. A microscopic view of a plant section reveals a greater than usual number of xylem cells. What might this suggest about the plant's ability to conduct and store water? In what kind of environment would a large number of xylem cells be beneficial to survival? Explain why.

6. Identifying and learning the names of plant tissues can be challenging, even for experienced plant biologists. Create a table which lists the three major tissue types, the specific cell types in each, and their special structures including the name, a description, and the function of each structure. Add a section to describe the location and importance of meristematic tissue.

13.3 Leaves

Green leaves are the major sites of photosynthesis. They contain chlorophyll, the green pigment necessary to capture light energy. They must also be able to obtain carbon dioxide from the air and water to use as the building blocks for sugars and starches, the products of photosynthesis. If maximizing photosynthesis were the only objective, we would expect leaves to be very wide to maximize their exposure to the light, and to have systems to readily obtain carbon dioxide and water. However, there are other equally important considerations. Leaves must not dry out—a difficult problem when faced with bright sunlight, hot, dry air, and high winds. There is also the problem of hungry herbivores. As a result, leaves occur in a great variety of shapes, sizes, and textures. Leaves also have a variety of internal structures. These characteristics have evolved very gradually. They are the features which allowed the plants to survive the **biotic** and **abiotic** factors of their habitats. Those plants with characteristics which did not promote survival simply died.

The blade is the flattened main body of the leaf. Leaves are positioned along the stem at points called **nodes**. The distance between successive nodes is called an **internode** (Figure 1). Typically, leaves have a network of veins or vascular bundles of conducting and supporting tissue. In many plants, each leaf is connected

biotic: describes anything related to living things. Biotic factors are all living things in an area and include interactions within and between species, such as competition and predation.

abiotic: describes anything related to non-living things. Abiotic factors include temperature, humidity, light availability, and soil conditions such as water content, texture, and mineral composition.

nodes: the locations where leaves are attached to the stem

internode: the space between two successive nodes on the same stem

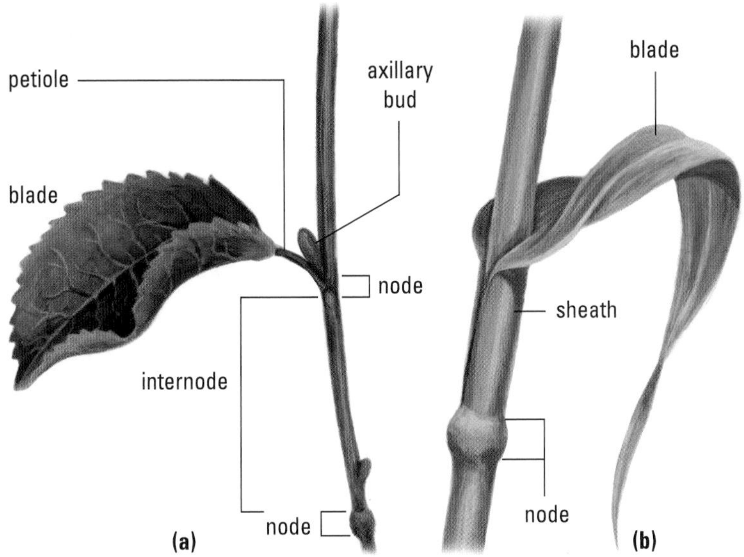

Figure 1
Typical leaf forms of (a) dicots and (b) monocots

simple leaf: a leaf that is not divided into leaflets

compound leaf: a leaf that is divided into two or more leaflets

to the stem by a leaf stalk called a petiole. The vascular tissue in the stem usually sends out one branch to the leaf through the petiole. Once in the leaf, the vascular tissue branches out. If these new veins branch and rebranch throughout the whole leaf, the leaf is said to have net venation. This is the normal pattern for dicots (**Figure 2(a)**). If these new leaf veins tend to run from the petiole to the leaf tip without joining one another, the leaf is said to have parallel venation. This is the normal pattern for monocots (**Figure 2(b)**). If the leaf has a single, undivided blade, it is called a **simple leaf**. A **compound leaf** has a blade divided into two or more leaflets (**Figure 3**).

Figure 2
(a) A small part of a dicot leaf (from a silver maple tree) showing net venation
(b) The monocot leaves of a plant called rose twisted stalk, showing parallel venation

(a)   (b)

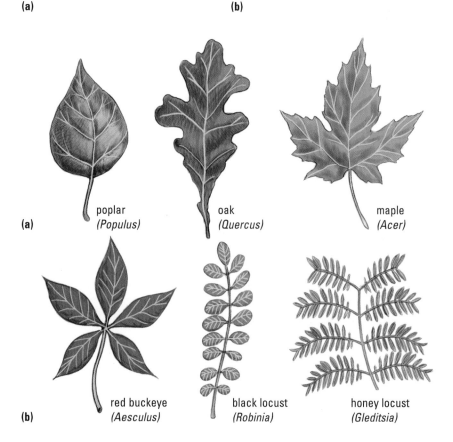

(a)
poplar
(*Populus*)

oak
(*Quercus*)

maple
(*Acer*)

(b)
red buckeye
(*Aesculus*)

black locust
(*Robinia*)

honey locust
(*Gleditsia*)

Figure 3
Examples of **(a)** simple leaves and **(b)** compound leaves

Stomata

During photosynthesis, the leaf must acquire a constant supply of carbon dioxide and be able to release the oxygen produced. The exchange of these gases is regulated by tiny pores called stomata (singular: stoma). Stomata are found in the epidermis of leaves or stems but are mostly found in the lower epidermis of leaves. As well as regulating carbon dioxide and oxygen diffusion, stomata also allow water vapour to escape from the leaf. The loss of water vapour in plants is called **transpiration**. The water diffuses and evaporates into the air spaces of the leaves and out to the atmosphere through the stomata.

When the stomata are open, the plant can obtain needed carbon dioxide; however, the plant also loses water, which can pose significant problems for the plant. When the stomata are closed, water is conserved, but carbon dioxide cannot be obtained. The opening and closing of each stoma is regulated by a pair of sausage-shaped **guard cells** (Figure 5).

transpiration: the loss of water through the surfaces of the plants. Most transpiration occurs through leaf stomata.

guard cells: the cells that occur in pairs around each stoma in the epidermis of a leaf or a stem. They regulate the opening and closing of the stoma.

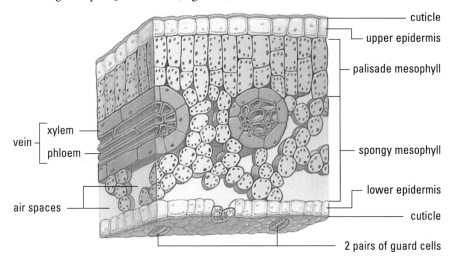

Figure 4
A three-dimensional drawing of a typical leaf showing internal and surface structures

Most epidermal cells do not contain chloroplasts. However, the guard cells around the stomata do contain many chloroplasts. When a pair of guard cells contains low levels of water, they are somewhat limp and rest against each other, closing the stoma. As water builds up in the leaf tissues, as it does most nights, the guard cells tend to swell. However, the portion of the cell wall of each guard cell that faces the other is thickened. As the cells enlarge, they swell less where their walls are thickened. The pairs of swollen guard cells look similar to kidney beans and the stomata are now open. At sunrise, photosynthesis begins in the chloroplasts. Carbon dioxide levels drop and oxygen levels increase in the leaves relative to the concentration of these gases in air. Because the stomata are open, gaseous exchange occurs by simple diffusion. Water vapour is also lost.

Throughout the day, as the water concentration in the guard cells drops, the cells begin to shrink. Gradually, as the pairs of guard cells become limp and collapse, the stomata close. The opening and closing of the stomata are also related to the concentration of carbon dioxide in the guard cells. This mechanism tends to allow the stomata to be open in the daytime for gaseous exchange and closed at night to conserve water.

During the hottest part of the day, plants may lose excessive water. If this loss occurs, the guard cells, along with all the other cells, lose water and become limp. The stomata close and as a result, gaseous exchange of carbon dioxide and oxygen is prevented and photosynthesis is slowed down or temporarily stopped. In addition, light levels, temperature, and abscisic acid concentrations play a key role in the opening and closing of the stomata.

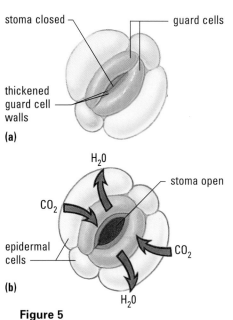

Figure 5
The closing **(a)** and opening **(b)** of stomata are regulated by the levels of water and carbon dioxide in the guard cells.

Counting Stomata

- Obtain a prepared slide of the epidermis of any leaf. Alternatively, ask your teacher how to make your own wet mount, or imprint.
- Observe the specimen under medium power and note the appearance and pattern of stomata among the epidermal cells. Note the shape of the guard cells.
- Count the number of stomata in a single field of view. If there are too many stomata, count the number in a one-quarter "pie section" of the field of view and then multiply your result by four.
- Your teacher will provide you with the field of view diameter for your microscope from which you can determine the radius. Determine the area (in square millimetres) of your field of view using this formula: area = πr^2. Show your calculations with your recorded result.
- Use your results to estimate the total number of stomata on a leaf. Explain how you obtained this number.

Mesophyll

mesophyll: the region of photosynthetic cells between the epidermal layers of leaves

palisade mesophyll: one or two layers of brick-shaped cells, rich in chloroplasts and found tightly packed beneath the upper epidermis of most leaves

spongy mesophyll: a layer of irregularly shaped cells containing chloroplasts between the palisade mesophyll and the lower epidermis of most leaves. Many air spaces are randomly distributed within this layer.

Between the upper and lower surfaces of a leaf is a photosynthetic region called **mesophyll** (Figure 4). The mesophyll consists of parenchyma cells containing lots of chloroplasts. In most plants, the mesophyll has two different areas based on the orientation and shape of the cells. The **palisade mesophyll** occurs under the upper epidermis. Here, the cells are shaped like bricks and are tightly packed together in one or two layers. The longer sides of the cells are at right angles to the upper epidermis. These palisade cells contain many chloroplasts and are the primary site for photosynthesis. The **spongy mesophyll** lies between the palisade mesophyll and the lower epidermis. These cells have fewer chloroplasts, are irregular in shape, and are randomly arranged with large air spaces scattered among them. These air spaces promote the rapid diffusion of carbon dioxide into cells and oxygen gas out of them. In the leaves of some plants, the mesophyll does not form two distinct areas.

As a result of photosynthesis, carbon dioxide levels drop in the mesophyll cells. Since carbon dioxide levels are lower within the cells than in the surrounding air spaces, carbon dioxide diffuses into the mesophyll cells, providing more reactants for photosynthesis. Similarly, as oxygen gas is produced within the cells during photosynthesis, the concentration rises, resulting in the diffusion of oxygen gas out of the cells into the surrounding air spaces. If the stomata are open, gaseous exchange occurs between the air spaces and the atmosphere. If the stomata are closed, the process of photosynthesis quickly consumes the available carbon dioxide within the air spaces in the mesophyll and further photosynthesis effectively stops.

Leaf Adaptations to Abiotic Factors

The extreme conditions of some terrestrial environments have made it difficult for many plants to survive. Diversity of species has resulted in the survival of some species in those locations but not others. In addition, diversity within a species has also allowed those individuals of a species which could somehow cope with the extreme conditions to survive, while others died. Plants with very broad leaves to trap low light energy will survive in shaded areas but die if they germinate in open, sunny fields. Plants whose leaves appear in the very early spring, before

DID YOU KNOW ?

The single most abundant protein on Earth, RUBISCO (acronym for *ribu*lose 1,5-*bis*phosphate *c*arboxylase/*o*xygenase), is the enzyme responsible for creating organic molecules containing carbon from the inorganic carbon dioxide in the air.

Try This Activity

Examining Water Loss

In this activity, you will examine water loss in leaves.

Materials: large fresh green leaf, scissors, water, two small binder clips or clothes pins, paper towel, balance, incubator or small fan

Procedure
- Trace the leaf on the paper towel. Cut out the leaf shape.
- Wet the paper leaf until it is saturated but not dripping.
- Determine and record the masses of the leaf and the wet paper cutout.
- If possible, place the leaf and the cutout in a drying oven or incubator overnight. Otherwise, hang them up to dry. You may wish to use a small fan to shorten the drying time.
- When the paper cutout looks noticeably drier, determine the final masses of both the leaf and the cutout.
- Calculate the percent loss of mass for both the leaf and the paper cutout.
- Comment on the effectiveness of the leaf cuticle in preventing drying out.

leaves of the surrounding trees emerge and create shaded conditions, will survive in a deciduous forest, but plants whose leaves appear late will not survive.

Most conifers are evergreen, which means they keep their leaves throughout the winter. This characteristic is especially beneficial in regions with a short growing season. These trees avoid expending the large amounts of time, energy, and nutrients required to grow a complete set of new leaves each year. The leaves of most conifers, such as pine and spruce, are modified as thin, long needles. The needles have a small surface area and a thick, waxy cuticle. Although these features make the leaves inefficient for photosynthesis, they greatly reduce water loss. This prevention is advantageous since, during the winter, lost water cannot be replaced by roots buried in frozen ground.

Plants that can survive in areas of low precipitation or high salt content in the soil usually have leaves with thick layers of water storage tissue or are covered with an extra thick, waxy cuticle to prevent water loss. These plants also tend to have fewer than the usual number of stomata. The spines of cacti are the remnants of leaves. Water loss is reduced because they have very few stomata and extremely small surface areas. Photosynthesis takes place in the fleshy, green stems. Cacti thrive in sunny desserts but would not survive in shaded forests (**Figure 6**).

(a)

(b)

Figure 6
(a) Conifer needles are modified leaves which reduce water loss but still perform photosynthesis.
(b) Cactus spines are more radically modified leaves. Most photosynthesis occurs in the fleshy stems.

Leaf Adaptations to Biotic Factors

Leaves are extremely vulnerable to herbivores. Herbivores are attracted to tender leaves with mild flavours. Any plants that happen to have tough, hairy, prickly, or bitter leaves are more likely to survive herbivore appetites (**Figure 7**). However, there is always a tradeoff. The very characteristics which help the plants survive herbivores reduce their photosynthetic efficiency. Also, diversity among the herbivores has allowed certain herbivores to cope with the plant features. Some herbivores have tough mouth tissues, efficient teeth, or special digestive enzymes; others have a poor sense of taste.

Diversity also means that while some plants continue to supply nutritious food for herbivores, other plants produce toxic chemicals in their tissues which actually control herbivore populations. The nicotine in tobacco leaves is an insecticide. An even more convoluted situation involves the common milkweed plant and the monarch butterfly. While milkweeds produce a **toxin** which has a horrible taste and is highly toxic to almost all insects and large herbivores, monarch butterfly caterpillars are immune to this poison. In fact, the milkweed toxins accumulate in the fatty tissues of these caterpillars. In this way, the caterpillars and adult monarchs are themselves toxic and unpalatable to their enemies.

toxin: a poison produced in the body of a living organism. It is not harmful to the organism itself but to other organisms.

(a)   **(b)**

Figure 7
Some leaf adaptations discourage hungry herbivores. Plants such as **(a)** woolly lamb's ears, which have hairy leaves, stems, and flower heads, and **(b)** common milkweed, which have a horrible taste, are avoided by herbivores. Look closely and you can see a monarch butterfly caterpillar among the leaves and seed pods of the milkweed.

Other Leaf Adaptations

In addition to performing photosynthesis, some plants use leaves to accomplish a number of other functions (**Figure 8**). In some plants, all the leaves are modified, but in others, regular leaves exist along with modified ones. Onion bulbs are modified leaves that are specialized for storage of water and nutrients. Some plants develop specialized leaves called tendrils for attachment to surfaces or objects for support. A variety of plants, such as cacti, produce sharp spines, which are actually modified leaves. Plants with such specialized leaves survive better than those without them when dealing with hungry herbivores. Even the petals of flowers are modified leaves which attract pollinators for reproductive purposes. The Venus fly-trap, the pitcher plant, and sundews are all examples of carnivorous plants, although the term carnivorous is misleading because they do not require animals as food; they all can photosynthesize. However, studies have shown that they thrive when animal protein is available. There are also some

(a)

(b)

(c)

(d)

(e)

(f)

plants which have lost not only their leaves but also their ability to photosynthesize. For example, Indian pipe leaves are reduced to tiny bits of tissue. There are no chloroplasts in any part of this plant. Thus, these plants are heterotrophic. They are saprophytes and obtain their nutrients from decaying organic material in the soil. Indian pipe has vascular tissue and complete flowers.

Practice

Understanding Concepts

1. State the primary function of leaves.
2. What two functions are served by vascular tissue within leaves?
3. Why are guard cells essential?
4. What two substances control the swelling or collapse of guard cells?
5. Describe four plant features which help protect them from hungry herbivores.

Figure 8
Leaf adaptations.
(a) An onion bulb consists of modified leaves. Note the dried real leaves at one end of the bulb and dried up roots at the other end.
(b) A very prickly plant growing along a roadside in southern Spain
(c) Colourful petals of common spring garden flowers
(d) Leaves of the pitcher plant attract, trap, and digest insects in a pool of water.
(e) Leaves of sundew with sticky projections that bend down to trap insects that land on them
(f) Indian pipe has no chlorophyll and is a heterotrophic plant.

Activity 13.3.1

Leaf Adaptations

Leaves are the site of photosynthesis and exhibit many adaptations to maximize their photosynthetic efficiency. Leaves are also the site of water loss through transpiration and evaporation. Up to 90% of the water taken in by roots is lost to the air through leaves. Plants called **xerophytes** can survive, or even thrive, in water-deficient environments and have evolved numerous features to conserve water. In contrast, plants which live in water are called **hydrophytes** and exhibit features

xerophytes: plants that survive or thrive in areas with very little moisture

hydrophytes: plants living on or in water

uniquely suited to their environment. Most plants thrive in environments with a moderate water supply and are referred to as **mesophytes**.

Questions

What are the features of a typical mesophytic leaf?
What leaf adaptations are found in xerophytes?
What special features do the leaves of hydrophytic plants have?

Materials

prepared slides of *Syringa*, *Yucca*, *Zea* (corn), *Potamogeton*, *Pinus* (pine), *Verbascum*, and *Oleander*
whole samples of *Aloe*, rubber plant leaf, and fig leaf
microscope
any potted mesophyte, such as *Geranium* or *Coleus*
any potted xerophyte, such as jade
cobalt chloride paper
paper clip or adhesive tape
plastic wrap

Procedure

Part 1: Observing Leaf Adaptations

This lab will consist of a series of numbered lab stations. Each station will have one or two specimens for you to examine. Take approximately 10 min to complete and record your observations at each station. Your teacher will instruct you on how and when to change stations.

1. Copy **Table 1** into your notebook but leave large spaces for your answers, about 15 lines for each row of the table. Your whole table may occupy 3 notebook pages. Record your answers as you complete the steps. At each lab station, use the information and questions provided here as guides to your observations. List and describe the most significant features of each leaf. Add drawings where appropriate.

Table 1

Station	Specimen(s)	Observed feature(s)	Adaptive significance
1	*Syringa*	?	?
2	*Yucca*	?	?
	Aloe	?	?
3	*Pinus*	?	?
	Zea	?	?
4	*Verbascum*	?	?
	Oleander	?	?
5	*Potamogeton*	?	?
6	*Philodendron*	?	?

Station 1: *Syringa*—A Mesophyte

2. The prepared *Syringa* slide has a cross section of a leaf blade. View the leaf under low power. This leaf represents a typical mesophyte dicot leaf. Observe and record the shape of the whole cross section of this leaf.

3. View the slide under medium and high powers. Look for the palisade mesophyll. Describe and draw what you see.

4. Look for the dark pairs of guard cells in the epidermal layers which open and close the stomata. Compare the number of guard cells seen in the upper and lower epidermal tissue.

Station 2: *Yucca* and *Aloe*—Succulent Xerophytes

5. View the prepared *Yucca* slide under medium power. Look for large numbers of vascular bundles and a thick cuticle.

6. Break off a small piece of one *Aloe* leaf. Observe and describe its interior.

Station 3: *Zea* and *Pinus*—Monocot and Gymnosperm Xerophytes

7. View the prepared slides of a *Zea* leaf and a *Pinus* needle using medium and high powers. Corn and pine are well adapted for conserving water during hot summer days. Note that most stomata are located on the upper epidermis of the corn leaf. The leaf also possesses unusually large epidermal cells. These cells collapse in hot, dry conditions, causing the leaf to curl up and inwards. Describe the position of the pine guard cells relative to the needle surface.

Station 4: *Verbascum* and *Oleander*—Xerophytes with Specialized Surface Features

8. View the prepared slides of *Verbascum* and *Oleander* under medium and high powers. Note the hairs extending from the surface of the *Verbascum* leaf.

9. Carefully note the location of stomata within the large pits on the lower surface of the *Oleander* leaf. These pits also contain tiny hairs. Look for a thick cuticle and multilayered epidermis in the *Oleander*.

Station 5: *Potamogeton*—A Hydrophyte

10. View the prepared slide of a leaf cross section under medium and high powers. Note that the palisade layer is near the upper leaf surface. Note the location of the stomata and the size of the spongy mesophyll layer. What effect would this specialized tissue have on leaves that grow in water?

Station 6: *Philodendron*—A Tropical Rain Forest Vine

11. *Philodendrons* are vines that live in tropical rain forests. They begin life on the ground in deep shade and grow up into the canopy along tree trunks. Examine the leaves, stems, and aerial roots of this plant. Note the "drip tip" on the leaf.

Part 2: Testing Transpiration in Leaves

12. Dry cobalt chloride paper is blue. In the presence of water, it turns pink. Select a healthy, large leaf on a potted mesophyte. Do not remove this leaf. While being careful not to damage the leaf or plant stem, fold a piece of cobalt chloride paper over the edge of the leaf, covering part of the upper and lower leaf surface. Cover the cobalt chloride paper with a slightly larger piece of plastic wrap. Fasten the paper and plastic to the leaf with a paper clip or tape, being careful not to damage the leaf (**Figure 9**).

13. After 15 min, remove and examine the indicator paper. Record your observations.

14. Repeat steps 12 and 13 using a xerophyte plant.

Figure 9
Without removing or damaging the leaf, apply the cobalt chloride paper as shown.

Analysis

(a) In the *Syringa* leaf, how does the location of the palisade layer enhance photosynthesis? Where are most of the stomata located?

(b) Compare the *Yucca* and *Aloe* leaves with the leaves of other common plants. How do the special features of these leaves contribute to their function?

(c) What is the function of the thick cuticle of the *Yucca* leaf?

(d) Comment on the size and shape of the *Pinus* needles. How does their overall size influence water loss?

(e) How does *Zea*'s ability to curl its leaves in hot weather influence water loss?

(f) How might the hairs on the *Verbascum* leaf influence surface air flow? How might the hairs influence water loss?

(g) How do the stomatal pits on the *Oleander* leaf influence water loss?

(h) How is the location of the stomata on the *Potamogeton* leaf well suited to a floating leaf?

(i) Discuss whether the large air spaces in the *Potamogeton* leaf are wasted space or if they serve a useful purpose.

(j) Describe the leaf surfaces of the *Philodendron* as smooth and waxy or rough and dull. How would the surface influence what happens to rainwater that falls on these leaves?

(k) How might the "drip tip" on *Philodendron* leaves influence what happens to rainwater that falls on these leaves?

(l) What is the advantage of having leaves that shed surface water quickly?

(m) How is water able to escape from the surfaces of leaves? What structures are involved?

(n) Compare the transpiration rates from top and bottom leaf surfaces. Account for any differences.

(o) Compare the overall transpiration rates from the mesophytes and xerophytes. Account for any differences.

(p) What advantages are there to having stomata in the lower leaf surface rather than the upper leaf surface?

(q) What features or structures of a xerophytic plant allow it to survive in hot, dry environments?

(r) If the leaves of xerophytes have far fewer leaf stomata than those of mesophytes, what can you infer about the potential rates of photosynthesis and growth in these plants? Explain your answer.

SUMMARY Leaves

1. Leaves are the site of photosynthesis.

2. Transpiration and evaporation of water and gaseous exchange of carbon dioxide and oxygen occur through the stomata..

3. When there is plenty of water in the guard cells, they swell in a way that opens the stomata, but when the water content decreases, the guard cells collapse, closing the stomata.

4. The mesophyll is a photosynthetic layer consisting of palisade and spongy mesophyll.

   • The palisade mesophyll cells, found just under the upper epidermis of leaves, are brick shaped, tightly packed together, and have many chloroplasts.

- The irregularly shaped, spongy mesophyll cells with interspersed large air spaces lie between the palisade mesophyll and the lower epidermis of leaves and have fewer chloroplasts.

5. Leaves have adaptations (e.g., cactus spines, evergreen needles, tough fibres, hairy leaves, thorns, bright colours, and toxic compounds) to various abiotic and biotic factors.

6. A xerophyte is a plant that survives or thrives in areas with very little moisture.

7. A hydrophyte is a plant living on or in water.

8. A mesophyte is a plant that thrives with moderate moisture.

Section 13.3 Questions

Understanding Concepts

1. What is the advantage of having air spaces within leaves?

2. List the features of xerophyte leaves that reduce water loss.

3. Describe a special adaptation found in each of the following leaves:
   (a) *Potamogeton*   (b) *Pinus*   (c) *Verbascum*
   (d) *Aloe*   (e) *Zea*

4. Photosynthesis requires water and carbon dioxide and produces sugars and oxygen. Using proper terminology, describe the pathways taken by these reactants and products as they enter and/or exit the leaf.

5. Describe the key problem that exists for plants that have many broad, thin leaves. How does the cuticle help overcome this problem?

6. Tropical rain forest plants have no shortage of water.
   (a) How have the leaves of rain forest plants evolved to cope with rainy conditions?
   (b) How might rain forest plants be harmed by having continuously damp or wet leaves?

7. During the winter, the ground is frozen. The days can be cold and sunny while the air is very dry. Given these conditions, suggest reasons why pine trees have evolved to retain their leaves during the winter. Also, suggest the benefits of their very narrow, waxy leaves.

Applying Inquiry Skills

8. A number of leaves are obtained from a rare plant collection and carefully examined. Based on the following information, classify each leaf type as a mesophyte, xerophyte, or hydrophyte.
   - Leaf A: broad and thin with a double palisade layer and many stomata on the lower leaf surface
   - Leaf B: no stomata on the lower surface but does have stomata on the upper surface; vascular bundles almost completely lacking in xylem
   - Leaf C: very fleshy with many vascular bundles but very few stomata

9. Examine a variety of houseplants. Name each one and describe any features it has which might allow it to survive dry, humid, cold, and hot climates. Look for and mention any plants with "drip tips." Speculate on the natural habitat of each plant described.

(continued)

10. Why is it not surprising that many hydrophytes have little or no xylem tissue?

11. The leaves of many underwater plants are finely divided, dramatically increasing the surface area that is in contact with water. Explain how this leaf structure might be an adaptation for obtaining carbon dioxide.

13.4 Roots

Although usually hidden below ground, a plant's root system is critically important and is often larger than the plant's entire shoot system (**Figure 2**, page 495). Roots absorb water and minerals from the soil, physically support and anchor plants, and store carbohydrates. Water and minerals that enter the roots are transported through vascular tissue up to the stem, leaves, and flowers. Many of the carbohydrates produced in the shoot system are transported down to the roots for storage. Roots are also responsible for producing a variety of compounds, such as hormones, which are used throughout the plant.

The **primary root** is the very first main root which develops from the seed. A **secondary root**, also called a lateral root, is smaller than the main root and branches from it. The two main types of plant roots are **taproots** and **fibrous roots**. If a young root increases in diameter, grows downward, and develops small lateral roots, it is called a taproot. Examples of plants with taproots are carrots, beets, dandelions, and oak trees. In grasses and other monocots, the primary root is short-lived and is replaced by **adventitious roots**. Adventitious roots and their lateral roots form a fibrous root system. Generally, these fibrous roots do not penetrate as deeply as taproots but their total combined length can be enormous. The fibrous roots of a single rye grass plant were estimated to have a total surface area of over 600 m^2 while occupying just 6 L of soil!

Each root has a protective **root cap** at the tip. Just behind the cap, the epidermal cells often have fine, microscopic **root hairs**. Root hairs increase the root's surface area for absorption. At the centre of each root is the vascular cylinder (stele), which contains the vascular tissues, xylem and phloem, and some ground tissue.

Surrounding the vascular cylinder is the **endodermis**, which regulates lateral movement of water and minerals. The endodermis is really the innermost layer of the cortex. Between the vascular cyclinder and the endodermis is the **pericycle**. The pericycle is important because it is made up of meristematic tissue, which gives rise to secondary roots.

Secondary growth in roots is associated with an increase in diameter of the vascular cylinder and, in much older roots, the formation of bark. Most monocot roots as well as some **annual** dicots do not exhibit secondary growth (**Figure 1**). It is, however, a definite characteristic of **perennial** dicot roots. Some of the ground tissue within the stele becomes meristematic and begins to divide. This tissue is the vascular cambium. Vascular cambium separates the xylem and phloem. When the cambium divides by mitosis, it produces phloem cells to the outside and xylem cells to the inside. As the xylem increases in size, the cambium is displaced outward, causing an increase in the diameter of the root. Primary xylem and phloem are produced in the first year of the plant's life. After that time, all xylem and phloem that are produced are called secondary (**Figure 2**). As

primary root: the first root developed from the seed

secondary root: smaller root branches growing sideways from a primary root

taproots: root systems where the primary root remains predominant, though very small secondary roots may be present

fibrous roots: root systems whose primary roots have disintegrated and have been replaced by adventitious roots

adventitious roots: roots that develop from a part of the plant other than a root. They often form huge tufts at the base of the stem. There is no main root because most are the same size as the others. However, smaller secondary roots do branch out from these roots.

root cap: a loose mass of cells forming a protective cap covering the apical meristems of most root tips

root hairs: microscopic extensions of the epidermal cells near the tip of a root. Root hairs function in the absorption of water and minerals.

endodermis: a layer of rectangular cells surrounding the vascular cylinder. It is the innermost layer of the cortex.

pericycle: a thin layer of lateral meristematic cells that surrounds the vascular cylinder

annual: describes plants that complete their entire life cycle, from seed to reproduction to death, in one year

perennial: describes plants that grow and reproduce repeatedly for many years

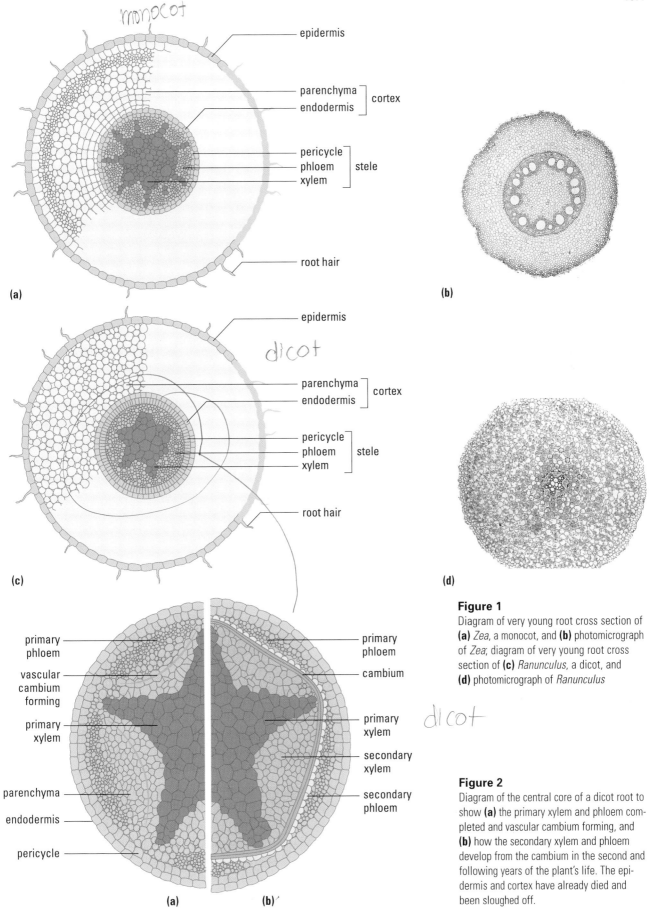

monocot

epidermis

parenchyma ⎤
endodermis ⎦ cortex

pericycle ⎤
phloem   ⎥ stele
xylem   ⎦

root hair

(a)

(b)

dicot

epidermis

parenchyma ⎤
endodermis ⎦ cortex

pericycle ⎤
phloem   ⎥ stele
xylem   ⎦

root hair

(c)

(d)

primary
phloem

vascular
cambium
forming

primary
xylem

parenchyma

endodermis

pericycle

(a)

primary
phloem

cambium

primary
xylem

secondary
xylem

secondary
phloem

(b)

dicot

Figure 1
Diagram of very young root cross section of
(a) *Zea*, a monocot, and **(b)** photomicrograph
of *Zea*; diagram of very young root cross
section of **(c)** *Ranunculus*, a dicot, and
(d) photomicrograph of *Ranunculus*

Figure 2
Diagram of the central core of a dicot root to
show **(a)** the primary xylem and phloem completed and vascular cambium forming, and
(b) how the secondary xylem and phloem
develop from the cambium in the second and
following years of the plant's life. The epidermis and cortex have already died and
been sloughed off.

the pericycle gets pushed further and further outwards, some of its cells become meristematic and develop into the **cork cambium**, which produces compact layers of **cork** cells. These cells become impervious to water. When the epidermis and remnants of the cortex die and get rubbed off, what remains will be the accumulating corky layers of the bark which help to protect the roots from excess moisture loss, bacterial invasion, and mechanical damage (**Figure 3**).

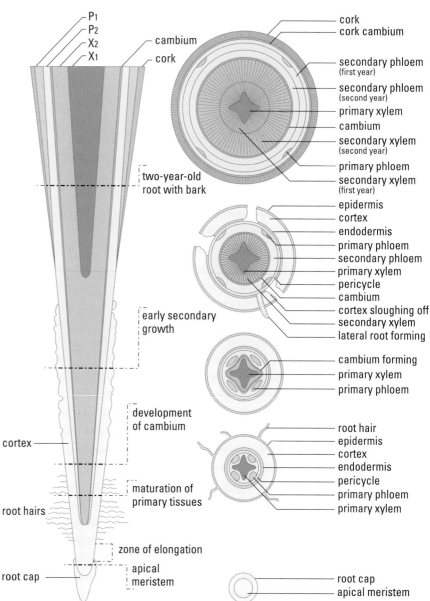

Figure 3

Diagrams to show a longitudinal section of a two-year-old dicot root and five cross sections through different regions of that root. Note the labels carefully. Be aware that in spite of the size of the drawing, this root could be very small.

Root Adaptations

Some plants have roots that have evolved special structures or habits that allow those plants to survive in special situations. In waterlogged, oxygen-poor soils, such as in mangrove swamps, roots may have specialized extensions called **pneumatophores** which grow up out of the water and function to supply oxygen to the root tissues below. In some plants, such as ivy or various types of *Philodendron*, small **aerial roots** grow from the leaf nodes along the stem. They also absorb oxygen from the air. In addition, if this part of the stem contacts the soil, these roots grow downwards as normal roots (**Figure 4**). The aerial roots of

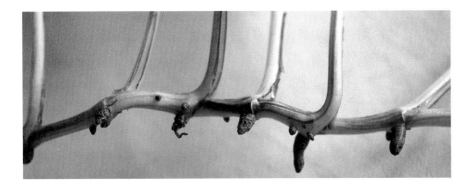

orchids never reach the ground. Many orchids are epiphytes; they grow on the stems and branches of other plants. They are not parasitic because they have retained their ability to photosynthesize. Strangler figs are also epiphytes but their roots do grow down to the ground (**Figure 5**). In many other plants, such as the carrot, roots are greatly expanded and specialized as a major carbohydrate-storage organ. The roots of some plants, such as walnut trees, release toxins into the soil from their roots, seeds, or fallen leaves. These toxins inhibit the germination of other plant seeds, thus reducing competition for light and soil nutrients. This strategy is called **allelopathy**.

allelopathy: the suppression of growth and development of neighbouring plants by a plant of a different species. This effect is caused by chemicals secreted by the roots or contained in the leaves of the allelopathic species. Not all surrounding plants will suffer the same effect.

(a)

(b)

Figure 5
(a) These drawings show the growth of a strangler fig plant over time. The strangler fig begins its life as a small epiphyte growing high up in a tree. The fig grows numerous aerial roots which, over time, grow down the tree trunk to reach the ground. These roots lengthen, thicken, and fuse together. The fig depends on the tree to support its bulky body. Eventually the roots completely engulf and strangle the tree. The tree's death results in greater light exposure for the strangler fig, which is then able to grow quickly, often to immense size.
(b) A photograph of a small section of tree trunk encased in large aerial roots of a strangler fig.

SUMMARY **Roots**

1. Roots absorb water and minerals, support and anchor the plant, and store carbohydrates.

2. Taproots increase in length and diameter and develop lateral roots. Carrots have taproots.

3. Fibrous roots are branching roots of similar size. They do not penetrate as deeply as taproots but have a very long combined length. Grasses usually have fibrous roots.

4. In the centre of the root is the vascular cylinder.

5. In perennial dicots, the vascular cambium separates xylem and phloem. The cells of the vascular cambium divide by mitosis, producing phloem

cells to the outside and xylem cells to the inside. These new cells cause an increase in the root's diameter.

6. Surrounding the vascular cylinder is the pericycle, which forms lateral roots.

7. The endodermis surrounds the vascular cylinder and pericycle, and regulates the lateral movement of water and minerals in the root.

8. Pneumatophores and aerial roots form on some plants.

9. Epiphytes have roots that seldom reach the soil.

10. Allelopathy, the release of toxins into the soil, is a survival mechanism of a few plants.

Practice

Understanding Concepts

1. State the major functions of plant roots.

2. Compare taproots and fibrous roots.

3. The small roots seen growing laterally from carrots are not root hairs. What are they?

4. Root tips are often covered by large numbers of fine root hairs. How do these hairs assist the root in performing one of its functions?

5. Vascular tissue includes both xylem and phloem cells. Which of these cell types is located closest to the centre of the root?

Activity 13.4.1

Root Anatomy

In this activity, you will identify and compare the tissues in roots from a variety of representative plants. You will base your comparison on examinations of whole specimens and microscopic sections.

Questions
What types of plant tissues and structures are found in the root?
How are these tissues and structures arranged?
How are monocot and dicot root features similar or different?

Materials
microscope slides and cover slips
single-edged razor blade
germinating radish seeds with well-developed root hairs
prepared slides of cross sections of *Ranunculus* (buttercup) and *Zea* (corn) roots
prepared slides of longitudinal sections of an onion root tip
dissecting microscope or compound microscope
aerial root of a *Philodendron* or carrot taproot
fresh or herbarium specimens of a taproot and a fibrous root system
Lugol's solution in dropper bottles

Procedure
Part 1: The Root Tip

Always be careful when using a sharp instrument.

1. Obtain a germinating radish seed. Using a single-edged razor blade, cut off and prepare a wet mount of the bottom 1 cm of the root tip.

2. Examine the root tip using a dissecting microscope or a compound microscope on low power. Locate a mass of cells covering the root tip. This is the root cap. It covers the apical meristem, the region of active cell division.

3. Note the root hairs extending out from the root just behind the root cap. Examine the root hairs carefully. Determine the number of cells which make up each root hair.

4. Obtain a prepared microscope slide of a longitudinal section of an onion root tip. Examine the slide under low and medium power. Locate the root cap and the apical meristem region. Note that the cells of the apical meristem are small. Now look for the region in which the cells appear to be much greater in length. This is the region of cell elongation which functions to push the root tip through the soil.

Part 2: Monocot and Dicot Root Tissues

5. Obtain a prepared slide of a *Zea* root cross section. Observe this monocot root under low and medium power of a microscope. Locate the epidermis and the cortex. Note the presence of starch grains in the cells of the cortex. Locate the endodermis, the large xylem cells, the phloem cells, the vascular cambium, the pericycle, and the pith. Draw a cross-sectional "pie wedge" of the *Zea* root and label each part you located.

6. Obtain a prepared slide of a *Ranunculus* root cross section. This plant is a dicotyledon. Examine the slide and observe the location and arrangement of tissues. Take note of any differences between the *Ranunculus* and the *Zea*, particularly the central vascular tissue with its star-shaped bundle of large xylem cells. Draw a cross-sectional "pie wedge" of the *Ranunculus* root and label the epidermis, cortex, endodermis, and phloem and xylem tissues.

7. Examine fresh or herbarium samples of a taproot and fibrous root system. Note and compare the arrangements and relative sizes of the roots of each of these systems.

Part 3: Testing for the Presence of Starch

8. Lugol's solution contains iodine and is an indicator for starch. Iodine turns blue-black in the presence of starch. Using a single-edged razor blade, cut a very thin cross section of a small fresh root; the aerial root of a *Philodendron* is ideal. Place the section on a slide. Add one drop of Lugol's solution and then place a cover slip on top.

 Lugol's solution is toxic and can cause an itchy rash. Avoid skin and eye contact. Wash all splashes off your skin and clothing thoroughly. If you get any chemical in your eyes, rinse for at least 15 min and inform your teacher.

9. Observe the stained root section under low power. Sketch the cross section and show the location of starch grains.

Analysis

(a) Based on your observations of both the live root and the prepared section, make a generalized sketch of a growing root tip. Label the root cap, root hairs, apical meristem, and zone of elongation.

(b) What differences did you observe between the actively dividing cells and those a little further back from the tip?

(c) Explain the function of a root cap based on its structure and location.

(d) What tissue gives rise to root hairs? How many cells does a single root hair consist of?

(e) What root function is demonstrated by the presence of starch grains within the cortex?

(f) Based on your observations, what feature(s) could be used to distinguish between monocot and dicot roots?

(g) Use the distinguishing criteria you described in your answer to (f) to determine if the plant you observed in step 8 was a monocot or a dicot.

Section 13.4 Questions

Understanding Concepts

1. What root cell activity pushes the root tip through the soil?

2. What tissue forms lateral roots?

3. What is the difference between aerial roots and pneumatophores?

4. What are the differences between annual and perennial plants?

Applying Inquiry Skills

5. As part of a certain field experiment, the root systems of various types of plants were examined.
   (a) It was noticed that cactus roots have very shallow, widely spreading root systems. Explain how such a feature would enhance survival in desert conditions.
   (b) Epiphyte roots were often found exposed to the air, green in colour, and with few or no root hairs. Hypothesize how each of these features might be beneficial to such plants. Hypothesize the climatic conditions in which epiphytes would be successful.
   (c) Grasses and other monocots often have fibrous root systems, while dicots often have taproot systems. Which root system would you expect to penetrate to greater depths, and what advantage would it provide to the plant? Explain which root system you predict would be better suited for reducing soil erosion.

Making Connections

6. A homeowner plants a vegetable garden next to a large walnut tree. The garden receives lots of sunlight and she ensures it is well fertilized and watered. However, the plants fail to thrive. What is the likely cause of this problem? What options does this gardener have?

13.5 Stems

Stems and leaves make up the plant shoot system. Stems provide support for the plant and serve as a transport link to and from leaves, roots, and reproductive parts such as flowers, fruits, and seeds. Stems may also serve to store water and carbohydrates. Some plants have **herbaceous** stems. Herbaceous stems are thin, soft, green, short-lived, and contain little or no wood. They carry out photosynthesis and thus produce some carbohydrates for the plant. Usually, herbaceous stems do not grow more than one metre tall. Exceptions include palm trees, which are not actually trees at all, but huge herbaceous monocots lacking true woody tissue. Stems that are not herbaceous are **woody**. Examples of plants with woody stems include grape vines, shrubs, conifers, and dicot trees (eg., oaks).

During the first year of growth, woody shoots and stems and herbaceous shoots and stems closely resemble each other (**Figures 1** and **2**), with growth at the apical meristems increasing the shoot length. However, monocot and dicot

herbaceous: describes the fleshy stems of annual plants. These stems usually do not survive more than one year, especially if there is a cold winter. They are also called nonwoody stems.

woody: describes stems of perennial plants. They increase in diameter each year as more and more vascular tissue is created. The xylem cells, even after they have died, create the hard, woody tissue called wood.

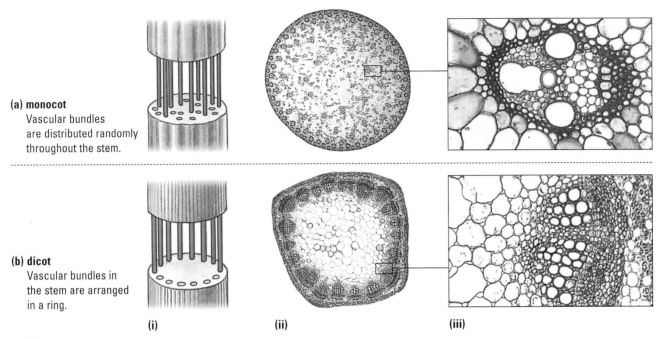

(a) monocot
Vascular bundles are distributed randomly throughout the stem.

(b) dicot
Vascular bundles in the stem are arranged in a ring.

(i)   (ii)   (iii)

Figure 1
The very young stem structure of **(a)** a monocot and **(b)** a dicot
(i)   Drawings to show the distribution of **vascular bundles** in longitudinal stem sections
(ii)   Photomicrographs of young stem cross sections
(iii) Highly magnified photomicrographs of vascular bundles as seen in cross sections

vascular bundles: collections of xylem and phloem tissue, separate from other collections, running longitudinally through stems

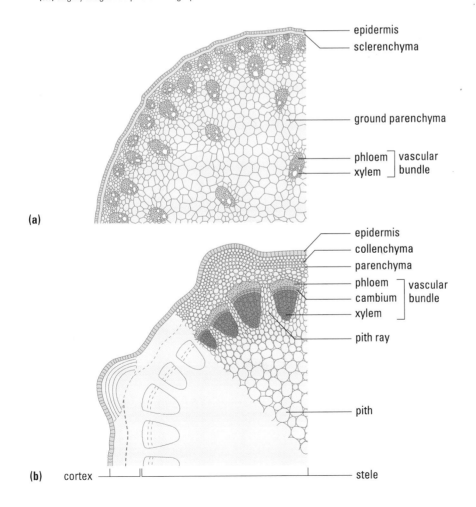

epidermis
sclerenchyma

ground parenchyma

phloem ⎤ vascular
xylem ⎦ bundle

(a)

epidermis
collenchyma
parenchyma
phloem ⎤ vascular
cambium ⎦ bundle
xylem
pith ray

pith

(b)   cortex ⎿⎾ ⎿⎾ stele

Figure 2
Diagrams of a small portion of the cross sections of very young stems of **(a)** *Zea*, a monocot, and **(b)** *Ranunculus*, a dicot

herbaceous plants do have different arrangements of vascular tissues within the stem. Remember, these stems seldom survive more than a year. Any lateral growth during that first year is still called primary growth.

However, just like their roots, the stems of many dicots, most notably trees, continue to grow year after year. Secondary growth occurs after year one and involves the vascular and cork cambia. As in roots, cells of the vascular cambium divide by mitosis to produce secondary phloem to the outside and secondary xylem to the inside. The secondary phloem cells tend to crush the somewhat fragile phloem cells of the previous years as pressure is exerted outwards. The secondary xylem cells do not crush earlier xylem cells because they have thick walls. Remember that the oldest xylem is nearest the centre of the stem; the youngest is next to the cambium (**Figure 3**).

The secondary xylem thickens and forms tissue known as wood. Each year a new layer of xylem is added to this thickening core. This adds size, hardness, and strength to the aging stem or trunk. Xylem cells produced and maturing in the spring tend to grow very large because of the moisture available. As the season gets

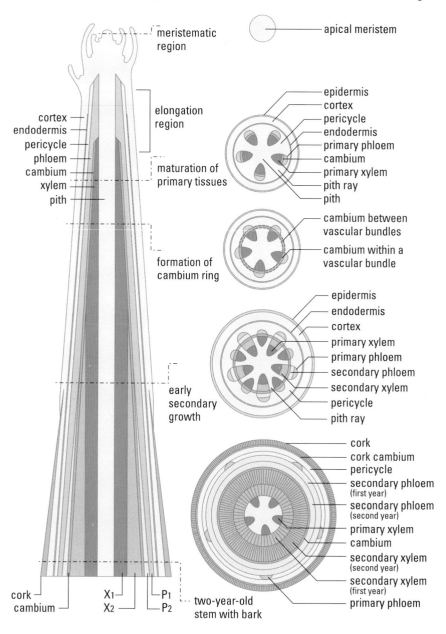

Figure 3

Diagrams showing a longitudinal section of a two-year-old dicot shoot and five cross sections through different regions of that shoot. Note the labels carefully. Be aware that in spite of the size of the drawing, this shoot could be very small.

later and later, conditions often get much drier. The result is that the xylem cells that get produced and mature later are much smaller. One zone of spring xylem plus one zone of summer-fall xylem form an **annual ring** (Figures 4 and 5). In very wet years, the rings are wider than in very dry years. By counting the number of rings, we can learn the age of a woody plant. In tropical zones, growth in wood is somewhat uniform all year and annual rings do not appear.

Additional changes take place as the tree ages. No new ground tissue is added to the tree, and the primary tissue may be displaced by secondary tissue. As in the root, the cork cambium is formed. The cork cells produced by this layer will accumulate in layers. The actual cells die but the cell walls remain to form cork, which offers protection against mechanical damage and damage

annual ring: the increase in the amount of secondary xylem during one year. The number of annual rings indicates the age of the woody plant.

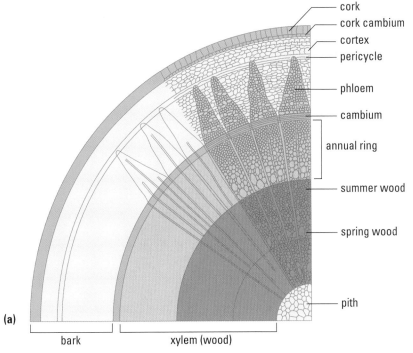

(a)

bark

xylem (wood)

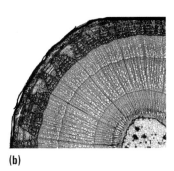

(b)

Figure 4
Drawing of part of a cross section of **(a)** a three-year-old woody plant, *Tilia* (basswood), and **(b)** photomicrograph of a *Tilia*

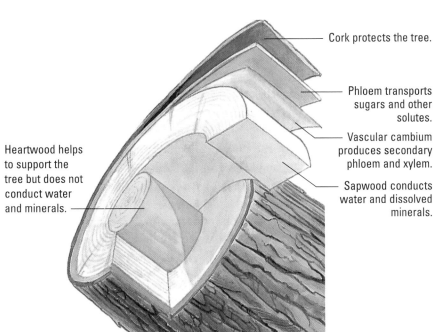

Cork protects the tree.

Phloem transports sugars and other solutes.

Vascular cambium produces secondary phloem and xylem.

Sapwood conducts water and dissolved minerals.

Heartwood helps to support the tree but does not conduct water and minerals.

Figure 5
The major tissues within a woody stem. Secondary growth causes an increase in diameter each year.

bark: the outer layers on older stems, branches, and trunks. Bark consists of every layer from the vascular cambium outwards: phloem, any remaining cortex, cork cambium, and cork.

heartwood: the older, harder, nonliving central wood in tree trunks. It is often darker due to the accumulation of oils and resins and its basic function is to provide support.

sapwood: the younger, softer, outer wood in tree trunks that is important for transporting water and dissolved materials as well as for support

sap: the fluid within any part of a plant, found mostly within the xylem and phloem tissues

caused by bacteria, fungi, and insects. As the stem expands, the cork cracks and is renewed from the inside.

Bark is composed of everything from the vascular cambium outwards: phloem, any remaining cortex, cork cambium, and cork. If the bark is removed, the phloem will be destroyed and no further carbohydrates will be transported to the root for storage. The tree will die, although it may be a slow death taking place over three or more years.

As trees age, old xylem cells become plugged with various substances such as oils and resins, which prevent water and dissolved materials from moving up or down. The oils and resins often cause the xylem tissue to darken. This darkened region is called the **heartwood**, because it lies at the heart or centre of the stem. The living xylem tissue surrounding the heartwood is called the **sapwood** and carries water and dissolved minerals. In sugar maples, the xylem also stores and transports carbohydrates. This area is often somewhat light in colour. Each year some sapwood is changed to heartwood. The fluid within any part of the plant is called **sap**. The trunks of trees are simply stems on a large scale (**Figure 6**).

(a)

(b)

(c)

Figure 6
Stems of trees are the largest structures produced by plants.
(a) Trunks of giant sequoia trees
(b) Evidence that very large trees once grew in the heart of Toronto
(c) Palm tree trunks

tubers: thick underground stems specialized for carbohydrate storage and asexual reproduction

runners: thin stems which grow along the ground producing roots and shoots at their nodes

Stem Adaptations

Just as leaves and roots are often modified to meet special needs, stems exhibit a variety of specializations (**Figure 7**). Cacti have thick, fleshy stems that perform photosynthesis and contain large amounts of parenchyma tissue for water storage. Sometimes stems are modified for asexual reproduction. For example, regular white potatoes are **tubers**, a form of underground stem, and the thin, reddish strawberry **runners** are stems, not roots. Some plants have stems which twine completely around other plants or structures. Members of the same species always twine in the same direction. The *Cecropia* tree of Central and South America has hollow stems which are home to symbiotic ants. The ants benefit from the protective habitat and, in turn, offer protection for the Cecropia by attacking leaf-eating insects. Have you ever wondered why birch trees have peeling bark? Many tropical rain forest trees such as the *Gumbo limbo* have shedding bark as well. Shedding bark helps get rid of epiphytes which grow on the tree trunk and branches.

(a)

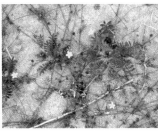
(b)

(c)

(d)

Figure 7
Modified stems
(a) The fleshy bodies of cacti
(b) Runners of silverweed
(c) Twining stems of black swallowwort wound around a branch of a large shrub
(d) Peeling white birch trunk

SUMMARY Stems

1. Stems provide support, serve as a transport link between roots and the rest of the plant, and store nutrients.
2. Both woody and herbaceous stems have primary growth.
3. Herbaceous stems are green, soft, short-lived, have little or no wood, and are usually short.
4. Primary growth occurs at apical meristems.
5. Stems of dicots, such as trees, undergo secondary as well as primary growth.
   - Secondary growth occurs in the vascular cambium.
   - Secondary xylem forms wood and each year a new layer is added.
   - Cork cambium produces cork cells.
   - Bark consists of all tissue from the vascular cambium outwards.
   - Heartwood is darker, dead, older xylem that is plugged with resins and oils.
   - Sapwood is lighter, live, younger xylem and functions in the transport of water and dissolved materials, and in some nutrient storage.
6. Stem adaptations include peeling bark, twining stems, tubers, runners, and hollow stems to house symbiotic insects.

Activity 13.5.1

Stem Anatomy

In this activity, you will identify and compare the plant tissues in stems from a variety of representative plants. You will examine whole specimens, microscopic sections, and models.

Questions

What types of tissues are found in monocot and dicot stems?
How are these tissues arranged?
What pattern of tissue growth results in the formation of a woody stem?

Materials

prepared slides of stem cross sections from *Zea* (corn), *Medicago* (alfalfa), and *Tilia* (basswood)
model of a mature tree cross section or an entire cross section of an actual tree trunk

single-edged razor blade
microscope slides and cover slips
phloroglucinol stain in dropper bottles

microscope
fresh *Coleus* stem

Procedure

Part 1: Comparing Monocot and Dicot Stems

1. Obtain a prepared slide of a *Zea* stem cross section. Observe this monocot stem under low power. Notice that the vascular bundles are scattered throughout the stem. Observe a single vascular bundle under medium or high power. Try to identify the large xylem vessel cells—they often look like the eyes and nose of a face. The thinner-walled phloem cells appear to the outside of the xylem cells and are clustered together. You may be able to distinguish the sieve tubes and their companion cells. Forming a fibrous sheath of ground tissue on the outside of the vascular bundle is a layer of thick-walled sclerenchyma cells. Draw a single vascular bundle and label the xylem, phloem, and sclerenchyma cells.

2. Obtain a prepared slide of a *Medicago* stem cross section. Examine this dicot under low power and observe the location and arrangement of tissues. Draw a cross-sectional "pie wedge" of the stem and label the epidermis, phloem, xylem, and vascular cambium tissue.

Part 2: Woody Stems

3. Observe a cross section of a young *Tilia* stem under low power. Note the wide inner region of xylem cells surrounding a smaller central pith. If the stem is two or three years old you will observe concentric layers of xylem tissue. Draw a simple "pie wedge" of this stem, labelling the cortex, phloem, vascular cambium, xylem, and pith regions.

4. Examine a model of a mature tree cross section or a cross section of an actual tree trunk. Attempt to count the growth rings in the xylem to estimate the age of the tree specimen. Record your estimate.

5. Each ring consists of a lighter band of larger cells produced during the spring and a dark band of smaller cells produced during the summer of the same year. Compare the width of annual rings from different years.

6. Look for evidence of heartwood in your specimen or model. Where is it located? Try to estimate how many years of heartwood exist.

Part 3: Examining a Living Stem

7. Using a single-edged razor blade, cut a very thin cross section of a fresh *Coleus* stem. Prepare a wet mount of this section and observe under low power. Note the location of the vascular bundles. If available, use a single drop of phloroglucinol stain for the wet mount medium. Phloroglucinol stains xylem a deep red.

Analysis

(a) Explain why there is a difference between spring and summer xylem. How does this difference affect the appearance of these two regions within a single annual ring?

(b) How old was the mature tree sample you examined? Why was the absolute age difficult to establish? How did you obtain your estimate of the age of the heartwood?

(c) How can you determine if *Coleus* is a monocot or a dicot?

Phloroglucinol stain is toxic and can cause an itchy rash. Avoid skin and eye contact. Wash all splashes off your skin and clothing thoroughly. If you get any chemical in your eyes, rinse for at least 15 min and inform your teacher.

Always be careful when using a sharp instrument.

Practice

Understanding Concepts

1. List the key functions of plant stems.
2. What is the function of the vascular cambium?
3. Compare the arrangement of vascular bundles in young monocot and dicot stems.
4. Which type of plant typically exhibits secondary growth of the stem?
5. Describe the purpose(s) served by the dead woody tissue in plants with very tall stems.
6. In which part(s) of a woody stem does transport of water and dissolved materials take place?
7. Beavers feed mainly on the phloem and vascular cambium of various trees. Why might this tissue be more nutritious than that of the xylem?
8. Describe how modifications of the stem have allowed cacti to survive desert environments.

Making Connections

9. How might a study of tree rings give clues about climatic conditions during the life of the tree?
10. Suggest a way to count and examine tree rings without cutting down, and thereby killing, the tree.

13.6 Transport in Plants

Small plants and animals use simple diffusion or very simple branching tubules to transport various materials throughout their bodies. Just as more complex animals depend on their circulatory systems for the transportation of materials, larger plants rely on a system of vascular tissue to perform similar functions. Plants transport nutrients and water as well as a variety of carbohydrates and hormones throughout their bodies. For a comparison of the transport systems of animals and plants, see **Table 1**.

Table 1: Internal Transport Systems of Animals and Plants

	Structures	Main components	Functions	Rate and direction of flow
Mammals	arteries, veins, capillaries, heart	complex mixture of whole blood cells and liquid plasma	transport of oxygen, water, food, wastes, hormones, and the immune (defence) system	relatively rapid; blood circulates in a "loop"; water is lost through excretion and evaporation from moist surfaces (e.g., lungs)
Vascular plants	xylem, phloem; no pumping organ	water and dissolved materials in xylem; complex chemical mixture in phloem	transport of water and minerals, various carbohydrates, hormones, and other chemical products	relatively slow; xylem carries water and minerals up; phloem distributes substances in all directions; water is lost through transpiration and evaporation from leaves

One of the primary materials required by plants is water. The height that water must move in tall trees can exceed 100 m; however, a large tree may lose thousands of litres of water a day through transpiration and evaporation. How do plants lift water to such great heights and how can they replace the huge volumes lost through the leaves?

Figure 1

In early spring, when the days are warm but the nights are below freezing, the sap flows upwards through the xylem bringing stored carbohydrates to the quickly developing leaf and flower buds. Traditionally, a special metal tube was driven into a sugar maple tree and a bucket was hung on the tube. The pails were emptied regularly. Now, plastic tubing is usually connected to the metal tubes and the sap drawn by a pump to a central location.

translocation: the process of moving products of photosynthesis throughout the plant body

Over the past several decades, scientists have been trying to answer these questions. Countless hypotheses have been made and large numbers of experiments performed. Numerous documents describe why each idea is the right one. Scientists have now reached a point where they believe that many theories contribute to the truth about transportation in plants and that we should avoid debating that one explanation answers all the questions.

It is easy to observe the conducting tissues throughout a plant, but how does water with its dissolved materials move through them? The best description for that movement is bulk, or mass, flow. Capillary action plus adhesion and cohesion theories are all related to the hydrogen bond attraction among water molecules and to the tubule walls. Although all of these mechanisms may have some role to play in keeping the water column intact as it moves through the plant, none can really account for all of the bulk flow which occurs. **Figure 1** shows how the sap of sugar maple trees is collected.

Several processes are involved. Water evaporates from the inner leaf cell walls into the air spaces and out through the stomata if they are open. This water loss creates a reduced pressure in the leaves and water flows upwards from the roots where there is a relatively higher pressure. This lost water is replaced by bulk flow from the roots. When water moves upwards from the roots, the cell membranes of the root cells allow more water and dissolved minerals to enter by osmosis, diffusion, and/or active transport.

You could sit for hours with a straw leading from your mouth into a glass of lemonade, but that lemonade is never going to come up to your mouth on its own. The suction action you must apply with your mouth is really just reducing the pressure at the end of the straw in your mouth. As soon as you reduce that pressure, the lemonade moves up the straw. Once the pressure in your mouth increases, no more lemonade comes up. Think back to the leaf. Evaporation from the leaves reduces the pressure at the top of the plant. As long as there is water available to the roots, it will move up, assisted by the forces mentioned earlier.

Carbohydrates produced by photosynthesis are moved from leaf cells to other plant parts through phloem tissue. This process is called **translocation**. Phloem, like xylem tissue, forms a continuous pipeline between leaves and roots. The most accepted explanation of translocation is called the pressure-flow hypothesis. Fluids will flow from an area of higher pressure toward an area of lower pressure. Carbohydrates will flow from their source (where they are made), which has a higher pressure, to where they are stored or used ("sink"), which has a lower pressure. The driving force is a positive pressure gradient from source to sink.

SUMMARY Transport in Plants

1. As water is lost through leaves by transpiration and evaporation, it is replaced by bulk flow of water from the roots.

2. The loss of water from the leaves reduces the pressure there, and water containing dissolved materials is drawn up from the roots.

3. The water column remains intact because of capillary action, plus adhesion and cohesion.

4. Translocation is the distribution of the products of photosynthesis to other plant parts. It occurs in phloem.

5. Sap flows from the source to a sink because of a pressure gradient.

Investigation 13.6.1

Water Movement in Stems and Leaves

INQUIRY SKILLS

○ Questioning   ◉ Recording
○ Hypothesizing   ◉ Analyzing
◉ Predicting   ◉ Evaluating
○ Planning   ◉ Communicating
◉ Conducting

Questions

In what parts of a stem is water conducted upwards?

How does the presence of leaves influence the rate of water uptake?

How does water escape from leaves, and how is the rate of loss influenced by the leaf surface or the leaf type?

Predictions

Predict whether the colouring will move further up the stalk with leaves or the stalk without leaves.

Predict whether the colouring will move through all the stalk tissues or just some.

Materials

2 fresh celery stalks	microscope
250-mL beaker	microscope slides and cover slips
metric ruler	single-edged razor blade
red food colouring	paper towel

Procedure

1. Cut approximately 1 cm from the bottom of two celery stalks that are similar in both diameter and length. Remove the leaves from one celery stalk.

2. Place 50 mL of red food colouring in a beaker. Place both celery stalks in the beaker. *Wait 30 min before continuing!*

3. Place both stalks on a paper towel, and using a single-edged razor blade, start at the bottom of both stalks and cut each stalk at 1 cm intervals. Keep the pieces in the order they were cut. Continue cutting until the colouring can no longer be seen.

4. Using a ruler, determine the distance in millimetres the colouring has moved in each stalk. Record your results.

5. Make a slide of a very thin section of the lower celery stalk to observe the coloured tissues under a microscope at low power.

Always be careful when using a sharp instrument.

Analysis

(a) Before step 3, what evidence was there that the colouring had moved up the stalks?

(b) In step 3, what evidence suggested that the colouring had moved up?

(c) Did your results from step 4 support your first prediction?

(d) Examine your celery cross sections. Did your observations support your second prediction?

(e) What type, or types, of cells are carrying the stain?

(f) Explain how the presence of leaves affects the rate of water uptake.

(g) Explain how water does or does not move up the celery stalk with no leaves.

Understanding Concepts

1. State two similarities and two differences between the transport systems found in mammals and those found in plants.
2. Describe the bulk flow of water from the roots to the leaves.
3. Name three different mechanisms that help keep the water column intact.
4. Explain how carbohydrates produced in leaves are transported throughout the plant.
5. When pine and spruce trees are damaged, a thick, sticky gum seeps from the wound. Suggest reasons why this process benefits the trees.

Making Connections

6. Maple syrup is produced from the sap moving in the xylem of sugar maple trees in early spring. Explain why it is important for syrup producers to take only a limited amount of sap from each tree.

13.7 Reproduction

Flowers

The variety of angiosperm flowers was discussed in Chapter 10. They exhibit a dramatic variety of colours, shapes, and sizes, and are probably the most complex and intricate of all plant structures. The role of the various intricate flower parts is to ensure successful pollination and fertilization. Indirectly, flowers are also responsible for protecting and adequately distributing the seeds because it is flower parts which form the seed coats and fruit.

Seed Growth and Development

In Chapter 10, it was shown that spermatophytes are divided into gymnosperms and angiosperms. Seed development in both these groups is very similar. However, the following text focuses on angiosperm seeds formed inside fruit. A seed consists of an embryo, tissue to provide nutrients for the embryo, and a protective coat. The growing embryo slowly forms a root and shoot structure. If the plant is a monocot, a single seed leaf (cotyledon) develops; if it is a dicot, two seed leaves form (**Figure 1**). The one or two cotyledons may contain all the nutrients for the embryo, or there may be an additional nutrient-rich material called the endosperm. Seeds with large cotyledons often have insignificant endosperm tissue. Seeds with a large quantity of endosperm often have insignificant cotyledons. Seeds, protected by the seed coat, may enter a **dormant** period, often lasting many years.

When temperature and moisture conditions are optimum, seeds begin to germinate. Once the seed and seed coat absorb water, the embryo begins to grow rapidly. The seed coat ruptures and the root and shoot emerge. The water- and nutrient-absorbing root begins to grow downward, and the shoot grows upward. This upward and downward growth occurs because of the presence of certain chemicals and the effect of gravity. In some plants, the cotyledons are raised out

dormant: describes a state of extremely slow biological activity. A dormant seed contains a living embryo but it does not grow; it remains protected by a seed coat and sometimes the fruit as well.

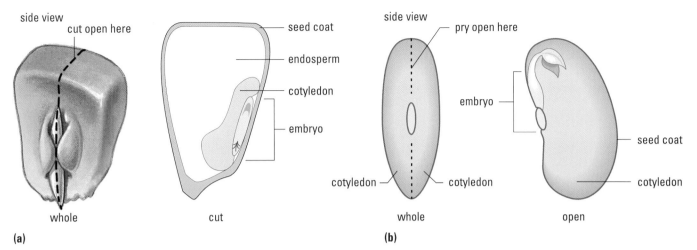

of the soil and some of these actually photosynthesize for a short time. In other plants, the cotyledons remain underground. Nutrients stored in the endosperm and/or the cotyledons support development until the new plant has sufficient root surface area and chlorophyll to be self-sustaining.

Seed Adaptations

Being stationary creates numerous challenges for plants, not the least of which is the dispersal of offspring. If all the seeds from a plant were to drop to the ground and germinate immediately below the parent plant, certain difficulties would arise. The developing seedlings would have to compete with the parent for light, water, and minerals, and the rate of survival would be low. Competition within a species is reduced through the evolution of various mechanisms for dispersing seeds. Plant species unable to disperse seeds adequately do not survive well. Also, individuals within a species that cannot disperse their seeds well are less likely to survive than individuals that can. Over time, all successful plants seem to have developed some means of ensuring survival of their young. This evolution is ongoing.

Fruit development represents a significant investment in resources on the part of the plant. This fruit is the "price" the plant pays for improved seed dispersal. Some fruits have special features such as hooks or spines that will attach to the fur of mammals, the feathers of birds, or the clothes of humans. Fleshy fruits (e.g., apples, strawberries, and tomatoes) are both attractive to and nutritious for animals. Once eaten by an animal, seeds pass unharmed through the digestive tract and are deposited along with the animal's waste, which is an ideal fertilizer. Birds are capable of carrying seeds tremendous distances in their digestive tract.

Some plant seeds will not germinate unless they have passed through the digestive tract of a specific animal. The dodo, a large flightless bird that lived on the island of Mauritius, went extinct in the 1600s. It ate the fruits of the tree *Calvaria* and was the only bird naturally capable of inducing these seeds to germinate. Following the dodo's extinction, not a single seed of this tree germinated. There are now fewer than 20 of these trees remaining. In the 1970s, scientists fed some of the large *Calvaria* seeds to turkeys. After passing through the complete digestive tract of the turkeys, some of these seeds germinated—the first of their kind to do so in over 300 years!

The common practice of squirrels or chipmunks hiding or burying acorns and other large seeds for their winter food supply is another example of seed dispersal (**Figure 2**). Rarely are all the acorns consumed; some of the remaining seeds grow into new oak trees.

Figure 1
(a) Corn has a typical monocot seed.
(b) Beans have typical dicot seeds. When the seeds are opened as shown, the embryo as well as the nutritive tissue can be seen.

Figure 2
Many seed-eating rodents, such as this chipmunk, help plants reproduce by dispersing and burying seeds.

Plants accomplish seed dispersal in a variety of ways (**Figure 3**). Many plants have capsule fruits or pods that explode when mature. The resulting explosion throws the seeds some distance away from the parent plant. Some plants, such as the orchid and poppy, produce small, lightweight seeds that can be carried great distances by the winds. Others, such as dandelions and milkweed, have fluffy, parachutelike structures attached to their seeds. Still others, such as the maple, sycamore, elm, and ash, have winglike structures attached to the seeds. For seeds that are dispersed by water, such as those of water lilies and coconuts, air is trapped in the seeds and fruits to enable them to float. Often, seeds of water plants are enclosed in a waxy, waterproof coating that protects the seed during travel over long distances in water. Later, as the coating wears away, water can penetrate the seed and trigger germination.

(a) **(b)** **(c)** **(d)**

(e) **(f)** **(g)** **(h)**

Figure 3

Variety of mechanisms for seed dispersal

(a) Milkweed pods releasing seeds with their own parachutes

(b) Wild cucumber fruit open at the bottom to release the seeds

(c) Thick squash fruit protecting the seeds

(d) A cluster of Jack-in-the-pulpit fruit, each containing one seed

(e) Kiwi fruit with many tiny black seeds

(f) Clusters of staghorn sumac fruit, each with its own seed

(g) A common burdock flower showing the form of its future fruit with many barbed seeds

(h) Fruit of white baneberry

Practice

Understanding Concepts

1. What is the functional role of flower parts?
2. What is a seed?
3. Where, specifically, would you expect to find starch in a seed? How would you prove your answer to be correct?
4. What basic condition is required for seed germination?
5. What influences the direction of growth of the new shoot and new root?
6. What are the possible locations of the cotyledons after germination?
7. Describe four specific and different methods of seed dispersal.

SUMMARY **Reproduction**

1. Spermatophytes include gymnosperms and angiosperms.
2. Flower parts of angiosperms are responsible for fruit and seed production.
3. A seed contains an embryo plus nutrient-rich material which is in one or two cotyledons and/or endosperm tissue. Seeds can remain dormant for years.
4. When conditions are ideal, the seed coat absorbs water and the embryo grows rapidly. The root and shoot grow quickly and break open the seed coat to allow germination.
5. Nutrients stored in the endosperm or cotyledon support development in the early stages.
6. Seeds must be dispersed away from the parent to reduce competition with the parent. Some seed dispersal adaptations are wings, parachutes, hooks, and fleshy edible fruit.

Activity 13.7.1

Monocot and Dicot Seeds

You will observe and compare the structure and germination of monocot and dicot seeds.

Questions

Do all seeds take the same time to germinate?
How do monocot and dicot seeds compare in structure?
What function is played by seed structures during germination and shoot development?
Where will the starch be found in opened corn and bean seeds?

Materials

masking tape
four 100-mL beakers or jars   paper towel
single-edged razor blade   probe
hand lens (magnifying glass)   Lugol's solution

For Part 1:

bean and corn seeds—soaked in water for 2 h
two other types of seeds such as grass, radish, carrot, poppy, etc.—soaked in water for 2 h

For Part 2:

bean and corn seeds—soaked in water for 24 h
dry bean seeds
four 250-mL beakers

Procedure

Part 1: Germination and Shoot Development

1. Use masking tape to label 4 beakers A, B, C, and D. Use beaker A for bean seeds and beaker B for corn seeds. Both these bean and corn seeds have been soaked for 2 h before use. Use beakers C and D for the 2 additional selected seeds.

2. Line each beaker with a double layer of wet paper towel and leave 2 cm of water in the beaker. It is important that the towels be kept wet throughout the activity.

3. Position 2 bean seeds so they are wedged between the wall of the beaker and the wet towels (**Figure 4**). Repeat for the other seeds in beakers B, C, and D.

4. Every day check that the towel remains moist and add water so that the depth is restored to 2 cm. Examine the germinating seeds daily with a hand lens until the root and shoot are established. Allow 2 weeks to complete the activity. Record your observations in a daily log. Once changes to the seeds occur, add labelled drawings to your written observations. You will need 1 set of notes and drawings for each beaker.

Part 2: Examining Embryos

5. Obtain 1 dry bean seed and 1 bean seed that has been soaked in water for 24 h. Compare the appearance of the soaked and dry bean seeds.

6. Use your fingers to gently rub the surface to remove the seed coat from the soaked bean seed. Use a probe to pry the 2 sections of the seed apart. Refer to **Figure 1**, page 529 as a guide. Find the embryo and examine it with a hand lens. Draw a diagram of the cotyledon and the attached embryo.

7. Use one drop of Lugol's solution to test various parts of the bean seed for the presence of starch. On your diagram indicate the regions containing starch.

8. Obtain a corn seed that has been soaked for 24 h. Carefully examine the seed. Note that one side of the corn seed is lighter. This indicates the location of the embryo. Lay the corn seed down with the embryo facing upward. Describe the appearance of the embryo. Refer to **Figure 1**, page 529 as a guide.

Figure 4

 Lugol's solution is toxic and can cause an itchy rash. Avoid skin and eye contact. Wash all splashes off your skin and clothing thoroughly. If you get any chemical in your eyes, rinse for at least 15 min and inform your teacher.

9. Using a razor blade, cut the seed in half lengthwise as shown in **Figure 1**, page 529. Test various areas of the corn seed for starch using the same technique you used to test the bean seed. Make a drawing to show which areas of the corn seed have the greatest amount of starch.

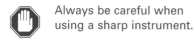

Always be careful when using a sharp instrument.

Analysis

(a) What are the observed differences in the germination process of the bean seed and that of the corn grain?

(b) Explain why two similar seeds might not germinate at the same time.

(c) Suggest a reason for at least one difference you observed in step 5.

(d) In what ways does the corn kernel differ from the bean seed?

Evaluation

(e) Why was it important to use soaked seeds in Part 1?

(f) Why was it important to keep the paper towel wet in Part 1?

(g) Why is it necessary to soak the seeds in Part 2 for a much greater length of time?

Section 13.7 Questions

Understanding Concepts

1. What function does the starch serve in the seed?

2. Eventually, seedlings no longer need cotyledons. Why not?

3. Why do seeds need to be dispersed so that they do not germinate directly underneath the parent plant?

4. Why might it also be undesireable for a seed to be dispersed very far away to an area where no other individuals of the same species exist?

5. The production of fruit represents a very high "cost" to plants in terms of nutrient energy. How does fruit improve chances of species survival?

6. Describe some specific mechanisms for seed dispersal.

Applying Inquiry Skills

7. (a) Alone or in a group of two or three, outline a simple step-by-step experimental design which could be used to test the effect of temperature on seed germination. Specify how you would set up your control. State your independent and dependent variables.

   (b) Exchange your answer to part (a) with another person or group. Record your questions or comments about their procedure and make suggestions to improve the design.

Making Connections

8. Most plants do not produce fruit unless their flowers are pollinated. How does this information influence decisions concerning if and when to spray insecticides near an orchard?

13.8 External Factors Affecting Plant Growth

Plants, like all organisms, must obtain chemical nutrients from their environment. However, unlike heterotrophic organisms, most plants are able to build their own carbohydrates using raw inorganic materials. They do require a source of light for energy, but their nutrient demands are quite simple. They require carbon dioxide and water, which are essential for photosynthesis, as well as a supply of other basic nutrients needed for the formation of energy-rich organic compounds. Like all aerobic organisms, plants also require oxygen for cellular respiration. They obtain carbon dioxide and oxygen from the air, and water and other nutrients from the soil.

Light Requirements

Earth is constantly being bombarded by sunlight. Each day, an immense amount of radiant energy reaches Earth from the Sun. Of this total, less than 1% is captured by plants in the process of photosynthesis. Green plants trap this light energy using a variety of pigments. Terrestrial plants contain chlorophyll as well as a variety of other pigments, including carotenoids. Chlorophyll absorbs light toward the red and blue area of the visible light spectrum, while carotenoids absorb light toward the blue–green area of the spectrum.

Both the quantity and quality of light influence a plant's ability to perform photosynthesis. The quantity of light is limited by natural environmental factors, such as latitude and competition from taller plants. The quality of light is influenced by shading by other plants, cloud cover, time of day, and angles of incidence during different seasons. The duration of natural lighting is dictated by the seasons. This variable **photoperiod**, with lengthening daylight hours in spring and shortening daylight hours in the fall, affects the productivity and reproductive life cycles of plants. Photoperiodism is the physiological responses of organisms to the varying photoperiods. Different plants react differently to photoperiod length. Short-day plants flower and reproduce when the photoperiods are shortening, while long-day plants flower and reproduce when the photoperiods are lengthening (**Figure 1**). An example of a short-day plant is the chrysanthemum; a long-day plant example is spinach.

photoperiod: the number of daylight hours

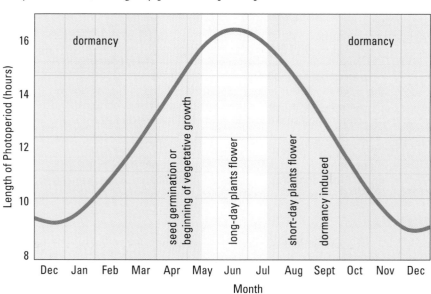

Figure 1

Relationship between number of hours of daylight and plant growth and development. Note that both short- and long-day seeds germinate at the same time, but flowering times differ. The data reflects the responses of plants growing in the upper regions of the Northern Hemisphere.

In artificial environments such as greenhouses or other indoor environments, both the quantity and quality of lighting may be dramatically altered relative to what is happening in the natural environment. The control of lighting and other factors such as temperature, humidity, and soil nutrients permits greenhouse operators to provide whatever growing conditions their plants require whenever the operators wish. By careful regulation of these external factors, plants like greenhouse tomatoes can be stimulated to produce flowers and fruit in the middle of winter in northern Ontario and even at Eureka on Ellesmere Island in the high eastern Arctic, where total darkness prevails several months of the year.

Investigation 13.8.1

The Effect of Light on Plant Growth

INQUIRY SKILLS

○ Questioning   ● Recording
○ Hypothesizing   ● Analyzing
● Predicting   ● Evaluating
● Planning   ● Communicating
● Conducting

In this investigation, you will design and conduct an experiment to examine the effects of one variable on the growth rate of seedlings. You will grow your plants in a standard potting soil mix under artificial light. Note: Time the planting of your seeds so that they are just beginning to emerge from the soil when you are ready to begin your experiment.

Questions

Choose one of the following:
How does light duration affect the growth rate of plants?
How does intensity of light affect the growth rate of plants?
How does quality of light affect the growth rate of plants?

Predictions

Choose one of the following that is related to the question you have chosen:
Predict the number of hours of light which will cause your seedlings to grow the most.
Predict the growth difference which will result from a two-hour difference of light per day.
Predict the effect of very bright light on plant growth compared with less intense light.
Predict the brightness which will result in the greatest growth of your seedlings.
Predict the effect of different colours of light on the growth of your seedlings.

Materials

potting soil and trays
adjustable lighting fixtures with timers
light meter
light bulbs of various intensities
bean, radish, or other seeds
assorted coloured cellophane sheets (red, green, yellow, etc.)

Design

(a) Choose only one independent variable for your work: photoperiod, light intensity, or light quality (colour).

(b) Determine how you will adjust your chosen independent variable.

(c) Decide on the method you will use to measure growth rate.
- What will you measure as a growth indicator? Will you measure more than one thing?
- How will you record your results?
- How will you report your results (e.g., in a graph or table format)?

(d) What variables will you need to control? Think carefully to make sure your list is complete and think about how you will control them.

(e) Describe your control setup and what and how you will measure the control.

(f) Decide how many seedlings you will use for each setup.

Procedure

1. Submit your experimental design to your teacher for approval before continuing with the experiment. Your submitted design must have each of the following:
   - title
   - question
   - prediction
   - materials list
   - procedure, including a description of your control setup, methods of varying the independent variable, methods of measuring and recording changes in the dependent variable, and a description of all the variables you will control and how you will accomplish these tasks

2. Conduct your experiment over a period of one to two weeks. Do not allow your plants to become too dry. Keep the soil moist but well drained.

Analysis

(g) How did your chosen variable influence the growth rate of your plants compared to your control?

(h) Describe the findings of students who chose to test a different variable.

(i) How might these results be useful to the operator of a greenhouse? How might some of these results also be useful to a farmer growing field crops?

Evaluation

(j) How did your results compare with those of other students in the class who chose the same variable? Suggest reasons for any differences.

(k) Why was it important to use more than one seedling for both your test group and your control group? What number of seedlings would you suggest if you were going to repeat the investigation? Why?

(l) Growth rates of plants can be measured using several features of the plant. Explain which feature would be the most convenient to measure. Explain which feature would give the most valid measurement of growth.

(m) Suggest ways that you could improve your experimental design.

Soil Nutrients

macronutrients: 9 nutrients required by plants in relatively large quantities (greater than 1000 mg/kg of dry mass)

micronutrients: 8 nutrients required by plants in relatively small quantities (less than 100 mg/kg of dry mass)

There are a minimum of 17 elements now considered by plant nutritionists to be essential for the healthy growth and reproduction of most vascular plants. The nine **macronutrients** are those elements needed in relatively large quantities (greater than 1000 mg/kg of dry mass), while the eight **micronutrients** are those needed in much smaller amounts (less than 100 mg/kg of dry mass). Of the nine macronutrients, carbon, oxygen, and hydrogen, which are the major components of most organic molecules, are obtained primarily from carbon dioxide and water. These elements make up over 95% of the dry mass of plants. The remaining macronutrients (**Table 1**) and all eight micronutrients are obtained as dissolved ions from the soil.

Table 1: Selected Macronutrients

Macronutrient	Functions	Deficiency symptoms	Comments
nitrogen	constituent of amino acids and thus all proteins as well as nucleic acids and chlorophyll; very important for leaf growth	chlorosis: the yellowing of old leaves due to a reduction in chlorophyll	obtained from nitrate or ammonium ions; nitrogen gas cannot be used directly by plants
potassium	involved in water balance, including the operation of the stomata; required for protein synthesis	retarded growth; weak stems; chlorosis of older leaves	present in large amounts in most soils but often in insoluble form
phosphorus	component of ATP, nucleic acids, phospholipids, and some proteins; critical for mitosis and cell division	lack of or poor seed and fruit development; leaves become dark and reddish; stunted growth	present mostly in fruit and seeds as well as in meristematic cells
calcium	constituent of cell walls; involved in membrane permeability	pronounced abnormalities; stunted growth, especially of roots; weakened condition	neutralizes harmful soil acids; presence can facilitate the uptake of potassium
magnesium	component of chlorophyll and **coenzymes**	chlorosis; sometimes reddening of leaves	enhances the uptake of phosphorus
sulfur	component of most proteins	stunted growth; yellow young leaves	usually plentiful in the soil

Studies have shown that for some plants, the nitrogen–phosphorus ratio is more important than the actual amounts of those nutrients. The eight micronutrients—iron, chlorine, boron, manganese, zinc, copper, molybdenum, and nickel—play vital roles in plant physiology. Some of these elements function mostly as **cofactors**, while others have a structural role in specific molecules. They are needed in only minute amounts, yet their absence can cause death or seriously weaken a plant.

Plant nutrient requirements are variable. Recent studies have found that a number of plant species may require elements such as sodium and cobalt, while dicots have been found to have a greater demand for calcium and boron than monocots.

Careful observation of specific symptoms can help in the diagnosis of mineral deficiencies, which can then be confirmed by soil tests. In natural soils, most nutrients are either present in the form of minerals or are locked up in dead organic matter. Nutrients are released in the form of inorganic ions through the physical and chemical weathering of rocks and the decomposition of organic matter by fungi and bacteria. Many nutrients which form positively charged ions, such as potassium ions (K^+), calcium ions (Ca^{2+}), and magnesium ions (Mg^{2+}), bind to negatively charged clay particles within the soil (**Figure 2**, page 538). For plants to absorb these **cations**, they must be released from the clay particles. This release is accomplished in part by hydrogen ions (H^+), which are released from root hairs into the soil. The hydrogen ions exchange places with the positive nutrient ions. The released nutrient ions are then available for uptake by plants. Negatively charged nutrient ions such as nitrate ions (NO_3^-) and sulphate ions (SO_4^{2-}), which are the main sources of nitrogen and sulfur, are not held by soil particles. As a result, they are more easily absorbed by plant roots. However, they are also more easily **leached** away by heavy rains or irrigation. Mycorrhizae, as discussed in Chapter 10, seem to play an important role in the absorption of phosphorus in the form of phosphate ions (PO_4^{3-}) for most plants.

coenzymes: organic molecules necessary for the activity of some enzymes

cofactors: substances necessary for the activity of another substance, usually an enzyme. Coenzymes are organic cofactors.

cations: ions with a positive charge

leached: washed away as a soluble substance by rainwater or a watering system

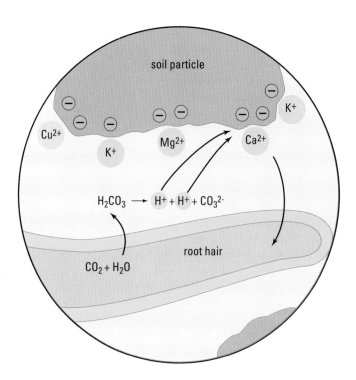

Figure 2
Cation exchange mechanism

When plants are eaten, all the nutrients which they incorporated into their bodies are cycled through herbivores and carnivores. When any living organism releases wastes or dies, decomposers in the environment eventually break down the wastes or bodies and return these nutrients to the soil.

Nitrogen: A Special Case

The nitrogen demands of plants are high, surpassed only by the need for water and carbon dioxide. As nitrogen is a primary component of proteins, those plants which contain a high percentage of protein have a correspondingly high demand for nitrogen. One main reservoir of nitrogen in the environment is the organic matter of living and dead organisms. Dead organic matter is broken down by various bacteria and fungi in a series of reactions which converts the nitrogen compounds first into ammonium ions (NH_4^+), then into nitrite ions (NO_2^-), and subsequently into nitrate ions (NO_3^-), which are readily absorbed by plants through their roots.

The other major reservoir of nitrogen is the atmosphere, which is 79% nitrogen. Unfortunately, the nitrogen gas in air is in a form that is not useful to most living organisms. Atmospheric nitrogen, however, is added to the soil through the action of **nitrogen-fixing bacteria**. These specialized species of bacteria are able to convert nitrogen gas (N_2) from the atmosphere into ammonium ions, which then convert into nitrite ions, and finally into nitrate ions. Although some of these special bacteria live free in the soil, the most important nitrogen-fixing bacteria form symbiotic relationships with plants. These bacteria grow within plant root cells, forming **nodules** (Figure 3, page 539). Although many plant species form such relationships, the best known are the **legumes**, such as peas, beans, clover, and alfalfa. The bacteria supply the legumes with nitrates and the bacteria receive a supply of carbohydrates from the plants. The legumes only receive a small portion of the nitrogen compounds manufactured in the nodules. Estimates for the amount of nitrogen added to Earth's soil each year by these bacteria are very large. Natural ecosystems are balanced; thus, other processes must remove nitrogen from the soil. These processes include the removal of plants (such as harvesting of food crops),

nitrogen-fixing bacteria: bacteria that can convert atmospheric nitrogen gas (N_2) into ammonium ions (NH_4^+). They tend to live in nodules on the roots of legumes and have a symbiotic relationship with the legumes.

nodules: swellings on the roots of legumes that contain symbiotic nitrogen-fixing bacteria

legumes: a group of angiosperms, including peas, beans, clover, and alfalfa, which tend to have nodules containing nitrogen-fixing bacteria on their roots erosion, leaching of nitrogen compounds from the soil, and the action of certain bacteria which convert the nitrate ions back to nitrogen gas.

Practice

Understanding Concepts

1. How do plant nutritional requirements differ fundamentally from those of almost all other organisms?
2. How can technology be used to manipulate photoperiod and why would we want to do this?
3. What are the six soil nutrients needed by plants in the largest quantities?
4. What vital role do nitrogen-fixing bacteria play in the environment?

Applying Inquiry Skills

5. Explain how one year of intensive farming can reduce soil nutrients more than several years of wild plant growth.
6. What might happen if there were no nitrogen-fixing bacteria in the soil in a given area?

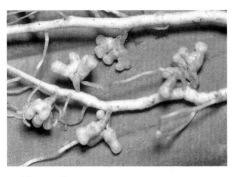

Figure 3
A legume root showing nodules containing nitrogen-fixing bacteria

SUMMARY External Factors Affecting Plant Growth

1. Unlike animals, plants can create their own carbohydrates from inorganic material.

2. Photosynthesis depends partly on the quantity, the quality (colour), and duration of light.

3. Macronutrients are required in large amounts, micronutrients in small amounts.
   - Carbon, oxygen, and hydrogen are primarily obtained from air and water.
   - All other nutrients are obtained as dissolved ions from the soil.
   - Observation of symptoms can indicate which nutrient is deficient.

4. A cation exchange system exchanges hydrogen ions (H^+) from the roots for cations such as potassium ions (K^+), calcium ions (Ca^{2+}), and magnesium ions (Mg^{2+}), which are bound to clay particles in the soil.

5. Negatively charged ions, such as nitrate ions (NO_3^-) and sulphate ions (SO_4^{2-}), are absorbed quite easily.

6. A plant's greatest demands are for water and carbon dioxide; nitrogen is the next greatest demand.

7. Nitrogen is made available when organic matter breaks down. Atmospheric nitrogen is added to soil through the action of nitrogen-fixing bacteria, which live in nodules on legume roots or free in the soil.

Section 13.8 Questions

Understanding Concepts

1. Explain how both the quantity and quality of light influence plant growth.

(continued)

2. Explain the cation exchange system. Why is this mechanism so important to plants?

3. In phosphorus-poor soils, growth slows in many plant tissues. Explain why.

4. Explain how nutrients can be lost from the soil.

Applying Inquiry Skills

5. When a dried plant is burned, the mass of the ash is very low relative to the initial mass of the plant. Account for this fact based on the chemical composition of plant tissue.

6. High soil pH interferes with the uptake of iron by plants. If a plant exhibits symptoms of iron deficiency, what action(s) should be taken to reverse the condition?

7. When magnesium is unavailable in the soil, plant leaves often fade from green to yellow. Suggest an explanation for this observation.

8. The changing of leaf colour in the autumn is triggered mainly by the changing photoperiod. If a tree that originates in Kentucky was moved 1100 km north to central Ontario, explain when its leaves would change colour in comparison to when they would change in its natural habitat.

13.9 Internal Factors Affecting Plant Growth

Plant development and growth are controlled by a combination of many external environmental factors. Growth and development are also influenced by internal chemical regulators—hormones, and pigments such as phytochrome.

Plant growth regulators are similar to animal hormones, but in addition to stimulating growth, they can inhibit growth. Also, plant growth regulators are usually produced in the growing tissue itself, not by some distant source, as animal hormones are. The five hormone groups which exist in most plants are auxins, gibberellins, cytokinins, ethylene, and abscisic acid. Environmental conditions determine the amounts of plant growth regulators that the plants synthesize as well as the sensitivity of plant tissues to these regulators. The relationships which exist are those which have helped the plants survive.

Auxins

Auxins are the best-known plant hormones. Although they both stimulate and inhibit plant growth, they are primarily involved in promoting plant cell elongation. Apical bud growth and the growth and ripening of fruit are stimulated by auxins, but lateral bud growth is inhibited by them. Auxins also regulate a wide range of other plant activities, including cell division in the vascular cambium and the dropping of fruit and leaves. Although auxins are present in very low quantities in plant tissues, they exert a strong influence. Synthetic auxins are widely used in a number of ways, especially as a component of some herbicides. They cause the unwanted plants to grow so fast that they exhaust their carbohydrate reserves and die.

Gibberellins

Gibberellins, like auxins, are best known for promoting cell division and elongation in plant shoots. In some plants they cause the stem to elongate just before the plant flowers. This process, called bolting, produces a long stem that raises the flower up to pollinators and the wind. Plants with this characteristic are more successful than plants which do not bolt. Dwarf varieties of some plants have a mutated gene which blocks the synthesis of gibberellin. If gibberellins are artificially applied to the plant stems of a dwarf variety, the stems will grow normally. Gibberellins are also used commercially to stimulate seed germination.

Cytokinins

Cytokinins are hormones that stimulate cell division and leaf mesophyll growth. Cytokinins are mostly concentrated in endosperm tissue and young fruit. These hormones are important commercially because of their use in the expanding field of biotechnology. They are used in tissue cultures to allow the production of many new plants which have the identical genotype as one original, genetically altered cell.

Ethylene

Ethylene has a significant role in fruit ripening. It also interacts with a number of other growth regulators. Studies have shown that many fruits begin to produce ethylene gas just before the rise in cellular respiration that corresponds to mass conversion of starch into sugars; these sugars increase the sweetness of the fruit. Ethylene also is related to both a colour change and a softening of the tissue of ripening fruit. When growers do not want fruit to ripen in storage or during transport, they keep the fruit in conditions which keep ethylene concentrations low. Ethylene is applied externally when they do want the fruit to ripen.

Abscisic acid

Abscisic acid is a growth regulator which usually acts as an inhibitor. It is important in promoting the closure of stomata, inducing seed and bud dormancy, and providing resistance to water stress.

SUMMARY Internal Factors Affecting Plant Growth

1. Plant growth regulators are usually produced in the growing tissue.
2. Plant hormones can inhibit as well as stimulate growth.
3. Auxins stimulate cell division and elongation in the apical bud but inhibit these processes in lateral buds. Auxins play a role in ripening and dropping of fruit. Synthetic auxins are used in herbicides.
4. Gibberellins play a major role in the elongation of stems.
5. Cytokinins stimulate cell division and leaf mesophyll growth. They play a significant role in the field of biotechnology.
6. Ethylene is produced by plants and seems to affect the colour, texture, and sugar content of ripening fruit.
7. Abscisic acid tends to inhibit various processes, resulting in dormancy, which allows the plants to survive harsh conditions.

- ◐ Define the Issue
- ○ Identify Alternatives
- ◐ Research
- ◐ Analyze the Issue
- ◐ Defend a Decision
- ◐ Evaluate

Explore an Issue

Debate: Plant Hormones

Just as doctors prescribe hormones in the treatment of diseases, farmers apply hormones to growing plants to spur flowering or to make the fruit develop without seeds.

Statement: The application of synthetic hormones is safe for humans and the environment, and is an ethical practice.

- First define whether the issue applies to both food plants and non-food plants. Would your answers be different for food plants than for nonfood plants?
- Consider whether your answer would be different if the issue were about using synthetic hormones in animals and why this would, or would not, make a difference.
- In your group, research the issue. Learn more about the hormones that are used, their effects on quantity and quality of the plant, their cost, and any effects on other organisms or on the environment.
- Search for information in newspapers, periodicals, CD-ROMs, and on the Internet.
  Follow the links for Nelson Biology 11, 13.9.

GO TO www.science.nelson.com

- Write a list of points and counterpoints which your group considered.
- Decide whether your group agrees or disagrees with the statement.
- Prepare to defend your group's position in a class discussion.

Practice

Understanding the Concepts

1. Where are plant growth hormones produced in a plant?
2. What problem would a plant have if it had deficient quantities of gibberellins? Why?
3. Explain the advantage that abscisic acid gives a plant. When is this advantage of particular value?
4. Briefly explain the role of cytokinins in tissue cultures in biotechnology labs.

Applying Inquiry Skills

5. Some people place fresh fruit in bags to increase the speed of ripening. Which growth regulator is being concentrated by this action and what three things seem to be influenced by this regulator?
6. Cold storage of fresh fruit often slows the ripening process. Explain what the colder temperatures must do in the plant.

Key Expectations

Throughout this chapter, you have had opportunities to do the following:

- Describe the structure and function of the components of each of the leaf, the stem, and the root of a representative vascular plant (13.1, 13.2, 13.3, 13.4, 13.5, 13.6).
- Express the result of any calculation involving experimental data to the appropriate number of decimal places or significant figures (13.3).
- Identify, using a microscope and models, the plant tissues in roots, stems, and leaves (13.3, 13.4, 13.5, 13.6).
- Select appropriate instruments and use them effectively and accurately in collecting observations and data (13.3, 13.4, 13.5, 13.6, 13.7, 13.8).
- Select and use appropriate modes of representation to communicate scientific ideas, plans, and experimental results (13.3, 13.4, 13.5, 13.6, 13.7, 13.8).
- Differentiate between monocot and dicot plants by observing and comparing the structure of their seeds and identifying vascular differences between plants (13.4, 13.5, 13.6, 13.7).
- Design and carry out an experiment to determine the factors that affect the growth of a population of plants, identifying and controlling major variables (13.8).
- Demonstrate the skills required to plan and carry out investigations, using laboratory equipment safely, effectively, and accurately (13.8).
- Describe the effects of growth regulators (13.9).
- Locate, select, analyze, and integrate information on topics under study, working independently and as part of a team, and using appropriate library and electronic research tools, including Internet sites (13.9).

Key Terms

abiotic
adventitious roots
aerial roots
allelopathy
annual ring
annual
apical meristems
bark
biotic
cations
coenzymes
cofactors
collenchyma
companion cells
compound leaf
cork
cork cambium
cortex
dormant
endodermis
epidermis
fibrous roots
guard cells
heartwood
herbaceous
hydrophytes
internode
lateral meristems
leached
legumes
macronutrients
meristems
mesophyll
mesophytes
micronutrients
nitrogen-fixing bacteria
nodes
nodules
palisade mesophyll
parenchyma
perennial
pericycle
periderm
photoperiod
pith
pneumatophores
primary growth
primary root
root cap
root hairs
runners
sap
sapwood
sclerenchyma
secondary growth
secondary root
sieve tubes
simple leaf
spongy mesophyll
succulents
taproots
toxin
tracheids
translocation
transpiration
tubers
vascular bundles
vascular cambium
vessels
woody
xerophytes

Make a Summary

In this chapter, you have studied plant forms and functions to better understand the relationship between the anatomy of plants and their internal and external environments. To summarize your learning, create a poster which illustrates the various structures and functions of plant tissues and organs. Because there are so many parts to consider, you might like to focus on one major plant part. You could begin by placing a diagram of a generalized vascular plant in the centre of the poster and drawing attention to the part you have chosen. Then, add images and information linked to that part, using the key terms above as necessary. Be creative and include significant environmental factors.

Reflect on your Learning

Revisit your answers to the Reflect on Your Learning questions at the beginning of this chapter.

- How has your thinking changed?
- What new questions do you have?

Understanding Concepts

1. What term corresponds to each of the following descriptions?
   (a) a region of plant cell division
   (b) a plant with a single seed leaf
   (c) tissue responsible for transporting carbohydrates
   (d) thin, waxy leaf coating
   (e) contains an angiosperm male haploid nucleus
   (f) a nonflowering plant capable of producing seeds
   (g) dead cells used for transporting water within a plant

2. Make simple sketches of the two types of xylem and phloem cells. Label as many parts as you can.

3. Provide an example of each of the following plant adaptations:
   (a) structure that reduces air flow over a leaf surface;
   (b) plant with a green photosynthetic stem;
   (c) poisons released into the environment to kill or damage competing plants;
   (d) roots for obtaining oxygen from above water level;
   (e) feature allowing leaves to float on the water surface.

4. What are the advantages of having woody stems?

5. Explain the mechanism by which guard cells open and close stomata.

6. Trace the pathway of water through a plant from entry to exit. Use the following terms in your description: xylem, stomata, root hair, epidermis, endodermis, mesophyll, cortex, and phloem.

7. Explain how plants have been successful at adapting to and competing under different environmental conditions.

8. Explain the difference between macronutrients and micronutrients, giving three examples of each.

9. If the nitrogen in air is useless to plants, explain how plants are able to make use of it eventually.

10. How are plant hormones similar to animal hormones and how are they different?

11. In the spring, trees transport high-energy nutrient reserves to new growing leaves and stems. This sap is harvested from sugar maple trees. In which type of tissue do you think the sap is flowing? Explain.

12. Discuss how greenhouse operators are able to provide poinsettias in December and lilies in March and April even though these are not the normal time of year for these plants to flower.

13. Describe three different plant abnormalities and suggest a mineral deficiency that could have caused them.

14. Suggest the advantages and disadvantages of being wind pollinated rather than animal pollinated. Consider the following:
   • Which type of plant needs to make the most pollen?
   • Which plant needs to produce the fanciest and most complex flower?
   • Which type of flower must "give away" high-energy food molecules to pollinators?
   • Would wind-pollinated plants do better if they were crowded or spread far apart?

15. Identify each plant growth regulator that is most likely responsible for the following effects:
   (a) When placed in a sealed bag, the fruit ripens quickly.
   (b) When exposed to this compound, the plant flower stalks elongate dramatically.
   (c) When this chemical is not present, the plants are dwarfs.

16. The epidermis on a peach produces many fine hairs. Suggest a function for this peach fuzz.

17. A number of plants adapted to hot, dry environments possess very fuzzy leaves. These leaves appear quite light in colour. What effect do you think this has on the internal temperature of the leaf and on water loss due to transpiration and evaporation?

18. The following is a list of foods eaten by a large deer. For each item, explain how the deer's actions affected the plant's life cycle.
   (a) beet root   (b) grass leaves
   (c) young clover flowers (d) ripe crab apples

19. The century plant is very slow growing, and flowers and produces seeds only once when it is about 150 years old! How might this slow rate of seed production actually increase the plant's reproductive success?

20. A dormant seed from the sacred lotus plant was carbon-dated at 1288 years old when it successfully germinated. Seeds from arctic lupins were even older when they successfully germinated. Discuss the advantages of seed dormancy.

21. Some tropical plant species, such as fig trees, produce flowers and fruits throughout the year and often when few other plants are fruiting. Many animals are dependent on these fruits to sustain them. How does this help the reproductive success of the fig plants?

22. How is slippery and/or peeling tree bark advantageous to trees growing in tropical rain forests?

Making Connections

23. Many commercial insecticides such as nicotine, rotenone, and pyrethrum come from plants. Rotenone is derived from the roots of several tropical plants, while pyrethrum is obtained from chrysanthemum flowers. Which plant organ(s) would you expect to contain the greatest variety and concentration of toxins? Support your reasoning.

24. Scientists studying a few plots of tropical rain forest, totalling only 6.6 ha, identified 711 different plant species. By contrast, the same area in Ontario might have fewer than 100 plant species. How should this influence decisions made concerning the cutting of tropical rain forests?

25. Explain how a knowledge of nitrogen-fixing bacteria and allelopathic relationships would be of use to a gardener, a farmer, and people working to regenerate a natural area.

26. Scientists are very interested in the complex relationships between plants, animals, and fungi. How might the extinction of each of the following affect plants in their environment?
    (a) a soil fungus species (b) a seed-eating bird
    (c) a fruit-eating monkey (d) a leaf-eating mammal

27. **Figure 1** shows a May-apple plant. It has two leaves and one or two white flowers at the bases of the leaves. It grows in deciduous forests and blooms in May.
    (a) What type of venation does the May-apple have?
    (b) On the basis of this venation, how many cotyledons would you expect its seeds to have?
    (c) What are two visible features of May-apple leaves that would allow it to thrive on the floor of a forest?
    (d) Explain why this plant has evolved flower and seed production in the spring rather than another time of year.

28. **Figure 2** shows a colourful plant growing on sand.
    (a) What type of plant is shown here?
    (b) What visible characteristic makes this plant well suited to growing on dry, windy sand dunes?
    (c) What name can be used for these specialized stems?
    (d) What advantage(s) do such stems provide in a sand dune habitat?
    (e) What type of tissue is probably not present in huge quantities in these stems? Explain how you know.

Figure 2

29. **Figure 3** is a legume.
    (a) Give three specific examples of legumes.
    (b) This legume is growing on a sandy lakeshore. What would you expect to find if you did a soil analysis near this plant? Explain why.
    (c) What would you expect to see if you carefully dug up the roots of this plant?
    (d) Why might you not see the feature you suggested in (c) if you pulled the plant out to see the roots?

Figure 1
A May-apple plant in early May

Figure 3
A legume growing along a sandy lakeshore in Ontario

14

The Importance of Plants

If asked why plants are important, people would suggest the economic value and medical potential of plant products (**Figure 1**) and would also mention habitat and food for various animals, the aesthetic value of plants, and the oxygen they produce. Some might even stress that plants help reduce the threat of global warming by removing huge amounts of carbon dioxide (CO_2) from the atmosphere. Yet we often cut trees to build more or wider roads to carry more and more cars which add carbon dioxide to the air.

BIOCAP (The Biosphere Implications of CO_2 Policy in Canada) is an organization consisting of over 120 researchers from 22 Canadian universities and 15 research institutes. It includes numerous biologists, foresters, environmental analysts, and chemical engineers. Their research goal is to better understand the complex carbon cycling within our global ecosystem. Knowledge about plant life is essential to BIOCAP's work.

In this chapter, you will be able to

- illustrate the process of succession and describe examples of plant competition;
- connect the role of plants to the maintenance of biological diversity and the survival of all organisms on the planet;
- design and conduct an experiment to study the effects of fertilizers on plant growth;
- describe and explain some uses of plants in food and industrial processes, in therapeutic products, and in chemical products;
- evaluate the choices and tradeoffs which often must be made when choosing how to develop, market, and buy new plant products;
- gather facts and make choices related to which plant research projects should receive funding, taking into consideration the desires of the public and societal and environmental concerns.

Try This Activity

Plant Product Potpourri

- In your notebook, create a table with five columns labeled as shown in **Table 1**.

Table 1: Plant Products

Product	Use(s)	Country or region of origin	Plant part from which it is obtained	Properties of plant which complement use(s)

- Examine the list below. Choose at least five products from the list and write them in the left column of the table you just drew. Allow large spaces for writing about your findings.

paper	plywood	cotton fabric	rice
turpentine	camphor	rubber band	chewing gum
vanilla	vegetable oil	cork	cinnamon
cocoa powder	coffee	tannin	aspirin

- Use a dictionary or encyclopedia, search the Internet, read labels in a local supermarket, and talk to family and friends to gather the information needed to fill in the rest of the table for each product. Discuss your results with the class.
  (a) Which of these plant products are obtained and used "as is" without any processing?
  (b) Which of these products require minimal processing?
  (c) Which of these products do you consider very important in your everyday life? What are your reasons for these choices?

Humans always have been, and always will be, dependent upon plant life. Cultivating, harvesting, and processing plants for various products necessitate an understanding of the growth requirements of plants, as well as the invention and application of technologies appropriate to the survival of our planet. These processes provide opportunities for employment, research, and innovation. The wise use of plant science and technology is critical for a prosperous and sustainable future. This chapter explores how plants are necessary for the health and well-being of humans and all the other organisms that share our planet.

Figure 1
Describe how all of these items, including the background, are connected to plant products.

Reflect on your Learning

1. Make a list of all the plant products you use, directly or indirectly, in a typical day. How important are plants in your daily life on a scale of 1 to 10, with 1 representing very unimportant and 10 representing very important?

2. Suggest some specific examples to show how science and technology are related to the growth, harvesting, and processing of plants.

3. Describe briefly some environmental impacts of current agricultural and forestry practices.

4. Suggest possible problems related to an increased dependency on technological advances.

5. What does "sustainable future" mean?

6. Suggest some ways that technology might help lead to a sustainable future.

14.1 Succession

Few events in nature appear as devastating as the destruction of a mature forest by a forest fire (**Figure 1**). All that remains is a blackened landscape with a few solitary tree trunks and stems starkly pointing to the sky. However, despite appearances, forest fires create opportunities for new life (**Figures 2** and **3**). The soil is enriched by the mineral content of the ash. Within a few weeks, seeds of annual and perennial plants will take advantage of the opportunity presented by the fertile soil, open ground, available sunlight, and lack of competition, and will germinate and grow. Within two or three years, shrubs and young trees are quite evident and growing rapidly to provide browse (young twigs, leaves, and shoots) for returning animals to eat. A decade or two later, an untrained observer would probably never know that the area had once been burned. In 100 to 150 years, the forest will again reach maturity. Once mature, the forest will remain in a steady state until another disturbance, caused by nature or humans, again disrupts the ecosystem.

Figure 1
Forest fires, however started, always seem like disasters. Nevertheless, fires play a major role in maintaining healthy ecosystems.

Figure 3
New life beginning after a devastating forest fire

Figure 2
The colourful, metre-tall fireweed is one of the first plants to germinate and thrive in a recently burned area, whether it was a whole forest or a farmer's barn.

succession: a series of gradual changes in the vegetation of an area followed by gradual changes in the animals in the area

climax community: the final, self-perpetuating stage of succession. The composition of the climax community depends on the abiotic factors of the area.

Succession is a series of very gradual changes occurring in the vegetation of an area, accompanied by changes in animal species. At all times, succession is influenced by the biotic and abiotic factors characteristic of the region. The stages or steps of succession are not distinct; they blend into one another in a continuous gradual sequence.

Vegetation always alters the soil and influences the microclimate in the immediate area. New conditions, often not favourable to existing plants, will be suitable to other types of vegetation. For example, the dappled shade created by poplar trees is not suitable for the germination of poplar seeds, but coniferous seeds are able to germinate and produce thriving trees. As the conifers grow, they eventually replace the poplars. In an area undergoing succession, the oldest plants are of different species than the youngest plants. Eventually, the stage is reached when the seeds produced by the mature plants thrive in the same area. In these areas, the oldest and youngest plants are the same species. This **climax community** is self-perpetuating, or renews itself, rather than preparing the way for other plants.

Although climax communities are considered to be stable, or in a steady state, they are not static; there are always numerous biotic and abiotic forces at work. Climax communities, once established, are able to tolerate routine stresses, such as a large tree falling.

Succession is a process that occurs in all aquatic and terrestrial environments, but the speed at which it occurs, the stages involved, and the eventual climax community that results are all dependent on the abiotic factors in the area. Succession in northern Ontario is different from that occurring in southern Ontario, which, in turn, varies from that occurring in southern Michigan or central Saskatchewan. The climax community is black spruce forests in northern Ontario; beech-maple forests in southern Ontario; oak forests in southern Michigan; and grasslands in central Saskatchewan.

Primary succession (Table 1, page 550) occurs in an area in which no living things existed previously and where no organic substrate exists. Examples of this process are the invasion of plant life on a newly formed volcanic island, on land released from a retreating glacier, on a sandy beach, or on a single bare rock. Soil is absent and must be developed to a stage where vegetation can take hold. Soil building is a long, slow process and depends on abiotic factors such as temperature, slope, and the composition of ground materials, and on biotic factors such as the presence of lichens.

primary succession: succession which begins without preexisting organic material or soil

Lichens are **pioneer organisms** and can survive under extremes of heat, cold, and drought. Lichens can live in barren locations because the fungal hyphae can grip the slightest irregularities in rock surfaces. Lichens can produce their own carbohydrates. They absorb water from rain and from the air, and they obtain minerals from the sand, rocky surfaces, and dust particles. Once the lichen is thick enough, its own tissues can act as a substrate for mosses. Bits of airborne debris can get caught by the lichens or mosses and may contribute to soil building. The death and decay of the lichens, mosses, or moss parts contribute to further soil development, paving the way for larger plants such as ferns or small angiosperms.

pioneer organisms: organisms capable of surviving harsh conditions and establishing themselves in bare, barren, or open areas to initiate the process of primary or secondary succession

Pioneer angiosperms typically have small seeds that are easily carried by the wind or within the digestive tracts of birds. Grasses meet these criteria. Other plant seeds can also travel long distances but need more organic material before they can survive. For example, dandelion seeds can travel great distances but need sufficient soil for their taproots.

Secondary succession occurs following the partial or complete destruction of an existing community. The regrowth of an area after a forest fire is an example of secondary succession, as are the changes that occur after a farmer stops cultivating a field. The pioneer species in secondary succession must be tolerant of harsh conditions, but since soil is already present, the lengthy process of soil formation is not necessary. Grasses, dandelions, milkweed, fireweed, and raspberry are examples of pioneers in secondary succession. In secondary succession, many plants will come from seeds already in the soil. Although there are likely to be many seeds of many different species already in the soil, the abiotic conditions, especially the availability of light and nitrogen, will determine which ones germinate and succeed.

secondary succession: succession which begins with organic material or soil already present

In both primary and secondary succession, the early communities are relatively simple, with a small number of species interacting in uncomplicated food webs. As succession progresses, the communities have larger numbers of species of many types of organisms, and the food webs are more complex. The final stages and the climax community have the greatest diversity of organisms interacting in convoluted food webs. If a simple community in the early stages of succession is disturbed—for example, by fire, flood, toxic spill, or pest infestation—the results will be severe and recovery will take a long time. Biologically diverse communities at later stages of succession will suffer less from similar disturbances and recovery will be faster.

Table 1: Primary Successional Stages Along Georgian Bay Shoreline

Stage	Dominant plants		Plant characteristics	Effects on environment
inorganic sand and different types of accumulated organic drift, wood bits, bird feathers, and dead insects get washed ashore	no plant life			decay of the organic matter adds various nutrients to the inorganic sand base
pioneer community	grasses with fibrous roots anchor themselves in the sand		heat resistant, shade intolerant; provides food for insects (e.g., grasshoppers)	establish soil; decrease soil temperature and evaporation; increase soil fertility
early stage	perennial herbs, shrubs, fast-growing trees (e.g., poplar)		shade intolerant, fast growing, and short-lived	stabilize and enrich soil; pioneers decrease because of shade and overcrowding; provide food for greater variety of consumers
middle stage	conifers such as pine and larger shrubs		tall, woody, slower growing, and sun tolerant	add to and stabilize soil; provide increased habitat and food for variety of animals
later stage leading to climax community	young deciduous hardwood trees, older conifers, shrubs, and herbaceous perennials		tall conifers; young, slow-growing, and shade-tolerant understorey	shade allows only shade-tolerant species to germinate and thrive; conifer seeds seldom successful; soil enriched as older trees fall and decay; more complex food webs
climax community	all ages of deciduous hardwood trees (e.g., oak, beech, and maple); understorey shrubs, saplings, and herbaceous perennials		tall, slow-growing, and long-lived trees and shade-tolerant understorey	shade allows only shade-tolerant species to germinate and thrive; their own seeds are successful; high biodiversity; complex food webs possible; cooler temperatures and higher humidity

Competition

Succession is a long-term process driven by various types of **competition**. Competition is a relationship in which two organisms place demands on the same environmental resource. Competition between organisms of different species is **interspecific** competition. Competition between organisms of the same species is **intraspecific** competition. The more closely related the two competitors are, the greater their competition. Plants mainly compete for water, space, nutrients, and light. For animals, competition also involves territory, food, nesting sites, and mates.

A species of plant that grows faster than a competing species will develop a root system more quickly, will claim a greater amount of water and soil nutrients, and will limit the physical space available for the growth of competitors (**Figure 4**). The shade created by the leaves of the rapidly growing plant will limit photosynthesis for other plants, thereby slowing the growth of competing plants even more. The more rapidly growing plant out-competes the other. The "winner" will thrive; the "loser" may survive, but will not thrive.

Competitive reproductive strategies in plants include the production of vast quantities of seeds and the production of seeds which have some structural means of being widely dispersed. Plants have also evolved many competitive strategies that, in effect, are ways to avoid interfering with other plants. For example, if two species of plants situated near each other grow, flower, and produce seeds at the same time, they will be in competition for resources at the same time. However, in a species that exhibits genetic variation, there may be individuals of each species that develop later or earlier than the others. These individuals will face less competition and, therefore, will more likely be successful and pass on their genes to future generations.

In some situations, a slower-growing species might have an advantage by producing a chemical that slows or inhibits the growth rate of its competitors. Allelopathy differs from general competition because it involves the addition of a chemical to the environment.

competition: a relationship in which two organisms place demands on the same environmental resource

interspecific: between two species

intraspecific: within one species

Figure 4
Relative to grass, dandelions are very successful in competing for water and minerals because they have more extensive root systems. Dandelions also form a rosette of leaves which spread outward to prevent the grass from getting adequate sunlight. Most people think that grassy lawns are more desirable than dandelions and ignore the economic and environmental costs of getting rid of the dandelions.

Try This
Activity

Fire Release

The jack pine is a fire release species with specialized cones which only open when exposed to the high temperatures of a fire.

- Obtain several mature jack pine cones (**Figure 5**) and place them in an oven at 200°C.
- Observe them to see if and when they open.
- Collect any released seeds after the cones have cooled.
- Put them in little pots of moist potting soil to see if they will germinate.

Figure 5
Jack pine cones

Human Intervention in Succession

Have you ever witnessed or heard reports about someone's farm pond gradually filling in with "weeds," or people complaining because their once pristine lakefront property now has a variety of cattails, water lilies, duckweed, and algae making their shoreline "dirty"? What these people are witnessing is aquatic succession, promoted by an increase of nutrients in the water. Humans are responsible, directly and indirectly, for most of this increased nutrient load.

Succession in various types of environments and the nutritional requirements of plants have been studied so thoroughly that scientists can now predict how the vegetation will change in a given area and what type of climax community will eventually exist there. Such information is of great value to individuals and groups that are trying to regenerate natural areas where a lot of disturbance has occurred. One rule of thumb when planning to rehabilitate a landscape is that we should only consider plants that would naturally grow successfully on the site.

Another point to remember is that we can assist succession but we cannot force it to jump several stages. It is not possible to convert a site to its climax community immediately, because as you have learned, climax forests require special conditions that must be allowed to develop gradually. However, it is possible to give natural succession a boost. For example, in southern Ontario, it would be foolish to plant beeches or sugar maples in a grassy area even though we know that they would eventually form the climax community in that area. They simply could not thrive in the exposure to winds and hot sun, and if they ever did produce any seeds, they certainly would not germinate. A more reasonable choice would be to plant trees from more intermediate successional stages, such as poplars and a few white pines. Another successful technique is to plant appropriate species at the edges of an existing forest (**Figure 6**). These areas would eventually seed themselves, but it could take several decades. A decade later when the pines and poplars become established, we could try planting a few oak and maple saplings among them.

Figure 6
Secondary school students giving natural succession a boost by planting suitable species in a grassy area next to a remnant forest

Practice

Understanding Concepts

1. In addition to providing food, what roles do plants have in the natural environment?
2. Distinguish between primary and secondary succession.
3. How does succession influence the overall diversity in an ecosystem?
4. What characteristics are typical of pioneer species in the following places?
   (a) on a pile of recently fallen rocks
   (b) on an unused railway bed

Fires

Forest fires due to lightning strikes are a natural phenomenon. Forest fires can also be due to human carelessness. Although the effects seem devastating, a fire may be beneficial by promoting the release of large amounts of nutrients into the environment. Many forest plants can tolerate periodic fires, and many animal and plant species actually benefit from the changes brought about by fires.

Fires are sometimes prescribed to achieve a specific goal in a specific area. For example, in Ontario, grasslands are not a normal climax community, with the exception of a small region near Windsor, but, for a variety of reasons, the maintenance of grasslands may be desired. The way to accomplish this is to burn areas carefully when saplings or other undesirable plants develop beyond a desired size. The desired herbaceous plants will recover by the next growing season. Prescribed fires may also be used to control the spread of dry bush fires. By burning a designated area of dry brush, we can prevent undesired fires from spreading.

The detrimental effects of forest fires include threats to wildlife, human lives, property, and the logging industry. They also release large amounts of various greenhouse gases into the atmosphere.

Debate: Forest Fire Suppression

You have learned about the environmental benefits and disadvantages of forest fires. Fire suppression is a topic that is under debate due to its environmental and economic impacts.

Statement: Ontario should commit more money to prevent and fight forest fires throughout the province.

- In your group, research the issue. Search for information in newspapers, periodicals, CD-ROMs, and on the Internet. Start by finding out what Ontario's current policy is concerning forest fire prevention and control.
  Follow the links for Nelson Biology 11, 14.1.

GO TO www.science.nelson.com

- As you carry out your research, write a list of points for and against the statement. You might consider these factors:
  - the extent of annual forest fire damage in Ontario
  - the type and amounts of greenhouse gases released
  - forest floor composition with and without fires
  - the effect of fire on forest insect pests
  - the effects on wildlife if fire does or does not occur
- Reflect on the information you have gathered and develop an opinion about the statement.
- Write a position paper summarizing your position.

 Succession

1. Succession is the series of gradual changes in the vegetation of an area as it develops toward a climax community.

2. Primary succession occurs in an area in which soil building is required.

3. Secondary succession occurs in an area in which soil building is not required. There will often be seeds lying dormant in the soil or there may already be some plants growing there.

4. Pioneer plants must be tolerant of open conditions, such as bright light, high soil temperature, and low humidity. Once established, they create microclimates that allow other species to establish themselves. This pattern is repeated again and again.

5. The composition of the climax community is determined by the abiotic factors of the area. It is self-perpetuating and in a state of dynamic equilibrium.

6. Competition may be interspecific or intraspecific. It involves common demands on one or more resources. Competition among plants often involves the timing of parts of their life cycles and their growth rates.

7. Knowledge of the special requirements for different species and the local succession patterns allows humans to make the right decisions regarding a regeneration project to increase the survivorship of the species planted.

Section 14.1 Questions

Understanding Concepts

1. During succession, plants are in direct competition for light, water, nutrients, and space. Suggest which plant structures are most involved in competition for the following:
   (a) light
   (b) water and mineral nutrients
   (c) space

2. If the climax community is dominated by tall trees, which factor in question 1 would be the most significant? Explain your answer.

3. If the climax community is dominated by prairie grasses, which factor in question 1 would be the most significant? Explain your answer.

4. Explain why the climax communities in two different areas might be the same or different.

Applying Inquiry Skills

5. In 1883, a massive volcanic eruption buried the Indonesian island of Krakatau and the surrounding islands in ash to a depth of over 30 m. All plant and animal life was wiped out. By 1934, more than 270 plant species had returned to Krakatau.
   (a) Would you call this an example of primary or secondary succession? Explain why.
   (b) Propose two ways in which plants might have reached these uninhabited Pacific islands.
   (c) Hypothesize what the pattern of animal recolonization might have been. What types of animals may have arrived first? Explain why some of these could survive while others probably could not.

14.2 Plants and Biodiversity

Earth's ecosystems are highly varied, containing perhaps 15 million species of animals dependent on about 300 000 species of plants. On average, each plant species supports 50 different animal species in a complex food web. Consequently, the greater the plant diversity in an ecosystem, the higher the total **biodiversity** of that ecosystem. This 50 to 1 ratio has significant ramifications. When a single species of plant becomes extinct or when a particular plant community is destroyed, many other organisms are threatened.

biodiversity: the number of different species of all living organisms in a given area (sometimes called biological diversity)

Animal species diversity is influenced not only by the variety of plant species but also by the physical structure of the plant community in which they live. Forests of equal-aged trees are quite uniform in height and are home to far fewer species than forests of mixed-aged trees (**Figure 1**).

It should not be surprising that biological diversity is highest in tropical rain forests. Rain forests are continuously warm and wet, ideal conditions for many

Figure 1
Although tropical rain forests are known for their variety of life forms living at different heights, mixed deciduous-coniferous forests also provide various layers of canopy, understorey, forest floor, and soil in which many nonplant organisms make their home or find food.

plants, and tend to have a complex physical structure. Biodiversity reaches a minimum in areas that experience extreme temperatures or lack water such as in deserts, on mountaintops, or in the polar regions. Where little or no plant life exists, very few other living organisms can survive.

Compare the biodiversity in a **monoculture** of corn, wheat, soybeans, or a plantation of pines (**Figure 2**) with that of a wild prairie or natural forest. Such monocultures are often sprayed with herbicides to kill all competing plants as well as with insecticides and fungicides to kill the organisms which have a negative impact on the crop. In natural ecosystems, countless different plant species live in the same area and support a food web involving thousands of different species. Humans benefit greatly from the products of monocultures, but we must not lose sight of the ecological consequences of our actions.

As mentioned in Chapter 10, genetic diversity is enhanced through sexual reproduction because genes from both parents are combined in a variety of ways. In any given situation, those individuals with the genetic traits that allow them to thrive will pass those characteristics on to their offspring. Those members of the same species that have traits that do not allow them to survive simply die. Diversity within a species is the key to survival of that species. Similarly, diversity among species within a community will also guarantee the relative stability of that community even if it is disrupted by a disturbance such as a fire, flood, or an insect infestation. Some species will suffer more than others but, in time, balance will be restored.

In general, there will be greater biological diversity in a community if there are no weather extremes, if the topography is varied, if each species of the community exhibits a wide variation of sizes and ages, and if a wide range of habitats are available to the organisms living there.

Biodiversity in many parts of the world is threatened by some **exotic** plant or animal species which have been introduced by intent or by accident. Many exotic species are **invasive**, which means that they are capable of out-competing **native** species. They often are not susceptible to native diseases or pests. Local herbivores frequently ignore exotic plant species as food. The new species may also succeed because they are hardy perennials with extensive root systems that give rise to more and more of the unwanted species. As native plant species fail to thrive, there is increasing danger that the invasive species will take over more

Figure 2
This equal-aged pine plantation supports little biodiversity.

monoculture: the cultivation or growth of a single species

exotic: describes species that are foreign or not native

invasive: capable of out-competing the other species in any given area

native: describes species that originate in a particular region

Figure 3

A 2001 publication focusing on maintaining biodiversity

and more of the area and that the biodiversity of all organisms will decrease (**Figure 3**). It should be noted that there are exotic species that are not invasive and a few native species which are considered invasive in certain circumstances.

SUMMARY ### Plants and Biodiversity

1. Each plant supports about 50 different animal species through food webs.
2. Changes in the plants in a community will have a great impact on the whole community.
3. Diversity within a species is important for the survival of that species.
4. Diversity among the species in any community is important for the survival of that community.
5. Biodiversity is increased when there are no extreme weather conditions, if the topography is varied, and if there is a wide range of available habitats.
6. Diverse communities can withstand disturbances more easily than less diverse communities.
7. In some areas, biodiversity is threatened by invasive exotic species.

Practice

Understanding Concepts

1. List three conditions that lead to greater biological diversity, and explain how each factor influences biodiversity.
2. What does it mean for a species to be invasive?

Reflecting

3. In what ways would your life and that of future generations be affected by a large loss in biological diversity? How important is the extinction of species to you personally?

Try This
Activity

Explore Ontario's Biodiversity and Species at Risk

The Royal Ontario Museum (ROM) is home to Ontario's Centre for Biodiversity and Conservation Biology. Scientists at the centre conduct research to gain understanding of the interrelationships of species, their genetic diversity, and their phylogenetic relationships. They estimate the risks of extinction and recommend appropriate strategies for resource management and conservation programs.

- Find the Web site for the Centre for Biodiversity and Conservation Biology. Find out about their research. You can follow links to virtual field guides to help you learn about wildlife in your part of Ontario. Find out which Ontario species are threatened with extinction. Discover which invasive exotic species are causing problems in Ontario and what methods are being suggested for controlling them.

Follow the links for Nelson Biology 11, 14.2.

GO TO www.science.nelson.com

Understanding Concepts

1. How does an increase in genetic diversity enhance the overall health of an ecosystem?
2. What factors cause exotic species to be invasive?

Making Connections

3. You decide to plant a decorative flower bed filled with hundreds of begonias using a variety of colours to create a design. When you describe your plans to a landscape architect at the nursery, she suggests that you should create the same colourful display using at least six different flowering plants.
   (a) Offer the scientific reasoning behind her advice.
   (b) Explain the environmental advantage of taking her advice.

14.3 Fertilizers

In Chapter 13, you learned about the 17 essential nutrients that plants must acquire. In most natural situations, plants are able to obtain adequate supplies of the essential elements from the surrounding soil. Nutrients taken up by plants are replaced through natural cycles. Human agricultural practices, however, can result in a steady depletion of soil nutrients as plants are harvested year after year. For example, the growing and harvesting of crops that are high in protein content, such as corn, will result in a rapid decline in available soil nitrogen, since protein-rich plants take up large amounts of nitrogen. Intensive agricultural practices result in reductions of one or more of three nutrients: nitrogen, phosphorus, and potassium (NPK). If soils are deficient in one or more nutrients, the answer is to add **fertilizers** that contain whatever nutrients are required in suitable amounts to make the soil fertile.

The nutritional demands of plants change as they grow and develop. Phosphorus is important in root formation and thus is needed most in early stages of development. Nitrogen is used in large quantities during the most active growing phase of the plant when vegetative growth is important. Too much nitrogen, however, can promote excessive leafy growth and delay or reduce flowering and fruit production. Potassium is also important for active growth as it is needed for strong stems. As plants enter the reproductive phase—when they produce flowers, seeds, and fruits—phosphorus and potassium are again needed in larger quantities while the nitrogen requirement decreases. For these reasons, fertilizers higher in nitrogen are often applied from early to mid season, while fertilizers with a reduced nitrogen content are used to encourage flower and fruit production later in the growing season, when vegetative growth is not encouraged. Fertilizers can be natural or synthetic. **Natural fertilizers** can be physically processed, but the chemical changes occur naturally. **Synthetic fertilizers** have been created through chemical processes directed by humans.

Natural Fertilizers

Manure and **compost** (Figures 1 and 2, page 558) are natural soil conditioners because they add a lot of humus to the soil. Humus is dark, spongy, decayed plant material that increases the soil's ability to hold water. It also contains minerals to increase the general soil fertility. Some municipal recycling programs accept

fertilizers: any minerals added to soil, usually to replace those removed by crops

natural fertilizers: fertilizers produced without human-directed chemical processes

synthetic fertilizers: fertilizers produced through human-directed chemical processes

manure: animal waste

compost: a mixture that consists largely of decayed organic matter and is used as a soil conditioner and source of minerals

Figure 1
Manure direct from livestock barns can be used on fields. Gardeners can purchase bagged manure for use on their gardens.

Figure 2
Home composting prevents valuable nutrients from being thrown away.

Figure 3
Municipal composting reduces the need for landfill space and produces valuable soil conditioner. Some programs use indoor digesters. The program shown here is outdoors and uses a large machine to regularly mix the yard waste with air.

sewage sludge: semisolid matter produced during sewage treatment

crop rotation: the agricultural practice of planting a field in successive years with various crops, each of which has a different nutrient requirement

Figure 4
In order to provide a nitrogen supply to rice crops, the small floating water fern *Azolla* (seen here covering the surface of the water) is often grown in the flooded rice paddies. Symbiotic nitrogen-fixing bacteria in the fern accumulate nitrogen. As the rice plants grow, they prevent sunlight from reaching the ferns. The small ferns die and decompose, releasing their nitrogen supplies.

kitchen scraps and/or organic yard waste. These programs offer the potential for significant contributions to waste reduction and nutrient recycling (**Figure 3**). Properly processed **sewage sludge** from large urban centres also has the potential to be a significant soil conditioner. This sludge must be processed to kill any potentially pathogenic bacteria and must be carefully tested to ensure that safety levels for heavy metals and other contaminants are not exceeded. All of these natural sources of nutrients are not very concentrated, have varying amounts of different minerals, and release their nutrients slowly over time.

Crop rotation is a natural way to maintain and improve soil fertility without the addition of any type of soil conditioner or fertilizer. With crop rotation, legumes are typically planted every second or third year. The nitrogen-fixing bacteria living in legume root nodules not only supply the legume with nitrogen but also add a lot of nitrogen to the soil for subsequent crops. The legumes may be harvested, leaving the roots in the ground to decompose, or the entire plant may be plowed under as "green manure" to maximize the addition of nitrogen to the soil. In the following one or two years, a crop with high nitrogen demand, such as corn, can be grown on the field with little or no additional input of nitrogen. During the cultivation of rice crops, a tiny fern is sometimes used to add nitrogen (**Figure 4**) instead of adding other fertilizers.

Synthetic Fertilizers

In 1913, a process for the synthesis of ammonia was discovered. Using ammonia, millions of tonnes of synthetic nitrogen fertilizers are produced annually. The production and widespread use of synthetic fertilizers began in the 1930s and revolutionized agriculture. All synthetic fertilizers (**Figure 5**) are more concentrated and release their nutrients into the soil much more rapidly than do most natural fertilizers. Global food production increased dramatically. The application of specific amounts of quick-release concentrated fertilizers has enabled farmers to increase crop yields by rapidly replacing those nutrients removed by harvesting and/or lost due to erosion and leaching.

Unfortunately, we have paid a high price for these convenient synthetic fertilizers. It was once thought that manufacturing fertilizers was very efficient. We now understand that making synthetic fertilizers requires large inputs of energy. Synthetic fertilizers are prone to leaching and runoff and may alter the natural chemical balance in groundwater and surface water, and may affect associated

Figure 5
Synthetic fertilizers have an NPK code indicating the percent content of available nitrogen, phosphorus, and potassium, always in that order. For example, a label reading 5-10-10 indicates 5% nitrogen, 10% phosphorus, and 10% potassium. The remaining material is usually filler.

Figure 6
Great blue herons build huge nests in large trees, but the nutrient concentration in their droppings kills the nesting trees. Naturalists have a dilemma because they want to protect both the herons and the trees. Homeowners notice a similar effect on their lawns if excess fertilizer is used; areas that are too rich in nutrients show brown spots of dead grass.

organisms. In addition, the use of synthetic fertilizers can lead to imbalances in natural soil composition and biological activity, resulting in a loss of organic matter and increased erosion.

Another danger of using any fertilizer is that if too much is added, it can actually kill plants (**Figure 6**).

Hydroponics

Growing plants without soil is now common practice in home and commercial greenhouses. In **hydroponics**, the soil is replaced by a sterile solution of aerated water and essential plant nutrients or sterile sand and the nutrient solution (**Figure 7**). Growing plants hydroponically can reduce or eliminate pathogens and pests that normally reside in soils. It permits the precise regulation of nutrient concentrations which can be tailored to the changing needs of the growing plants. Although such technology increases the costs of production, the result is rapid growth and high yields.

hydroponics: a system of growing plants without soil but instead with a sterile medium and a solution containing all required nutrients

SUMMARY   Fertilizers

1. Synthetic fertilizers can be added to soil to replace nitrogen, phosphorus, and potassium lost as a result of current agricultural practices.

2. Nutritional requirements change over the plant's life cycle, for example, nitrogen promotes vegetative growth but can inhibit reproductive growth.

3. Natural fertilizers tend to release nutrients more slowly than synthetic fertilizers but condition the soil by increasing its ability to hold water.

4. Crop rotation increases soil fertility; legumes are planted every few years to replenish nitrogen.

5. Synthetic fertilizer labels show the percent content of nitrogen, phosphorus, and potassium.

Figure 7
Hydroponically grown lettuce has a much larger root system than lettuce grown in soil.

6. Synthetic fertilizers have allowed increased food production but their manufacture is energy intensive and they often cause chemical imbalances in soils, groundwater, and surface water.

7. Hydroponics is a system of growing plants without soil using only a solution of essential plant nutrients. It allows exact regulation of nutrients but is expensive to operate.

Practice

Understanding Concepts

1. Describe how plant nutrient requirements change as a plant goes through its life cycle.
2. What is the significance of 20-20-10 on a package of synthetic fertilizer?
3. What sort of impact have synthetic fertilizers had on food productivity?
4. Why do hydroponic greenhouses have high yields?

INQUIRY SKILLS

○ Questioning
● Hypothesizing
● Predicting
● Planning
● Conducting
● Recording
● Analyzing
● Evaluating
● Communicating

 Some fertilizers could be hazardous to your health. Wash your hands after handling them.

Investigation 14.3.1

Assessing Synthetic Fertilizers

In this investigation, you will design and conduct an experiment to assess the effects of only one variable related to synthetic fertilizers on the growth rate of seedlings. You will grow your plants in a standard potting soil mix under artificial light.

Question

Choose one of the following questions:

How do three synthetic fertilizers with different NPK values affect the growth rates of plants?

How do three different quantities of the same synthetic fertilizer affect the growth rate of plants?

How do three different brand name fertilizers with the same NPK values affect the growth rate of plants?

Hypothesis/Prediction

From your question, create a hypothesis/prediction that will be tested in your investigation.

Materials

potting soil
potting trays
adjustable lighting fixtures with timers
light meter
bean, radish, or other seeds
three types of synthetic fertilizer with different NPK values
three different brands of synthetic fertilizers with identical NPK values

Design

Depending on your hypothesis/prediction, state the independent variable and the dependent variable.

In designing your investigation, consider the following:

- How will you adjust your chosen independent variable?
- What method will you use to measure growth rate?
  - What will you measure as a growth indicator? Will you measure more than one thing?
- How will you record your results?
- How will you report your results—in a graph or table format?
- What variables will you need to control? (Think carefully to make sure your list is complete.)
  - How will you control these variables?
  - How will you set up your control?
- How many seedlings will you use for each setup?

Procedure

1. Submit your experimental design to your teacher for approval before continuing with the experiment.
2. Conduct your experiment over a period of three to four weeks.
3. Submit a written lab report. Be sure to include a conclusion that addresses your question and your prediction.

Analysis

(a) How did your chosen variable influence the growth rate of your plants?
(b) How might these results be useful to the operator of a greenhouse and to a farmer growing field crops?

Evaluation

(c) How did your results compare with those of other students in the class who chose the same variable? Suggest reasons for any differences.
(d) Describe the findings of students who chose to test a different variable.
(e) Justify the number of seedlings that you used for each test. What number of seedlings would you suggest if you were going to repeat the work? Explain your answer.
(f) Growth rates of plants can be determined by measuring several features of the plant. Explain which feature would be the most convenient to measure and which would give the most valid measurement of growth.
(g) Suggest ways that you could improve your experimental design.

Section 14.3 Questions

Understanding Concepts

1. Which kinds of plants have the greatest need for nitrogen?
2. What effect does too much nitrogen have on flower and fruit production?
3. What are the advantages of using compost as a soil conditioner?
4. Why should many farmers regularly rotate a legume crop with a nonlegume crop?

(continued)

14.4 Food Plants

With over 6 billion people to feed each day, the production of food is one of the world's most important challenges. Although Earth is home to over 300 000 plant species, over 80% of all food calories consumed by humans are supplied by only a few species (**Table 1**). The two most important are wheat (**Figure 1**) and rice (**Figure 2**).

About 150 species of plants are intensively cultivated for food while at least 5000 species are known to be edible. It should also be noted that corn (**Figure 3**) and barley (**Figure 4**) are used mainly for livestock feed. Animal products only account for 8% of the world's food supply.

Table 1: Major Food Crops of the World (1998/1999)

Plant	World production (millions of tonnes/year)	Description
wheat	589	monocot seed crop
rice	577	monocot seed crop
potatoes	299	dicot tuber crop
manioc (cassava)	162	dicot root crop
soybeans	158	dicot seed crop
sugar cane/sugar beets	152	monocot root and stem crop
sweet potatoes	139	dicot seed crop
sorghum/millet	89	monocot fruit crop
corn (maize)	613	dicot root crop
barley	139	monocot seed crop

Figure 1
Whole wheat kernels

Figure 2
Basmati (upper left), brown (upper right), and wild rice

Figure 3
Corn

Figure 4
Whole barley kernels

Domestication of Plants

The domestication of plants began with wheat, peas, and olives about 10 000 years ago in southwest Asia and dramatically increased the amount of food produced, which helped support human population growth. The seeds from indigenous wild plants were actively collected, planted, and cultivated. As the food was harvested, seeds or tubers from the best plants were selected for the planting of subsequent crops. **Artificial selection** led to plant varieties which look quite unlike the wild plants from which they originated. The wild ancestors of corn and tomato, for example, are unrecognizable to all but the most expert plant biologists. The dramatic results of plant breeding can be seen in *Brassica*; six different vegetables—kale, Brussels sprouts, broccoli, kohlrabi, cabbage, and cauliflower—all have a single common ancestor, the sea cabbage, *Brassica oleracea* (**Figure 5**).

artificial selection: the intentional choosing of individuals of a species for the purpose of reproduction. Choices are based on the presence or absence of certain traits with the result that the desired characteristics will appear in subsequent generations.

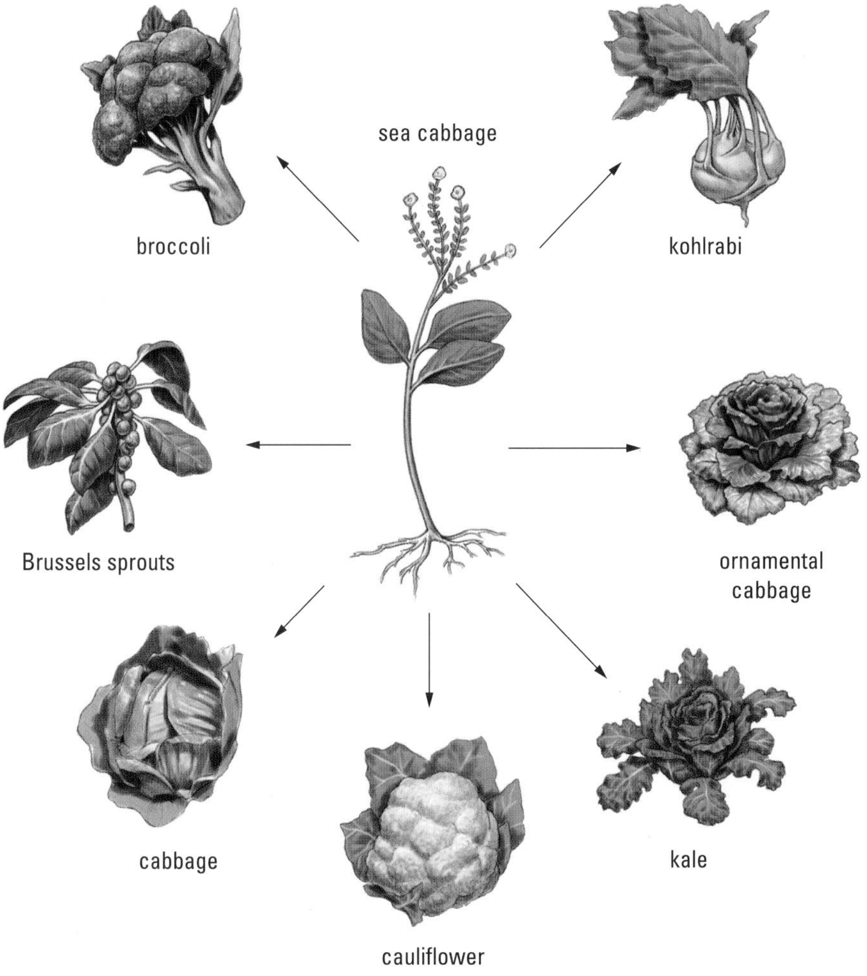

Figure 5
The cabbage family tree

Ontario has a large agricultural sector which produces a wide range of foods. Among Ontario's most valuable food crops are vegetables such as tomatoes, carrots, onions, and potatoes; grains such as wheat, oats, and barley; corn (mostly for use as livestock feed); canola (for oil); soybeans; and fruits such as apples, pears, peaches, grapes, and strawberries.

A significant number of plants are not grown for their nutritional value but rather for their flavour or novelty. Popular examples are listed in **Table 2** and shown in **Figures 6** to **12** on page 564.

Figure 6
Chocolate is used in many treats.

Figure 7
Cola drinks

Figure 8
Fresh and dried rosemary (left); and fresh and dried thyme

Figure 9
Spices: cinnamon sticks, several whole cloves, and two whole nutmegs

Table 2: Specialty Food Plant Products

Food	Description
chocolate	one of the world's favourite foods derived from the seeds of the tropical cacao plant
cola	drink flavouring extracted from the seeds of the coca plant
herbs	delicate flavourings usually derived from various parts of herbaceous plants (e.g., basil, rosemary, thyme)
spices	strong flavourings usually derived from various parts (often the fruit or seeds) of woody plants (e.g., pepper, vanilla, nutmeg, cinnamon)
teas	prepared from young dried leaves of the East Asian *Thea* shrub; herbal teas are not true teas and are prepared from many different plants (e.g., camomile)
coffee	the dried and roasted bean of the coffee tree, a shrub native to Africa; second only to petroleum in dollar value as an internationally traded commodity
hops	the dried flowers of a woody vine provide beer's bitter taste

Figure 10
Real teas: green (lower left) and black (lower right); herbal tea: camomile (top)

Figure 11
Roasted coffee beans

Figure 12
Hops in a compressed and pelletized form ready for beer making

Activity 14.4.1

Planning a Vegetable Garden

Before planting, farmers and gardeners must choose from a wide selection of plants. Each plant **variety** (**Figure 13**) has particular characteristics that make it more or less suitable for growing in a particular region and for satisfying a particular market demand. In this activity, you will select one plant species. Then you will examine the types of food plant characteristics that are preferred and the tradeoffs that result when one trait is selected over another.

variety: a subspecies of a plant species. In humans we refer to subspecies as *races*. For domesticated animals, the term *breed* applies.

Questions

Once you have chosen your plant species, what are the most important considerations when selecting a specific variety for planting?

What tradeoffs must be made when selecting one characteristic over another?

How do the demands of consumers influence the plant selection process?

Materials

seed catalogues and seed packages
information about seed companies or about gardening

Figure 13
Seed packages

Procedure

1. Choose only one food plant. The following are some possibilities:

tomato	lettuce	carrot	cucumber
pumpkin	squash	strawberry	pea
bean	melon	sweet corn	onion

2. Using information sources, investigate the range of traits that are available for your particular species of food plant. The information should reveal details about each variety within the chosen species, including colour, flavour, size, disease resistance, growing season, soil conditions required, and planting instructions.

   Compile information for at least five different varieties. Choose varieties that best emphasize the range of characteristics within the plant species you have chosen. Compare their individual characteristics in a table. You will need six columns. The heading for the first column will be Characteristic. The headings of the other five columns will be the specific names of the five varieties of the plant you chose. List the following characteristics in the first column leaving two to three lines between each one: colour, flavour, size, disease resistance (if any), growing season (number of days required), whether it is a hybrid, soil requirements, special care needed, nutritional value, and cost.

Analysis

(a) Of the varieties of the species of food plant that you examined, which one requires the longest growing season?

(b) Which one produces the largest crop? How was that amount described? How do you think it was calculated?

(c) What advantages, if any, did the hybrid varieties have?

(d) How are the research and development costs reflected in the price of hybrid varieties?

(e) If you had to pick only one variety of your species to grow, which would you choose? Explain.

(f) What tradeoffs did you have to make when you made your selection in (e)?

(g) What tradeoffs are most likely to occur for people growing food crops in northern climates?

(h) Comment on how each of the following technologies improves our ability to consume low-cost, high-quality food during the winter months:

   (i) plant breeding

   (ii) hydroponics/greenhouse growing

   (iii) refrigeration

   (iv) transportation technologies

   (v) preserving/canning of foods

The Green Revolution

Traditionally, farmers grew only varieties of native plants that had evolved to thrive under local conditions. The wide variety of environmental conditions meant that there was a large range of varieties within native plant species representing a great diversity of genotypes and phenotypes.

Beginning in the 1930s, experiments based on a knowledge of Mendelian genetics led to the development of a number of high-yield hybrid dwarf varieties of wheat, rice, and corn. When grown with appropriate fertilizer inputs, these shorter and stronger varieties produced up to five times as much grain as traditional native varieties. In contrast, the native plant varieties did not respond well to synthetic fertilizers. Applying fertilizer did increase seed production, but larger seed heads were too heavy for the tall, slender stems and the plants collapsed under their own weight. The hybrids resulted in large increases in food production and became very attractive to farmers. Farmers started to rely more and more on the high-yielding hybrid varieties instead of the native varieties.

The new hybrids made possible the **green revolution** during which certain developing countries, especially Mexico, India, Pakistan, and the Philippines, dramatically increased their food production. Within 7 years of receiving 100 kg of the hybrid wheat seed in 1963, India became an exporter of wheat and officially announced it would be able to end hunger and poverty for the first time in its history—although that predicted result has never happened. Similar green revolutions occurred in Mexico with wheat, and in the Philippines with rice. Even the United States experienced its own green revolution with corn.

However, the green revolution also had many disadvantages. The high-yielding varieties of food crops were less disease resistant and required increased use of both fertilizers and pesticides, widespread use of irrigation, plus more fuel for the intensive mechanical farming methods. In many poor regions, farmers were unable to afford the chemicals and machinery needed to take advantage of the new crop varieties. At the same time, the increased global grain supply resulted in a drop in the market value of the crops. Also, the decreased cultivation of local varieties of crops has resulted in a loss in the genetic diversity of the world's food crops. The name "green revolution," based on the pigment intricately involved with photosynthesis, brought a new hope for the future. However, considering the increase in human numbers and all the negative aspects of these new hybrids, some people think that what was really achieved was not so green after all.

green revolution: the large increase in food production experienced from the late 1960s into the 1980s

Practice

Understanding Concepts

1. What are the six most important food crops on Earth?
2. List five foods that are successfully produced on Ontario farms. List five foods that are not likely to be successfully grown on Ontario farms.
3. In what ways does technology play a role in providing us with foods from foreign countries?
4. How valid is it to refer to humans as grass and seed eaters?
5. Corn is a very important crop but is not a major component of our diet. What major role does corn play in human food production?
6. When did the domestication of the world's major food crops begin?
7. What major technologies were associated with the green revolution?
8. What factors caused the green revolution to have a negative impact on poor farmers?

Protecting and Utilizing Genetic Diversity

Until recently, the importance of maintaining genetic diversity was not well understood. It is now well documented that reduced genetic diversity renders plant crops very susceptible to insect attacks and disease. Fewer genes in the **gene pool** mean a smaller range of possible responses to environmental stress. For example, the Irish potato famine of 1846 and 1847 was caused by the outbreak of a fungus, which the genetically uniform potato crop was unable to resist. If the potato crop had had more genetic diversity, some plants might have been more resistant to the fungus and been able to survive. The crop failure resulted in the death of over half a million people from starvation and related diseases, while another 1.5 million people emigrated.

We once selectively bred plants in order to create uniform types with desirable characteristics. Now the many problems associated with uniformity are recognized and a number of methods are being used to ensure that important plants will not be wiped out by one drastic stress such as a disease, pest, or environmental condition. Not only is there renewed interest in cultivating native wild varieties, but biotechnology is used to incorporate special genes from the wild gene pool into plants that already have other desired traits.

Modern tomatoes have been given important disease resistance by crossbreeding them with wild Peruvian tomatoes, while a single gene from an Ethiopian wild grass now protects the world's barley crop from yellow dwarf disease. By maintaining a large gene pool, there is a greater likelihood that resistance to a wide variety of stresses will be available for further crossbreeding. Just as there are blood banks to store blood for various medical uses, we also now have seed banks which store seeds from vast numbers of plants. Scientists working with seed banks conduct research using the genetic material from samples of the seeds (**Figure 14**).

Protection of genetic resources occurs not only through seed banks, but also through conservation of the areas where the plants grow wild. Maintaining a diversity of plants is important both because of its usefulness to humans and because plants provide habitat and food for all other forms of life. Many organizations are working to protect threatened habitats while others are collecting seeds from the wild for seed banks. This rush to collect genetic information has led to disputes between companies and countries over the ownership of genetic resources.

gene pool: all the genes of all the individuals in a specific population

Figure 14
Researchers in seed banks collect and study seeds and their genetic material.

SUMMARY Food Plants

1. Six species of plants account for over 80% of all food calories: wheat, rice, corn, potatoes, sweet potatoes, and manioc.

2. Many plant products are produced for their special flavour.

3. Only about 150 of the 5000 plant species known to be edible are cultivated for food.

4. Artificial selection and crossbreeding are used extensively in agriculture. Through artificial selection, the plants with the desired traits of each generation were selected to reproduce, creating the next generation. Through crossbreeding of different varieties and species, most of today's hybrid food plants were created.

5. The green revolution occurred from the late 1960s into the 1980s and resulted in increases in food production throughout the world. The hybrid crops grown depended on irrigation, fertilizers, and pesticides, which were not readily available to poor farmers and which created environmental problems.

6. Genetic diversity within a species is important for the survival of that species by promoting resistance to disease and environmental stresses.

7. Seed banks and other conservation methods are being implemented to try to stop the loss of genetic diversity.

Section 14.4 Questions

Understanding Concepts

1. Give examples of six foods which you eat on a regular basis that are produced largely from seed crops.

2. List the modern food plants that are related to the sea cabbage. What technology permitted them to be developed?

3. Artificial selection is an early form of biotechnology. Explain how it works.

Applying Inquiry Skills

4. Create a pie graph using the data in **Table 1** (page 562), which displays the relationship between the total world production with the following four categories: monocot seeds; dicot seeds; monocot roots, stems, and fruits; and dicot roots, stems, and tubers.

5. Consider the biological roles of leaves, stems, roots, and seeds. Explain why none of our major food crops are leaves.

6. Why is the calculation of the world's supply of food per capita more important than the total world production of food?

Making Connections

7. Tapioca is obtained from the manioc, or cassava, plant. What part of the plant do you think is the source of tapioca, the starchy material used to make tapioca pudding? Explain the reasons for your choice.

8. Explain how the successful breeding of dwarf grass varieties led to increased food production.

9. List three of the world's most important food crops that grow in Ontario.

10. How might the goals of a commercial plant grower differ from those of a backyard vegetable gardener? Give examples to illustrate your answer.

11. Coffee is a multibillion-dollar crop.
    (a) How might the demand for coffee influence efforts to protect tropical rain forests?
    (b) How might the demand for coffee influence efforts to help nations feed themselves?

12. Research and development of plant products is extremely expensive.
    (a) What are specific examples of plant research that you feel are important? What are your reasons?
    (b) What should be the source of funding for that research?
    (c) What increase in costs of these products would you tolerate in order to pay for that research?
    (d) What are specific examples of plant research that you feel

14.5 Fuels, Fibres, and Wood Products

Plants are the source of many products in addition to food. These include fuels and fibres. To date, however, only a small fraction of the plant species on earth have been tested for their commercial value to humans. There is extremely active scientific research in this area.

Fuels

For most of human history, plants have been the primary source of fuel for cooking and heating. Woody plants were burned as probably the first and only fuel for centuries. Firewood is still the main source of energy in many regions of the world. Although today hydro and nuclear power do meet some energy demands in some parts of the world, our dependency on fossil fuels seems to be greater than ever for heating, electricity, and transportation. What most people forget is that our fossil fuels began as plants. These fuels are reserves of reduced carbon resulting from photosynthesis in earlier geological times. Dead plant material, under conditions of high pressure and temperature over long periods of time, become coal, oil, and gas.

Wood consists mainly of the cellulose walls of dead xylem tissue. The most significant factor in determining the energy content of a piece of wood is its dry mass. For this reason, denser hardwood tree species contain more energy for their volume than softwoods (**Figure 1**). In general, hardwoods are trees such as oak, maple, beech, and birch, while trees such as pine, spruce, and cedar are known as softwoods.

At present, almost half of the world's annual tree harvest is used as fuel (**Figure 2**). This exploitation is leading to severe firewood shortages in many parts of the world. As available firewood is cut down at alarming rates, the planet's ability to absorb carbon dioxide is rapidly dropping. As forests and other vegetated areas are destroyed, habitat loss follows. In sub-Saharan Africa, firewood gathering is causing **desertification** because plant roots that used to hold the soil to prevent erosion are gone.

Another plant-based fuel is peat, discussed in Chapter 10, section 10.7. Peat moss that has accumulated over thousands of years in bog environments is harvested, dried, and burned in ovens and furnaces.

Ethanol is becoming an increasingly important fuel. This alcohol is produced from organic matter either through fermentation or industrial chemical processing. The large ethanol industry in Brazil uses sugar cane and cassava, widely available plants, as the raw materials for fermentation, allowing Brazil to become self-sufficient in energy production and to avoid the high costs of foreign oil. In Ontario, corn, sugar beets, hemp, and wood have all been proposed as crops for establishing an ethanol industry. Theoretically, any plant material could be used as a source of raw material for ethanol production. Ethanol production is promising as a fuel, especially in the transportation sector. However, people are now asking whether using potential food crops to produce a fuel is environmentally sound or ethical.

Fibres

For thousands of years, humans have used plant fibres to produce woven materials. We use these products as fabrics for clothing and furniture, as ropes, string, and thread, and as canvases for our art (**Table 1** and **Figures 3–8**, page 570).

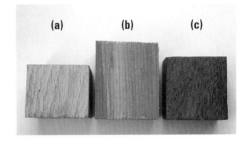

Figure 1
The block of oak wood **(a)** is 3.3 cm x 4.3 cm x 4.1 cm and has a mass of 48 g. The block of cedar wood **(b)** is 4.2 cm x 3.8 cm x 5.0 cm and has a mass of 31 g. The block of teak wood **(c)** is 3.3 cm x 4.0 cm x 4.1 cm and has a mass of 60.5 g. Which one will release the most energy when burned? Explain.

desertification: the process of becoming desertlike. One major cause of desertification is the loss of topsoil through erosion after the removal of vegetation.

Figure 2
Wood-burning fireplaces create an enjoyable atmosphere and release a lot of heat energy. Unfortunately, a lot of this heat goes up the chimney and not into the room.

Plant fibres are pulped, not woven, to make paper (**Figure 9**). The world demand for these materials is enormous and continuing to increase. The advent of photocopiers and desktop printers has resulted in a huge jump in paper consumption. Since 1984, the annual per capita use of paper has increased from about 150 kg in North America to well over 400 kg. This amount dwarfs the current annual 15 kg per capita consumed by people in other regions of the world. When e-mail and the Internet became popular, some people forecasted a reduction in paper use, but use of paper has actually exploded.

Plant fibres are almost entirely composed of cell wall material or cellulose. A plant's genes and its environment dictate the chemical composition of these cell walls. Each plant species produces unique combinations of various chemicals in its cell walls, resulting in fibres with a wide variety of physical and chemical properties.

Figure 3
Three very different-looking materials but all 100% cotton

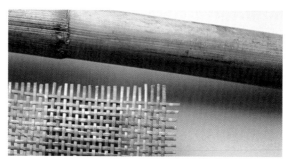

Figure 4
Natural or painted rattan is used to make wicker furniture. The caning is usually left natural.

Figure 5
Sisal rope

Figure 6
Jute rope, twine, and fabric

Table 1: Important Plant Fibre Products

Fibre	Source and use
cotton	seed hairs from cotton plants; the most important fabric for clothes, towels, etc.
rattan	the stem of a climbing palm used whole or in strips to make wicker and cane furniture
jute	a bark fibre used for making rope and a coarse fabric called burlap
sisal	leaf fibres used to make rope
flax	a stem fibre made into linen fabric and paper
ramie	a bark fibre used in knitted or woven fabrics
wood pulp	long softwood fibres used for making various paper and cardboard products

Figure 7
Linen fabric wears well and absorbs moisture, but it wrinkles readily.

Figure 8
Knitted fabric with added strength and lustre because of its 55% ramie content

Figure 9
New paper sheets being cut and packaged

Practice

Hemp Production

Fibres produced by agricultural means were the original fibres in the pulp and paper industry. In the last century, technological innovations resulted in the establishment of wood as the dominant fibre used. Recently, however, cheap wood fibre has become more scarce, prompting a reevaluation of agricultural sources of fibre in both North America and Europe. Various plants have been proposed as candidates for reestablishing an agricultural fibre industry. They include flax, kenaf, straw (from grain crop residue), and hemp (**Figure 10**).

Hemp (*Cannabis sativa*) has proven to be one of the most promising and controversial possibilities. It is one of the oldest cultivated plants in the world and has been grown to produce high-quality fibres for ropes and clothing (**Figure 11**). For example, the original Levi's jeans were made of hemp cloth. In addition to its valuable fibres, hemp produces seeds rich in oil. The oil can be used in the manufacture of cosmetics and has the potential to be used in foods.

However, hemp has been little used in the recent past because *Cannabis* is also known as marijuana and can produce the psychoactive compound THC (delta-9 tetra-hydrocannabinol). In the early 1930s, when concern mounted over the use of marijuana as a drug, Canada banned its use, possession, and cultivation. The renewed interest in growing hemp has led to the breeding and testing of new varieties. The first of these to be approved by Health Canada was a new variety which has an extremely low THC content (less than 0.3%), therefore posing no drug-related concern.

At the present time, the production of high-quality hemp fibres is not economically viable in Ontario. Current world production relies on low-cost, labour-intensive processes which are not available in Canada. To create a commercially viable manufacturing industry in Ontario, new low-cost processing technologies will have to be developed. To ensure large quantities of strong fibres, fertilizers high in potassium must be applied. Hemp crops seem to have few weed, insect, or fungus problems, an environmental advantage because no herbicides, insecticides, or fungicides are needed.

There is also renewed medical interest in *Cannabis sativa*. Marijuana has been considered for a variety of medicinal uses, including as a painkiller, in the treatment of glaucoma, and to increase appetite and reduce nausea in cancer and AIDS patients. Despite marijuana's potential usefulness, it remains generally illegal in Canada. However, growth of medicinal marijuana is allowable for a few special projects which have been granted a licence by the government.

Figure 10
A mature hemp plant

Figure 11
Hemp fibre products

Wood

Wood is the world's most popular construction material. Its light weight, strength, and durability make it ideally suited for structural materials. Hardwoods are widely used for fine furniture and floors, while softwoods are well suited to floors, ceilings, and framing within walls. Lumber (**Figure 12**, page 572), plywood (**Figure 13**, page 572), veneer (**Figure 14**, page 572), particleboard, chipboard, and fibreboard (**Figure 15**) are all made of wood. The versatility of wood makes it valuable for different uses (**Figure 16**).

In the early 1900s, much of Ontario was covered in forests. With steadily increasing demand for wood products and wood pulp, and with technology helping us remove more trees faster, our forests have steadily disappeared. The wood available now is in forests much further from the markets and it has to be transported long distances. Timber production in Ontario is big business; Ontario's annual harvest of wood is worth $12 billion. Conifer species account for the majority of Ontario's tree harvest.

Figure 12
Unfinished and unstained pieces of cedar (lower left), spruce (upper right), and (top to bottom) teak, walnut, oak, and pine lumber

Figure 13
Plywood is made by gluing together multiple thin layers of wood with the grain in each layer running crosswise to the layers next to it. Plywood is stronger than solid wood. Shown are pieces of plywood with various thicknesses and different numbers of layers.

Figure 14
Veneer is a thin layer of fine wood usually applied to the surface of wood of lesser quality. Shown are unfinished and unstained pieces of (left to right) oak, walnut, maple, cherry, rosewood, and mahogany veneer.

Figure 15
Chipboard, particleboard, and fibreboard are all made by compressing wood chips and/or sawdust together with glue and resin. They are all structural materials. The largest piece shown is chipboard, which is composed of large wood chips. The two pieces of particleboard, made of wood chips and sawdust, show different degrees of particle compression. The single piece of medium density fibreboard, MDF, appears uniform in texture because the very fine wood particles are extremely compressed.

Figure 16
Boxes and other decorative items

Canada's Future Forests

Approximately 40% of the world's northern forests are found in Canada. Despite warnings from ecologists about the dangers of deforestation, many Canadians assume that our forests provide an inexhaustible supply of lumber because nature takes care of reforestation. Although legislation establishing the principle of sustained yield was first enacted in 1929, little real forest management was practised until years later. By then, the valuable white pine forests of eastern Canada had already been seriously depleted. Timber businesses are worried about our present and future forests.

Forest regeneration projects are not new. Many people are concerned because these initiatives have created "junk forests," in which new growth has failed to become properly established. The cause may be poor planting and/or harvesting techniques. Clear-cutting along slopes drastically increases erosion and decreases soil nutrients. Long-term decreased soil fertility has been linked with the depletion of mycorrhizae, which are important to the survival of trees. Planning and implementing sustainable forest management practices are of paramount importance. Trees must be harvested using methods which will not hamper reforestation efforts, and areas must be replanted with care and planning so that timber may be constantly harvested. Critics point out that planted monocultures are really just tree farms or plantations and cannot replace diverse forests as habitats for the plants, animals, and other life forms in a natural ecosystem. Some forests must be left unharvested to provide continuous habitats.

While it would be beneficial to the environment to stop the harvesting of trees, it would pose problems for society. Forest products account for a major share of Canada's exports and translate into over a million jobs directly in the forest industry and indirectly in all the secondary industries which use wood or wood pulp.

A part of the solution lies somewhere in between the two extremes of stopping harvesting and continuing at the present rates. A better understanding of forest ecosystems will be required before Canadians can be confident that our forest management practices are truly sustainable.

SUMMARY Fuels, Fibres, and Wood Products

Fuels

1. Presently, almost half of the world annual tree harvest is used as fuel for heating and cooking.
2. When burned, hardwoods release more energy than softwoods.
3. Peat, formed from compressed moss, is a fuel in many regions of the world.
4. Ethanol is produced from organic matter through fermentation or chemical processing and is being promoted increasingly as a fuel.

Fibres

1. Fibres are used for fabrics, rope, canvases, and paper products.
2. Plant fibre is composed almost entirely of cellulose.
3. Agricultural fibres are being explored as alternatives to wood fibres.
4. Hemp fibre is a very promising alternative but is also very controversial because of its connection to marijuana.

Wood Products

1. Wood is our most common construction material.
2. Lumber, plywood, veneer, particleboard, chipboard, and fibreboard are all valuable wood products.
3. Forest regeneration efforts have not been greatly successful for a variety of reasons, including poor harvesting and planting techniques.

4. Clear-cutting on slopes results in soil erosion, loss of soil nutrients, and disturbed mycorrhizae.

5. Replanted monocultures do not offer the ecological benefits of diverse natural forests.

Section 14.5 Questions

Understanding Concepts

1. What is the specific plant material which releases energy when wood is burned?

2. Name three fossil fuels and explain their connection to plants.

3. How is ethanol produced?

4. Considering economic and environmental issues, what advantages and disadvantages does ethanol fuel have compared to gasoline?

5. What is an environmental advantage of growing hemp as a fibre crop?

6. Hemp is a very fast-growing plant which can easily reach a height of 2 to 4 m. How does this characteristic influence its ability to compete with weeds?

Making Connections

7. Why do some people think that ethanol fuel production is unethical?

8. How has plant-breeding technology led to a reconsideration of hemp farming?

9. If more and more hemp is planted, how might this influence hemp pest problems? Explain.

10. In what ways does our increasing rate of paper consumption have an impact on the environment?

11. Years ago, all the branches were trimmed off and left to rot where trees were cut down. Now, they are brought out and used in various products. Explain why this change could be considered both good and bad.

12. Why is proper forest regeneration very important economically as well as environmentally?

14.6 Medicinal and Nonmedicinal Chemical Plant Products

Long before the advent of modern science and technology, people around the world were using plants or substances derived from them as medicines. In tropical rain forests, where plant diversity is greatest, hundreds of different plants have been used for a multitude of purposes including the production of medicines, antidotes, poisons, dyes, gums, soaps, and perfumes.

Medicinal Chemical Plant Products

Countless lives have been saved by the appropriate use of plant medicines. Today, about one quarter of our prescription drugs contain at least one chemical obtained from plants (**Figure 1**). A partial list is found in **Table 1**. Scientists are continually testing new plants and traditional remedies in hopes of discovering valuable medicines. Destruction of rain forests and the loss of native cultures make this research urgent as the ancient knowledge of indigenous peoples, and the plants themselves, are being lost forever at an alarming rate.

(a)

(b)

(c)

Figure 1
Plants such as **(a)** opium poppy, **(b)** foxglove, and **(c)** rosy periwinkle are the sources of many medicines.

Table 1: Important Medicinal Chemical Plant Products

Drug	Plant source	Uses
atropine	belladonna	treatment of excessive muscle contractions
codeine, morphine	opium poppy	painkiller
colchicine	autumn crocus	antitumour agent
digitoxin	foxglove	treatment of heart disorders
diosgenin	yams	birth control pills
L-Dopa	velvet bean	treatment of Parkinson's disease
quinine	cinchona tree	antimalarial agent
aspirin	willow	painkiller; anti-inflammatory
taxol	Pacific yew	treatment of ovarian cancer
vincristine, vinblastine	rosy periwinkle	treatment of childhood leukemia

> **DID YOU KNOW ?**
>
> With the discovery of the drugs vincristine and vinblastine, the survival rate of people with Hodgkin's disease, a form of lymphoma, has jumped from 20% to over 90%. The rosy periwinkle is native only to the tropical rain forests of Madagascar, a region experiencing rapid deforestation and species extinction.

Although many plants have beneficial medical effects, not all plant medicines are effective or safe. Recently, the medical community has become concerned about the increased popularity of herbal remedies. Unlike pharmaceutical drugs, which must undergo rigorous scientific testing for both safety and effectiveness, herbal remedies are largely unregulated. Although many of these substances are probably harmless and some may have benefits, others could pose a threat to your health. For example, gingko may reduce your blood platelet count and interfere with proper blood clotting, while ginseng has been associated with increased blood pressure and rapid heart rates. Taking such herbs may be particularly dangerous in combination with prescribed drugs or prior to undergoing surgery. We must not assume a product is safe just because it is natural. Much more research and good scientific studies need to be carried out before the safety and effectiveness of these herbal remedies can be determined with any confidence (**Figure 2**).

Figure 2
Echinacea is one of the most popular herbal remedies even though some studies indicate that it is medically ineffective.

Nonmedicinal Chemical Plant Products

A wide variety of nonmedicinal chemicals are also derived from plants. The most familiar of these is rubber, which is obtained from the white sap of tropical rain forest trees (**Figure 3**). Examples of other chemicals are shown in **Table 2** and **Figures 4** and **5**.

Figure 3
When done carefully, tapping rubber trees does not kill them.

Table 2: Important Nonmedicinal Chemical Plant Products

Chemical products	Uses
dyes: henna, bayberry, logwood, safflower, woad, alkanet, and indigo	colour or stain for hair, wood, fabric, margarine, and microscopic specimens
latex products: chicle, natural rubber, and gutta-percha	chewing gum, rubber gloves, hoses, tires, golf balls, and cables
essential oils: sandalwood, rosewood, lavender, and eucalyptus	cosmetics, perfumes, soaps, and candles
poisons and pesticides: strychnine, rotenone, pyrethrum, curare, and nicotine	hunting poisons or paralyzers, agricultural pest control, and insecticides
resins: rosin, lacquer, varnish, and turpentine	roofing compounds, wood finishes, binding agents in wood products, and paint solvent and thinner
tannins: from tree barks, oak galls, and tea	used in the tanning of leather and as dyes
waxes: carnauba and jojoba	commercial wax products, cosmetics, foods, and high pressure lubricants (e.g., for automobile transmissions)

Figure 4
The leather tanning process gets its name from the tannins used. Tanned leather is often dyed.

Human Abuse of Plant Chemicals

Ethanol—the alcohol in beer, wine, and liquor—and tobacco are examples of harmful but legal drugs that are derived from plants. Plants are also the source of a wide range of illegal psychoactive drugs including opium, cocaine, marijuana, and mescaline. The number of deaths and the harm to health that result from the abuse of these substances is greater than most people imagine or will acknowledge.

The tobacco plant contains nicotine, one of the most addictive chemicals known. Millions of people die, including more than 40 000 Canadians, every year from tobacco-related illness. It has been proven conclusively that smoking dramatically increases the risks of premature death and disability. This situation is by far the most dramatic example of a legal activity that results in unnecessary human suffering and loss of life. Still, tobacco ranks as one of the most commercially valuable plant crops in the world.

Figure 5
Once considered a desert weed, the jojoba shrub is proving to be very valuable. The wax obtained from the seeds is an excellent substitute for sperm whale oil and ambergris used in perfumes and other cosmetics and has removed the excuse for hunting the sperm whale. Jojoba thrives in very dry and even salty conditions and may provide an excellent source of income in countries with large desert regions unable to support food crops.

Practice

Understanding Concepts

1. Name two plant-derived medicinal chemicals and describe what they do or are used to treat. If you wish, describe how one of the chemicals in **Table 1** has helped you, a family member, or a friend.

2. What are some advantages of herbal remedies? What are some disadvantages?

3. List three important nonmedicinal plant products and their uses.

Making Connections

4. Why are new drugs so expensive? Is the high price justified? Explain.

5. Explain why our future ability to discover new medicines is threatened by the loss of natural tropical environments.

Conservation Ecologist

Conservation ecologists have a wide variety of potential jobs. They might be sent out to confirm recent sightings of a snowy owl in southern Ontario during a particularly hard winter in the Arctic. They also might work with universities, municipalities, businesses, or conservation groups. They might analyze the current biological diversity in an area and how to maintain or increase it, or they might be interested in how to preserve a specific species or habitat. Plant and animal conservation ecologists often work together to understand the complex relationships within a natural community and what threatens that balance. Many types of university programs are relevant to conservation ecology.

Forest Technicians and Scientists

Forest technicians and scientists work in many fields, including the logging industry and in conservation. Their work might include detailed examinations of tree bark for evidence of insect damage. They may also be involved in planning which species of trees to grow in a certain area of land given the environmental conditions and the goals of a logging company. They might also work with conservation biologists in deciding which species will best achieve certain environmental goals, such as maintaining soil fertility or providing habitat for a particular species. Forest technicians and scientists are trained at community colleges or complete a university degree in forestry or a related subject. Many go on to take additional courses.

Urban Planner

Urban planners help decide how a municipality's land will be distributed among business, residential, recreational, and other uses. They often carefully study topographical maps and aerial photos before recommending which areas should be developed and which should be regenerated as green areas. When planning an urban park, they need to know which plant species will survive in city conditions, which ones might contribute to air quality in a city, and which ones have ornamental value. They must also consider the needs and wants of the people living in the area. Their job often requires making compromises. Urban planners must have a university degree and many also do postgraduate studies.

Practice

Making Connections

6. On the Internet, follow the links for Nelson Biology 11, Chapter 14 Careers. Identify several careers that require knowledge about plants. Select a career that interests you from the list you made. Read about or interview someone with that type of job.

GO TO www.science.nelson.com

7. Describe a typical day at the job you chose. Explain what the duties are and how some of the topics covered in Chapters 10, 13, and 14 relate to that work.

8. Explain what qualifications are needed in order to get such a job. What university or college programs are needed? What extra training would be useful?

Medicinal and Nonmedicinal Chemical Plant Products

1. Many plant chemicals, both medicinal and nonmedicinal, are derived from plants found in tropical rain forests.

2. About one quarter of all prescription drugs contain at least one plant-derived ingredient.

3. Although most herbal remedies are probably harmless, and some may actually help a given ailment, there is no scientific consensus about their effectiveness or safety.

4. Other important plant-derived chemicals are rubber, dyes, poisons, oils, and waxes.

5. Some plant products, both legal and illegal, are addictive and cause health problems.

Section 14.6 Questions

Understanding Concepts

1. Use at least two examples to demonstrate whether or not plant-derived medicines are products of modern technology.

2. What plant products might be useful in treating each of the following conditions?
   (a) cancer
   (b) heart disease
   (c) severe pain
   (d) muscle spasms
   (e) Parkinson's disease

3. List five different chemical plant products that you have used lately.

4. What nonmedicinal uses are made of plant compounds that are poisonous?

5. Nicotine is toxic to insects. How does its presence benefit tobacco plants? In what plant part would you expect to find the highest concentration of nicotine?

6. How might the jojoba plant help establish an agricultural economy in countries with a hot, dry climate?

7. (a) What role does modern science play in determining whether a plant product is classified as a herbal remedy or a pharmaceutical?
   (b) Explain why you trust, or do not trust, medicines that have been approved by the established scientific community.
   (c) Explain why you trust, or do not trust, remedies that have not been approved by the established scientific community.

8. Many plant poisons are protein compounds which are destroyed by high temperatures. How is this fact put to use by native peoples who use these poisons for hunting with blow guns and arrows?

9. From which part of a rubber tree do you think natural rubber is extracted: the mesophyll, phloem, meristematic tissue, or xylem? Explain why.

Applying Inquiry Skills

10 How has plant research indirectly influenced the survival of some species of whales?

Making Connections

11. There are many diseases which cause human suffering and death.
    (a) What is different about the suffering and death caused by smoking?
    (b) How effective or ineffective has modern science been at controlling or reducing these health problems? Explain your answer.

12. The laws governing tobacco smoking are lenient in comparison to laws that ban other substances. Why has this situation occurred? What political and economic factors influence tobacco laws?

13. Consider your opinions on research funding for plant-derived medicinal and nonmedicinal chemicals.
    (a) Discuss whether you think the efforts of the past have provided enough benefits to warrant our continued search for new plant chemical products.
    (b) What should be the source of funding for additional plant science research?
    (c) How much would you, personally, pay per month to support such research? Why did you choose the amount you did?
    (d) Describe a particular research subject you would support.
    (e) Describe a particular research subject you would not support.

14.7 Pest Management

Humans are in direct competition with many other species for plant products. In both cultivated fields and natural forests, a host of fungi, insects, and other organisms feed on and in plants (**Figure 1**). At least one third of the world's food production is lost to pre- and post-harvest pests. In addition, the plants we utilize must compete for water, light, and nutrients with many unwanted plants.

For thousands of years, humans have attempted to eliminate or control competitors. As agriculture expanded rapidly to meet the demands of an increasing human population, so did the need for effective pest control. Pesticides have a major role in the agriculture and forestry industries and are widely used around homes and gardens (**Figure 2**). Recently, some cities and towns have begun restricting the use of pesticides on public property and/or have banned homeowners and private companies from using pesticides for purely aesthetic purposes (**Figure 3**).

Figure 1
The devastating effects of gypsy moths in some Canadian forests

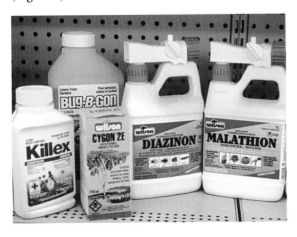

Figure 2
Reading the fine print on the pesticide containers is important. Does it harm pets? If so, what about the other mammals in your yard? Also note that the pesticide may kill beneficial insects.

Figure 3
Which sign would you like to see posted in your neighbours' yards?

The Importance of Plants **579**

Figure 4
In Canada, the peregrine falcon and other birds of prey have been victims of persistent chemicals, especially DDT. The high levels of DDT and related residues interfered with calcium metabolism and prevented the peregrine falcons from producing hard egg shells. The thin shells broke quickly before the chicks could develop. Those that managed to develop and hatch often had deformities which prevented them from thriving. As a direct result, the eastern race of the peregrine falcon is now extinct.

biological accumulation: the process by which persistent chemicals accumulate in the body of an individual organism throughout its entire life

persistent chemicals: fat-soluble chemicals that are not easily excreted from animal bodies and, thus, tend to accumulate

biological magnification: the process by which persistent chemicals seem to get multiplied as they pass along a food chain

DDT

The first important pesticide to be synthesized was DDT (**d**ichloro**d**iphenyl-**t**richlorethane). DDT was extremely toxic to insects, seemed relatively harmless to humans and other mammals, and was very economical to produce and apply. The widespread use of DDT not only dramatically increased crop yields but also was responsible for saving millions of human lives by controlling the mosquitoes responsible for the spread of malaria. The discoverer of DDT, Swiss chemist Paul Mueller, was awarded the Nobel prize in 1948.

However, DDT had many negative ecological effects. For example, when DDT was sprayed in Borneo to control mosquitoes and the spread of malaria, the lizards that fed on sprayed insects began to die. Then the pet cats, which fed on dead lizards, began to die. Without cats, the rat population soon escalated, increasing the risk of plagues spread by the rodents. The disruption of the food web became even more evident when certain caterpillar populations began to increase dramatically. Apparently, DDT affected wasps and other predators of the caterpillar, but had little effect on the caterpillars themselves. Eventually, the caterpillars, searching for new food sources, devastated food crops and even began eating the leaves that were used to thatch roofs. Although the ecosystem eventually stabilized, this example illustrates how a change in one part of a food web can affect a number of interrelationships.

In addition to many other negative environmental impacts, such as the decimation of populations of eagles, peregrine falcons (**Figure 4**), osprey, pelicans, and condors, DDT has been linked to cancer and other human health problems. DDT is now banned in many countries, including Canada. However, chemical factories in some countries that banned the application of DDT and other similar chemicals still produce them. They sell them to poorer countries that cannot afford more expensive pesticides.

Bioaccumulation and Biomagnification

Biological accumulation, often shortened to bioaccumulation, describes what happens to **persistent chemicals** in the body of any one individual organism. Because persistent chemicals are soluble in fats, not water, they do not tend to be released from the individual's body along with water carrying water-soluble wastes. These chemicals are stored in the organism's fatty tissue. As the organism gets older, these chemicals continue to accumulate, hence the name bioaccumulation. As a result, people who eat fish from a river where there might be persistent chemicals should choose to eat a fish that is two years old instead of the larger five-year-old catch of the same species.

Another important term is **biological magnification**, also known as biological amplification, or by the short forms biomagnification and bioamplification. Biomagnification is the apparent multiplying effect that occurs as the persistent chemicals pass through a food chain (**Figure 5**).

At each stage in the food chain, the concentration gets greater, not because of any chemical change, but because a predator further along on the food chain feeds on many organisms that have already accumulated chemicals in their bodies. When a predator fish eats a prey fish, it consumes in that meal all the persistent chemicals that the prey fish accumulated over its whole lifetime. As a result, people who eat fish from a river where there might be persistent chemicals should choose to eat fish species that are near the beginning of the food chain, not the end.

The Borneo story and the peregrine falcon case provide examples of animals that have suffered from the effects of bioaccumulation and biomagnification.

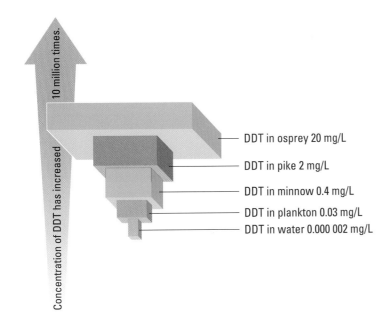

Figure 5
Biomagnification of DDT in a food chain

Humans are also at risk. During the 1950s and 1960s, fat-soluble DDT turned up in human breast milk throughout North America. DDT levels were especially high in humans who lived in areas where DDT was used for spraying crops. However, DDT was also found in mothers' milk in the Canadian north. DDT was banned from use in Canada and the US in the late 1970s but DDT residues continued to be found everywhere. Migratory birds, like the mallard duck and the Canada goose, winter in Central America and Mexico, where DDT is still used. A similar scenario exists for the migrating fish of the Atlantic and Pacific Oceans. Remember that, once applied, pesticides know no boundaries and can travel the world. Is it any wonder that DDT is present in the fat of Canada's seals and polar bears and in the milk of Inuit mothers?

Ironically, many of the pests that pesticides are intended to kill have a highly diverse gene pool and extremely efficient reproductive cycles. As a result, there always seem to be some individuals that are more or less resistant to the chemical. Although the great majority die, those that do survive can reproduce and pass their characteristic of resistance on to the next generation. This next generation thrives not only because they are resistant but also because they have less competition for food. Just when there were signs that malaria might be completely eradicated by killing mosquitoes with DDT, mosquitoes evolved resistance to the pesticide. Hundreds of species of insects now exist with known resistance to a wide range of pesticides. There are worries that science will not be able to continually develop new chemicals to stay a few steps ahead of the insects.

Alternatives to Pesticides

In the past few decades, strong efforts have been made to find ways to avoid excessive use of pesticides. These efforts include using chemicals that do not persist in the environment, and the development through selective breeding and genetic engineering of plant varieties that are resistant to disease and pests.

Another alternative, biological controls, can be very effective. In many greenhouses, ladybugs (**Figure 6**) and parasitic wasps are used to control insect pest infestations. In some countries that cannot afford the most modern pesticides, a switch to biological controls has been very effective.

Figure 6
Ladybugs are harmless to plants but are an effective means of controlling insect pests.

An increasingly popular approach is integrated pest management (IPM), which employs careful monitoring of pest levels, the use of natural biological controls, and the judicious application of pesticides only when necessary.

As the risks and benefits of pesticide use continue to be hotly debated, many consumers are turning to organically grown food as an alternative. This type of food is grown without the use of any synthetic pesticides or fertilizers. However, many doubt that organic farming can produce enough food to feed the world.

Table 1: Pesticides Used on Surveyed Crops in Ontario 1998 (Tonnes of Active Ingredient)

Crop	Herbicides	Insecticides	Fungicides	Nematocides	Pest growth regulators	Total pesticides
field corn	1982.9	8.1	-	-	0.2	1991.2
soybeans	1274.6	5.0	0.2	8.2	-	1288.0
grains	356.6	-	2.1	-	-	358.7
tobacco	7.7	31.9	8.9	262.4	201.1	512.0
canola	12.3	0.1	-	-	-	12.4
fruits	13.7	70.1	459.2	-	0.1	543.1
vegetables	124.0	33.3	153.6	20.4	-	331.3
other	147.4	4.5	25.7	-	-	177.6
total	3919.2	153.0	649.7	291.0	201.4	5214.3

Practice

Understanding Concepts

1. What are the most important competitors for our food and plant crops?
2. How do pesticides increase our crop yields?
3. What are persistent chemicals?
4. What is bioaccumulation?
5. What is biomagnification?
6. What do the letters IPM mean?
7. What does "organically grown" mean?

SUMMARY Pest Management

1. Synthetic pesticides have a major role in agriculture and forestry. Pesticides are also used by private businesses and citizens.

2. Pesticides account for much of the increased agricultural production of the past few decades.

3. DDT was the first important pesticide.
   - It dramatically increased crop production and controlled malarial mosquitoes.
   - It also had severe negative environmental effects and has been linked to human health problems.

4. Biological accumulation is the buildup of persistent chemicals in the body of one individual organism throughout its lifetime. Persistent chemicals are fat soluble and do not get excreted from animal bodies.

5. Biological magnification is the apparent multiplication of levels of persistent chemicals as they get passed along a food chain.

6. Alternatives to traditional pesticides include practices such as the development of safer pesticides, biological controls, and integrated pest management.

Section 14.7 Questions

Understanding Concepts

1. Briefly summarize your knowledge about DDT in each area:
   (a) effectiveness
   (b) safety
   (c) cost
   (d) long-term impact

2. Give one example to illustrate how the elimination of one species can lead to an increase or decrease in the number of another species.

3. Why would a water-soluble pesticide have a reduced tendency to bioaccumulate?

4. What two Ontario crops require very large herbicide applications?

5. How does genetic diversity help insects fight back against pesticides?

Applying Inquiry Skills

6. Many organic farmers use routine crop rotation and mixed-crop farming as strategies to reduce pest problems. Speculate about how these strategies work to decrease pest population numbers.

7. Refer to **Table 1** in section 14.7 to determine what type of Ontario plant crop is most susceptible to a combination of insect and fungal attack. When you have purchased this type of food, what evidence of insects or fungus growth have you seen?

8. Refer to **Table 1** in section 14.7 to determine whether the greatest competition for food crops comes from animals, fungi, or weeds. Explain your answer. What assumption must you make?

Making Connections

9. What is IPM and what are its various components? In what ways is this concept beneficial?

10. Explain how the use of pesticides can have short-term benefits but lead to long-term harm.

11. (a) What herbicides are used in your area for lawn weed control? Consider the lawns in front of houses, apartment buildings, or office buildings.
    (b) Who uses these herbicides and how much is used?
    (c) What other options are available?
    (d) What change of attitude might be beneficial?

12. What evidence indicates that the herbicides used in your area are safe when used properly?

13. Speculate which groups of people or which other species face the most risk due to pesticide use.

14. How do people in your area feel about the use of herbicides? Survey your friends and family and collect their opinions.

15. Most of the pesticide which is sprayed on a field or forest never reaches its intended target organism. Instead, well over 90% of it gets either blown or washed away. For alternative solutions, many biotechnology companies are genetically engineering plants to produce their own pesticides.
    (a) How could this help the environment?
    (b) How might it affect the cost of farming?
    (c) How might you be affected?

14.8 A Sustainable Future

A sustainable future is a future involving a wide range of processes and activities that can take place without impairing our ability to continue in the same way. Although humans have been cultivating and harvesting plants for millennia, our future depends on increasing our understanding and utilization of plants. During this century, the demand for nutritious food will escalate dramatically as the human population passes 8 billion. The world's petroleum reserves will continue to be slowly depleted and become more expensive, and the demand will rise for alternative and renewable sources of industrial materials and energy. Global warming and climate change may negatively impact our most important food-growing regions and our valuable forests.

To address these concerns, wiser use must be made of our planet's resources. We will need to discover and cultivate new plants for foods and chemical products. We must manage farmland and forests better and protect the genetic diversity of the world's endangered ecosystems. If the promises of genetically engineered plants are to be realized they must be proven safe for both human consumers and the environment.

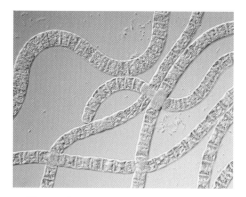

Figure 1
Spirulina is a rich source of protein.

New Kinds of Plants

There exists the possibility of finding and breeding new plants to supplement the limited number of plants now under cultivation. The winged bean is a nitrogen-fixing, fast-growing vine native to tropical New Guinea. Such plants do not require inputs of nitrogen fertilizer and are, therefore, less expensive to grow and better for the environment. *Amaranthus* was once grown extensively in South America but the invading Spaniards stopped its cultivation to prevent its use in "pagan" ceremonies. There is now renewed interest in this fast-growing pseudo-grain, which produces large quantities of seeds rich in protein that contains lysine. The amino acid lysine is essential to our diets but is rare in plant proteins. Mixing even small amounts of amaranth flour with wheat flour would greatly enhance its nutritional value. Guayule, a desert shrub native to northern Mexico and the southern United States, is an excellent source of high-quality natural rubber and is now under cultivation in 30 countries.

Many more plants are worthy of investigation. The wax gourd is a large, edible, melonlike vegetable which can be stored without refrigeration for up to a year. Palm oil is used in hundreds of different products, from lipsticks to jet engines, yet hardly any of the 28 000 species of palm have been studied by scientists. An unusual and promising source of protein is *Spirulina* (**Figure 1**). Not a plant but a marine cyanobacterium, *Spirulina* is photosynthetic and nitrogen-fixing and can be grown in saltwater ponds. Such ponds with *Spirulina* are more productive than wheat fields, and the harvested product has a higher protein content than soybeans.

The Land Base

There is a limited amount of land available for the production of food. Most land that is well suited to agriculture is already under cultivation. In many tropical regions, increasing pressure to plant cash crops, such as coffee and bananas for foreign markets, has resulted in the clearing of rain forest and the replacement of important local food crops with these export crops. In many parts of the world, poor cultivation practices and changing climate patterns are leading to soil erosion and desertification. These trends threaten to reduce both genetic biodiversity and the land base fit for cultivation.

In addition, the irrigation of new land may prove difficult, as large freshwater resources are scarce. In many areas where irrigation has already been used, groundwater supplies are dwindling and the soil is becoming increasingly salty. This process of **salinization** occurs when dissolved salts in the groundwater are brought to the surface during ongoing irrigation. The water is lost through transpiration and evaporation, leaving the salts behind to accumulate in the soil (**Figure 2**). Every year an estimated 200 000 ha of irrigated cropland become useless for food production due to salinization. It is not discussed much and does not make headlines, but salinization is a serious problem in central Canada and the United States. As a result, **halophytes** are gaining the attention of researchers. These plants grow naturally in deserts, salt marshes, and coastal regions. Examples of some halophytes growing along highways or in roadside ditches are *Atriplux patula,* a herbaceous plant, *Hippophae rhamnoides,* a shrub, and the Russian olive tree. Although these examples will not likely be used for human food, they illustrate that the millions of kilometres of coastline and millions of hectares of desert that exist may be able to support crops of halophytes, using salt water for irrigation.

New plants and cultivation methods must be investigated but care must also be taken to examine the real causes of our problems and deal with them, not just the symptoms. When acid rain lowered the pH of many lakes, researchers found a fish which could tolerate acidic conditions. Many assumed that this discovery meant we really did not have to worry about the acid rain problem. It soon became clear that although the fish could tolerate the acidic waters, species needed by the fish as food could not. In addition, other problems such as heavy metal leaching persisted.

Even if we could manage to produce adequate human food sources on the planet, we should not ignore the impact of a rapid growth in human numbers. Where would we all live, what would fulfill our energy needs, and what would happen to all our waste?

Preserving Genetic Diversity

The value of maintaining genetic diversity cannot be overstated. Future improvements in plant technology ultimately depend on the availability of a range of genetic characteristics in the plants themselves. Their genetic diversity will provide the characteristics necessary to allow us to develop new food crops, new medicines, and other commercially valuable plant products. At present, a very large percentage of the genetic diversity of plant species has yet to be identified and preserved. One vital step which needs to be taken is the collection and preservation of the genetic material of the world's cultivated varieties. These varieties represent the thousands of traditionally cultivated plant types which contain much of the total genetic diversity of our most important crops. A second step will be to collect and preserve the genetic material of wild plants which are the closest relatives of these valuable cultivated plants. Third and most important is the preservation of the many thousands of species of plants that are found in the world's threatened ecosystems. These plant communities house the majority of the world's "library" of genetic information. To allow these libraries to disappear without ever having studied their contents would be an enormous error.

salinization: the process in which salts in the groundwater are brought to the surface by irrigation and left to accumulate there as the water leaves the soil and plants through transpiration and evaporation

halophytes: plants that can tolerate living in salty habitats because they have the ability to take water from concentrated solutions

Figure 2
Salinization is a serious problem for croplands. Plants with salt tolerance are being sought.

SUMMARY **A Sustainable Future**

1. The human population will soon surpass 8 billion.

2. Global warming and climate change may reduce the amount of land fit for cultivation.

3. Farmland and forests must be used more wisely.

4. There are some promising investigations regarding plants which might yield alternative food and other resources.

5. We are already using most of the land which is suitable for growing crops.

6. Salinization is a problem when irrigation is used excessively.

7. Halophytes may be helpful but we cannot rely on them to replace all currently useful plants.

8. Efforts must be made to reduce the causes of all our problems.

9. Maintaining genetic diversity is a critical element in creating a sustainable future for our planet.

Section 14.8 Questions

Understanding Concepts

1. Why are new varieties of nitrogen-fixing plants so eagerly sought?

2. How does the chemical composition of amaranth seeds make them particularly valuable?

3. List some factors which limit the availability of more land for the cultivation of crops.

4. (a) What are halophytes?
   (b) What would be the advantage of finding a highly nutritious halophyte?

5. How can the routine irrigation of soil with normal well water lead to salinization?

Making Connections

6. How can a world demand for tropical cash crops such as bananas and coffee negatively impact food production in the tropical areas?

7. List and discuss some of the problems we must try to solve other than finding new food sources.

8. **Figure 3** shows a tree diagram that can be used for a concept map.
   (a) Draw a similar diagram in your notebook. Your map should have four major branches, each with three smaller branches.
   (b) Label the major branches: Science, Technology, Society, Environment.
   (c) Complete your map by placing the following labels on the smaller branches: chemistry, ecology, genetics, agriculture, forestry, pharmaceuticals, medicines, food, economy, climate change, soil, habitats.

Figure 3

Key Expectations

Throughout this chapter, you have had opportunities to do the following:

- Illustrate the process of succession and the role of plants in the maintenance of diversity and the survival of organisms (14.1, 14.2).
- Locate, select, analyze, and integrate information on topics under study, working independently and as part of a team, and using appropriate library and electronic research tools, including Internet sites (14.1, 14.4, 14.6).
- Express opinions supported by your own research about the case for funding certain projects in plant science or technology (14.2, 14.7, 14.8).
- Design and carry out an experiment to determine the factors that affect the growth of a population of plants, identifying and controlling major variables (14.3).
- Select appropriate instruments and use them effectively and accurately in collecting observations and data (14.3).
- Select and use appropriate modes of representation to communicate scientific ideas, plans, and experimental results (14.3, 14.4).
- Describe and explain ways in which society supports and influences plant science and technology (14.3, 14.4, 14.5, 14.6, 14.7).
- Describe how a technology relates to plant function and evaluate it on the basis of identified criteria such as safety, cost, availability, and impact on everyday life and the environment (14.3, 14.7).
- Identify various factors that influence result in trade-offs in the development of food technologies (14.4).
- Express the result of any calculation involving experimental data to the appropriate number of decimal places or significant figures (14.4, 14.5).
- Describe and explain some of the uses of plants in food and therapeutic products and in industrial processes (14.4, 14.5, 14.6).
- Identify and describe science- and technology-based careers related to the subject area under study (14.6).
- Compile information about the chemical products derived from plants and, either by hand or computer, display the information in a variety of formats, including diagrams, flow charts, tables, graphs, and scatter plots (14.6).

Key Terms

artificial selection	interspecific
biodiversity	intraspecific
biological accumulation	invasive
	manure
biological magnification	monoculture
	native
climax community	natural fertilizers
competition	persistent chemicals
compost	pioneer organisms
crop rotation	primary succession
desertification	salinization
exotic	secondary succession
fertilizers	sewage sludge
gene pool	succession
green revolution	synthetic fertilizers
halophytes	variety
hydroponics	

Make a Summary

In this chapter, you focused on the importance of plants and their products. To summarize these essential aspects of plant science and technology, create a concept map, which may be modelled on the simplified example in **Figure 3**, page 586. Your concept map should show the general types of plant products and the subcategories of these general types. In most cases, you will also be able to include specific examples. Also, add the "things" (not limited to three as shown) required by the plants before they can yield these products.

Reflect on your Learning

Revisit your answers to the Reflect on Your Learning questions at the beginning of this chapter.

- How has your thinking changed?
- What new questions do you have?

Understanding Concepts

1. What is succession? Explain why it happens. Do not merely list specific stages.

2. What is the difference between primary succession and secondary succession?

3. What is the relationship between plants and animals in succession?

4. What are some factors that influence global plant distribution and abundance?

5. Explain what is meant if a location is said to have low biodiversity.

6. What are the environmental consequences of an area with low biodiversity?

7. Describe one specific situation which would illustrate how diversity is important to the survival of a specific species.

8. Explain, with examples, the difference between interspecies and intraspecies competition.

9. What plant tissues would you expect to produce the most toxic chemicals: leaves, stems, roots, or fruits? Justify your answer.

10. What major discovery led to the creation of the first synthetic fertilizers?

11. Indicate how nutrient demands for the elements N, P, and K change from the seedling to the fruiting stage of plant growth.

12. Describe one advantage and one disadvantage of using compost instead of synthetic fertilizer on your garden.

13. What environmental damage can be caused by excessive fertilizer applications?

14. List at least five factors you would investigate before choosing one of four kinds of cucumber to plant in your garden.

15. In addition to food, what other general types of products are derived from plants?

16. Describe the role of artificial selection in the breeding and development of food crops.

17. Describe the key technological advances required by the green revolution.

18. Over hundreds of years, plant breeders have managed to modify the sea cabbage dramatically through the process of artificial selection. Match the food plant on the right with the special characteristic that would have

been selected in order to produce it. See **Figure 5** in section 14.4 to help you.

(a) an enlarged terminal bud	1. cauliflower
(b) a thickened and swollen stem	2. cabbage
(c) an enlarged flower	3. kohlrabi
(d) enlarged lateral buds	4. broccoli
(e) thickened stems and flowers	5. Brussels sprouts
(f) enlarged leaves	6. kale and collards

19. What type of plant product (food, medicine, fibre, etc.) is provided by each of the following?

(a) tannins	(f) rosy periwinkle
(b) flax	(g) hemp
(c) jojoba	(h) rattan
(d) logwood	(i) turpentine
(e) sisal	(j) salicylic acid

20. Refer to **Figure 1**.
    (a) Pieces x and y are both softwood trees. In general, what types of trees are these?
    (b) Give a specific reason to explain which piece of wood would yield the larger amount of energy when burned.
    (c) Suggest two uses for these woods other than burning.

Figure 1
The piece of pine **(x)** is 2.0 cm x 24.2 cm x 2.9 cm and has a mass of 75.1 g. The piece of maple **(y)** is 5.7 cm x 22.2 cm x 0.85 cm and has a mass of 63.0 g.

21. What are the three key steps that must be taken to protect the genetic diversity of the world's plants?

Applying Inquiry Skills

22. You perform some soil tests in your garden. The results indicate that the levels of all three major nutrients are

low. The nutrient requirements for each type of plant you are planning to grow are provided. You will be planting them in separate garden beds. Suggest an NPK number for a fertilizer that would be appropriate for each garden bed early in the season.

(a) sweet potato N-low   P-medium   K-high
(b) strawberry   N-medium   P-medium   K-low
(c) leaf lettuce N-high   P-very high K-very high
(d) watermelons   N-medium   P-medium   K-medium

23. Use the data in **Table 1** in section 14.7 to draw a pie chart that displays pesticide use by type on all crops in Ontario. Each segment should represent the percentage of the total for each type of pesticide.

24. Use the data in **Table 1** in section 14.7 to do both parts of this question. Examine also **Figures 2** and **3**, which show insecticide and fungicide use on Ontario crops. For each pie chart, determine and record in your notebook the percentages and their corresponding crops.

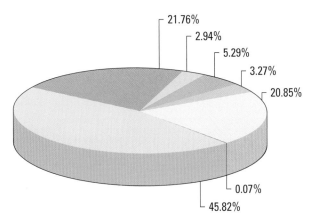

Figure 2
Insecticide use

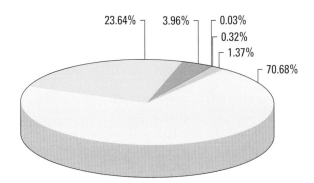

Figure 3
Fungicide use

Making Connections

25. It takes approximately 15 kg of grain to produce 1 kg of meat. Explain one obvious environmental benefit of a vegetarian diet.

26. Describe the method(s) you think early humans used to discover the medicinal properties of plants before the advent of the modern scientific method.

27. What impacts has the long-term use of fertilizers had on society and the environment?

28. How have environmental, economic, and societal concerns influenced the use of pesticides?

29. What are some questions you should ask before trying any herbal remedy? A perscription drug? Why should you ask these questions?

30. Some personal products advertise that they contain all natural ingredients.
    (a) Name three examples of specific products which make that claim.
    (b) What do many people assume when they read that something is "all natural"?
    (c) Explain why that assumption could be dangerous.

31. Trees are considered a renewable resource. Why?

32. Some feel that we really only need to conserve non-renewable resources such as fossil fuels. Why should resources such as trees also be conserved?

33. In the mid 1990s, people predicted that we would soon have a paperless society.
    (a) What was meant by that prediction?
    (b) Why did those people think they were right?
    (c) Comment on how correct or incorrect they were.

Plants That Changed the World

In this unit task, you will explore the relationships between a particular plant and human society. Your challenge will be to gather, synthesize, analyze, and present your findings in an informed and creative manner. Your final product will include a written report, a visual display, and/or a presentation to the class as determined in consultation with your teacher.

Your task is divided into six distinct components. These components and your corresponding specific expectations are as follows:

Component	Specific expectations and suggestions
plant biology	• Research and report on both the taxonomy and natural history of a plant. Include its size, appearance, growth habits, where it grows (geographic distribution), and its soil and climatic requirements. • Use a variety of illustrations, maps, and diagrams.
plant product(s)	• Identify the useful product(s) obtained from this plant. Identify and describe the plant structures and functions associated with the product. • Determine which physical and chemical properties of the product make it valuable. If possible, obtain a sample of the product and evaluate these properties directly (e.g., determine the strength of a plant fibre or the caloric content of a food plant). • Include images, drawings, and/or samples of the plant product.
historical/ societal connections	• Research and then place the use of this plant product in a historical context and illustrate with a short chronology (time line). • Identify various influences this product has had on quality of life, the economy, and the environment. • Evaluate the significance of this plant product on a global scale. • Support your evaluation with quantitative data.
technology and research	• Use a flow chart and diagrams to outline and illustrate the technologies used to culture, grow, harvest, and process the plant and plant product, include planting, fertilizing, irrigating, and pest management techniques, where applicable. • Identify any economic, social, or environmental costs associated with the production of this plant product. • Describe any current research initiatives associated with this plant product and state what questions you would like answered.
career connections	• Identify one or more occupations associated with each step in your technology flow chart and place them in a "careers list." • Select one career and determine the educational background required. Try to be very specific.
conclusion and synthesis	• Based on your research and analysis, express an informed opinion which supports the following statement: This plant product has had a significant impact on both human society and on individuals in their daily lives. The properties and value of this product can be explained through an understanding of plant science. These same properties and values may be further improved through the application of appropriate technology.

Assessment

Your completed task will be assessed according to the following criteria:

Process
- Use of a variety of sources, including print and electronic research tools.
- Demonstrate skills in conducting a thorough search of available reference materials.
- Locate broad-ranging information pertinent to the plant and plant product.
- Compile and summarize the information.
- Select appropriate methods for the analysis and presentation of your findings.
- Evaluate the quality of data and information obtained.

Product
- Compile, organize, and present the data using a variety of formats, including tables, illustrations, flow charts, and diagrams.
- Clearly demonstrate an understanding of the connections between the anatomy and growth of a plant and its potential uses.
- Express and present your findings in a creative and effective manner.
- Use both scientific terminology and the English language appropriately.

Understanding Concepts

1. Determine whether each statement is true or false. Rewrite the false statements to make them true.
   (a) Palisade cells are found in the leaf and produce a waxy cuticle.
   (b) Guard cells tend to open stomata during the night and close them during the day.
   (c) Root hairs are extensions of xylem cells which increase surface area for absorption.
   (d) Water lilies and geraniums are examples of xerophytes.
   (e) Monocotyledons typically have parallel venation in their leaves.
   (f) The pericycle is a meristematic region of the root responsible for secondary growth.
   (g) Secondary xylem growth forms wood and aids in support.
   (h) Sieve tubes and companion cells are responsible for conducting water throughout the plant body.
   (i) Succession eventually forms a climax forest.

2. Briefly describe the significance of each of the following:
   (a) allelopathy
   (b) mycorrhizae
   (c) aerial roots
   (d) cactus spines
   (e) nodular nitrogen-fixing bacteria
   (f) succulent leaves
   (g) cotyledons
   (h) vessels and tracheids
   (i) seed coat
   (j) plant competition

3. Look carefully at the photograph in **Figure 1**.

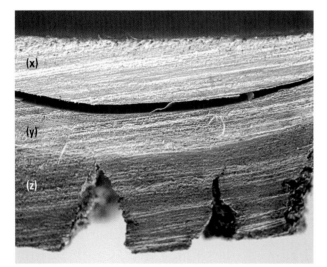

Figure 1
The edge of a cross section of a mature tree

(a) What specific type of cells are found in area x?
(b) What is the function of these cells?
(c) What specific type of cells are found in area y?
(d) What is area z?
(e) What thin but special layer of tissue lies between x and y? Ignore the dark crack in this area.
(f) What is the significance of this special thin layer?
(g) Of the areas seen, which is/are part of the tree's bark?

4. Create a table in your notebook to compare the following features of monocots and dicots: leaf venation, seed structure, and arrangement of vascular tissue in stems and roots.

5. Name the growth regulator that is primarily responsible for each of the following:
   (a) rapid stem elongation
   (b) conversion of stored starches into sugars, such as when an apple ripens
   (c) leaf growth
   (d) elongation of root tips
   (e) guard cell activity

6. Seeds can remain dormant for extended periods of time.
   (a) Why is seed dormancy a survival strategy?
   (b) What factors trigger them to germinate?

7. List the important biological roles of the following nutrients in plants:
   (a) nitrogen
   (b) potassium
   (c) phosphorus
   (d) magnesium
   (e) calcium
   (f) sulfur

8. Most soil particles possess a negative electrical charge. Explain how this charge influences the availability of valuable ions for uptake by plant roots. Include a discussion of cation exchange mechanisms.

9. (a) Why is it not surprising that seeds and roots contain proteins and molecules with large quantities of chemical energy?
   (b) Explain which plant products would be easier to store for long periods of time: seeds or roots.

10. Give two or three examples of plant products which are used for each of the following purposes. Identify the plant part from which the product is derived.
    (a) flavourings
    (b) therapeutic products
    (c) wood finishes
    (d) car or furniture waxes
    (e) cosmetics

11. Humans use a wide variety of plant fibres and wood.
    (a) What specific cell component makes up the bulk of fibre material?
    (b) What plant part is harvested to produce cotton fibre?
    (c) What are the advantages of plywood over solid lumber?

12. The production of ethanol fuel is a major industry in Brazil and has potential to grow in Ontario.
    (a) What plants are being used in Brazil to supply this industry?
    (b) What plants are the best candidates for this use in Ontario?
    (c) Ethanol fuel is considered a renewable resource. Explain.
    (d) What arguments are being used against this industry?

Applying Inquiry Skills

13. Refer to the tree cross section shown in **Figure 2**.
    (a) Estimate the age of this tree when it was cut. Explain how you arrived at your answer.
    (b) What name applies to the dark area, x?
    (c) What name applies to the lighter area, y?
    (d) What is the reason for the colour difference?

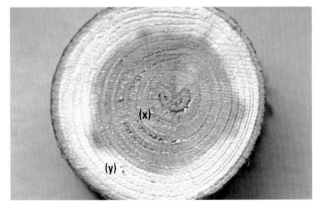

Figure 2
A cross section of a young mulberry tree

14. Present the following data in an appropriate graphical form using percentage values.

Table 1: Global Use of Wood (millions of m³)

	Industrial wood consumption	Firewood consumption
Developed countries	1106	221
Developing countries	348	1485
1980 world totals	1454	1706

15. Describe the type of investigation that scientists must have used to gather evidence about what nutrients plants need in order to grow.

16. Refer to the two partial cross sections in **Figure 3**. One piece of oak (y) is slightly darker than the other (x) because it has been exposed to air for longer. As a forest technician, you are asked to analyze these two tree sections.
    (a) Compare the growth rates of trees x and y.
    (b) Explain the evidence for your previous answer.
    (c) If both trees, x and y, grew from acorns from the same parent tree, what factors could account for the different growth rates?

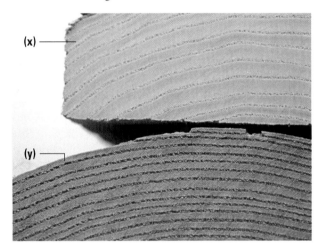

Figure 3
Small portions from near the outer edge of cross sections of two different oak trees, bark removed

17. Refer to **Figure 4**. As a plant biologist, you have been asked to answer the following questions posed by a tree plantation operator.
    (a) Explain whether the oldest xylem is closer to x or y.
    (b) Compare the changes in tree ring diameters as the tree ages. Suggest ways of accounting for these changes.

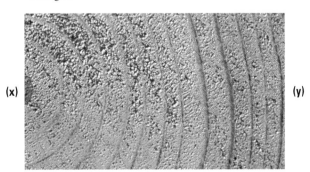

Figure 4
A small portion of the cross section of a pine tree

18. A plant biologist performs an investigation to compare the growth rates of conifer seedlings in both sterilized and unsterilized soil. She finds that even though both soils contain adequate quantities of nutrients, the plants in the sterile soil demonstrate very stunted growth. What is the most likely explanation?

19. Two farmers are contemplating planting a tomato crop. One farmer lives in southern Ontario and sells his tomatoes fresh in the late summer and fall to small local grocers. The second farmer lives in California. She sells most of her crop fresh during the winter months to Ontario grocery store chains.
    (a) Contrast the important characteristics and trade-offs that each might consider when choosing a tomato variety to plant.
    (b) How might the expectations and preferences of the consumer change from fall to winter when choosing tomatoes?

20. You are a young scientist who is not yet able to identify specific plant species. You walk through a forest and note that there seem to be four kinds of trees: A, B, C, and D. You can tell there are four different species even if you do not know exactly what they are. The specimens of A and B seem to be all quite old, though still healthy. Of the specimens of C and D, some are mature, but not old, while others are at various levels of maturity. Among the plants making up the understorey are seedlings and saplings of C and D. No seedlings or saplings of species A and B are found.
    (a) In terms of the process of succession, describe what stage this forest has reached. Explain.
    (b) Describe specifically what this area will look like after 150 years.
    (c) Explain why it will look different than it does now.

Making Connections

21. Describe one specific situation that illustrates how diversity is important to the survival of an entire community.

22. Outline the impact the following technologies have had on both the quantity and quality of global food production:
    (a) synthetic fertilizers
    (b) modern pesticide use
    (c) artificial selection and plant breeding

23. Based on your own research and classroom discussions during this unit, express your opinions on society's willingness to fund and use new technologies to make advances in plant cultivation.

24. Suppose you have been given the responsibility of choosing which types of plant research should be funded. Rank the following research proposals from most to least important and briefly comment on what criteria you used to judge them:
    (a) continued research into finding or creating valuable food crops with increased pest resistance and nutritional value
    (b) a global effort to collect seed samples from plants for preservation in an international "gene pool bank"
    (c) development of new strategies to use natural biological controls and organic farming techniques to eliminate the use of synthetic fertilizers and pesticides
    (d) an intensive research program aimed at discovering new plant products including foods, medicines, and other chemicals

25. Refer to **Figure 5** for the following questions.
    (a) Common nightshade tends to grow up and around any other plants or structures near it. Which part of the plant is specialized to grow this way?
    (b) What advantages does the plant obtain by having this growth pattern?
    (c) What type of leaf venation does this plant have?
    (d) Suggest why the leaves of this plant never grow very large.

Figure 5
Common nightshade growing along the top of a farm fence

26. Refer to **Figure 6.**
    (a) Elderberries are often picked by humans to make jelly or wine. They are also sought by birds, who eat hundreds at a time. Explain which of these actions assists the elderberry plant.
    (b) If gibberellins were absent from these plants, how would their structure be different? Explain why.
    (c) If ethylene were not produced by this plant, what visible feature would likely not be present? Explain why not.

Figure 6
A red elderberry shrub with fruit clusters, growing at the edge of a forest

Figure 8
Pinesap

27. Note the various plants in **Figure 7.**
    (a) What is the common name and biological name for the type of plant that is in the centre? Explain how you know at such a distance.
    (b) Using the biological name, identify what type of plant the others in the photo are.
    (c) What two major problems are encountered by very tall plants?
    (d) What type of tissue solves both these problems? Explain how.
    (e) What leaf features of the centre plant allow the leaves to stay on the plant all winter?

29. **Figure 9** shows marsh marigolds in early summer.
    (a) In what type of habitat would you expect these plants to live?
    (b) What is the name for plants that live in this habitat?
    (c) What hormone activity is obvious from this photo?
    (d) The stalked structures are the fruit clusters. What used to be found at that location? What would you expect to find inside the fruit?
    (e) These plants almost disappear by late autumn. What two biological categories would these plants belong to?
    (f) Explain why you would not expect to find bark on these plants.

Figure 7
A variety of trees

Figure 9
Marsh marigolds

28. **Figure 8** shows pinesap and a separate green vine on a forest floor.
    (a) Why might you think pinesap is not a plant?
    (b) What might you think it is? Why?
    (c) If you were told it is a flowering plant, what would you look for in addition to flowers, assuming you had access to a microscope?
    (d) What plant structures seem to be missing?

Appendixes

APPENDIX A

Skills Handbook   598

A1: Scientific Inquiry   598
    Planning an Investigation   598
A2: Decision Making   604
    A Risk–Benefit Analysis Model   607
A3: Technological Problem Solving   608
A4: Lab Reports   610
A5: Graphic Organizers   614
A6: Math Skills   615
    Scientific Notation   615
    Uncertainty in Measurements   615
    Probability   617
    Graphs   618

APPENDIX B

Safety Skills   620

B1: Safety Conventions and Symbols   620
B2: Safety in the Laboratory   622
B3: Care and Use of the Microscope   627

APPENDIX C

Reference   630

C1: Numerical Prefixes and Units   630
C2: Greek and Latin Prefixes and Suffixes   632
C3: Periodic Table of the Elements   634

APPENDIX D

Answers   636

Scientific Inquiry

Planning an Investigation

In our attempts to further our understanding of the natural world, we encounter questions, mysteries, or events that are not readily explainable. We can use controlled experiments, correlational studies, or observational studies to help us look for answers or explanations. The methods used in scientific inquiry depend, to a large degree, on the purpose of the inquiry.

Controlled Experiments

Controlled experiments are performed when the purpose of the inquiry is to create or test a scientific concept. In a controlled experiment, an independent variable is purposefully and steadily changed to determine its effect on a second, dependent variable. All other variables are controlled or kept constant.

The common components for controlled experiments are outlined in the flow chart below. *Even though the sequence is presented as linear, there are normally many cycles through the steps during the actual experiment.*

Process Description

Choose a topic that interests you. Determine whether you are going to create, test, or use a scientific concept and whether you are going to carry out a given procedure or develop a new experimental design. Indicate your decision in a statement of the purpose.

Your question forms the basis for your investigation. Controlled experiments are about relationships, so the question could be about the effects on variable A when variable B is changed. The question may also be about what causes the change in variable A. In this case, you might speculate about possible variables and determine which variable causes the change.

A hypothesis is a tentative explanation. You must be able to test your hypothesis, which can range in certainty from an educated guess to a concept that is widely accepted in the scientific community. A prediction is based upon a hypothesis or a more established scientific explanation, such as a theory. In the prediction you state what outcome you expect from your experiment.

The design of a controlled experiment identifies how you plan to manipulate the independent variable, measure the response of the dependent variable, and control all the other variables.

| Stating the purpose | → | Asking the question | → | Hypothesizing/ predicting | → | Designing the investigation |

Example: Diffusion and Osmosis

The purpose of this investigation is to examine the conditions under which water moves into and out of cells.

Under what conditions does water move in and out of cells?

If a model cell is placed in a hypotonic solution (having lower concentrations of solute than the cell), water will move into the cell. If a cell is placed in a hypertonic solution (having higher concentrations of solute than the cell), water will move out of the cell.

Three model cells are constructed from dialysis tubing. Two are filled with distilled water and one with a starch suspension. One of the cells with distilled water is placed in a beaker of starch suspension. The other two cells are placed in beakers of distilled water. Iodine is added to the distilled water in the beakers. (Iodine is used as an indicator to detect the presence of starch.) The mass of each cell is recorded and other observations are made at the beginning and after 10 and 20 min.

SKILLS HANDBOOK

Table 1: Observations of Model Cells

Model cell	Initial mass (g)	Mass after 10 min (g)	Mass after 20 min (g)	Other observations
cell A—dialysis tube with distilled water in beaker of distilled water				
cell B—dialysis tube with starch suspension in beaker of distilled water				
cell C—dialysis tube with distilled water in beaker of starch suspension				

There are many ways to gather and record observations during an investigation. It is helpful to plan ahead and think about what data you will need and how best to record them. This helps to clarify your thinking about the question posed at the beginning, the variables, the number of trials, the procedure, and your skills. It will also help you organize your evidence for easier analysis later.

After thoroughly analyzing your observations, you may have sufficient and appropriate evidence to enable you to answer the question posed at the beginning of the investigation.

At this stage of the investigation, you will evaluate the processes that you followed to plan and perform the investigation. Evaluating the processes includes reviewing the design and the procedure. You will also evaluate the outcome of the investigation, which involves assessing the evidence—whether it supports the hypothesis or not—and the hypothesis itself.

In preparing your report, your objectives should be to describe your design and procedure accurately, and to report your observations accurately and honestly.

→ **Gathering, recording, and organizing observations** → **Analyzing the observations** → **Evaluating the evidence and the hypothesis** → **Reporting on the investigation**

The measurements and observations will be recorded in a table like **Table 1**.

The results will be analyzed to determine if there were any changes in mass and if there were any other observable changes. Changes in mass and/or colour should enable us to determine what is happening with the cells.

To evaluate this investigation, we have to ask ourselves several questions. Is our model cell an appropriate model? Do the results allow us to determine how water moves into and out of cells? Does the evidence gathered support our hypothesis? Are there possible sources of error that may invalidate the evidence?

For the format of a typical lab report, see the sample Lab Report in Appendix A4.

Correlational Studies

When the purpose of scientific inquiry is to test a suspected relationship (hypothesis) between two different variables, but a controlled experiment is not possible, a correlational inquiry is conducted. In a correlational study, the investigator tries to determine whether one variable is affecting another without purposefully changing or controlling any of the variables. Instead, variables are allowed to change naturally. It is often difficult to isolate cause and effect in correlational studies. A correlational inquiry requires very large sample numbers and many replications to increase the certainty of the results.

The flow chart below outlines the components/processes that are important in designing a correlational study. The investigator can conduct the study without doing experiments or fieldwork, for example, by using databases prepared

Process Description

Choose a topic that interests you. Determine whether you are going to replicate or revise a previous study, or create a new one. Indicate your decision in a statement of the purpose.

In planning a correlational study, it is important to pose a question about a possible statistical relationship between variable A and variable B.

A hypothesis is a tentative explanation. In a correlational study, a hypothesis can range in certainty from an educated guess to a concept that is widely accepted in the scientific community.

The design of a correlational study identifies how you will gather data on the variables under study and also identifies the potential source. There are two possible sources—observations made by the investigator and existing data.

| Stating the purpose | → | Asking the question | → | Hypothesizing/ predicting | → | Designing the investigation |

Example: Colourblindness in Males and Females

The purpose of this investigation is to determine if there is a statistical relationship between the sex of an individual and colourblindness.

Is there a greater percentage of males than females who are colourblind?

Red-green colour blindness is a sex-linked recessive disorder. In a large sample, the incidence of colourblindness among males will be significantly higher than the incidence among females.

An equal number (two hundred each) of males and females in the school will be surveyed for colourblindness. The data will be recorded in a table and the percentage of colourblind individuals will be calculated. The actual, or observed, percentage will be compared to the percentage we would expect if there were no relationship between the two variables.

by other researchers to find relationships between two or more variables. The investigator can also make his or her own observations and measurements through fieldwork, interviews, and surveys.

Even though the sequence is presented as linear, there are normally many cycles through the steps during the actual study.

There are many ways to gather and record observations during your investigation. It is helpful to plan ahead and think about what data you will need and how best to record them. This is an important step because it helps to clarify your thinking about the question posed at the beginning, the variables, the number of trials, the procedure, and so on. It will also help you organize your information for easier analysis later.

After thoroughly analyzing your observations, you may have sufficient and appropriate evidence to enable you to answer the question posed at the beginning of the investigation.

At this stage of the investigation, you will evaluate the processes that you followed to plan and perform the investigation. Evaluating the processes includes reviewing the design and the procedure. You will also evaluate the outcome of the investigation, which involves assessing the evidence—whether it supports the hypothesis or not—and the hypothesis itself. The results of your investigation may be used to create new related studies.

In preparing your report, your objectives should be to describe your design and procedure accurately, and to report your observations accurately and honestly.

Gathering, recording, and organizing observations → **Analyzing the observations** → **Evaluating the evidence and the hypothesis** → **Reporting on the investigation**

Individual and group results will be recorded in tables like **Tables 2** and **3**.

Table 2: Individual Colourblind Test Results

Participant #	Colour plate #	Identified correctly (Y/N)
1	1	
	2	
	3	
	4	
2	1	
	2	
	3	
	4	

Once we have totalled the number of individuals who are colourblind, we will calculate a percentage for each sex. If there was no relationship between the sex of the individual and the incidence of colourblindness, we would expect an equal percentage of males and females to be colourblind.

Table 3: Percentage of Colourblind and Normal by Sex

	Colourblind	Normal
Male		
Female		

Was the sample size adequate? If the survey was completed on 10 other samples of this size, what is the likelihood that the results would be different? Do the survey results indicate that the incidence of colourblindness is significantly higher among males than females or can any difference be attributed to random variation in the sample?

For the format of a typical lab report, see the sample Lab Report in Appendix A4.

Observational Studies

Often the purpose of inquiry is simply to study a natural phenomenon with the intention of gaining scientifically significant information to answer a question. Observational studies involve observing a subject or phenomenon in an unobtrusive or unstructured manner, often with no specific hypothesis. A hypothesis to describe or explain the observations may, however, be generated after repeated observations, and modified as new information is collected over time.

The flow chart below summarizes the stages and processes of scientific inquiry through observational studies. *Even though the sequence is presented as linear, there are normally many cycles through the steps during the actual study.*

Process Description

Choose a topic that interests you. Determine whether you are going to replicate or revise a previous study, or create a new one. Indicate your decision in a statement of the purpose.

In planning an observational study, it is important to pose a general question about the natural world. You may or may not follow the question with the creation of a hypothesis.

A hypothesis is a tentative explanation. In an observational study, a hypothesis can be formed after observations have been made and information gathered on a topic. A hypothesis may be created in the analysis.

The design of an observational study describes how you will make observations relevant to the question.

Stating the purpose → **Asking the question** → **Hypothesizing/ predicting** → **Designing the investigation**

Example: Local Vegetation

The purpose of our investigation is to conduct an inventory of the plants in our local area to determine the most common trees and to determine if the vegetation in our local area is typical of that found in the boreal forest biome.

What are the most common species of trees in our local region? Is the vegetation in our region representative of the vegetation in the boreal forest biome?

There is considerable variation in the vegetation found throughout any biome. However, one would expect that certain species would be more common than others and would be found throughout the biome. Since we are geographically located in what is defined as the boreal forest biome, we would expect that the most common trees in this area would be those that define the vegetation of the boreal forest (i.e., spruce, fir, pine).

We will conduct an inventory of trees in 10 sample areas of our local region. Each sample area will be 10 000 m^2 (100 m $\times$ 100 m). All species of trees will be identified (using the common and scientific names) and counted in each sample area. The sample areas will be selected by placing a scaled grid over a map of the region and then randomly selecting 10 cells of the grid.

Table 4: Inventory of Trees in Sample Areas of Local Region

Tree species (common and scientific name)	Sample area										
	1	2	3	4	5	6	7	8	9	10	Total

There are many ways to gather and record observations during an investigation. During your observational study, you should quantify your observations where possible. All observations should be objective and unambiguous. Consider ways to organize your information for easier analysis.

After thoroughly analyzing your observations, you may have sufficient and appropriate evidence to enable you to answer the question posed at the beginning of the investigation. You may also have enough observations and information to form a hypothesis.

At this stage of the investigation, you will evaluate the processes used to plan and perform the investigation. Evaluating the processes includes evaluating the materials, the design, the procedure, and your skills. The results of most such investigations will suggest further studies, perhaps correlational studies or controlled experiments to explore tentative hypotheses you may have developed.

In preparing your report, your objectives should be to describe your design and procedure accurately, and to report your observations accurately and honestly.

Gathering, recording, and organizing observations

Analyzing the observations

Evaluating the evidence and the hypothesis

Reporting on the investigation

The data will be recorded in a table like **Table 4**.

The data from the table will be plotted on a bar graph to show the frequency distribution of the various species. This will make it easier to identify the most common species.

In the evaluation of this investigation, we must first decide whether our sample areas collectively represent the forested areas of our local region. We can then compare the frequencies of the different species in the local area to the frequencies of those species in the typical boreal forest. The results of this investigation might lead us to speculate about the reasons (e.g., climate, topography, soil) for any differences in vegetation between our local area and the typical boreal forest.

For the format of a typical lab report, see the sample Lab Report in Appendix A4.

Decision Making

Modern life is filled with environmental and social issues that have scientific and technological dimensions. An issue is defined as a problem that has at least two possible solutions rather than a single answer. There can be many positions, generally determined by the values that an individual or a society holds, on a single issue. Which solution is "best" is a matter of opinion; ideally, the solution that is implemented is the one that is most appropriate for society as a whole.

The common processes involved in the decision-making process are outlined in the graphic below.

Even though the sequence is presented as linear, you may go through several cycles before deciding you are ready to defend a decision.

Process Description

The first step in understanding an issue is to explain why it is an issue, describe the problems associated with the issue, and identify the individuals or groups, called stakeholders, involved in the issue. You could brainstorm the following questions to research the issue: Who? What? Where? When? Why? How? Develop background information on the issue by clarifying facts and concepts, and identifying relevant attributes, features, or characteristics of the problem.

Examine the issue and think of as many alternative solutions as you can. At this point, it does not matter if the solutions seem unrealistic. To analyze the alternatives, you should examine the issue from a variety of perspectives. Stakeholders may bring different viewpoints to an issue and these may influence their position on the issue. Brainstorm or hypothesize how different stakeholders would feel about your alternatives.

Formulate a research question that helps to limit, narrow, or define the issue. Then develop a plan to find reliable and relevant sources of information. Outline the stages of your information search: gathering, sorting, evaluating, selecting, and integrating relevant information. You may consider using a flow chart, concept map, or other graphic organizer to outline the stages of your information search. Gather information from many sources including newspapers, magazines, scientific journals, the Internet, and the library.

Defining the issue → **Identifying alternatives/positions** → **Researching the issue**

Example: Genetically Modified Foods

Genetically modified (GM) foods are foods that have been altered by the insertion of genes from a different species. Crops can be genetically modified to grow quickly, to be resistant to diseases and pests, or to be efficient at absorbing nutrients from the soil.

There is growing public debate about such foods. Have they been tested enough to assure us that there will be no long-term effects from eating these foods? What will be the impact of genetically engineered species on natural species?

The issue is the overall safety of foods produced from genetically engineered organisms. In this debate, there are basically two positions: you either support genetic engineering in agriculture or you do not support it. **Table 1**, page 606, lists the potential groups or stakeholders who may be positively or negatively affected by the issue. Develop your knowledge of the background information on the issue by clarifying facts and concepts, and by identifying relevant attributes, features, or characteristics of the problem.

One possible solution for people concerned about GM food is to ban its production. Since farmers who grow GM foods would lose income, it might be necessary to offer subsidies for farmers who grow GM crops.

Think about how different stakeholders might feel about the alternatives. What would be the perspective of a consumer? a local politician? a farmer? an agronomist? a geneticist? an economist? a nutritionist? **Table 2**, page 606, lists the possible perspectives on an issue.

Remember that one person could have more than one perspective, or two people looking at an issue from the same perspective might disagree. For example, geneticists might disagree about the impact of genetically engineered species on natural species.

A possible research question is: What are the advantages and disadvantages of genetically modified foods?

In this stage, you will analyze the issue and clarify where you stand. First, you should establish criteria for evaluating your information to determine its relevance and significance. You can then evaluate your sources, determine what assumptions may have been made, and assess whether you have enough information to make your decision.

Once the issue has been analyzed, you can begin to evaluate the alternative solutions. You may decide to carry out a risk–benefit analysis—a tool that enables you to look at each possible result of a proposed action and helps you make a decision.

There are five steps that must be completed to effectively analyze the issue:

1. Establish criteria for determining the relevance and significance of the data you have gathered.

2. Evaluate the sources of information.

3. Identify and determine what assumptions have been made. Challenge unsupported evidence.

4. Determine any causal, sequential, or structural relationships associated with the issue.

5. Evaluate the alternative solutions, possibly by conducting a risk–benefit analysis.

After analyzing your information, you can answer your research question and take an informed position on the issue. You should be able to defend your solution in an appropriate format—debate, class discussion, speech, position paper, multimedia presentation (e.g., computer slide show), brochure, poster, video.

Your position on the issue must be justified using supporting information that you have researched. You should be able to defend your position to people with different perspectives. Ask yourself the following questions:

• Do I have supporting evidence from a variety of sources?
• Can I state my position clearly?
• Can I show why this issue is relevant and important to society?
• Do I have solid arguments (with solid evidence) supporting my position?
• Have I considered arguments against my position, and identified their faults?
• Have I analyzed the strong and weak points of each perspective?

The final phase of decision making includes evaluating the decision itself and the process used to reach the decision. After you have made a decision, carefully examine the thinking that led to your decision.

Some questions to guide your evaluation include:

• What was my initial perspective on the issue? How has my perspective changed since I first began to explore the issue?
• How did we make our decision? What process did we use? What steps did we follow?
• In what ways does our decision resolve the issue?
• What are the likely short- and long-term effects of the decision?
• To what extent am I satisfied with the final decision?
• What reasons would I give to explain our decision?
• If we had to make this decision again, what would I do differently?

Analyzing the issue	Defending the decision	Evaluating the process

The causal, sequential, or structural relationships related to the issue include the following: GM food has arisen from the need to produce more and better food for the world's population. A consequence may be that consumers will refuse to buy the GM products, thereby increasing the demand for traditionally produced food and increasing prices accordingly. Another consequence is that public funds are being spent to support research in genetically modified food crops.

Table 3, page 607, shows a risk–benefit analysis of producing GM foods.

By reviewing the research and performing a risk–benefit analysis on the information, we conclude that the benefits of using GM foods are greater than the risks. The production of GM foods will solve an immediate problem of world hunger by producing more food faster, producing food with a higher nutritional content, and by reducing the need for chemical use in food production. The use of GM foods provides an immediate solution to world hunger, which we consider to be a major benefit for their production. There are, however, calculated risks involved in full-scale GM food production. We may produce novel organisms whose long-term effects on an ecosystem cannot be known ahead of time. However, our evidence indicates that the probability of serious damage is minimal when compared with the overall benefits that GM foods will bring to society.

This issue came to my attention through the many newspaper articles I had read that warned of a growing danger to people and the environment with the production and sale of GM foods. After researching the issue, I found the vast majority of recent scientific reports found that, while there are some long-term unknowns, the production and consumption of GM products are as safe as those of non-GM foods. In fact, some studies indicated that the production of some organically grown foods could be even more dangerous. While every effort was made to obtain the most current, scientific information on GM foods and their effects on human, other species, and the environment, the number and quality of studies focusing on the long-term environmental impact of GM foods were relatively low. These studies need to be carried out. By examining such research, we may be able to make better-informed decisions regarding the use of these foods.

Table 1: Stakeholders in the GM Food Debate

Stakeholders	Viewpoint
scientists	GM foods may end world hunger.
farmers	GM crops use less pesticides and herbicides.
doctors	GM foods may be created to solve health problems. For example, children whose major diet staple is rice often have a vitamin A deficiency. Rice can be genetically engineered to produce vitamin A.
environmentalists	Genes from GM crops may spread to create superweeds and superbugs.
politicians	GM foods can end world hunger.
health critics	Genetic alterations may produce substances that are poisonous or that trigger allergies and disease.

Table 2: Perspectives on an Issue

cultural	customs and practices of a particular group
ecological	interactions among organisms and their natural habitat
economic	the production, distribution, and consumption of wealth
educational	the effects on learning
emotional	feelings and emotions
environmental	the effects on physical surroundings
esthetic	artistic, tasteful, beautiful
moral/ethical	what is good/bad, right/wrong
legal	the rights and responsibilities of individuals and groups
spiritual	the effects on personal beliefs
political	the effects on the aims of a political group or party
scientific	logical or research based
social	the effects on human relationships, the community, or society
technological	machines and industrial processes

A Risk–Benefit Analysis Model

Risk–benefit analysis is a tool used to organize and analyze information gathered in research. A thorough analysis of the risks and benefits associated with each alternative solution can help you decide on the best alternative.

- Research as many aspects of the proposal as possible. Look at it from different perspectives.
- Collect as much evidence as you can, including reasonable projections of likely outcomes if the proposal is adopted.
- Classify every individual potential result as being either a benefit or a risk.
- Quantify the size of the potential benefit or risk (perhaps as a dollar figure, or a number of lives affected, or on a scale of 1 to 5).
- Estimate the probability (percentage) of that event occurring.
- By multiplying the size of a benefit (or risk) by the probability of its happening, you can calculate a probability value for each potential result.

- Total the probability values of all the potential risks, and all the potential benefits.
- Compare the sums to help you decide whether to accept the proposed action.

Table 3 shows an incomplete risk–benefit analysis of the issue. Note that although you should try to be objective in your assessment, the beliefs of the person making the risk–benefit analysis will have an effect on the final sums. The possible outcomes considered for analysis, the assessment of the relative importance of a cost or benefit, and the probability of the cost or benefit actually arising will vary according to who does the analysis. For example, would you agree completely with the values placed in the "Cost" and "Benefit" columns of the analysis in **Table 3**?

Table 3: Risk–Benefit Analysis of Producing Foods from Genetically Modified Organisms

Risks				Benefits			
Possible result	Cost of result (scale of 1 to 5)	Probability of result occurring (%)	Cost × probability	Possible result	Benefit of result (scale of 1 to 5)	Probability of result occurring (%)	Benefit × probability
GM foods increase human health risks.	very serious 5	research is inconclusive (50%)	250	Food supplies increase and food becomes cheaper.	great 5	very likely (90%)	450
GM crops are more competitive and eliminate natural species.	serious 4	likely (80%)	320	GM foods have higher nutritional value than their natural counterparts	great 5	likely (75%)	375
GM species negatively affect other species in the food chain.	serious 4	somewhat likely (60%)	240	GM crops require less chemical fertilizers.	high 4	somewhat likely (60%)	240
Total risk value			**810**	**Total benefit value**			**1065**

Technological Problem Solving

There is a difference between science and technology. The goal of science is to understand the natural world. The goal of technological problem solving is to develop or revise a product or a process in response to a human need. The product or process must fulfill its function but, in contrast with scientific problem solving, it is not essential to under-stand why or how it works. Technological solutions are evaluated based on such criteria as simplicity, reliability, efficiency, cost, and ecological and political ramifications.

Even though the sequence presented in the graphic below is linear, there are normally many cycles through the steps in any problem-solving attempt.

Process Description

This process involves recognizing and identifying the need for a technological solution. You need to clearly state the question(s) that you want to investigate to solve the problem and the criteria you will use as guidelines and to evaluate your solution. In any design, some criteria may be more important than others. For example, if the product solution measures accurately and is economical, but is not safe, then it is clearly unacceptable.

Use your prior knowledge and experience to propose possible solutions. Creativity is also important in suggesting novel solutions.

You should generate as many ideas as possible about the functioning of your solution and about potential designs. During brainstorming, the goal is to generate many ideas without judging them. They can be evaluated and accepted or rejected later.

To visualize the possible solutions, it is helpful to draw sketches. Sketches are often better than verbal descriptions to communicate an idea.

Planning is the heart of the entire process. Your plan will outline your processes, identify potential sources of information and materials, define your resource parameters, and establish evaluation criteria.

Seven types of resources are generally used in developing technological solutions to problems—people, information, materials, tools, energy, capital, and time.

Defining the problem → **Identifying possible solutions** → **Planning**

Example: Aquarium Activity and Fish Death

You need to conduct an investigation to determine how temperature affects respiration rate in mammals. You need an instrument to measure the consumption of oxygen. A respirometer measures oxygen consumption during respiration. Respirometers are commercially available but may be inaccessible because of cost or other limitations. Your challenge then is to design a respirometer that will enable you to measure oxygen consumption with a degree of accuracy sufficient to answer your scientific question.

Generate as many ideas as possible about the functioning of a respirometer and about potential designs.

In this problem, you may need, and be restricted by, the following resources:
People—yourself, your partner(s), and your biology teacher
Information—You already understand the biochemical process of respiration. You need to understand how a respirometer works. What chemical reactions occur? What variable is measured as an indicator of oxygen consumption? In what units do you measure oxygen consumption?
Materials—Your choice of materials is limited by the proposed design of your respirometer, cost, availability, and time. You are restricted to materials that can be obtained in school or at home. Materials to consider include a large tray, beaker or flask, rubber stopper, rubber tubing, glass tubing, pipette, graduated cylinder, plastic bottles, carbon dioxide (CO_2) absorbent such as potassium hydroxide (KOH) solution or pellets, plasticene, silicone caulking.
Tools—The tools you will need will depend on your proposed design. For this problem, you are likely to require only a knife and scissors. Your design should not require any specialized tools or machines.
Energy—The only energy requirement for this problem is your own energy.
Capital—The capital resources must be minimal; otherwise, it may be more efficient to purchase a commercial model.
Time—Because of the time limit on the scientific investigation, there is an even shorter time limit on the construction of a respirometer that will be used to collect data in the investigation. You should be able to construct your respirometer within 30 min.
The following are design criteria:
- large enough to contain a small mammal (e.g., mouse, gerbil)
- accurate to at least + or − 5 mL/min
- must incorporate a thermometer
- can be reused without significant maintenance
- safe for animal subject

In this phase, you will construct and test your prototype using systematic trial and error. Try to manipulate only one variable at a time. Use failures to inform the decisions you make before your next trial. You may also complete a cost–benefit analysis on the prototype.

To help you decide on the best solution, you can rate each potential solution on each of the design criteria using a five-point rating scale, with 1 being poor, 2 fair, 3 good, 4 very good, and 5 excellent. You can then compare your proposed solutions by totalling the scores.

Once you have made the choice among the possible solutions, you need to produce and test a prototype. While making the prototype, you may need to experiment with the characteristics of different components. A model, on a smaller scale, might help you decide whether the product will be functional. The test of your prototype should answer three basic questions:

- Does the prototype solve the problem?
- Does it satisfy the design criteria?
- Are there any unanticipated problems with the design?

If these questions cannot be answered satisfactorily, you may have to modify the design or select another potential solution.

In presenting your solution, you will communicate your solution, identify potential applications, and put your solution to use.

Once the prototype has been produced and tested, the best presentation of the solution is a demonstration of its use—a test under actual conditions. This demonstration can also serve as a further test of the design. Any feedback should be considered for future redesign. Remember that no solution should be considered the absolute final solution.

The technological problem-solving process is cyclical. At this stage, evaluating your solution and the process you used to arrive at your solution may lead to a revision of the solution.

Evaluation is not restricted to the final step. However, it is important to evaluate the final product using the criteria established earlier, and to evaluate the processes used while arriving at the solution. Consider the following questions:

- To what degree does the final product meet the design criteria?
- Did you have to make any compromises in the design? If so, are there ways to minimize the effects of the compromises?
- Did you exceed any of the resource parameters?
- Are there other possible solutions that deserve future consideration?
- How did your group work as a team?

Constructing/testing solutions → Presenting the preferred solution → Evaluating the solution and process

Table 1 illustrates a rating for two different respirometer designs. Note that although Design 1 came out with the highest rating, there is one factor (subject safety) that suggests that we should go with Design 2. This is what is referred to as a tradeoff. We have to compromise on criteria such as size and accuracy in order to ensure that the apparatus is safe for the subject specimen. By reviewing or evaluating the product and the process so far, we may be able to modify Design 2 to optimize its performance in meeting the other criteria.

Table 1: Rating for Respirometer Designs

Design criterion	Design 1	Design 2
size	4	3
accuracy	4	3
thermometer	5	3
ease of reuse	4	3
subject safety	2	5
total score	**19**	**17**

1—poor, 2—fair, 3—good, 4—very good, 5—excellent

Design 1 scored higher than Design 2; however, we feel that the safety of the subject is a very important factor that must be addressed in our design. While our best effort went into designing and creating a respirometer, a comparison between our instrument and a commercial model is recommended. By comparing the results between our instrument and a commercial model, an estimation of our instrument's accuracy can be determined. As a final step, our design could be improved, thus justifying the creation and use of our instrument over the use and purchase of a commercial model.

Lab Reports

When carrying out investigations, it is important that scientists keep records of their plans and results, and share their findings. In order to have their investigations repeated (replicated) and accepted by the scientific community, scientists generally share their work by publishing papers in which details of their design, materials, procedure, evidence, analysis, and evaluation are given.

Lab reports are prepared after an investigation is completed. To ensure that you can accurately describe the investigation, it is important to keep thorough and accurate records of your activities as you carry out the investigation.

Investigators use a similar format in their final reports or lab books, although the headings and order may vary. Your lab book or report should reflect the type of scientific inquiry that you used in the investigation and should be based on the following headings, as appropriate. (See **Figure 1** for a sample lab report.)

Title

At the beginning of your report, write the number and title of your investigation. In this course the title is usually given, but if you are designing your own investigation, create a title that suggests what the investigation is about. Include the date the investigation was conducted and the names of all lab partners (if you worked as a team).

Purpose

State the purpose of the investigation. Why are you doing this investigation?

Question

This is the question that you attempted to answer in the investigation. If it is appropriate to do so, state the question in terms of independent and dependent variables.

Hypothesis/Prediction

Based on your reasoning or on a concept that you have studied, formulate an explanation of what should happen (a hypothesis). From your hypothesis you may make a prediction, a statement of what you expect to observe, before carrying out the investigation. Depending on the nature of your investigation, you may or may not have a hypothesis or a prediction.

Design

This is a brief general overview (one to three sentences) of what was done. If your investigation involved independent, dependent, and controlled variables, list them. Identify any control or control group that was used in the investigation.

Materials

This is a detailed list of all materials used, including sizes and quantities where appropriate. Be sure to include safety equipment such as goggles, lab apron, latex gloves, and tongs, where needed. Draw a diagram to show any complicated setup of apparatus.

Procedure

Describe, in detailed, numbered, step-by-step format, the procedure you followed in carrying out your investigation. Include steps to clean up and dispose of waste.

Observations

This includes all qualitative and quantitative observations that you made. Be as precise as appropriate when describing quantitative observations, include any unexpected observations, and present your information in a form that is easily understood. If you have only a few observations, this could be a list; for controlled experiments and for many observations, a table will be more appropriate.

Analysis

Interpret your observations and present the evidence in the form of tables, graphs, or illustrations, each with a title. Include any calculations, the results of which can be shown in a table. Make statements about any patterns or trends you observed. Conclude the analysis with a statement based only on the evidence you have gathered, answering the question that initiated the investigation.

Evaluation

The evaluation is your judgment about the quality of evidence obtained and about the validity of the prediction and hypothesis (if present). This section can be divided into two parts:

- Did your observations provide reliable and valid evidence to enable you to answer the question? Are you confident enough in the evidence to use it to evaluate any prediction and/or hypothesis you made?
- Was the prediction you made before the investigation supported or falsified by the evidence? Based on your evaluation of the evidence or prediction, is the hypothesis supported or should it be rejected?

Investigation 2.5—Movement of Water Into and Out of Cells
April 15, 2015

By Barry La Drew and Eileen Jong

Purpose

The purpose of this investigation is to examine the conditions under which water moves into and out of cells, using dialysis tubing as a model of a cell membrane.

Question

Under what conditions does water move in and out of cells?

Hypothesis/Prediction

If a model cell is placed in a hypotonic solution (having lower concentrations of solute than the cell), water will move into the cell. If a cell is placed in a hypertonic solution (having higher concentrations of solute than the cell), water will move out of the cell.

Design

Three model cells are constructed from dialysis tubing. Two are filled with distilled water and one with starch suspension. One of the cells with distilled water is placed in a beaker of starch suspension. The other two cells are placed in beakers of distilled water. Iodine is added to the distilled water in the beakers. (Iodine is used as an indicator to detect the presence of starch.) The mass of each cell is recorded and other observations are made at the beginning and after 10 and 20 min.

Materials

lab apron	latex gloves
medicine dropper	dialysis tubing
funnel	paper towels
scissors	distilled water
100 mL graduated cylinder	4% starch suspension
three 250-mL beakers	iodine
triple-beam balance	

Procedure

1. A lab apron and latex gloves were put on.

2. Three strips of dialysis tubing (each about 25 cm long) were cut. The strips were then soaked in a beaker of tap water for approximately 2 min.

3. To find the opening, one end of the dialysis tube was rubbed between our fingers. A knot was tied near the other end of the tube. These steps were repeated for the other tubes.

4. Using a graduated cylinder, 15 mL of 4% starch suspension was measured. The suspension was poured through a funnel into the open end of one of the dialysis tubes.

5. The second and third dialysis tubes were filled with 15 mL of distilled water. In one tube, 20 drops of iodine were added.

Figure 1
Sample Lab Report

6. A knot was tied in the open end of each tube to close it.

7. The outside of all the tubes was rinsed with distilled water to remove any fluids that may have leaked out during the tying process.

8. Excess water was gently blotted from the dialysis tubes and the mass of each tube was measured.

9. An observation table similar to **Table 1** was constructed and the measurements recorded.

10. The dialysis tube with the starch suspension was placed into a beaker containing 100 mL of distilled water. The tube with only distilled water was placed in a second beaker containing 100 mL of distilled water. The tube with distilled water and iodine was placed in a beaker containing 100 mL of starch suspension.

11. Twenty drops of iodine were added to each of the beakers containing distilled water. All three tubes were observed closely for any colour change at the beginning.

12. After 10 min, the dialysis tubes were removed from the beakers. Any excess liquid from the tubes was gently blotted off and the mass of each tube measured and recorded.

13. The dialysis tubes were then returned to the appropriate beakers.

14. After another 10 min, the tubes were removed again, blotted dry, and the mass of each measured and recorded.

15. The contents of the tubes were poured down the sink and the sink rinsed. The tubes were placed in the regular garbage disposal.

Observations

Table 1: Observation Table

Model cell	Initial mass (g)	Mass after 10 min (g)	Mass after 20 min (g)	Other observations
dialysis tube with distilled water in beaker of distilled water	17.5	17.5	17.5	no observable change in the solution inside the tube
dialysis tube with starch suspension in beaker of distilled water	24.7	25.3	28.4	suspension inside the dialysis tube turned dark
dialysis tube with distilled water in beaker of starch suspension	18.5	17.8	15.7	starch suspension in the beaker turned dark

Analysis

- The tube with the distilled water in the beaker containing distilled water showed no change in mass.
- The tube with the starch suspension increased in mass during each 10-minute period. The solution in this tube turned dark.
- The tube with the distilled water in the beaker with the starch suspension decreased in mass during each 10-minute period. The starch suspension in the beaker turned dark.

The mass of the cell with the starch suspension inside increased because water moved into the cell. The suspension in this tube turned dark, which suggests that iodine must have also moved into the tube along with the water molecules.

The mass of the cell with the starch suspension outside decreased because water moved out of the cell. The suspension outside the cell turned dark, which suggests that iodine also must have moved outside the tube along with the water molecules.

The tube with the distilled water acted as a control. It had the same solution inside as outside. There was no observable change in it.

From our observations, we can conclude that water will move out of a cell in a hypertonic solution and into a cell in a hypotonic solution.

Evaluation

The design is adequate because the observations allowed us to reach a conclusion about the conditions under which water moves in and out of cells. The procedure is easy to follow and does not require any special skills.

Water or starch suspension remaining on the tubes after they were taken out of the beakers may have affected the measurement of the mass of the tubes. However, this should not make a difference since this error would have affected the mass of all tubes. Errors in measurement would probably not have accounted for the increase in the mass of the tube with the starch or the decrease in the mass of the tube with distilled water. So, while there may have been errors in measurement, it should not have affected the quality of the evidence.

The evidence supports the prediction. To verify our conclusion, we could repeat the investigation or do a similar investigation with different solutions.

Since we were working with model cells, we cannot be sure that this would happen with real cells.

Graphic Organizers

When you are describing situations, issues, or events, it is sometimes helpful to record your ideas so that you can see them and compare them with those of other people. A graphic organizer such as a table can be used to record your observations. These visual representations of ideas indicate your understanding of a topic and can take many forms.

PMI Chart

A PMI chart is used to examine both sides of an issue. Positive aspects of a topic or issue are recorded in the P (plus) column. Negative aspects are recorded in the M (minus) column. Interesting or controversial questions are recorded in the I (interesting) column (**Table 1**).

Table 1: A PMI Chart

P	M	I

Table 2: A KWL Chart

K	W	L

KWL Chart

A KWL chart can help you identify prior knowledge and experience, decide what new information you want to learn about, and reflect on your learning. Before you begin a new concept, lesson, or unit, list what you know about a topic in the K column and what you want to know in the W column. After studying the new topic, list what you learned in the L column (**Table 2**).

Venn Diagram

A Venn diagram is used to show similarities and differences in two or more concepts. Write all similarities between the concepts in the overlapping section of the circles and all unique traits of each concept in the nonoverlapping parts of the appropriate circles (**Figure 1**).

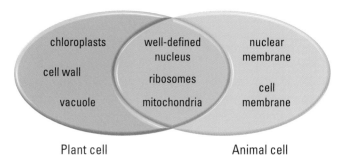

Figure 1
Venn diagram: plant and animal cells

Fishbone Diagram

A fishbone diagram is used to identify separate causes and effects. In the head of the fish, identify the effect, topic, or

result. At the end of each major bone, identify the major subtopics or categories. On the minor bones that attach to each major bone, add details about the subtopics or possible causes of each effect or result (**Figure 2**).

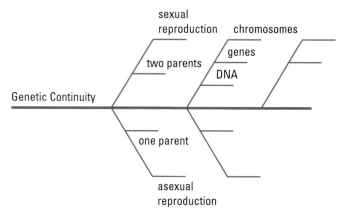

Figure 2
Fishbone diagram: genetic continuity

The Concept Map

Concept maps are used to show connections between ideas and concepts, using words or visuals. Put the central idea in the middle of a sheet of paper. Organize the ideas most closely related to each other around the centre. Draw arrows between the ideas that are related. On each arrow, write a short description of how the terms are related to each other. Expand and add ideas or relationships as you think of them (**Figure 3**).

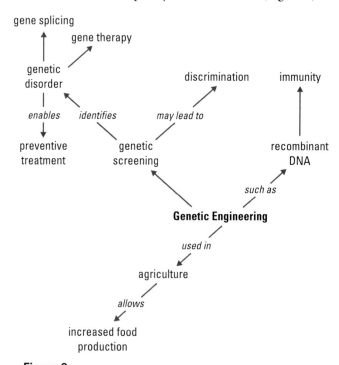

Figure 3
Concept map: genetic engineering

Math Skills

Scientific Notation

It is difficult to work with very large or very small numbers when they are written in common decimal notation. Usually it is possible to accommodate such numbers by changing the SI prefix so that the number falls between 0.1 and 1000; for example, 237 000 000 mm can be expressed as 237 km and 0.000 000 895 kg can be expressed as 0.895 mg. However, this prefix change is not always possible, either because an appropriate prefix does not exist or because it is essential to use a particular unit of measurement. In these cases, the best method of dealing with very large and very small numbers is to write them using scientific notation. Scientific notation expresses a number by writing it in the form $a \times 10^n$, where $1 < |a| < 10$ and the digits in the coefficient a are all significant. **Table 1** shows situations where scientific notation would be used.

Table 1: Examples of Scientific Notation

Expression	Common decimal notation	Scientific notation
124.5 million kilometres	124 500 000 km	1.245×10^8 km
154 thousand picometres	154 000 pm	1.54×10^{-5} pm
602 sextillion /mol	602 000 000 000 000 000 000 000 /mol	6.02×10^{23}/mol

To multiply numbers in scientific notation, multiply the coefficients and add the exponents; the answer is expressed in scientific notation. Note that when writing a number in scientific notation, the coefficient should be between 1 and 10 and should be rounded to the same certainty (number of significant digits) as the measurement with the least certainty (fewest number of significant digits). Look at the following examples:

$$(4.73 \times 10^5 \text{ m})(5.82 \times 10^7 \text{ m}) = 27.5 \times 10^{12} \text{ m}^2 = 2.75 \times 10^{13} \text{ m}^2$$

$$(3.9 \times 10^4 \text{ N})(5.3 \times 10^{-3} \text{ m}) = 20.7 \times 10^7 \text{ N·m} = 2.07 \times 10^8 \text{ N·m}$$

On many calculators, scientific notation is entered using a special key, labelled EXP or EE. This key includes "× 10" from the scientific notation; you need to enter only the exponent. For example, to enter

7.5×10^4	press	7.5 EXP 4
3.6×10^{-3}	press	3.6 EXP +/−3

Uncertainty in Measurements

There are two types of quantities that are used in science: exact values and measurements. Exact values include defined quantities (1 m = 100 cm) and counted values (5 cars in a parking lot). Measurements, however, are not exact because there is some uncertainty or error associated with every measurement.

There are two types of measurement error. **Random error** results when an estimate is made to obtain the last significant figure for any measurement. The size of the random error is determined by the precision of the measuring instrument. For example, when measuring length, it is necessary to estimate between the marks on the measuring tape. If these marks are 1 cm apart, the random error will be greater and the precision will be less than if the marks are 1 mm apart.

Systematic error is associated with an inherent problem with the measuring system, such as the presence of an interfering substance, incorrect calibration, or room conditions. For example, if the balance is not zeroed at the beginning, all measurements will have a systematic error; if using a metre stick that has been worn slightly, all measurements will contain an error.

The precision of measurements depends upon the gradations of the measuring device. **Precision** is the place value of the last measurable digit. For example, a measurement of 12.74 cm is more precise than a measurement of 127.4 cm because the first value was measured to hundredths of a centimetre whereas the latter was measured to tenths of a centimetre.

When adding or subtracting measurements of different precision, the answer is rounded to the same precision as the least precise measurement. For example, using a calculator, add

$$11.7 \text{ cm} + 3.29 \text{ cm} + 0.542 \text{ cm} = 15.532 \text{ cm}$$

The answer must be rounded to 15.5 cm because the first measurement limits the precision to a tenth of a centimetre.

No matter how precise a measurement is, it still may not be accurate. Accuracy refers to how close a value is to its true value. The comparison of the two values can be expressed as a percentage difference. The percentage difference is calculated as:

$$\% \text{ difference} = \frac{|\text{experimental value} - \text{predicted value}|}{\text{predicted value}} \times 100$$

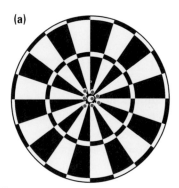
(a)

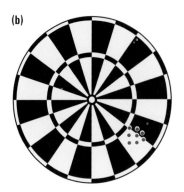
(b)

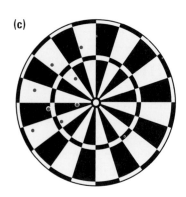
(c)

Figure 1
The positions of the darts in each of these figures are analogous to measured or calculated results in a laboratory setting.
The results in **(a)** are precise and accurate, in **(b)** they are precise but not accurate, and in **(c)** they are neither precise nor accurate.

Figure 1 shows an analogy between precision and accuracy, and the positions of darts thrown at a dartboard.

How certain you are about a measurement depends on two factors: the precision of the instrument used and the size of the measured quantity. More precise instruments give more certain values. For example, a mass measurement of 13 g is less precise than a measurement of 12.76 g; you are more certain about the second measurement than the first. Certainty also depends on the measurement. For example, consider the measurements 0.4 cm and 15.9 cm; both have the same precision. However, if the measuring instrument is precise to ± 0.1 cm, the first measurement is 0.4 ± 0.1 cm (0.3 cm or 0.5 cm) or an error of 25%, whereas the second measurement could be 15.9 ± 0.1 cm (15.8 cm or 16.0 cm) for an error of 0.6%. For both factors—the precision of the instrument used and the value of the measured quantity— the more digits there are in a measurement, the more certain you are about the measurement.

Significant Digits

The certainty of any measurement is communicated by the number of significant digits in the measurement. In a measured or calculated value, significant digits are the digits that are certain plus one estimated (uncertain) digit. Significant digits include all digits correctly reported from a measurement.

Follow these rules to decide if a digit is significant:

1. If a decimal point is present, zeros to the left of the first non-zero digit (leading zeros) are not significant.

2. If a decimal point is not present, zeros to the right of the last non-zero digit (trailing zeros) are not significant.

3. All other digits are significant.

4. When a measurement is written in scientific notation, all digits in the coefficient are significant.

5. All digits in counted and defined values are significant.

Table 2 shows some examples of significant digits.

An answer obtained by multiplying and/or dividing measurements is rounded to the same number of significant digits

Table 2: Certainty in Significant Digits

Measurement	Number of significant digits
32.07 m	4
0.0041 g	2
5×10^5 kg	1
6400 s	2
100 people (counted)	infinite
78.0 cm	3
204.0 cm	4
10.0 kJ	3

as the measurement with the fewest number of significant digits. For example, we could use a calculator to solve the following equation:

$$77.8 \text{ km/h} \times 0.8967 \text{ h} = 69.76326 \text{ km}$$

However, the certainty of the answer is limited to three significant digits, so the answer is rounded up to 69.8 km.

Rounding Off

The following rules should be used when rounding answers to calculations.

1. When the first digit discarded is less than five, the last digit retained should not be changed.
   3.141 326 rounded to 4 digits is 3.141

2. When the first digit discarded is greater than five, or if it is a five followed by at least one digit other than zero, the last digit retained is increased by 1 unit.
   2.221 672 rounded to 4 digits is 2.222
   4.168 501 rounded to 4 digits is 4.169

3. When the first digit discarded is five followed by only zeros, the last digit retained is increased by 1 if it is odd, but not changed if it is even.
   2.35 rounded to 2 digits is 2.4
   2.45 rounded to 2 digits is 2.4
   −6.35 rounded to 2 digits is −6.4

Measuring and Estimating

Many people believe that all measurements are reliable (consistent over many trials), precise (to as many decimal places as possible), and accurate (representing the actual value). But there are many things that can go wrong when measuring.

- There may be limitations that make the instrument or its use unreliable (inconsistent).
- The investigator may make a mistake or fail to follow the correct techniques when reading the measurement to the available precision (number of decimal places).
- The instrument may be faulty or inaccurate; a similar instrument may give different readings.

For example, when measuring the temperature of a liquid, it is important to keep the thermometer at the proper depth and the bulb of the thermometer away from the bottom and sides of the container. If you sit a thermometer with its bulb at the bottom of a liquid-filled container, you will be measuring the temperature of the bottom of the container and not the temperature of the liquid. There are similar concerns with other measurements.

To be sure that you have measured correctly, you should repeat your measurements at least three times. If your measurements appear to be reliable, calculate the mean and use that value. To be more certain about the accuracy, repeat the measurements with a different instrument.

Every measurement is a best estimate of the actual value. The measuring instrument and the skill of the investigator determine the certainty and the precision of the measurement. The usual rule is to make a measurement that estimates between the smallest divisions on the scale of the instrument.

Probability

In scientific investigations, probability is a measure of the likelihood of a specific event occurring and is usually expressed as a number between 0 and 1. A probability of 0 means the event will not occur; a probability of 1 means the event will definitely occur. Probabilities may also be expressed as fractions or as percents.

There are two types of probability: theoretical probability and experimental probability. Theoretical probability is the likelihood of an event occurring based on the information known about certain conditions. This is an expectation.

$$\text{theoretical probability} = \frac{\text{number of desired outcomes}}{\text{total number of possible outcomes}}$$

Example

Black fur colour is a dominant trait in guinea pigs, while white fur colour is a recessive trait. What is the theoretical probability of a pair of heterozygous black guinea pigs (Bb) producing offspring with white fur (bb)?

Using a Punnett square (**Figure 2**), we show that if four offspring were produced, it is expected that three would have black fur and one would have white fur.

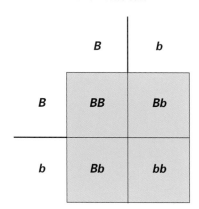

Figure 2
Punnett square showing a Bb x Bb cross

$$\text{theoretical probability} = \frac{\text{number of desired outcomes}}{\text{total number of possible outcomes}}$$
$$= \frac{\text{number of offspring with white fur}}{\text{total number of possible outcomes}}$$
$$= \frac{1}{4}$$
$$= 0.25$$
$$= 25\%$$

The theoretical probability of producing white offspring is $\frac{1}{4}$ or 25%. So if a litter had eight offspring, then you could expect two to have white fur.

Experimental probability is based on the recorded outcomes or events of an investigation. The more often an experiment is repeated or the more observations made, the closer the experimental probability will be to the theoretical probability.

$$\text{experimental probability} = \frac{\text{number of desired outcomes observed}}{\text{total number of observations}}$$

Example

Black fur colour is a dominant trait in guinea pigs, while white fur colour is a recessive trait. Two heterozygous black guinea pigs (Bb) were crossed. The litter contained six offspring with black fur and one with white fur. What is the experimental probability of producing offspring with white fur?

$$\text{experimental probability} = \frac{\text{number of desired outcomes observed}}{\text{total number of observations}}$$
$$= \frac{\text{number of offspring with white fur}}{\text{total number of possible offspring}}$$
$$= \frac{1}{7}$$
$$\doteq 0.14$$
$$= 14\%$$

The experimental probability of having offspring with white fur is 14%. If you performed the same analysis on a large number of litters, you would expect the experimental probability to be the same as (or very close to) the theoretical probability.

Graphs

There are many types of graphs that you can use to organize your data. You need to identify which type of graph is best for your data before you begin graphing. Three of the most useful kinds are bar graphs, circle (pie) graphs, and point-and-line graphs.

Bar Graphs

When at least one of the variables is qualitative, use a bar graph to organize your data (**Figure 3**). For example, a bar graph would be a good way to present the data collected from a study of the number of plants (quantitative) and the type of plants found (qualitative) in a local nursery. In this graph, each bar stands for a different category, in this case a type of plant.

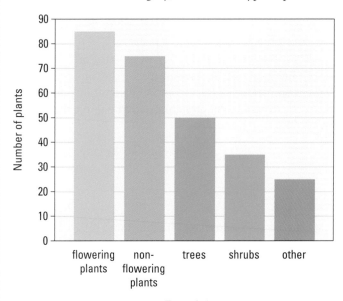

Figure 3
Bar graph

Circle Graphs

Circle graphs and bar graphs are used for similar types of data. A circle graph is used if the quantitative variable can be changed to a percentage of a total quantity (**Figure 4**). For example, if you surveyed a local nursery to determine the types of plants found and the number of each, you could make a circle graph. Each piece in the graph stands for a different category (e.g., the type of plant). The size of each piece is determined by the percentage of the total that belongs in each category (e.g., the percentage of plants of a particular type).

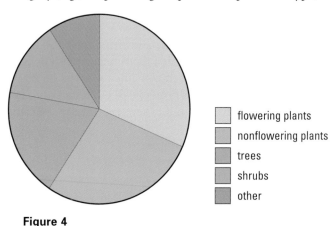

Figure 4
Circle graph

Point-and-Line Graphs

When both variables are quantitative, use a point-and-line graph. For example, we can use the following guidelines and the data in **Table 3** to construct the point-and-line graph shown in **Figure 5**.

Table 3: Number of Brine Shrimp Eggs Hatched in Salt Solutions of Various Concentrations

Day	2% salt	4% salt	6% salt	8% salt
1	0	0	0	1
2	0	11	2	3
3	0	14	8	5
4	2	20	17	8
5	5	37	25	15
6	6	51	37	31

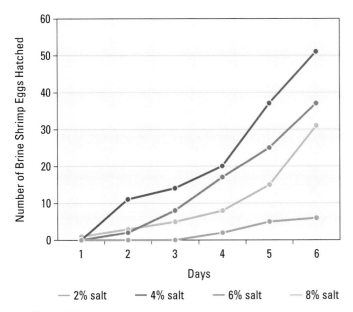

Figure 5
Point-and-line graph

1. Use graph paper and construct your graph on a grid. The horizontal edge on the bottom of this grid is the *x*-axis and the vertical edge on the left is the *y*-axis. Do not be too thrifty with graph paper—a larger graph is easier to interpret.

2. Decide which variable goes on which axis and label each axis, including the units of measurement. The independent variable is generally plotted along the *x*-axis and the dependent variable along the *y*-axis. The exception to this is when you plot a variable against time: regardless of which is the independent or dependent variable, always plot time on the *x*-axis. This convention ensures that the slope of the graph always represents a rate.

3. Title your graph. The title should be a concise description of the data contained in the graph.

4. Determine the range of values for each variable. The range is the difference between the largest and smallest values. Graphs often include a little extra length on each axis, to make them appear less cramped.

5. Choose a scale for each axis. This will depend on how much space you have and the range of values for each axis. Each line on the grid usually increases steadily in value by a convenient number, such as 1, 2, 5, 10, 50, or 100.

6. Plot the points. Start with the first pair of values, which may or may not be at the origin of the graph.

7. After all the points are plotted, draw a line through the points to show the relationship between the variables, if possible. Not all points may lie exactly on a line; small errors in each measurement may have occurred and moved the points away from the perfect line. Draw a line that comes closest to most of the points. This is called the line of best fit—a smooth line that passes through or between the points so that there are about the same number of points on each side of the line. The line of best fit may be straight or curved.

8. If you are plotting more than one set of data on one graph, use different colours or symbols to indicate the different sets, and include a legend (**Figure 5**).

Safety Conventions and Symbols

Although every effort is undertaken to make the science experience a safe one, there are some inherent risks. These risks are generally associated with the materials and equipment used, and with disregard of safety instructions when conducting investigations. There may also be risks associated with the location of the investigation. Most of these risks pose no more danger than what is normally experienced in everyday life. With an awareness of the possible hazards, knowledge of the rules, appropriate behaviour, and common sense, these risks can be practically eliminated.

Remember, you share the responsibility not only for your own safety, but also for the safety of those around you. Always alert the teacher in case of an accident.

In this text, chemicals, equipment, and procedures that are hazardous are highlighted in red and are preceded by the appropriate Workplace Hazardous Materials Information System (WHMIS) symbol or by ✋ .

WHMIS Symbols and HHPS

The Workplace Hazardous Materials Information System (WHMIS) provides workers and students with complete and accurate information regarding hazardous products. All chemical products supplied to schools, businesses, and industries must contain standardized labels and be accompanied by a Material Safety Data Sheet (MSDS) providing detailed information about the product. Clear and standardized labelling is an important component of WHMIS (**Table 1**). These labels must be present on the product's original container and added to other containers if the product is transferred.

The Canadian Hazardous Products Act requires manufacturers of consumer products containing chemicals to include a symbol specifying both the nature of the primary hazard and the degree of this hazard. In addition, any secondary hazards, first-aid treatment, storage, and disposal must be noted. Household Hazardous Product Symbols (HHPS) are used to show the hazard, and the degree of the hazard is indicated by the type of border surrounding the illustration (**Figure 1**).

	CORROSIVE
	This material can burn your skin and eyes. If you swallow it, it will damage your throat and stomach.
	FLAMMABLE
	This product or the gas (or vapour) from it can catch fire quickly. Keep this product away from heat, flames, and sparks.
	EXPLOSIVE
	Container will explode if it is heated or if a hole is punched in it. Metal or plastic can fly out and hurt your eyes and other parts of your body.
	POISON
	If you swallow or lick this product, you could become very sick or die. Some products with this symbol on the label can hurt you even if you breathe (or inhale) them.

Danger

Warning

Caution

Figure 1
Household Hazardous Product Symbols

Table 1: Workplace Hazardous Materials Information System (WHMIS)

Class and type of compounds	WHMIS symbol	Risks	Precautions
Class A: Compressed Gas Material that is normally gaseous and kept in a pressurized container		• could explode due to pressure • could explode if heated or dropped • possible hazard from both the force of explosion and the release of contents	• ensure container is always secured • store in designated areas • do not drop or allow to fall
Class B: Flammable and Combustible Materials Materials that will continue to burn after being exposed to a flame or other ignition source		• may ignite spontaneously • may release flammable products if allowed to degrade or when exposed to water	• store in properly designated areas • work in well-ventilated areas • avoid heating • avoid sparks and flames • ensure that electrical sources are safe
Class C: Oxidizing Materials Materials that can cause other materials to burn or support combustion		• can cause skin and eye burns • increase fire and explosion hazard • may cause combustibles to explode or react violently	• store away from combustibles • wear body, hand, face, and eye protection • store in proper container that will not rust or oxidize
Class D: Toxic Materials— Immediate and Severe Poisons and potentially fatal materials that cause immediate and severe harm		• may be fatal if ingested or inhaled • may be absorbed through the skin • small volumes have a toxic effect	• avoid breathing dust or vapours • avoid contact with skin and eyes • wear protective clothing, and face and eye protection • work in well-ventilated areas and wear breathing protection
Class D: Toxic Materials— Long-Term Concealed Materials that have a harmful effect after repeated exposures or over a long period		• may cause death or permanent injury • may cause birth defects or sterility • may cause cancer • may be sensitizers causing allergies	• wear appropriate personal protection • work in a well-ventilated area • store in appropriate designated areas • avoid direct contact • use hand, body, face, and eye protection • ensure respiratory and body protection is appropriate for the specific hazard
Class D: Biohazardous Infectious Materials Infectious agents or a biological toxin causing a serious disease or death		• may cause anaphylactic shock • includes viruses, yeasts, moulds, bacteria, and parasites that affect humans • includes fluids containing toxic products • includes cellular components	• special training is required to handle materials • work in designated biological areas with appropriate engineering controls • avoid forming aerosols • avoid breathing vapours • avoid contamination of people/area • store in special designated areas
Class E: Corrosive Materials Materials that react with metals and living tissue		• eye and skin irritation on exposure • severe burns/tissue damage on longer exposure • lung damage if inhaled • may cause blindness if contacts eyes • environmental damage from fumes	• wear body, hand, face, and eye protection • use breathing apparatus • ensure protective equipment is appropriate • work in a well-ventilated area • avoid all direct body contact • use appropriate storage containers and ensure proper nonventing closures
Class F: Dangerously Reactive Materials Materials that may have unexpected reactions		• may react with water • may be chemically unstable • may explode if exposed to shock or heat • may release toxic or flammable vapours • may vigorously polymerize • may burn unexpectedly	• handle with care, avoiding vibration, shocks, and sudden temperature changes • store in appropriate containers • ensure storage containers are sealed • store and work in designated areas

Safety in the Laboratory

General Safety Rules

Safety in the laboratory is an attitude and a habit more than it is a set of rules. It is easier to prevent accidents than to deal with the consequences of an accident. Most of the following rules are common sense:

- Do not enter a laboratory unless a teacher or other supervisor is present, or you have permission to do so.
- Familiarize yourself with your school's safety regulations.
- Make your teacher aware of any allergies and other health problems you may have.
- Wear eye protection, lab aprons or coats, and gloves when appropriate.
- Wear closed shoes (not sandals) when working in the laboratory.
- Place your books and bags away from the work area. Keep your work area clear of all materials except those that you will use in the investigation.
- Do not chew gum, eat, or drink in the laboratory. Food should not be stored in refrigerators in laboratories.
- Know the location of MSDS information, exits, and all safety equipment, such as the fire blanket, fire extinguisher, and eyewash station.
- Avoid sudden or rapid motion in the laboratory that may interfere with someone carrying or working with chemicals or using sharp instruments.
- Never engage in horseplay or practical jokes in the laboratory.
- Ask for assistance when you are not sure how to do a procedural step.
- Never attempt any unauthorized experiments.
- Never work in a crowded area or alone in the laboratory.
- Always wash your hands with soap and water before and after you leave the laboratory. Definitely wash your hands before you touch any food.
- Use stands, clamps, and holders to secure any potentially dangerous or fragile equipment that could be tipped over.
- Do not taste any substance in a laboratory.
- Never smell chemicals unless specifically instructed to do so by the teacher. Do not inhale the vapours, or gas, directly from the container. Take a deep breath to fill your lungs with air, then waft or fan the vapours toward your nose.
- Clean up all spills, even water spills, immediately.
- If you are using a microscope with a mirror, never direct the mirror to sunlight. The concentrated reflected light could hurt your eyes badly.
- Do not forget safety procedures when you leave the laboratory. Accidents can also occur outdoors, at home, and at work.

Eye and Face Safety

- Always wear approved eye protection in a laboratory, no matter how simple or safe the task appears to be. Keep the safety glasses over your eyes, not on top of your head. For certain experiments, full face protection may be necessary.
- If you must wear contact lenses in the laboratory, be extra careful; whether or not you wear contact lenses, do not touch your eyes without first washing your hands. If you do wear contact lenses, make sure that your teacher is aware of it. Carry your lens case and a pair of glasses with you.
- Do not stare directly at any bright source of light (e.g., a burning magnesium ribbon, lasers, the Sun). You will not feel any pain if your retina is being damaged by intense radiation. You cannot rely on the sensation of pain to protect you.
- Never look directly into the opening of flasks or test tubes.

Handling Glassware Safely

- Never use glassware that is cracked or chipped. Give such glassware to your teacher or dispose of it as directed. Do not put the item back into circulation.
- Never pick up broken glassware with your fingers. Use a broom and dustpan.
- Do not put broken glassware into garbage containers. Dispose of glass fragments in special containers marked "Broken Glass."
- Heat glassware only if it is approved for heating. Check with your teacher before heating any glassware.
- Be very careful when cleaning glassware. There is an increased risk of breakage from dropping when the glassware is wet and slippery.
- If you need to insert glass tubing or a thermometer into a rubber stopper, get a cork borer of a suitable size. Insert the borer in the hole of the rubber stopper, starting from the small end of the stopper. Once the borer is pushed all the way through the hole, insert the tubing or thermometer through the borer. Ease the borer out of the hole, leaving the tubing or thermometer inside. To remove the tubing or thermometer from the stopper, push the borer from the small end through the stopper until it shows at the other end. Ease the tubing or thermometer out of the borer.
- Protect your hands with heavy gloves or several layers of cloth before inserting glass into rubber stoppers.

Using Sharp Instruments Safely

- Make sure your instruments are sharp. Surprisingly, one of the main causes of accidents with cutting instruments is using a dull instrument. Dull cutting instruments require more pressure than sharp instruments and are, therefore, much more likely to slip.
- Always transport a scalpel in a dissection case or box. Never carry the scalpel from one area of the laboratory to another with an exposed blade.
- Select the appropriate instrument for the task. Never use a knife when scissors would work best.
- Always cut away from yourself and others.

Fire Safety

- Immediately inform your teacher of any fires. Very small fires in a container may be extinguished by covering the container with a wet paper towel or a ceramic square to cut off the supply of air. Alternatively, sand may be used to smother small fires. A bucket of sand with a scoop should be available in the laboratory.
- If anyone's clothes or hair catch fire, tell the person to drop to the floor and roll. Then use a fire blanket to help smother the flames. Never wrap the blanket around a person on fire; the chimney effect will burn the lungs. For larger fires, immediately evacuate the area. Call the office or sound the fire alarm if close by. Do not try to extinguish larger fires. Your prime concern is to save lives. As you leave the classroom, make sure that the windows and doors are closed.
- If you use a fire extinguisher, direct the extinguisher at the base of the fire and use a sweeping motion, moving the extinguisher nozzle back and forth across the front of the fire's base. Different extinguishers are effective for different classes of fires. The fire classes are outlined below. Fire extinguishers in the laboratory are 2A10BC. They extinguish classes A, B, and C fires.
- Class A fires involve ordinary combustible materials that leave coals or ashes, such as wood, paper, or cloth. Use water or dry chemical extinguishers on class A fires.
- Class B fires involve flammable liquids such as gasoline or solvents. Carbon dioxide or dry chemical extinguishers are effective on class B fires.
- Class C fires involve live electrical equipment, such as appliances, photocopiers, computers, or laboratory electrical apparatus. Carbon dioxide or dry chemical extinguishers are recommended for class C fires. Do not use water on live electrical devices as this can result in severe electrical shock.
- Class D fires involve burning metals, such as sodium, potassium, magnesium, or aluminum. Sand, salt, or graphite can be used to put out class D fires. Do not use water on a metal fire as this can cause a violent reaction.
- Class E fires involve a radioactive substance. These require special consideration at each site.

Heat Safety

- Keep a clear workplace when performing experiments with heat.
- Make sure that heating equipment, such as the burner, hot plate, or electric heater, is secure on the bench and clamped in place when necessary.
- Do not use a laboratory burner near wooden shelves, flammable liquids, or any other item that is combustible.
- Take care that the heat developed by the heat source does not cause any material close by to get hot enough to burst into flame. Do not allow overheating if you are performing an experiment in a closed area. For example, if you are using a light source in a large cardboard box, be sure you have enough holes at the top of the box and on the sides to dissipate heat.
- Before using a laboratory burner, make sure that long hair is always tied back. Do not wear loose clothing (wide long sleeves should be tied back or rolled up).
- Always assume that hot plates and electric heaters are hot and use protective gloves when handling.
- Do not touch a light source that has been on for some time. It may be hot and cause burns.
- In a laboratory where burners or hot plates are being used, never pick up a glass object without first checking the temperature by lightly and quickly touching the item, or by placing your hand near but not touching it. Glass items that have been heated stay hot for a long time, even if they do not appear to be hot. Metal items such as ring stands and hot plates can also cause burns; take care when touching them.
- Never look down the barrel of a laboratory burner.
- Always pick up a burner by the base, never by the barrel.
- Never leave a lighted burner unattended.
- Any metal powder can be explosive. Do not put these in a flame.
- When heating a test tube over a laboratory burner, use a test-tube holder and a spurt cap. Holding the test tube at an angle, with the open end pointed away from you and others, gently move the test tube back and forth through the flame.

- To heat a beaker, put it on the hot plate and secure with a ring support attached to a utility stand. (A wire gauze under the beaker is optional.)
- Remember to include a cooling time in your experiment plan; do not put away hot equipment.

To use a burner:

- Tie back long hair and tie back or roll up wide long sleeves.
- Secure the burner to a stand using a metal clamp.
- Check that the rubber hose is properly connected to the gas valve.
- Close the air vents on the burner. Use a sparker to light the burner.
- Open the air vents just enough to get a blue flame.
- Control the size of the flame using the gas valve.

Electrical Safety

- Water or wet hands should never be used near electrical equipment such as a hotplate, a light source, or a microscope.
- Do not use the equipment if the cord is frayed or if the third pin on the plug is missing. If the teacher allows this, then make sure the equipment has a double-insulated cord.
- Do not operate electrical equipment near running water or a large container of water.
- Check the condition of electrical equipment. Do not use if wires or plugs are damaged.
- If using a light source, check that the wires of the light fixture are not frayed, and that the bulb socket is in good shape and well secured to a stand.
- Make sure that electrical cords are not placed where someone could trip over them.
- When unplugging equipment, remove the plug gently from the socket. Do not pull on the cord.

Handling Chemicals Safely

Many chemicals are hazardous to some degree. When using chemicals, operate under the following principles:

- Never underestimate the risks associated with chemicals. Assume that any unknown chemicals are hazardous.
- Use a less hazardous chemical wherever possible.
- Reduce exposure to chemicals as much as possible. Avoid direct skin contact if possible.
- Ensure that there is adequate ventilation when using chemicals.

The following guidelines do not address every possible situation but, used with common sense, are appropriate for situations in the high school laboratory.

- Obtain an MSDS for each chemical and consult the MSDS before you use the chemical.
- Know the emergency procedures for the building, the department, and the chemicals being used.
- Wear a lab coat and/or other protective clothing (e.g., apron, gloves), as well as appropriate eye protection at all times in areas where chemicals are used or stored.
- Never use the contents from a bottle that has no label or has an illegible label. Give any containers with illegible labels to your teacher. When leaving chemicals in containers, ensure that the containers are labelled. Always double-check the label, once, when you pick it up, and a second time when you are about to use it.
- Carry chemicals carefully using two hands, one around the container and one underneath.
- Always pour from the side opposite the label on a reagent bottle; your hands and the label are protected as previous drips are always on the side of the bottle opposite of the label.
- Do not let the chemicals touch your skin. Use a laboratory scoop or spatula for handling solids.
- Pour chemicals carefully (down the side of the receiving container or down a stirring rod) to ensure that they do not splash.
- Always pour volatile chemicals in a fume hood or in a well-ventilated area.
- Never pipet or start a siphon by mouth. Always use a pipet suction device (such as a bulb or a pump).
- If you spill a chemical, use a chemical spill kit to clean up.
- Return chemicals to their proper storage place according to your teacher's instructions.
- Do not return surplus chemicals to stock bottles. Dispose of excess chemicals in an appropriate manner as instructed by your teacher.
- Clean up your work area, the fume hood, and any other area where chemicals were used.
- Wash hands immediately after handling chemicals and before and after leaving the lab, even if you wore gloves. Definitely wash your hands before you touch any food.

Handling Animals, Plants, and Other Organisms Safely

- Do not perform any investigation on any animal that might cause suffering or pain, or that might pose a health hazard to you or anyone else in the school.

- Animals that live in the classroom should be treated with care and respect, and be kept in a clean, healthy environment.
- Ensure that your teacher is aware of any plant or animal allergies that you may have.
- Never bring a plant, animal, or other organism to school without receiving prior permission from the teacher.
- Keep cages and tanks clean—both for your health and the health of the organism. Most jurisdictions recommend no live mammals or birds in the laboratory. Reptiles often carry Salmonella.
- Wear gloves and wash your hands before and after feeding or handling an animal, touching materials from the animal's cage or tank, or handling bacterial cultures.
- Do not grow any microorganisms other than those that occur naturally on mouldy bread, cheese, and mildewed objects. Anaerobic bacteria should not be grown.
- Cultures should be grown at room temperature or in the range of 25°C to 32°C. Incubation at 37°C may encourage the growth of microorganisms that are capable of living in the human body.
- Bacteria from soils should not be grown because of the possibility of culturing tetanus-causing organisms.
- Spores collected from household locations, such as telephones or bathrooms, should not be cultured in the laboratory. The body can destroy small numbers of these bacteria, but may not be able to cope with large numbers.
- All surfaces and equipment used in culturing microorganisms should be washed down with a disinfectant (e.g., a solution of bleach).
- Apparatus used in microbiology should be autoclaved because liquid disinfectants and germicidal agents generally cannot guarantee complete sterilization. The oven of an ordinary kitchen stove may be used.
- Wild or sick animals should never be brought into the lab. Dead animals, wild or tame, that have died from unknown causes should also not be brought into the lab.
- Preserved specimens should be removed from the preservative with gloves or tongs, and rinsed thoroughly in running water.
- Before going on field trips, become familiar with any dangerous plants and animals that may be common in the area (e.g., stinging nettles and poisonous plants).

Waste Disposal

Waste disposal at school, at home, and at work is a societal issue. To protect the environment, federal and provincial governments have regulations to control wastes, especially chemical wastes. For example, the WHMIS program applies to controlled products that are being handled. Most laboratory waste can be washed down the drain or, if it is in solid form, placed in ordinary garbage containers. However, some waste must be treated more carefully. It is your responsibility to follow procedures and to dispose of waste in the safest possible manner according to the teacher's instructions.

Flammable Substances

Flammable liquids should not be washed down the drain. Special fire-resistant containers are used to store flammable liquid waste. Waste solids that pose a fire hazard should be stored in fireproof containers. Care must be taken not to allow flammable waste to come into contact with any sparks, flames, other ignition sources, or oxidizing materials. The method of disposal depends on the nature of the substance.

Corrosive Solutions

Solutions that are corrosive but not toxic, such as acids, bases, and oxidizing agents, should be disposed of in a container provided by the teacher, preferably kept on the teacher's desk. Do not pour corrosive solutions down the drain.

Heavy Metal Solutions

Heavy metal compounds (e.g., lead, mercury, and cadmium compounds) should not be poured down the drain. These substances are cumulative poisons and should be kept out of the environment. A special container should be kept in the laboratory for heavy metal solutions. Pour any heavy metal waste into this container. Remember that paper towels used to wipe up solutions of heavy metals, as well as filter papers with heavy metal compounds embedded in them, should be treated as solid toxic waste.

Toxic Substances

Solutions of toxic substances, such as oxalic acid, should not be poured down the drain, but should be disposed of in the same manner as heavy metal solutions, but in a separate container.

Organic Material

Remains of plants and animals can generally be disposed of in school garbage containers. Before disposal, organic material should be rinsed thoroughly to rid it of any excess preservative. Fungi and bacterial cultures should be autoclaved or treated with a fungicide or antibacterial soap before disposal.

First Aid

The following guidelines apply in case of an injury, such as a burn, cut, chemical spill, ingestion, inhalation, or splash in the eyes.
- Always inform your teacher immediately of any injury.
- Know the location of the first-aid kit, fire blanket, eyewash station, and shower, and be familiar with the contents and operation of them.

- If the injury is a minor cut or abrasion, wash the area thoroughly. Using a compress, apply pressure to the cut to stop the bleeding. When bleeding has stopped, replace the compress with a sterile bandage. If the cut is serious, apply pressure and seek medical attention immediately.
- If the injury is the result of chemicals, drench the affected area with a continuous flow of water for 15 min. Clothing should be removed as necessary. Retrieve the Material Safety Data Sheet (MSDS) for the chemical; this sheet provides information about the first-aid requirements for the chemical.
- If you get a solution in your eye, quickly use the eyewash or nearest running water. Continue to rinse the eye with water for at least 15 min. This is a very long time—have someone time you. Unless you have a plumbed eyewash system, you will also need assistance in refilling the eyewash container. Have another student inform your teacher of the accident. The injured eye should be examined by a doctor.

- If you have ingested or inhaled a hazardous substance, inform your teacher immediately. The MSDS provides information about the first-aid requirements for the substance. Contact the Poison Control Centre in your area.
- If the injury is from a burn, immediately immerse the affected area in cold water or run cold water gently over the burned area. This will reduce the temperature and prevent further tissue damage.
- In case of electric shock, unplug the appliance and do not touch it or the victim. Inform your teacher immediately.
- If a classmate's injury has rendered him/her unconscious, notify the teacher immediately. The teacher will perform CPR if necessary. Do not administer CPR unless under specific instructions from the teacher. You can assist by keeping the person warm and by reassuring him/her once conscious.

Care and Use of the Microscope

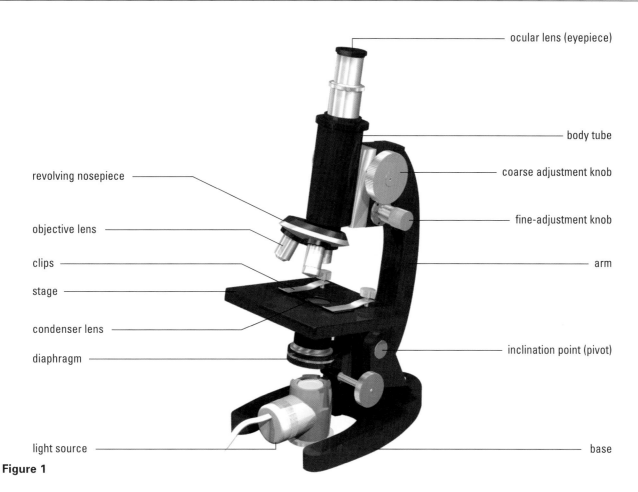

ocular lens (eyepiece)

body tube

coarse adjustment knob

fine-adjustment knob

arm

inclination point (pivot)

base

revolving nosepiece

objective lens

clips

stage

condenser lens

diaphragm

light source

Figure 1

The microscope (**Figure 1**) is a useful tool in making observations and collecting data during scientific investigations in biology.

The eyepiece lens usually magnifies 10 times. This information is printed on the side of the eyepiece. The microscopes that you will likely use will have three objective lenses: scanning (magnifies 4 times), low (magnifies 10 times), and high (magnifies 40 times). To determine the total magnification, multiply the magnification of the eyepiece lens by the magnification of the objective lens. For example, the magnification obtained using the low-power objective lens is 10 times multiplied by 10 times, or 100 times. In other words, a specimen viewed with the low power will appear 100 times larger than it actually is.

To view very small specimens, an oil immersion lens can be attached to the revolving nosepiece. This lens magnifies 100 times. A drop of immersion oil is placed between the slide and this objective lens.

Use of the Compound Microscope

1. Obtain a microscope from the storage area. Grasp the arm with one hand and use the other to support the base of the microscope.

2. If the microscope has a built-in light supply, plug it in. Place the cord so that it will not be hooked accidentally.

3. Rotate the revolving nosepiece until the shortest (scanning) objective lens clicks into place.

4. Place a prepared slide of the letter *f* on the stage and centre it. Hold the slide in place with the stage clips.

5. Turn the coarse-adjustment knob away from you to lower the lens down as far as possible. Watch from the side to ensure that the lens does not contact the slide.

6. Keeping both eyes open, look through the eyepiece and turn the coarse-adjustment knob toward you until the object comes into view. Use the fine-adjustment knob to focus the image.

7. Adjust the diaphragm to control the amount of light and fix the contrast. If the microscope has a mirror in the base, adjust it to receive the appropriate amount of light.

8. Compare the orientation of the letter *f* as viewed on the slide with the orientation as viewed through the eyepiece.

9. Compare the movement of the object on the slide with the movement of the object as viewed through the eyepiece.

10. With the image in focus and centred in the field of view, rotate the nosepiece until the next longer objective lens clicks into place.

11. Use the fine-adjustment knob to refocus the image if necessary.

12. Readjust the diaphragm or the mirror to regulate the amount of light. The higher the lens power, the more light is necessary.

13. Repeat the previous three steps with the high-power lens.

14. View other prepared slides to practise locating objects within the field of view and focusing the microscope.

15. When you have finished viewing a specimen, always rotate the nosepiece so that the shortest (scanning) lens is centred before making any adjustments with the focusing knobs.

16. Remove and clean the slide and cover slip and return them to their appropriate location.

17. Return the microscope to the storage area.

Investigating Depth of Field

The depth of field is the amount of an image that is in sharp focus when it is viewed under a microscope.

1. Cut two pieces of thread of different colours.

2. Make a temporary dry mount by placing one thread over the other in the form of an X in the centre of a microscope slide. Cover the threads with a cover slip.

3. Place the slide on the microscope stage and turn on the light.

4. Position the scanning objective lens close to, but not touching, the slide.

5. View the crossed threads through the ocular lens (eyepiece). Slowly rotate the coarse-adjustment knob until the threads come into focus.

6. Rotate the nosepiece to the low-power objective lens. Focus on the upper thread by using the fine adjustment knob. You will probably notice that you cannot focus on the lower thread at the same time. The depth of the object that is in focus at any one time represents the depth of field.

7. Repeat step 6 for the high-power objective lens. The stronger the magnification, the shallower the depth of field.

Determining Field of View

It is often necessary to measure the size of objects viewed through the microscope. The field of view is the circle of light seen while looking through the eyepiece. Once the size of the field of view is determined, it is possible to estimate the size of a specimen by comparing it to the size of the field of view.

1. With the scanning lens in place, put a transparent ruler on the stage. Position the millimetre marks on the ruler immediately below the objective lens.

2. Using the coarse-adjustment knob, focus on the marks on the ruler.

3. Move the ruler so that one of the millimetre markings is just at the edge of the field of view. Note the diameter of the field of view in millimetres under the scanning objective lens.

4. Using the same procedure, measure the field of view for the low-power objective lens.

5. Most high-power lenses provide a field of view that is less than one millimetre in diameter, so it cannot be measured with a ruler. The following steps can be followed to calculate the field of view of the high-power lens.
   - Calculate the ratio of the magnification of the high-power lens to that of the low-power lens.

$$\text{ratio} = \frac{\text{magnification of high-power lens}}{\text{magnification of low-power lens}}$$

   - Use the ratio to determine the diameter of the field of view under high-power magnification.

$$\text{field of view diameter} \atop \text{(high-power)} = \frac{\text{field of view diameter (low-power)}}{\text{ratio}}$$

6. Using a table like **Table 1**, determine the magnification of each of the lenses on your microscope and record the diameter of the field of view.

Table 1

Lens	Magnification	Eyepiece magnification	Total magnification	Diameter (mm)
scanning		10X		
low		10X		
high		10X		

Estimating Size

- Determine the size of the field of view you are viewing.
- Estimate the length and width of the specimen by comparing it to the diameter of the field of view.
- **Figure 2** shows a skin cell viewed under high power. One might estimate it to be 0.2 mm wide.

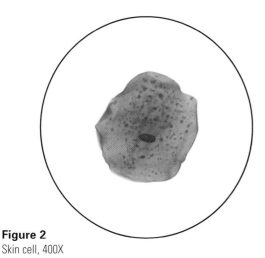

Figure 2
Skin cell, 400X

Preparing a Wet-Mount Slide

Specimens of all types can be mounted in a fluid medium before they are examined. Water is the most convenient, but 30% glycerol can also be used. Glycerol helps make the material more transparent and it does not dry out as fast as water. A cover slip is used so that there are no reflecting surfaces.

1. Clean a slide and cover slip, holding them by the edges so that you do not leave fingerprints on them. Lay them on a clean, dry surface.

2. Place a drop of water or glycerol in the centre of the slide.

3. Transfer the specimen into the water.

4. Hold the edges of the cover slip between thumb and forefinger. Place the cover slip in an almost vertical position on the slide so that the cover slip just touches the edge of the drop of water or glycerol (**Figure 3**).

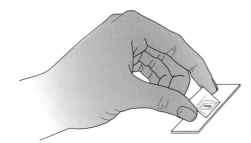

Figure 3

5. The water or glycerol will spread along the edge of the cover slip. Supporting the cover slip with the point of a needle, gently lower the cover slip on to the slide.

6. Use blotting paper to dry off any excess water or glycerol around the cover slip. Dry any water that may accidentally drop on the stage of the microscope or on the objective lenses.

Staining Techniques

Staining procedures are used to increase the contrast between cells of various tissues and the mounting medium, and make the features more visible. Biological stains may be temporary or permanent. In this course, we are not concerned with creating permanent slides, so temporary staining is adequate.

Stains react differently with different components of cells; they may affect the cell walls, they may stain the cytoplasm, or they may be used as a test for a specific substance (e.g., iodine for starch). There are many stains available but those listed below are more common.

In most cases, the staining technique is simple. After a wet-mount slide has been prepared, a drop of the stain is placed beside and under one corner of the cover slip. A piece of paper towel or filter paper touched to the water on the opposite side of the cover slip will draw the stain through the entire specimen in a few seconds. The stained specimen may then be observed.

Lugol's (Iodine) Solution

This is an especially good stain for plant cells like onion cells. A weak solution of iodine in potassium iodide is a specific colour indicator for starch; it turns blue-black. Starch grains in tissues show up very clearly. Cellulose cell walls remain unstained. Sections may be mounted directly in iodine as described above.

Aniline Sulfate

Lignified walls of plant cells turn bright yellow. Specimens may be left in aniline sulphate or transferred to glycerol.

Sudan IV Indicator

A saturated solution of Sudan IV in 70% alcohol will stain all fatty substances orange-pink or red. Fresh sections should be left in the stain for 20 min. Wash with 50% alcohol to remove excess stain and transfer to a drop of glycerol. Fats can be detected as droplets within cells.

Toluidine Blue

This is an excellent water-soluble double stain. Lignified cell walls stain green, greenish blue, or bright blue. Lignified phloem fibres stain bright blue, and xylem elements stain a greenish colour. Most nonlignified cell walls (e.g., parenchyma) stain pinkish purple. Nuclei become visible and chloroplasts keep their distinctive green colour. Specimens should be left in stain for approximately 30 s, rinsed with water, and then dry mounted.

Numerical Prefixes and Units

Throughout *Nelson Biology 11* and in this reference section, we have attempted to be consistent in the presentation and usage of units. As far as possible, *Nelson Biology 11* uses the International System (SI) of Units. However, some other units have been included because of their practical importance, wide usage, or use in specialized fields. For example, Health Canada and the medical profession continue to use millimetres of mercury (mm Hg) as the units for measurement of blood pressure, although the Metric Practice Guide indicates that this unit is not to be used with the SI.

The most recent *Canadian Metric Practice Guide* (CAN/CSA-Z234.1-89) was published in 1989 and reaffirmed in 1995 by the Canadian Standards Association.

Other data in this reference section has been taken largely from *Lange's Handbook of Chemistry*, Fifteenth Edition, McGraw-Hill, 1999.

Numerical Prefixes

Prefix	Power	Symbol
deca–	10^1	*da*
hecto–	10^2	*h*
kilo–	10^3	*k**
mega–	10^6	*M**
giga–	10^9	*G**
tera–	10^{12}	*T*
peta–	10^{15}	*P*
exa–	10^{18}	*E*
deci–	10^{-1}	*d*
centi–	10^{-2}	*c**
milli–	10^{-3}	*m**
micro–	10^{-6}	*μ**
nano–	10^{-9}	*n**
pico–	10^{-12}	*p*
femto–	10^{-15}	*f*
atto–	10^{-18}	*a*

* commonly used

Common Multiples

Multiple	Prefix
0.5	hemi-
1	mono-
1.5	sesqui-
2	bi-, di-
2.5	hemipenta-
3	tri-
4	tetra-
5	penta-
6	hexa-
7	hepta-
8	octa-
9	nona-
10	deca-

Some Examples of Prefix Use

0.0034 mol = 3.4×10^{-3} mol = 3.4 **milli**moles or 3.4 mmol

1530 L = 1.53×10^3 L = 1.53 **kilo**litres or 1.53 kL

SI Base Units

Quantity	Symbol	Unit name	Symbol
amount of substance	n	mole	mol
electric current	I	ampere	A
length	L, l, h, d, w	metre	m
luminous intensity	I_v	candela	cd
mass	M	kilogram	kg
temperature	T	kelvin	K
time	t	second	s

Some SI Derived Units

Quantity	Symbol	Unit	Unit symbol	Expression in SI base unit
area	A	square metre	m^2	m^2
density	ρ, D	kilogram per cubic metre	kg/m^3	kg/m^3
displacement	$\vec{d}$	metre	m	m
energy	E, E_k, E_p	joule	J	$kg \cdot m^2/s^2$
pressure	P p	pascal newton per square metre	Pa N/m^2	$kg/(m \cdot s^2)$
volume	V	cubic metre	m^3	m^3

Greek and Latin Prefixes and Suffixes

Greek and Latin Prefixes

Prefix	Meaning	Prefix	Meaning	Prefix	Meaning
a-	not, without	em-	inside	micr-, micro-	small
ab-	away from	en-	in	mono-	one
abd-	led away	end-, endo-	within	morpho-	form, shape
acro-	end, tip	epi-	at, on, over	muc-, muco-	slime
adip-	fat	equi-	equal	multi-	many
aer-, aero-	air	erythro-	red	myo-	muscle
agg-	to clump	ex-, exo-	away, out	nas-	nose
agro-	land	flag-	whip	necro-	corpse
alb-	white	gamet-, gamo-	marriage, united	neo-	new
allo-	other	gastr-, gastro-	stomach	neur-, neuro-	nerve
ameb-	change	geo-	earth	noct-	night
amphi-	around, both	glyc-	sweet	odont-, odonto-	tooth
amyl-	starch	halo-	salt	oligo-	few
an-	without	haplo-	single	oo-	egg
ana-	up	hem-, hema-, hemato-	blood	orni-	bird
andro-	man	hemi-	half	oss-, osseo-, osteo-	bone
ant-, anti-	opposite	hepat-, hepa-	liver	ovi-	egg
anth-	flower	hetero-	different	pale-, paleo-	ancient
archae-, archaeo-	ancient	histo-	web	patho-	disease
archi-	primitive	holo-	whole	peri-	around
astr-, astro-	star	homeo-	same	petro-	rock
aut-, auto-	self	hydro-	water	phag-, phago-	eat
baro-	weight (pressure)	hyper-	above	pharmaco-	drug
bi-	twice	hypo-	below	phono-	sound
bio-	life	infra-	under	photo-	light
blast-, blasto-	sprout (budding)	inter-	between	pneum-	air
carcin-	cancer	intra-	inside of, within	pod-	foot
cardio-, cardia-	heart	intro-	inward	poly-	many
chlor-, chloro-	green	iso-	equal	pseud-, pseudo-	false
chrom-, chromo-	colour	lact-, lacti-, lacto-	milk	pyr-, pyro-	fire
co-	with	leuc-, leuco-	white	radio-	ray
cosmo-	order, world	lip-, lipo-	fat	ren-	kidney
cut-	skin	lymph-, lympho-	clear water	rhizo-	root
cyan-	blue	lys-, lyso-	break up	sacchar-, saccharo-	sugar
cyt-, cyto-	cell	macro-	large	sapr-, sapro-	rotten
dendr-, dendri-, dendro-	tree	mamm-	breast	soma-	body
dent-, denti-	tooth	meg-, mega-	great	spermato-	seed
derm-	skin	melan-	black	sporo-	seed
di-	two	meningo-	membrane	squam-	scale
dors-	back	mes-, meso-	middle	sub-	beneath
ec-, ecto-	outside	meta-	after, transition	super-, supra-	above

Greek and Latin Prefixes

Prefix	Meaning	Prefix	Meaning	Prefix	Meaning
sym-, syn-	with, together	ultra-	beyond	xanth-, xantho-	yellow
telo-	end	uro-	tail, urine	xer-, xero-	dry
therm-, thermo-	temperature, heat	vas-, vaso-	vessel	xyl-	wood
tox-	poison	vita-	life	zoo-	animal
trans-	across	vitro-	glass	zygo-	yoke
trich-	hair	vivi-	alive		

Greek and Latin Suffixes

Suffix	Meaning	Suffix	Meaning	Suffix	Meaning
-aceous	like	-lysis	loosening	-phyll	leaf
-blast	budding	-lyt	dissolvable	-phyte	plant
-cide	kill	-mere	share	-pod	foot
-crin	secrete	-metry	measure	-sis	a condition
-cut	skin	-mnesia	memory	-some	body
-cyte	cell	-oid	like	-stas, -stasis	halt
-emia	blood	-ol	alcohol	-stat	to stand, stabilize
-gen	born, agent	-ole	oil	-tone, -tonic	strength
-genesis	formation	-oma	tumour	-troph	nourishment
-graph, -graphy	to write	-osis	a condition	-ty	state of
-gynous	woman	-pathy	suffering	-vorous	eat
-itis	inflammation	-ped	foot	-yl	wood
-logy	the study of	-phage	eat	-zyme	ferment

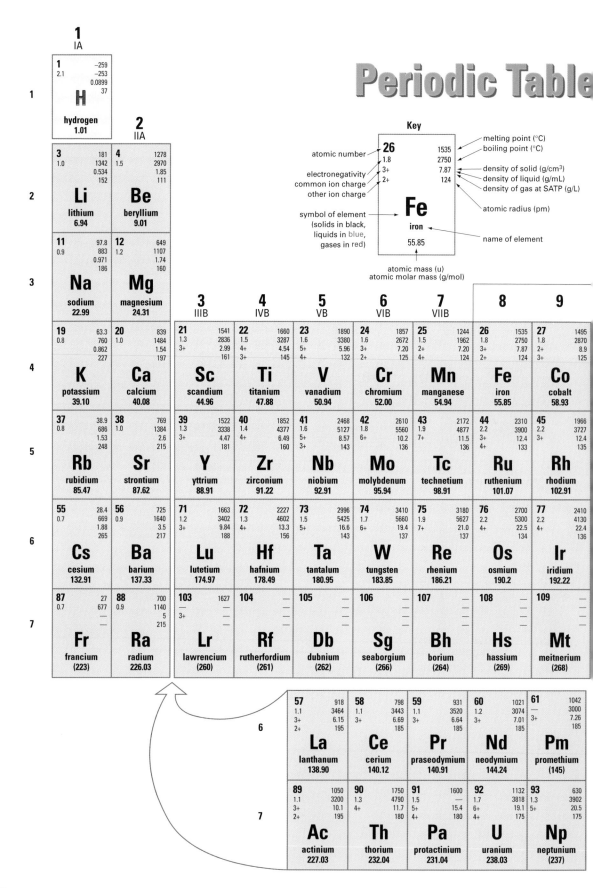

Periodic Table

Key

atomic number	**26**	1535 → melting point (°C)
	1.8	2750 → boiling point (°C)
electronegativity	3+	7.87 → density of solid (g/cm³)
common ion charge	2+	124 → density of liquid (g/mL)
other ion charge		→ density of gas at SATP (g/L)
	Fe	→ atomic radius (pm)
symbol of element	iron	
(solids in black,		
liquids in blue,	55.85	→ name of element
gases in red)		

atomic mass (u)
atomic molar mass (g/mol)

Group 1 / IA

1 | −259
2.1 | −253
 | 0.0899
 | 37
H
hydrogen
1.01

Group 2 / IIA

3	181	4	1278
1.0	1342	1.5	2970
	0.534		1.85
	152		111
Li		**Be**	
lithium		beryllium	
6.94		9.01	

11	97.8	12	649
0.9	883	1.2	1107
	0.971		1.74
	186		160
Na		**Mg**	
sodium		magnesium	
22.99		24.31	

Group 3 / IIIB — Group 9

19	63.3	20	839	21	1541	22	1660	23	1890	24	1857	25	1244	26	1535	27	1495
0.8	760	1.0	1484	1.3	2836	1.5	3287	1.6	3380	1.6	2672	1.5	1962	1.8	2750	1.8	2870
	0.862		1.54	3+	2.99	4+	4.54	5+	5.96	3+	7.20	2+	7.20	3+	7.87	2+	8.9
	227		197		161	3+	145	4+	132	2+	125	4+	124	2+	124	3+	125
K		**Ca**		**Sc**		**Ti**		**V**		**Cr**		**Mn**		**Fe**		**Co**	
potassium		calcium		scandium		titanium		vanadium		chromium		manganese		iron		cobalt	
39.10		40.08		44.96		47.88		50.94		52.00		54.94		55.85		58.93	

37	38.9	38	769	39	1522	40	1852	41	2468	42	2610	43	2172	44	2310	45	1966
0.8	686	1.0	1384	1.3	3338	1.4	4377	1.6	5127	1.8	5560	1.9	4877	2.2	3900	2.2	3727
	1.53		2.6	3+	4.47	4+	6.49	5+	8.57	6+	10.2	7+	11.5	3+	12.4	3+	12.4
	248		215		181		160	3+	143		136		136	4+	133		135
Rb		**Sr**		**Y**		**Zr**		**Nb**		**Mo**		**Tc**		**Ru**		**Rh**	
rubidium		strontium		yttrium		zirconium		niobium		molybdenum		technetium		ruthenium		rhodium	
85.47		87.62		88.91		91.22		92.91		95.94		98.91		101.07		102.91	

55	28.4	56	725	71	1663	72	2227	73	2996	74	3410	75	3180	76	2700	77	2410
0.7	669	0.9	1640	1.2	3402	1.3	4602	1.5	5425	1.7	5660	1.9	5627	2.2	5300	2.2	4130
	1.88		3.5	3+	9.84	4+	13.3	5+	16.6	6+	19.4	7+	21.0	4+	22.5	4+	22.4
	265		217		188		156		143		137		137		134		136
Cs		**Ba**		**Lu**		**Hf**		**Ta**		**W**		**Re**		**Os**		**Ir**	
cesium		barium		lutetium		hafnium		tantalum		tungsten		rhenium		osmium		iridium	
132.91		137.33		174.97		178.49		180.95		183.85		186.21		190.2		192.22	

87	27	88	700	103	1627	104	—	105	—	106	—	107	—	108	—	109	—
0.7	677	0.9	1140	—	—		—		—		—		—		—		—
	—		5	3+	—		—		—		—		—		—		—		—
	—		215		—		—		—		—		—		—		—		—
Fr		**Ra**		**Lr**		**Rf**		**Db**		**Sg**		**Bh**		**Hs**		**Mt**	
francium		radium		lawrencium		rutherfordium		dubnium		seaborgium		borium		hassium		meitnerium	
(223)		226.03		(260)		(261)		(262)		(266)		(264)		(269)		(268)	

Period 6 (lanthanides)

57	918	58	798	59	931	60	1021	61	1042
1.1	3464	1.1	3443	1.1	3520	1.2	3074	—	3000
3+	6.15	3+	6.69	3+	6.64	3+	7.01	3+	7.26
2+	195		185		185		185		185
La		**Ce**		**Pr**		**Nd**		**Pm**	
lanthanum		cerium		praseodymium		neodymium		promethium	
138.90		140.12		140.91		144.24		(145)	

Period 7 (actinides)

89	1050	90	1750	91	1600	92	1132	93	630
1.1	3200	1.3	4790	1.5	—	1.7	3818	1.3	3902
3+	10.1	4+	11.7	5+	15.4	6+	19.1	5+	20.5
2+	195		180	4+	180	4+	175		175
Ac		**Th**		**Pa**		**U**		**Np**	
actinium		thorium		protactinium		uranium		neptunium	
227.03		232.04		231.04		238.03		(237)	

of the Elements

18 VIIIA

Period							18 VIIIA

Group headings (top): 13 IIIA · 14 IVA · 15 VA · 16 VIA · 17 VIIA · 18 VIIIA

Period 1

2	−272 / −269 / X / 0.179 / 50 — **He** helium 4.00

Period 2

| 5 — 2300 / 2.0 / 2550 / X / 2.34 / 88 — **B** boron 10.81 |
| 6 — 3550 / 2.5 / 4827 / X / 2.26 / 77 — **C** carbon 12.01 |
| 7 — −210 / 3.0 / −196 / 1.25 / 70 — **N** nitrogen 14.01 |
| 8 — −218 / 3.5 / −183 / 1.43 / 66 — **O** oxygen 16.00 |
| 9 — −220 / 4.0 / −188 / 1.70 / 64 — **F** fluorine 19.00 |
| 10 — −249 / X / −246 / 0.900 / 62 — **Ne** neon 20.18 |

Period 3

| 13 — 660 / 1.5 / 2467 / X / 2.70 / 143 — **Al** aluminum 26.98 |
| 14 — 1410 / 1.8 / 2355 / X / 2.33 / 117 — **Si** silicon 28.09 |
| 15 — 44.1 / 2.1 / 280 / 1.82 / 110 — **P** phosphorus 30.97 |
| 16 — 113 / 2.5 / 445 / 2.07 / 104 — **S** sulfur 32.06 |
| 17 — −101 / 3.0 / −34.6 / 3.21 / 99 — **Cl** chlorine 35.45 |
| 18 — −189 / X / −186 / 1.78 / 95 — **Ar** argon 39.95 |

Groups (lower): 10 · 11 IB · 12 IIB

Period 4

| 28 — 1455 / 1.8 / 2730 / 2+ / 8.90 / 3+ / 124 — **Ni** nickel 58.69 |
| 29 — 1083 / 1.9 / 2567 / 2+ / 8.92 / 1+ / 128 — **Cu** copper 63.55 |
| 30 — 420 / 1.6 / 907 / 2+ / 7.14 / 133 — **Zn** zinc 65.38 |
| 31 — 29.8 / 1.6 / 2403 / 3+ / 5.90 / 122 — **Ga** gallium 69.72 |
| 32 — 937 / 1.8 / 2830 / 4+ / 5.35 / 123 — **Ge** germanium 72.61 |
| 33 — 817 / 2.0 / 613 / 5.73 / 121 — **As** arsenic 74.92 |
| 34 — 217 / 2.4 / 684 / 4.81 / 117 — **Se** selenium 78.96 |
| 35 — −7.2 / 2.8 / 58.8 / 3.12 / 114 — **Br** bromine 79.90 |
| 36 — −157 / — / −152 / X / 3.74 / 112 — **Kr** krypton 83.80 |

Period 5

| 46 — 1554 / 2.2 / 2970 / 2+ / 12.0 / 4+ / 138 — **Pd** palladium 106.42 |
| 47 — 962 / 1.9 / 2212 / 1+ / 10.5 / 144 — **Ag** silver 107.87 |
| 48 — 321 / 1.7 / 765 / 2+ / 8.64 / 149 — **Cd** cadmium 112.41 |
| 49 — 157 / 1.7 / 2080 / 3+ / 7.30 / 163 — **In** indium 114.82 |
| 50 — 232 / 1.8 / 2270 / 4+ / 7.31 / 2+ / 140 — **Sn** tin 118.69 |
| 51 — 631 / 1.9 / 1750 / 3+ / 6.68 / 5+ / 141 — **Sb** antimony 121.75 |
| 52 — 450 / 2.1 / 990 / 6.2 / 137 — **Te** tellurium 127.60 |
| 53 — 114 / 2.5 / 184 / 4.93 / 133 — **I** iodine 126.90 |
| 54 — −112 / X / −107 / 5.89 / 130 — **Xe** xenon 131.29 |

Period 6

| 78 — 1772 / 2.2 / 3827 / 4+ / 21.5 / 2+ / 138 — **Pt** platinum 195.08 |
| 79 — 1064 / 2.4 / 2808 / 3+ / 19.3 / 1+ / 144 — **Au** gold 196.97 |
| 80 — −39.0 / 1.9 / 357 / 2+ / 13.5 / 1+ / 160 — **Hg** mercury 200.59 |
| 81 — 304 / 1.8 / 1457 / 1+ / 11.85 / 3+ / 170 — **Tl** thallium 204.38 |
| 82 — 328 / 1.8 / 1740 / 2+ / 11.3 / 4+ / 175 — **Pb** lead 207.20 |
| 83 — 271 / 1.9 / 1560 / 3+ / 9.80 / 5+ / 155 — **Bi** bismuth 209.98 |
| 84 — 254 / 2.0 / 962 / 2+ / 9.40 / 4+ / 167 — **Po** polonium (209) |
| 85 — 302 / 2.2 / 337 / — / 142 — **At** astatine (210) |
| 86 — −71 / X / −61.8 / 9.73 / 140 — **Rn** radon (222) |

Period 7

| 110 — **Uun** ununnilium (269, 271) |
| 111 — **Uuu** unununium (272) |
| 112 — **Uub** ununbium (277) |
| 113 |
| 114 — **Uuq** ununquadium (285) |
| 115 |
| 116 — **Uuh** ununhexium (289) |
| 117 |
| 118 — **Uuo** ununoctium (293) |

Lanthanide series (Period 6)

| 62 — 1074 / 1.2 / 1794 / 3+ / 7.52 / 2+ / 185 — **Sm** samarium 150.36 |
| 63 — 822 / — / 1527 / 3+ / 5.24 / 2+ / 185 — **Eu** europium 151.96 |
| 64 — 1313 / 1.1 / 3273 / 3+ / 7.90 / 180 — **Gd** gadolinium 157.25 |
| 65 — 1356 / 1.2 / 3230 / 3+ / 8.23 / 175 — **Tb** terbium 158.92 |
| 66 — 1412 / — / 2567 / 3+ / 8.55 / 175 — **Dy** dysprosium 162.50 |
| 67 — 1474 / 1.2 / 2700 / 3+ / 8.80 / 175 — **Ho** holmium 164.93 |
| 68 — 1529 / 1.2 / 2868 / 3+ / 9.07 / 175 — **Er** erbium 167.26 |
| 69 — 1545 / 1.2 / 1950 / 3+ / 9.32 / 175 — **Tm** thulium 168.93 |
| 70 — 819 / 1.1 / 1196 / 3+ / 6.97 / 2+ / 175 — **Yb** ytterbium 173.04 |

Actinide series (Period 7)

| 94 — 641 / 1.3 / 3232 / 4+ / 19.8 / 6+ / 175 — **Pu** plutonium (244) |
| 95 — 994 / 1.3 / 2607 / 3+ / 13.7 / 4+ / 175 — **Am** americium (243) |
| 96 — 1340 / — / 3110 / 3+ / 13.5 / 4+ / — — **Cm** curium (247) |
| 97 — 986 / — / 3+ / 14 / 4+ / — — **Bk** berkelium (247) |
| 98 — 900 / — / 3+ / — — **Cf** californium (251) |
| 99 — 860 / — / 3+ / — — **Es** einsteinium (252) |
| 100 — 1527 / — / 3+ / — — **Fm** fermium (257) |
| 101 — 1021 / — / 3074 / 2+ / 3+ / — — **Md** mendelevium (258) |
| 102 — 863 / — / 2+ / 3+ / — — **No** nobelium (259) |

Answers

The following answers are for selected questions from the practice, chapter review, and unit review sections. The answers provided are short and do not represent the complete answer to a question.

Chapter 1

1.4 Practice (page 20)

7. mitochondria, ribosomes, endoplasmic reticulum, Golgi apparatus, lysosomes, microfilaments and microtubules

Chapter 1 Review (pages 28–29)

6. (a) ribosomes
   (b) mitochondria
   (c) lysosome, vacuole
14. 750 μm
16. No, she is not correct.
17. (a) transmission electron microscope
    (b) light microscope
    (c) scanning electron microscope

Chapter 2

2.2 Practice (pages 36–37)

2. Water is not organic. Carbon dioxide is not organic.
5. fructose, glucose, galactose

2.11 Practice (page 67)

3. lactic acid

Chapter 2 Review (pages 70–71)

10. no
23. water soluble: glucose, starch, amino acids, proteins, phospholipids, sucrose
    fat soluble: phospholipids, triglycerides, waxes and butter

Unit 1 Review (pages 74–77)

12. (a) vesicle, lysosome
    (b) cell membrane, cytoplasm, endoplasmic reticulum
    (c) cell membrane, cytoplasm
    (d) cell membrane, cytoplasm, Golgi apparatus, vesicles
42. (a) maltose
    (b) sucrose
46. A—blue
    B—eventually will turn yellow
    C—yellow at night, blue in the day
    D—yellow at night, blue in the day

Chapter 3

3.2 Practice (page 90)

1. prophase, metaphase, anaphase, telophase
7. Cytokinesis produces two distinct and separate cells.
8. cell division or cell death

3.4 Practice (page 98)

3. no

3.5 Practice (pages 101–102)

2. Cancer cells grown in tissue culture can divide every 24 hours. This rate is unlikely to occur in the body.
3. The division phase of the life cycle is proportionately longer.

Chapter 3 Review (pages 124–127)

2. (a) A and B are plant cells, C and D are animal cells.
12. (b) no
16. (a)

Cell phase	Area #1	Area #2	Total cell count	Time spent in phase (min)
interphase	91	70	161	533
prophase	10	14	24	79.6
metaphase	2	1	3	10
anaphase	2	1	3	10
telophase	4	4	8	26.5

(b)

interphase	300°
prophase	35°
metaphase	5°
anaphase	5°
telophase	15°

19. (a) graph A
22. (a) No

Chapter 4

4.2 Practice (pages 137–138)

2. Purebred refers to parents that are homozygous for a trait.
3. (a) $\frac{3}{4}$ tall, $\frac{1}{4}$ dwarf
   (b) $\frac{1}{2}$ tall, $\frac{1}{2}$ dwarf
   (c) $\frac{3}{4}$ purple, $\frac{1}{4}$ white
   (d) All have green pods.
   (e) All have round seeds.
4. (b) 5

4.3 Practice (pages 142–143)

2. possible genotypes: SS, Ss, ss; probability of shortsightedness is $\frac{3}{4}$, probability of normal vision is $\frac{1}{4}$

4.4 Practice (page 144)

2. $E^4E^4 \times E^3E^4 = \frac{1}{2}$ honey, $\frac{1}{2}$ white
   $E^4E^4 \times E^3E^3 =$ all honey
3. $\frac{1}{2}$ apricot, $\frac{1}{4}$ honey, $\frac{1}{4}$ white

4.5 Practice (pages 145–146)

2. $\frac{1}{2}$ white, $\frac{1}{2}$ pink

3. $\frac{1}{2}$ red, $\frac{1}{2}$ roan

4. $\frac{1}{4}$ *LL* (long), $\frac{1}{2}$ *LR* (oval), $\frac{1}{4}$ *RR* (round)

4.6 Practice (page 154)

1. (a) ratio in F_1 generation: $\frac{1}{2}$ *BbHh* (black, short), $\frac{1}{2}$ *Bbhh* (black, long)

   (b) ratio in F_1 generation: $\frac{1}{4}$ *BbHh* (black, short), $\frac{1}{4}$ *Bbhh* (black, long), $\frac{1}{4}$ *bbHh* (white, short), $\frac{1}{4}$ *bbhh* (white, long)

   (c) ratio in F_1 generation: $\frac{1}{4}$ *BBHh* (black, short), $\frac{1}{4}$ *BbHh* (black, short), $\frac{1}{4}$ *BBhh* (black, long), $\frac{1}{4}$ *Bbhh* (black, long)

2. male = *BbHh*, female A = *bbHH*, female B = *BbHh*, female C = *bbhh*

3. the child could be phenotype A, Rh+; A, Rh-; O, Rh+; or O, Rh–

Chapter 4 Review (pages 158–159)

8. codominance

9. $\frac{1}{2}$ long stem, $\frac{1}{2}$ short stem

11. A man with blood type O and a woman with blood type AB can only have children with blood types A or B (about $\frac{1}{2}$ each). A woman with blood type AB could have children with AB blood, depending on the blood type of the father. However, she could never have children with blood type I^O, because all of her gametes have either an I^A or I^B allele, both of which are dominant over I^O.

12. (b) no

14. F_1 generation: all are black, short. F_2 generation: $\frac{9}{16}$ black, long; $\frac{3}{16}$ black, short; $\frac{3}{16}$ white, long; $\frac{1}{16}$ white, short

17. (a) no

    (b) yes

    (c) no

Chapter 5

5.3 Practice (page 167)

6. (b) unlikely

7. (a) F_1: all children have colour vision

   (b) F_1: 50% of the males colour blind, all females have colour vision

   (c) father: X^cY, mother: X^CX^c or X^cX^c

Chapter 5 Review (pages 188–189)

9. $X^RX^r \times X^RY$; F_1: all females have wild-type eyes, and half of the males have white eyes.

10. *XX*, +/*tra* × *XY*, *tra/tra*; F_1: all males are normal, but half of the females transform; F_2: all males are normal, and $\frac{1}{3}$ of the females are normal.

11. (b) $\frac{3}{4}$ normal, $\frac{1}{4}$ hemophilic

    (d) II-4 = X^HX^h; II-5 = X^HY

Unit 2 Review (pages 192–193)

18. (a) dominant

19. (a) F_1: all yellow and round; F_2: $\frac{9}{16}$ yellow, round; $\frac{3}{16}$ yellow, wrinkled; $\frac{3}{16}$ green, round; $\frac{1}{16}$ green, wrinkled

(b) Observed F_2 Phenotypes

Phenotype	Theoretical	Actual
round, yellow seeds	56.25%	56.65%
round, green seeds	18.75%	19.42%
wrinkled, yellow seeds	18.75%	18.16%
wrinkled, green seeds	6.25%	5.76%

20. (a) $\frac{1}{2}$ wild-type, $\frac{1}{4}$ apricot, $\frac{1}{4}$ honey

    (b) $E^1E^4 \times E^3E^4$

21. (a) $\frac{1}{2}$ green, $\frac{1}{2}$ yellow

    (b) $\frac{3}{16}$

22. (c) $\frac{1}{2}$

    (d) 100%

23. baby A—couple 4; baby B—couple 1, baby C—couple 2, baby D—couple 3

Chapter 6

6.4 Practice (page 216)

2. mucus, hydrochloric acid, pepsinogens, and other substances

6.6 Practice (page 224)

1. pigments and bile salts

6.9 Practice (page 233)

6. 11 040 kJ

7. 16.24 h

Chapter 6 Review (pages 238–239)

12. (a) 6300 kJ

    (c) 77 min

15. test tube A = pink; test tube B = pink; test tube C = clear

Chapter 7

7.2 Practice (page 246)

3. Albumins maintain osmotic pressure in capillaries; Globulins produce antibodies to provide protection against invading microbes and parasites; Fibrinogens are important in blood clotting.

7.5 Practice (page 258)

3. Chest pains, or angina, occur when too little oxygen reaches the heart.

4. Coronary bypass operation diverts blood around an area of blockage.

7.8 Practice (page 266)

6. Beta 1 blockers

7.10 Practice (page 274)

1. in the arterioles

Chapter 7 Review (pages 278–279)

3. False, deoxygenated blood is found in the pulmonary artery.
22. (a) Subject C, the hemoglobin level is higher than normal. People living at higher altitude have higher RBC and hemoglobin to compensate for air with less oxygen. The oxygen content is also increased because of an elevated hematocrit.
    (b) Subject B has a lower hemoglobin level. Iron is a component of hemoglobin.

Chapter 8

8.4 Practice (page 297)

4. Patient A 4.8 L
   Patient B 1.5 L
5. One cannot voluntarily exhale residual air into a respirometer in order to measure its volume.
6. vital capacity

Chapter 8 Review (pages 314–315)

1. Figure 1 shows the components of the human respiratory system.
   (a) w—alveolus, x—trachea, y—bronchiole
10. (a) time = approx. 2.2 units
    (b) time = approx. 1.0 unit
    (c) time = approx. 3.0 units
11. It would be beneficial.
13. fetal hemoglobin
14. (b) the first flask with the limewater

Unit 3 Review (pages 318–321)

10. (b) 2 and 3
13. (a) test tube 3
15. (a) $x = 4300$, $y = 2000$, $z = 600$
    (b) subject #1

Chapter 9

9.2 Practice (page 338)

5. DNA examples: smallpox, chickenpox
   RNA examples: measles, mumps

9.3 Practice (page 344)

7. sexual reproduction in which genetic material is exchanged between two cells

Chapter 9 Review (page 361)

11. Examples: *Volvox, Spirogyra, Paramecium*
13. kill both harmful and beneficial bacteria

Chapter 10

10.4 Practice (page 375)

6. e.g., pine tree
   e.g., apple tree

10.8 Practice (pages 389–390)

1. (a) bryophytes have no vascular system
2. one of the first plants to colonize a region

10.9 Practice (page 399)

1. tracheophytes; pteridophytes and spermatophytes
7. monocotyledons and dicotyledons

Chapter 10 Review (pages 404–405)

22. (a) sample B
    (b) a lichen
23. (b) a fern
30. (a) moss and fern
    (c) crustose, foliose, and fruticose

Chapter 11

11.1 Practice (pages 411–412)

2. three
3. The coelom is a fluid-filled space inside the body between the body wall and the gut.
4. bilateral symmetry

11.3 Practice (page 420)

2. ectoderm, mesoderm, endoderm

11.4 Practice (page 434)

4. Mollusks have gills; segmented worms exchange gases by direct diffusion through their skin.

11.5 Practice (pages 440–441)

4. a network of open tubes that reaches throughout the body and is connected directly to the atmosphere

Chapter 11 Review (pages 444–445)

7. "coel" means hollow
9. during moulting
19. sponges, hydra and flatworms, and some arthropods, such as crayfish
25. arthropod, nematode, arthropod (crustacean), cnidarian
33. parasitic flatworms, roundworms, annelids, land snails, arthropods

Chapter 12

12.1 Practice (page 449)

1. Urochordata, Cephalochordata, Vertebrata

12.4 Practice (page 460)

2. eggs laid in water, external fertilization, three-chambered heart, herbivorous as tadpoles, carnivorous as adults, gills when immature, lungs when adult

12.6 Practice (page 467)

4. a tail that can be used for grasping

Chapter 12 Review (page 481)

13. (d) scoop them from the air

Unit 4 Review (pages 484-487)

8. Eubactieria: (g) cyanobacteria, (i) *E. coli*; Protista: (c) amoeba, (f) euglena; Fungi: (d) mushroom, (h) yeast; Plantae: (a) pine tree, (b) moss, (l) fern; Animalia: (e) ant, (j) sponge, (k) worm

10. (a), (d), (f), and (g) are true

11. (b), (f), and (g) are true

35. (a) disagree
    (b) agree
    (c) disagree
    (d) agree

Chapter 13

13.1 Practice (page 497)

2. chemical potential energy and chemical building blocks

13.7 Practice (page 531)

3. in the cotyledon and/or endosperm

13.8 Practice (page 539)

2. artifical light, light blockers, and timers

Chapter 13 Review (pages 544–545)

11. xylem
28. (e) xylem
29. (b) high nitrogen content

Chapter 14

14.2 Practice (page 556)

1. no weather extremes, varied topography, each species has wide variation in sizes and ages, wide range of habitats available

14.5 Practice (page 571)

2. hardwood
3. process of becoming desertlike

Chapter 14 Review (pages 588–589)

1. gradual series of changes in vegetation in an area followed by changes in animals in the area

23. insecticides—2.9%; growth regulators—3.9%; nematocides—5.6%; fungicides—12.5%; herbicides—75.3%

Unit 5 Review (pages 592–595)

1. (f) and (h) are true
13. (a) 10–12 years
16. (a) x grew faster
17. (a) closer to x
20. (a) beginning of the climax community
27. (a) evergreen and conifer/gymnosperm
    (d) vascular tissue
28. (b) fungus

Glossary

abiotic: describes anything related to nonliving things. Abiotic factors include temperature, humidity, light availability, and soil conditions such as water content, texture, and mineral composition.

active transport: involves the use of cell energy to move materials across a cell membrane against the concentration gradient

addiction: a compulsive need for a harmful substance. Addiction is characterized by tolerance to the substance and physiological withdrawal symptoms.

adenosine triphosphate (ATP): a compound that temporarily stores chemical energy

adventitious roots: roots that develop from a part of the plant other than a root. They often form huge tufts at the base of the stem. There is no main root because most are the same size as the others. However, smaller secondary roots do branch out from these roots.

aerial roots: adventitious roots that grow from leaf nodes along stems

alcoholic fermentation: chemical decomposition of a carbohydrate in the absence of oxygen. It usually involves the conversion of carbohydrate into alcohol and carbon dioxide gas.

allantois: the embryonic membrane that functions in respiration and in the storage of metabolic wastes in reptiles, birds, and some mammals

alleles: two or more alternate forms of a gene. The alleles are located at the same position on one of the pairs of homologous chromosomes.

allelopathy: the suppression of growth and development of neighbouring plants by a plant of a different species. This effect is caused by chemicals secreted by the roots or contained in the leaves of the allelopathic species. Not all surrounding plants will suffer the same effect.

alternation of generations: refers to the complete life cycle of a plant, where the haploid stage produces gametes, and the diploid stage produces spores

alveoli: sacs of the lung in which the exchange of gases between the atmosphere and the blood occurs

amino acids: organic chemicals that can be linked together to form proteins

amnion: a membrane that surrounds the embryo and holds the amniotic fluid

amniotic eggs: eggs that have leathery or calcified shells that surround internal membranes, including the amnion

amylase: an enzyme that breaks down complex carbohydrates

amyloplasts: colourless plastids that store starch

anaerobic respiration: respiration that takes place without oxygen

anemia: the reduction in blood oxygen due to low levels of hemoglobin or poor red blood cell production

aneurysm: a fluid-filled bulge found in the weakened wall of an artery

angiosperms: spermatophytes that produce seeds enclosed in fruit formed by certain flower parts

annual: describes plants which complete their entire life cycle, from seed to reproduction to death, in one year

annual ring: the increase in the amount of secondary xylem during one year. The number of annual rings indicates the age of the woody plant.

antheridium: the sex organ that produces male gametes in mosses and ferns

antibiotics: chemicals produced synthetically or by microorganisms that inhibit the growth of or destroy certain other microorganisms

antibodies: proteins formed within the blood that react with antigens

antigen: a substance, usually protein in nature, that stimulates the formation of antibodies

antioxidants: chemicals that reduce the danger of oxygen-free radicals. Vitamin C is a common antioxidant.

aorta: the largest artery in the body. The aorta carries oxygenated blood to the tissues of the body.

apical meristems: regions at the tips of all roots and shoots. They are responsible for the primary growth, which lengthens shoots and roots throughout the life of the plant.

apomorphic: in cladistics, this term describes a recent characteristic that is derived from, but no longer the same as, an ancestral characteristic. Derived characteristics arose at some time after the first "splitting" of a member from the group, and therefore derived characteristics differ among the members of the group.

archaebacteria: in a six-kingdom system, a group of prokaryotic microorganisms distinct from eubacteria that possess a cell wall not containing peptidoglycan and that live in harsh environments such as salt lakes and thermal vents

archegonium: the sex organ that produces female gametes in mosses and ferns

arteriosclerosis: a group of disorders that cause the blood vessels to thicken, harden, become winding, and lose their elasticity

artery: a blood vessel that carries blood from the heart to the body

artificial selection: the intentional choosing of individuals of a species for the purpose of reproduction. Choices are based on the presence or absence of certain traits with the result that the desired characteristics will appear in subsequent generations.

asexual reproduction: the production of offspring from a single parent; offspring inherit the genes of that parent only

atherosclerosis: a degeneration of the blood vessel caused by the accumulation of fat deposits along the inner wall

atria: thin-walled chambers of the heart that receive blood from veins

atrioventricular (AV) node: a small mass of tissue in the right atrioventricular region through which impulses from the sinoatrial node are passed to the ventricles

atrioventricular (AV) valves: heart valves that prevent the backflow of blood from the ventricles into the atria

autonomic nervous system: the part of the nervous system that controls the motor nerves that regulate homeostasis. Autonomic nerves are not under conscious control.

autosomes: the chromosomes not involved in sex determination

autotrophs: organisms that use energy, usually light, to synthesize their own food from inorganic compounds

bacteriophages: a category of viruses that infect and destroy bacterial cells

bark: the outer layers on older stems, branches, and trunks. Bark consists of every layer from the vascular cambium outwards: phloem, any remaining cortex, cork cambium, and cork.

Barr body: a small, dark spot of chromatin located in the nucleus of a female mammalian cell

basal metabolic rate: the minimum amount of energy that a resting animal requires to maintain life processes

bile salts: the components of bile that break down fat globules

binomial nomenclature: a method of naming organisms by using two names—the genus and the species name. Scientific names are italicized.

biodiversity: the number of different species of all living organisms in a given area (sometimes called biological diversity)

biological accumulation: the process by which persistent chemicals accumulate in the body of an individual organism throughout its entire life

biological magnification: the process by which persistent chemicals seem to get multiplied as they pass along a food chain

biotechnology: the use of living things in industrial or manufacturing applications

biotic: describes anything related to living things. Biotic factors are all living things in an area and include interactions within and between species, such as competition and predation.

blastula: an embryonic stage consisting of a ball of cells produced by cell division following the fertilization of an egg

breathing: the movement of gases between the respiratory membrane of living things and their external environment

bronchi: the passage from the trachea to either the left or right lung

bronchial asthma: a respiratory disorder characterized by a reversible narrowing of the bronchial passage

bronchioles: the smallest passageways of the respiratory tract

bronchitis: an inflammation of the bronchial tubes

Brownian motion: the random movement of molecules

buffer: a substance capable of neutralizing acids and bases, thus maintaining the original pH of the solution

capillary: a blood vessel that connects arteries and veins. Capillaries are the sites of fluid and gas exchange.

capsid: the protective protein coat of viruses

carbohydrates: nutrients made up of a single sugar molecule or many sugar molecules. Carbohydrates contain only carbon, hydrogen, and oxygen.

carbonic anhydrase: an enzyme found in red blood cells. The enzyme speeds the conversion of carbon dioxide and water to form carbonic acid.

cardiac output: the amount of blood pumped from the heart each minute. Cardiac output is determined by multiplying the heart rate by the stroke volume (the quantity of blood pumped with each heartbeat).

cations: ions with a positive charge

cell fractionation: the process by which cell components are separated by centrifugation

cell membrane: a structure that surrounds the cytoplasm of the cell and regulates the movement of materials in and out of the cell

cellular respiration: the total of a series of chemical processes by which nutrients are broken down to provide energy

cellulose: the carbohydrate that forms the cell walls of plant cells

centrioles: small protein bodies that are found in the cytoplasm of animal cells

centromere: the structure that holds chromatids together

cephalization: the concentration of nerve tissue and receptors at the anterior end of an animal's body

chitin: a nitrogenous polysaccharide of long fibrous molecules

chloroplasts: plastids that contain the green pigment chlorophyll and specialize in photosynthesis

chorion: the outer membrane around a developing embryo

chromatid: one of two chromosome strands resulting from the duplication of a chromosome. The pair are called sister chromatids and they remain attached at their centromere until separated during mitosis.

chromatin: the tangled fibrous complex of DNA and protein within a eukaryotic nucleus

chromoplasts: plastids that store orange and yellow pigments

chromosomes: long threads of genetic material found in the nucleus of cells

cilia: tiny hairlike protein structures found on some cells that sweep foreign debris from the respiratory tract

cirrhosis: a chronic inflammation of liver tissue characterized by an increase of nonfunctioning fibrous tissue and fat

cladistics: a system which classifies organisms based on the presence or absence of shared derived characteristics

climax community: the final, self-perpetuating stage of succession. The composition of the climax community depends on the abiotic factors of the area.

clitellum: a smooth swollen band found about one third of the way along the body of some annelids. It secretes a protective covering for the eggs.

coagulation: a permanent change in a protein's shape due to amino acid bond disruption

coelom: the fluid-filled space inside the body, lined with a layer of cells called the peritoneum

coenzymes: organic molecules necessary for the activity of some enzymes

cofactors: substances necessary for the activity of another substance, usually an enzyme. Coenzymes are organic cofactors.

collenchyma: a living ground tissue that offers flexible support for primary growth

colon: the largest segment of the large intestine, where water reabsorption occurs

companion cells: small cells lying next to the sieve elements and directing their activities

competition: a relationship in which two organisms place demands on the same environmental resource

complementary base pair: a pair that forms between nitrogenous bases in the DNA molecule. The pairings are adenine and thymine (A–T) and guanine and cytosine (G–C).

compost: a mixture that consists largely of decayed organic matter and is used as a soil conditioner and source of minerals

compound leaf: a leaf which is divided into two or more leaflets

computerized axial tomography (CAT) scan: a procedure in which an X-ray machine takes many pictures of an object from different angles; a computer then reassembles the images to allow viewing of the object in cross section and in 3-D

concentration gradient: a difference in the number of molecules or ions of a substance between adjoining regions. Without the addition of energy, molecules tend to diffuse from the area of higher concentration to the area of lower concentration.

conjugation: a form of sexual reproduction in which genetic material is exchanged between two cells

cork: describes the cells produced by the cork cambium that eventually form a layer of dead cells which provide a protective covering for roots over two years old. Cork also describes the protective layers.

cork cambium: a lateral meristem formed by the pericycle in dicots over two years old

coronary arteries: arteries that supply the cardiac muscle with oxygen and other nutrients

cortex: the parenchyma tissue, usually with slightly thicker cell walls, surrounding the vascular tissue in roots and stems

cotyledon: a seed leaf that stores carbohydrates for the seedling and often is the first photosynthetic organ of a young seedling

crop: a receptacle for storing undigested food

crop rotation: the agricultural practice of planting a field in successive years with various crops, each of which has a different nutrient requirement

crossing over: the exchange of genetic material between two homologous chromosomes

cuticle: in plants, a layer of noncellular material secreted by epidermal cells designed to protect cells from drying out

cysts: cells that have a hardened protective covering on top of the cell membrane

cytokinesis: the division of cytoplasm

cytoplasm: a fluid that contains all cellular parts enclosed inside the plasma membrane except the nucleus

Dalton's law of partial pressure: the total pressure of a mixture of nonreactive gases is equal to the sum of the partial pressures of the individual gases.

dehydration synthesis: a series of chemical reactions that allow two molecules to bond by the formation of a water molecule

denature: to disrupt amino acid bonds (that hold a protein molecule together) by physical or chemical means, changing the protein's shape, which may or may not be temporary

deoxyribonucleic acid (DNA): a molecule that carries genetic information in cells

depressant: a drug that slows down the action of the central nervous system, often causing a decrease in heart and breathing rate

desertification: the process of becoming desertlike. One major cause of desertification is the loss of topsoil through erosion after the removal of the vegetation.

detoxify: to remove the effects of a poison

diaphragm: a sheet of muscle that separates the organs of the chest cavity from those of the abdominal cavity

diastole: relaxation (dilatation) of the heart, during which the cavities of the heart fill with blood

dichotomous key: a two-part key used to identify living things. *Di* means two.

diffusion: the movement of molecules from an area of higher concentration to an area of lower concentration

dihybrid cross: a type of cross that involves two genes, each consisting of nonidentical alleles

dikaryotic: describes cells that contain two haploid nuclei, each of which came from a separate parent

dioecious: describes organisms in which the male and female reproductive organs or gonads are carried by separate individuals

diploid: refers to twice the number of chromosomes in a gamete. Every cell of the body, with the exception of sex cells, contains a diploid chromosome number.

dominant: alleles of this type determine the expression of the genetic trait in offspring

dormant: describes a state of extremely slow biological activity. A dormant seed contains a living embryo but it does not grow; it remains protected by a seed coat and sometimes the fruit as well.

Down syndrome: a trisomic disorder in which a zygote receives three homologous chromosomes for chromosome pair number 21

duodenum: the first segment of the small intestine

ectoderm: the outermost tissue layer of an animal embryo. In more complex animals, this layer will become the skin, the outermost parts of the nervous system, and various other outer organs, structures, and systems, depending on the organism.

ectoplasm: the thin, semi-rigid (gelled) layer of the cytoplasm under the plasma membrane

ectotherms: organisms that are not able to maintain a constant body temperature. They are also described as poikilothermic.

electron transport system: a series of progressively stronger electron acceptors, with energy release at each step

emphysema: a respiratory disorder characterized by an overinflation of the alveoli

endergonic: any process that requires (consumes) energy

endocytosis: a process by which the cell membrane wraps around a particle and pinches off a vesicle inside the cell

endoderm: the innermost tissue layer of an animal embryo. In more complex animals, this layer will become the digestive tract, the respiratory tract, and various other inner organs, structures, and systems, depending on the organism.

endodermis: a layer of rectangular cells surrounding the vascular cylinder. It is the innermost layer of the cortex.

endoplasm: the fluid part of the cytoplasm that fills the inside of the cell. The endoplasm is responsible for an amoeba's shape as it moves.

endoplasmic reticulum: a series of canals that transport materials throughout the cytoplasm

endoscope: an instrument to view the interior of the body

endoskeleton: an internal skeleton

endospores: dormant cells of bacilli bacteria that contain genetic material encapsulated by a thick, resistant cell wall. These forms of cells develop when environmental conditions become unfavourable.

endotherms: organisms that are able to maintain a constant body temperature. They are also described as homeothermic.

enterokinase: an enzyme of the small intestine that converts trypsinogen to trypsin

enucleated: the condition where a cell does not contain a nucleus

enzymes: protein molecules that increase the rate at which biochemical reactions proceed

epidermis: the outermost cell layer of a multicellular plant experiencing primary growth

epiglottis: a structure that covers the glottis (opening of the trachea) during swallowing

epiphytes: plants that grow on the stems and branches of other plants but continue to photosynthesize

equilibrium: a condition in which all acting influences are balanced, resulting in a stable environment

erepsins: enzymes that complete protein digestion by converting small-chain peptides to amino acids

erythrocytes: red blood cells that contain hemoglobin and carry oxygen

esophagus: a tube that carries food from the mouth to the stomach

eubacteria: in a six-kingdom system, a group of prokaryotic microorganisms that possess a peptidoglycan cell wall and reproduce by binary fission

eukaryotic: a type of cell that has a true nucleus. The nuclear membrane surrounds a well-defined nucleus.

exergonic: any process that gives off (releases) energy

exocytosis: a process by which particles are released from a cell by fusing a particle-filled vesicle with the cell membrane

exoskeleton: a tough outer covering or cuticle that provides protection and support to an organism

exotic: describes species that are foreign or not native

expiratory reserve volume: the amount of air that can be forcibly exhaled after a normal exhalation

external intercostal muscles: muscles that raise the rib cage, decreasing pressure inside the chest cavity

extracellular fluid (ECF): fluid that occupies the spaces between cells and tissues; includes plasma and interstitial fluid

facultative anaerobes: bacteria that prefer environments with oxygen, but can live without oxygen

fatty acids: long chains of carbon and hydrogen joined together. The end of the chain has an acid group (–COOH)

fertilization: the union of male and female sex cells

fertilizers: any minerals added to soil, usually to replace those removed by crops

fibrous roots: root systems whose primary roots have disintegrated and have been replaced by adventitious roots

filtration: the selective movement of materials through capillary walls by a pressure gradient

frond: the leaf portion of a fern

gallstones: crystals of bile salts that form in the gall bladder

gametes: sex cells that have a haploid chromosome number

gametogenesis: the formation of gametes (sex cells) in animals

gametophyte: refers to a stage in a plant's life cycle in which cells have haploid nuclei. This stage begins with the haploid spores. During this stage, the sex cells (gametes) are produced by mitosis.

gastrin: a digestive hormone secreted by the stomach that stimulates the release of gastric juices to digest proteins

gastrovascular cavity: a digestive compartment, usually with a single opening that functions as both mouth and anus

gemmae: the small clumps of haploid photosynthetic cells produced in little cup-shaped structures on the gametophtye plant. They are dispersed by splashes of rain to grow into other gametophyte plants.

gene pool: all the genes of all the individuals in a specific population

gene therapy: a procedure by which defective genes are replaced with normal genes in order to cure genetic disorders

genes: sections of a chromosome, each of which contains one set of instructions

genetic engineering: intentional production of new genetic material by substituting or altering existing material

genome: the complete set of instructions contained within the DNA of an organism

genotype: the alleles an organism contains

germ layers: layers of cells in the embryo that give rise to specific tissues in the adult

gizzard: a muscular chamber designed to physically break down food

glycerol: a three-carbon chain molecule containing three hydroxyl (–OH) groups

glycogen: the form of carbohydrate storage in animals

Golgi apparatus: a protein-packaging organelle composed of membranous sacs

Green Revolution: the large increase in food production experienced from the late 1960s into the 1980s

guard cells: the cells that occur in pairs around each stoma in the epidermis of a leaf or a stem. They regulate the opening and closing of the stoma.

gymnosperms: spermatophytes that produce "naked" seeds, usually inside cones

halophytes: plants that can tolerate living in salty habitats because they have the ability to take water from concentrated solutions

haploid: refers to the number of chromosomes in a gamete

heartwood: the older, harder, nonliving central wood in tree trunks. It is often darker due to the accumulation of oils and resins and its basic function is to provide support.

hemocoel: an open space or body cavity filled with blood; part of an open circulatory system

herbaceous: describes the fleshy stems of annual plants. These stems usually do not survive more than one year, especially if there is a cold winter. They are also called nonwoody stems.

heredity: the passing of traits from parents to offspring

hermaphroditic: sharing both male and female sex cells or organs

heterozygous: a genotype in which the alleles of a pair are different

homeostasis: a process by which a constant internal environment is maintained despite changes in the internal and external environment

hominid: a family consisting of existing and extinct human species and close relatives

homologous chromosomes: paired chromosomes similar in shape, size, gene arrangement, and gene information

homozygous: a genotype in which both alleles of a pair are the same

host range: the limited number of host species, tissues, or cells that a virus or other parasite can infect

hybridization: the mating of two different parents to produce offspring with desirable characteristics of both parents

hybridomas: cells that result from the fusion of two different cells

hybrids: offspring that differ from their parents in one or more traits. Interspecific hybrids result from the union of two different species.

hydrolytic enzymes: enzymes that use water to break down molecules

hydrophytes: plants living on or in water

hydroponics: a system of growing plants without soil but instead a sterile medium and a solution containing all required nutrients

hypertonic solution: a solution where the concentration of solutes outside a cell is higher than that found inside the cell

hypha: one of the filaments of the mycelium

hypotonic solution: a solution where the concentration of solutes outside a cell is lower than that found inside the cell

inbreeding: the process by which breeding is limited to a number of desirable phenotypes

inspiratory reserve volume: the amount of air that can be forcibly inhaled after a normal inhalation

internal intercostal muscles: muscles that pull the rib cage downward, increasing pressure inside the chest cavity

internode: the space between two successive nodes on the same stem

interphase: the time interval between nuclear divisions. During this phase, a cell increases in mass, roughly doubles the cytoplasmic components, and duplicates its chromosomes.

interspecific: between two species

interstitial: refers to the space between cells

intraspecific: within one species

invasive: capable of out-competing the other species in any given area

invertebrates: multicellular, eukaryotic heterotrophs that do not have a notochord

isotonic solution: a solution where the concentration of solute molecules outside a cell is equal to the concentration of solute molecules inside the cell

jaundice: the yellowish discoloration of the skin and other tissues brought about by the collection of bile pigments in the blood

karyotype chart: a picture of chromosomes arranged in homologous pairs

Klinefelter syndrome: a trisomic disorder in which a male carries an XXY condition

lacteals: small vessels that provide the products of fat digestion access to your circulatory system

lactic acid: an organic molecule that is half of a glucose molecule with the molecular formula $C_3H_6O_3$

larva: an intermediate form that an organism goes through to achieve its adult form. Tadpoles are larval frogs; caterpillars are larval butterflies or moths.

larynx: the voice box

lateral line: a line of sensory cells along each side of a fish's body

lateral meristems: cylindrical regions in roots and stems. They are responsible for all increases in diameters of roots and stems.

law of independent assortment: if genes are located on separate chromosomes, they are inherited independently of each other

leached: washed away as a soluble substance by rain water or a watering system

legumes: a group of angiosperms, including peas, beans, clover, and alfalfa, which tend to have nodules containing nitrogen-fixing bacteria on their roots

leukocytes: white blood cells

lichen: a combination of an alga or cyanobacterium and a fungus growing together in a symbiotic relationship

lipases: lipid-digesting enzymes

lipids: a chemical group which includes fats and oils

liposomes: artificial lipid vesicles

lymph nodes: round masses of tissue that supply lymphocytes to the bloodstream and remove bacteria and foreign particles from the lymph

lymph: the fluid found outside capillaries. Most often, the lymph contains some small proteins that have leaked through capillary walls.

lymphocytes: white blood cells that produce antibodies

lysis: the destruction or bursting open of a cell, e.g., when an invading virus replicates in a bacterium and many viruses are released

lysogeny: the dormant state of a virus

lysosomes: vesicles that contain a variety of enzymes able to break down large molecules

macromolecules: large molecules that are made by joining several separate units, such as joining several sugar units to form a starch molecule

macronutrients: 9 nutrients required by plants in relatively large quantities (>1000 mg/kg of dry mass)

manure: animal waste

marsupials: mammals that give birth to partially developed embryos that continue further development in the mother's pouch

meiosis: two-stage cell division in which the chromosome number of the parental cell is reduced by half. Meiosis is the process by which gametes are formed.

meristems: regions of the plant where some cells retain the ability to divide repeatedly by mitosis

mesoderm: the middle tissue layer of an animal embryo. In more complex animals, this later will become the muscles and other connective tissues, the blood vessels and blood cells, and various other organs, structures, and systems, depending on the organism.

mesophyll: the region of photosynthetic cells between the epidermal layers of leaves

mesophytes: plants that thrive with moderate moisture

metamorphosis: a series of stages that an organism goes through, from egg to adult. The intermediate forms are quite often different from the final form.

metastasis: the event where a cancer cell breaks free from the tumour and moves into another tissue

micronutrients: 8 nutrients required by plants in relatively small quantities (<100 mg/kg of dry mass)

microvilli: microscopic fingerlike outward projections of the cell membrane

minerals: elements (such as copper, iron, calcium, potassium, etc.) required by the body, often in trace amounts. Minerals are inorganic.

mitochondria: organelles that provide cells with a form of stored chemical energy

mitosis: a type of cell division in which a daughter cell receives the same number of chromosomes as the parent cell

Monera: in a five-kingdom system, a kingdom that includes organisms that lack a true nucleus

monoculture: the cultivation or growth of a single species

monohybrid cross: a cross that involves one allele pair of contrasting traits

monosomy: the condition where there is a single chromosome in place of a homologous pair

monotremes: mammals that reproduce by laying eggs

motile: capable of movement. Motile animals are able to move from place to place by expending cellular energy.

mucus: a protein produced by a layer of epithelial cells known as a mucous membrane

mutations: a heritable change in the molecular structure of DNA. Many mutations change the appearance of the organism.

mycelium: a collective term for the branching filaments that make up the part of a fungus not involved in sexual reproduction

mycorrhizae: symbiotic relationships between the hyphae of certain fungi and the roots of many specific plants

myogenic muscle: muscle that contracts without external nerve stimulation

native: describes species that originate in a particular region

natural fertilizers: fertilizers produced without human-directed chemical processes

nematocysts: stinging capsules that aid in the capture of prey

nephridia: open-ended tubules that function in excretion

nitrogen-fixing bacteria: bacteria that can convert atmospheric nitrogen gas (N_2) into ammonium ions (NH_4^+). They tend to live in nodules on the roots of legumes and have a symbiotic relationship with the legumes.

nodes: the locations where leaves are attached to the stem

nodules: swellings on the roots of legumes that contain symbiotic nitrogen-fixing bacteria

nonvascular: to be without the conductive tissues found in vascular plants. Nonvascular plants are referred to as bryophytes.

notochord: a skeletal rod of connective tissue that runs lengthwise along the dorsal surface and beneath the nerve cord. Notochords are present at some time during vertebrate development.

nuclear imaging: a medical imaging technique that uses radionuclides to view organs and tissues of the body

nuclear magnetic resonance (NMR) technology: a technique to determine the behaviour of the nucleus of an atom. In magnetic resonance imaging, NMR technology is used to produce a picture of the internal structures of the human body.

nucleolus: a small, spherical structure located inside the nucleus

nucleotides: the basic structural units of nucleic acid. Each unit is composed of a five-carbon sugar, a phosphate, and a nitrogenous base.

nucleus: the control centre for the cell, which contains hereditary information. The nucleus is bound by a double membrane.

nutrients: the raw materials needed for cell metabolism

obligate aerobes: bacteria that require oxygen for respiration

obligate anaerobes: bacteria that conduct respiration processes in the absence of oxygen

oocytes: immature eggs

ootid: an unfertilized ovum

operculum: also called a gill cover. It is a bony plate covering the gill chamber.

organ system: a group of organs that have related functions. Organ systems often interact.

organs: structures composed of different tissues specialized to carry out specific functions

osmosis: the diffusion of water molecules across a selectively permeable membrane

ovules: the plant structures that contain the megaspore mother cell and, later, the single haploid megaspore, which is the female gametophyte

oxidize: the loss of electrons from an atom or molecule

palisade mesophyll: one or two layers of brick-shaped cells, rich in chloroplasts and found tightly packed beneath the upper epidermis of most leaves

parasympathetic nervous system: a system of nerves that forms a division of the autonomic nervous system and that returns the body to normal resting levels following adjustments to stress

parenchyma: a living ground tissue that makes up the bulk of the plant body. Parenchyma tissues take part in several tasks, including photosynthesis, storage, and regeneration.

passive transport: the movement of materials across a cell membrane without the use of energy from the cell

pathogen: a disease-causing organism

pedigree chart: a graphic presentation of a family tree that permits patterns of inheritance to be followed for a single gene

pepsin: a protein-digesting enzyme produced by the stomach

perennial: describes plants which grow and reproduce repeatedly for many years

pericycle: a thin layer of lateral meristematic cells that surrounds the vascular cylinder

periderm: a protective covering that replaces the epidermis in plants that show extensive secondary growth

peristalsis: rhythmic, wavelike contraction of smooth muscle that moves food along the gastrointestinal tract

peritoneum: a covering membrane that lines the body cavity and covers the internal organs

persistent chemicals: fat-soluble chemicals that are not easily excreted from animal bodies and, thus, tend to accumulate

phagocytosis: a form of endocytosis in which solid particles are engulfed by cells

pharynx: a muscular section of the digestive tract. Air and/or food passes through this muscular tube.

phenotype: the observable traits of an organism that arise because of the interaction between genes and the environment

phloem: a vascular tissue that transports sugars, which were synthesized in the leaves, throughout the plant and down to the root for storage

phospholipids: the main components of cell membranes. They are composed of a phosphate group and two fatty acids attached to the glycerol backbone.

phosphorylation: the addition of one or more phosphate groups to a molecule

photoperiod: the number of daylight hours

photoreceptors: sensory receptors that detect light energy

photosynthesis: the process by which plants use chlorophyll to trap sunlight energy and use it to produce carbohydrates

phylogeny: the history of the evolution of a species or a group of organisms

pinocytosis: a form of endocytosis in which liquid droplets are engulfed by cells

pioneer organisms: organisms capable of surviving harsh conditions and establishing themselves in bare, barren, or open areas to initiate the process of primary or secondary succession

pith: the parenchyma tissue at the very centre of roots and stems

placental: a type of mammal that has all of the embryo development within the uterus of the female

placoid scales: toothlike scales composed of dentine, within a pulp cavity

plasma: the fluid portion of the blood

plasmid: a small ring of genetic material

plastids: organelles that function as factories for the production of sugars or as storehouses for starch and some pigments

platelets: component of blood responsible for initiating blood clotting

plesiomorphic: in cladistics, this term describes a primitive characteristic that is thought to be ancestral to all members of the group under consideration. Primitive characteristics are widespread and cannot be used to distinguish between members of such a group.

pleural membrane: a thin, fluid-filled membrane that surrounds the outer surface of the lungs and lines the inner wall of the chest cavity

pneumatophores: roots which grow upwards into the air to take in oxygen

polar bodies: cells that contain all the genetic information of a haploid ovum but lack sufficient cytoplasm to survive; formed during meiosis in females

pollen: the grains that contain the haploid male gametophyte in seed plants

pollination: the transfer of pollen from the pollen-producing organs to the organs containing the female gametophyte

polypeptide: a chain of amino acids held together by peptide bonds

prehensile: capable of grasping

primary growth: all plant growth originating at apical meristems resulting in increases in length, as well as growth originating at the lateral meristems in the first year of a plant's life

primary root: the first root developed from the seed

primary succession: succession that begins without preexisting organic material or soil

proglottids: the segmentlike divisions of a tapeworm's body

prokaryotic: a type of cell that does not have its chromosomes surrounded by a nuclear membrane

prothallus: the gametophyte plant of ferns

Protista: a kingdom originally proposed for all unicellular organisms such as the amoeba. More recently, multicellular algae have been added to the kingdom.

protonema: the young gametophyte of a moss in the early stages after the germination of the spore

protoplasm: the entire contents of a cell. The protoplasm includes the nucleus and the cytoplasm.

pseudocoelom: a fluid-filled cavity that lacks the mesodermal lining of a true coelom

psychoactive drugs: drugs that affect the nervous system and often result in changes to behaviour

pulmonary circulatory system: the system of blood vessels that carries deoxygenated blood to the lungs and oxygenated blood back to the heart

pulse: a change in the diameter of the arteries following heart contractions

Punnett square: a chart used by geneticists to show the possible combinations of alleles in offspring

Purkinje fibres: nerve fibres that branch and carry electrical impulses throughout the ventricles

pus: protein fragments that remain when white blood cells engulf and destroy invading microbes

radioisotopes: unstable chemicals that emit bursts of energy as they break down

radionuclides: the nucleii of unstable atoms that emit rays of energy. In nuclear imaging techniques, the energy emitted by radionuclides injected into the body is scanned to produce a picture.

recessive: alleles of this type are overruled by dominant alleles, which determine the genetic trait

recessive lethal: a trait that, when both recessive alleles are present, results in death or severe malformation of the offspring. Usually, recessive traits occur more frequently in males.

recombinant DNA: DNA that is created when fragments of DNA from two or more different organisms are spliced together

respiration: all processes involved in the exchange of oxygen and carbon dioxide between cells and the environment. Respiration includes breathing, gas exchange, and cellular respiration.

respiratory membrane: the living membrane where the diffusion of oxygen and other gases occurs. Gases are exchanged between the living cells of the body and the external environment (the atmosphere or water).

restriction enzyme: an enzyme that attacks a specific sequence of nitrogenous bases on a DNA molecule to create fragments of DNA with unpaired bases

rhizoids: hairlike structures that function like tiny roots and are found on the lower surfaces of certain parts of mosses, ferns, and other small organisms. They probably evolved before true roots with vascular tissue.

rhizomes: the barely visible stems of ferns

ribonucleic acid (RNA): a single-stranded genetic messenger that carries genetic information from the nucleus to the cytoplasm

ribosomes: structures within the cell, where protein synthesis occurs

root cap: a loose mass of cells forming a protective cap covering the apical meristems of most root tips

root hairs: microscopic extensions of the epidermal cells near the tip of a root. Root hairs function in the absorption of water and minerals.

runners: thin stems which grow along the ground producing roots and shoots at their nodes

S

salinization: the process in which salts in the groundwater are brought to the surface by irrigation and left to accumulate there as the water leaves the soil and plants through transpiration and evaporation

sap: the fluid within any part of a plant; mostly found within the xylem and phloem tissues

saprophytes: organisms that obtain nutrients from dead or nonliving organic matter

sapwood: the younger, softer, outer wood in tree trunks that is important for transporting water and dissolved materials as well as for support

sclerenchyma: a ground tissue whose mature cells are dead. These cells have thick walls composed of cellulose and lignin. Sclerenchyma supports mature plants and often protects seeds.

scolex: the knoblike head of a tapeworm

secondary growth: plant growth originating at lateral meristems that results in increased diameters of roots and stems in the second and all subsequent years of a plant's life

secondary root: smaller root branches growing sideways from a primary root

secondary succession: succession which begins with organic material or soil already present

secretin: a hormone that stimulates pancreatic and bile secretions

sedentary: used to sitting still much of the time; moving little and rarely

seed: an ovule after fertilization, containing an embryo which developed from the zygote

segmentation: the repetition of body units that contain some similar structures

segregation: the separation of paired alleles during meiosis

selective breeding: the crossing of desired traits from plants or animals to produce offspring that have one or several of the favoured characteristics

selectively permeable membrane: a barrier that allows some molecules to pass through, but prevents other molecules from penetrating

semilunar valves: valves that prevent the backflow of blood from arteries into the ventricles

septum: a wall of muscle that separates the right heart pump from the left

serum: the liquid that remains after the solid and liquid components of blood have been separated

sessile: not capable of independent movement. Sessile animals remain fixed in one place throughout their adult lives.

sewage sludge: semisolid matter produced during sewage treatment

sex chromosomes: the pair of chromosomes that have a role in the sex of an individual

sex-linked traits: traits that are controlled by genes located on the sex chromosomes

sexual reproduction: the production of offspring from the union of two sex cells, one from each different parent; the genetic makeup of the offspring is different from that of either parent

sieve tubes: long tubes formed by many sieve elements to allow easy passage of water and dissolved materials

simple leaf: a leaf which is not divided into leaflets

sinoatrial (SA) node: a small mass of tissue in the right atrium that originates the impulses stimulating the heartbeat

sinus: a body cavity or air space surrounding an internal organ

somatic cells: all the cells of an organism other than the sex cells

sorus: a cluster of sporangia on a fern frond

species: a group of organisms that look alike and can interbreed under natural conditions to produce fertile offspring

spermatocytes: sperm-producing sex cells

sphincters: constrictor muscles that surround a tubelike structure

sphygmomanometer: a device used to measure blood pressure. Blood pressure is traditionally measured in millimetres of mercury (mm Hg).

spindle fibres: protein structures that guide chromosomes during cell division

spleen: a lymphoid organ that acts as a reservoir for blood and a filtering site for lymph

spongy mesophyll: a layer of irregularly shaped cells containing chloroplasts between the palisade mesophyll and the lower epidermis of most leaves. Many air spaces are randomly distributed within this layer.

sporangia: the reproductive structures in which spores are produced

spore mother cells: the last sporophyte cells. They undergo meiosis to produce the haploid spores.

spores: reproductive cells that can produce a new organism without fertilization. Spores have a haploid number of chromosomes. Fungal spores have thick, resistant outer coverings to protect them.

sporophyte: refers to a stage in a plant's life cycle in which cells have diploid nuclei. This stage arises from the union of two haploid gametes. During this stage, spores are produced by meiosis.

starch: a large carbohydrate molecule used by plants to store energy

sternum: the breast bone found at the front of the chest in birds and mammals

stimulant: a drug that speeds the action of the central nervous system, often causing an increase in heart and breathing rate

stomata: pores in the epidermis of plants, particularly in leaves. They permit the exchange of gases between the plant and atmosphere while at the same time helping to prevent excessive water loss.

substrate: a surface in or on which an organism grows or is attached

succession: a series of gradual changes in the vegetation of an area followed by gradual changes in the animals in the area

succulents: plants which have thick, fleshy parts due to the presence of large amounts of parenchyma for water storage

swim bladder: a small, balloonlike structure filled with gases which helps fish maintain buoyancy and allows them to float at different depths. It is also called an air bladder.

symbiotic relationship: a relationship between two organisms in which both partners benefit from the interaction. Some of these relationships may be necessary for the survival of the partners while others benefit both partners but are not necessary.

sympathetic nervous system: a system of nerves that forms a division of the autonomic nervous system and that prepares the body for stress

synapsis: the pairing of homologous chromosomes

synthetic fertilizers: fertilizers produced through human-directed chemical processes

systemic circulatory system: the system of blood vessels that carries oxygenated blood to the tissues of the body and deoxygenated blood back to the heart

systole: contraction of the heart, during which blood is pushed out of the heart

taproots: root systems where the primary root remains predominant, though very small secondary roots may be present

taxa: categories used to classify organisms

taxonomy: the science of classifying organisms

tetrad: a pair of homologous chromosomes, each with two chromatids

thymus gland: a lymphoid organ in which lymphocytes mature, multiply, and differentiate

tidal volume: the amount of air inhaled and exhaled in a normal breath

tissues: groups of cells that work together to perform specialized tasks

totipotent: having the ability to support the development of an egg to an adult

toxin: a poison produced in the body of a living organism. It is not harmful to the organism itself but to other organisms.

trachea: the windpipe

tracheids: xylem cells with tapered, overlapping ends and pits in their cell walls for conducting water and dissolved materials in plants

transcription: the process by which an mRNA molecule is built using the sequence of nucleotides in DNA as a template

translation: the process by which polypeptides (proteins) are produced at the ribosomes. Transfer RNA (tRNA) positions amino acids according to the sequence of nucleotides in mRNA.

translocation: the process of moving the products of photosynthesis throughout the plant body through phloem

transpiration: the loss of water through the surfaces of a plant. Most transpiration occurs through the leaf stomata.

transposons: specific segments of DNA that can move along the chromosome

triglyceride: a lipid composed of glycerol and three fatty acids which are bonded together

trisomy: the condition where there are three homologous chromosomes in place of a homologous pair

trypsin: a protein-digesting enzyme

tubers: thick underground stems specialized for carbohydrate storage and asexual reproduction

turgor pressure: the pressure exerted by water against the cell membrane and the cell walls of plant cells

Turner syndrome: a monosomic disorder in which a female has a single X chromosome

ulcer: a lesion along the surface of an organ

vaccines: solutions that are prepared from viral components or inactivated viruses

vacuole: a large, fluid-filled compartment in the cytoplasm of a plant cell that stores sugars, minerals, proteins, and water and is important in maintaining turgor pressure

variety: a subspecies of a plant species. In humans we refer to subspecies as *races*. For domesticated animals, the term *breed* applies.

vascular bundles: collections of xylem and phloem tissue, separate from other collections, running longitudinally through stems

vascular cambium: a lateral meristem which is responsible for creating new xylem and phloem tissue

vascular: describes the system of conductive tissue (xylem and phloem) found in plants to transport water and dissolved minerals throughout a plant. Vascular plants are referred to as tracheophytes.

vasoconstriction: the narrowing of blood vessels. Less blood goes to the tissues when the arterioles constrict.

vasodilation: the widening of blood vessels. More blood moves to tissues when arterioles dilate.

vegetative: describes any part of a fungus or plant that is not involved with sexual reproduction

veins: blood vessels that carry blood from the body to the heart

ventricles: muscular, thick-walled chambers of the heart that deliver blood to the arteries

vertebrates: multicellular, eukaryotic heterotrophs that have a notochord at some stage in their life

vesicles: small sacs or packets that are released by the Golgi apparatus. Vesicles are important in the processes of exocytosis and endocytosis.

vessels: long tubes of vessel elements for conducting water and dissolved materials in plants

villi: small fingerlike projections that extend into the small intestine which increase surface area for absorption

viruses: microscopic particles capable of reproducing only within living cells

vital capacity: the maximum amount of air that can be exhaled

vitamins: organic molecules needed in trace amounts for normal growth and metabolic processes

viviparous: describes animals that give birth to live young

woody: describes stems of perennial plants. They increase in diameter each year as more and more vascular tissue is created. The xylem cells, even after they have died, create the hard, woody tissue called wood.

xerophytes: plants that survive or thrive in areas with very little moisture

xylem: a vascular tissue in plants that carries water and dissolved materials up from the roots to the other plant parts

zoology: the study of animal life

zygote: a cell resulting from the union of a male and a female sex cell until it divides, and then it is called an embryo

Index

A

Abiotic, 501
Abscisic acid, 541
Absorption, in digestion, 225–26
Acorns, 529
Active transport, 56–57
Adam's apple, 286
Addiction, 306
Adenosine deaminase (ADA), 185
    deficiency, 338
Adenosine diphosphate (ADP), 62
Adenosine triphosphate (ATP), 14
    production of, 62
    in respiration, 64–65
Adipose, 15
Adventitious roots, 512
Aerial roots, 514
Aerobic respiration, 60, 64
Agar, 352
Agnatha, 452
Agranulocytes, 245, 246
Agriculture
    alternative methods of, 400–401
    biotechnology in, 99, 362
Albumins, 244
Alcohol, 223, 265, 305–306
Alcoholic fermentation, 65–66
Algae, 350, 351–53
Alimentary canal, 209
Allantois, 458
Alleles, 132
    multiple, 143
Allelopathy, 515, 551
Allen, Dr. Theresa, 49
Allergic reactions, 273
Alternation of generations, 378–79
Alveoli, 287
Amino acids, 15, 41
    and protein, 45–47
Ammonia, 558
Amniocentesis, 121
Amnion, 458
Amniotic eggs, 458
Amoeba, 208–209, 354
Amoebic dysentery, 355
Amphibians, 450, 455–56
Amylase, 210
Amyloplasts, 20
Anaerobic respiration, 64–66
Anaphase, 88–89
Anaphylactic reactions, 273
Anderson, Dr. Gail, 442
Anemia, 245
Aneurysm, 250
Angioplasty, 258
Angiosperms, 391, 394–99
Animal breeder, 184
Animal kingdom, 408–11
Animal-like protists, 353–55
Animals
    cloning, 96–97
    and humans, 475
    reproduction in, 113–14
Annelida (phylum), 421–22
Annual, 512
Annual ring, 521
Antheridium, 382
Anthocyanins, 72
Antibiotics, 178, 345–46
Antibodies, 23, 245, 247
Antigen, 247
Antioxidants, 235
Aorta, 257
Apes, 467
Apical meristems, 494
Apomorphic, 450
Archaebacteria, 329, 330, 340–41, 348
Archegonium, 382
Arctic fox, 470
Aristotle, 162, 326
Arrhythmia, 267
Arteries, 241, 250–51
Arteriolar resistance, 264
Arterioles, 250
Arteriosclerosis, 251
Arthritis, 24
Arthropods, 435–40
Artificial selection, 563
Artificial sweeteners, 30
Ascaris, 420
Asexual reproduction, 82
Aspartame, 30
Asthma, 294–95
Atherosclerosis, 251
Athletes, 280–81
Athlete's foot, 8
Atmosphere, composition of, 282
Atria, 256
Atrioventricular (AV) node, 259
Atrioventricular (AV) valves, 256
Autonomic nervous system, 250
Autosomes, 115
Autotrophs, 58
Auxins, 540
Avery, Oswald, 173
Aves, 462–63. See also Birds

B

Bacteria, 8, 340–45. See also Archaebacteria
    and Eubacteria kingdoms
    beneficial effects of, 343
    harmful effects of, 344–45
    nitrogen-fixing, 538
    poison-eating, 343–44
    resistance to antibiotics, 345–46
Bacterial conjugation, 178
Bacteriophages, 335
Bailey, Dr. B.W.D., 25
Bangham, Alec, 49
Bark, 522
Barr, Dr. Murray, 166
Barr bodies, 166
Basal metabolic rate (BMR), 230
Bats, 474
Bear, Dr. Christine, 181
Bears, 475
Benden, Edouard van, 162
Bennett, Dr. G.F., 356
Beta-blockers, 266
Bile, 222–23
Bile salts, 222
Binary fission, 342
Binomial nomenclature, 328
BIOCAP, 546
Biodiversity, 554–56
Biological accumulation, 580
Biological controls, 582
Biological magnification, 580
Biosurgery, 423
Biotechnology, in agriculture, 99
Biotic, 501
Birds, 450, 461–65
    digestion in, 210
Bivalves, 430
Blastula, 91
Blood
    artificial, 247–48
    components of, 244–46
Blood cells, artificial, 18
Blood clots, 246
Blood flow, one-way, 256–57
Blood groups, 246–47
Blood pressure, 264–65, 267
Blood sugar level, 227
Blood transfusions, 246–48
Blood vessels, 250–53
Bolting, 541
Bone marrow, 245, 246, 275
Boonstra, Dr. Rudy, 323
Botanist, 61
Boveri, Theodor, 163
Bread mould, 369
Breathing, 283
Breathing movements, 287–89
Briggs, Robert, 96
Bronchi, 287
Bronchial asthma, 294
Bronchioles, 287
Bronchitis, 294–95
Brown, Robert, 9
Brownian motion, 50
Buffer, 292

C

Cacti, 498, 505, 522
Calorie, 230
Cambium, 496
Cancer
    causes of, 102
    cells, 15, 91, 100–101
    lung, 295, 299–302
Cannon, Walter, 206
Canola, 141

Capillaries, 214, 251–52
Capillary fluid exchange, 272
Capsid, 335
Carbohydrates, 34–37, 62
    in plants, 526
Carbon dioxide
    and aerobic respiration, 64–65
    and global warming, 68, 546
    partial pressure of, 290–91
    and photosynthesis, 503, 504
Carbon dioxide transport, 292
Carbonic acid, 292
Carbonic anhydrase, 292
Cardiac catheterization, 257–58
Cardiac muscle, 259
Cardiac output, 264
Cardiovascular disease, 266–67
Carnivorous plants, 506–507
Carotenoids, 534
Cations, 537
Cell clock, 90–91
Cell cycle, 86–89
Cell division
    principles of, 84–85
    and reproduction, 112–15
Cell fractionation, 14
Cell membrane, 9, 10, 11, 47–49
Cell organization, levels of, 205
Cell research, 23–25
Cells
    aging of, 90–91
    artificial, 18
    looking at, 12–14
    and proteins, 41
Cell structure, overview, 10–12
Cell theory, 8–9, 84
Cellular respiration, 14, 60, 62–63
Cellulose, 36, 225
Cell walls, of plants, 20
Centipedes, 439
Centrifuge, 14
Centrioles, 87
Centromere, 86
Cephalization, 410
Cephalopods, 431
Cetaceans, 474
Chang, Dr. Thomas, 18
Chelicerates, 438
Chemicals of life, 32
Chitin, 36, 364
Chlorophyll, 501, 534
Chloroplasts, 19, 60, 503
Cholecystokinin (CCK), 222–23
Cholesterol, 39
Chondrichthyes, 452–53
Chordata (phylum), 448–50
Chorion, 458
Chorionic villus sampling (CVS), 121
Chromatid, 86
Chromatin, 86
Chromoplasts, 20
Chromosomal theory, 163
Chromosomes, 10, 84–85
    double-stranded, 86

homologous, 104
    looking inside, 172–73
    in meiosis, 103–107
    sex, 115
Cicada, 438
Cilia, 12, 286
Ciliophora, 354–55
Circulatory system
    and exercise, 268
    illustrated, 254
    importance of, 242–43
Cirrhosis, 223, 305–306
Cladistics, 450–51
Classification, levels of, 328–30
Climax community, 548
Clitellum, 422
Cloning, 94–99
Cnidaria (phylum), 413–15
Coagulation, 42
Codominance, 144–45
Coelacanth, 454
Coelom, 408
Coenzymes, 537
Cofactors, 537
Collenchyma, 498
Colon, 224
Companion cells, 499
Competition, 551
Complementary base pair, 175
Complex carbohydrates, 36
Compost, 557–58
Compound eyes, 437
Compound leaf, 502
Computerized axial tomography (CAT) scan, 206
Concentration, 50
Concentration gradient, 52
Concept map, 614
Conifers, 391–92, 505
Conjugation, 178, 343
Conservation ecologist, 577
Contractile vacuole, 53
Controlled experiments, 598
Coral, 413, 415
Cork, 514
Cork cambium, 514
Corn, 155
Cornea, 24
Coronary arteries, 257
Coronary bypass, 257
Correlational studies, 600
Cortex, 498
Cotyledon, 397
Coyotes, 475
Crayfish, 436
Crick, Francis, 174–75
Cristae, 15
Crop (stomach), 209
Crop rotation, 558
Cross-fertilization, 134
Crossing over, 105
Crustaceans, 439
Cuticle, 377
Cyclamate, 30

Cystic fibrosis, 42, 79, 181
Cysts, 354
Cytokinesis, 84, 89
Cytokinins, 541
Cytology, 162
Cytoplasm, 10, 11
Cytoplasmic organelles, 14–18
Cytoskeleton, 372–73

D

Dalton's law of partial pressure, 290
Dandelions, 551
Daphnia, 82–83
Daughter cells, 84
DDT, 580–82
Deber, Dr. Charles, 42
Decision making, 604–607
Dehydration synthesis, 35
Denaturation, 42
Deoxyribonucleic acid (DNA), 10–11, 41, 45–47, 172
    discovering structure of, 174–75
    fingerprinting, 182–83
    recombinant, 99, 178
    structure of, 46, 175–77
Deoxyribose, 35
Depressants, 304
Dermal tissue system, 497
Desertification, 569
Detoxify, 223
Diabetes, 185
Diaphragm, 288
Diastole, 261
Dichotomous key, 329, 330
Dicots, 397
Dietician, 61
Dieting, 232
Diffusion, 50–51
Digestion, 208–10, 214–15
Digestive enzymes, 208
Digestive system, illustrated, 211
Digitalis, 266
Dihybrid crosses, 150–52
Dikaryotic, 366
Dinosaurs, 459, 460
Dioecious, 431
Diploid, 103
Disaccharides, 35
DNA fingerprinting, 182–83
Dodo bird, 529
Dolly (sheep), 97–98
Doogie mice, 160
Dominant traits, 132
Dormant, 528
Down, John, 120
Downey, Richard, 141
Down syndrome, 117–18, 120–21
Duchenne muscular dystrophy, 182
Duodenum, 216

E

Earthworms, 421–22. *See also* Worms
    circulatory system in, 243
    digestion in, 209
    dissection, 424–26
    respiration in, 283
Echinoderms, 432–33
Ectoderm, 242, 408
Ectoplasm, 354
Ectotherms, 449
Egg cells, 114
Electrocardiographs, 260–61
Electron microscope, 6–7
Electron transport system, 63
Emphysema, 294, 295
Endocytosis, 17, 56
Endoderm, 242, 408
Endodermis, 512
Endoplasm, 354
Endoplasmic reticulum, 16
Endoscope, 215
Endoskeleton, 432
Endosperm, 528
Endospores, 343
Endotherms, 449
Energy, food, 230–32
Energy flow, in photosynthesis/cellular
    respiration, 58–60
Energy requirements, calculating, 232
Enterokinase, 217
Enucleated cells, 96
Environment, and phenotype, 160
Enzymes, 15, 41
    digestive, 208
    restriction, 178
Epidermis, 497
Epiglottis, 286
Epinephrine, 266, 268
Epiphytes, 387, 515
Equatorial plate, 88
Equilibrium, 52
Erepsins, 217, 218
Erythrocytes, 84, 244–45. *See also* Red blood
    cells
Escherichia coli, 8
Esophagus, 209, 212, 285
Essential amino acids, 41, 234
Essential nutrients, 234–35
Estimating, 617
Ethanol, 569
Ethylene, 541
Eubacteria, 329, 330, 340–41, 342, 348
Euglena, 351
Eukaryotic cells, 10. *See also* Protista
    kingdom
Exercise, 268
Exergonic chemical reactions, 62
Exocytosis, 17, 56
Exoskeleton, 436
Exotic species, 555
Expiratory reserve volume, 296
Extracellular fluid (ECF), 272
Exxon Valdez, 430

F

Facilitated diffusion, 53
Facultative anaerobes, 342
Farming, alternative, 400–401
Fats, 30. *See also* Lipids
    and atherosclerosis, 251
    in diet, 234
    saturated, 37–38
    unsaturated, 37–38
Fatty acids, 37
Fermentation, 342
Ferns, 386–89
Fertilization, 82
Fertilizers, 557–59
Fibre, in diets, 225
Fibres, 569–70
Fibrinogens, 244
Fibrous roots, 512
Field mushroom, 366–67
Filtration, 272
Fish, 450, 452–54
Fishbone diagram, 614
Six-kingdom system, 328–30
Flagella, 12
Flatworms, 417–18
Fleming, Walter, 162
Flowers, 131, 396–99, 528
Fluid pressure, 272–73
Fluosol, 248
Food energy, 230–32
Food plants, 562–67
Forensic entomology, 442
Forensic scientist, 184
Forest fires, 548, 552–53
Forests, 572–73
Forest technician/scientist, 577
Fox, Terry, 101
Foxglove, 266
Franklin, Rosalind, 174, 175
Free radicals, 235
Frogs, respiration in, 285
Frond, 386
Fructose, 35
Fruit, 399, 529
Fruit fly, 143, 164
Fuels, 569
Fungal mimicry, 371–72
Fungi, 8, 329
    classification of, 365
    importance of, 369–73
    life cycle of, 366–67
    vs. plants, 364
Fungilike protists, 356

G

Galactose, 35
Gall, 102
Gall bladder, 222–24
Gallstones, 223
Gametes, 103
Gametogenesis, 113
Gametophyte, 113, 379

Garden peas, characteristics of, 130–31
Garneau, Dr. Marc, 492
Gas exchange and transport, 290–93
Gastrin, 227–28
Gastropods, 430
Gastrovascular cavity, 208
Gastrula, 96
Gemmae, 382
Gene pool, 567
Genes, 10, 45
    jumping, 177–78
    in plants, 131–32
Gene therapy, 49, 185–86
    and viral vectors, 338–40
Genetic diversity, 567, 585
Genetic engineering, 95
Genetic research, and technologies, 180–82
Genetics
    early development in, 162
    terms used in, 134–35
Genetics counsellor, 184
Genetic screening, 146
Genome, 180
Genotype, 134
Genus, 328
Germinate, 366
Germ layers, 408
Gibberellins, 91, 541
Gills, 283
Gizzard, 209
Global warming, 68, 546
Globulins, 244
Glucose, 15, 35, 62, 65
Glycerol, 37
Glycogen, 36, 62
Glycoproteins, 48
Golgi, Camillo, 16
Golgi apparatus, 16–17
Govind, Dr. C.K., 197
Grant, Dr. David, 304
Granulocytes, 245, 246
Graphic organizers, 614
Graphs, 618–19
Grasshopper, 437
Grbic, Dr. Vojislava, 489
Great blue herons, 559
Greek prefixes/suffixes, 632–33
Green revolution, 566
Griffiths, Frederick, 173
Ground tissue system, 498
Guard cells, 503
Gymnosperms, 391–93

H

Hematocrit, 244
Hemoglobin, 244–45, 291–92
Hemophilia, 169
Hagfish, 452
Hair, 470
Hall, Dr. Judith, 181
Halophytes, 585
Haploid, 103
Harvey, William, 252

Hayden, Dr. Michael, 181
Heart
    artificial, 262–63
    diseases, 266–67
    structure of, 256–57
    tempo of, 259–61
Heart drugs, 266
Heart murmur, 261–62
Heart sounds, 261–62
Heart transplants, 240–41
Heartwood, 522
Hemocoel, 438
Hemp, 571
Hepatitis B, 337
Herbaceous stems, 518
Heredity, 128
    early beliefs, 130–34
    searching for chemical of, 172–73
Hermaphroditic, 412
Heterotrophs, 60
Heterozygous, 135
Hiccup, 287
Hillier, James, 6
Histamine, 273
Homeostasis, 52, 226–27
    and digestion, 227–28
    and psychoactive drugs, 304–306
Hominid, 467
Homologous chromosomes, 104
Homozygous, 135
Hooke, Robert, 8–9
Hormones, 48–49
Horn-Miller, Waneek, 280
Horticulturalist, 184
Host range, 336
Human Genome Project, 46, 180
Humus, 557
Huntington's chorea, 146, 181
Hybridization, 140
Hybridomas, 23–24
Hydra, 94, 208–209, 414, 415
Hydrochloric acid, 214
Hydrogen bond, 175
Hydrogen ions, 292
Hydrolytic enzymes, 208
Hydrophytes, 507
Hydroponics, 559
Hypersensitivity, 273
Hypertension, 49, 267
Hypertonic solution, 53
Hypha, 365
Hypotonic solution, 52

Immune system, 242
Inbreeding, 140
Incomplete dominance, 144–45
Inflammation, 273
Ingestion, 210–12
Insects, 439
    respiration in, 284
Inspiratory reserve volume, 296
Insulin, 48

Integrated pest management (IPM), 582
Intercostal muscles, 288
Intermediate inheritance, 144
Internode, 501
Interphase, 86
Interspecific competition, 551
Interstitial fluid, 243
Intestine
    large, 224–26
    small, 216–18
Intraspecific competition, 551
Invasive species, 555
Invertebrates, 408
    classifying, 410–11
Iron, 245
Isotonic solutions, 52

Jaundice, 223
Jeffreys, Alec, 183
Jojoba, 576
Joule, James, 230
Joules, 230
Jumping gene theory, 177

Karyotype charts, 117, 119
Karyotyping, 121
Kennedy, Dr. Brian, 185
King, Thomas, 96
Klinefelter syndrome, 118–19
Koala bear, 466
Krill, 440
KWL chart, 614

L

Lab reports, 610–13
Lab technician, 61
Lacteals, 226
Lactic acid fermentation, 64–65
Lactobacillus bulgaricus, 8
Lactose, 35, 218
Ladybugs, 582
Laennec, Rene, 262
Lamprey, 452
Land base, 584–85
Landsteiner, Karl, 246
Larva, 412
Larynx, 286
Lasers, 215
Lateral line, 453
Lateral meristems, 495
Latin prefixes/suffixes, 632–33
Law of independent assortment, 151
Leached, 537
Leaves, 501–507
Leeches, 422, 423
Legumes, 538
Leonardo da Vinci, 162
Leukemia, 91
Leukocytes, 245, 246. See also White blood cells

Lichens, 370–71, 549
Limbs, in mammals, 471–72
Linnaeus, Carl, 328
Lipase, 218
Lipids, 11, 37–39, 62, 218. See also Fats
Liposomes, 49
Liver, 222–24, 305–306
Liver tissue transplants, 98
Lobe-finned fish, 454
Lung cancer, 295, 299–302
Lung volume, 296
Lymph, 274
Lymphatic system, 274–76
Lymph nodes, 274
Lymphocytes, 274
Lymphoid organs, 275–76
Lysis, 336
Lysogeny, 336
Lysosomes, 17–18, 24, 208

M

MacLeod, Colin, 173
Macromolecules, 32
Macronutrients, 536
Maggots, 423
Magnetic resonance imaging (MRI), 206
Maiman, Theodore, 215
Malaria, 355, 574, 580
Maltose, 35, 62
Mammals, 450, 465–77
Manatees, 474
Manure, 557–58
Margarine, 38
Marijuana, 571
Marshall, Dr. Barry, 215
Marsupials, 466
Mastigophora, 354
Math skills, 615–19
McCarthy, Senator Joseph, 175
McCarty, MacLyn, 173
McClintock, Barbara, 177
Measuring, 617
Medicinal plant products, 574–75
Medusa, 413, 415
Meiosis, 103
    abnormal, 116–19
    compared with mitosis, 108–109
    stages of, 104–107
Mendel, Gregor, 130–34, 162
Meristems, 494–96
Mesoderm, 242, 408
Mesophyll, 504
Mesophytes, 508
Messenger RNA (mRNA), 11, 41
Metabolism, factors affecting, 231–32
Metamorphosis, 412
Metaphase, 88
Metastasis, 101
Microfilaments, 18
Micronutrients, 537
Micropipette, 96
Microscopes, 6–7, 162
    care and use of, 627–29

Microspheres, 9
Microtubules, 18
Microvilli, 225
Middle lamella, 20
Milkweed, 506
Minerals, 235
Mitochondria, 14–15, 32, 60
Mitosis, 84, 86–89
    compared with meiosis, 108–109
Models, 460
Mollusks, 428–32
Monera kingdom, 328, 329. *See also*
    Archaebacteria *and* Eubacteria
Monocots, 397
Monoculture, 555
Monohybrid cross, 135
Monosaccharides, 35
Monosomy, 116
Monotremes, 465–66
Morgan, Thomas Hunt, 163–66
Mosses, 381–85
Motile, 410
Mucus, 214
Mueller, Paul, 580
Mullins, Kary, 180
Multiple alleles, 143
Multiple marker screen, 121
Multiples, common, 630
Murrin, Dr. Faye, 372–73
Muscle cells, 15
Muscular dystrophy, 181–82
Mushroom, 366–67
Mussels, and glue, 431–32
Mussivand, Dr. Tofy, 262
Mutations, 164
Mycelium, 364
Mycorrhizae, 370
Myogenic muscle, 259

N

Nanotechnology, 3
Native species, 555
Natural fertilizers, 557–58
Nematocysts, 414
Nematoda (phylum), 419–20
Nephridia, 422
Nicotine, 306
Nitrogen, in plants, 538–39, 557–58
Nitrogen-fixing bacteria, 538
Nitrogenous bases, 175
Nitroglycerine, 266
Nodes, 501
Nodules, 538
Nondisjunction, 116–19
    disorders, 117–18
Nonvascular, 374
Notochord, 408
Nuclear imaging, 206
Nuclear magnetic resonance (NMR)
    technology, 206
Nuclear medicine, 206
Nucleic acids, 45–47
Nucleolus, 11

Nucleotides, 41
Nucleus, 9, 10–11
Nurse, 298
Nutrients, 32–34, 208
    essential, 234–35
Nutritionist, 61

O

Obligate aerobes, 342
Obligate anaerobes, 342
Observational studies, 602
Oils, 37
Olestra, 30
Oncologist, 298
Oocytes, 114
Ootid, 114
Operculum, 453
Opossum, 466
Orchids, 515
Organelles, 13
Organic farming, 400–401
Organs, 204
    monitoring, 205–206
Organ systems, 204
Organ transplants, 303–304
Osmosis, 51–53
Osmotic pressure, 272–73
Osteichthyes, 453–54
Ostrich, 463
Ovules, 392
Oxidation, 282
Oxidization, 65
Oxygen
    challenge of getting, 283–85
    and circulation, 244–45, 252
    partial pressure of, 290–91
    and photosynthesis, 504
    and respiration, 64–65
Oxygen transport, 291

P

Pacific yew, 392
Palisade mesophyll, 504
Pancreas, 216–18
Papillae, 211
Paramecium, 354–55
Parasympathetic nervous system, 260
Parenchyma, 498
Parent cells, 84
Parkinson's disease, 24–25
Partial pressure, 290–91
Passive transport, 50–53
Pasteur, Louis, 9
Pathogen, 173
Pauling, Linus, 175
Peat, 384–85, 569
Pectin, 20
Pedigree charts, 141–42
Penguin, 463
Penrose, L.S., 120
Pepsin, 214
Peregrine falcon, 580

Perennial, 512
Pericycle, 512
Periderm, 497
Periodic table, 634–35
Peristalsis, 212
Peritoneum, 409
Permutation, 47
Persistent chemicals, 580
Pest control officer, 479
Pest management, 579–82
Petiole, 502
Phagocytosis, 56
Pharynx, 209, 285
Phenotype, 135
    effect of environment on, 160
Phloem, 377, 499, 500, 512
Phospholipid bilayer, 48
Phospholipids, 39
    molecule of, 48
Phosphorylation, 62
Photoperiod, 534
Photoreceptors, 437
Photosynthesis, 19, 494, 501, 504
    energy flow in, 58–60
Phylogenetic tree, 329, 330
Phylogeny, 329
Pine tree, 392–94
Pinocytosis, 56
Pioneer organisms, 549
Pith, 498
Placental mammals, 466
    major orders of, 468–69
Placoid scales, 453
Planaria, 417–18
Plantae kingdom, 374
Plant breeders, Canadian, 141
Plant cells, special structures of, 19–20
Plant hormones, 540–41
Plantlike protists, 350–53
Plants
    and biodiversity, 554–56
    carnivorous, 506–507
    chemical products of, 574–75
    cloning, 94–95
    domestication of, 563
    evolution of, 375–77
    external factors affecting growth of,
        534–39
    and fertilizers, 557–59
    and food, 562–67
    vs. fungi, 364
    internal factors affecting growth of,
        540–41
    life cycle of, 378–79
    light requirements of, 534–35
    Mendel's experiments with, 130–34
    new kinds of, 584
    nonfood products of, 569–73
    and pest management, 579–82
    reproduction in, 112–13, 528–30
    seed, 390–401
    and soil nutrients, 536–39, 557–59
    structure and function, 494–96
    and succession, 548–53

and sustainability, 584–85
transport in, 525–26
Plant tissues, 497–99
Plant variety, 564
Plasma, 244
Plasmid, 178
Plastids, 19–20
Platelets, 245–46
Platyhelminthes, 417–18
Platypus, 446, 447, 465
Plesiomorphic, 450
Pleural membrane, 287
Pleurisy, 287
PMI chart, 614
Pneumatophores, 514
Pneumothorax, 289
Polar bodies, 114
Politics, and science, 175
Pollen, 392
Pollination, 131, 393, 398
Polychaetes, 422
Polyp, 413
Polypeptides, 41
Polysaccharides, 36
Porifera (phylum), 412
Prebus, Albert, 6
Prefixes
    Greek and Latin, 632–33
    numerical, 630
Prehensile, 466
Prenatal testing, 121, 146
Pressure-flow hypothesis, 526
Primary growth, 496
Primary root, 512
Primary succession, 549, 550
Primates, 466–67
Prions, 338
Probability, 617–18
Progeria, 127
Proglottids, 418
Prokaryotes, 328
Prokaryotic cell, 10
Prophase, 87–88
Proteins, 15, 41–42, 62, 172–73
    and amino acids, 45–47
    in digestion, 217
    research, 42
Prothallus, 387
Protista kingdom, 328, 349–56
Protonema, 382
Protoplasm, 10
Protozoa, 353–55
Pseudocoelom, 409
Psychoactive drugs, 304–306
Pulmonary circulatory system, 256
Pulse, 241
Punnett, Reginald, 135
Punnett square, 135
Purines, 175
Purkinje fibres, 260
Pus, 245
Pyrimidines, 175

R

Radioisotopes, 14
Radionuclides, 206
Recessive lethal trait, 166
Recessive traits, 132
Recombinant DNA, 99, 178
Red blood cells, 84, 244–45. *See also*
    Erythrocytes
Red bone marrow, 275
Rennin, 214
Replication, viral, 336
Reproduction
    and cell division, 112–15
    in plants, 528–30
Reptiles, 450, 457–59
Research scientist, 479
Residual air, 291
Respiration
    process, 282–83
    cellular, types of, 64–66
Respiratory membrane, 283
Respiratory stress syndrome, 287
Respiratory system
    disorders of, 294–95
    illustrated, 286
    importance of, 282–85
    in mammals, 285–89
Respiratory therapist, 298
Restriction enzymes, 178
Rhesus factor, 247
Rhizoids, 376
Rhizomes, 386
Ribonucleic acid (RNA), 11, 45
Ribosomal RNA (rRNA), 11
Ribosomes, 11, 15–16
Risk–benefit analysis model, 607
Roentgen, Wilhelm, 205
Root cap, 512
Root hairs, 512
Roots, 512–15
Rough endoplasmic reticulum, 16
Roundworms, 419–20
Rubber, 576
RUBISCO, 504
Runners, 522

S

Saccharin, 30
Safety conventions and symbols, 620–21
Safety in the laboratory, 622–26
Salinization, 585
Saliva, 210–11
Salt, on roads, 53
Sanger, Frederick, 180
Sap, 522, 526
Saprophytes, 365
Sapwood, 522
Sarcodina, 354
Saturated fats, 37–38
Saunders, Sir Charles, 141
Scanning electron microscope (SEM), 6–7

Scanning tunnelling microscope (STM), 3,
    6–7
Schleiden, Mathias, 9
Schwann, Theodor, 9
Science, and politics, 175
Scientific inquiry skills, 598–603
Scientific notation, 615
Sclerenchyma, 498
Scolex, 418
Sea cucumbers, 432
Sea urchins, 432
Secondary growth, 496
Secondary root, 512
Secondary succession, 549
Secretin, 217, 227
Sedentary, 430
Seed plants, 390–401
Seeds, 393, 528–30
Sefton, Dr. Michael, 241
Segmentation, 421
Segregation, 106, 134
Selective breeding, 140–42
Selectively permeable membrane, 11
Semilunar valves, 256
Septum, 256
Serum, 247
Sessile, 410
Sewage sludge, 558
Sex chromosomes, 115, 167
Sex linkage, 163–66
Sex-linked traits, 165
Sexual reproduction, 82
Sharks, 452–53
Sieve tubes, 499
Significant digits, 616
Simple leaf, 502
Singh, Dr. Gurmit, 15
Single-trait inheritance, 134–35
Sinoatrial (SA) node, 259
Sinus, 243
SI units, 631
Skates, 453
Skin, artificial, 24
Slime moulds, 356
Smallpox, 338
Smith, Alyssa, 24, 98
Smoking, 299–302, 306
Smooth endoplasmic reticulum, 16
Snail, 243
Soap, 39
Soil nutrients, 536–39, 557–59
Somatic cells, 163
Sorus, 387
Species, 328
Spermatocytes, 91
Spermatophytes, 390
Sperm cells, 12, 91, 114
Sphincters, 214
Sphygmomanometer, 264
Spider silk, 438
Spindle fibres, 87
Spiny anteater, 465
Spleen, 275

Sponges, 243, 412
Spongy mesophyll, 504
Sporangia, 366
Spore mother cells, 379
Spores, 112, 355
Sporophyte, 112, 379
Sporozoa, 355
Squids, 431
Staphylococcus epidermis, 8
Starch, 36, 62
Steffasson, Baldur, 141
Stele, 512
Stems, 518–22
Sternum, 463
Steroids, 39
Stethoscope, 262
Stewart, Fredrick, 94
Stimulants, 304
Stomach
    and digestion, 214
    structure of, 204
Stomata, 377, 503
Strangler fig, 515
Streptococcus lactis, 8
Substrate, 364
Succession, 548–53
Succulents, 498
Sucrose, 35
Suffixes, Greek and Latin, 633
Sugars, 14–15, 30
    disaccharides, 35
    monosaccharides, 35
    polysaccharides, 36
Sulfamates, 30
Sustainability, 584–85
Sutton, Walter S., 163
Sweating, 227
Swim bladder, 453–54
Symbiotic relationship, 370
Symmetry, 406, 410
Sympathetic nervous system, 260
Synapsis, 105
Synthetic fertilizers, 557, 558–59
Systemic circulatory system, 256
Systole, 261

Tachycardia, 260
Tapeworms, 418–19
Taproots, 512
Taste buds, 210–11
Taxa, 328
Taxol, 392
Taxonomic systems, 328–30
Taxonomy, defined, 326
Tay-Sachs disease, 17–18
Technological problem solving, 608–609
Technologies, and genetic research, 180–82
Teeth, 211–12, 471
Telomeres, 90
Telophase, 89
Test cross, 138
Tetrad, 105

Thermostat, 227
Thymus gland, 275–76
Thyroid gland, 231
Tidal volume, 296
Tissue cultures, 24
Tissues, 244
T lymphocytes (T cells), 275–76, 340
Tobacco, 576
Tomatoes, 492, 567
Totipotent nucleus, 96
Toxin, 506, 515
Trachea, 285
Tracheids, 499
Transcription, 41
Transfer RNA (tRNA), 41
Translation, 41
Translocation, 526
Transmission electron microscope (TEM),
    6–7, 13
Transpiration, 503
Transport, in plants, 525–26
Transposons, 177
Trees, 520–22
Tremblay, Michel, 185
Trichinosis, 420
Triglycerides, 37
Trisomy, 116
Tropical rain forests, 554–55
Trypsin, 217
Tsui, Dr. Lap-Chee, 79, 181
Tubers, 522
Tumour cells, 15
Turgor pressure, 19, 53
Turner syndrome, 118
Twins, 96
Tyrosine phosphatase, 185

Uchida, Dr. Irene, 120–21
Ulcers, 214–15
Ultrasound, 121
Uncertainty, in measurements, 615–16
Uniramia, 439
Unsaturated fats, 37–38
Urban planner, 577

Vaccines, 337
Vacuole, 19
Van Leeuwenhoek, Anton, 9
Varicose veins, 253
Variety, plant, 564
Vascular, 374
Vascular bundles, 519
Vascular cambium, 496, 512
Vascular plant structure and function,
    494–96
Vascular tissue system, 499
Vasoconstriction, 250
Vasodilation, 250
Vegetarianism, 202, 234
Vegetative, 366

Veins, 241, 252–53
Venation, 502
Venn diagram, 614
Ventricles, 256
Venules, 252
Vertebrates, 408
    ecological role of, 449–50
Vesicles, 17
Vessels, 499
Veterinary technician, 479
Villi, 225
Viral vectors, 338–40
Virchow, Rudolph, 9
Viroids, 338
Viruses, 334–40
    human, examples, 337
Vital capacity, 296
Vitamins, 234–35
Viviparous, 470
Vocal cords, 286

Water buttercup, 160
Water transport, in plants, 525–26
Watson, James, 174–75, 180
Waxes, 39
Webber, Kristine, 475
Weisman, August, 162
Whales, 474
Wheat, 141
White blood cells, 17, 91, 245, 246. *See also*
    Leukocytes
Wilkins, Maurice, 174
Wilmut, Dr. Ian, 97–98
Wolkow, Dr. Bob, 3
Wolves, 475
Wood, 569, 571–72
Woody stems, 518
Workplace Hazardous Materials Information
    System (WHMIS), 620, 621
Worms, 417–23. *See also* Earthworms
Worton, Dr. Ron, 181

X chromosomes, 115
Xerophytes, 507
X-ray diffraction, 174
X-ray photographs, 205–206
X-ray technician, 298
Xylem, 376, 499, 512

Y chromosomes, 115
Yeast, 65–66
Yogurt, 8

Zoology, 408
Zygote, 104

Credits

(Fig 4, 5) Daniel W. Gotshall/Visuals Unlimited; (Fig 6) Gerald & Buff Corsi/Visuals Unlimited; (Fig 7) David Fleetham/Visuals Unlimited; **331** (Fig 3) Visuals Unlimited; (Fig 4) K.G. Murti/Visuals Unlimited; **332** (Fig 6) Gerard Lacz/Peter Arnold; (Fig 7) C. Allan Morgan/Peter Arnold; (Fig 8) David Fleetham/Visuals Unlimited; (Fig 9) Science VU/Visuals Unlimited; (Fig 10) Shane Moore/Animals Animals; (Fig 11) Still Pictures/Peter Arnold (Fig 12) David Barron/Animals Animals; **333** (from left to right) Rob & Ann Simpson/Visuals Unlimited; Richard Walters/Visuals Unlimited; Thomas Gula/Visuals Unlimited; Leroy Simon/Visuals Unlimited; Kevin & Betty Collins/Visuals Unlimited; **337** Science VU/Visuals Unlimited; **341** (top) David M. Phillips/Visuals Unlimited; **341** (bottom) Dr. Hans Reichenbach; **343** George P. Chapman/Visuals Unlimited; **344** Arthur M. Stegelman/Visuals Unlimited; **350** (Fig 1) Cabisco/VU; **350** (Table 2 from top to bottom) Peter Parks/Oxford Scientific Films; Cabisco/VU; Peter Parks/Oxford Scientific Films; L.L. Sims/Visuals Unlimited; Sinclair Stammers/Visuals Unlimited; Biophoto Associates/Photo Researchers; **351** (Fig 2a) Biophoto Associates/Photo Researchers; **351** (Fig 3a) R. Kessel-G. Shih/Visuals Unlimited; (Fig 3b) Cabisco/Visuals Unlimited; **352** (Fig 4a) Simon Fraser/Science Photo Library; (Fig 4b) L.L. Sims/Visuals Unlimited; (Fig 4c) Link/Visuals Unlimited; (Fig 5) David Hoffman/Stone; (Fig 6) D. Ditchburn/Visuals Unlimited; **353** (Table 3 from top to bottom) Andrew Syred/Science Photo Library; Manfred Kage/Oxford Scientific Films; Eric Grave/Science Photo Library; CNRI/Science Photo Library; Roy Ficken/Memorial University; S. Flegert/Visuals Unlimited; **354** Cabisco/Visuals Unlimited; **356** (top) R. Calentene/ Visuals Unlimited; **356** (bottom) Onderstepoort Veterinary Research Institute. **TEN: 363** Mark E. Gibson/Visuals Unlimited; **364** (Table 1 left) Janice Palmer; Table 1 (right, from top to bottom) Jana R. Jurak/Visuals Unlimited; Kevin & Betty Collins/Visuals Unlimited; Walter H. Hodge/Peter Arnold; Janice Palmer; **365** (Table 2 from top to bottom) Dr. Tony Brain/Science Photo Library; Dr. Jeremy Burgess/Science Photo Library; J.L. Lepore/Photo Researchers; J. Forsdyke/Science Photo Library; CNRI/Science Photo Library; **366** G.A. Maclean/Oxford Scientific Films; **367** CNRI/Science Photo Library; **369** C. Reeves/CPG/Masterfile; 370, 371 Janice Palmer; 375 Smithsonian Institution; 376 Janice Palmer; 377 Bob Semple; **381, 383** Janice Palmer; **384** John G. Roberts/First Light; **386** (Fig 1) Joe McDonald/Visuals Unlimited; **386** (Fig 2) Janice Palmer; **389** (from left to right) Kjell B. Sandved/Photo Researchers; Scott Camazine/Photo Researchers; Biophoto Associates/Photo Researchers; Kjell B. Sandved/Photo Researchers; **390** Janice Palmer; **391** (left) Janice Palmer; **391** (right) First Light; **392** Thomas Kitchin/First Light; **394-399** Janice Palmer; **400** (left) Inga Spence/Visuals Unlimited; **400** (right) Ivy Images; **402, 405** Janice Palmer. **ELEVEN: 406** R. Williamson/Visuals Unlimited; **407** (clockwise from top left) Hans Pfletshinger/Peter Arnold; Ken Lucas/Visuals Unlimited; Rudy Kuiter/Oxford Scientific Films; Michael Fogden/Oxford Scientific Films; Visuals Unlimited; Norbert Wu/Oxford Scientific Films; **413** David Fleetham/Visuals Unlimited; **414** (Fig 4a) J.A.L. Cooke/Oxford Scientific Films; (Fig 4b) Peter Parks/Oxford Scientific Films; (Fig 4c) Glenn Oliver/Visuals Unlimited; **421** David Thompson/Oxford Scientific Films; **422** (Fig 6) Kathie Atkinson/Oxford Scientific Films; **422** (Fig 7) Martin Dohrm/Science Photo Library; **428** (Fig 1b) Biophoto Associates/Photo Researchers; (Fig 1c) Gregory Dimijian/Photo Researchers; **430** Howard Hall/Oxford Scientific Films; **430** (Fig 4a)

Jason Puddifoot/First Light; **430** (Fig 4b, c) G.I. Bernard/Oxford Scientific Films; **431** (left) Peter Parks/Oxford Scientific Films; **431** (right) Douglas Faulkner/Photo Researchers; **432** (Fig 6) David Fleetham/Visuals Unlimited; (Fig 7) Fredrik Ehrenstrom/Oxford Scientific Films; (Fig 8) Colin Mullens/Oxford Scientific Films; **436** Cosmos Blank/Photo Researchers; **437** Dr. Jeremy Burgess/ Science Photo Library; **438** (Fig 5) Dick Poe/Visuals Unlimited; (Fig 6) Ken Lucas/Visuals Unlimited; (Fig 7) Ken M. Highfall/Photo Researchers; **439** (Fig 8a) Bob Semple; (Fig 8b) Marilyn Kazmers/Peter Arnold; (Fig 9) Tom McHugh/Photo Researchers; **439** (Fig 11a) Michael Fogden/Oxford Scientific Films; (Fig 11b) Phil Devries/Oxford Scientific Films; (Fig 11c) Michael Fogden/ Oxford Scientific Films; (Fig 11d) Bob Semple; **440** Doug Cheeseman/ Peter Arnold; **442** Courtesy of Gail Anderson; **444** (from top to bottom) Bruce Davidson/Animals Animals; K. Atkinson/Oxford Scientific Films/Animals Animals; Joe Clark/Animals Animals; **445** (Fig 3) J.A.L. Cooke/Oxford Scientific Films. **TWELVE: 447** (Fig 1) Fritz Prenzel/Animals Animals; (Fig 2) J-P Varin/Jacana/Photo Researchers; **448** G.I. Bernard/Oxford Scientific Films/Animals Animals; **450** Gilbert S. Grant/Photo Researchers; **452** (top) Patrice Ceisel/Visuals Unlimited; **452** (bottom) David B. Fleetham/Visuals Unlimited; **453** (Fig 3) Ken Lucas/Visuals Unlimited; **453** (Fig 4a) Tom McHugh/Photo Researchers; (Fig 4b) John Paling/Oxford Scientific Films; (Fig 4c) Kjell B. Sandved/Visuals Unlimited; (Fig 4d) Daniel W. Gotshall/Visuals Unlimited; **454** Peter Scoones/Planet Earth Pictures; **455** (from top to bottom) Nathan W. Cohen/Visuals Unlimited; Joseph T. Collins/Photo Researchers; G.I. Bernard/Oxford Scientific Films; Stephen Dalton/Oxford Scientific Films; Michael Fogden/Oxford Scientific Films; **457** (from top to bottom) Richard Packwood/Oxford Scientific Films; Michael Fogden/Oxford Scientific Films; Beth Davidow/Visuals Unlimited; **460** A.J. Copley/Visuals Unlimited; **462** James L. Amos/Photo Researchers; **465** (top) E.C. Williams/Visuals Unlimited; **465** (bottom) John Alcock/Visuals Unlimited; **466** (top) W.A. Banaszewski/Visuals Unlimited; **466** (bottom) Gerald & Buff Corsi/Visuals Unlimited; **470** Cheryl A. Ertelt/Visuals Unlimited; **471** (Fig 6a) William Bacon/Photo Researchers; (Fig 6b) Renee Lynn/Stone; **473** L.S. Stepanowicz/Visuals Unlimited; **474** (Fig 15) Stephen Dalton/Oxford Scientific Films; (Fig 16) François Gohier/Photo Researchers; **479** (top left) Jeffrey Howe/Visuals Unlimited; **479** (top right) Jeff Greenberg/Visuals Unlimited; **479** (bottom right) Science VU/Visuals Unlimited; **481** (from left to right) Joe McDonald/Visuals Unlimited; Tom Ulrich/Visuals Unlimited; Arthur Morris/Visuals Unlimited; Maslowski/Visuals Unlimited; Gerard Fuehrer/Visuals Unlimited; **485** Cabisco/Visuals Unlimited. **THIRTEEN: 489** Courtesy of Dr. Vojislava Grbic/University of Western Ontario; **493** Chris O'Meara/ AP/CP Picture Archive; **494, 497** Janice Palmer; **499, 500** Mike Summner/University of Manitoba; **502-507** Janice Palmer; **513** (Fig 1b) © John D. Cunningham/Visuals Unlimited; (Fig 1d) © Stan Elems/ Visuals Unlimited; **515** (Fig 4) Janice Palmer; (Fig 5) Doug Fraser; **519** Mike Sumner/University of Manitoba; **522** (Fig 6a) Science VU/Visuals Unlimited; (Fig 6b) Janice Palmer; (Fig 6c) Carol & Don Spencer/Visuals Unlimited; **521** (Fig 4b) © Biodisc/Visuals Unlimited; **523, 526** Janice Palmer; **529** Tim Fitzharris/Masterfile; **530** Janice Palmer; **539** Cabisco/Visuals Unlimited; **545** Janice Palmer. **FOURTEEN: 547** Janice Palmer; **548** (Fig 1) Arnold J. Karpoff/Visuals Unlimited; **548** (Fig 2) Janice Palmer; (Fig 3) Steve McCutcheon/Visuals Unlimited; **550-552** Janice Palmer; **555** (Fig 1) Janice Palmer; (Fig 2)